Physical Constants

Symbol	Name	Value
q	Magnitude of electronic charge	1.60218×10^{-19} C
m_o	Electron mass in free space	9.10938×10^{-31} kg
q/m_o	Charge/mass ratio (electron)	1.75882×10^{11} C/kg
c	Speed of light in vacuum	2.99792×10^8 m/s
ε_o	Permittivity of vacuum	8.8542×10^{-12} F/m
k	Boltzmann's constant	1.38065×10^{-23} J/K
		8.61734×10^{-5} eV/K
h	Planck's constant	6.62607×10^{-34} J · s
		4.13567×10^{-15} eV · s
A_o	Avogadro's constant	6.02214×10^{26} molecules/kg · mole
kT	Thermal energy	0.02586 eV ($T = 27°C$)
		0.02526 eV ($T = 20°C$)

Source: http://physics.nist.gov/constants

Conversion Factors

1 Å	$= 0.1$ nm
	$= 10^{-4}$ μm
	$= 10^{-8}$ cm
	$= 10^{-10}$ m
1 μm	$= 10^4$ Å
	$= 10^3$ nm
	$= 10^{-4}$ cm
	$= 10^{-6}$ m
1 mil	$= 10^{-3}$ in
	$= 25.4$ μm
1 eV	$= 1.60218 \times 10^{-19}$ J
λ (μm)	$= 1.2398/E$ (eV)
λ (Å)	$= 1.2398 \times 10^4/E$ (eV)
	$= 12.398/E$ (keV)

SEMICONDUCTOR MATERIAL AND DEVICE CHARACTERIZATION

SEMICONDUCTOR MATERIAL AND DEVICE CHARACTERIZATION

Third Edition

DIETER K. SCHRODER
Arizona State University
Tempe, AZ

IEEE PRESS

A JOHN WILEY & SONS, INC., PUBLICATION

Published by John Wiley & Sons, Inc., Hoboken, New Jersey.
Published simultaneously in Canada.

For general information on our other products and services or for technical support, please contact our Customer Care Department within the United States at (800) 762-2974, outside the United States at (317) 572-3993 or fax (317) 572-4002.

Wiley also publishes its books in a variety of electronic formats. Some content that appears in print may not be available in electronic formats. For more information about Wiley products, visit our web site at www.wiley.com.

Library of Congress Cataloging-in-Publication Data:

Schroder, Dieter K.
    Semiconductor material and device characterization / by Dieter K. Schroder.
        p. cm.
    "A Wiley-Interscience Publication."
    Includes bibliographical references and index.
    ISBN-13: 978-0-471-73906-7 (acid-free paper)
    ISBN-10: 0-471-73906-5 (acid-free paper)
    1. Semiconductors. 2. Semiconductors–Testing. I. Title.
    QC611.S335 2005
    621.3815'2—dc22

                                                    2005048514

Printed in the United States of America.

10 9 8 7 6 5 4 3 2 1

Chapter 12

This is a new chapter, dealing with *Failure Analysis and Reliability*. I have taken some sections from other chapters in the second edition and expanded them. I introduce failure times and distribution functions here, then discuss electromigration; hot carriers; gate oxide integrity; negative bias temperature instability; stress induced leakage current; electrostatic discharge that are of concern for device reliability. The rest of this chapter deals with the more common failure analysis techniques: quiescent drain current; mechanical probes; emission microscopy; fluorescent microthermography; infrared thermography; voltage contrast; laser voltage probe; liquid crystals; optical beam induced resistance change and noise.

Several people have supplied experimental data and several concepts were clarified by discussions with experts in the semiconductor industry. I acknowledge their contributions in the figure captions. Tom Shaffner from the National Institute of Standards and Technology has continued to be an excellent source of knowledge and a good friend and Steve Kilgore from Freescale Semiconductor has helped with electromigration concepts. The recent book *Handbook of Silicon Semiconductor Metrology*, edited by Alain Diebold, is an excellent companion volume as it gives many of the practical details of semiconductor metrology missing here. I thank executive editor G. Telecki, R. Witmer and M. Yanuzzi from John Wiley & Sons for editorial assistance in bringing this edition to print.

DIETER K. SCHRODER

Tempe, AZ

Chapter 2
Contactless $C-V$ added; integral capacitance augmented; series capacitance added/augmented; free carrier absorption augmented; new lateral profiling section; added Appendix 2—equivalent circuit derivations.

Chapter 3
Augmented circular contact resistance section; added considerations of parasitic resistance in TLM method; expanded barrier height section by adding BEEM; added Appendix dealing with parasitic resistance effects.

Chapter 4
Added section of pseudo MOSFETs for silicon-on-insulator characterization; added several MOSFET effective channel length measurement methods and deleted some of the older methods.

Chapter 5
Added Laplace DLTS; added a section to the time constant extraction portion in Appendix 5.2.

Chapter 6
Expanded the section on oxide thickness measurements; added considerations for the effect of leaky gate oxides on conductance and charge pumping; added the $DC-IV$ method; expanded the section on gate oxide leakage currents; added Appendix 6.2 considering the effects of wafer chuck parasitic capacitance and leakage current.

Chapter 7
Clarified the optical lifetime section; added *Quasi-steady-state Photoconductance*; augmented the free carrier absorption and diode current lifetime method; added leaky oxide current considerations to the pulsed MOS capacitor technique.

Chapter 8
Added the effects of gate depletion, channel location, gate current, interface traps, and inversion charge frequency response to the extraction of the effective mobility. I also added a section on contactless mobility measurements.

Chapter 9
This chapter is new and introduces charge-based measurement and Kelvin probes. I have also included probe-based measurements here and expanded these by including scanning capacitance, scanning Kelvin force, scanning spreading resistance, and ballistic electron emission microscopy.

Chapter 10
Expanded confocal optical microscopy, photoluminescence, and line width measurement.

Chapter 11
Made some small changes.

PREFACE TO THIRD EDITION

Semiconductor characterization has continued its relentless advance since the publication of the second edition. New techniques have been developed, others have been refined. In the second edition preface I mentioned that techniques such as scanning probe, total-reflection X-ray fluorescence and contactless lifetime/diffusion length measurements had become routine. In the intervening years, probe techniques have further expanded, charge-based techniques have become routine, as has transmission electron microscopy through the use of focused ion beam sample preparation. Line width measurements have become more difficult since lines have become very narrow and the traditional SEM and electrical measurements have been augmented by optical techniques like scatterometry and spectroscopic ellipsometry. In addition to new measurement techniques, the interpretation of existing techniques has changed. For example, the high leakage currents of thin oxides make it necessary to alter existing techniques/theories for many MOS-based techniques.

I have rewritten parts of each chapter and added two new chapters, deleted some outdated material, clarified some obscure/confusing parts that have been pointed out to me. I have redone most of the figures, deleted some outdated ones or replaced them with more recent data. The third edition is further enhanced through additional problems and review questions at the end of each chapter and examples throughout the book, to make it a more attractive textbook. I have added 260 new references to bring the book as up-to-date as possible. I have also changed the symbol for sheet resistance from ρ_s to R_{sh}, to bring it in line with more accepted use.

I list the main additional or expanded material here briefly by chapter. There are many other smaller changes throughout the book.

Chapter 1
New sheet resistance explanation; new 4-point probe derivation; use of 4-point probe for shallow junctions and high sheet resistance sample; added the *Carrier Illumination* method.

9   Charge-based and Probe Characterization   523

10   Optical Characterization   563

CONTENTS

1

RESISTIVITY

1.1 INTRODUCTION

The *resistivity* ρ of a semiconductor is important for starting material as well as for semiconductor devices. Although carefully controlled during crystal growth, it is not truly uniform in the grown ingot due to variability during growth and segregation coefficients less than unity for the common dopant atoms. The resistivity of epitaxially grown layers is generally very uniform. Resistivity is important for devices because it contributes to the device series resistance, capacitance, threshold voltage, hot carrier degradation of MOS devices, latch up of CMOS circuits, and other parameters. The wafers resistivity is usually modified locally during device processing by diffusion and ion implantation, for example.

The resistivity depends on the free electron and hole densities n and p, and the electron and hole mobilities μ_n and μ_p according to the relationship

$$\rho = \frac{1}{q(n\mu_n + p\mu_p)} \tag{1.1}$$

ρ can be calculated from the measured carrier densities and mobilities. For extrinsic materials in which the majority carrier density is much higher than the minority carrier density, it is generally sufficient to know the majority carrier density and the majority carrier mobility. The carrier densities and mobilities are generally not known, however. Hence we must look for alternative measurement techniques, ranging from *contactless*, through *temporary contact* to *permanent contact* techniques.

Semiconductor Material and Device Characterization, Third Edition, by Dieter K. Schroder
Copyright © 2006 John Wiley & Sons, Inc.

1.2 TWO-POINT VERSUS FOUR-POINT PROBE

The *four-point probe* is commonly used to measure the semiconductor resistivity. It is an absolute measurement without recourse to calibrated standards and is sometimes used to provide standards for other resistivity measurements. Two-point probe methods would appear to be easier to implement, because only two probes need to be manipulated. But the interpretation of the measured data is more difficult. Consider the two-point probe or two-contact arrangement of Fig. 1.1(a). Each contact serves as a current *and* as a voltage probe. We wish to determine the resistance of the device under test (DUT). The total resistance R_T is given by

$$R_T = V/I = 2R_W + 2R_C + R_{DUT} \qquad (1.2)$$

where R_W is the wire or probe resistance, R_C the contact resistance, and R_{DUT} the resistance of the device under test. Clearly it is impossible to determine R_{DUT} with this measurement arrangement. The remedy is the four-point probe or four-contact arrangement in Fig. 1.1(b). The current path is identical to that in Fig. 1.1(a). However, the voltage is now measured with two additional contacts. Although the voltage path contains R_W and R_C as well, the current flowing through the voltage path is very low due to the high input impedance of the voltmeter (around 10^{12} ohms or higher). Hence, the voltage drops across R_W and R_C are negligibly small and can be neglected and the measured voltage is essentially the voltage drop across the DUT. By using four rather than two probes, we have eliminated parasitic voltage drops, even though the voltage probes contact the device on the same contact pads as the current probes. Such four contact measurements are frequently referred to as *Kelvin measurements*, after Lord Kelvin.

An example of the effect of two versus four contacts is shown in Fig. 1.2. The drain current–gate voltage characteristics of a metal-oxide-semiconductor field-effect transistor were measured with one contact on source and drain (no Kelvin), one contact on source and two contacts on drain (Kelvin-Drain), two contacts on source and one on drain (Kelvin-Source), and two contacts on source and drain (Full Kelvin). It is quite obvious that eliminating contact and probe resistances in the "Full Kelvin" has a significant effect on the measured current. The probe, contact, and spreading resistances of a two-point probe arrangement on a semiconductor are illustrated in Fig. 1.3.

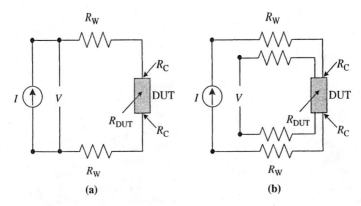

Fig. 1.1 Two-terminal and four-terminal resistance measurement arrangements.

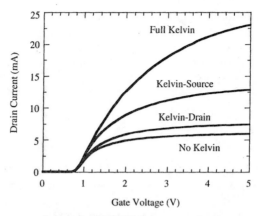

Fig. 1.2 Effect of contact resistance on MOSFET drain current. Data courtesy of J. Wang, Arizona State University.

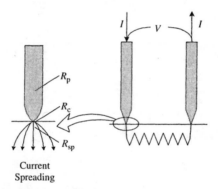

Fig. 1.3 Two-point probe arrangement showing the probe resistance R_p, the contact resistance R_c, and the spreading resistance R_{sp}.

The four-point probe was originally proposed by Wenner[1] in 1916 to measure the earth's resistivity. The four-point probe measurement technique is referred to in Geophysics as *Wenner's method*. Valdes adopted it for semiconductor wafer resistivity measurements in 1954.[2] The probes are generally collinear, *i.e.*, arranged in-line with equal probe spacing, but other probe configurations are possible.[3]

Exercise 1.1

Problem: This exercise deals with data presentation. Frequently non-linear behavior is encountered in presenting data of semiconductor materials or devices, where one parameter may be proportional to another parameter to some power, *e.g.*, $y = Kx^b$, where both the prefactor K and exponent b are constant. One parameter may vary exponentially with another parameter, *e.g.*, $I = I_o \exp(\beta V)$. What is the best way to present the information to be able to extract "b" and "β"?

Solution: Consider the relationship $y = Kx^b = 8x^5$. Plots of y versus x on a linear scale, shown in Fig. E1.1(a) and (b), do not allow "b" do be determined, regardless what scale is used because the curves are non-linear. However, when the same data are plotted on a log-log plot as in (c), "b" is simply the slope of such a plot. In this case the slope is 5, because

$$\log(y) = \log(Kx^b) = \log(K) + \log(x^b) = \log(K) + b\log(x)$$

and the slope m is

$$m = \frac{d[\log(y)]}{d[\log(x)]} = b = 5$$

If the data are plotted as in (d), which is also a log-log plot, the data must first be converted to "log" before the slope is taken. When that is done, the slope is again $m = 5$.

Let us now consider the relationship $y = y_o\exp(\beta x) = 10^{-14}\exp(40x)$. Obviously, a linear-linear plot, shown in (e), allows neither y_o nor β to be extracted. When, however, the data are plotted on a semilog plot, as in (f), we have

$$\ln(y) = \ln(y_o) + \beta x \Rightarrow \log(y) = \log(y_o) + \beta x/\ln(10)$$

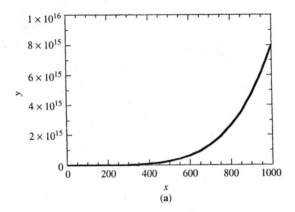

(a)

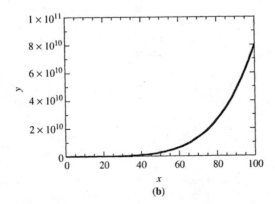

(b)

Fig. E1.1

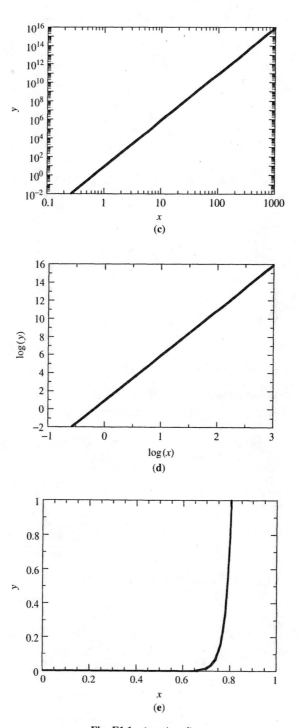

Fig. E1.1   (*continued*)

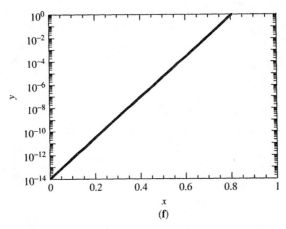

(f)

Fig. E1.1 (*continued*)

The slope m is

$$m = \frac{d[\log(y)]}{dx} = \frac{\beta}{\ln(10)} = \frac{\beta}{2.3036} = \frac{14}{2.3036 \times 0.8}$$

and the intercept at $x = 0$ is $y_o = 10^{-14}$.

To derive the four-point probe resistivity expression, we start with the sample geometry in Fig. 1.4(a). The electric field $\mathscr{E}$ is related to the current density J, the resistivity ρ, and the voltage V through the relationship[2]

$$\mathscr{E} = J\rho = -\frac{dV}{dr}; \quad J = \frac{I}{2\pi r^2} \tag{1.3}$$

The voltage at point P at a distance r from the probe, is then

$$\int_0^V dV = -\frac{I\rho}{2\pi} \int_0^r \frac{dr}{r^2} \Rightarrow V = \frac{I\rho}{2\pi r} \tag{1.4}$$

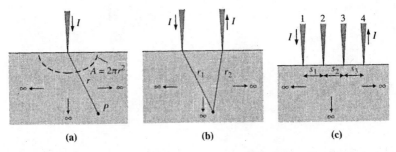

(a)   (b)   (c)

Fig. 1.4 (a) one-point probe, (b) two-point, and (c) collinear four-point probe showing current flow and voltage measurement.

For the configuration in Fig. 1.4(b), the voltage is

$$V = \frac{I\rho}{2\pi r_1} - \frac{I\rho}{2\pi r_2} = \frac{I\rho}{2\pi}\left(\frac{1}{r_1} - \frac{1}{r_2}\right) \tag{1.5}$$

where r_1 and r_2 are the distances from probes 1 and 2, respectively. The minus sign accounts for current leaving through probe 2. For probe spacings s_1, s_2, and s_3, as in Fig. 1.4(c), the voltage at probe 2 is

$$V_2 = \frac{I\rho}{2\pi}\left(\frac{1}{s_1} - \frac{1}{s_2 + s_3}\right) \tag{1.6}$$

and at probe 3 it is

$$V_3 = \frac{I\rho}{2\pi}\left(\frac{1}{s_1 + s_2} - \frac{1}{s_3}\right) \tag{1.7}$$

The total measured voltage $V = V_{23} = V_2 - V_3$ becomes

$$V = \frac{I\rho}{2\pi}\left(\frac{1}{s_1} - \frac{1}{s_2 + s_3} - \frac{1}{s_1 + s_2} + \frac{1}{s_3}\right) \tag{1.8}$$

The resistivity ρ is given by

$$\rho = \frac{2\pi}{(1/s_1 - 1/(s_1 + s_2) - 1/(s_1 + s_2) + 1/s_3)}\frac{V}{I} \tag{1.9}$$

usually expressed in units of ohm · cm, with V measured in volts, I in amperes, and s in cm. The current is usually such that the resulting voltage is approximately 10 mV. For most four-point probes the probe spacings are equal. With $s = s_1 = s_2 = s_3$, Eq. (1.9) reduces to

$$\rho = 2\pi s \frac{V}{I} \tag{1.10}$$

Typical probe radii are 30 to 500 μm and probe spacings range from 0.5 to 1.5 mm. The spacings vary for different sample diameter and thickness.[4] For $s = 0.1588$ cm, $2\pi s$ is unity, and ρ becomes simply $\rho = V/I$. Smaller probe spacings allow measurements closer to wafer edges, an important consideration during wafer mapping. Probes to measure metal films should not be mixed with probes to measure semiconductors. For some applications, *e.g.* magnetic tunnel junctions, polymer films, and semiconductor defects, microscopic four-point probes with probe spacings of 1.5 μm have been used.[5]

Semiconductor wafers are not semi-infinite in extent in either the lateral or the vertical dimension and Eq. (1.10) must be corrected for finite geometries. For an arbitrarily shaped sample the resistivity is given by

$$\boxed{\rho = 2\pi s F \frac{V}{I}} \tag{1.11}$$

where F corrects for probe location near sample edges, for sample thickness, sample diameter, probe placement, and sample temperature. It is usually a product of several independent correction factors. For samples thicker than the probe spacing, the simple, independent correction factors contained in F of Eq. (1.11) are no longer adequate due to interactions between thickness and edge effects. Fortunately the samples are generally thinner than the probe spacings, and the correction factors can be independently calculated.

1.2.1   Correction Factors

Four-point probe correction factors have been calculated by the method of images,[2, 6] complex variable theory,[7] the method of Corbino sources,[8] Poisson's equation,[9] Green's functions,[10] and conformal mapping.[11-12] We will give the most appropriate factors here and refer the reader to others where appropriate.

The following correction factors are for *collinear* or *in-line probes* with equal probe spacing, s. We write F as a product of three separate correction factors

$$F = F_1 F_2 F_3 \tag{1.12}$$

Each of these factors can be further subdivided. F_1 corrects for sample thickness, F_2 for lateral sample dimensions, and F_3 for placement of the probes relative to the sample edges. Other correction factors are discussed later in the chapter.

Sample thickness must be corrected for most measurements since semiconductor wafers are not infinitely thick. A detailed derivation of thickness correction factors is given by Weller.[13] Sample thicknesses are usually on the order of the probe spacing or less introducing the correction factor[14]

$$F_{11} = \frac{t/s}{2\ln\{[\sinh(t/s)]/[\sinh(t/2s)]\}} \tag{1.13}$$

for a *non-conducting* bottom wafer surface boundary, where t is the wafer or layer thickness. If the sample consists of a semiconducting layer on a semiconductor substrate, it is important that the layer be electrically isolated from the substrate. The simplest way to do this is for the two regions to be of opposite conductivity, *i.e.*, n-layer on a p-substrate or p-layer on an n-substrate. The space-charge region is usually sufficiently insulating to confine the current to the layer.

For a *conducting* bottom surface the correction factor becomes

$$F_{12} = \frac{t/s}{2\ln\{[\cosh(t/s)]/[\cosh(t/2s)]\}} \tag{1.14}$$

F_{11} and F_{12} are plotted in Fig. 1.5. Conducting bottom boundaries are difficult to achieve. Even a metal deposited on the wafer back surface does not ensure a conducting contact. There is always a contact resistance. Most four-point probe measurements are made with insulating bottom boundaries.

For thin samples Eq. (1.13) reduces to

$$F_{11} = \frac{t/s}{2\ln(2)} \tag{1.15}$$

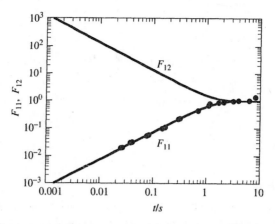

Fig. 1.5 Wafer thickness correction factors versus normalized wafer thickness; t is the wafer thickness, s the probe spacing. The data points are taken from ref. 15.

using the approximation $\sinh(x) \approx x$ for $x \ll 1$. Eq. (1.15) is valid for $t \leq s/2$. For very thin samples that satisfy the conditions for F_2 and F_3 to be approximately unity, we find from Eqs. (1.11), (1.12), and (1.15)

$$\rho = \frac{\pi}{\ln(2)} t \frac{V}{I} = 4.532 t \frac{V}{I} \tag{1.16}$$

Thin layers are often characterized by their *sheet resistance* R_{sh} expressed in units of ohms per square. The sheet resistance of uniformly doped samples is given by

$$R_{sh} = \frac{\rho}{t} = \frac{\pi}{\ln(2)} \frac{V}{I} = 4.532 \frac{V}{I} \tag{1.17}$$

subject to the constraint $t \leq s/2$. The sheet resistance characterizes thin semiconductor sheets or layers, such as diffused or ion-implanted layers, epitaxial films, polycrystalline layers, and metallic conductors.

The sheet resistance is a measure of the resistivity averaged over the sample thickness. The sheet resistance is the inverse of the sheet conductance G_{sh}. For *uniformly-doped* samples we find

$$R_{sh} = \frac{1}{G_{sh}} = \frac{1}{\sigma t} \tag{1.18}$$

where σ is the conductivity and t the sample thickness. For *non uniformly-doped* samples

$$R_{sh} = \frac{1}{\int_0^t [1/\rho(x)] \, dx} = \frac{1}{\int_0^t \sigma(x) \, dx} = \frac{1}{q \int_0^t [n(x)\mu_n(x) + p(x)\mu_p(x)] \, dx} \tag{1.19}$$

Exercise 1.2

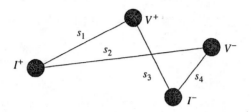

Fig. E1.2

Problem: Is there another way to derive the sheet resistance expression?

Solution: Consider a sample of thickness t and resistivity ρ. The four probes are arranged as in Fig. E1.2. Current I is injected at probe I^+ and spreads out cylindrically symmetric. By symmetry and current conservation, the current density at distance r from the probe is

$$J = \frac{I}{2\pi r t}$$

The electric field is

$$\mathscr{E} = J\rho = \frac{I\rho}{2\pi r t} = -\frac{dV}{dr}$$

Integrating this expression gives the voltage drop between probes V^+ and V^-, located at distances s_1 and s_2 from I^+ as

$$\int_{V_{s1}}^{V_{s2}} dV = -\frac{I\rho}{2\pi t}\int_{s_1}^{s_2}\frac{dr}{r} \Rightarrow V_{s1} - V_{s2} = V_{12} = \frac{I\rho}{2\pi t}\ln\left(\frac{s_2}{s_1}\right)$$

By the principle of superposition, the voltage drop due to current injected at I^- is

$$V_{34} = -\frac{I\rho}{2\pi t}\ln\left(\frac{s_3}{s_4}\right)$$

leading to

$$V = V_{12} - V_{34} = \frac{I\rho}{2\pi t}\ln\left(\frac{s_2 s_3}{s_1 s_4}\right)$$

For a collinear arrangement with $s_1 = s_4 = s$ and $s_2 = s_3 = 2s$

$$\rho = \frac{\pi t}{\ln(2)}\frac{V}{I}; R_{sh} = \frac{\pi}{\ln(2)}\frac{V}{I}$$

Exercise 1.3

Problem: What does *sheet resistance* mean and why does it have such strange units?

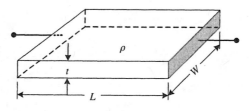

Fig. E1.3

Solution: To understand the concept of sheet resistance, consider the sample in Fig. E1.3. The resistance between the two ends is given by

$$R = \rho\frac{L}{A} = \rho\frac{L}{Wt} = \frac{\rho}{t}\frac{L}{W} \text{ ohms}$$

Since L/W has no units, ρ/t should have units of ohms. But ρ/t is not the sample resistance. To distinguish between R and ρ/t, the ratio ρ/t is given the units of ohms/square and is named sheet resistance, R_{sh}. Hence the sample resistance can be written as

$$R = R_{sh}\frac{L}{W} \text{ ohms}$$

The sample is sometimes divided into squares, as in Fig. E1.4. The resistance is then given as

$$R = R_{sh} \text{ (ohms/square)} \times \text{Number of squares} = 5R_{sh} \text{ ohms}$$

Looking at it this way, the "square" cancels.

The sheet resistance of a semiconductor sample is commonly used to characterize ion implanted and diffused layers, metal films, etc. The depth variation of the dopant atoms need not be known, as is evident from Eq. (1.19). The sheet resistance can be thought of as the depth integral of the dopant atom density in the sample, regardless of its vertical spatial doping density variation. A few sheet resistances are plotted in Fig. E1.5 versus sample thickness as a function of sample resistivity. Also shown are typical values for Al, Cu and heavily-doped Si.

Exercise 1.4

Problem: For the carrier density profiles in Fig. E1.6, do the sheet resistances of the three layers differ?

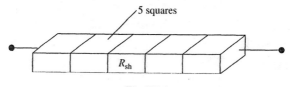

Fig. E1.4

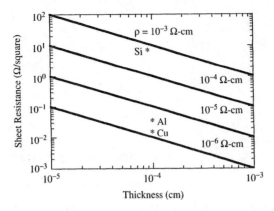

Fig. E1.5

Solution: Eq. (1.19) shows the sheet resistance to be inversely proportional to the conductivity-thickness product. For constant mobility, R_{sh} is inversely proportional to the area under the curves in Fig. E1.6. Since the three areas are equal, this implies that R_{sh} is the same for all three cases. In other words, it does not matter what the carrier distribution is, only the integrated distribution matters for R_{sh}.

Four-point probe measurements are subject to further sample size correction factors. For circular wafers of diameter D, the correction factor F_2 in Eq. (1.12) is given by[16]

$$F_2 = \frac{\ln(2)}{\ln(2) + \ln\{[(D/s)^2 + 3]/[(D/s)^2 - 3]\}} \tag{1.20}$$

F_2 is plotted in Fig. 1.6 for circular wafers. The sample must have a diameter $D \geq 40\ s$ for F_2 to be unity. For a probe spacing of 0.1588 cm, this implies that the wafer must be at least 6.5 cm in diameter. Also shown in Fig. 1.6 is the correction factor for rectangular samples.[6]

The correction factor 4.532 in Eq. (1.17) is for collinear probes with the current flowing into probe 1, out of probe 4, and with the voltage sensed across probes 2 and 3. For the current applied to and the voltage sensed across other probes, different correction factors obtain.[17] For probes *perpendicular* to and a distance d from a *non-conducting boundary*, the correction factors, for infinitely thick samples, are shown in Fig. 1.7.[2] It is obvious from the figures that as long as the probe distance from the wafer boundary is at least

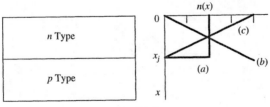

Fig. E1.6

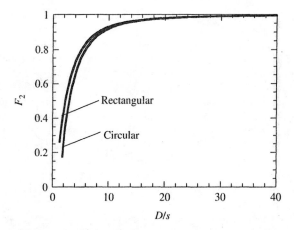

Fig. 1.6   Wafer diameter correction factors versus normalized wafer diameter. For circular wafers: D = wafer diameter; for rectangular samples: D = sample width, s = probe spacing.

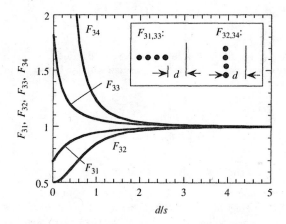

Fig. 1.7   Boundary proximity correction factors versus normalized distance d (s = probe spacing) from the boundary. F_{31} and F_{32} are for non-conducting boundaries, F_{33} and F_{34} are for conducting boundaries.

three to four probe spacings, the correction factors F_{31} to F_{34} reduce to unity. For most four-point probe measurements this condition is easily satisfied. Correction factors F_{31} to F_{34} only become important for small samples in which the probe is, of necessity, close to the sample boundary.

Other corrections must be applied when the probe is not centered even in a wafer of substantial diameter.[16] For rectangular samples it has been found that the sensitivity of the geometrical correction factor to positional error is minimized by orienting the probe with its electrodes within about 10% of the center.[11] For square arrays the error is minimized by orienting the probe array with its electrodes equidistant from the midpoints of the sides. There is also an angular dependence of the placement of a square array on the rectangular sample.[9, 11] We should mention that if the probe spacings are not exactly identical, there is a further small correction.[18]

The key to high precision four-point probe measurements, including reduced geometric effects associated with proximity of the probe to a non-conducting boundary, is the use of two measurement configurations at each probe location.[19–21] This technique is known as the "dual configuration" or as the "configuration switched" method. The first configuration is usually with current into probe 1 and out of probe 4 and with the voltage sensed across probes 2 and 3. The second measurement is made with current driven through probes 1 and 3 and voltage measured across probes 2 and 4. The advantages are: (i) the probe no longer needs to be in a high symmetry orientation (being perpendicular or parallel to the wafer radius of a circular wafer or to the length or width of a rectangular sample), (ii) the lateral dimensions of the specimen do not have to be known since the geometric correction factor results directly from the two measurements, and (iii) the two measurements self-correct for the actual probe spacings.

The sheet resistance in the dual configuration is given by[21]

$$R_{sh} = -14.696 + 25.173(R_a/R_b) - 7.872(R_a/R_b)^2 \qquad (1.21)$$

where

$$R_a = \frac{V_{f23}/I_{f14} + V_{r23}/I_{r14}}{2}; \quad R_b = \frac{V_{f24}/I_{f13} + V_{r24}/I_{r13}}{2} \qquad (1.22)$$

V_{f23}/I_{f14} is the voltage/current across terminals 2,3 and 1,4 with the current in the forward direction and V_{r23}/I_{r14} with the current in the reverse direction.

The resistivity of semiconductor ingots, measured with the four-point probe, is given by

$$\rho = 2\pi s \frac{V}{I} \qquad (1.23)$$

only if the ingot diameter D satisfies the relationship $D \geq 10$ s.[10, 22, 23]

1.2.2   Resistivity of Arbitrarily Shaped Samples

The collinear probe configuration is the most common four-point probe arrangement. Arrangement of the points in a square has the advantage of occupying a smaller area since the spacing between points is only s or $2^{1/2}$s, whereas in a collinear configuration the spacing between the outer two probes is 3s. The square arrangement is more commonly used, not as an array of four mechanical probes, but rather as contacts to square semiconductor samples.

The theoretical foundation of measurements on irregularly shaped samples is based on conformal mapping developed by van der Pauw.[24, 26] He showed how the specific resistivity of a flat sample of arbitrary shape can be measured without knowing the current pattern, if the following conditions are met: (1) the contacts are at the circumference of the sample, (2) the contacts are sufficiently small, (3) the sample is uniformly thick, and (4) the surface of the sample is singly connected, i.e., the sample does not contain any isolated holes.

Consider the flat sample of a conducting material of arbitrary shape, with contacts 1, 2, 3, and 4 along the periphery as shown in Fig. 1.8 to satisfy the conditions above. The resistance $R_{12,34}$ is defined as

$$R_{12,34} = \frac{V_{34}}{I_{12}} \qquad (1.24)$$

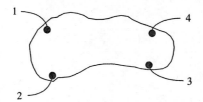

Fig. 1.8 Arbitrarily shaped sample with four contacts.

where the current I_{12} enters the sample through contact 1 and leaves through contact 2 and $V_{34} = V_3 - V_4$ is the voltage difference between the contacts 3 and 4. $R_{23,41}$ is defined similarly.

The resistivity is given by[24]

$$\rho = \frac{\pi}{\ln(2)} t \frac{(R_{12,34} + R_{23,41})}{2} F \qquad (1.25)$$

where F is a function only of the ratio $R_r = R_{12,34}/R_{23,41}$, satisfying the relation

$$\frac{R_r - 1}{R_r + 1} = \frac{F}{\ln(2)} \operatorname{ar cosh}\left(\frac{\exp[\ln(2)/F]}{2}\right) \qquad (1.26)$$

The dependence of F on R_r is shown in Fig. 1.9.

For a *symmetrical* sample such as the circle or the square in Fig. 1.10, $R_r = 1$ and $F = 1$. This allows Eq. (1.25) to be simplified to

$$\rho = \frac{\pi}{\ln(2)} t R_{12,34} = 4.532 t R_{12,34} \qquad (1.27)$$

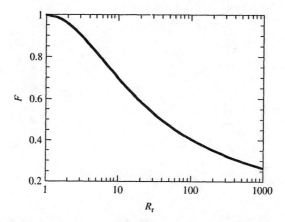

Fig. 1.9 The van der Pauw correction factor F versus R_r.

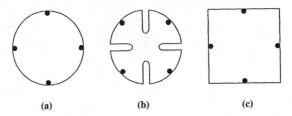

Fig. 1.10 Typical symmetrical circular and square sample geometries.

The sheet resistance becomes

$$R_{sh} = \frac{\pi R_{12,34}}{\ln(2)} = 4.532 R_{12,34} \qquad (1.28)$$

similar to the four-point probe expression in Eq. (1.17).

The van der Pauw equations are based on the assumption of negligibly small contacts located on the sample periphery. Real contacts have finite dimensions and may not be exactly on the periphery of the sample. The influence of non-ideal peripheral contacts is shown in Fig. 1.11. The correction factor C is plotted as a function of the ratio of contact size to sample side length d/l. C is defined as

$$\rho = Ct R_{12,34}; \quad R_{sh} = C R_{12,34} \qquad (1.29)$$

Figure 1.11 shows that corner contacts introduce less error than contacts placed in the center of the sample sides. However, if the contact length is less than about 10% of the side length, the correction is negligible for either contact placement.

The error introduced by non-ideal contacts can be eliminated by the cloverleaf configuration of Fig. 1.10(b). Such configurations make sample preparation more complicated and are undesirable, so square samples are generally used. One of the advantages of the van der Pauw structure is the small sample size compared with the area required for four-point probe measurements. For simple processing it is preferable to use the circular or

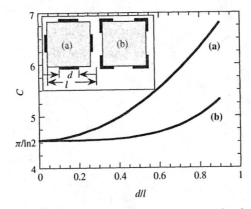

Fig. 1.11 Correction factor C versus d/l for contacts at the center and at the corners of the square. Data after ref. 25.

square sample geometries shown in Fig. 1.10. For such structures it is not always possible to align the contacts exactly.

Geometries other than those in Fig. 1.10 are also used. One of these is the *Greek cross* in Fig. 1.12. Using photolithographic techniques, it is possible to make such structures very small and place many of them on a wafer for uniformity characterization. The sheet resistance of the shaded area is determined in such measurements. For structures with $L = W$, the contacts should be placed so that $d \leq L/6$ from the edge of the cross, where d is the distance of the contact from the edge.[27] Surface leakage can introduce errors if L is too large.[28] A variety of cross sheet resistor structures have been investigated and their performance compares well with conventional bridge-type structures.[29] The measured voltages in cross and van der Pauw structures are lower than those in conventional bridge structures.

The cross and the bridge structures are combined in the cross-bridge structure in Fig. 1.13, allowing the *sheet resistance* and the *line width* to be determined. The sheet resistance, determined in the shaded cross area, is

$$R_{sh} = \frac{\pi}{\ln(2)} \frac{V_{34}}{I_{12}} \tag{1.30}$$

where $V_{34} = V_3 - V_4$ and I_{12} is the current flowing into contact I_1 and out of contact I_2.

The left part of Fig. 1.13 is a bridge resistor to determine the line width W. We mention the line width measurement feature only briefly here. Line width measurements are more fully discussed in Chapter 10. The voltage along the bridge resistor is

$$V_{45} = \frac{R_{sh} L I_{26}}{W} \tag{1.31}$$

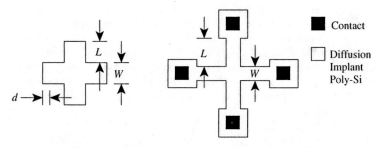

Fig. 1.12   A Greek cross sheet resistance test structure. d = distance of contact from edge.

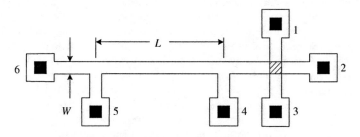

Fig. 1.13   A cross bridge sheet resistance and line width test structure.

where $V_{45} = V_4 - V_5$ and I_{26} is the current flowing from contact 2 to contact 6. From Eq. (1.31) the line width is

$$W = \frac{R_{sh} L I_{26}}{V_{45}} \qquad (1.32)$$

with R_{sh} determined from the cross structure and Eq. (1.30). A key assumption in this measurement is that the sheet resistance be identical for the entire test structure.

Since the bridge structure in Fig. 1.13 is suitable for resistance measurements, it can be used to characterize "dishing" during chemical-mechanical polishing of semiconductor wafers, where soft metal lines tend polish thinner in the central portion than at the edges leading to non-uniform thickness. This is particularly important for soft metals such as copper. With the resistance inversely proportional to metal thickness, resistance measurements can be used to determine the amount of dishing.[30]

1.2.3 Measurement Circuits

Four-point probe measurement circuits are given in various ASTM Standards. For example, ASTM F84[18] and F76[31] give detailed circuit diagrams. Today's equipment is supplied with computers to provide the current stimulus, measure the voltage and apply appropriate correction factors as well as provide the signals for the probe station stepping for wafer mapping.

1.2.4 Measurement Errors and Precautions

For four-point probe measurements to be successful a number of precautions must be taken and appropriate correction factors must be applied to the measured data.

Sample Size: As mentioned earlier, a number of corrections must be applied, depending on the location of the probe as well as sample thickness and size. For those cases where the wafer is uniformly doped in the lateral direction and its diameter is appreciably larger than the probe spacing, the wafer thickness is the chief correction. If the wafer or the layer to be measured is appreciably thinner than the probe spacing, the calculated resistivity varies directly with thickness. It is therefore very important to determine the thickness accurately for resistivity determination. For sheet resistance measurements the thickness need not be known.

Minority/Majority Carrier Injection: It is often stated that metal-semiconductor contacts do not inject minority carriers. That is not strictly true. Metal-semiconductor contacts do inject minority carriers, but their injection efficiency is low. However, under high current conditions it may not be negligible. Minority carrier injection causes *conductivity modulation* because increased minority carrier density leads to increased majority carrier density (to maintain charge neutrality) and subsequent enhanced conductivity. To reduce minority carrier injection, the surface should have a high recombination rate for minority carriers. This is best achieved by using lapped surfaces. For a highly polished wafer it may not be possible to achieve the necessary high surface recombination. Injected minority carriers will have decayed by recombination and cause very little error for voltage probes 3–4 minority carrier diffusion lengths from the injecting current probe. However, for high lifetime material the diffusion length may be longer than the probe spacing, and the measured resistivity will be in error. Another possible source of error is the *probe pressure-induced band gap narrowing* leading to enhanced minority carrier injection.

Minority carrier injection may be important for high resistivity materials. For silicon this applies for $\rho \geq 100$ ohm $\cdot$ cm. An error of less than 2% is introduced by minority carrier injection if the voltage across the two voltage-sensing probes is held to less than 100 mV for 1 mm probe spacings for samples with lapped surfaces. If the current density exceeds the value $J = qnv$, where $n \approx N_D$ for n-type samples and v is the thermal velocity, excess *majority* carriers can be injected into the sample, causing the resistivity to change. Majority carrier injection is usually of little concern if the four-point probe voltage does not exceed 10 mV.

Probe Spacing: A mechanical four-point probe exhibits small random probe spacing variations. Such variations give erroneous values of resistivity or sheet resistance, especially when evaluating uniformly doped wafers. In such cases it is very important to know whether any non-uniformities are due to the wafer, due to process variations, or due to measurement errors. An example is the evaluation of ion-implanted layers. It is known that ion-implanted layers can have sheet resistance uniformities better than 1%. For small probe spacing variations the correction factor[18]

$$F_S \approx 1 + 1.082(1 - s_2/s_m) \tag{1.33}$$

must be applied, where s_2 is the spacing between the inner two probes and s_m is the mean value of the probe spacings. Errors due to probe wander can be reduced by averaging several independent readings.

Current: Additional sources of error are the current amplitude and surface leakage current. The current can affect the measured resistivity in two ways: by an apparent resistivity increase produced by wafer heating and by an apparent resistivity decrease due to minority and/or majority carrier injection. The suggested four-point probe measurement current for silicon wafers is shown in Fig. 1.14 as a function of resistivity and sheet resistance.[18] The data were obtained by measuring the four-point probe resistivity as a function of current for a given sample. Such resistivity-current curves show typically a flat region bounded by non-linearities at both low and high currents. The flat region gives the appropriate current. Surface leakage is reduced or eliminated by enclosing the probe in a shielded enclosure held at a potential equal to the inner probe potential.

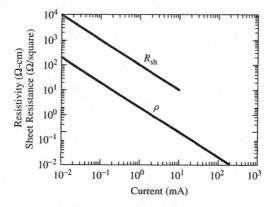

Fig. 1.14   Recommended four-point probe current versus Si sheet resistance and resistivity.

Temperature: It is important that the sample temperature be uniform in order not to introduce thermoelectric voltages. Temperature gradients can be caused by ambient effects but are more likely due to sample heating by the probe current. Current heating is most likely to occur in low resistivity samples where high currents are required to obtain readily measurable voltages.

Even if temperature variations are not caused by the measurement apparatus and there are no temperature gradients, there may still be temperature variations due to temperature fluctuations in the measurement room. Since semiconductors have relatively large temperature coefficients of resistivity, errors are easily introduced by failing to compensate for such temperature variations (*n*- and *p*-Si[18] and for *n*- and *p*-Ge).[32] For resistivities of 10 ohm · cm or higher, the Si coefficient is on the order of 1%/°C. Temperature corrections are made by using the correction factor[18]

$$F_T = 1 - C_T(T - 23) \tag{1.34}$$

where C_T is the temperature coefficient of resistivity and T is the temperature in °C.

Surface Preparation: Proper surface preparation is important for high sheet resistance Si measurements. For example, positive charge on the surface of a *p*-type layer on an *n*-type wafer, leads to a surface charge-induced space-charge region leaving only a portion of the layer in its neutral state. This, of course, increases the thickness-dependent sheet resistance. Similarly, a positive surface charge on an *n*-type implanted layer, leads to surface accumulation and a sheet resistance reduction. An example of this effect is shown in Fig. 1.15. Wafers dipped into boiling water or into H_2SO_4 or H_2O_2 exhibit stabilized surfaces while those etched in HF exhibit a time-dependent sheet resistance.[33]

High Resistivity, High Sheet Resistance Materials: Materials of very high resistivity are more difficult to measure by four-point probe or van der Pauw methods. Moderately doped wafers can become highly resistive at low temperatures and are similarly difficult

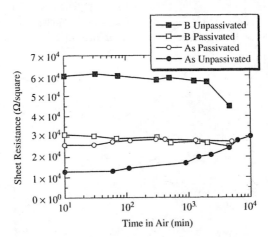

Fig. 1.15   Sheet resistance versus time in room temperature air. *B* implant: 8×10^{11} cm^{-2}, 70 keV through 59 nm oxide into *n*-Si substrate, annealed 1050°C, 15 s; As implant: 8×10^{11} cm^{-2} into bare *p*-Si substrate, annealed 1000°C, 30 min. Both passivated in boiling water for 10 min. After ref. 33.

to measure. Special measurement precautions must be observed. Thin semiconductor films usually have high sheet resistance. These include lightly doped layers, polycrystalline Si films, amorphous Si films, silicon-on-insulator, *etc*. It is possible to make four-point probe measurements with sheet resistances up to about $10^{10}-10^{11}$ ohms/square, provided one uses a stable low current as low as picoamperes. A further consideration is penetration of the probes through shallow implanted layers. One solution to this problem is to use mercury four-point probes instead of metal "needles".

A measurement for high-resistivity bulk wafers relies on providing the wafer with a large contact on one side and a small contact on the other side. A current is passed through the contacts and the voltage is measured. This arrangement, by itself, can suffer from surface leakage currents. By surrounding the small contact with a guard ring and holding the guard ring at the same or nearly the same potential as the small contact, surface currents are essentially suppressed.[34] It is of course necessary to ensure that the contacts are ohmic or as close to ohmic as possible so that the bulk resistivity and not the contact resistance is measured.

Two-terminal measurements are notorious for being complicated by contact effects and the true sample resistivity is not easy to determine as indicated by Eq. (1.2). Conventional van der Pauw measurements suitable for moderate or low resistivity materials are suspect for high resistance samples unless care is taken to eliminate current leakage paths and sample loading by the voltmeter. One approach around this problem is the "guarded" approach using high input impedance, unity gain amplifiers between each probe on the sample, and the external circuitry.[35] The unity gain amplifiers drive the shields on the leads between the amplifier and the sample, thereby effectively eliminating the stray capacitance in the leads. This reduces leakage currents and the system time constant. Measurements of resistances up to 10^{12} ohms have been made with such a system. The "guarded" approach can also be automated.[36]

1.3   WAFER MAPPING

Wafer mapping, originally developed to characterize ion implantation uniformity, has become a powerful process monitoring tool. Manual wafer mapping originated in the 1970s.[37] Today, highly automated systems are used. During wafer mapping the sheet resistance or some other parameter proportional to ion implant dose is measured at many locations across a sample. The data are then converted to two-dimensional or three-dimensional contour maps. Contour maps are a more powerful display of process uniformity than displaying the same data in tabular form. A well-designed contour map gives instant information about ion implant uniformity, flow patterns during diffusion, epitaxial reactor non-uniformities, *etc*. If desired, line scans along one line across the sample can also be displayed to show the uniformity along that line.

The most common sheet resistance wafer mapping techniques are: four-point probe sheet resistance, modulated photoreflectance, and optical densitometry.[38] Of these, the configuration-switched four-point probe method is commonly used. It allows for rapid comparison between samples and has been used for ion implantation, diffusion, poly-Si films, and metal uniformity characterization.[39] Example wafer maps are shown in Fig. 1.16.

1.3.1   Double Implant

Precaution needs to be taken to measure the sheet resistance of low-dose, single implanted layers by the four-point probe technique, because (1) it is difficult to make good electrical

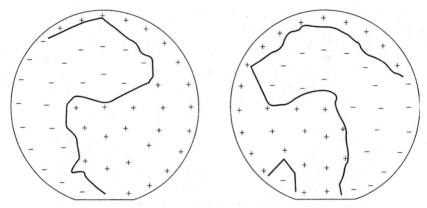

Fig. 1.16 Four-point probe contour maps; (a) boron, 10^{15} cm^{-2}, 40 keV, R_{sh}(average) = 98.5 ohms/square; (a) arsenic, 10^{15} cm^{-2}, 80 keV, R_{sh}(average) = 98.7 ohms/square; 1% intervals. 200 mm diameter Si wafers. Data courtesy of Marylou Meloni, Varian Ion Implant Systems.

contact from the probe to the semiconductor, (2) low doses give low carrier densities and low conductivity, and (3) the surface leakage current can be comparable to the measurement current. The conventional four-point probe method can be used provided the starting wafers are of high resistivity, and they are oxidized before the implant to stabilize the surface resistance and to prevent ion channeling. The wafer is implanted and annealed, the oxide is stripped, and the surface is stabilized in a hot sulfuric acid and hydrogen peroxide solution (piranha etch).

A modified four-point probe method, the *double implant technique*, is sometimes used for sheet resistance measurements of such layers.[20, 40] It is implemented as follows: A p-type (n-type) impurity is implanted into an n-type (p-type) substrate at a dose Φ_1 and energy E_1. For example, boron is implanted at a dose of $\Phi_1 = 10^{14}$ cm^{-2} and energy $E_1 = 120$ keV. The wafer is annealed to activate the implanted ions electrically. The sheet resistance R_{sh1} is measured and the data are stored. Next the desired low-dose impurity is implanted at dose Φ_2 and energy E_2, with $\Phi_2 < \Phi_1$. E_2 should be less than E_1 to prevent penetration through the first implant layer. The first implant energy is typically at least 10–20% higher, and the first implant dose is at least two orders of magnitude higher than the second implant. The second implant conditions might be $\Phi_2 = 10^{11}$ cm^{-2} and $E_2 = 100$ keV. The sheet resistance R_{sh2} after the second implant is measured and compared to R_{sh1} *without* annealing the second implant.

The second sheet resistance measurement relies on the implant damage of the second implant being proportional to the implant dose. This is true for low implant doses. Implanted, but not activated ions, do not contribute to electrical conduction. Furthermore, due to implant damage, the mobility is reduced making $R_{sh2} > R_{sh1}$. The impurity atomic mass of the first implant should be approximately the same mass as the second implant. It has also been found that (111)-oriented Si wafers are preferred over (100)-oriented wafers to reduce channeling effects. The double-implant method allows measurements immediately after the second implant. Implant doses as low as 10^{10} cm^{-2} can be measured by this technique. Test wafers can be annealed and reused, provided the anneal temperature is kept sufficiently low to prevent impurity redistribution. The method is also applicable for electrically inactive species, such as oxygen, argon, or nitrogen implants. A more detailed discussion is given in Smith et al.[40]

The double-implant technique suffers from several problems. Any sheet resistance non-uniformities resulting from the first implant and its activation cycle alter the low-dose measurement. Additionally, since this method derives its low-dose sensitivity from ion-implant *damage*, it is sensitive to post-implant relaxation, where implant damage anneals itself over a period of hours to days following the implant. If the measurement is made immediately after the second implant, damage relaxation has little effect. However, if the measurement is made several hours or days after the implant, damage relaxation can reduce the measured resistance by 10–20% for the types of implant doses and energies typical for low-dose implants. The measurement stability is improved by a 200°C, dry N_2 anneal for 45 min before making the measurement.[40]

1.3.2 Modulated Photoreflectance

Modulated photoreflectance is the modulation of the optical reflectance of a sample in response to waves generated when a semiconductor sample is subjected to periodic heat stimuli. In the *modulated photoreflectance* or *thermal wave* method an Ar^+ ion laser beam, incident on the semiconductor sample, is modulated at a frequency of 0.1 to 10 MHz, creating transient thermal waves near the surface that propagate at different speeds in damaged and crystalline regions. Hence, signals from regions with various damages differ, leading to a measure of crystal damage. The thermal wave diffusion length at a 1 MHz modulation frequency is 2 to 3 μm.[41] The small temperature variations cause small volume changes of the wafer near the surface and the surface expands slightly.[42] These changes include both thermoelastic and optical effects,[43] and they are detected with a second laser—the probe beam—by measuring the reflectivity change. The apparatus is illustrated in Fig. 1.17. Both pump and probe laser beams are focused to approximately 1 μm diameter spots, allowing measurements not only on uniformly implanted wafers but also on patterned wafers.

Modulated photoreflectance is commonly used to determine the implant dose of ion implanted wafers. Conversion from thermal wave signal to implant dose requires calibrated standards with known implant doses. The ability to determine ion-implant doses by thermal waves depends on the conversion of the single crystal substrate to a partially

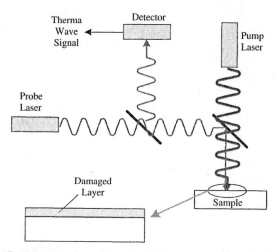

Fig. 1.17 Schematic diagram of the modulated photoreflectance apparatus.

disordered layer by the implant process. The thermal wave-induced thermoelastic and optical effects are changed in proportion to the number of implanted ions. Modulated photoreflectance implant monitoring is subject to post-implant damage relaxation. However, the laser detection scheme accelerates the damage relaxation process, and the sample stabilizes within a few minutes.

The technique is contactless and non-destructive and has been used to measure implant doses from 10^{11} to 10^{15} cm^{-2}.[44] Measurements can be made on bare and on oxidized wafers. The ability to characterize oxidized samples has the advantage of allowing measurements of implants through an oxide. The technique can discriminate between implant species since the lattice damage increases with implant atom size and the thermal wave signal depends on the lattice damage. It has been used for ion implantation monitoring, wafer polish damage, and reactive and plasma etch damage studies. Its chief strength lies in the ability to detect low-dose implants contactless and to display the information as contour maps. Example contour maps are shown in Fig. 1.18.

1.3.3   Carrier Illumination (CI)

Somewhat similar to modulated photoreflectance is *carrier illumination*, to determine junction depth. Optical characterization of activated shallow junctions requires high contrast between the active implant and the underlying layer. The index of refraction of the doped layer is slightly higher than the underlying silicon by virtue of its higher conductivity. However, this is insufficient to enable measurement using conventional methods. In carrier illumination, a focused laser ($\lambda = 830$ nm) injects excess carriers into the semiconductor, forming a dc excess carrier distribution and a $\lambda = 980$ nm probe beam measures the reflectance.[45] The carrier distribution is deduced from the reflected signal. The carrier density in the substrate is flat, and falls rapidly at the junction edge. This creates a steep gradient in the index of refraction at the edge of the doping profile. The index of refraction change Δn relates to the excess carrier density ΔN as

$$\Delta n = \frac{q^2 \Delta N}{2K_s \varepsilon_o m^* \omega^2} \tag{1.35}$$

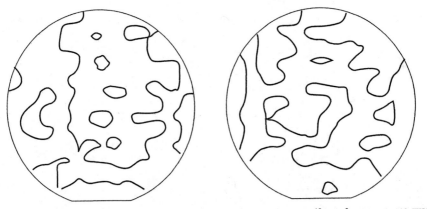

Fig. 1.18   Modulated photoreflectance contour maps; (a) boron, 6.5×10^{12} cm^{-2}, 70 keV, 648 TW units; (a) boron, 5×10^{12} cm^{-2}, 30 keV, 600 TW units; 0.5% intervals. 200 mm diameter Si wafers. Data courtesy of Marylou Meloni, Varian Ion Implant Systems.

where ω is the radial frequency of the light. Light is reflected from this distribution and interference with a reference leads to an interference signal correlating directly to the junction depth. By slowly modulating the laser generating the excess carriers, thereby maintaining the static distribution conditions, it is possible to use sensitive phase-locked methods to obtain a reflection signal with several orders of magnitude gain over a dc measurement.

The method works best for layers with active doping densities in excess of 10^{19} cm^{-3} to avoid high-level injection conditions in the active implanted region. High depth resolution is achieved because of the high index of refraction of the semiconductor. The measurement wavelength in silicon is about 270 nm, and a full 2π phase shift occurs in 135 nm. With a noise-limited phase resolution better than 0.5°, the depth resolution is about 0.2 nm. In addition to junction depth measurements, CI has been shown to be sensitive to the active dopant density and the profile abruptness and can also measure the thickness of the amorphous depth after a pre-amorphizing implant, making the CI method very sensitive for monitoring as-implanted low-dose ion implants.[46]

1.3.4   Optical Densitometry

In *optical densitometry* the doping density is determined by a technique entirely different from any of the methods discussed in this chapter. The method was developed for ion implantation uniformity and dose monitoring and does not use semiconductor wafers. A transparent substrate, typically glass, is coated with a thin film consisting of a polymer carrier and an implant sensitive radiochromic dye. During implant, the dye molecule undergoes heterolytic cleavage, resulting in positive ions with a peak light absorption at a wavelength of 600 nm.[47] When this polymer-coated glass wafer is ion implanted, the film darkens. The amount of darkening depends on the implant energy, dose, and species.

The optical densitometer, using a sensitive microdensitometer, detects the transparency of the entire wafer before and after implant and compares the final-to-initial difference in optical transparency with internal calibration tables. The optical transparency is measured over the entire implanted wafer and then displayed as a contour map. Calibration curves of optical density as a function of implant dose have been developed for implant doses from 10^{11} to 10^{13} cm^{-2}.

The method requires no implantation activation anneal and the results can be displayed within a few minutes of the implantation. The optical density is measured with about 1 mm resolution and lends itself well to ion doses as low 10^{11} cm^{-2}. As discussed earlier in this chapter, the doping density of low-dose implants is not easy to measure electrically, and this optical method is a viable alternate technique. It is also very stable. Table 1.1 compares three mapping techniques.[38]

1.4   RESISTIVITY PROFILING

A four-point probe measures the sheet resistance. The resistivity is obtained by multiplying by the sample thickness with the correct resistivity obtained only for *uniformly*-doped substrates. For non-uniformly doped samples, the sheet resistance measurement averages the resistivity over the sample thickness according to Eq. (1.19). The resistivity *profile* of a non-uniformly doped layer cannot be determined from a single sheet resistance measurement. Furthermore it is usually the dopant density profile that is desired, not the resistivity profile.

TABLE 1.1   Mapping Techniques for Ion Implantation Uniformity Measurements.

	Four-Point Probe	Double Implant	Spreading Resistance	Modulated Photoreflectance	Optical Densitometry
Type	Electrical	Electrical	Electrical	Optical	Optical
Measurement	Sheet Resistance	Crystal Damage	Spreading Resistance	Crystal Damage	Polymer Damage
Resolution (μm)	3000	3000	5	1	3000
Species	Active	Active, Inactive	Active	Inactive	Inactive
Dose Range (cm^{-2})	$10^{12}-10^{15}$	$10^{11}-10^{14}$	$10^{11}-10^{15}$	$10^{11}-10^{15}$	$10^{11}-10^{13}$
Results	Direct	Calibration	Calibration	Calibration	Calibration
Relaxation	Minor	Serious	Minor	Serious	Serious
Requires	Anneal	Initial Implant	Anneal		Measure before and after

Suitable techniques for determining dopant density profiles include the differential Hall effect, spreading resistance, capacitance-voltage, MOSFET threshold voltage, and secondary ion mass spectrometry. We will discuss the first two methods in this chapter and defer discussion of the others to Chapter 2.

1.4.1   Differential Hall Effect (DHE)

To determine a resistivity or dopant density depth profile, depth information must be provided. It is possible to measure the resistivity profile of a non-uniformly doped sample by measuring the resistivity, removing a thin layer of the sample, measuring the resistivity, removing, measuring, *etc.* The *differential Hall effect* is such a measurement procedure. The sheet resistance of a layer of thickness $(t - x)$ is given by

$$R_{sh} = \frac{1}{q \int_x^t [n(x)\mu_n(x) + p(x)\mu_p(x)] \, dx} \qquad (1.36)$$

where x is the coordinate from the surface into the sample as illustrated in Fig. 1.19. If the sample is a thin layer, it must be separated from the substrate by an insulating layer to confine the four-point probe current to the layer. For example, an n-type implant into a p-substrate is suitable, with the space-charge region of the resulting np junction acting as an "insulating" boundary. An n-type implant into an n-substrate is not suitable as the measuring current is no longer confined to the n-layer.

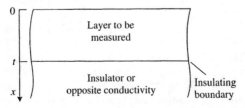

Fig. 1.19   Sample geometry with measurement proceeding from the surface into the sample.

The sheet resistance of a uniformly doped layer with constant carrier densities and mobilities is

$$R_{sh} = \frac{1}{q(n\mu_n + p\mu_p)t} \tag{1.37}$$

The sheet resistance is a meaningful descriptor not only for uniformly doped layers but also for non-uniformly doped layers, where both carrier densities and mobilities are depth dependent. In Eq. (1.36) R_{sh} represents an averaged value over the sample thickness $(t - x)$. Obviously, for $x = 0$, the sheet resistance is given by Eq. (1.19).

The sheet resistance is measured by the Hall effect or with a four-point probe as a function of depth by incremental layer removal. A plot of $1/R_{sh}(x)$ versus x leads to the sample conductivity $\sigma(x)$ according to the equation[48]

$$\frac{d[1/R_{sh}(x)]}{dx} = -q[n(x)\mu_n(x) + p(x)\mu_p(x)] = -\sigma(x) \tag{1.38}$$

Equation (1.38) is derived from (1.36) using Leibniz's theorem

$$\frac{d}{dc}\int_{a(c)}^{b(c)} f(x,c)\,dx = \int_{a(c)}^{b(c)} \frac{\partial}{\partial c}[f(x,c)]\,dx + f(b,c)\frac{\partial b}{\partial c} - f(a,c)\frac{\partial a}{\partial c} \tag{1.39}$$

The resistivity is determined from Eq. (1.38) and from the identity $\rho(x) = 1/\sigma(x)$ as

$$\rho(x) = -\frac{1}{d[1/R_{sh}(x)]/dx} = \frac{R_{sh}^2(x)}{dR_{sh}(x)/dx} = \frac{R_{sh}(x)}{d[\ln(R_{sh}(x))]/dx} \tag{1.40}$$

The dopant density determined by this method is illustrated in Exercise 1.5. Dopant density profiles determined by DHE, spreading resistance profiling, and secondary ion mass spectrometry are shown in Fig. 1.20.

Exercise 1.5

Problem: Given the sheet resistance versus depth plot of an *n*-Si layer on a *p*-Si substrate in Fig. E1.7(a), determine the resistivity and the doping density as a function of depth.

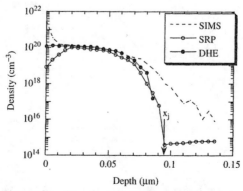

Fig. 1.20 Dopant density profiles determined by DHE, spreading resistance profiling, and secondary ion mass spectrometry. Data after ref. 49. Reprinted from the Jan. 1993 edition of *Solid State Technology*.

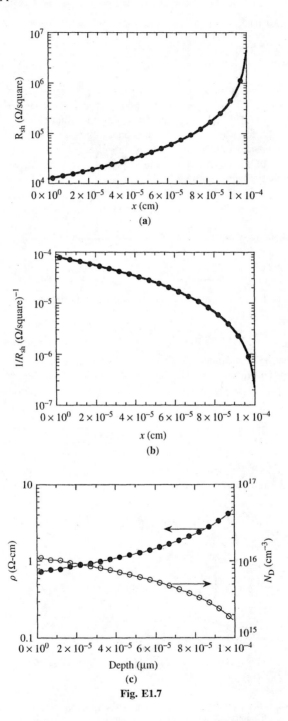

Fig. E1.7

Solution: Determine the slope of this plot as a function of x. Then determine $\rho(x)$ versus x using Eq. (1.40). Remember, in problems where the data are given in terms of "log" as in the figure above, you need to use the conversion "$\ln(10)\ln(x) = \log(x)$". The resistivity and doping density data so derived are shown in Figs. E1.7(b) and (c). Conversion of "ρ to N_D" used a mobility of 800 cm^2/V·s.

A word of caution regarding sheet resistance measurements of thin layers is in order here. Surface charges can induce space-charge regions at the sample surface. If that happens, then the neutral layer that governs the sheet resistance is thinner than the physical layer, introducing an error into the measurement. It is generally not a problem for Si, but can be a problem for GaAs, where surface charge-induced space-charge regions are very common. Corrections need to be applied then.[50-51]

Repeated removal of well-controlled thin layers from a heavily-doped semiconductor is difficult to do by chemical etching. It can, however, be done with *anodic oxidation*. During anodic oxidation a semiconductor is immersed in a suitable electrolyte in an anodization cell. A current is passed from an electrode to the semiconductor sample through the electrolyte, causing an oxide to grow at room temperature. The oxide grows by consuming a portion of the semiconductor. By subsequently etching the oxide, that portion of the semiconductor consumed during the oxidation is removed as well. This can be done very reproducibly.

Two anodization methods are possible. In the *constant voltage* method, the anodization current is allowed to fall from an initial to a final predetermined value. In the *constant current* method, the voltage is allowed to rise until a preset value is attained. The oxide thickness is directly proportional to the net forming voltage in the constant current anodization method, where the net forming voltage is the final cell voltage minus the initial cell voltage.

A variety of anodization solutions have been used. The non-aqueous solutions *N*-methylacetamide, tetrahydrofurfuryl alcohol and ethylene glycol are suitable for silicon.[52] Ethylene glycol containing 0.04N KNO$_3$ and 1-5% water produces uniform, reproducible oxides at current densities of 2 to 10 mA/cm^2. For the ethylene glycol mixture 2.2 Å of Si are removed per volt.[52] A forming voltage of 100 V removes 220 Å of Si. Ge[53], InSb[54], and GaAs[55] have all been anodically oxidized.

The laborious nature of the differential conductivity profiling technique limits its applicability if the entire process is done manually. The measurement time can be substantially reduced by automating the method. Computer-controlled experimental methods have been developed in which the sample is anodized, etched and then the resistivity and the mobility are measured *in situ*.[49, 56-57]

1.4.2 Spreading Resistance Profiling (SRP)

The *spreading resistance probe* technique has been in use since the 1960s. Although originally used for lateral resistivity variation determination, it is mainly used today to generate resistivity and dopant density *depth profiles*. It has very high dynamic range ($10^{12}-10^{21}$ cm^{-3}) and is capable of profiling very shallow junctions into the nm regime. Substantial progress has been made in data collection and treatment. The latter relates to improved sample preparation and probe conditioning procedures, specialized constrained cubic spline smoothing schemes, universally applicable Schumann-Gardner-based correction factors with appropriate radius calibration procedures, and the development of physically based Poisson schemes for the correction of the carrier diffusion (spilling) phenomenon. Reproducibility is sometimes mentioned as an SRP problem. Reproducibility of 10% can be obtained routinely by "qualified" SRP systems, provided qualification procedures are rigorously implemented.[58]

The spreading resistance concept is illustrated in Fig. 1.21. The instrument consists of two carefully aligned probes that are stepped along the beveled semiconductor surface. The resistance between the probes is given by

$$R = 2R_p + 2R_c + 2R_{sp} \qquad (1.41)$$

where R_p is the probe resistance, R_c the contact resistance and R_{sp} the spreading resistance. The resistance is measured at each location.[59]

The sample is prepared by mounting it on a bevel block with melted wax. Bevel angles less than 1° can be readily prepared. The bevel block is inserted into a well-fitting cylinder, and the sample is lapped using a diamond paste or other polishing compound. Sample preparation is very important for successful SRP measurements.[60–61] Next the sample is positioned in the measurement apparatus with the bevel edge perpendicular to the probe stepping direction. It is very useful to provide the sample with an insulating (oxide or nitride) coating. The oxide provides a sharp corner at the bevel and also clearly defines the start of the beveled surface because the spreading resistance of the insulator is very high. Spreading resistance measurements should be made in the dark to avoid photoconductance effects and are primarily used for silicon.

A good discussion of sample preparation is given by Clarysse et al.[58] The bevel angle should be measured with a well-calibrated profilometer. In the absence of a top oxide, the measurement should be started at least 10–20 points before the bevel edge. The actual starting point can then be determined from a micrograph (dark field illumination, magnification 500×). The error on the starting point should not be larger than a few points (maximum 3). Typically, the raw resistance profile shows a transition at the starting point position. The probe imprints must be visible to be able to count them and to determine the starting point. The bevel edge must be sharp enough to reduce the uncertainty of the starting point as much as possible. Good bevel surfaces require a 0.1 or 0.05 μm, high-quality, diamond paste. The rotating glass plate, used for polishing the bevel, should have a peak-to-peak roughness of 0.13 μm. The probe separation must be below 30–40 μm. Typically, 100–150 data points are used for sub micrometer implants or epitaxial layers. For sub-100 nm structures, one should try to obtain 20–25 data points.

To understand spreading resistance, consider a metallic probe contacting a semiconductor surface as in Fig. 1.22. The current I flows from the probe of diameter $2r$ into a

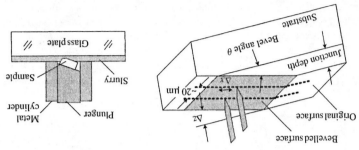

Fig. 1.21   Spreading resistance bevel block and the beveled sample with probes and the probe path shown by the dashed line.

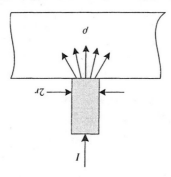

Fig. 1.22 A cylindrical contact of diameter $2r$ to a semiconductor. The arrows represent the current flow.

semiconductor of resistivity ρ. The current is concentrated at the probe tip and spreads out radially from the tip. Hence the name *spreading resistance*. For a non-indenting, cylindrical contact with a planar, circular interface and a highly conductive probe, the spreading resistance for a semi-infinite sample is[62]

$$R_{sp} = \frac{\rho}{4r} \quad \text{ohms} \qquad (1.42a)$$

For a hemispherical, indenting probe tip of radius r, the spreading resistance is

$$R_{sp} = \frac{\rho}{2\pi r} \quad \text{ohms} \qquad (1.42b)$$

Equation (1.42a) has been verified by comparing spreading resistance with four-point probe measurements. The spreading resistance can be expressed as[63]

$$R_{meas} = R_{cont} + R_{spread} = R_{cont} + \frac{\rho}{2r}C \qquad (1.43)$$

where C is a correction factor that depends on sample resistivity, probe radius, current distribution and probe spacing. It should be noted that the radius r is not necessarily the physical radius. The contact resistance also depends on wafer resistivity and probe pressure and on the surface state density. These surface states dominate the Schottky barrier height of the metal/semiconductor contact. The surface state density and energy distribution are expected to be different for polished and beveled surfaces. High surface state densities induce Fermi level pinning.[64] On beveled SRP p-type material the contact is expected to be surrounded by a depleted region while n-type material has an inversion layer near the surface.

A weight of approximately 5 g is applied and the probes have to be conditioned to form an area of small microcontacts, believed to be necessary to break through the thin native oxide on the bevel surface. Despite the relatively low weight very high local pressures result. Assuming a 1 μm radius, a straightforward division by the contact area leads to an estimate of the contact pressure of approximately 16 GPa.

About 80% of the potential drop due to current spreading occurs within a distance of about five times the contact radius. The probe penetration is about 10 nm for probe loads of 10 to 12 g.[65] The relationship between SRP measured resistance and Si resistivity is

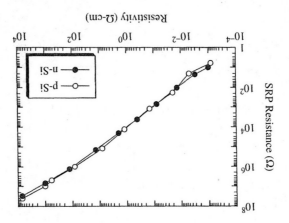

Fig. 1.23 Calibration curves for conventional SRP measurements. After ref. 63.

shown in Fig. 1.23.[63] For a contact radius of 1 μm, Eq. (1.42a) predicts $R_{sp} \approx 2500\rho$.

The fact that the spreading resistance is about 10^4 times higher than ρ is the reason that R_{sp} dominates over R_p and R_c in Eq. (1.41). However, if the metal-semiconductor barrier height is significant, then the measured resistance does include a non-negligible contact resistance, as in GaAs, for example.

The tungsten-osmium alloy probes, are mounted in gravity-loaded probe arms. The probe tips are shaped so that they can be positioned very close together, often with less than 20 μm spacing. The probe arms are supported by a kinematic bearing system with five contacts giving the arms only one degree of freedom, which is a rotation around the horizontal axis. This virtually eliminates lateral probe motion during contact to the sample minimizing probe wear and damage to the semiconductor. The probes deform only slightly elastically upon contacting the semiconductor, thus making very reproducible contacts. The probes are "conditioned" using the "Gorey-Schneider technique"[63] for the contact area of the probe to consist of a large number of microscopic protrusions to penetrate the thin oxide layer on silicon surfaces. An example SRP plot and the resulting dopant density profile is shown in Fig. 1.24.

The conversion of spreading resistance data to a carrier density profile and subsequently to a doping density profile is a complicated task that involves data smoothing to reduce measurement noise, a deconvolution algorithm, and a correct model for the contact.[67] An important aspect of SRP is the fact that spreading resistance measures a carrier distribution along a *beveled* surface. It has often been assumed that this profile is identical to the vertical *carrier* profile. Furthermore, the vertical *carrier* profile is often assumed to be identical to the vertical *doping* profile. This is not true for shallow junctions where the redistribution of mobile carriers, referred to as carrier spilling, distorts the measured SR profiles. For example, electrons from the highly doped n^+ layer in an n^+p junction spill into the lowly p-doped substrate. Hence, an SRP plot, that is expected to show a resistance maximum at the metallurgical junction due to the space-charge region with few carriers, may not show such a maximum at all.[67] The actual plot suggests the absence of a junction leading to the conclusion that the junction may be an n^+n junction. Carrier spilling accounts for SRP determined junction depths being usually less than those measured by SIMS.[68]

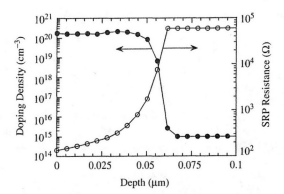

Fig. 1.24 High-resolution spreading resistance and dopant density profiles. Data courtesy of S. Weinzierl, Solid State Measurements, Inc.

The voltage between the probes during measurement is kept at around 5 mV to reduce the effect of contact resistance. The probe-semiconductor contact is a metal-semiconductor contact with the non-linear current–voltage characteristic

$$I = I_0(e^{qV/kT} - 1) \approx I_0 qV/kT \qquad (1.44)$$

for voltages less than $kT/q \approx 25$ mV.

The spreading resistance profiling technique is a comparative technique. Calibration curves are generated for a particular set of probes at a particular time using samples of known resistivity. Such calibration samples are commercially available for silicon. Comparison of the spreading resistance data to the calibration samples is necessary and sufficient for uniformly doped samples. For samples containing pn or high-low junctions, additional corrections are necessary. These multilayer corrections have evolved over the years where today very sophisticated correction schemes are used.[67–72] A different approach calculates the spreading resistance profile from an assumed doping profile.[73] The calculated profile is then compared to the measured profile and adjusted until they agree.

The bevel angle θ is typically $1°$–$5°$ for junction depths of 1–2 μm and $\theta \leq 0.5°$ for junction depths less than 0.5 μm. The equivalent depth, Δz, for each Δx step along the surface beveled at angle θ, is

$$\Delta z = \Delta x \sin(\theta) \qquad (1.45)$$

For a step of 5 μm and an angle of $1°$, the equivalent step height or measurement resolution is 0.87 nm. A plot of dopant density profiles determined by differential Hall effect, spreading resistance profiling, and secondary ion mass spectrometry (SIMS) is shown in Fig. 1.20. Note the good agreement between DHE and SRP for this sample. SIMS profiling is discussed in Chapter 2. The small SRP angles are determined by measuring a small slit of light that is reflected from the beveled and the unbeveled surfaces so that two images are detected. When the slit is rotated, the two images rotate also, and the rotation angle is measured and related to the bevel angle.[74] Surface profilometers can also be used for angle determination.[61]

Limitations in SRP profiling for very shallow junctions arise due to the large sampling volume induced by the large contact and probe spacing necessitating correction factors which can be as large as 2000. Moreover, additional correction factors have been identified to correct for carrier spilling, surface damage, microcontact distribution, and three-dimensional current flow. Unfortunately, all these corrections become increasingly important for very shallow profiles and scale with probe radius and probe separation. Probe penetration and bevel roughness also limit the depth resolution. In order to cope with the limited thickness of the layers, very shallow bevels are required.[75]

Almost all spreading resistance measurements are made with two probes, but three-probe arrangements have been used.[69] In the three-probe configuration one probe serves as the common point to both voltage and current circuits and is the only probe contributing to the measured resistance. The three-probe system is more difficult to keep aligned. Since probe alignment parallel to the bevel intersection with the top surface is crucial for depth profiling, the three-point spreading resistance probe is rarely used. *Micro spreading resistance*, known as *scanning spreading resistance microscopy* is discussed in Chapter 9.

1.5   CONTACTLESS METHODS

Contactless resistivity measurement techniques have become popular in line with the general trend toward other contactless semiconductor measurements. Contactless resistivity measurement methods fall into two broad categories: electrical and non-electrical measurements. Commercial equipment is available for both. Electrical contactless measurement techniques fall into several categories. (1) the sample is placed into a microwave circuit and perturbs the transmission or reflection characteristics of a waveguide or cavity[76], (2) the sample is capacitively coupled to the measuring apparatus[77], and (3) the sample is inductively coupled to the apparatus.[78-79]

1.5.1   Eddy Current

To be a viable commercial instrument, the apparatus should be simple with no special sample requirements. This rules out special sample configurations to fit microwave cavities, for example, and led to a variation of the inductively coupled approach. The eddy current measurement technique is based on the parallel resonant tank circuit of Fig. 1.25. The quality factor Q of such a circuit is reduced when a conducting material is brought close to the coil due to the power absorbed by the conducting material. An implementation of this concept is shown in Fig. 1.25(a), where the LC circuit is replaced by dual coils on ferrite cores separated to provide a gap for the wafer that is coupled to the circuit via the high permeability ferrite cores. The oscillating magnetic field sets up eddy currents in the semiconductor leading to Joule heating of the material.

The absorbed power P_a is[80]

$$P_a = K(V_T/n)^2 \int_0^t \sigma(x)\, dx \qquad (1.46)$$

where K is a constant involving the coupling parameters of the core, V_T the rms primary rf voltage, n the number of primary turns of the coil, σ the semiconductor conductivity, and t the thickness. With power given by $P_a = V_T I_T$, where I_T is the in-phase drive current

$$I_T = \frac{KV_T}{n^2} \int_0^t \sigma(x)\, dx = \frac{KV_T}{n^2}\frac{1}{R_{sh}} \qquad (1.47)$$

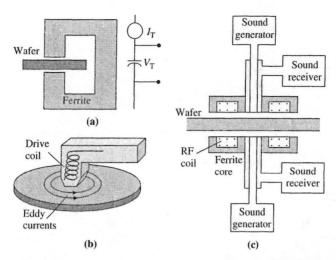

Fig. 1.25 (a) Schematic eddy current experimental arrangement, (b) practical implementation after Johnson[81], and (c) schematic showing the eddy current coils and the thickness sound generator.

If V_T is held constant through a feedback circuit, the current is proportional to the sample conductivity-thickness product, or it is inversely proportional to the sample sheet resistance. A more recent implementation is shown in Fig. 1.25(b).[81] Eddy current and other contactless techniques are discussed further in Chapter 7 in reference to lifetime measurements.

When an alternating current is induced in a conductor, the current is not uniformly distributed, but is displaced toward the surface. For high frequencies most of the current is concentrated in a layer near the surface known as the *skin depth*. Equation (1.46) is valid provided the sample is thinner than the skin depth δ given by

$$\delta = \sqrt{\rho/\pi f \mu_o} = 5.03 \times 10^3 \sqrt{\rho/f} \text{ cm} \tag{1.48}$$

where ρ is the resistivity ($\Omega \cdot$ cm), f the frequency (Hz), and μ_o the permeability of free space ($4\pi \times 10^{-9}$ H/cm). Equation (1.48) is plotted in Fig. 1.26 as a function of frequency. Comparison of four-point probe and eddy current wafer maps are shown in Fig. 1.27 for Al and Ti layers. Note the excellent agreement in the contours and the average sheet resistances.

To determine the wafer resistivity, its thickness must be known. In contactless measurements provision must be made to measure the wafer thickness without contact. Two methods are used: differential capacitance probe and ultrasound.[82] In the ultrasound method sound waves are reflected from the upper and lower wafer surfaces located between the two probes shown in Fig. 1.25(c). The phase shift of the reflected sound caused by the impedance variation of the air gap is detected by the sonic receiver. The phase shift is proportional to the distance from each probe to each surface. With known probe spacing, the wafer thickness can be determined.

One system to determine sample thickness by capacitance measurements is illustrated in Fig. 1.28.[83] Two capacitive probes of area A are separated by a distance s. The semiconductor wafer is held between the two capacitance probes. Each probe forms one plate of the capacitor, the wafer the other. The capacitance is $C_1 = \varepsilon_o A/d_1$ between the upper

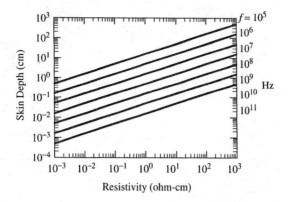

Fig. 1.26 Skin depth versus resistivity as a function of frequency.

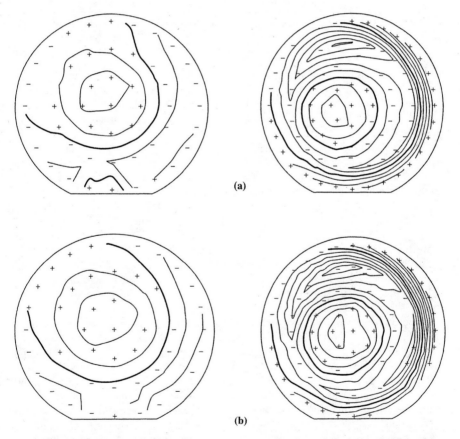

Fig. 1.27 (a) Four-point probe and (b) eddy current contour maps. *Left*: 1 μm aluminum layer, $R_{\text{sh,av}}(4\ \text{pt}) = 3.023 \times 10^{-2}$ ohms/square, $R_{\text{sh,av}}(\text{eddy}) = 3.023 \times 10^{-2}$ ohms/square, *right*: 20 nm titanium layer, $R_{\text{sh,av}}(4\ \text{pt}) = 62.90$ ohms/square, $R_{\text{sh,av}}(\text{eddy}) = 62.56$ ohms/square. Data courtesy of W.H. Johnson, KLA-Tencor.

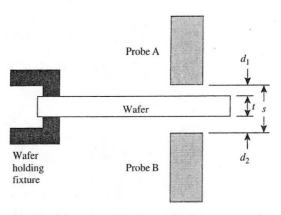

Fig. 1.28   Capacitive wafer thickness and flatness measurement system.

probe and the wafer and $C_2 = \varepsilon_o A/d_2$ between the lower probe and the wafer. From Fig. 1.28, the thickness t is

$$t = s - (d_1 + d_2) = s - \varepsilon_o A(C_1^{-1} + C_2^{-1}) \tag{1.49}$$

To determine t we only need to know the probe separation s and the capacitances C_1 and C_2.

The wafer thickness measurement is independent of the vertical wafer position in the gap. As the wafer moves in the vertical direction, both d_1 and d_2 change by equal and opposite amounts leaving the thickness reading unchanged. The median surface is determined by $d_1 + d_2$. By measuring the capacitance at many points on the wafer, the thickness and shape of the entire wafer can be determined. Bow and warpage, due to stress in the wafer, are determined from the median surface reading allowing the stress to be determined.[84] The flatness obtained by this capacitive technique is a function of only the wafer, not the mechanical support used in the instrument.

Resistivity measurements based on the eddy current technique are useful for uniformly-doped wafers. The technique has also found use for the measurement of highly conductive layers on less conductive substrates. The sheet resistance of the layer should be at least a hundred times lower than the sheet resistance of the substrate to measure the layer and not the substrate. This rules out measurements of diffused or ion-implanted layers on con-ducting substrates, which generally do not satisfy this rule. For example, sheet resistances of diffused or ion-implanted layers are typically 10 to 100 ohms/square, and the sheet resistance of a 10 ohm · cm, 650 µm thick Si wafer is 154 ohms/square. However, the sheet resistance of implanted or epitaxial layers on semi-insulating substrates (*e.g.*, GaAs) or of metal layers on semiconductor substrates can be measured. The sheet resistance of a 5000 Å Al layer is typically 0.06 to 0.1 ohms/square, making such layers 2000 times less resistive than the Si substrate. The layer *thickness* is determined from a sheet resistance measurement according to

$$t = R_{sh}/\rho \tag{1.50}$$

The layer resistivity must be determined from an independent measurement. Contactless resistance measurements are routinely used to determine sheet resistances and thicknesses of conducting layers.

Eddy current measurements require calibrated standards. Radial resistivity variations or other ρ non-uniformities under the transducer are averaged and may be different from that of other ρ or R_{sh} measurement techniques. The measurement frequency should be such that the skin depth is at least five times the sample thickness to be measured.

1.6 CONDUCTIVITY TYPE

The semiconductor *conductivity type* can be determined by wafer flat location, thermal emf, rectification, optically, and Hall effect. The Hall effect is discussed in Chapter 2. The simplest method utilizes the shape of the wafer flats for those wafers following a standard pattern. Silicon wafers are usually circular. They may have characteristic flats, illustrated in Fig. 1.29, provided for alignment and identification purposes. The primary flat (usually along the $\langle 110 \rangle$ direction) and secondary flats identify the conductivity type and orientation. Wafers of diameter ≤ 150 mm usually have the standard flats of Fig. 1.29. Larger wafers usually do not have flats; instead they are provided with notches that do not provide conductivity type information.

In the *hot* or *thermoelectric probe* method the conductivity type is determined by the sign of the thermal emf or Seebeck voltage generated by a temperature gradient. Two probes contact the sample surface: one is hot the other is cold as illustrated in Fig. 1.30(a). Thermal gradients generate currents in a semiconductor; the majority carrier currents for n and p-type materials are[85]

$$J_n = -qn\mu_n \mathcal{P}_n \, dT/dx; \quad J_p = -qp\mu_p \mathcal{P}_p \, dT/dx \tag{1.51}$$

where $\mathcal{P}_n < 0$ and $\mathcal{P}_p > 0$ are the *differential thermoelectric power.*

Consider the experimental arrangement of Fig. 1.30(a). The right probe is hot, the left probe is cold. $dT/dx > 0$ and the electron current in an n-type sample flows from left to right. The thermoelectric power can be thought of as a current generator. Some of the

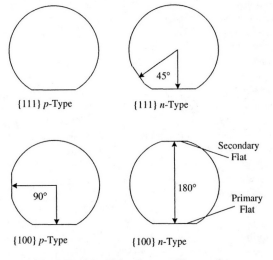

{111} *p*-Type   {111} *n*-Type

{100} *p*-Type   {100} *n*-Type

Fig. 1.29   Identifying flats on silicon wafers.

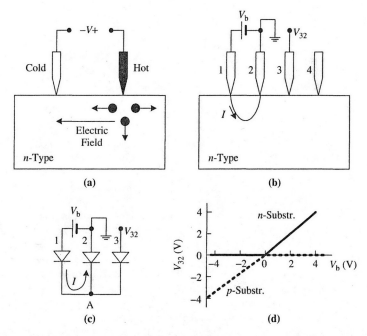

Fig. 1.30 Conductivity type measurements. (a) Hot probe; (b) rectifying probe, (c) equivalent circuit for (b), and (d) experimental data adapted from ref. 88.

current flows through the voltmeter causing the hot probe to develop a positive potential with respect to the cold probe.[86-87] There is a simple alternative view. Electrons diffuse from the hot to the cold region setting up an electric field that opposes the diffusion. The electric field produces a potential detected by the voltmeter with the hot probe positive with respect to the cold probe. Analogous reasoning leads to the opposite potential for p-type samples.

Hot probes are effective over the 10^{-3} to 10^3 ohm-cm resistivity range. The voltmeter tends to indicate n-type for high resistivity material even if the sample is weakly p-type because the method actually determines the $n\mu_n$ or the $p\mu_p$ product. With $\mu_n > \mu_p$ intrinsic or high resistivity material is measured n-type if $n \approx p$. In semiconductors with $n_i > n$ or $n_i > p$ at room temperature (narrow band gap semiconductors, for example), it may be necessary to cool one of the probes and let the room temperature probe be the "hot" probe.

In the *rectification* method, the sign of the conductivity is determined by the polarity of a rectified ac signal at a point contact to the semiconductor.[86-87] When two probes are used, one should be rectifying and the other should be ohmic. Current flows through a rectifying contact to n-type material if the metal is positive and for p-type if it is negative. Rectifying and ohmic contacts are difficult to implement with two-point contacts. Fortunately four-point probes can be used with appropriate connections. A dc voltage is applied and current flows between probes 1 and 2, and the resulting potential is measured between probes 3 and 2 in Fig. 1.30(b). For an n-substrate with positive V_b, the probe 1 metal-semiconductor diode is forward biased and probe 2 diode is reverse biased. Hence the current I is the leakage current of the reverse-biased diode and diode 1 in Fig. 1.30(c) has very low forward bias. The voltage at point A is

$$V_A = V_b + V_{D1} \approx V_b \tag{1.52}$$

The voltage is measured with a high-input impedance voltmeter with very low current between points A and 3. Hence, there is negligible voltage drop across diode 3 and $V_{32} \approx V_A$.

$$V_{32} \approx V_A \approx V_b \tag{1.53}$$

For p-substrates and the same bias arrangement as in Fig. 1.30(c) diode 1 is reverse and diode 2 forward biased. Consequently,

$$V_{32} \approx V_A \approx 0 \tag{1.54}$$

Equations (1.53) and (1.54) show how this probe arrangement can be used for semiconductor *type* determination. The voltage dependence is shown in Fig. 1.30(d). For thin semiconductor films, *e.g.*, silicon-on-insulator or polysilicon films, the metallic needle probes have been replaced with mercury probes.[88] This method of conductivity type measurement is built into some commercial four-point probe instruments.

In the *optical* method, an incident modulated laser beam creates a time-varying surface photovoltage (SPV) in the sample, detected with a non-vibrating, optically transparent Kelvin probe held up to several cm from the sample surface. The principle is the surface photovoltage method discussed in Section 7.4.5. The SPV is negative for p-type and positive for n-type semiconductors.

1.7 STRENGTHS AND WEAKNESSES

Four-Point Probe: The weakness of the four-point probe technique is the surface damage it produces and the metal it deposits on the sample. The damage is not very severe but sufficient not to make measurements on wafers to be used for device fabrication. The probe also samples a relatively large volume of the wafer, preventing high-resolution measurements. The method's strength lies in its established use and the fact that it is an absolute measurement without recourse to calibrated standards. It has been used for many years in the semiconductor industry and is well understood. With the advent of wafer mapping, the four-point probe has become a very powerful process-monitoring tool. This is where its major strength lies today.

Differential Hall Effect: The weakness of this method is its tediousness. The layer removal by anodic oxidation is well controlled, but it is also slow, limiting the method to relatively few data points per profile when done manually. That restriction is lifted when the technique is automated. The sheet resistance can be measured by four-point probe or Hall effect. Repeated four-point probe measurements on the same area create damage, rendering the measurements questionable. That problem does not exist for Hall samples. The method is destructive. The method's strength lies in its inexpensive equipment when using "home assembled" equipment. For those dopant profiles that cannot be profiled by capacitance-voltage measurements, only secondary ion mass spectrometry and spreading resistance methods are the alternatives. Equipment for those measurements is significantly more expensive, leaving anodic oxidation/four-point probe as a viable, inexpensive alternative.

Spreading Resistance: The weakness of the spreading resistance profiling technique is the necessity of a skilled operator to obtain reliable profiles. The system must be periodically calibrated against known standards, and the probes must be periodically reconditioned. It does not work well for semiconductors other than Si and Ge. The sample preparation is not trivial, and the measurement is destructive. The conversion of the measured spreading resistance data to doping density profiles depends very much on the algorithm. Several algorithms are in use, and others are being developed. The strengths of SRP lie in the ability to profile practically any combination of layers with very high resolution and no depth limitation and no doping density limitations. Very high resistivity material must be carefully measured and interpreted. The equipment is commercially available and it is used extensively. Hence there is a large background of knowledge related to this method, which has been in use over the past 40 years.

Contactless Techniques: The weakness of the eddy current technique is its inability to determine the sheet resistance of thin diffused or ion-implanted layers. In order to detect such sheet resistances, it is necessary for the sheet resistance of the layer to be on the order of a hundred times lower than the sheet resistance of the substrate. This is only attainable when the sheet consists of a metal on a semiconductor or a highly doped layer on an insulating substrate. The eddy current technique is often used to measure the sheet resistance of metal layers on semiconductor substrates to determine their thickness. The strength of the eddy current method lies in its non-contacting nature and the availability of commercial equipment. This is ideal for measuring the resistivity of semiconductor wafers and the layer thickness.

Optical Techniques: The weakness of optical techniques is that the measurements are qualitative with quantitative doping measurements requiring calibrated standards. Profiling is generally not possible, and only average values are obtained. The optical densitometry and modulated photoreflectance techniques have become commercially available methods. They are mainly used for wafer mapping of ion-implanted wafers. Their strength lies in their ability to measure the implants non-destructively, with small spot size, and rapidly and in displaying the information in the form of contour plots. The modulated photoreflectance technique is able to measure through an oxide and is routinely used for ion implantation monitoring. Disadvantages are possible laser drift and post-implant damage relaxation. Disadvantages of optical densitometry are the Al backing plate that must be affixed to the wafer rear surface before implantation and removed for optical sensing and the film's UV sensitivity. Without the backing plate the optical sensors in the ion implanter will register a loading error.

APPENDIX 1.1

Resistivity as a Function of Doping Density

Figures A1.1(a) and (b) show the resistivity for boron- and phosphorus-doped Si. For boron-doped Si, the boron density is related to the resistivity by[89]

$$N_B = \frac{1.33 \times 10^{16}}{\rho} + \frac{1.082 \times 10^{17}}{\rho[1 + (54.56\rho)^{1.105}]} \ [\text{cm}^{-3}]$$

$$\rho = \frac{1.305 \times 10^{16}}{N_B} + \frac{1.133 \times 10^{17}}{N_B[1 + (2.58 \times 10^{-19}N_B)^{-0.737}]} \ [\Omega\text{-cm}] \qquad (A1.1)$$

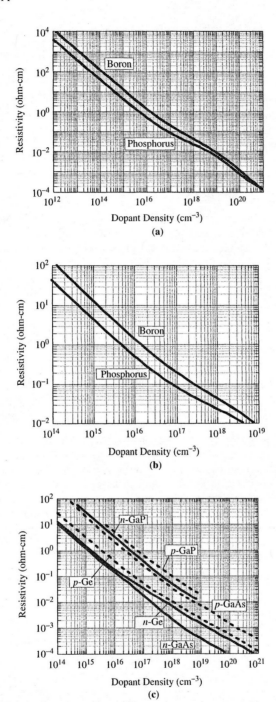

Fig. A1.1 (a) and (b) Doping density versus resistivity for *p*-type (boron-doped) and *n*-type (phosphorus-doped) silicon at 23°C. Data from ASTM F723; (c) for Ge, GaAs, and GaP. Data from Ref. 95 and 96.

For phosphorus-doped Si, the phosphorus density is related to the resistivity by[89]

$$N_P = \frac{6.242 \times 10^{18} 10^Z}{\rho} \ [\text{cm}^{-3}], \quad \text{where } Z = \frac{A_0 + A_1 x + A_2 x^2 + A_3 x^3}{1 + B_1 x + B_2 x^2 + B_3 x^3} \quad \text{(A1.2a)}$$

where $x = \log_{10}(\rho)$, $A_0 = -3.1083$, $A_1 = -3.2626$, $A_2 = -1.2196$, $A_3 = -0.13923$, $B_1 = 1.0265$, $B_2 = 0.38755$, and $B_3 = 0.041833$. The resistivity is

$$\rho = \frac{6.242 \times 10^{18} 10^Z}{N_P} \ [\Omega\text{-cm}], \quad \text{where } Z = \frac{C_0 + C_1 y + C_2 y^2 + C_3 y^3}{1 + D_1 y + D_2 y^2 + D_3 y^3} \quad \text{(A1.2b)}$$

and $y = \log_{10}(N_P) - 16$, $C_0 = -3.0769$, $C_1 = 2.2108$, $C_2 = -0.62272$, $C_3 = 0.057501$, $D_1 = -0.68157$, $D_2 = 0.19833$, and $D_3 = -0.018376$.

Resistivity plots for Ge, GaAs, and GaP are shown in Fig. A1.1(c).

APPENDIX 1.2

Intrinsic Carrier Density

The intrinsic carrier density n_i for Si has been described by a number of equations over the years. The most recent and most accurate expressions are[90-91]

$$n_i = 5.29 \times 10^{19} (T/300)^{2.54} \exp(-6726/T) \qquad \text{(A2.1a)}$$

$$n_i = 2.91 \times 10^{15} T^{1.6} \exp(-E_G(T)/2kT) \qquad \text{(A2.1b)}$$

where the temperature-dependent band gap is given by[92]

$$E_G(T) = 1.17 + 1.059 \times 10^{-5} T - 6.05 \times 10^{-7} T^2 \qquad (0 \le T \le 190 \ K) \quad \text{(A2.2a)}$$

$$E_G(T) = 1.1785 - 9.025 \times 10^{-5} T - 3.05 \times 10^{-7} T^2 \qquad (150 \le T \le 300 \ K) \quad \text{(A2.2b)}$$

T is in Kelvin. n_i and E_G are plotted in Figs. A2.1 and A2.2. Eq. (A2.1a) is based on experiments over the 78–340 K temperature range.[92] Equation (A2.1a) has been rewritten

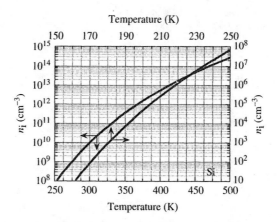

Fig. A2.1   Silicon intrinsic carrier density versus temperature.

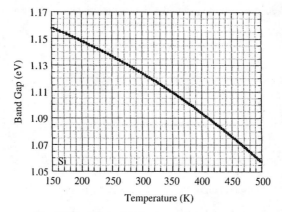

Fig. A2.2 Silicon band gap versus temperature.

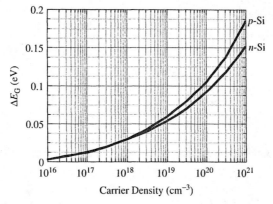

Fig. A2.3 Silicon band gap narrowing versus carrier density.

as Eq. (A2.1b) by Trupke et al.[91] At $T = 300$ K, $n_i = 9.7 \times 10^9$ cm^{-3}. This is slightly lower than the earlier value by Sproul and Green[93] due to band gap narrowing. Band gap narrowing is expressed by

$$n_{i,eff} = n_i \; \exp(\Delta E_G / kT) \tag{A2.3}$$

where the band gap narrowing energy, ΔE_G, is shown in Fig. A2.3.[94]

REFERENCES

1. F. Wenner, "A Method of Measuring Earth Resistivity," *Bulletin of the Bureau of Standards* **12**, 469–478, 1915.

2. L.B. Valdes, "Resistivity Measurements on Germanium for Transistors," *Proc. IRE* **42**, 420–427, Feb. 1954.

3. H.H. Wieder, "Four Terminal Nondestructive Electrical and Galvanomagnetic Measurements," in *Nondestructive Evaluation of Semiconductor Materials and Devices* (J.N. Zemel, ed.), Plenum Press, New York, 1979, 67–104.

4. R. Hall, "Minimizing Errors of Four-Point Probe Measurements on Circular Wafers," *J. Sci. Instrum.* **44**, 53–54, Jan. 1967.

5. D.C. Worledge, "Reduction of Positional Errors in a Four-point Probe Resistance Measurement," *Appl. Phys. Lett.* **84**, 1695–1697, March 2004.

6. A. Uhlir, Jr., "The Potentials of Infinite Systems of Sources and Numerical Solutions of Problems in Semiconductor Engineering," *Bell Syst. Tech. J.* **34**, 105–128, Jan. 1955; F.M. Smits, "Measurement of Sheet Resistivities with the Four-Point Probe," *Bell Syst. Tech. J.* **37**, 711–718, May 1958.

7. M.G. Buehler, "A Hall Four-Point Probe on Thin Plates," *Solid-State Electron.* **10**, 801–812, Aug. 1967.

8. M.G. Buehler, "Measurement of the Resistivity of a Thin Square Sample with a Square Four-Probe Array," *Solid-State Electron.* **20**, 403–406, May 1977.

9. M. Yamashita, "Geometrical Correction Factor for Resistivity of Semiconductors by the Square Four-Point Probe Method," *Japan. J. Appl. Phys.* **25**, 563–567, April 1986.

10. S. Murashima and F. Ishibashi, "Correction Devisors for the Four-Point Probe Resistivity Measurement on Cylindrical Semiconductors II," *Japan. J. Appl. Phys.* **9**, 1340–1346, Nov. 1970.

11. D.S. Perloff, "Four-Point Probe Correction Factors for Use in Measuring Large Diameter Doped Semiconductor Wafers," *J. Electrochem. Soc.* **123**, 1745–1750, Nov. 1976; D.S. Perloff, "Four-Point Probe Sheet Resistance Correction Factors for Thin Rectangular Samples," *Solid-State Electron.* **20**, 681–687, Aug. 1977.

12. M. Yamashita and M. Agu, "Geometrical Correction Factor for Semiconductor Resistivity Measurements by Four-Point Probe Method," *Japan. J. Appl. Phys.* **23**, 1499–1504, Nov. 1984.

13. R.A. Weller, "An Algorithm for Computing Linear Four-point Probe Thickness Correction Factors," *Rev. Sci. Instrument.* **72**, 3580–3586, Sept. 2001.

14. J. Albers and H.L. Berkowitz, "An Alternative Approach to the Calculation of Four-Probe Resistances on Nonuniform Structures," *J. Electrochem. Soc.* **132**, 2453–2456, Oct. 1985.

15. J.J. Kopanski, J. Albers, G.P. Carver, and J.R. Ehrstein, "Verification of the Relation Between Two-Probe and Four-Probe Resistances as Measured on Silicon Wafers," *J. Electrochem. Soc.* **137**, 3935–3941, Dec. 1990.

16. M.P. Albert and J.F. Combs, "Correction Factors for Radial Resistivity Gradient Evaluation of Semiconductor Slices," *IEEE Trans. Electron Dev.* **ED-11**, 148–151, April 1964.

17. R. Rymaszewski, "Relationship Between the Correction Factor of the Four-Point Probe Value and the Selection of Potential and Current Electrodes," *J. Sci. Instrum.* **2**, 170–174, Feb. 1969.

18. ASTM Standard F84-93, "Standard Method for Measuring Resistivity of Silicon Slices With a Collinear Four-Point Probe," *1996 Annual Book of ASTM Standards*, Am. Soc. Test. Mat., West Conshohocken, PA, 1996.

19. D.S. Perloff, J.N. Gan and F.E. Wahl, "Dose Accuracy and Doping Uniformity of Ion Implantation Equipment," *Solid State Technol.* **24**, 112–120, Feb. 1981.

20. A.K. Smith, D.S. Perloff, R. Edwards, R. Kleppinger and M.D. Rigik, "The Use of Four-Point Probe Sheet Resistance Measurements for Characterizing Low Dose Ion Implantation," *Nucl. Instrum. and Meth.* **B6**, 382–388, Jan. 1985.

21. ASTM Standard F1529-94, "Standard Method for Sheet Resistance Uniformity by In-Line Four-Point Probe With the Dual-Configuration Procedure," *1996 Annual Book of ASTM Standards*, Am. Soc. Test. Mat., West Conshohocken, PA, 1996.

22. H.H. Gegenwarth, "Correction Factors for the Four-Point Probe Resistivity Measurement on Cylindrical Semiconductors," *Solid-State Electron.* **11**, 787–789, Aug. 1968.

23. S. Murashima, H. Kanamori and F. Ishibashi, "Correction Devisors for the Four-Point Probe Resistivity Measurement on Cylindrical Semiconductors," *Japan. J. Appl. Phys.* **9**, 58–67, Jan. 1970.

24. L.J. van der Pauw, "A Method of Measuring Specific Resistivity and Hall Effect of Discs of Arbitrary Shape," *Phil. Res. Rep.* **13**, 1–9, Feb. 1958.

25. W. Versnel, "Analysis of Symmetrical van der Pauw Structures With Finite Contacts," *Solid-State Electron.* **21**, 1261–1268, Oct. 1978.

26. L.J. van der Pauw, "A Method of Measuring the Resistivity and Hall Coefficient on Lamellae of Arbitrary Shape," *Phil. Tech. Rev.* **20**, 220–224, Aug. 1958; R. Chwang, B.J. Smith and C.R. Crowell, "Contact Size Effects on the van der Pauw Method for Resistivity and Hall Coefficient Measurement," *Solid-State Electron.* **17**, 1217–1227, Dec. 1974.

27. Y. Sun, J. Shi, and Q. Meng, "Measurement of Sheet Resistance of Cross Microareas Using a Modified van der Pauw Method," *Semic. Sci. Technol.* **11**, 805–811, May 1996.

28. M.G. Buehler and W.R. Thurber, "An Experimental Study of Various Cross Sheet Resistor Test Structures," *J. Electrochem. Soc.* **125**, 645–650, April 1978.

29. M.G. Buehler, S.D. Grant and W.R. Thurber, "Bridge and van der Pauw Sheet Resistors for Characterizing the Line Width of Conducting Layers," *J. Electrochem. Soc.* **125**, 650–654, April 1978.

30. R. Chang, Y. Cao, and C.J. Spanos, "Modeling the Electrical Effects of Metal Dishing Due to CMP for On-Chip Interconnect Optimization," *IEEE Trans. Electron Dev.* **51**, 1577–1583, Oct. 2004.

31. ASTM Standard F76-02, "Standard Test Method for Measuring Resistivity and Hall Coefficient and Determining Hall Mobility in Single Crystal Semiconductors," *1996 Annual Book of ASTM Standards*, Am. Soc. Test. Mat., West Conshohocken, PA, 1996.

32. DIN Standard 50430-1980, "Testing of Semiconducting Inorganic Materials: Measurement of the Specific Electrical Resistivity of Si or Ge Single Crystals in Bars Using the Two-Probe Direct-Current Method," *1995 Annual Book of ASTM Standards*, Am. Soc. Test. Mat., Philadelphia, 1995.

33. J.T.C. Chen, "Monitoring Low Dose Single Implanted Layers With Four-Point Probe Technology," *Nucl. Instrum. and Meth.* **B21**, 526–528, 1987.

34. T. Matsumara, T. Obokata and T. Fukuda, "Two-Dimensional Microscopic Uniformity of Resistivity in Semi-Insulating GaAs," *J. Appl. Phys.* **57**, 1182–1185, Feb. 1985.

35. P.M. Hemenger, "Measurement of High Resistivity Semiconductors Using the van der Pauw Method," *Rev. Sci. Instrum.* **44**, 698–700, June 1973.

36. L. Forbes, J. Tillinghast, B. Hughes and C. Li, "Automated System for the Characterization of High Resistivity Semiconductors by the van der Pauw Method," *Rev. Sci. Instrum.* **52**, 1047–1050, July 1981.

37. P.A. Crossley and W.E. Ham, "Use of Test Structures and Results of Electrical Test for Silicon-On-Sapphire Integrated Circuit Processes," *J. Electron. Mat.* **2**, 465–483, Aug. 1973; D.S. Perloff, F.E. Wahl and J. Conragan, "Four-Point Sheet Resistance Measurements of Semiconductor Doping Uniformity," *J. Electrochem. Soc.* **124**, 582–590, April 1977.

38. C.B. Yarling, W.H. Johnson, W.A. Keenan, and L.A. Larson, "Uniformity Mapping in Ion Implantation," *Solid State Technol.* **34/35**, 57–62, Dec. 1991; 29–32, March 1992.

39. J.N. Gan and D.S. Perloff, "Post-Implant Methods for Characterizing the Doping Uniformity and Dose Accuracy of Ion Implantation Equipment," *Nucl. Instrum. and Meth.* **189**, 265–274, Nov. 1981; M.I. Current, N.L. Turner, T.C. Smith and D. Crane, "Planar Channelling Effects in Si (100)," *Nucl. Instrum. and Meth.* **B6**, 336–348, Jan. 1985.

40. A.K. Smith, W.H. Johnson, W.A. Keenan, M. Rigik and R. Kleppinger, "Sheet Resistance Monitoring of Low Dose Ion Implants Using the Double Implant Technique," *Nucl. Instrum. and Meth.* **B21**, 529–536, March 1987; S.L. Sundaram and A.C. Carlson, "Double Implant Low Dose Technique in Analog IC Fabrication," *IEEE Trans. Semicond. Manuf.* **4**, 146–150, Nov. 1989.

41. A. Rosencwaig, "Thermal-wave Imaging," *Science* **218**, 223–228, Oct. 1982.

42. N.M. Amer and M.A. Olmstead, "A Novel Method for the Study of Optical Properties of Surfaces," *Surf. Sci.* **132**, 68–72, Sept. 1983; N.M. Amer, A. Skumanich, and D. Ripple, "Photothermal Modulation of the Gap Distance in Scanning Tunneling Microscopy," *Appl. Phys. Lett.* **49**, 137–139, July 1986.

43. A. Rosencwaig, J. Opsal, W.L. Smith and D.L. Willenborg, "Detection of Thermal Waves Through Optical Reflectance," *Appl. Phys. Lett.* **46**, 1013–1015, June 1985.

44. W.L. Smith, A. Rosencwaig and D.L. Willenborg, "Ion Implant Monitoring with Thermal Wave Technology," *Appl. Phys. Lett.* **47**, 584–586, Sept. 1985; W.L. Smith, A. Rosencwaig, D.L. Willenborg, J. Opsal and M.W. Taylor, "Ion Impiant Monitoring with Thermal Wave Technology," *Solid State Technol.* **29**, 85–92, Jan. 1986.

45. P. Borden, "Junction Depth Measurement Using Carrier Illumination," in *Characterization and Metrology For ULSI Technology 2000* (D.G. Seiler, A.C. Diebold, T.J. Shaffner, R. McDonald, W.M. Bullis, P.J. Smith, and E.M. Sekula, eds.) *Am. Inst. Phys.* **550**, 175–180, 2001; P. Borden, L. Bechtler, K. Lingel, and R. Nijmeijer, "Carrier Illumination Characterization of Ultra-Shallow Implants," in *Handbook of Silicon Semiconductor Metrology* (A.C. Diebold, ed.), Marcel Dekker, New York, 2001, Ch. 5.

46. W. Vandervorst, T. Clarysse, B. Brijs, R. Loo, Y. Peytier, B.J. Pawlak, E. Budiarto, and P. Borden, "Carrier Illumination as a Tool to Probe Implant Dose and Electrical Activation," in *Characterization and Metrology for ULSI Technology 2003* (D.G. Seiler, A.C. Diebold, T.J. Shaffner, R. McDonald, S. Zollner, R.P. Khosla, and E.M. Sekula, eds.) *Am. Inst. Phys.* **683**, 758–763, 2003.

47. J.P. Esteves and M.J. Rendon, "Optical Densitometry Applications for Ion Implantation," in *Characterization and Metrology for ULSI Technology 1998* (D.G. Seiler, A.C. Diebold, W.M. Bullis, T.J. Shaffner, R. McDonald, and E.J. Walters, eds.) *Am. Inst. Phys.* **449**, 369–373, 1998.

48. R.A. Evans and R.P. Donovan, "Alternative Relationship for Converting Incremental Sheet Resistivity Measurements into Profiles of Impurity Concentration," *Solid-State Electron.* **10**, 155–157, Feb. 1967.

49. S.B. Felch, R. Brennan, S.F. Corcoran, and G. Webster, "A Comparison of Three Techniques for Profiling Ultrashallow p^+n Junctions," *Solid State Technol.* **36**, 45–51, Jan. 1993.

50. R.S. Huang and P.H. Ladbrooke, "The Use of a Four-Point Probe for Profiling Sub-Micron Layers," *Solid-State Electron.* **21**, 1123–1128, Sept. 1978.

51. D.C. Look, "Hall Effect Depletion Correction in Ion-Implanted Samples: Si^{29} in GaAs," *J. Appl. Phys.* **66**, 2420–2424, Sept. 1989.

52. H.D. Barber, H.B. Lo and J.E. Jones, "Repeated Removal of Thin Layers of Silicon by Anodic Oxidation," *J. Electrochem Soc.* **123**, 1404–1409, Sept. 1976, and references therein.

53. S. Zwerdling and S. Sheff, "The Growth of Anodic Oxide Films on Germanium," *J. Electrochem Soc.* **107**, 338–342, April 1960.

54. J.F. Dewald, "The Kinetics and Mechanism of Formation of Anode Films on Single-Crystal InSb," *J. Electrochem Soc.* **104**, 244–251, April 1957.

55. B. Bayraktaroglu and H.L. Hartnagel, "Anodic Oxides on GaAs: I Anodic Native Oxides on GaAs," *Int. J. Electron.* **45**, 337–352, Oct. 1978; "II Anodic Al_2O_3 and Composite Oxides on GaAs," *Int. J. Electron.* **45**, 449–463, Nov. 1978; "III Electrical Properties," *Int. J. Electron.* **45**, 561–571, Dec. 1978; "IV Thin Anodic Oxides on GaAs," *Int. J. Electron.* **46**, 1–11, Jan. 1979; H. Müller, F.H. Eisen and J.W. Mayer, "Anodic Oxidation of GaAs as a Technique to Evaluate Electrical Carrier Concentration Profiles," *J. Electrochem. Soc.* **122**, 651–655, May 1975.

56. R. Galloni and A. Sardo, "Fully Automatic Apparatus for the Determination of Doping Profiles in Si by Electrical Measurements and Anodic Stripping," *Rev. Sci. Instrum.* **54**, 369–373, March 1983.

57. L. Bouro and D. Tsoukalas, "Determination of Doping and Mobility Profiles by Automatic Electrical Measurements and Anodic Stripping," *J. Phys. E: Sci. Instrum.* **20**, 541–544, May 1987.

58. T. Clarysse, W. Vandervorst, E.J.H. Collart, and A.J. Murrell, "Electrical Characterization of Ultrashallow Dopant Profiles," *J. Electrochem. Soc.* **147**, 3569–3574, Sept. 2000.

59. R.G. Mazur and D.H. Dickey, "A Spreading Resistance Technique for Resistivity Measurements in Si," *J. Electrochem Soc.* **113**, 255–259, March 1966; T. Clarysse, D. Vanhaeren, I. Hoflijk, and W. Vandervorst, "Characterization of Electrically Active Dopant Profiles with the Spreading Resistance Probe," *Mat. Sci. Engineer.* **R47**, 123–206, 2004.

60. M. Pawlik, "Spreading Resistance: A Quantitative Tool for Process Control and Development," *J. Vac. Sci. Technol.* **B10**, 388–396, Jan/Feb. 1992.

61. ASTM Standard F672-88, "Standard Method for Measuring Resistivity Profile Perpendicular to the Surface of a Silicon Wafer Using a Spreading Resistance Probe," *1996 Annual Book of ASTM Standards*, Am. Soc. Test. Mat., West Conshohocken, PA, 1996.

62. R. Holm, *Electric Contacts Theory and Application*, Springer Verlag, New York, 1967.

63. T. Clarysse, M. Caymax, P. De Wolf, T. Trenkler, W. Vandervorst, J.S. McMurray, J. Kim, and C.C. Williams, J.G. Clark and G. Neubauer, "Epitaxial Staircase Structure for the Calibration of Electrical Characterization Techniques," *J. Vac. Sci. Technol.* **B16**, 394–400, Jan./Feb. 1998.

64. T. Clarysse, P. De Wolf, H. Bender, and W. Vandervorst, "Recent Insights into the Physical Modeling of the Spreading Resistance Point Contact," *J. Vac. Sci. Technol.* **B14**, 358–368, Jan./Feb. 1996.

65. W.B. Vandervorst and H.E. Maes, "Probe Penetration in Spreading Resistance Measurements," *J. Appl. Phys.* **56**, 1583–1590, Sept. 1984.

66. J.R. Ehrstein, "Two-Probe (Spreading Resistance) Measurements for Evaluation of Semiconductor Materials and Devices," in *Nondestructive Evaluation of Semiconductor Materials and Devices* (J.N. Zemel, ed.), Plenum Press, New York, 1979, 1–66.

67. R.G. Mazur and G.A. Gruber, "Dopant Profiling on Thin Layer Silicon Structures with the Spreading Resistance Technique," *Solid State Technol.* **24**, 64–70, Nov. 1981.

68. W. Vandervorst and T. Clarysse, "Recent Developments in the Interpretation of Spreading Resistance Profiles for VLSI-Technology," *J. Electrochem. Soc.* **137**, 679–683, Feb. 1990; W. Vandervorst and T. Clarysse, "On the Determination of Dopant/Carrier Distributions," *J. Vac. Sci. Technol.* **B10**, 302–315, Jan/Feb. 1992.

69. P.A. Schumann, Jr. and E.E. Gardner, "Application of Multilayer Potential Distribution to Spreading Resistance Correction Factors," *J. Electrochem Soc.* **116**, 87–91, Jan. 1969.

70. S.C. Choo, M.S. Leong, H.L. Hong, L. Li and L.S. Tan, "Spreading Resistance Calculations by the Use of Gauss-Laguerre Quadrature," *Solid-State Electron.* **21**, 769–774, May 1978.

71. H.L. Berkowitz and R.A. Lux, "An Efficient Integration Technique for Use in the Multilayer Analysis of Spreading Resistance Profiles," *J. Electrochem Soc.* **128**, 1137–1141, May 1981.

72. R. Piessens, W.B. Vandervorst and H.E. Maes, "Incorporation of a Resistivity-Dependent Contact Radius in an Accurate Integration Algorithm for Spreading Resistance Calculations," *J. Electrochem Soc.* **130**, 468–474, Feb. 1983.

73. R.G. Mazur, "Poisson-Based Analysis of Spreading Resistance Profiles," *J. Vac. Sci. Technol.* **B10**, 397–407, Jan/Feb. 1992.

74. A.H. Tong, E.F. Gorey and C.P. Schneider, "Apparatus for the Measurement of Small Angles," *Rev. Sci. Instrum.* **43**, 320–323, Feb. 1972.

75. W. Vandervorst, T. Clarysse and P. Eyben, "Spreading Resistance Roadmap Towards and Beyond the 70 nm Technology Node," *J. Vac. Sci. Technol.* **B20**, 451–458, Jan./Feb. 2002.

76. J.A. Naber and D.P. Snowden, "Application of Microwave Reflection Technique to the Measurement of Transient and Quiescent Electrical Conductivity of Silicon," *Rev. Sci. Instrum.* **40**, 1137–1141, Sept. 1969; G.P. Srivastava and A.K. Jain, "Conductivity Measurements of Semiconductors by Microwave Transmission Technique," *Rev. Sci. Instrum.* **42**, 1793–1796, Dec. 1971.

77. C.A. Bryant and J.B. Gunn, "Noncontact Technique for the Local Measurement of Semiconductor Resistivity," *Rev. Sci. Instrum.* **36**, 1614–1617, Nov. 1965; N. Miyamoto and J.I. Nishizawa, "Contactless Measurement of Resistivity of Slices of Semiconductor Materials," *Rev. Sci. Instrum.* **38**, 360–367, March 1967.

78. H.K. Henisch and J. Zucker, "Contactless Method for the Estimation of Resistivity and Lifetime of Semiconductors," *Rev. Sci. Instrum.* **27**, 409–410, June 1956.

79. J.C. Brice and P. Moore, "Contactless Resistivity Meter for Semiconductors," *J. Sci. Instrum.* **38**, 307, July 1961.

80. G.L. Miller, D.A.H. Robinson and J.D. Wiley, "Contactless Measurement of Semiconductor Conductivity by Radio Frequency-Free Carrier Power Absorption," *Rev. Sci. Instrum.* **47**, 799–805, July 1976.

81. W.H. Johnson, "Sheet Resistance Measurements of Interconnect Films," in *Handbook of Silicon Semiconductor Metrology* (A.C. Diebold, ed.), Marcel Dekker, New York, 2001, Ch. 11.

82. P.S. Burggraaf, "Resistivity Measurement Systems," *Semicond. Int.* **3**, 37–44, June 1980.

83. J.L. Kawski and J. Flood, *IEEE/SEMI Adv. Man. Conf.*, 106 (1993); ASTM Standard F1530-94, "Standard Method for Measuring Flatness, Thickness, and Thickness Variation on Silicon Wafers by Automated Noncontact Scanning," *1996 Annual Book of ASTM Standards*, Am. Soc. Test. Mat., West Conshohocken, PA, 1996.

84. ADE Flatness Stations Semiconductor Systems Manual.

85. S.M. Sze, *Physics of Semiconductor Devices*, 2nd ed., Wiley, New York, 1981.

86. W.A. Keenan, C.P. Schneider and C.A. Pillus, "Type-All System for Determining Semiconductor Conductivity Type," *Solid State Technol.* **14**, 51–56, March 1971.

87. ASTM Standard F42-93, "Standard Test Methods for Conductivity Type of Extrinsic Semiconducting Materials," *1996 Annual Book of ASTM Standards*, Am. Soc. Test. Mat., West Conshohocken, PA, 1996.

88. S. Hénaux, F. Mondon, F. Gusella, I. Kling, and G. Reimbold, "Doping Measurements in Thin Silicon-on-Insulator Films," *J. Electrochem. Soc.* **146**, 2737–2743, July 1999.

89. ASTM Standard F723-88, "Standard Practice for Conversion Between Resistivity and Dopant Density for Boron-Doped and Phosphorus-Doped Silicon," *1996 Annual Book of ASTM Standards*, Am. Soc. Test. Mat., West Conshohocken, PA, 1996.

90. K. Misiakos and D. Tsamakis, "Accurate Measurements of the Silicon Intrinsic Carrier Density from 78 to 340 K," *J. Appl. Phys.* **74**, 3293–3297, Sept. 1993.

91. T. Trupke, M.A. Green, P. Würfel, P.P. Altermatt, A. Wang, J. Zhao, and R. Corkish, "Temperature Dependence of the Radiative Recombination Coefficient of Intrinsic Crystalline Silicon," *J. Appl. Phys.* **94**, 4930–4937, Oct. 2003.

92. W. Bludau, A. Onton, and W. Heinke, "Temperature Dependence of the Band Gap of Silicon," *J. Appl. Phys.* **45**, 1846–1848, April 1974.

93. A.B. Sproul and M.A. Green, "Improved Value for the Silicon Intrinsic Carrier Concentration from 275 to 375 K," *J. Appl. Phys.* **70**, 846–854, July 1991.

94. A. Schenk, "Finite-temperature Full Random-phase Approximation Model of Band Gap Narrowing for Silicon Device Simulation," *J. Appl. Phys.* **84**, 3684–3695, Oct. 1998; P.P. Altermatt, A. Schenk, F. Geelhaar, and G. Heiser, "Reassessment of the Intrinsic Carrier Density in Crystalline Silicon in View of Band-gap Narrowing," *J. Appl. Phys.* **93**, 1598–1604, Feb. 2003.

95. D.B. Cuttriss, "Relation Between Surface Concentration and Average Conductivity in Diffused Layers in Germanium," *Bell Syst. Tech. J.* **40**, 509–521, March 1961.

96. S.M. Sze and J.C. Irvin, "Resistivity, Mobility, and Impurity Levels in GaAs, Ge, and Si at 300 K," *Solid-State Electron.* **11**, 599–602, June 1968.

PROBLEMS

1.1 The function $y = x^n$ is plotted in Fig. P1.1. Determine n.

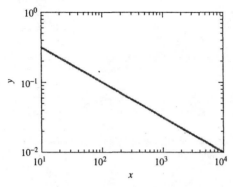

Fig. P1.1

1.2 Determine y_o and x_1 in the equation $y = y_o \exp((x/x_1) - 1)$ plotted in Fig. P1.2.

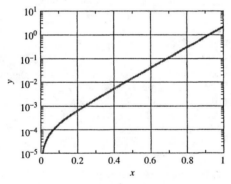

Fig. P1.2

1.3 Plot the $\log(y) - x$ data of Fig. P1.3(a) on the $x - y$ figure in Fig. P1.3(b). Write numeric values on the y axis of Fig. P1.3(b).

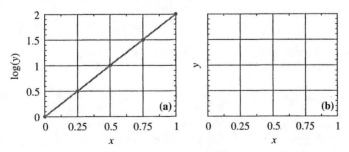

Fig. P1.3

1.4 Derive an expression for the resistivity of a semiconductor sample infinite in extent laterally and vertically measured with a *square* four-point probe with the probes spaced a distance s shown in Fig. P1.4. Current I enters probe 1 and leaves probe 4; voltage V is measured between probes 2 and 3.

Fig. P1.4

1.5 Derive an expression for the resistivity of a semiconductor sample infinite in extent laterally and vertically measured with a four-point probe with the probes spaced as shown in Fig. P1.5. Current I enters probe 1 and leaves probe 4; voltage V is measured between probes 2 and 3.

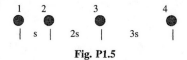

Fig. P1.5

1.6 Consider an n-type wafer containing small n^+ regions. A four-point probe is placed on this wafer so that probe 1 of a conventional *in-line* four-point probe, is placed on one of those n^+ regions. The other three probes are on the n-portion of the wafer. In this four-point probe, current enters probe 1 and leaves probe 4; the voltage is measured across probes 2 and 3. There are no n^+ regions between probes 2 and 4. Is the correct sheet resistance measured in this case?

1.7 The resistance of the semiconductor sample in Fig. P1.7 is measured between the two contacts as a function of wafer thickness t. The results are:

t (μm)	200	400	600	800	1000
R (Ω)	318.3	623.9	929.5	1235.1	1540.7

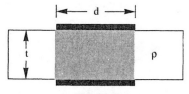

Fig. P1.7

Determine the resistivity ρ in $\Omega \cdot cm$ and the specific contact resistance ρ_c in $\Omega \cdot cm^2$ for $d = 0.01$ cm. Assume the current is confined to the area of the contact, shown by the shaded region. The contact is circular with the contact resistance given by $R_c = \rho_c/A$, where A is the contact area.

1.8 From the $I-V$ curve in Fig. P1.8 determine the conductance $g = dI/dV$ at $I = 10^{-7}$ A.

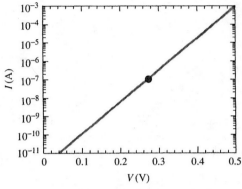

Fig. P1.8

1.9 The resistance R of a semiconductor sample in Fig. P1.9(a) is measured between the two contacts as a function of circular contact of radius $r = d/2$. R is shown as a function of r and $1/r$ in P1.9(b) and (c). Derive an expression for the resistance in terms of the resistivity ρ, radius r and thickness t. Neglect contact resistance and assume the current follows the shaded region. Determine the resistivity ρ (in $\Omega \cdot$ cm) for $t = 400$ μm.

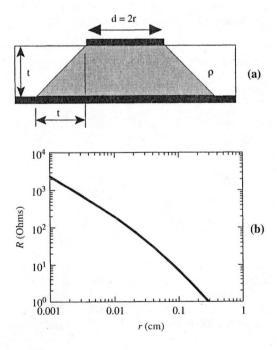

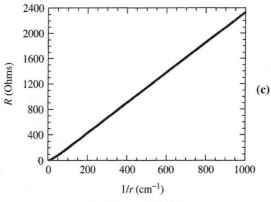

Fig. P1.9   (*continued*)

1.10 The conducting region in Fig. P1.10 of thickness $t = 0.1$ μm and resistivity $\rho = 0.1$ Ω · cm, is deposited on an insulating substrate. $L = 1$ mm, $W = 100$ μm. Determine the resistance between contacts A and B.

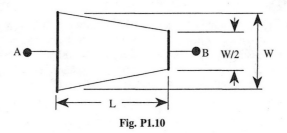

Fig. P1.10

1.11 The semiconductor structure in Fig. P1.11 has thickness t, inside and outside radii r_1 and r_2, and resistivity ρ. Determine the resistance R (in Ω) between the inner ring and the outer ring, *i.e.*, for the doughnut-shaped sample, for $\rho = 15$ Ω · cm, $t = 500$ μm and $r_2/r_1 = 100$. Current flows radially as indicated by the bold arrows. *Hint*: $R = \rho L/A$ becomes $dR = \rho dr/A(r)$.

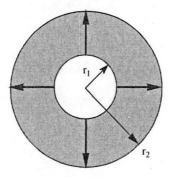

Fig. P1.11

1.12 The sheet resistance is measured in an anodic oxidation experiment. The results are shown in Fig. P1.12. Determine and plot the *resistivity*, ρ (in $\Omega \cdot$ cm), and the *carrier density*, n (in cm^{-3}), versus x for this sample. To determine $n(x)$, use $\mu_n = 1180$ cm^2/V $\cdot$ s.

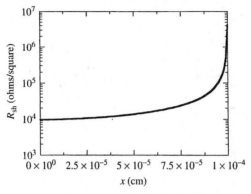

Fig. P1.12

1.13 The semiconductor structure in Fig. P1.13 consists of two films of width $W = 20$ μm, lengths $L_1 = 150$ μm and $L_2 = 100$ μm, thicknesses $t_1 = 0.6$ μm and $t_2 = 0.3$ μm, and resistivities $\rho_1 = 10$ ohm-cm and $\rho_2 = 1$ ohm-cm. Determine the sheet resistance of each film (in ohms/square) and the resistance between points A and B (in ohms). The dark regions at points A (not seen) and B are ideal ohmic contacts with zero resistance. The boundary between the two films has zero resistance.

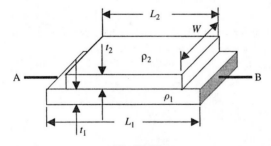

Fig. P1.13

1.14 The resistivity of a semiconductor layer of thickness t varies according to $\rho = \rho_o(1 - kx/t)$, where k is a constant. L is the sample length, W is the sample width and x is the dimension along the sample thickness. Derive an expression for the *sheet resistance* of this sample.

1.15 For the *n*-type layers on a *p*-type substrate in Fig. P1.15:

(a) determine R_{sh}

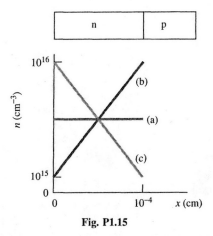

Fig. P1.15

(b) calculate and plot: σ versus x (linear-linear plot), ρ versus x (linear-linear plot), R_{sh} versus x (log-linear plot), and $1/R_{sh}$ versus x (log-linear plot) for the three cases. Use $\mu_n = 1250$ cm^2/V · s.

1.16 An arbitrarily shaped van der Pauw sample of thickness $t = 500$ μm was measured. The measured resistances were: $R_{12,34} = 74\ \Omega$ and $R_{23,41} = 6\ \Omega$. Determine the *resistivity* and *sheet resistance* of this sample.

1.17 An arbitrarily shaped van der Pauw sample of thickness $t = 350$ μm was measured. The measured resistances were: $R_{12,34} = 59\ \Omega$ and $R_{23,41} = 11\ \Omega$. Determine the *resistivity* and *sheet resistance* of this sample.

1.18 An arbitrarily shaped, uniformly doped van der Pauw sample has a thickness of 500 μm. The measured resistances are $R_{12,34} = 90\ \Omega$ and $R_{23,41} = 9\ \Omega$. Determine the *resistivity* and the *sheet resistance* of this sample.

1.19 In the cross bridge test structure in Fig. 1.13, consisting of a uniformly-doped layer on an insulating substrate, the following parameters are determined: $V_{34} = 58$ mV, $I_{12} = 1$ mA, $V_{45} = 1.75$ V, $I_{26} = 0.1$ mA. An independent measurement has given the resistivity of the film as $\rho = 0.0184\ \Omega \cdot$ cm and $L = 500$ μm. Determine the film sheet resistance R_{sh} (in Ω/square), the film thickness t (in μm), and the line width W (in μm).

1.20 The doping profile $N_D(x)$ of an ion implanted layer is given by

$$N_D(x) = \frac{\phi}{\Delta R_p \sqrt{2\pi}} \exp\left[-0.5\left(\frac{x - R_p}{\Delta R_p}\right)^2\right],$$

where ϕ is the implant dose, R_p is the range, and ΔR_p the straggle. Determine the sheet resistance for an arsenic layer implanted ($E = 100$ keV) into p-type Si doped to $N_A = 10^{15}$ cm^{-3}. Use $\phi = 10^{15}$ cm^{-2}, $R_p = 577$ Å, $\Delta R_p = 204$ Å, and $\mu_n = 100$ cm^2/V · s. Assume $N_D(x) = n(x)$.

Hint: First you have to find the junction depth.

1.21 The doping profile $N_D(x)$ of an ion-implanted layer is given by

$$N_D(x) = \frac{\phi}{\Delta R_p \sqrt{2\pi}} \exp\left[-0.5\left(\frac{x - R_p}{\Delta R_p}\right)^2\right],$$

where ϕ is the implant dose, R_p the range, and ΔR_p the straggle. Determine the sheet resistance for an n-type dopant layer (arsenic) implanted at an energy of 60 keV into a p-type Si wafer doped to $N_A = 10^{16}$ cm^{-3}. Use $\phi = 5 \times 10^{15}$ cm^{-2}, $R_p = 368$ Å, $\Delta R_p = 133$ Å, and $\mu_n = 50$ cm^2/V $\cdot$ s. Assume $N_D(x) = n(x)$.

   Hint: At the junction depth x_j: $N_A = N_D$.

1.22 **(a)** In a cross bridge test structure in Fig. 1.13 of a semiconductor layer on an insulating substrate, the following parameters are determined: $V_{34} = 18$ mV, $I_{12} = 1$ mA, $V_{45} = 1.6$ V, $I_{26} = 1$ mA. An independent measurement has given the resistivity of the film as $\rho = 4 \times 10^{-3}$ $\Omega \cdot$ cm and $L = 1$ mm. Determine the film sheet resistance R_{sh} (Ω/square), the film thickness t (μm), and the line width W (μm).

   (b) In one particular cross bridge test structure, the leg between contacts V_4 and V_5 is overetched. For this particular structure $V_{45} = 3.02$ V for $I_{26} = 1$ mA; it is known that half of the length L has a reduced W, *i.e.*, W', due to a fault during pattern etching. Determine the width W'.

1.23 In a cross bridge test structure in Fig. 1.13 consisting of a uniformly-doped layer on an insulating substrate, measurements give: $V_{34} = 58$ mV, $I_{12} = 1$ mA, $V_{45} = 1.75$ V, $I_{26} = 0.1$ mA. An independent measurement has given $\rho = 1.84 \times 10^{-2}$ $\Omega \cdot$ cm and $L = 500$ μm.

   (a) Determine the film sheet resistance R_{sh} (in Ω/square), the film thickness t (in μm), and the line width W (in μm).
   It is usually assumed that the sheet resistance R_{sh}, measured in region A in Fig. P1.23, is the same in the entire structure. Suppose that is not the case. What effect will that have on the line width measurement?

   (b) Determine the widths $W(a)$ and $W(b)$ in Fig. P1.23 that are calculated if the sheet resistance in the cross hatched region is R_{sh1} and in the white region it is R_{sh} (as determined in (i)), where $R_{sh} = 0.5 R_{sh1}$, but you assumed it was R_{sh} everywhere. Give your answer as $W(a)/W$ and $W(b)/W$, where W is the width for uniform sheet resistance.

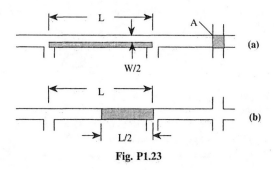

Fig. P1.23

1.24 Consider a p-type semiconductor cross bridge test structure on an insulating substrate. The layer, of thickness t, is non-uniformly doped according to $N_A = 10^{19} \exp(-kx)$, where k is a constant and x is the dimension along the sample thickness. Determine R_{sh}, V_{34} and V_{45}. Use $I_{12} = I_{26} = 1$ mA, $\mu_p = 100$ cm^2/V · s, $t = 1$ µm, $k = 10^5$ cm^{-1}, $L = 500$ µm, and $W = 10$ µm. Neglect the electron contribution to the layer resistivity and assume $N_A = p$.

1.25 (a) In the cross bridge test structure in Fig. P1.25, consisting of a uniformly doped layer on an insulating substrate, measurements give: $V_{34} = 11$ mV, $I_{12} = 0.5$ mA, $V_{45} = 50$ mV, $I_{26} = 1$ µA. The resistivity of the film is $\rho = 5 \times 10^{-3}$ Ω · cm and $L = 100$ µm. Determine the film *sheet resistance* R_{sh} (in Ω/square), the film *thickness* t (in µm), and the *line width* W (in µm).

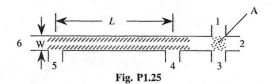

Fig. P1.25

(b) It is usually assumed that the resistivity in the "L" region is uniform. Suppose that is not the case. Determine the *effective line width* W_{eff} if the resistivity in the shaded "L" region varies linearly from 5×10^{-3} Ω · cm at terminal 5 to 10^{-2} Ω · cm at terminal 4. The resistivity in region "A", I_{12}, I_{26} and the physical width W are the same as in (a).

1.26 A sample with doped regions as shown in Fig. P1.26 is characterized by a spreading resistance probe. The minimum lateral step (along the beveled direction) that the probe can be moved is 2 µm. Determine the maximum bevel angle θ (in degrees) to ensure a minimum of 20 measurement points per doped region?

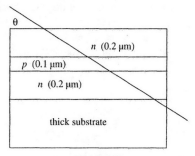

Fig. P1.26

1.27 Draw the spreading resistance plots, R_{sp} versus depth, for a p^+n and an n^+n junction on the same plot. The n-substrates are the same and the resistivity of the n^+ region is the same that of the p^+ region. These are *qualitative curves*, without numerical values.

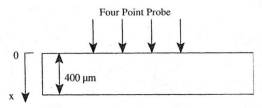

Fig. P1.28

1.28 Determine the sheet resistance R_{sh} for a Si wafer of thickness $t = 400$ μm, shown in Fig. P1.28, for:

(a) $N_A(x) = N_A(0)\exp(-x/L)$; $N_D = 0$, *i.e.*, no donors.

(b) $N_A(x) = N_A(0)\exp(-x/L)$; $N_D = 10^{16}$ cm^{-3}, *i.e.*, uniformly doped with donors.

Use $p(x) - n(x) - N_A(x) + N_D(x) = 0$, $p(x)n(x) = n_i^2$, $n_i = 10^{10}$ cm^{-3}, $N_A(0) = 10^{17}$ cm^{-3}, $L = 5$ μm,

$$\mu_p(x) = 54.3 + \frac{406.9}{1 + \left(\dfrac{N_A(x) + N_D}{2.35 \times 10^{17}}\right)^{0.88}}; \mu_n(x) = 92 + \frac{1268}{1 + \left(\dfrac{N_A(x) + N_D}{1.3 \times 10^{17}}\right)^{0.91}}.$$

The sheet resistance is measured on the top surface. Assume the pn junction in (b) is an insulating boundary. Neglect the width of the pn junction space-charge region. Assume that the four-point probe spacing s is much larger than the wafer thickness t.

1.29 Consider the sample in Fig. P1.29(a). Give a value for the sheet resistance R_{sh}. To convert from N_A to ρ, use Fig. A1.1. Then positive charge of density 10^{12} cm^{-2} is deposited on the upper surface, as in (b), and the charge remains there. This charge sheet does not change the measurement condition, *i.e.*, no surface current flows, but it does change the sample configuration, by causing the p-layer to be partially depleted and this depleted region can be considered an insulating region. Give a value for the new sheet resistance R_{sh}.

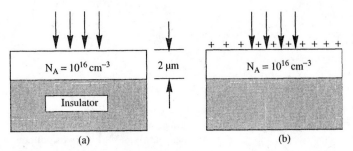

Fig. P1.29

1.30 The hot probe is used to determine the semiconductor *type, i.e. n*-type or *p*-type. For the arrangement in Fig. P1.30, determine the conductivity type and draw the band diagram. The sample is uniformly doped and in the dark, *i.e.*, it is not illuminated.

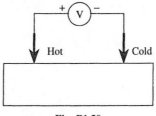

Fig. P1.30

Hint: The electron current density is $J_n = n\mu_n\, dE_F/dx - qn\mu_n P_n\, dT/dx$, where $P_n < 0$ is the differential thermoelectric power.

REVIEW QUESTIONS

- What is the best way to plot power law data?
- What is the best way to plot exponential data?
- Why is a four-point probe better than a two-point probe?
- Why is resistivity inversely proportional to doping density?
- What is an important application of wafer mapping?
- What is sheet resistance and why does it have such strange units?
- Why is sheet resistance commonly used to describe thin films?
- What are *van der Pauw* measurements?
- What is the main advantage of *Eddy current* measurements?
- What are advantages and disadvantages of the *modulated photoreflectance* (therma wave) technique?
- What is *carrier illumination* and what material parameters does it provide?
- How is *spreading resistance* profiling implemented?
- How can *conductivity type* be determined?

2

CARRIER AND DOPING DENSITY

2.1 INTRODUCTION

The carrier density is related to the resistivity, as shown in Chapter 1. It is, however, usually not derived from resistivity measurements but is measured independently. The carrier density and doping density are frequently assumed to be identical. While that is true for uniformly doped materials, the two may differ substantially for non-uniformly doped materials.

We discuss in this chapter methods for determining the carrier and the doping density. Among the electrical methods capacitance-voltage, spreading resistance, and Hall effect techniques are most commonly used. Being current-voltage or capacitance-voltage techniques, they determine the *carrier* density. Secondary ion mass spectrometry, an ion beam technique, has also found wide application for measuring the *doping* density. Optical methods, such as free carrier absorption, infrared spectroscopy, and photoluminescence, are sparingly employed. Infrared spectroscopy and photoluminescence have the advantage of very high sensitivity and the ability to *identify* the doping impurities.

2.2 CAPACITANCE-VOLTAGE (C-V)

2.2.1 Differential Capacitance

The *capacitance-voltage* technique relies on the fact that the width of a reverse-biased space-charge region (scr) of a semiconductor junction device depends on the applied voltage. This scr width dependence on voltage lies at the heart of the $C-V$ technique. The $C-V$ profiling method has been used with Schottky barrier diodes using deposited

Semiconductor Material and Device Characterization, Third Edition, by Dieter K. Schroder
Copyright © 2006 John Wiley & Sons, Inc.

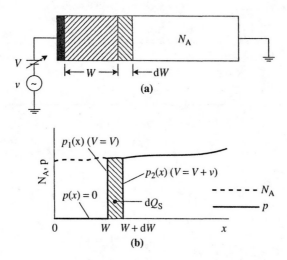

Fig. 2.1   (a) A reverse-biased Schottky diode, and (b) the doping density and majority carrier density profiles in the depletion approximation.

metal, mercury, and liquid electrolyte contacts, pn junctions, MOS capacitors, MOSFETs, and metal-air-semiconductor structures.

We consider the Schottky barrier diode of Fig. 2.1(a). The semiconductor is p-type with doping density N_A. A dc bias V produces a space-charge region of width W. The differential or small signal capacitance is defined by

$$C = \frac{dQ_m}{dV} = -\frac{dQ_s}{dV} \qquad (2.1)$$

where Q_m and Q_s are the metal and semiconductor charges. The negative sign accounts for *negative* charge in the semiconductor scr (negatively charged ionized acceptors) for *positive* voltage on the metal for reverse bias. The capacitance is determined by superimposing a small-amplitude ac voltage v on the dc voltage V. The ac voltage frequency is typically 10 kHz to 1 MHz with 10 to 20 mV amplitude, but other frequencies and other voltages can be used.

Let us consider the diode to be biased to dc voltage V plus a sinusoidal ac voltage v. Imagine the ac voltage increasing from zero to a small positive voltage adding a charge increment dQ_m to the metal contact. The charge increment dQ_m must be balanced by an equal semiconductor charge increment dQ_s for overall charge neutrality.

The semiconductor charge is given by

$$Q_s = qA \int_0^W (p - n + N_D^+ - N_A^-) \, dx \approx -qA \int_0^W N_A \, dx \qquad (2.2)$$

where the approximation obtains for $N_D = 0$ and $p \approx n \approx 0$ in the depletion approximation. Another assumption is that all acceptors are ionized. For acceptors or donors with energy levels deep within the band gap, the true dopant density profile may not be measured, as discussed further in Section 2.4.6.

The charge increment dQ_s in Fig. 2.1(b) comes about through a slight increase in the scr width. From Eqs. (2.1) and (2.2)

$$C = -\frac{dQ_s}{dV} = qA\frac{d}{dV}\int_0^W N_A\,dx = qAN_A(W)\frac{dW}{dV} \tag{2.3}$$

In going from Eq. (2.2) to (2.3), we have neglected the term $dN_A(W)/dV$, assuming N_A does not vary over the distance dW, or variations of N_A over a distance dW cannot be obtained with the $C-V$ technique. The capacitance in these equations is given in units of F *not* F/cm^2.

The capacitance of a reverse-biased junction, when considered as a parallel plate capacitor, is

$$C = \frac{K_s\varepsilon_o A}{W} \tag{2.4}$$

Differentiating Eq. (2.4) with respect to voltage and substituting dW/dV into Eq. (2.3) gives

$$N_A(W) = -\frac{C^3}{qK_s\varepsilon_o A^2\,dC/dV} = \frac{2}{qK_s\varepsilon_o A^2 d(1/C^2)/dV} \tag{2.5}$$

using the identity $d(1/C^2)/dV = -(2/C^3)\,dC/dV$. Note the *area dependence* in these expressions. Since the area appears as A^2, it is very important that the device area be precisely known for accurate doping profiling. From Eq. (2.4) we find the scr width dependence on capacitance as

$$W = \frac{K_s\varepsilon_o A}{C} \tag{2.6}$$

Equations (2.5) and (2.6) are the key equations for doping profiling.[1-2] The doping density is obtained from the slope dC/dV of a $C-V$ curve or from the slope $d(1/C^2)/dV$ of a $1/C^2-V$ curve. The depth at which the doping density is evaluated is obtained from Eq. (2.6). For a Schottky barrier diode there is no ambiguity in the scr width since it only spreads into the substrate. Space-charge region spreading into the metal is totally negligible. The doping density profile equations are equally well applicable for asymmetrical pn junctions, *i.e.*, p$^+$n or n$^+$p junctions, with one side of the junction more highly doped than the other side. If the doping density of the heavily doped side is 100 or more times higher than that of the lowly doped side, then the scr spreading into the heavily doped region can be neglected, and Eqs. (2.5) and (2.6) hold. If that condition is not met, the equations must be modified or both doping density and depth will be in error.[3] The correction, however, is fraught with difficulty. It has been proposed that no unique doping density profile can be derived from $C-V$ measurements under those conditions.[4] If the doping density profile of one side of the junction is known, then the profile on the other side can be derived from the measurements.[5] Fortunately, most pn junctions for doping density profiling, are of the p$^+$n or n$^+$p type, and corrections due to doping asymmetries are not necessary.

MOS capacitors (MOS-C) and MOSFETs can also be used for profiling.[6] For an MOS-C, the measurement is slightly more complicated because the device must remain in deep depletion during the measurement, ensured with a rapidly varying dc ramp voltage or by using pulsed gate voltages. In the latter case, the gate voltage is pulsed from $V_G = 0$ to $V_G = V_{G1}$, then from $V_G = 0$ to $V_G = V_{G2}$, where $V_{G2} > V_{G1}$, etc. The capacitance is measured immediately after the pulse before minority carriers have had time to be generated. MOS-C doping density profile measurements are influenced by interface traps and minority carrier generation, discussed in more detail in Section 2.4.3. Equation (2.5) applies directly to MOS-Cs when both interface states and minority carriers can be neglected, but the scr width expression becomes[7-8]

$$W = K_s \varepsilon_o A \left(\frac{1}{C} - \frac{1}{C_{ox}} \right)$$ (2.7)

Equation (2.7) differs from Eq. (2.6) by the oxide capacitance C_{ox}, because part of the gate voltage is dropped across the oxide. The MOS-C profiling technique has also been implemented by driving the device into deep depletion and measuring the current instead of the capacitance.[9-10] The interference of minority carrier generation with differential capacitance profile measurements can be avoided by providing a minority carrier sink, such as a reverse-biased pn junction, adjacent to the MOS-C. A MOSFET provides such minority carrier collecting junctions. Minority carriers are drained from the channel region of the MOSFET provided the source/drain voltages are equal to or higher than the gate voltage. Since there are no minority carriers in this case, the measurement can be made in steady-state, *i.e.*, no need for pulsed gate voltage.

A *contactless* capacitance and doping profiling version uses a contact held in close proximity to the semiconductor wafer. The sensor electrode, 1 mm diameter and coated with high dielectric strength thin film, is surrounded by an independently biased guard electrode. The sensor electrode is held above the wafer by a porous ceramic air bearing which provides for a very stable distance from the wafer as long as the load on the air bearing does not change, shown in Fig. 2.2. The controlled load is provided by pressurizing a bellows. As air escapes through the porous surface, a cushion of air forms on the wafer that acts like a spring and prevents the porous surface from touching the wafer. The porosity and air pressure are designed such that the disk floats approximately 0.5 μm above the wafer surface. A stainless steel bellows acts to constrain the pressurized air and to raise the porous disk when the air pressure is reduced. If the air pressure fails, the disk moves up, rather than falling down and damaging the wafer.[11]

To prepare the wafer, it is placed in a low-concentration ozone environment at a temperature of about 450°C, reducing the surface charge on the wafer, especially critical for *n*-Si, makes it more uniform, reduces the surface generation velocity and allows deeper depletion.[12] A recent comparison of epitaxial resistivity profiles by the contactless with Hg-probe $C-V$ measurements compared very favorably.[13] The capacitance of the air gap is measured by biasing the semiconductor surface in accumulation. Light is used

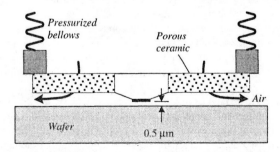

Fig. 2.2   Contactless doping profiling arrangement. Pressurized air maintains the electrode at approximately 0.5 μm above the sample surface.

to collapse any possible space-charge region due to surface charge while the sensor is lowered and while the air gap modulation due to the electrostatic attraction is determined to eliminate any series space-charge capacitance. Assuming that the air gap does not vary with changing electrode voltage, the capacitance of the air gap is the measured capacitance at its maximum value. The doping density profile is determined from Eqs. (2.5) and (2.7) with C_{ox} in Eq. (2.7) replaced by C_{air}.

For the derivation of Eq. (2.5) we used the *depletion approximation*, which neglects minority carriers and assumes total depletion of majority carriers in the space-charge region to a depth W and perfect charge neutrality beyond W, as illustrated in Fig. 2.1(b). This is a reasonably good approximation when the scr is reverse biased and when the substrate is uniformly doped. Furthermore, we used as the incremental charge variation the acceptor ion density at the edge of the space-charge region. The ac probe voltage exposes more or less ionized acceptors at the scr edge, as shown in Fig. 2.1. The charges that actually move in response to the ac voltage are the mobile holes, not the acceptor ions. Hence, the differential capacitance-voltage profiling technique determines the *carrier density*, not the *doping density*. What is actually measured is an *apparent or effective carrier density*, which is neither the true carrier density nor the doping density. Fortunately, the apparent density is approximately the majority carrier density and the relevant equations become

$$p(W) = -\frac{C^3}{q K_s \varepsilon_o A^2 \, dC/dV} = \frac{2}{q K_s \varepsilon_o A^2 d(1/C^2)/dV} \tag{2.8}$$

$$W = \frac{K_s \varepsilon_o A}{C} \tag{2.9}$$

$$W = K_s \varepsilon_o A \left(\frac{1}{C} - \frac{1}{C_{ox}} \right) \tag{2.10}$$

The equations for the *majority carrier density* rather than *doping* density can be derived from majority carrier currents in diodes[14] or from surface potentials in MOS capacitors.[15]

It is worthwhile to say a few words about the $C-V$ interpretation of Eq. (2.8). Both dC/dV and $d(1/C^2)/dV$ methods are used, with the $d(1/C^2)/dV$ the preferred method. We demonstrate this in Fig. 2.3. $C-V$ and $1/C^2-V$ curves of a Si pp$^+$ junction are shown in Fig. 2.3(a). It is difficult to tell from the $C-V$ curve whether the doping density of this sample is constant or not. When the $C-V$ curve is converted to a $1/C^2-V$ curve, it is immediately obvious that the carrier density is not uniform with a discontinuity at around 3 V. The carrier density profile determined with Eqs. (2.8) and (2.9) is shown in Fig. 2.3(b).

The use of the majority carrier density rather than the doping density in the profile equations is an important point and has been the subject of much discussion.[16-28] We demonstrate the concept for a *non-uniform acceptor doping density* profile by the heavy curve in Fig. 2.4(a). The majority hole density profile shown by the light line differs from the doping density profile even in thermal equilibrium. Some of the holes diffuse from the highly doped region to the lowly doped region and an equilibrium profile is established as a result of both diffusion and drift. The steeper the doping gradient, the more p and N_A

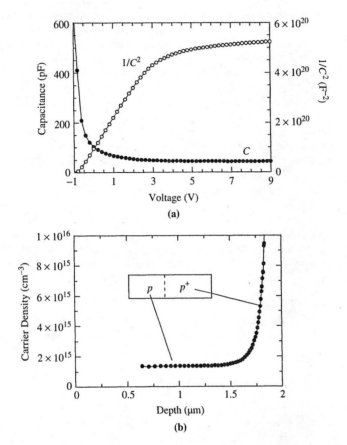

Fig. 2.3 (a) $C-V$ and $1/C^2-V$ curves of a Si n^+p diode, (b) $p(\text{x})\text{-}W$ profile.

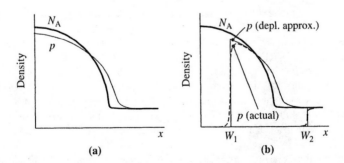

Fig. 2.4 A schematic representation of the doping and majority carrier density profiles of a non-uniformly doped layer. (a) zero-biased junction, (b) reverse-biased junction showing the doping density profile, the majority carrier profiles in the depletion approximation and the actual majority carrier profiles for two reverse-bias voltages.

differ from one another. The majority carrier density deviation from the doping density is governed by the *extrinsic Debye length* L_D, more generally called the Debye length

$$L_D = \sqrt{\frac{kT K_s \varepsilon_o}{q^2 (p + n)}} \tag{2.11}$$

L_D is a measure of the distance over which a charge imbalance is neutralized by majority carriers under steady-state or equilibrium conditions.

When a scr forms as a result of a reverse biased Schottky diode, for example, the carrier distribution becomes that in Fig. 2.4(b). We show the majority carrier distribution expected from the depletion approximation for scr widths W_1 and W_2, corresponding to two different reverse-bias voltages. The actual majority carrier distribution is also shown. The two differ appreciably and it is quite obvious from these curves that the doping density profile is not what is measured by differential capacitance profiling. It is also not clear that it is the majority carrier distribution that is measured. It has been shown by detailed computer calculations that what is actually measured is an *effective* or *apparent* majority carrier density profile, that is closer to the true majority carrier density profile than to the doping density profile.[18] The doping density profile, the majority carrier density profile, and the effective majority carrier density profile are identical for uniformly doped substrates, but not for non-uniformly doped substrates.

The Debye length sets a limit to the *spatial resolution* of the measured profile. This Debye length problem arises because the capacitance is determined by the movement of majority carriers and the majority carrier distribution cannot follow abrupt spatial changes in doping density profiles. Detailed calculations show that if a doping density step occurs within one Debye length, the majority carrier and the apparent densities agree fairly well with one another, but both differ appreciably from the true doping density profile.[18] For a more gradual transition, the majority carrier density agrees quite well with the apparent densities with depletion occurring from the lowly doped or from the highly doped side. The agreement with the doping density profile is also quite reasonable.

A relationship between the measured majority carrier density and the doping density is[16]

$$N_A(x) = p(x) - \frac{kT K_s \varepsilon_o}{q^2} \frac{d}{dx} \left(\frac{1}{p(x)} \frac{dp(x)}{dx} \right) \tag{2.12}$$

Extensive computer simulations have shown that Eq. (2.12) is too much of a simplification.[17-18, 26] For low-high junctions, *e.g.*, a p-p$^+$ junction, the results depend on whether the junction is profiled from the p-side or from the p$^+$-side. The simulations show that a step profile cannot be resolved accurately to less than about $2-3L_D$, with the Debye length determined by the carrier density on the highly doped side of the junction. A doping density ramp profile, for example, cannot be distinguished accurately from a step unless it is appreciably wider than a Debye length.

Equations (2.4) to (2.9) are derived in the *depletion approximation*, which assumes zero mobile carrier density in the space-charge region. This is a reasonably good approximation for reverse bias. However, for zero- or forward-biased Schottky and pn junctions, the approximation loses its validity, and majority carrier profiling becomes inaccurate. Under forward bias an additional capacitance due to excess minority carrier storage in the quasi-neutral regions is introduced, rendering the method still less accurate. The concept of a zero- or forward-biased junction does not apply for an MOS-C. However, the role of mobile carriers is clearly just as important as it is for junction devices.

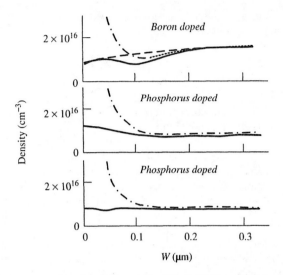

Fig. 2.5 Doping density profiles for three samples. The solid lines are experimental data. The dashed lines indicate the profiles in the absence of interface states. The dot-dash lines show the profiles when the depletion approximation is used. Reprinted with permission after ref. 28.

Neglect of *majority carriers* has been shown to lead to errors in pulsed MOS-C doping density profile determinations for surface potentials below 0.1 V,[7, 19, 27] corresponding to a distance of approximately $2-3L_D$ from the SiO_2-Si interface. It has been suggested that profiling below this limit is possible by accounting for majority carriers.[28] Fairly complex equations are necessary for this correction, but they apply only to uniformly doped substrates. Nevertheless, they are useful, and results of such a modified analysis are shown in Fig. 2.5, where the dash-dot lines show the profile under the usual Debye length limitation and the corrected experimental data points show the profile all the way to the surface. Other considerations to be observed during profile measurements are discussed in the ASTM standard F419.[29] As with all ASTM methods this is a good source of practical information and precautions to observe during measurement. One more caution: a common technique for the preparation of metal-semiconductor contacts uses chemically etched, hydrogen-terminated Si. Hydrogen can diffuse several microns into Si at room temperature and compensate boron acceptors,[30] leading to erroneous carrier density profiles. The B-H complex dissociates for $T \geq 180°C$ anneals.

2.2.2 Band Offsets

When two semiconductors with different band gaps are joined, the conduction and valence bands cannot both be aligned, as illustrated in Fig. 2.6. Band offsets may exist in the conduction band, ΔE_c, the valence band, ΔE_v, or both. Band offsets can be determined with various techniques. One of the earliest was the infrared absorption method.[31] A widely used method is photoemission spectroscopy, where photons incident on a sample eject electrons.[32] The electron energy is related to the band gap and band offset and the band offset is measured directly.

An electrical technique is based on C–V measurements. It is easiest to determine band offsets on n-N or p-P isotype heterojunctions. Here the lower-case letters n, p

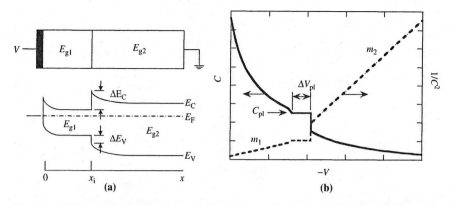

Fig. 2.6 (a) Cross-sectional and band diagram of two semiconductors with different band gaps, (b) schematic $C-V$ and $1/C^2-V$ plots. Real plots are smeared out and do not exhibit the sharp features shown here.

refer to the narrow band gap, and the upper case letters N, P to the wide band gap semiconductor. Schottky barrier diodes are formed on the structure, as in Fig. 2.6(a). The $C-V$ and $1/C^2-V$ curves of such a structure are shown schematically in Fig. 2.6(b). The doping density profiles of the two materials are determined from the slopes m_1 and m_2. The plateau capacitance C_{pl} is related to the thickness of the narrow band semiconductor and the plateau voltage ΔV_{pl} is related to the band offset. The $C-V$ curve yields an apparent or effective electron density, n^*, that differs from the true electron density and from the doping density.

We follow the theory of Kroemer et al.[33] The method was originally shown to be applicable to abrupt junctions, but was later shown to be applicable to graded junctions as well.[34] There may be an interfacial charge Q_i at the heterointerface, given by

$$Q_i = -q \int_0^\infty [N_D(x) - n^*(x)] \, dx \tag{2.13}$$

where $N_D(x)$ is the doping density. The conduction band discontinuity is

$$\Delta E_c = \frac{q^2}{K_s \varepsilon_o} \int_0^\infty [N_D(x) - n^*(x)](x - x_i) \, dx - kT \ln\left[\frac{n_2/N_{c2}}{n_1/N_{c1}}\right] \tag{2.14}$$

where n_1, n_2 are the free electron densities in the layer and the substrate, N_{c1}, N_{c2} the effective density of states in the conduction band in the layer and the substrate, and x_i the location of the heterojunction interface. Knowledge of the position of x_i is important. Any error in x_i translates into an error in the band offset and it can be determined self-consistently by comparing the measured apparent carrier density with the calculated carrier density.[35] A plot of apparent carrier density of an n-GaAs/N-AlGaAs heterojunction is shown in Fig. 2.7. The experimental data are shown by the data points. From this plot $Q_i/q = 2.74 \times 10^{10}$ cm^{-2} and $\Delta E_c = 0.248$ eV were extracted.

MOS capacitor measurements have also been used to determine band offsets. These measurements rely on a good oxide/semiconductor interface and hence are more applicable to Si-based structures. The technique has been used for determining the band offset of SiGe/Si heterojunctions with the band offset almost entirely in the valence band.[36] The

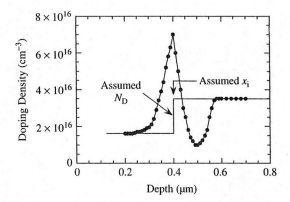

Fig. 2.7   Doping density plot of n-GaAs/N-Al$_{0.3}$Ga$_{0.7}$As heterojunction; the points are experimental data, the straight line is the assumed donor density. Data adapted from ref. 33.

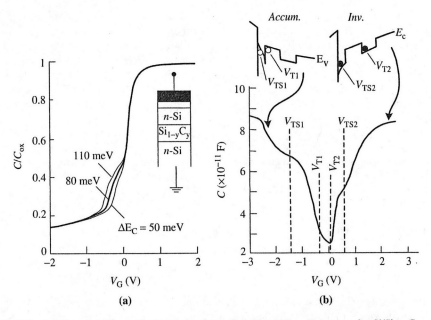

Fig. 2.8   (a) Measured (heavy line) and simulated (light lines) C_{hf}–V_G curves for Si/Si$_{0.98}$C$_{0.013}$ MOS capacitor (b) C_{lf}–V_G characteristics of Si/Si$_{0.7}$Ge$_{0.3}$ MOS capacitor showing threshold voltages and carrier confinement in accumulation and inversion and band diagrams in accumulation and inversion. Data adapted from refs. 37 and 38.

low-frequency C–V curve exhibits two threshold voltages associated with the SiO$_2$/Si and the heterojunction interfaces. It also shows a plateau with a width dependent on the band offset. Example MOS-C C–V_G curves for Si/SiC and Si/SiGe are shown in Fig. 2.8.[36-37] In both cases, the plateaus due to band offsets are clearly seen. Fig. 2.8(a) shows the high-frequency C–V_G offset of the Si/SiC heterojunction.

The valence band and conduction band alignment of the heterostructure in Fig. 2.8(b) show hole confinement in accumulation and electron confinement in inversion.[38] The low-frequency $C-V_G$ characteristic shows the plateaus in accumulation and inversion, due to carrier confinement. The characteristic exhibits two threshold voltages in accumulation V_{T1} and V_{TS1} and two threshold voltages in inversion V_{T2} and V_{TS2}. V_{T1} corresponds to hole accumulation at the top strained Si/SiGe heterojunction and V_{TS1} is related to the Si/SiO$_2$ interface. Similarly, V_{T2} and V_{TS2} correspond to the electron build up in inversion at the SiGe/(buried) strained Si heterojunction and Si/SiO$_2$ interface, respectively.

Current-voltage measurements are generally less reliable for band offset determination. Usually rectification of pn heterojunctions is interpreted for band offset determination. In principle, n-N and p-P heterojunctions should also show rectification. When they do not, that has been falsely interpreted as no band offset. Deep-level transient spectroscopy has also been used to determine band offsets.[39] Kroemer gives a good discussion and critique of band offset measurements.[40]

Internal photoemission and core-level X-ray photoemission spectroscopy provide more direct band gap offsets. In internal photoemission (discussed more fully in Section 3.5.4) electrons are excited from the valence and/or conduction band of the narrow band gap semiconductor to the wide band gap semiconductor.[41] If the conduction band of the right-hand semiconductor is populated by electrons at the interface, then there is a lower photoemission threshold energy which characterizes the conduction-band discontinuity ΔE_c. If the narrow band gap semiconductor is p-type, then the valence-band offset ΔE_v is determined. Valence band offsets are most reliably determined from the energy positions of core level lines in X-ray photoelectron spectra recorded with bulk samples of the two semiconductors in contact.[42] Since the escape depths of the photoelectrons are on the order of 2 nm, one of the two semiconductors must be sufficiently thin.

2.2.3 Maximum-Minimum MOS-C Capacitance

Equations (2.8) and (2.10) hold for the depletion portion of the equilibrium and the deep-depletion portion of non-equilibrium MOS-C $C-V_G$ curves but not for strong inversion. The deep-depletion $C-V_G$ curve C_{dd} is shown in Fig. 2.9. A simple method to determine the doping density of an equilibrium MOS-C is to measure the maximum high-frequency

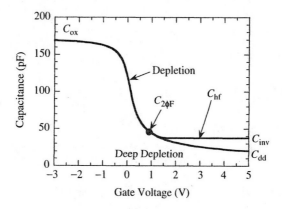

Fig. 2.9 $C-V_G$ curve for an SiO$_2$/Si MOS capacitor. $N_A = 10^{17}$ cm^{-3}, $t_{ox} = 10$ nm, $A = 5 \times 10^{-4}$ cm^2.

capacitance of an MOS-C in strong accumulation C_{ox} and the minimum high-frequency capacitance in strong inversion C_{inv}.[43] Interface traps play no role in this measurement if the gate voltage is sufficiently high for the device to be in strong inversion. Minority carrier generation does not exist with the device in equilibrium. The max-min capacitance method yields the average doping density over the scr width with the device in strong inversion.

Such a measurement is sufficient for uniformly doped substrates but not accurate for non-uniform doping densities. Information about non-uniformly doped substrates can also be extracted from such *equilibrium* MOS-C *C–V* curves by linearizing a non-uniformly doped layer on a uniformly doped substrate.[44] The measurement requires a knowledge of the substrate doping density and extracts the surface density and layer depth from the measured capacitance-voltage curves by iteration.

The *maximum-minimum capacitance* technique relies on the dependence of the scr width of a strongly inverted MOS capacitor on the substrate doping density. The general MOS-C capacitance is

$$C = \frac{C_{ox} C_s}{C_{ox} + C_s} \tag{2.15}$$

where $C_s = K_s \varepsilon_o A / W$ is the semiconductor capacitance. The capacitance C_{inv} is the strong inversion or *minimum* capacitance for which the space-charge region width is

$$W = W_{inv} = \sqrt{\frac{2 K_s \varepsilon_o \phi_{s,inv}}{q N_A}} \tag{2.16a}$$

where $\phi_{s,inv}$ is the surface potential in strong inversion. The surface potential $\phi_{s,inv}$ is frequently approximated by $\phi_{s,inv} \approx 2\phi_F$.[45] But $\phi_{s,inv}$ is actually slightly higher than $2\phi_F$, i.e., $\phi_{s,inv} \approx 2\phi_F + 4kT/q$.[46] For the approximate case of $\phi_{s,inv} \approx 2\phi_F = 2(kT/q) \ln(N_A/n_i)$

$$W = W_{2\phi F} = \sqrt{\frac{2 K_s \varepsilon_o 2\phi_F}{q N_A}} \tag{2.16b}$$

Equations (2.15) and (2.16b) lead to

$$N_A = \frac{4\phi_F}{q K_s \varepsilon_o A^2} \frac{C_{2\phi F}^2}{(1 - C_{2\phi F}/C_{ox})^2} \tag{2.17}$$

where $C_{2\phi F}$ is indicated on Fig. 2.9. $C_{2\phi F}$ is, of course, not known for a given C–V_G curve. Consequently, Eq. (2.17) is usually given as

$$N_A = \frac{4\phi_F}{q K_s \varepsilon_o A^2} \frac{C_{inv}^2}{(1 - C_{inv}/C_{ox})^2} = \frac{4\phi_F}{q K_s \varepsilon_o A^2} \frac{R^2 C_{ox}^2}{(1 - R)^2} \tag{2.18}$$

where $R = C_{inv}/C_{ox}$. C_{inv} and C_{ox} are shown on Fig. 2.9. A small inconsistency in Eq. (2.18) is the use of $\phi_{s,inv} = 2\phi_F$ in conjunction with C_{inv}. It should be $\phi_{s,inv} \approx 2\phi_F + 4kT/q$. This is a small error, however.

An empirical relationship between C_{inv} and N_A for silicon at room temperature is[47]

$$\log(N_A) = 30.38759 + 1.68278 \log(C_1) - 0.03177[\log(C_1)]^2 \tag{2.19}$$

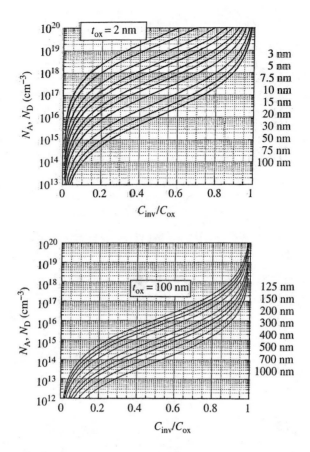

Fig. 2.10 Doping density versus C_{inv}/C_{ox} as a function of oxide thickness for the SiO_2/Si system at $T = 300$ K.

where "log" is the logarithm to base 10, $C_1 = RC_{ox}/[A(1-R)]$ and the capacitances are in units of F, the area is in units of cm^2, and N_A is in units of cm^{-3}. The equation is identical for n-type substrates with N_D substituted for N_A.

We show in Fig. 2.10 curves calculated from Eq. (2.18), giving the doping density as a function of C_{inv}/C_{ox}. These curves are useful for a first order estimate of the doping density, but they may hide depth-dependent features for spatially varying doping densities. Depth-dependent doping density profiles may be measured by gradually immersing the wafer in an etch bath so that the surface becomes a slightly sloped plane along which the impurity gradient is gradually changing. MOS capacitors formed on the etched and oxidized surface can be used to determine the doping density under each MOS-C as determined from its C_{inv}/C_{ox} ratio.[48]

The doping density of a poly-Si gate can be determined by the C_{inv}/C_{max} method using the connection of Fig. 2.11(a).[49] With source, drain, and substrate connected together and a gate voltage above the threshold voltage, the source/drain/substrate form one continuous n-layer, representing the "gate" of the MOS capacitor. The "substrate" of this capacitor

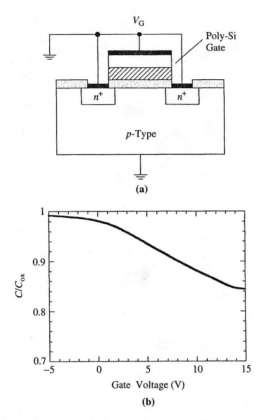

Fig. 2.11   (a) MOSFET connection to determine the doping density of the gate, (b) resulting $C-V$ curve calculated, $N_D = 5 \times 10^{19}$ cm^{-3}, $t_{ox} = 10$ nm.

is the poly-Si gate shown as depleted in Fig. 2.11(a). The resulting $C-V_G$ curve has the shape of Fig. 2.11(b). Although C_{inv} is not much lower than C_{ox}, it is, nevertheless, possible to extract the doping density using Fig. 2.10. However, it takes a significant gate voltage to invert the gate and the gate oxide may break down before inversion is reached. In that case one could match the depletion part of the $C-V_G$ curve with theory to extract N_D.

Exercise 2.1

Problem: For a p-type Si MOS capacitor, $C_{inv}/C_{ox} = 0.22$ and $t_{ox} = 15$ nm.

  (a) Determine the doping density for this device using $K_{ox} = 3.9$, $K_s = 11.7$, $n_i = 10^{10}$ cm^{-3}, $A = 5 \times 10^{-4}$ cm^2, and $T = 27°$C.
  (b) Determine C_{inv}/C_{ox} when $N_A = 5 \times 10^{15}$ cm^{-3}. Use the approach that leads to Eq. (2.18) for this problem.
  (c) Use Eq. (2.19) to determine N_A instead of Eq. (2.18).

Solution:

$$N_A = \frac{4\phi_F}{q K_s \varepsilon_o A^2} \frac{C_{inv}^2}{(1 - C_{inv}/C_{ox})^2} = \frac{4\phi_F}{q K_s \varepsilon_o A^2} \frac{R^2 C_{ox}^2}{(1 - R)^2}$$

(a) With $R = 0.22$, $K_{ox} = 3.9$ and $t_{ox} = 15$ nm, we find $C_{ox} = 1.15 \times 10^{-10}$ F and $C_{inv} = 2.53 \times 10^{-11}$ F. Solving the above equation gives: $N_A = 4 \times 10^{16}$ cm^{-3}

(b) for $N_A = 5 \times 10^{15}$ cm^{-3}: $C_{inv}/C_{ox} = 0.097$

(c) Using Eq. (2.19): $N_A = 4.48 \times 10^{16}$ cm^{-3}

Note the 10% difference between N_A determined with Eqs. (2.18) and (2.19)

2.2.4 Integral Capacitance

The differential capacitance technique has some limitations when used as a process monitor where accuracy and measurement time are important.[50] In particular, the required differentiation often results in noisy profiles. The integral capacitance technique is based on *integrating* a portion of the pulsed MOS-C $C-V$ curve to obtain a partial implant dose P_Φ, with the partial dose proportional to the implanted dose. The chosen dose includes the doping density between $x = x_1$ and $x = x_2$ and contains most of the implanted layer, but does not extend into the region where the doping density equals the uniform background doping density nor into the region within 2 to 3 Debye lengths from the surface. The partial dose is given by[50]

$$P_\Phi = \int_{x_1}^{x_2} N_A(x)\, dx = \frac{1}{qA} \int_{V_1}^{V_2} C\, dV \tag{2.20}$$

Note the linear dependence on device area rather than the square dependence of the $C-V$ method. The second parameter that is measured is related to the projected range R or implant depth at the density peak. It is defined by[50]

$$R = t_{ox} + \frac{1}{P_\Phi} \int_{x_1}^{x_2} x N_A(x)\, dx = \frac{K_s \varepsilon_o}{q P_\Phi}(V_2 - V_1) + (1 - K_s/K_{ox})t_{ox} \tag{2.21}$$

This expression for R incorporates P_Φ with only one integration. The repeatability of this technique for a given device was accurate to 0.1%, and the authors claim that the repeatability in partial dose measurement has been improved by over a factor of ten by going to the integral capacitance technique.[50]

A different MOS capacitor integral approach gives the implanted dose.[51] Example $C-V_G$ curves for various implant doses are shown in Fig. 2.12(a). The doping density profiles (symbols) in Fig. 2.12(b) are extracted from measured deep depletion CV-curves using the method of Fig. 2.5. The solid lines represent the simulated implanted doping densities. The deviation of the two profiles illustrates that the simple integration of the $C-V_G$ profile does not yield the true doping densities. The proposed technique relies on measuring the deep-depletion MOS capacitor $C-V_G$ curve. The depleted majority carrier charge at a certain space-charge region width is determined from the value of the majority carrier charge in strong accumulation and the corresponding change of majority carrier charge ΔQ when the MOS-C is driven into deep depletion. ΔQ is obtained by integrating the deep depletion $C-V_G$ curve. A second approach is measuring the depletion $C-V_G$ curves of the implanted sample and a reference sample. The implant dose is obtained

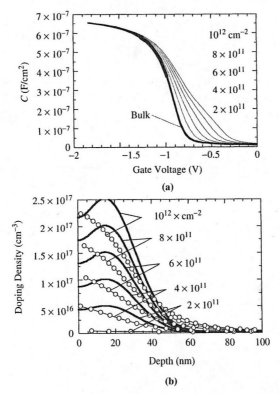

Fig. 2.12 (a) Deep depletion $C-V_G$ curves as a function of boron ion implant dose at 40 keV into p-Si substrates, $t_{ox} = 4.1$ nm, (b) doping profiles determined by conventional $C-V$ profiling (symbols) and simulation (lines). The "bulk" curve in (a) is for the unimplanted substrate. After ref. 51.

from the difference charge ΔQ by integrating the two $C-V_G$ curves starting at the same accumulation capacitances to the same deep-depletion capacitance.

2.2.5   Mercury Probe Contacts

Capacitance profiling requires junction devices. At times it is desirable to use a device whose junction can be fabricated without subjecting the material to high temperature treatments. Conventional Schottky barrier device fabrication is done near room temperature, but a metal must be deposited on the wafer. When a temporary contact is needed, as in evaluating epitaxial layers for example, a mercury probe is frequently used, where mercury contacts the sample through a well-defined orifice. Mercury probes can make contact either to the sample bottom or to the top. The contact area is sufficiently well defined to be useful for profile measurements. A mercury probe with a probe diameter as small as 7 μm has been used for $C-V$ measurements, allowing lateral capacitance profiles by continuously dragging the probe across the wafer.[52]

The mercury contact does not damage the wafer nor leave mercury on the surface.[53] The semiconductor surface should be treated before the Hg contacts the surface for reproducible measurements. Current leakage and junction breakdown of the mercury Schottky contact, usually at its edge, are the most important limiting factors for accurate doping profiling. To minimize current leakage and maximize junction breakdown voltage, a thin oxide layer is usually grown on n-Si surfaces by dipping the wafers in hot nitric acid or hot sulphuric acid. This oxide is about 3 nm thick. Dip p-Si wafers in HF for 30 s, rinse in flowing DI water and dry the wafer,[53] giving an oxide-free surface which is desirable for most reproducible results. The mercury should be very pure, so periodic mercury changes are recommended. It is also helpful to reduce the junction leakage by applying a wetting agent, *e.g.*, Kodak Photo-Flo, on the wafer surface to reduce moisture accumulation, before making the mercury contact.

2.2.6   Electrochemical $C-V$ Profiler (ECV)

The *electrochemical capacitance-voltage* profiling technique is based on the measurement of the capacitance of an electrolyte-semiconductor Schottky contact at a *constant* dc bias voltage. Depth profiling is achieved by electrolytically etching the semiconductor between capacitance measurements with no depth limitation. However, the method is destructive because it etches a hole into the sample. Early measurements divided the measurement and etch processes; later they were combined into one operation.[54] The present technique uses a combined process in which both etching and measurement are performed with the same apparatus. An excellent review is given by Blood.[55]

The electrochemical method is schematically shown in Fig. 2.13. The semiconductor wafer is pressed against a sealing ring in the electrochemical cell containing an electrolyte. The ring opening defines the contact area by means of spring-loaded back contacts pressing the wafer against the sealing ring. The etching and measuring conditions are controlled by the potential across the cell by passing a dc current between the semiconductor and the carbon electrode to maintain the required overpotential measured with respect to the saturated calomel electrode. To reduce series resistance, the ac voltages are measured with a platinum electrode located near the sample.

With a small reverse dc bias applied between the electrolyte and the semiconductor, two low-voltage signals of different frequencies are applied to the electrolyte. The carrier density measurement is based on Eq. (2.8) or on the relationship

$$p(W) = \frac{2K_s\varepsilon_o}{q}\frac{\Delta V}{\Delta(W)^2} \tag{2.22}$$

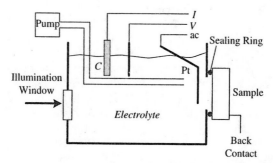

Fig. 2.13 Schematic diagram of the electrochemical cell showing the Pt, saturated calomel and carbon electrodes and the pump to agitate the electrolyte and disperse bubbles on the semiconductor surface. Reprinted with permission after Blood.[55]

where ΔV is the modulation component of the applied ac voltage (typically 100–300 mV at 30–40 Hz) and $\Delta(W)^2$ is the resulting scr width modulation. W is determined by measuring the imaginary component of the current with a phase-sensitive amplifier using typically a 50 mV, 1–5 kHz signal and Eq. (2.9). W and $p(W)$ are obtained through appropriate electronic circuits.[54] The 1–5 kHz frequency is significantly lower than the 0.1–1 MHz frequency typically used for conventional differential capacitance profiling to reduce the $r_s C$ time constant, where r_s is the series resistance of the electrolyte and C the device capacitance. The resistance-capacitance product must meet certain criteria for the measurements to be valid as discussed in Section 2.4.2 on *Series Resistance*. ECV profiling is more sensitive to deep traps due to the low frequencies, but for most materials this is rarely a problem.

Equations (2.9) and (2.22) provide the density at depth W. Depth profiling is achieved by dissolving the semiconductor electrolytically, which depends on the presence of holes. For p-type semiconductors, holes are plentiful and dissolution is readily achieved by forward biasing the electrolyte-semiconductor junction. For n-type material, holes are generated by illuminating and reverse biasing the junction. The depth W_R depends on the dissolution current I_{dis} according to the relationship[54]

$$W_R = \frac{M}{zF\rho A} \int_0^t I_{dis} \, dt \tag{2.23}$$

where M is the semiconductor molecular weight, z the dissolution valency (number of charge carriers required to dissolve one semiconductor atom), F the Faraday constant $(9.64 \times 10^4$ C), ρ the semiconductor density, and A the contact area. W_R is determined by integrating the dissolution current electronically. The measurement depth of the carrier density is

$$x = W + W_R \tag{2.24}$$

A key advantage of ECV over conventional $C-V$ profiling is the unlimited profile depth, since the semiconductor can be etched to any desirable depth. The electrolyte must be chosen appropriately for each semiconductor and suitable electrolytes are for InP: 0.5 M HCl in H_2O,[56] Pear etch (37% HCl:70% HNO_3:methanol (36:24:1000)),[57] FAP (48% HF:99% CH_3COOH:30% H_2O_2:H_2O (5:1:0.5:100)), and UNIEL A:B:C (1:4:1) where A:48% HF:99% CH_3COOH:85% o-H_3PO_4:H_2O (5:1:2:100), B: 0.1 M N-n-butylpyridinium chloride ($C_9H_{14}ClN$), C: 1 M NH_3F_2; for GaAs Tiron (1,2-dihydroxybenzene-3.5-disulfonic acid disodium salt $C_6H_2(OH)_2(SO_3Na)_2 \cdot H_2O$),[58] EDTA ($Na_2 \cdot$ EDTA (0.1 M) basified with ethylenediamine to pH of 9.1,[55, 59] UNIEL, and ammonium tartrate (($NH_4)_2C_4O_6$, FW184.15, basified with NH_4OH to pH of 11.5 or higher); for Si NaF/H_2SO_4 and 0.1 M NH_4HF_2.[60–62] One of the most successful electrolytes for GaAs:AlGaAs and InP based alloys is Na_2 EDTA (0.1 M) basified with ethylenediamine to pH 9–10.[63] The chemical nature of the electrolyte determines the quality of the etch well and the tendency to avoid film formation, both of which affect the carrier density.

The technique is eminently suitable for III–V materials because the dissolution valency, $z = 6$, is well defined and the electrolyte etches the semiconductor very controllably. The dissolution valency is not well defined for Si where it can vary between 2 and 5, affected by electrolyte concentration, dopant type and density, electrode potential, and illumination intensity. Furthermore hydrogen bubbles generated during the dissolution process hinder the uniformity and degrade the depth resolution. The hydrogen bubble problem is overcome by using a pulsed jet of the electrolyte.[61–62] Electrochemical profiling of silicon

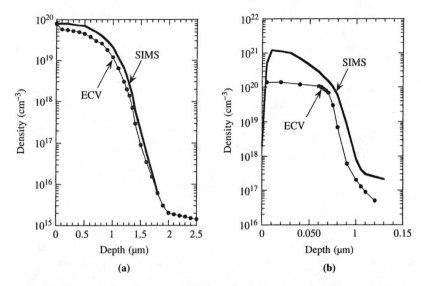

Fig. 2.14 Profiles obtained with the ECV profiler and with SIMS. (a) $p^+(B)/p(B)$ Si and (b) n^+ (As)/$p(B)$ Si. Reprinted after Peiner et al., Ref. 64 by permission of the publisher, the Electrochemical Society, Inc.

has in the past been limited to thin layers. However, 0.1M NH$_4$HF$_2$ with one drop of Triton X-100 added to 100 ml of solution electrolyte, for which $z = 3.7 \pm 0.1$, gives good results for Si. Example density profiles are shown in Fig. 2.14. The etch rate is typically a few microns/hour and depths to 20 µm are readily obtained in III–V materials. The etch rate for Si is on the order of 1 µm/hr.

The accuracy and reproducibility of ECS profiling is discussed in detail.[65] The cell and sample preparation are the greatest source of error with the most likely causes for variability being the condition of the sealing ring, the difference in the way the sample is mounted and the way trapped air bubbles are cleared from the sample surface. The ring areas should be measured at least three times a week. Ideally, the area of the etch well at the end of each run should be measured and inspected for signs of sealing ring wear or damage and for non-uniform etching, due to bubbles, etc. Sealing rings typically last for 150 measurements, with the wetted area getting progressively larger.

Problems may arise due to highly doped surface layers, high contact resistance or poor etching. A highly-doped surface layer, particularly for n-type material, creates difficulties for measuring underlying lowly doped layers, due to seepage at the edge of the ring. Complications arise if the sample exhibits significant parallel conduction, as the device can no longer be modeled by a simple two-element series or parallel model. The presence of crystalline defects in the sample can also lead to uneven etching.

2.3 CURRENT-VOLTAGE (I-V)

2.3.1 MOSFET Substrate Voltage—Gate Voltage

Differential capacitance profile measurements are typically made at frequencies of 0.1–1 MHz on large-diameter devices to reduce stray capacitances and increase the signal/noise ratio. These constraints make measurements on small-geometry MOSFETs difficult because the capacitance is extremely small. To overcome this limitation, several methods have been developed allowing the doping density profile to be extracted from MOSFET current-voltage measurements.

In the MOSFET substrate voltage-gate voltage method the MOSFET is biased in its *linear* region by a low drain-source voltage V_{DS} and an appropriate gate-source voltage V_{GS}. A source-substrate or body potential V_{SB} forces the space-charge region under the gate to extend into the substrate, allowing the doping density profile to be obtained. The inversion charge density is held constant, approximated by a constant drain current, by adjusting V_{GS} whenever V_{SB} is changed. The relevant equations are[66–67]

$$p(W) = \frac{K_{ox}\varepsilon_o}{q\,K_s t_{ox}^2} \frac{d^2 V_{SB}}{d V_{GS}^2} \tag{2.25}$$

$$W = \frac{K_s\varepsilon_o}{C_{ox}} \frac{d V_{SB}}{d V_{GS}} \tag{2.26}$$

A feedback circuit to implement this technique is shown in Fig. 2.15(a). V_{DS} is held constant, V_{GS} is varied, and a constant current I_1 is applied between the input terminals of the operational amplifier connected between the source (S) and ground. With the operational amplifier differential input voltage and input current nearly zero, current I_1 is forced through the MOSFET and the drain current is $I_D = I_1$. When V_{GS} is changed, the op amp adjusts its output voltage, *i.e.*, the voltage V_{SB} between the source and substrate (B), to maintain $I_D = I_1$.[68] A modified version of this technique, where the restriction to slowly varying doping density profiles is overcome by approximating the substrate doping density as a simple analytic function, has been proposed.[69]

The assumption that constant drain current corresponds to constant inversion charge is only true to a first approximation. It is known that in a MOSFET the effective mobility varies with gate voltage (see Chapter 8), requiring a correction in the analysis.[67, 70] However, for the commonly used mobility expression $\mu_{eff} = \mu_o/[1 + \theta(V_{GS} - V_T)]$, the mobility dependence on gate voltage does not affect the profile.[71] The drain-source voltage should be maintained below about 100 mV to ensure linear MOSFET operation, and the profile is affected by short-channel effects.[67, 72]

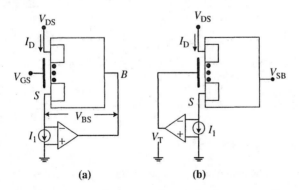

Fig. 2.15 Operational amplifier circuit for (a) the MOSFET substrate/gate voltage method, (b) the MOSFET threshold voltage method.

2.3.2   MOSFET Threshold Voltage

In the MOSFET threshold voltage profiling technique, the *threshold voltage* is measured as a function of substrate bias.[73-75] The threshold voltage of a MOSFET is

$$V_T = V_{FB} + 2\phi_F + \frac{\sqrt{2qK_s\varepsilon_o N_A(2\phi_F + V_{SB})}}{C_{ox}} = V_{FB} + 2\phi_F + \gamma\sqrt{2\phi_F + V_{SB}} \quad (2.27)$$

where $\gamma = (2qK_s\varepsilon_o N_A)^{1/2}/C_{ox}$ and the substrate bias $V_{SB} = V_S - V_B$ is positive for n-channel devices. The doping density profile is obtained by measuring V_T as a function of V_{SB}, plotting V_T against $(2\phi_F + V_{SB})^{1/2}$ and measuring the slope $\gamma = dV_T/d(2\phi_F + V_{SB})^{1/2}$ of this plot. The doping density is from Eq. (2.27)

$$N_A = \frac{\gamma^2 C_{ox}^2}{2qK_s\varepsilon_o} \quad (2.28)$$

assuming we can neglect variations of $d(2\phi_F)/d(2\phi_F + V_{SB})^{1/2}$. The profile depth is

$$W = \sqrt{\frac{2K_s\varepsilon_o(2\phi_F + V_{SB})}{qN_A}} \quad (2.29)$$

In Eq. (2.28) ϕ_F depends on N_A [$\phi_F = (kT/q)\ln(N_A/n_i)$], but N_A is not known *a priori*. A suitable approach is to plot V_T versus $(2\phi_F + V_{SB})^{1/2}$ using $2\phi_F = 0.6$ V. Then take the slope and find N_A. With this value of N_A find a new ϕ_F, replot V_T versus $(2\phi_F + V_{SB})^{1/2}$, repeating the procedure until a profile is obtained. One or two iterations usually suffice. In Fig. 2.16 we show doping density profiles obtained from MOSFET threshold voltage, spreading resistance, and pulsed MOS-C $C-V_G$ measurements. The pulsed MOS-C measurements were made on a test MOS-C structure processed identically to the MOSFET. The data are compared to a SUPREM3 calculated profile. The threshold voltage technique can also be used for depletion-mode devices.[74-75]

The threshold voltage is measured as a function of substrate bias with the circuit in Fig. 2.15(b). This method is discussed in more detail in Chapter 4 Section 4.8, as the

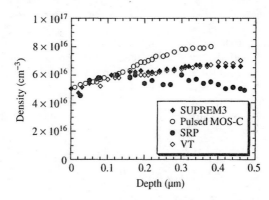

Fig. 2.16 Dopant density profiles determined by MOSFET threshold voltage, SRP, pulsed $C-V$, and SUPREM3. Reprinted after ref. 73 by permission of IEEE (© 1991, IEEE).

constant drain current method. The current I_1 is chosen as typically $I_1 \approx 1\ \mu\text{A}$. The output of the op amp gives the threshold voltage directly.

2.3.3   Spreading Resistance

Spreading resistance profiling is commonly used for Si. The sample is beveled, and two spreading resistance probes are stepped along the beveled surface. The spreading resistance is measured as a function of sample depth, and the doping density profile is calculated from the measured resistance profile. This technique is discussed in Section 1.4.2. Very high resolution profiles can be generated by using shallow bevel angles. An application of SRP to very thin MBE Si layers is given by Jorke and Herzog, who also discuss carrier spilling and low-high and high-low transitions.[76]

2.4   MEASUREMENT ERRORS AND PRECAUTIONS

Many $C-V$ measurements are made with no corrections of any kind because such corrections often only produce small changes in the measured doping density profile. Sometimes corrections are not made because the experimenter is unaware of possible corrections or they are too difficult to make. Nevertheless, one should be aware of possible measurement errors and means of correcting them.

2.4.1   Debye Length and Voltage Breakdown

The Debye length limitation is discussed in Section 2.2.1 and in numerous papers.[14-28, 77] To summarize briefly, mobile majority carriers do not follow the profile of the dopant atoms if the dopant density profile varies spatially over distances less than the Debye length. The majority carriers are more smeared out than the dopant atoms and a measured profile of steep dopant gradients (abrupt high-low junctions and steep-gradient ion implanted samples) will result in neither the doping nor the majority carrier density profile. Instead an effective or apparent carrier density profile is obtained, which is closer to the majority carrier density profile than to the doping density profile. It is possible to correct the measured profile by iterative calculations,[23] but due to the mathematical complexity this is rarely done.

Another consequence of the Debye length limitation is the inability to profile closer than about $3L_D$ from the surface using MOS devices. Although corrections are possible to calculate the profile to the surface, this is not routinely done. Even considering the Debye length limitation, it is possible to profile closer to the surface with MOS-Cs and MOSFETs than it is with Schottky barrier diodes or pn junctions. For MOS devices the limit is approximately $3L_D$, for Schottky diodes it is approximately the zero-bias scr width W_{0V}, and for pn junctions it is the junction depth plus the zero-bias scr width. The $3L_D$ limit is shown as the lower profile depth limit in Fig. 2.17.

For degenerately doped semiconductors the resolution is limited by the Thomas-Fermi screening length L_{TF} rather than the Debye length.[78] L_{TF} is given by

$$L_{TF} = \left(\frac{\pi}{3(p+n)}\right)^{1/6} \sqrt{\frac{\pi K_s \varepsilon_o \hbar^2}{q^2 m^*}} \tag{2.30}$$

where $\hbar$ is Planck's constant and m^* is the effective mass. For semiconductors with quantum confinement, *i.e.*, δ-doped semiconductors as well as compositional quantum

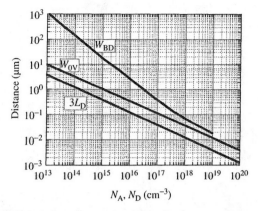

Fig. 2.17 Spatial profiling limits. The "3 L_D" line is the lower limit for conventional MOS-C profiling, the zero bias "W_{0V}" line is the lower limit for pn and Schottky diode profiling, and the "W_{BD}" line is the upper profile limit governed by bulk breakdown.

wells, the resolution is limited by the spatial extent of the ground state wave function given by[78]

$$L_\delta = 2\sqrt{\frac{7}{5}} \left(\frac{4K_s\varepsilon_o\hbar^2}{9q^2N^{2D}m^*} \right)^{1/3} \tag{2.31}$$

where N^{2D} is the two-dimensional doping density in units of cm^{-2}, for example. This equation shows the resolution of high effective mass materials to be better than for low m^* materials. For example, the resolution for p-GaAs is better than that for n-GaAs.

When the profile is generated by sweeping a reverse-bias voltage, the upper profile depth limit is determined by semiconductor breakdown. The space-charge region obviously no longer increases beyond breakdown. The breakdown limit is also shown on Fig. 2.17 as W_{BD}. Breakdown considerations do not apply to the electrochemical profiler. A theoretical study incorporating Debye length and breakdown limitation as well as majority carrier diffusion in steep-gradient profiles gives the dose and energy limits of Si and GaAs ion-implanted layers that can be profiled by the differential capacitance technique.[26]

2.4.2 Series Resistance

A pn or Schottky diode consists of a junction capacitance C, a junction conductance G, and a series resistance r_s, as shown in Fig. 2.18(a). The conductance governs the junction leakage current and depends on processing conditions. The series resistance depends on the bulk wafer resistivity and on the contact resistances. Capacitance meters assume the device to be represented by either the parallel equivalent circuit in Fig. 2.18(b) or the series equivalent circuit in Fig. 2.18(c). Comparing the two circuits to the original Fig. 2.18(a) circuit, allows C_P, G_P, C_S, and R_S to be written as (see Appendix 2.2)[79]

$$C_P = \frac{C}{(1 + r_sG)^2 + (\omega r_sC)^2}; G_P = \frac{G(1 + r_sG) + r_s(\omega C)^2}{(1 + r_sG)^2 + (\omega r_sC)^2} \tag{2.32}$$

$$C_S = C[1 + (G/\omega C)^2]; R_S = r_s + \frac{1}{G[1 + (\omega C/G)^2]} \tag{2.33}$$

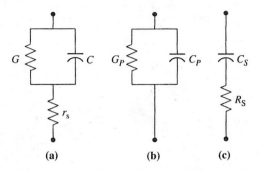

Fig. 2.18 (a) Actual circuit, (b) parallel equivalent circuit, and (c) series equivalent circuit for a pn or Schottky diode.

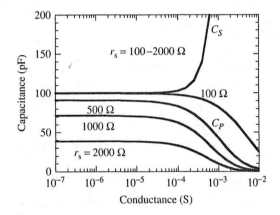

Fig. 2.19 C_S and C_P versus G as a function of r_s. $C = 100$ pF, $f = 1$ MHz.

where $\omega = 2\pi f$. To determine C from series connected measurements at two different frequencies, C_S in Eq. (2.33) can be written as

$$C = \frac{\omega_2^2 C_{S2} - \omega_1^2 C_{S1}}{\omega_2^2 - \omega_1^2} \tag{2.34}$$

where C_{S1} and C_{S2} are the measured capacitances at frequencies ω_1 and ω_2, respectively.

The capacitances C_P and C_S are plotted in Fig. 2.19. C_S is independent of the series resistance r_s, whereas C_P depends strongly on r_s. Both capacitances deviate from C at high G. With the quality factor Q for a parallel circuit defined by $Q = \omega C/G$, we find the true capacitance to be measured for $Q \geq 5$. Figure 2.19 clearly shows that for junction devices with $Q \geq 5$, the series equivalent circuit is the one to use for capacitance measurements if series resistance is suspected.

A real device may have series resistance and capacitance as parasitic elements, shown in Fig. 2.20. This is the case if the back contact is an evaporated metal contact without ohmic contact formation. For example, if a metal is deposited on the wafer front to form a Schottky diode for $C-V$ measurements, the same metal deposited on the wafer back also

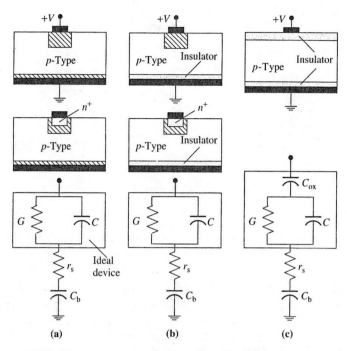

Fig. 2.20 Equivalent circuits with series resistance and capacitance for (a) front and rear Schottky contacts, (b) front Schottky and rear oxide contact, and (c) front and rear oxide contacts. The elements within the rectangles represent the intrinsic device.

forms a Schottky diode, as in Fig. 2.20(a). Fortunately, the back contact usually has much higher capacitance because it has larger area than the front contact and the back Schottky diode is forward biased when the front Schottky diode is reverse biased. Having two back-to-back Schottky diodes allows the necessary current to flow to bias the front diode. If the back contact consists of an insulator, as in 2.20(b), the front Schottky or pn diode is always zero biased, since there is no dc current flow. Hence this configuration does not work for dc doping profiling. On the other hand, the arrangement in 2.20(c), consisting of MOS contacts on the front and the back will work, since MOS C–V measurements do not require dc current to flow.

One of the problems with the configuration in 2.20(a) is the voltage distribution between front and back contacts. Although most of the applied voltage drops across the front reverse-biased junction, a portion drops across the back forward-biased rear junction. The measured voltage is, of course, the total voltage. The effect of this is illustrated in Fig. 2.21,[80] showing $1/C^2$–V plots of an n-Si wafer with front and back Schottky and with front Schottky and back ohmic contacts. An interesting feature is the negative voltage intercept, attributed to the distribution of the applied bias voltage between the front and the back contact diodes for small voltages. Since $1/C^2$–V curves are also used to determine the junction built-in potential V_{bi}, this curve will obviously yield an incorrect V_{bi}. To determine the correct built-in potential the curve must be shifted to the right. The $1/C^2$–V curve becomes "normal" when the back Schottky contact is a sintered Au/Sb ohmic contact.

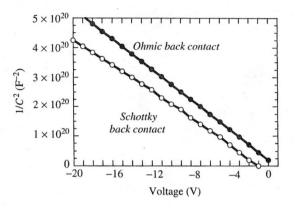

Fig. 2.21 $1/C^2$ versus voltage curves for n-Si wafers with $A = 3.14 \times 10^{-2}$ cm^2, $t = 640$ μm, $N_D \sim 5 \times 10^{14}$ cm^{-3}. Curve (a): front and back Al Schottky contacts, (b): front Au/Pd Schottky and back Au/Sb ohmic contacts. After Mallik et al., ref. 80.

Care must be exercised when preparing samples for capacitance measurements, especially if the device is at the wafer stage and measurements are made on a probe station. If the wafer is provided with a metallic back contact, there is usually no problem, provided the wafer resistivity itself does not contribute significant series resistance. However, wafers placed on a probe station without any back metallization can have appreciable contact resistance. This can be checked by reducing the measurement frequency. If C_P increases it is likely a series resistance problem. Measurement of C_S does not have this problem. It is important that a vacuum be pulled for all probe capacitive measurements to reduce the resistance between the wafer and the probe chuck. If an MOS device, *e.g.*, MOS capacitor or MOSFET is measured, and if the back contact resistance is a problem, it may be advantageous to leave the oxide on the back surface and place the wafer on the probe station making a large-area *capacitive* back contact (Fig. 2.20(c)). The contact capacitance C_b, much larger than the device capacitance because its area is usually the area of the entire sample, approximates a short circuit.

Series resistance also interferes with dopant profile measurements. For a wafer with negligible series resistance, there is zero phase shift between the rf voltage applied to the device and the rf current flowing through it when the *conductance* is measured. For the *capacitance* measurement there is a 90° phase shift, which is the basis of phase-sensitive capacitance measurements. When series resistance is not negligible, an additional phase shift ϕ is introduced into the measurement. This must be taken into account or the measured dopant profile determined from Eqs. (2.5) and (2.6) will be in error.[81]

An approximate way to consider series resistance is from Eqs. (2.5a), (2.6) and (2.32) with $r_s G \ll 1$. It can be shown that the measured density, $N_{A,meas}(W)$, and depth, W_{meas}, are related to N_A and W by the relationships

$$N_{A,meas} = \frac{N_A}{1 - (\omega r_s C)^4} \tag{2.35}$$

$$W_{meas} = W[1 + (\omega r_s C)^2] \tag{2.36}$$

Clearly, both density and depth increase with series resistance.

Exercise 2.2

Problem: The parallel circuit (Fig. 2.18(b)) $C_P - V$ curve of an n$^+$p junction, measured at a frequency of 1 MHz, is shown in Fig. E2.1. It is suspected that series resistance is significant in this device. An additional measurement at $f = 10$ kHz and lower frequencies confirmed this because $C(10$ kHz$) = 200$ pF at zero volts. The effect of series resistance is negligible at 10 kHz. $A = 4.25 \times 10^{-3}$ cm^2.

Determine the series resistance r_s and the carrier density profile. The conductance G of this device is negligibly small.

Solution: Solving Eq. (2.32) for r_s, neglecting the $r_s G$ term, gives $r_s = (1/\omega C)$ $\sqrt{C/C_P - 1}$.

With $C_P = 94$ pF and $C = 200$ pF, we find $r_s = 845\ \Omega$.

Now solving Eq. (2.32) for C gives $C = \dfrac{1 - \sqrt{1 - 4(\omega r_s C_P)^2}}{2C_P(\omega r_s)^2}$

Substituting $r_s = 845\ \Omega$ and the C_P from Fig. E2.1, gives the plot in Fig. E2.2(a). Replotting as $1/C^2$ is also shown as is the slope $d(1/C^2)/dV$ in Fig. E2.2(b). From

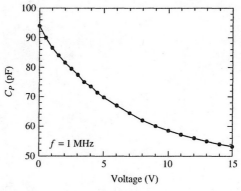

Fig. E2.1

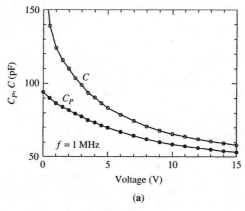

(a)

Fig. E2.2

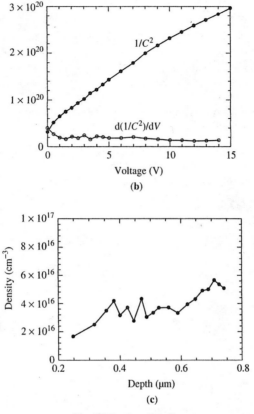

Fig. E2.2   (*continued*)

Eq. (2.5(b)) we find $N_A = 6.7 \times 10^{37}/[d(1/C^2)/dV]$. Using the slope $d(1/C^2)/dV$ and Eq. (2.6) gives the carrier density profile in Fig. E2.2(c).

Another approach is to write C_P in Eq. (2.32) as

$$\frac{1}{C_P} = \frac{(1 + r_s G)^2 + (2\pi f r_s C)^2}{C} \approx \frac{1 + (2\pi f r_s C)^2}{C}$$

Then plot $1/C_P$ versus f^2. The slope is $(2\pi r_s)^2 C$ and the intercept is $1/C$, allowing both r_s and C to be determined.

The effect of series resistance on a dopant profile of an epitaxial GaAs layer grown on a semi-insulating substrate is illustrated in Fig. 2.22. The correct profile is the one labeled $r_s = 0$. To obtain the other curves, external resistors were placed in series with the device to demonstrate the effect. Semiconducting layers on insulating or semi-insulating substrates are particularly prone to series resistance effects since both contacts are made on the top surface and lateral series resistance can be substantial.[82] For more details of capacitance measurements for devices with leaky junctions, wafer chuck parasitic capacitance and other considerations see Appendix A6.2.

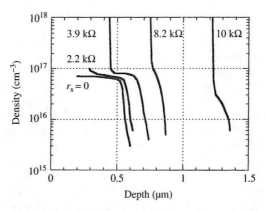

Fig. 2.22 Measured dopant profiles for a GaAs epitaxial layer on a semi-insulating substrate. The series resistance was obtained by placing resistors in series with the device. Reprinted after ref. 81 by permission of IEEE (© 1975, IEEE).

2.4.3 Minority Carriers and Interface Traps

In a reverse-biased Schottky barrier or pn junction diode, the scr width remains constant as a function of time because thermally generated electron-hole pairs are swept out of the scr and leave through the ohmic contacts of the device. Thermally generated minority carriers in a deep-depleted MOS capacitor (MOS-C), on the other hand, drift to the SiO_2-Si interface to form an inversion layer and the device is unable to remain in deep depletion, leading to errors in doping density profile measurements. For a more complete discussion of the behavior of MOS capacitors in their non-equilibrium or deep-depletion state see Section 7.6.2. Minority carriers can be neglected when the MOS-C is driven rapidly into deep depletion by applying a high ramp rate gate voltage. Alternately, a pulse train of successively higher gate voltage pulses can be applied with the device being cycled between accumulation and deep depletion.

The effect of minority carriers is shown on Fig. 2.23. When the MOS-C is driven into deep depletion by a rapidly varying ramp voltage, curve (i) in Fig. 2.23(a) results.

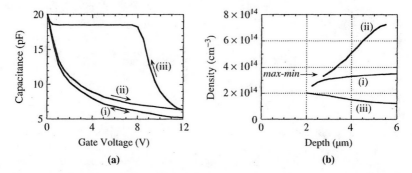

Fig. 2.23 (a) Equilibrium C–V_G curve of an MOS-C, (b) deep-depletion curves for (i) 5 V/s and (ii), (iii) 0.1 V/s sweep rates, (c) the carrier density profiles determined from (b). $C_{ox} = 98$ pF, $t_{ox} = 120$ nm. Courtesy of J.S. Kang, Arizona State University.

For negligible minority carrier generation the curve is identical for the gate voltage being swept from left to right or from right to left as indicated by the arrows. The doping density profile obtained from this curve is shown in Fig. 2.23(b) by (i). If the curve is swept very slowly, then the equilibrium high-frequency curve is obtained. For an intermediate sweep rate curve (ii) results. This curve lies above curve (i) and the extracted doping density profile, shown in Fig. 2.23(b) by (ii), is in error because dC/dV for (ii) is lower than dC/dV for (i). If curve (ii) is swept from right to left, resulting in curve (iii), its doping density profile is lower for similar reasons, as shown in Fig. 2.23(b) by curve (iii). It is possible to correct for these effects but corrections are not necessary for high sweep rates.[83]

Using the *max-min MOS-C capacitance* method to determine N_A, we find for equilibrium $C_{min}/C_{ox} = 0.19$, coupled with $t_{ox} = 120$ nm $N_A \approx 3.5 \times 10^{14}$ cm^{-3}. This value is very close to curve (i) in Fig. 2.23(b). Of course, the C_{min}/C_{ox} approach does not give a doping density profile, but considering its simplicity, it yields a density that compares favorably with the differential capacitance derived value.

The effects of *minority carrier* generation are a problem for high carrier generation rates in devices with low generation lifetimes. It is more difficult to drive the MOS-C into deep depletion under those conditions. Cooling to liquid nitrogen temperatures for high generation rates works well to reduce the effects of minority carrier generation.[84] Providing a collecting junction is another way to reduce the effect of minority carriers. As soon as minority carriers are generated, they are collected by the reverse-biased junction as in MOSFETs with source and drain reverse biased and in gate-controlled diodes.

A further complication is introduced by interface traps invariably present in all MOS capacitors. The interface trap density is usually negligibly low for properly annealed, high quality SiO_2-Si interfaces. When interface states do play a role, they cause the $C-V$ curves to be stretched out. Their effect on doping profiling can be corrected by measuring the high-frequency capacitance C_{hf} and the low-frequency capacitance C_{lf} according to[85]

$$N_{A,corr} = \frac{1 - C_{lf}/C_{ox}}{1 - C_{hf}/C_{ox}} N_{A,uncorr} \tag{2.37}$$

The effects of interface traps are considerably reduced in the pulsed MOS-C doping density profile technique when the modulation frequency is increased. Modulation frequencies of 30 MHz have been suggested,[19] but most measurements are made at 1 MHz or lower. Interface trap effects are also reduced when the device is cooled. Interface traps or interfacial layers can also give errors in Schottky barrier capacitance profiling. It has been found that if the diode ideality factor n is larger than 1.1, erroneous profiles are obtained.[86] Ideality factors $n \leq 1.1$ are satisfactory for profiling.

2.4.4 Diode Edge and Stray Capacitance

$C-V$ profiling relies on an accurate knowledge of the capacitance and of the device area. While the capacitance can be accurately measured, the area cannot always be accurately determined. Furthermore the capacitance may contain stray capacitance components. The device contact area can be measured but the effective area differs from the contact area due to lateral space-charge region spreading. The effective capacitance is[87]

$$C_{eff} = C(1 + bW/r) \tag{2.38}$$

where $C = K_s \varepsilon_o A/W$, $A = \pi r^2$, r is the contact radius, $b \approx 1.5$ for Si and GaAs, and $b \approx 1.46$ for Ge. Eq. (2.38) assumes the lateral extent of the space-charge region to be identical to the vertical extent. The lateral scr effect diminishes as the contact radius increases and $r \geq 100$ bW ensures for the second term in the bracket to contribute no more than 1% to the effective capacitance. For $W = 1$ μm, $r \geq 150$ μm whereas for $W = 10$ μm, $r \geq 1500$ μm. This is not a particularly severe limitation. It should be considered, however, because the effective doping density is related to the actual doping density by

$$N_{A,eff} = (1 + bW/r)^3 N_A \tag{2.39}$$

Equation (2.38) shows the edge capacitance to be a constant, and it can be nulled prior to differential profile measurements by using a dummy capacitor of an appropriate value. For mercury-probe profiling it has been proposed to make the contact sufficiently large that the edge capacitance effects can be neglected. The minimum recommended contact radius depends on the substrate doping density and should be[53]

$$r_{min} = 0.037(N/10^{16})^{-0.35} \text{ cm} \tag{2.40}$$

where N is the doping density. Equation (2.40) is valid for the doping range of 10^{13} to 10^{16} cm^{-3}. The minimum radius is about 8.3×10^{-2} cm for $N = 10^{15}$ cm^{-3}.

A diode junction capacitance consists of the true capacitance, C, the perimeter capacitance, C_{per} and the corner capacitance C_{cor}. The effective capacitance can be approximated by[88]

$$C_{eff} = AC + PC_{per} + NC_{cor} \tag{2.41}$$

where A is the area, P the perimeter, and N the number of corners. By using diodes with various areas and perimeters, it is possible to separate the various components and extract the true diode capacitance.[88]

Stray capacitance is more difficult to determine. It includes cable and probe capacitances, bonding pads, and gate protection diodes in MOSFETs. Cable and probe capacitances can be eliminated by nulling the capacitance meter without contact to the diode. Bonding pad capacitance can usually be calculated. Since the diode, MOS-C, or MOS-FET can be made much smaller than the bonding pad, it becomes important to know the bonding pad capacitance contribution accurately.

2.4.5 Excess Leakage Current

Junction devices occasionally show excessively high reverse-biased leakage currents leading to erroneous doping density profiles, especially for Schottky barrier devices. The assumption in the conventional profile equations is that the voltage is measured across the reverse-biased space-charge region only. For most devices that is a good assumption since the resistance of the reverse-biased scr is much higher than the semiconductor quasi-neutral region resistance. For excessive leakage currents, however, an appreciable voltage can be developed across the quasi-neutral regions. This voltage is automatically included in the recorded voltage introducing errors in the measured profiles.[89]

2.4.6 Deep Level Dopants/Traps

Capacitance measurements, being a measure of charge responding to an applied time-varying voltage, will detect any charge that responds to the applied voltage. We have already considered the contribution of interface traps to the capacitance. Deep level impurities or traps in the semiconductor bulk can also produce errors in capacitance profiles.[90–92] The contribution of traps is a complicated function of the density and energy level of the traps as well as the sample temperature and the frequency of the ac voltage. The ac voltage frequency is often assumed to be sufficiently high for the traps to be unable to follow it. Even if that is true, there is still cause for concern because the reverse bias dc voltage usually changes sufficiently slowly for the traps to be able to respond. This can give rise to profile errors that are both time and depth dependent. Fortunately, for trap densities much less than the doping density, say 1% or less, the contribution of traps is usually negligible. Capacitance measurements of traps are discussed in Chapter 5.

A potential problem arises for deep-lying dopant atoms not fully ionized at the measurement temperature. For the common dopants, *e.g.*, P, As, and B in Si and Si in GaAs, this is of no concern. However, for SiC, for example, some dopant energy levels can lie deep in the band gap. Consider the reverse-biased Schottky contact on a p-type substrate illustrated in Fig. 2.24. The dopant impurity has an energy level $E_A = E_v + \Delta E$. In the quasi-neutral region (qnr) the impurities are only partially ionized. The unionized, neutral atoms are indicated by N_A^o. Obviously $p \neq N_A$ in the qnr and the resistivity ρ is not uniquely related to N_A since $\rho \sim 1/p$. The degree of ionization depends on ΔE, N_A, and the temperature. However, in the space-charge region (scr) the situation is different. Let us assume the reverse bias V_1 has been applied for a sufficiently long time that all

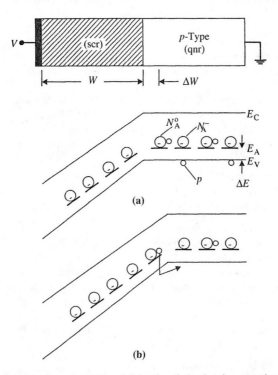

Fig. 2.24 Band diagram of a reverse-biased Schottky diode showing complete ionization in the space-charge region (scr) but only partial ionization in the quasi-neutral region (qnr). (a) $V = V_1$, (b) $V = V_1 + \Delta V$.

holes have been emitted from neutral acceptors. The emission time constant, discussed in Chapter 5, is

$$\tau_e = \frac{\exp(\Delta E/kT)}{\sigma_p v_{th} N_v} \tag{2.42}$$

where σ_p is the capture cross section, v_{th} the thermal velocity, and N_v the effective density of states in the valence band.

Now consider an ac voltage superimposed on the dc voltage with the ac voltage swinging positively causing the scr width to increase from W to $W + \Delta W$. Some of the neutral acceptors originally in the qnr now find themselves in the scr. If $\tau_e < 1/\omega$, where $\omega = 2\pi f$, then those holes trapped on acceptors will be emitted during the ac half cycle and the device behaves as a normal, shallow-level acceptor device. However, for $\tau_e > 1/\omega$ there is insufficient time for hole emission and the device will behave abnormally. The premise that it is p or N_A that is measured in a uniformly doped sample is no longer true. What is measured is an effective carrier density related to the doping density in an unknown way. During the negative ac voltage swing, the scr narrows and holes are captured rather than emitted. Capture is usually very fast and does not constitute a limit. It is emission that is the limit since τ_e depends exponentially on ΔE. Whether the true carrier or dopant density profile can be determined depends on the energy level of the dopant, the temperature, and the measurement frequency. The case of In in Si, whose energy level is at $E_V + 0.16$ eV, has been discussed by Schroder et al.[90] A more general treatment directed at traps in a semiconductor containing shallow level dopants is given by Kimerling.[91]

2.4.7 Semi-Insulating Substrates

Epitaxial or implanted layers on semi-insulating or insulating substrates present unique profiling problems. Examples include silicon-on-insulator and GaAs implanted layers on semi-insulating substrates. Due to the high resistance of the substrate, both contacts must be made to the top surface, introducing series resistance, especially when the reverse-biased scr extends close to the substrate as illustrated in Fig. 2.25. The remaining neutral region of the layer, indicated by the thickness t, becomes very thin and appreciable series resistance r_s results. Similar problems occur when an n-type (p-type) layer is formed on a p-type (n-type) substrate. The measured density profiles sometimes exhibit minima

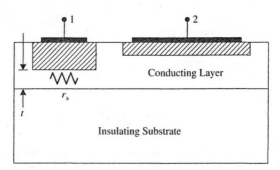

Fig. 2.25 Conducting layer on an insulating substrate showing the increasing series resistance with increasing back bias on contact 1.

near the interface between the two. Such minima are usually not real, but are artificially introduced by the sample geometry.[93]

An additional word of caution. Contact 1 in Fig. 2.25 should be rectifying and contact 2 should be ohmic. That is usually not possible when the conducting layer is lightly doped. In that case one should make contact 2, which is forward biased when contact 1 is reverse biased, much larger than contact 1. This ensures the C_2 to be much higher than C_1 because $A_2 \gg A_1$ and contact 2 is forward biased. As a first approximation, C_2 can be treated as a short circuit and C_1 is measured.

2.4.8 Instrumental Limitations

Capacitance meters determine the accuracy with which $p(x)$ and W are measured. The depth resolution should be limited by the Debye length rather than by the instrument. The overriding influence on the accuracy of $p(x)$ is the precision with which ΔC is measured.[94] There is a temptation to make ΔC large, but this introduces errors in the determination of the local value of $\Delta C / \Delta V$ because $C-V$ curves are not linear. It also degrades the depth resolution by increasing the modulation of W. It is common practice in analog profilers to keep ΔV constant by using a modulation voltage of constant amplitude. According to Eqs. (2.9) and (2.19)

$$\Delta V = \frac{q W p(W) \Delta W}{K_s \varepsilon_o} \quad \text{and} \quad \frac{\Delta W}{W} = -\frac{\Delta C}{C} \tag{2.43}$$

so that

$$\Delta C = -\frac{K_s \varepsilon_o C \Delta V}{q W^2 p(W)} \tag{2.44}$$

For constant $p(W)$ and constant ΔV, ΔC decreases as the sample is profiled because w increases and C decreases. Consequently profiles become noisier as the profile is measured deeper into the sample. Constant electric field increment feedback profilers alleviate this problem somewhat. An excellent discussion of instrumental limitations is given by Blood.[55]

2.5 HALL EFFECT

Those aspects of the *Hall effect* pertaining to carrier density measurements are discussed here. A more complete treatment of the Hall effect, including a derivation of the appropriate equations, is given in Chapter 8. The key feature of Hall measurements is the ability to determine the *carrier density*, the *carrier type*, and the *mobility*.

Hall theory predicts the Hall coefficient R_H as[95]

$$R_H = \frac{r(p - b^2 n)}{q(p + bn)^2} \tag{2.45}$$

where $b = \mu_n / \mu_p$ and r is the scattering factor whose value lies between 1 and 2, depending on the scattering mechanism in the semiconductor.[95] The scattering factor is also a function of magnetic field and temperature. In the high magnetic field limit $r \to 1$. The scattering factor can be determined by measuring R_H in the high magnetic field limit, *i.e.*,

$r = R_H(B)/R_H(B = \infty)$ where B is the magnetic field. The scattering factor in n-type GaAs was found to vary from 1.17 at $B = 0.1$ kG to 1.006 at $B = 83$ kG.[96] The high fields necessary for r to approach unity are not achievable in most laboratories. Typical magnetic fields are 0.5 to 10 kG, making $r > 1$ for typical Hall measurements. Since r is usually not known, it is frequently assumed to be unity.

The Hall coefficient is determined experimentally as

$$R_H = \frac{t V_H}{B I} \tag{2.46}$$

where t is the sample thickness, V_H the Hall voltage, B the magnetic field, and I the current. The thickness is well defined for uniformly doped wafers. However, the active layer thickness is not necessarily the total layer thickness for thin epitaxial or implanted layers on substrates of opposite conductivity type or on semi-insulating substrates. If depletion effects caused by Fermi level pinned band bending at the surface and by band bending at the layer-substrate interface are not considered, the Hall coefficient will be in error as will those semiconductor parameters derived from it.[97] Even the temperature dependence of the surface and interface space-charge regions should be considered for unambiguous measurements.[98]

For extrinsic p-type material with $p \gg n$, Eq. (2.45) reduces to

$$R_H = \frac{r}{qp} \tag{2.47}$$

and for extrinsic n-type it becomes

$$R_H = -\frac{r}{qn} \tag{2.48}$$

A knowledge of the Hall coefficient leads to a determination of the *carrier type* as well as the *carrier density*, according to Eq. (2.47) and (2.48). Usually r is assumed to be unity—an assumption generally introducing an error of less than 30%.[99]

The Hall effect is used to measure the carrier density, resistivity and mobility at a given temperature, and the carrier density as a function of temperature to extract additional information. For a p-type semiconductor of doping density N_A compensated with donors of density N_D, the hole density is determined from the equation[100]

$$\frac{p(p + N_D) - n_i^2}{N_A - N_D - p + n_i^2/p} = \frac{N_v}{g} \exp(-E_A/kT) \tag{2.49}$$

where N_v is the effective density of states in the valence band, g the degeneracy factor for acceptors (usually taken as 4), and E_A the energy level of the acceptors above the valence band with the top of the valence band as the reference energy. Equation (2.49) can be simplified for certain conditions.

1. At low temperatures where $p \ll N_D$, $p \ll (N_A - N_D)$, and $n_i^2/p \approx 0$

$$p \approx \frac{(N_A - N_D)N_v}{gN_D} \exp(-E_A/kT) \tag{2.50}$$

2. When N_D is negligibly small,

$$p \approx \sqrt{\frac{(N_A - N_D)N_v}{g}} \exp(-E_A/2kT) \tag{2.51}$$

3. At higher temperatures where $p \gg n_i$,

$$p \approx N_A - N_D \tag{2.52}$$

4. At still higher temperatures, where $n_i \gg p$

$$p \approx n_i \tag{2.53}$$

According to Eqs. (2.50) and (2.51), the slope of a $\log(p)$ versus $1/T$ plot gives an activation energy of either E_A or $E_A/2$, depending on whether there is a compensating donor density in the material or not. At higher temperatures, typically room temperature, the net majority carrier density is obtained with zero activation energy. At still higher temperatures the activation energy is that of n_i.

The experimental $\log(p)$ versus $1/T$ data can be fitted with an appropriate model, and a wealth of information can be extracted. Figure 2.26 shows the Hall carrier density data for an indium-doped silicon sample.[101] In addition to In, the sample contains Al, B, and P. For the acceptors (B, Al, and In) both the densities and the energy levels were extracted from the data. This figure demonstrates the powerful nature of Hall measurements.

Hall measurements are generally made on samples from which an average carrier density is derived. For uniformly doped samples the true density is obtained, but for non-uniformly doped samples an average value is determined. Occasionally one wants to measure spatially varying carrier density profiles. The Hall technique is suitable through differential Hall effect (DHE) measurements. Layers can be stripped reliably by anodic oxidation and subsequent oxide etch. Anodic oxidation consumes a certain fraction of the

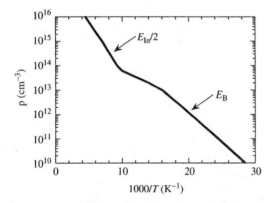

Fig. 2.26 Carrier density vs. reciprocal temperature for Si:In with Al and B contamination. $N_{In} = 4.5 \times 10^{16}$ cm^{-3}, $E_{In} = 0.164$ eV, $N_{Al} = 6.4 \times 10^{13}$ cm^{-3}, $E_{Al} = 0.07$ eV, $N_B = 1.6 \times 10^{13}$ cm^{-3}, $N_D = 2 \times 10^{13}$ cm^{-3}. Reprinted after ref. 101 by permission of IEEE (© 1980, IEEE).

semiconductor that is removed during the oxide etch. Layers can be removed in increments as small as 2.5 nm.[102] For a further discussion of DHE, see Section 1.4.1.

The interpretation of the differential Hall data becomes more complex when successive measurements are made. In order to generate a carrier density profile, the *sheet* Hall coefficient R_{Hsh}, given by $R_{Hsh} = V_H/BI$, and the sheet conductance G_{Hsh} must be measured repeatedly. The carrier density profile is obtained from Hall coefficient versus depth and from sheet conductance versus depth curves according to the relationship[103]

$$p(x) = \frac{r(dG_{Hsh}/dx)^2}{qd(R_{Hsh}G_{Hsh}^2)/dx} \tag{2.54}$$

where $G_{Hsh} = 1/R_{Hsh}$.

Occasionally the Hall sample consists of an n or p-film on a p or n-substrate. For film and substrate of opposite conductivity, the pn junction between them is usually assumed to be an insulating boundary. If that is not true, then the Hall data must be corrected.[104] This correction must also be made if the sample consist of a layer on an oppositely doped substrate and the junction separating the two is a good insulator, but the ohmic contact to the Hall sample is alloyed through the top layer, shorting it to the substrate. This can happen if the upper layer is an unintentional type conversion as has been observed in HgCdTe.[105]

For a simple two-layer structure with an upper layer of thickness t_1 and conductivity σ_1 and a substrate of thickness t_2 and conductivity σ_2 the Hall constant is[105–106]

$$R_H = R_{H1}\frac{t_1}{t}\left(\frac{\sigma_1}{\sigma}\right)^2 + R_{H2}\frac{t_2}{t}\left(\frac{\sigma_2}{\sigma}\right)^2 \tag{2.55}$$

where R_{H1} is the layer 1 Hall constant, R_{H2} is the substrate 2 Hall constant, $t = t_1 + t_2$, and σ is

$$\sigma = \frac{t_1\sigma_1}{t} + \frac{t_2\sigma_2}{t} \tag{2.56}$$

For $t_1 = 0$ we have $t = t_2$, $\sigma = \sigma_2$, and $R_H = R_{H2}$, with the substrate being characterized. If the upper layer is more heavily doped than the substrate or if it is formed by inversion through surface charges, for example, and $\sigma_2 \ll \sigma_1$, then

$$\sigma \approx \frac{t_1\sigma_1}{t}; R_H \approx \frac{tR_{H1}}{t_1} \tag{2.57}$$

and the Hall measurement characterizes the surface layer. This can be especially serious if the existence of the upper layer is not suspected.[105]

2.6   OPTICAL TECHNIQUES

2.6.1   Plasma Resonance

The optical reflection coefficient of a semiconductor is given by

$$R = \frac{(n-1)^2 + k^2}{(n+1)^2 + k^2} \tag{2.58}$$

where n is the refractive index and $k = \alpha\lambda/4\pi$ is the extinction coefficient, with α the absorption coefficient and λ the photon wavelength. The reflection coefficient of semiconductors is high at short wavelengths, tends to a constant, and then shows an anomaly at higher wavelengths. First, it decreases toward a minimum and then rises rapidly toward unity. R approaches unity when the photon frequency v, related to the wavelength through the relation $v = c/\lambda$, approaches the *plasma resonance frequency* v_p. The *plasma resonance wavelength* λ_p is given by[107]

$$\lambda_p = \frac{2\pi c}{q}\sqrt{\frac{K_s \varepsilon_o m^*}{p}} \qquad (2.59)$$

where p is the free carrier density in the semiconductor and m^* the effective mass. It is, in principle, possible to determine p from λ_p.

The plasma resonance wavelength is difficult to determine because it is not well defined. It is for this reason that the carrier density is determined not from the plasma resonance wavelength but from the wavelength λ_{min} at the *reflectivity minimum*, where $\lambda_{min} < \lambda_p$. The minimum wavelength is related to the carrier density through the empirical relationship

$$p = (A\lambda_{min} + C)^B \qquad (2.60)$$

where the constants A, B, and C are tabulated in ref. 108. The technique is useful only for carrier densities higher than 10^{18} to 10^{19} cm^{-3}.

The carrier densities determined with this technique are for uniformly doped substrates or for uniformly doped layers with layer thicknesses at least equal to $1/\alpha$. For diffused or implanted layers with varying carrier density profiles, a determination of the surface density is only possible if the shape of the profile and the junction depth are known.[109] A further complication for thin epitaxial layers is introduced by the phase shift at the epitaxial layer–substrate interface, adding an oscillatory component to the R-λ curve, making it more difficult to extract λ_{min}.[110]

2.6.2   Free Carrier Absorption

Photons of energy $hv > E_G$, absorbed in a semiconductor, generate electron-hole pairs. Photons of energy $hv < E_G$ can excite trapped electrons from the ground state of shallow-level impurities onto excited states as discussed in Section 2.6.3. It is also possible that photons of energy $hv < E_G$ excite free electrons (holes) in the conduction (valence) band to higher energy states in the band, *i.e.*, photons are absorbed by free carriers. This is the basis of *free carrier absorption*.

The free carrier absorption coefficient for holes is given by[95]

$$\alpha_{fc} = \frac{q^3\lambda^2 p}{4\pi^2\varepsilon_o c^3 nm^{*2}\mu_p} = 5.27 \times 10^{-17}\frac{\lambda^2 p}{n(m^*/m)^2\mu_p} \qquad (2.61)$$

where λ is the wavelength, c the velocity of light, n the refractive index, m^* the effective mass, and μ_p the hole mobility. However, care should be taken during the measurement not to use wavelengths that coincide with impurity or lattice absorption lines. For example, there is an absorption line in silicon due to interstitial oxygen at $\lambda = 9.05$ μm and substitutional carbon at $\lambda = 16.47$ μm. Lattice absorption lines are found near $\lambda = 16$ μm.

By fitting curves to experimental Si data good agreement is observed for[111]

$$\alpha_{fc,n} \approx 10^{-18}\lambda^2 n; \alpha_{fc,p} \approx 2.7 \times 10^{-18}\lambda^2 p \tag{2.62}$$

where n and p are the free carrier densities in cm^{-3} for n-Si and p-Si, respectively, and the wavelength is given in units of μm. Carrier densities of 10^{17} cm^{-3} or higher can be measured by this technique. The measurement becomes difficult for lower densities because the absorption coefficient is too low to be reliably determined. A modified expression has recently been published providing better agreement between sheet resistance and free carrier absorption measurements.[112] An expression for n-GaAs is[113]

$$\alpha_{fc}(\lambda = 1.5\ \mu m) = 0.81 + 4 \times 10^{-18}n; \alpha_{fc}(\lambda = 0.9\ \mu m) = 61 - 6.5 \times 10^{-18}n \tag{2.63}$$

Free carrier absorption also lends itself to sheet resistance measurements. Good agreement with experiment has been found in transmission using the expression[111]

$$T \approx (1 - R)^2 \exp(-k\lambda^2/R_{sh}) \tag{2.64}$$

with $k = 0.15$ for n-type Si and $k = 0.3375$ for p-type Si layers, where T is the transmittance. λ is in μm and R_{sh} in ohms/square. Free carrier density maps have been generated by scanning the infrared light beam. Carrier densities as low as 10^{16} cm^{-3} have been determined with a 1 mm resolution using $\lambda = 10.6\ \mu$m.[114]

2.6.3   Infrared Spectroscopy

Infrared spectroscopy relies on optical excitation of electrons (holes) from their respective donors (acceptors) into excited states. Consider the n-type semiconductor, shown in Fig. 2.27(a). At *low temperatures* most of the electrons are "frozen" onto the donors, and the free carrier density in the conduction band is very low. The electrons are mainly located on the lowest energy level or donor ground state in Fig. 2.27(b). With photons of energy $h\nu \leq (E_C - E_D)$ incident on the sample, two optical absorption processes can occur: electrons can be excited from the ground state to the conduction band giving a broad absorption continuum, and electrons can be excited from the ground state to one

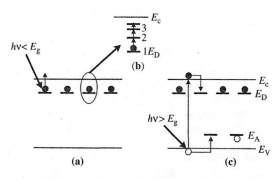

Fig. 2.27 (a) Energy band diagram for a semiconductor containing donors at low temperature, (b) energy band diagram showing the donor energy levels, (c) band diagram when both donors and acceptors are present. The "above-band gap" light fills donors and acceptors.

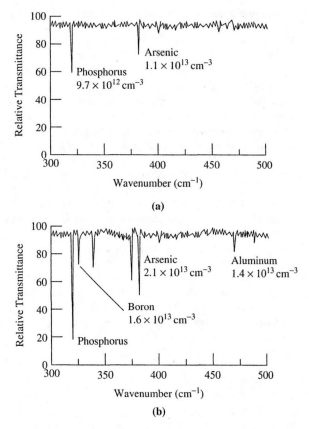

Fig. 2.28 (a) Donor impurity spectrum for 265 Ω-cm n-Si at $T \approx 12$ K, (b) spectrum for the sample in (a) with "above-band gap" illumination. Reprinted with permission after ref. 117.

of several excited states producing sharp absorption lines in the transmission spectrum, characteristic of the shallow-level impurities.[115–116] Such a transmittance curve is shown in Fig. 2.28(a) for phosphorus- and arsenic-containing silicon.[117] Additional information can be obtained by splitting the energy levels with a magnetic field.[118]

Through the use of Fourier transform techniques (Fourier transform infrared spectroscopy is discussed in Chapter 10), high sensitivity is obtained, and the detection limits are extremely low. Doping densities as low 5×10^{11} cm^{-3} have been measured in Si.[117] Such low densities can also be determined by Hall measurements, but contactless optical techniques are simpler, but require low temperatures.

Most electrical carrier density measurement methods determine the net carrier density $n = N_D - N_A$ in an n-type sample. The infrared spectroscopy technique as discussed so far also measures $N_D - N_A$, because there are only $n = N_D - N_A$ electrons frozen onto the donors at the low temperatures. Compensating acceptors are empty of holes because the holes are compensated by electrons. To measure N_D and N_A, the sample is illuminated with background light of energy $h\nu > E_G$.[117, 119–120] Some of the excess electron-hole pairs generated by the background light are captured by the ionized donors and acceptors. Virtually all donors and acceptors are neutralized, as shown in Fig. 2.27(c).

The long wavelength infrared radiation now can excite electrons into excited donor states *and* holes into excited acceptor states.

A spectrum for a Si sample without and with background light is shown in Fig. 2.28. The upper curve is without and the lower with background light. Two features distinguish Fig. 2.28(a) from 2.28(b): the *P* and As signals are increased, and the compensating *B* and Al impurities appear in the spectrum. It is possible to determine the *density* of all impurities and to *identify* them because each impurity has unique absorption peaks. The strongest absorption lines for Si are given by Baber.[117]

The infrared spectroscopy technique is very quantitative in identifying the *impurity type* but is qualitative in determining the *impurity density*. In order to determine the relationship between the absorption peak height and the impurity density, calibration data must be established using samples with known doping density. For uncompensated material this is fairly unambiguous. For compensated samples the procedure is more complex.[117]

The optical transmittance through a semiconductor wafer of thickness t is approximately

$$T \approx (1 - R)^2 \exp(-\alpha t) \tag{2.65}$$

For reasonable measurement sensitivity, αt should be on the order of unity or $t \approx 1/\alpha$. For $\alpha \approx 1$ to 10 cm^{-1}, applicable for shallow impurity absorption at low densities, the sample must be 1 to 10 mm thick. Samples of this thickness are convenient for bulk wafers but not for epitaxial layers, making IR spectroscopy of thin layers impractical.

A variation of this technique is *photothermal ionization spectroscopy* (PTIS) or *photoelectric spectroscopy*. Bound donor electrons are optically excited from the ground state to one of the excited states. At $T \approx 5$ to 10 K the sample phonon population is sufficiently high for carriers in the excited state to be transferred into the conduction band thermally leading to a change in sample conductivity. It is this photoconductivity change that is detected as a function of wavelength.[121-123] Doping densities as low as 10^9 cm^{-3} boron and gallium acceptors in Ge have been measured by the technique.[124] A disadvantage of PTIS is the need for ohmic contacts, but the advantage is its sensitivity for thin films. PTIS has been combined with magnetic fields for easier identification of impurities in GaAs and InP.[121]

2.6.4 Photoluminescence (PL)

Photoluminescence is a technique to detect and identify impurities in semiconductor materials, described in Chapter 10. PL relies on the creation of electron-hole pairs by incident radiation and subsequent *radiative* recombination photon emission. The radiative emission intensity is proportional to the impurity density. We discuss here briefly the application of PL to the measurement of doping densities in semiconductors.

Impurity identification by PL is very precise because the energy resolution is very high. It is the density measurement that is more difficult because it is not easy to draw a correlation between the intensity of a given impurity spectral line and the density of that impurity, due to non-radiative recombination through deep-level bulk or surface recombination centers.[125] Since the density of recombination centers can vary from sample to sample, even for constant shallow level densities, the photoluminescence signal can vary greatly.

This problem has been overcome by measuring both the intrinsic and the extrinsic PL peaks and using their ratio. It has been determined that the ratio $X_{TO}(\text{BE})/I_{TO}(\text{FE})$ is proportional to the doping density.[126] $X_{TO}(\text{BE})$ is the transverse optical phonon PL intensity

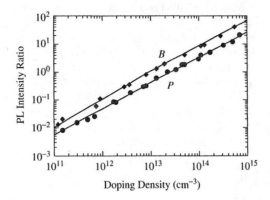

Fig. 2.29 PL intensity ratio versus doping density for B and P in Si. Reprinted with permission after ref. 128.

peak of bound excitons for element $X = B$ or P, and $I_{TO}(\text{FE})$ is the transverse optical phonon intrinsic PL intensity peak of free excitons. Good agreement is found between the resistivity measured electrically and the resistivity determined from photoluminescence for Si with the PL intensity ratio shown in Fig. 2.29 as a function of doping density. In InP the donor density as well as the compensation ratio was determined.[127]

2.7 SECONDARY ION MASS SPECTROMETRY (SIMS)

Secondary ion mass spectrometry is a very powerful technique for the analysis of impurities in solids. The details of SIMS are discussed in Chapter 11. In this section we briefly discuss the application of SIMS to semiconductor dopant profiling. The technique relies on removal of material from a solid by sputtering and on analysis of the sputtered ionized species. Most of the sputtered material consists of neutral atoms and cannot be analyzed. Only the *ionized* atoms can be analyzed by passing them through an energy filter and a mass spectrometer. It can detect all elements.

SIMS has good detection sensitivity for many elements, but its sensitivity is not as high as electrical or optical methods. Among the beam techniques it has the highest sensitivity and can detect dopant densities as low as 10^{14} cm^{-3}. It allows simultaneous detection of different elements, has a depth resolution of 1 to 5 nm, and can give lateral surface characterization on a scale of several microns. It is a destructive method since the very act of removing material by sputtering leaves a crater in the sample.

A SIMS doping density plot is produced by sputtering the sample and monitoring the secondary ion signal of a given element as a function of time. Such an "ion signal versus time" plot contains the necessary information for a dopant density profile. The *time axis* is converted to a *depth axis* by measuring the depth of the crater at the end of the measurement assuming a constant sputtering rate. This should be done for each sample, since the sputter rate varies with spot focus and ion current.[129] The *secondary ion signal* is converted to *impurity density* through standards of known dopant profile. The proportionality between ion signal and density is strictly true only if the matrix which contains the impurity is uniform. The ion yield of a given element is highly dependent on the matrix. For example, boron is implanted into Si at a given energy and dose to create

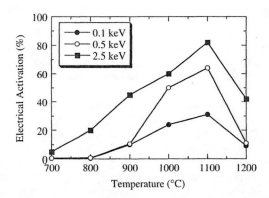

Fig. 2.30 Electrical activation of a 5×10^{14} cm^{-2} boron implantation for energies ranging from 100 eV to 5 keV after a 10 s RTA for different temperatures. Adapted from ref. 132.

a standard. The secondary ion signal is calibrated by assuming the total amount of boron in the sample to be equal to the implanted boron. The unknown sample of *B* implanted into Si is then compared to the standard.

SIMS determines the *total*, not the *electrically active* impurity density. For example, implanted, non-annealed samples give SIMS profiles very close to the predicted Gaussian distribution. Electrical measurements give very different results, with the ions not yet electrically activated. SIMS and electrical measurement agree quite well for activated samples as shown in Figs. 1.22 and 2.14.

Comparisons of SIMS dopant profiles with profiles measured by spreading resistance sometimes show a discrepancy in the lowly doped portions of the profile giving deeper junctions than those obtained by other methods (see Fig. 1.22).[130-131] The SIMS tail is likely caused by cascade mixing and knock-on of dopant atoms by the sputtering beam contributing to slightly deeper junctions or by the limited dynamic range of the SIMS instrument. When sputtering from a highly doped region near the surface to a lowly doped region deeper within the sample, the crater walls contain the entire doping density profile. Any stray signal from the crater walls adds to the signal from the lowly doped region in the central sputtered area giving the appearance of a higher dopant density and hence a deeper profile. Electronic or optical gating can suppress this signal. However, the ultimate limitation is material from the crater edges deposited on the crater floor adding to the crater floor signal. Another reason for the discrepancy is the nature of the species measured. Current is measured in SRP and the density of the electrons/holes depends on the activation of the implanted ions. SIMS, on the other hand, measures the total dopant density, regardless of activation. Figure 2.30 illustrates this point by showing the dependence of electrical activation of boron implanted into silicon on implant dose and activation anneal temperature.[132]

2.8 RUTHERFORD BACKSCATTERING (RBS)

Rutherford backscattering, discussed in Chapter 11, is a non-destructive, quantitative technique requiring no standards. It is based on backscattering of light ions from a sample. Usually monoenergetic He ions of 1–3 MeV energy are incident on and scattered from a sample and detected with a surface barrier detector. RBS is most useful for heavy elements in a light matrix. For example, As in Si or Te in GaAs are suitable, whereas *B* in Si and Si in GaAs are difficult to quantify, because in the interaction of a light ion (*e.g.*, He) with a heavy ion (*e.g.*, As), He loses less energy than if it interacted with a light ion (*e.g.*, *B*). No backscattering occurs from ions lighter than the probe ions.

The sensitivity of RBS is low compared to SRP and SIMS. The lowest detection limit is on the order of 10^{14} cm^{-2} atoms. For a layer 10^{-5} cm thick this corresponds to $10^{14}/10^{-5} = 10^{19}$ cm^{-3}. The sensitivity can be improved by using ions heavier than He, such as carbon. But heavy ions impair the depth resolution. Depth resolution can be improved by target tilting. Resolutions as low as 2–5 nm have been achieved.[133] RBS has an additional advantage and that is the ability to determine doping activation of implanted samples through *ion channeling* where the incoming ions are aligned with a crystal direction. Ions are channeled down the open channels and few ions are backscattered. Implanted, but non-activated atoms, typically occupy interstitial sites in the lattice causing increased backscattering. Analysis of the backscattered data allows a determination of the degree of electrical activation. None of the previous techniques give this type of information. RBS has also been used in the development of silicides and the effect of silicide formation on dopant distributions of impurities in semiconductors. This is an ideal application where no other technique is suitable. As a silicide forms, its formation is followed by RBS and by measuring the As distribution in the Si below the silicide, one can follow the As "snowplowing" ahead of the silicide front.[134]

2.9 LATERAL PROFILING

As semiconductor device dimensions shrink it becomes important to know the *vertical* as well the horizontal or *lateral* dopant profiles. The lateral profile is required as an input to computer aided design models. However, as device dimensions shrink so does the junction depth. Consequently, the lateral extent of a junction, which is typically assumed to be 0.6–0.7 of the vertical dimension, is very small. It has been proposed that a 10% doping density sensitivity down to 2×10^{17} cm^{-3} with sub 10 nm resolution is required for measured profiles to be useful for prediction of device characteristics.[135] What techniques are suitable for profile measurements at this scale?

It is important to distinguish between atomic and electrically active dopant profiles. Many techniques have been attempted but few have given quantitative results. We give a brief summary here of these techniques. More detailed discussions can be found in refs. 136–137. In scanning tunneling microscopy (STM) (discussed in Chapter 9) the probe is moved along the lateral portion of the junction. The tunneling current depends on doping density due to tip induced band bending at the semiconductor surface. A modification of STM is the measurement of the tunneling barrier height. The barrier height is obtained by measuring the tunneling current as a function of probe-sample distance. Changes of barrier height correspond to changes of dopant density. STM tips are very sharp and the technique has claimed a 1 to 5 nm spatial resolution. However, surface preparation is a significant issue.

Scanning or atomic force microscopy (AFM) has been combined with etching.[138] A cross-section of the device is first prepared by careful polishing. The sample is then etched in a suitable etch and the resulting topographical feature is determined with a technique capable of high-resolution imaging, *e.g.*, AFM, scanning electron microscopy or transmission electron microscopy. The technique relies on a dopant density dependent etch rate. Heavily doped regions etch faster than lightly doped regions. After etching, the surface is profiled and the physical profile is correlated with the dopant profile. Reference samples of known dopant density are required to calibrate the etch rate. Suitable etch solutions are for *p*-Si: $HF:HNO_3:CH_3 COOH$ (1:3:8) for a few seconds under strong illumination and for *n*-Si: $HF:HNO_3:H_2O$ (1:100:25).[139] A limitation of this technique is the limited sensitivity of $\sim 5 \times 10^{17}$ cm^{-3} for *p*-Si and *n*-Si.

The two main techniques that have emerged for lateral doping density profiling are *scanning capacitance microscopy* (SCM)[140] and *scanning spreading resistance microscopy* (SSRM).[141] In SCM a small-area capacitive probe measures the capacitance of a metal/semiconductor or an MOS contact, similar to techniques described earlier in this chapter.[142] If the capacitance measurement circuit is sufficiently sensitive, it is possible to measure the small capacitances of these probes. A problem is the non-planar nature of the contact. SCM is discussed in Chapter 9.

SSRM, based on the atomic force microscope (discussed in Chapter 9), measures the local spreading resistance between a sharp conductive tip and a large back surface contact. A precisely controlled force is used while the tip is stepped across the sample. SSRM sensitivity and dynamic range are similar to conventional spreading resistance (SRP discussed in Chapter 1). The small contact size and small stepping distance allows measurements on the device cross section with no probe conditioning. The high spatial resolution allows direct two-dimensional nano-SRP measurements, without the need for special test structures.

2.10   STRENGTHS AND WEAKNESSES

Differential Capacitance: The major weakness of the differential capacitance profiling method is its limited profile depth, limited at the surface by the zero-bias space-charge region width and in depth by voltage breakdown. The latter limitation is particularly serious for heavily doped regions. Further limitations are due to the Debye limit, which applies to all carrier profiling techniques. A minor weakness is the data differentiation introducing noise into the profile data.

The method's strength lies in its ability to give the carrier density profile with little data processing. A simple differentiation of the $C-V$ data suffices. It is an ideal method for moderately doped materials and is non-destructive when a mercury probe is used. It is well established with available commercial equipment. Its depth profiling capability is extended significantly for the electrochemical profiling method.

Max-Min MOS-C Capacitance: The weakness of this technique lies in its inability to provide a density profile. It determines only an average doping density in the space-charge region width of an MOS-C in equilibrium. Its strength lies in its simplicity. It merely requires a high-frequency $C-V$ measurement.

Integral Capacitance: The integral capacitance technique also does not provide a profile, which limits its usefulness. It does, however, provide, a value for an implant dose and depth, and its major strength lies in its accuracy. This is very important when monitoring ion implants with uniformities of 1%.

MOSFET Current-Voltage: The substrate/gate voltage technique requires two differentiations and has not found wide application. The threshold voltage method needs a proper definition of threshold voltage in its interpretation. The advantage of both methods is the fact that a MOSFET is measured directly. No special, large-area test structures are required. This is especially important when such test structures are not available. It is, however, subject to short- and narrow-channel effects.

Spreading Resistance: The weakness of SRP is the complexity of sample preparation as well as the interpretation of the measured spreading resistance profile. The measured data must be deconvolved, and either the mobility must be known or well calibrated standards must be used to extract the dopant profile. Its strength lies in being a well-known method that is routinely used by the semiconductor industry for Si profiling. It has no depth limit and can profile through an arbitrary number of pn junctions; it spans a very large doping density range from about 10^{13} cm^{-3} to 10^{21} cm^{-3}.

Hall Effect: The Hall effect is limited in its profiling ability through the inconvenience of providing repeated layer removal. This has been simplified with commercial equipment. Although it is utilized for profiling, it is not a routine method for generating profiles. Its advantage lies in providing average values of carrier density and mobility. For that it is used a great deal, as discussed in Chapter 8.

Optical Techniques: Optical techniques require specialized equipment with quantitative doping measurements requiring known standards. Profiling is generally not possible, and only average values are obtained. The major advantage of optical methods is their unprecedented sensitivity and accuracy in impurity identification. Furthermore, optical methods are, as a rule, contactless—a major advantage.

Secondary Ion Mass Spectrometry: The weakness of SIMS lies in the complexity of the equipment. It does not have the sensitivity of electrical and optical techniques. It is most sensitive for *B* in Si, for all other impurities it has reduced sensitivity. It is useless for semiconductors with stoichiometric dopant species. Reference standards must be used for quantitative interpretation of the raw SIMS data, and matrix effects can render measurement interpretation difficult. The strength of SIMS lies in its accepted use for dopant density profiling. It is the most commonly used method. It measures the dopant density profile not the carrier density profile and can be used for implanted samples before any activation anneals. That is not possible with electrical methods. It has high spatial resolution and can be used for any semiconductor.

Rutherford Backscattering: The weakness of RBS is its low sensitivity and the requirement of specialized equipment not readily available in most semiconductor laboratories. It is difficult to measure light elements. Its strength lies in its non-destructive and quantitative nature without recourse to standards. It is also capable of detecting activation effectiveness of implanted ions through ion channeling.

Lateral Profiling: Lateral profiling, although potentially very important, has not been developed to the point where it is a routine, accurate method. Many techniques are being evaluated, but none stands out as the most dominant method at this time, but capacitance and spreading resistance profiling look promising.

APPENDIX 2.1

Parallel or Series Connection?

Some capacitance meters have provision for parallel or series connection measurements, *i.e.*, the meter assumes the device under test to consist of either a parallel connection as shown in Fig. A2.1(a) or a series connection as in Fig. A2.1(b). The admittance Y_P of the parallel circuit and the impedance Z_S of the series circuit are

$$Y_P = G_P + j\omega C_P; Z_S = R_S + 1/j\omega C_S \qquad (A2.1)$$

where $\omega = 2\pi f$. Equating these two expressions as $Y_P = 1/Z_S$ gives

$$C_P = \frac{1}{1 + D_S^2} C_S; G_P = \frac{D_S^2}{1 + D_S^2} \frac{1}{R_S} \qquad (A2.2)$$

with the *dissipation factor* D_S:

$$D_S = \omega C_S R_S \qquad (A2.3)$$

Similarly, we can write

$$C_S = (1 + D_P^2) C_P; R_S = \frac{D_P^2}{1 + D_P^2} \frac{1}{G_P} \qquad (A2.4)$$

with the *dissipation factor* D_P

$$D_P = \frac{G_P}{\omega C_P} \qquad (A2.5)$$

The dissipation factor is sometimes expressed in terms of the *quality factor* Q. For the series and the parallel circuits, Q is given by

$$Q_S = \frac{1}{D_S} = \frac{1}{\omega C_S R_S}; Q_P = \frac{1}{D_P} = \frac{\omega C_P}{G_P} \qquad (A2.6)$$

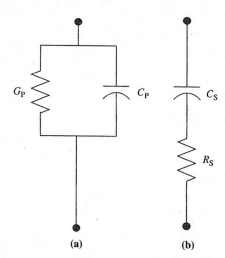

(a)   (b)

Fig. A2.1   (a) Parallel and (b) series connection of a capacitance having parallel conductance or series resistance.

For an ideal capacitor, $G_P = 0$ and $R_S = 0$, leading to $C_S = C_P$. Usually, however, $G_P \neq 0$ and $R_S \neq 0$. Unfortunately there is no unique criterion to select the appropriate measurement circuit. Series measurement circuit for low-impedance and parallel circuit for high-impedance samples are often used. The approximate instrumentation error for high dissipation values is given by

$$\% \ error = 0.1\sqrt{1 + D^2} \tag{A2.7}$$

Occasionally these concepts are expressed in terms of the *loss tangent*, $\tan(\delta)$, defined as

$$\tan(\delta) = \frac{\sigma}{K_s \varepsilon_o \omega} = \frac{1}{K_s \varepsilon_o \omega \rho} \tag{A2.8}$$

APPENDIX 2.2

Circuit Conversion

Let us consider the circuits in Figs. A2.2(a) and (b). The easiest way to convert from (a) to (b) is to consider the admittances of both circuits and to equate them. The admittance Y for (a) is

$$Y(a) = \frac{1}{Z(a)} = \frac{1}{r_s + 1/(G + j\omega C)} = \frac{G + j\omega C}{1 + r_s(G + j\omega C)}$$
$$= \frac{(G + j\omega C)(1 + r_s G - j\omega C)}{(1 + r_s G + j\omega r_s C)(1 + r_s G - j\omega r_s C)} \tag{A2.9}$$

where Z is the impedance. $Y(a)$ can be written as

$$Y(a) = \frac{G + r_s G^2 + r_s(\omega C)^2}{(1 + r_s G)^2 + (\omega r_s C)^2} + \frac{j\omega C}{(1 + r_s G)^2 + (\omega r_s C)^2} \tag{A2.10}$$

The admittance for (b) is simply

$$Y(b) = G_P + j\omega C_P. \tag{A2.11}$$

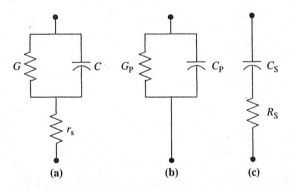

Fig. A2.2   (a) actual circuit, (b) parallel equivalent circuit, (c) series equivalent circuit.

Equating the real and imaginary parts of Eqs. (A2.10) and (A2.11) gives

$$C_P = \frac{C}{(1+r_sG)^2 + (\omega r_s C)^2}; G_P = \frac{G(1+r_sG) + r_s(\omega C)^2}{(1+r_sG)^2 + (\omega r_s C)^2} \qquad \text{(A2.12)}$$

For the circuits in Figs. A2.2(a) and (c), it is best to consider the impedances of both circuits and to equate them. The impedance of (a) is

$$Z(a) = r_s + \frac{1}{G+j\omega C} = \frac{(r_s(G+j\omega C)+1)(G-j\omega C)}{(G+j\omega C)(G-j\omega C)}$$

$$= \frac{r_s(G^2 + (\omega C)^2) + G}{G^2 + (\omega C)^2} - \frac{j\omega rC}{G^2 + (\omega C)^2} \qquad \text{(A2.13)}$$

and for (c) it is

$$Z(c) = R_S + \frac{1}{j\omega C_S} = R_S - \frac{j\omega C_S}{(\omega C_S)^2} \qquad \text{(A2.14)}$$

Equating real and imaginary parts of Eqs. (A2.13) and (A2.14) gives

$$C_S = C(1 + (G/\omega C)^2); R_S = \frac{r_s(G^2 + (\omega C)^2) + G}{G^2 + (\omega C)^2} = r_s + \frac{1}{G(1 + (\omega C/G)^2}$$

$$\text{(A2.15)}$$

REFERENCES

1. W. Schottky, "Simplified and Expanded Theory of Boundary Layer Rectifiers (in German)," *Z. Phys.* **118**, 539–592, Feb. 1942.

2. J. Hilibrand and R.D. Gold, "Determination of the Impurity Distribution in Junction Diodes from Capacitance-Voltage Measurements," *RCA Rev.* **21**, 245–252, June 1960.

3. R. Decker, "Measurement of Epitaxial Doping Density vs. Depth," *J. Electrochem. Soc.* **115**, 1085–1089, Oct. 1968.

4. L.E. Coerver, "Note on the Interpretation of $C-V$ Data in Semiconductor Junctions," *IEEE Trans. Electron Dev.* **ED-17**, 436, May 1970.

5. H.J.J. DeMan, "On the Calculation of Doping Profiles from C(V) Measurements on Two-Sided Junctions," *IEEE Trans. Electron Dev.* **ED-17**, 1087–1088, Dec. 1970.

6. W. van Gelder and E.H. Nicollian, "Silicon Impurity Distribution as Revealed by Pulsed MOS $C-V$ Measurements," *J. Electrochem. Soc.* **118**, 138–141, Jan. 1971.

7. Y. Zohta, "Rapid Determination of Semiconductor Doping Profiles in MOS Structures," *Solid-State Electron.* **16**, 124–126, Jan. 1973.

8. D.K. Schroder, *Advanced MOS Devices*, Addison-Wesley, Reading, MA, 1987, 64–71.

9. C.D. Bulucea, "Investigation of Deep-Depletion Regime of MOS Structures Using Ramp-Response Method" *Electron. Lett.* **6**, 479–481, July 1970.

10. G. Baccarani, S. Solmi and G. Soncini, "The Silicon Impurity Profile as Revealed by High-Frequency Non-Equilibrium MOS $C-V$ Characteristics," *Alta Frequ.* **16**, 113–115, Feb. 1972.

11. G.G. Barna, B. Van Eck, and J.W. Hosch, "In situ Metrology," in *Handbook of Silicon Semiconductor Technology* (A.C. Diebold, ed.), Dekker, New York, 2001.

12. M. Rommel, Semitest Inc., private correspondence.

13. K. Woolford, L. Newfield, and C. Panczyk, "Monitoring Epitaxial Resistivity Profiles Without Wafer Damage," *Micro*, July/Aug. 2002 (www.micromagazine.com).

14. D.P. Kennedy, P.C. Murley and W. Kleinfelder, "On the Measurement of Impurity Atom Distributions in Silicon by the Differential Capacitance Technique," *IBM J. Res. Develop.* **12**, 399–409, Sept. 1968.

15. J.R. Brews, "Threshold Shifts Due to Nonuniform Doping Profiles in Surface Channel MOS-FET's," *IEEE Trans. Electron Dev.* **ED-26**, 1696–1710, Nov. 1979.

16. D.P. Kennedy and R.R. O'Brien, "On the Measurement of Impurity Atom Distributions by the Differential Capacitance Technique," *IBM J. Res. Develop.* **13**, 212–214, March 1969.

17. W.E Carter, H.K. Gummel and B.R. Chawla, "Interpretation of Capacitance vs. Voltage Measurements of PN Junctions," *Solid-State Electron.* **15**, 195–201, Feb. 1972.

18. W.C. Johnson and P.T. Panousis, "The Influence of Debye Length on the $C-V$ Measurement of Doping Profiles," *IEEE Trans. Electron Dev.* **ED-18**, 965–973, Oct. 1971.

19. E.H. Nicollian, M.H. Hanes and J.R. Brews, "Using the MIS Capacitor for Doping Profile Measurements with Minimal Interface State Error," *IEEE Trans. Electron Dev.* **ED-20**, 380–389, April 1973.

20. C.P. Wu, E.C. Douglas and C.W. Mueller, "Limitations of the CV Technique for Ion-Implanted Profiles," *IEEE Trans. Electron Dev.* **ED-22**, 319–329, June 1975.

21. M. Nishida, "Depletion Approximation Analysis of the Differential Capacitance-Voltage Characteristics of an MOS Structure with Nonuniformly Doped Semiconductors," *IEEE Trans. Electron Dev.* **ED-26**, 1081–1085, July 1979.

22. G. Baccarani, M. Rudan, G. Spadini, H. Maes, W. Vandervorst and R. Van Overstraeten, "Interpretation of $C-V$ Measurements for Determining the Doping Profile in Semiconductors," *Solid-State Electron.* **23**, 65–71, Jan. 1980.

23. C.L. Wilson, "Correction of Differential Capacitance Profiles for Debye-Length Effects," *IEEE Trans. Electron Dev.* **ED-27**, 2262–2267, Dec. 1980.

24. D.J. Bartelink, "Limits of Applicability of the Depletion Approximation and Its Recent Augmentation," *Appl. Phys. Lett.* **38**, 461–463, March 1981.

25. H. Kroemer and W.Y. Chien, "On the Theory of Debye Averaging in the $C-V$ Profiling of Semiconductors," *Solid-State Electron.* **24**, 655–660, July 1981.

26. J. Voves, V. Rybka and V. Trestikova, "$C-V$ Technique on Schottky Contacts—Limitation of Implanted Profiles," *Appl. Phys.* **A37**, 225–229, Aug. 1985.

27. A.R. LeBlanc, D.D. Kleppinger and J.P. Walsh, "A Limitation of the Pulsed Capacitance Technique of Measuring Impurity Profiles," *J. Electrochem. Soc.* **119**, 1068–1071, Aug. 1972.

28. K. Ziegler, E. Klausmann and S. Kar, "Determination of the Semiconductor Doping Profile Right Up to Its Surface Using the MIS Capacitor," *Solid-State Electron.* **18**, 189–198, Feb. 1975.

29. ASTM Standard F419-94, "Standard Test Method for Net Carrier Density in Silicon Epitaxial Layers by Voltage-Capacitance of Gated and Ungated Diodes," *1996 Annual Book of ASTM Standards*, Am. Soc. Test. Mat., West Conshohocken, PA, 1996.

30. J.P. Sullivan, W.R. Graham, R.T. Tung, and F. Schrey, "Pitfalls in the Measurement of Metal/*p*-Si Contacts: The Effect of Hydrogen Passivation," *Appl. Phys. Lett.* **62**, 2804–2806, May 1993; A.S. Vercaemst, R.L. Van Meirhaeghe, W.H. Laflere, and F. Cardon, "Hydrogen Passivation Caused by "Soft" Sputter Etch Cleaning of Si," *Solid-State Electron.* **38**, 983–987, May 1995.

31. R. Dingle, "Confined Carrier Quantum States in Ultrathin Semiconductor Heterostructures," in *Festkörperprobleme/Advances in Solid State Physics* (H.J. Queisser, ed.), Vieweg, Braunschweig Germany, **15**, 1975, 21–48.

32. E.A. Kraut, R.W. Grant, J.R. Waldrop, and S.P. Kowalczyk, "Precise Determination of the Valence-Band Edge in X-Ray Photoemission Spectra: Application to Measurement of Semiconductor Interface Potentials," *Phys. Rev. Lett.* **44**, 1620–1623, June 1980.

33. H. Kroemer, W.Y. Chien, J.S. Harris, Jr., and D.D. Edwall, "Measurement of Isotype Heterojunction Barriers by $C-V$ Profiling," *Appl. Phys. Lett.* **36**, 295–297, Feb. 1980; M.A. Rao, E.J. Caine, H. Kroemer, S.I. Long, and D.I. Babic, "Determination of Valence and Conduction Band Discontinuities at the (Ga,In)P/GaAs Heterojunction by $C-V$ Profiling," *J. Appl. Phys.* **61**, 643–649, Jan. 1987; D.N. Bychkovskii, O.V. Konstantinov, and M.M. Panakhov, "Method for Determination of the Band Offset at a Heterojunction from Capacitance-Voltage Characteristics of an M-S Heterostructure," *Sov. Phys. Semicond.* **26**, 368–376, April 1992.

34. H. Kroemer, "Determination of Heterojunction Band Offsets by Capacitance-Voltage Profiling Through Nonabrupt Isotype Heterojunctions", *Appl. Phys. Lett.* **46**, 504–505, March 1985.

35. A. Morii, H. Okagawa, K. Hara, J. Yoshino, and H. Kukimoto, "Band Discontinuity at $Al_xGa_{1-x}P/GaP$ Heterointerfaces Studied by Capacitance-Voltage Measurements," *Japan. J. Appl. Phys.* **31**, L1161–L1163, Aug. 1992.

36. S.P. Voinigescu, K. Iniewski, R. Lisak, C.A.T. Salama, J.P. Noel, and D.C. Houghton, "New Technique for the Characterization of Si/SiGe Layers Using Heterostructure MOS Capacitors," *Solid-State Electron.* **37**, 1491–1501, Aug. 1994.

37. D.V. Singh, K. Rim, T.O. Mitchell, J.L. Hoyt, and J.F. Gibbons, "Measurement of the Conduction Band Offsets in $Si/Si_{1-x-y}Ge_xC_y$ and $Si/Si_{1-y}C_y$ Heterostructures Using Metal-Oxide-Semiconductor Capacitors," *J. Appl. Phys.* **85**, 985–993, Jan. 1999.

38. S. Chattopadhyay, K.S.K. Kwa, S.H. Olsen, L.S. Driscoll and A.G. O'Neill, "$C-V$ Characterization of Strained Si/SiGe Multiple Heterojunction Capacitors as a Tool for Heterojunction MOSFET Channel Design," *Semicond. Sci. Technol.* **18**, 738–744, Aug. 2003.

39. D.V. Lang, "Measurement of Band Offsets by Space Charge Spectroscopy," in *Heterojunctions and Band Discontinuities* (F. Capasso and G. Margaritondo, eds.), North Holland, Amsterdam, 1987, Ch. 9.

40. H. Kroemer, "Heterostructure Devices: A Device Physicist Looks at Interfaces," *Surf. Sci.* **132**, 543–576, Sept. 1983.

41. W. Mönch, *Electronic Properties of Semiconductor Interfaces*, Springer, Berlin 2004, 79–82.

42. E.A. Kraut, R.W. Grant, J.R. Waldrop, and S.P. Kowalczyk, "Precise Determination of the Valence-Band Edge in X-Ray Photoemission Spectra: Application to Measurement of Semiconductor Interface Potentials," *Phys. Rev. Lett.* **44**, 1620–1623, June 1980.

43. B.E. Deal, A.S. Grove, E.H. Snow and C.T. Sah, "Observation of Impurity Redistribution During Thermal Oxidation of Silicon Using the MOS Structure," *J. Electrochem. Soc.* **112**, 308–314, March 1965.

44. K. Iniewski and A. Jakubowski, "Procedure for Determination of a Linear Approximation Doping Profile in a MOS Structure," *Solid-State Electron.* **30**, 295–298, March 1987.

45. A.S. Grove, *Physics and Technology of Semiconductor Devices*, Wiley, New York, 1967.

46. E.H. Nicollian and J.R. Brews, *MOS Physics and Technology*, Wiley, New York, 1982.

47. W.E. Beadle, J.C.C. Tsai and R.D. Plummer, *Quick Reference Manual for Silicon Integrated Circuit Technology*, Wiley-Interscience, New York, 1985, Ch. 14.

48. J. Shappir, A. Kolodny and Y. Shacham-Diamand, "Diffusion Profiling Using the Graded C(V) Method," *IEEE Trans. Electron Dev.* **ED-27**, 993–995, May 1980.

49. W.W. Lin, "A Simple Method for Extracting Average Doping Concentration in the Polysilicon and Silicon Surface Layer Near the Oxide in Polysilicon Gate MOS Structures," *IEEE Electron Dev. Lett.* **15**, 51–53, Feb. 1994.

50. R.O. Deming and W.A. Keenan, "$C-V$ Uniformity Measurements," *Nucl. Instrum. and Meth.* **B6**, 349–356, Jan. 1985; "Low Dose Ion Implant Monitoring," *Solid State Technol.* **28**, 163–167, Sept. 1985.

51. R. Sorge, "Implant Dose Monitoring by MOS $C-V$ Measurement," *Microelectron. Rel.* **43**, 167–171, Jan. 2003.

52. R.S. Nakhmanson and S.B. Sevastianov, "Investigations of Metal-Insulator-Semiconductor Structure Inhomogeneities Using a Small-Size Mercury Probe," *Solid-State Electron.* **27**, 881–891, Oct. 1984.

53. J.T.C. Chen, Four Dimensions, private communication; P.S. Schaffer and T.R. Lally, "Silicon Epitaxial Wafer Profiling Using the Mercury-Silicon Schottky Diode Differential Capacitance Method," *Solid State Technol.* **26**, 229–233, April 1983.

54. T. Ambridge and M.M. Faktor, "An Automatic Carrier Concentration Profile Plotter Using an Electrochemical Technique," *J. Appl. Electrochem.* **5**, 319–328, Nov. 1975.

55. P. Blood, "Capacitance-Voltage Profiling and the Characterisation of III–V Semiconductors Using Electrolyte Barriers," *Semicond. Sci. Technol.* **1**, 7–27, 1986.

56. T. Ambridge and D.J. Ashen, "Automatic Electrochemical Profiling of Carrier Concentration in Indium Phosphide," *Electron. Lett.* **15**, 647–648, Sept. 1979.

57. R.T. Green, D.K. Walker, and C.M. Wolfe, "An Improved Method for the Electrochemical $C-V$ Profiling of InP," *J. Electrochem. Soc.* **133**, 2278–2283, Nov. 1986.

58. M.M. Faktor and J.L. Stevenson, "The Detection of Structural Defects in GaAs by Electrochemical Etching," *J. Electrochem. Soc.* **125**, 621–629, April 1978.

59. T. Ambridge, J.L. Stevenson and R.M. Redstall, "Applications of Electrochemical Methods for Semiconductor Characterization: I. Highly Reproducible Carrier Concentration Profiling of VPE "Hi-Lo" n-GaAs," *J. Electrochem. Soc.* **127**, 222–228, Jan. 1980; A.C. Seabaugh, W.R. Frensley, R.J. Matyi and G.E. Cabaniss, "Electrochemical $C-V$ Profiling of Heterojunction Device Structures," *IEEE Trans. Electron Dev.* **ED-36**, 309–313, Feb. 1989.

60. C.D. Sharpe and P. Lilley, "The Electrolyte-Silicon Interface; Anodic Dissolution and Carrier Concentration Profiling," *J. Electrochem. Soc.* **127**, 1918–1922, Sept. 1980.

61. W.Y. Leong, R.A.A. Kubiak and E.H.C. Parker, "Dopant Profiling of Si-MBE Material Using the Electrochemical CV Technique," in *Proc. First Int. Symp. on Silicon MBE*, Electrochem. Soc., Pennington, NJ, 1985, pp. 140–148.

62. M. Pawlik, R.D. Groves, R.A. Kubiak, W.Y. Leong and E.H.C. Parker, "A Comparative Study of Carrier Concentration Profiling Techniques in Silicon: Spreading Resistance and Electrochemical CV," in *Emerging Semiconductor Technology* (D.C. Gupta and R.P. Langer, eds.), **STP 960**, Am. Soc. Test. Mat., Philadelphia, 1987, 558–572.

63. A.C. Seabaugh, W.R. Frensley, R.J. Matyi, G.E. Cabaniss, "Electrochemical $C-V$ Profiling of Heterojunction Device Structures," *IEEE Trans. Electron Dev.* **36** 309–313, Feb. 1989.

64. E. Peiner, A. Schlachetzki, and D. Krüger, "Doping Profile Analysis in Si by Electrochemical Capacitance-Voltage Measurements," *J. Electrochem. Soc.* **142**, 576–580, Feb. 1995.

65. I. Mayes, "Accuracy and Reproducibility of the Electrochemical Profiler," *Mat. Sci. Eng.* **B80**, 160–163, March 2001.

66. J.M. Shannon, "DC Measurement of the Space Charge Capacitance and Impurity Profile Beneath the Gate of an MOST," *Solid-State Electron.* **14**, 1099–1106, Nov. 1971.

67. M.G. Buehler, "The D-C MOSFET Dopant Profile Method," *J. Electrochem. Soc.* **127**, 701–704, March 1980; M.G. Buehler, "Effect of the Drain-Source Voltage on Dopant Profiles Obtained from the DC MOSFET Profile Method," *IEEE Trans. Electron Dev.* **ED-27**, 2273–2277, Dec. 1980.

68. H.G. Lee, S.Y. Oh, and G. Fuller, "A Simple and Accurate Method to Measure the Threshold Voltage of an Enhancement-Mode MOSFET," *IEEE Trans. Electron Dev.* **ED-29**, 346–348, Feb. 1982.

69. H.J. Mattausch, M. Suetake, D. Kitamaru, M. Miura-Mattausch, S. Kumashiro, N. Shigyo, S. Odanaka, and N. Nakayama, "Simple Nondestructive Extraction of the Vertical Channel-Impurity Profile of Small-Size Metal–Oxide–Semiconductor Field-Effect Transistors," *Appl. Phys. Lett.* **80**, 2994–2996, April 2002.

70. M. Chi and C. Hu, "Errors in Threshold-Voltage Measurements of MOS Transistors for Dopant-Profile Determinations," *Solid-State Electron.* **24**, 313–316, April 1981.

71. G.S. Gildenblat, "On the Accuracy of a Particular Implementation of the Shannon-Buehler Method," *IEEE Trans. Electron Dev.* **36**, 1857–1858, Sept. 1989.

72. G.P. Carver, "Influence of Short-Channel Effects on Dopant Profiles Obtained from the DC MOSFET Profile Method," *IEEE Trans. Electron Dev.* **ED-30**, 948–954, Aug. 1983.

73. D.W. Feldbaumer and D.K. Schroder, "MOSFET Doping Profiling," *IEEE Trans. Electron Dev.* **38**, 135–140, Jan. 1991.

74. D.S. Wu, "Extraction of Average Doping Density and Junction Depth in an Ion-Implanted Deep-Depletion Transistor," *IEEE Trans. Electron Dev.* **ED-27**, 995–997, May 1980.

75. R.A. Burghard and Y.A. El-Mansy, "Depletion Transistor Threshold Voltage as a Process Monitor," *IEEE Trans. Electron Dev.* **ED-34**, 940–942, April 1987.

76. H. Jorke and H.J. Herzog, "Carrier Spilling in Spreading Resistance Analysis of Si Layers Grown by Molecular-Beam Epitaxy," *J. Appl. Phys.* **60**, 1735–1739, Sept. 1986.

77. H. Maes, W. Vandervorst and R. Van Overstraeten, "Impurity Profile of Implanted Ions in Silicon," in *Impurity Doping Processes in Silicon* (F.F.Y. Wang, ed.) North-Holland, Amsterdam, 1981, 443–638.

78. E.F. Schubert, J.M. Kuo, and R.F. Kopf, "Theory and Experiment of Capacitance-Voltage Profiling on Semiconductors with Quantum Confinement," *J. Electron. Mat.* **19**, 521–531, June 1990.

79. A.M. Goodman, "Metal-Semiconductor Barrier Height Measurement by the Differential Capacitance Method—One Carrier System," *J. Appl. Phys.* **34**, 329–338, Feb. 1963.

80. K. Mallik, R.J. Falster, and P.R. Wilshaw, "Schottky Diode Back Contacts for High Frequency Capacitance Studies on Semiconductors," *Solid-State Electron.* **48**, 231–238, Feb. 2004.

81. J.D. Wiley and G.L. Miller, "Series Resistance Effects in Semiconductor CV Profiling," *IEEE Trans. Electron Dev.* **ED-22**, 265–272, May 1975.

82. J.D. Wiley, "$C-V$ Profiling of GaAs FET Films," *IEEE Trans. Electron Dev.* **ED-25**, 1317–1324, Nov. 1978.

83. S.T. Lin and J. Reuter, "The Complete Doping Profile Using MOS CV Technique," *Solid-State Electron.* **26**, 343–351, April 1983.

84. D.K. Schroder and P. Rai Choudhury, "Silicon-on-Sapphire with Microsecond Carrier Lifetimes," *Appl. Phys. Lett.* **22**, 455–457, May 1973.

85. J.R. Brews, "Correcting Interface-State Errors in MOS Doping Profile Determinations," *J. Appl. Phys.* **44**, 3228–3231, July 1973; Y. Zohta, "Frequency Dependence of $\Delta V / \Delta(C^{-2})$ of MOS Capacitors," *Solid-State Electron.* **17**, 1299–1309, Dec. 1974.

86. B.L. Smith and E.H. Rhoderick, "Possible Sources of Error in the Deduction of Semiconductor Impurity Concentrations from Schottky-Barrier (C, V) Characteristics," *Brit. J. Appl. Phys.* **D2**, 465–467, March 1969.

87. J.A. Copeland, "Diode Edge Effect on Doping-Profile Measurements," *IEEE Trans. Electron. Dev.* **ED-17**, 404–407, May 1970; W. Tantraporn and G.H. Glover, "Extension of the $C-V$ Doping Profile Technique to Study the Movements of Alloyed Junction and Substrate Out-Diffusion, the Separation of Junctions, and Device Area Trimming," *IEEE Trans. Electron Dev.* **ED-35**, 525–529, April 1988.

88. E. Simoen, C. Claeys, A. Czerwinski, and J. Katcki, "Accurate Extraction of the Diffusion Current in Silicon p-n Junction Diodes," *Appl. Phys. Lett.* **72**, 1054–1056, March 1998.

89. P. Kramer, C. deVries and L.J. van Ruyven, "The Influence of Leakage Current on Concentration Profile Measurements," *J. Electrochem. Soc.* **122**, 314–316, Feb. 1975.

90. D.K. Schroder, T.T. Braggins, and H.M. Hobgood, "The Doping Concentrations of Indium-Doped Silicon Measured by Hall, $C-V$, and Junction Breakdown Techniques," *J. Appl. Phys.* **49**, 5256–5259, Oct. 1978.

91. L.C. Kimerling, "Influence of Deep Traps on the Measurement of Free-Carrier Distributions in Semiconductors by Junction Capacitance Techniques," *J. Appl. Phys.* **45**, 1839–1845, April 1974.

92. G. Goto, S. Yanagisawa, O. Wada and H. Takanashi, "An Improved Method of Determining Deep Impurity Levels and Profiles in Semiconductors," *Japan. J. Appl. Phys.* **13**, 1127–1133, July 1974.

93. K. Lehovec, "$C-V$ Analysis of a Partially Depleted Semiconducting Channel," *Appl. Phys. Lett.* **26**, 82–84, Feb. 1975.

94. I. Amron, "Errors in Dopant Concentration Profiles Determined by Differential Capacitance Measurements," *Electrochem. Technol.* **5**, 94–97, March/April 1967.

95. R.A. Smith, *Semiconductors*, Cambridge University Press, Cambridge, 1959, Ch. 5.

96. D.L. Rode, C.M. Wolfe and G.E. Stillman, "Magnetic-Field Dependence of the Hall Factor of Gallium Arsenide," in *GaAs and Related Compounds* (G.E. Stillman, ed.) Conf. Ser. No. 65, Inst. Phys., Bristol, 1983, pp. 569–572.

97. A. Chandra, C.E.C. Wood, D.W. Woodard and L.F. Eastman, "Surface and Interface Depletion Corrections to Free Carrier-Density Determinations by Hall Measurements," *Solid-State Electron.* **22**, 645–650, July 1979.

98. T.R. Lepkowski, R.Y. DeJule, N.C. Tien, M.H. Kim and G.E. Stillman, "Depletion Corrections in Variable Temperature Hall Measurements," *J. Appl. Phys.* **61**, 4808–4811, May 1987.

99. E.H. Putley, *The Hall Effect and Related Phenomena*, Butterworths, London, 1960, p. 106.

100. G.E. Stillman and C.M. Wolfe, "Electrical Characterization of Epitaxial Layers," *Thin Solid Films* **31**, 69–88, Jan. 1976.

101. T.T. Braggins, H.M. Hobgood, J.C. Swartz and R.N. Thomas, "High Infrared Responsivity Indium-Doped Silicon Detector Material Compensated by Neutron Transmutation," *IEEE Trans. Electron Dev.* **ED-27**, 2–10, Jan. 1980.

102. N.D. Young and M.J. Hight, "Automated Hall Effect Profiler for Electrical Characterisation of Semiconductors," *Electron. Lett.* **21**, 1044–1046, Oct. 1985.

103. R. Baron, G.A. Shifrin, O.J. Marsh, and J.W. Mayer, "Electrical Behavior of Group III and V Implanted Dopants in Silicon," *J. Appl. Phys.* **40**, 3702–3719, Aug. 1969.

104. R.D. Larrabee and W.R. Thurber, "Theory and Application of a Two-Layer Hall Technique," *IEEE Trans. Electron Dev.* **ED-27**, 32–36, Jan. 1980.

105. L.F. Lou and W.H. Frye, "Hall Effect and Resistivity in Liquid-Phase-Epitaxial Layers of HgCdTe," *J. Appl. Phys.* **56**, 2253–2267, Oct. 1984.

106. R.L. Petritz, "Theory of an Experiment for Measuring the Mobility and Density of Carriers in the Space-Charge Region of a Semiconductor Surface," *Phys. Rev.* **110**, 1254–1262, June 1958.

107. T.S. Moss, G.J. Burrell and B. Ellis, *Semiconductor Opto-Electronics*, Wiley, New York, 1973, 42–46.

108. ASTM Standard F398-82, "Standard Method for Majority Carrier Concentration in Semiconductors by Measurement of Wavelength of the Plasma Resonance Minimum," *1996 Annual Book of ASTM Standards*, Am. Soc. Test. Mat., West Conshohocken, PA, 1996.

109. T. Abe and Y. Nishi, "Non-Destructive Measurement of Surface Concentrations and Junction Depths of Diffused Semiconductor Layers," *Japan. J. Appl. Phys.* **7**, 397–403, April 1968.

110. A.H. Tong, P.A. Schumann, Jr. and W.A. Keenan, "Epitaxial Substrate Carrier Concentration Measurement by the Infrared Interference Envelope (IRIE) Technique," *J. Electrochem. Soc.* **119**, 1381–1384, Oct. 1972.

111. P.A. Schumann, Jr., W.A. Keenan, A.H. Tong, H.H. Gegenwarth and C.P. Schneider, "Silicon Optical Constants in the Infrared," *J. Electrochem. Soc.* **118**, 145–148, Jan. 1971 and references therein; D.K. Schroder, R.N. Thomas and J.C. Swartz, "Free Carrier Absorption in Silicon," *IEEE Trans. Electron Dev.* **ED-25**, 254–261, Feb. 1978.

112. J. Isenberg and W. Warta, "Free Carrier Absorption in Heavily-Doped Silicon Layers," *Appl. Phys. Lett.* **84**, 2265–2267, March 2004.

113. D.C. Look, D.C. Walters, M.G. Mier, and J.R. Sizelove, "Nondestructive Mapping of Carrier Concentration and Dislocation Density in n^+-type GaAs," *Appl. Phys. Lett.* **65**, 2188–2190, Oct. 1994.

114. J.L. Boone, M.D. Shaw, G. Cantwell and W.C. Harsh, "Free Carrier Density Profiling by Scanning Infrared Absorption," *Rev. Sci. Instrum.* **59**, 591–595, April 1988.

115. E. Burstein, G. Picus, B. Henvis and R. Wallis, "Absorption Spectra of Impurities in Silicon; I. Group III Acceptors," *J. Phys. Chem Solids* **1**, 65–74, Sept/Oct. 1956; G. Picus, E. Burstein and B. Henvis, "Absorption Spectra of Impurities in Silicon; II. Group V Donors," *J. Phys. Chem Solids* **1**, 75–81, Sept/Oct. 1956.

116. H.J. Hrostowski and R.H. Kaiser, "Infrared Spectra of Group III Acceptors in Silicon," *J. Phys. Chem. Solids* **4**, 148–153, 1958.

117. S.C. Baber, "Net and Total Shallow Impurity Analysis of Silicon by Low Temperature Fourier Transform Infrared Spectroscopy," *Thin Solid Films* **72**, 201–210, Sept. 1980; ASTM Standard F1630-95, "Standard Test Method for Low Temperature FT-IR Analysis of Single Crystal Silicon for III–V Impurities," *1996 Annual Book of ASTM Standards,* Am. Soc. Test. Mat., West Conshohocken, PA, 1996.

118. T.S. Low, M.H. Kim, B. Lee, B.J. Skromme, T.R. Lepkowski and G.E. Stillman, "Neutron Transmutation Doping of High Purity GaAs," *J. Electron. Mat.* **14**, 477–511, Sept. 1985.

119. J.J. White, "Effects of External and Internal Electric Fields on the Boron Acceptor States in Silicon," *Can. J. Phys.* **45**, 2695–2718, Aug. 1967; "Absorption-Line Broadening in Boron-Doped Silicon," *Can. J. Phys.* **45**, 2797–2804, Aug. 1967.

120. B.O. Kolbesen, "Simultaneous Determination of the Total Content of Boron and Phosphorus in High-Resistivity Silicon by IR Spectroscopy at Low Temperatures," *Appl. Phys. Lett.* **27**, 353–355, Sept. 1975.

121. G.E. Stillman, C.M. Wolfe and J.O. Dimmock, "Far Infrared Photoconductivity in High Purity GaAs," in *Semiconductors and Semimetals* (R.K. Willardson and A.C. Beer, eds.) Academic Press, New York, **12**, 169–290, 1977.

122. M.J.H. van de Steeg, H.W.H.M. Jongbloets, J.W. Gerritsen and P. Wyder, "Far Infrared Photothermal Ionization Spectroscopy of Semiconductors in the Presence of Intrinsic Light," *J. Appl. Phys.* **54**, 3464–3474, June 1983.

123. E.E. Haller, "Semiconductor Physics in Ultra-Pure Germanium," in *Festkörperprobleme 26* (P. Grosse, ed.), Vieweg, Braunschweig, 1986, 203–229.

124. S.M. Kogan and T.M. Lifshits, "Photoelectric Spectroscopy—A New Method of Analysis of Impurities in Semiconductors," *Phys. Stat. Sol.* (a) **39**, 11–39, Jan. 1977.

125. K.K. Smith, "Photoluminescence of Semiconductor Materials," *Thin Solid Films* **84**, 171–182, Oct. 1981.

126. M. Tajima, "Determination of Boron and Phosphorus Concentration in Silicon by Photoluminescence Analysis," *Appl. Phys. Lett.* **32**, 719–721, June 1978.

127. G. Pickering, P.R. Tapster, P.J. Dean and D.J. Ashen, "Determination of Impurity Concentration in n-Type InP by a Photoluminescence Technique," in *GaAs and Related Compounds* (G.E. Stillman, ed.) Conf. Ser. No. 65, Inst. Phys., Bristol, 1983, 469–476.

128. M. Tajima, T. Masui, T. Abe and T. Iizuka, "Photoluminescence Analysis of Silicon Crystals," in *Semiconductor Silicon 1981* (H.R. Huff, R.J. Kriegler and Y. Takeishi, eds.) Electrochem. Soc., Pennington, NJ, 1981, 72–89.

129. M. Pawlik, "Dopant Profiling in Silicon," in *Semiconductor Processing*, ASTM **STP 850** (D.C. Gupta, ed.) Am. Soc. Test. Mat., Philadelphia, PA, 1984, 391–408.

130. S.B. Felch, R. Brennan, S.F. Corcoran, and G. Webster, "A Comparison of Three Techniques for Profiling Ultrashallow p⁺n Junctions," *Solid State Technol.* **36**, 45–51, Jan. 1993.

131. E. Ishida and S.B. Felch, "Study of Electrical Measurement Techniques for Ultra-Shallow Dopant Profiling," *J. Vac. Sci. Technol.* **B14**, 397–403, Jan./Feb. 1996; S.B. Felch, D.L. Chapek, S.M. Malik, P. Maillot, E. Ishida, and C.W. Mageee, "Comparison of Different Analytical Techniques in Measuring the Surface Region of Ultrashallow Doping Profiles," *J. Vac. Sci. Technol.* **B14**, 336–340, Jan./Feb.1996.

132. E.J.H. Collart, K. Weemers, D.J. Gravesteijn, and J.G.M. van Berkum, "Characterization of Low-energy (100 eV–10 keV) Boron Ion Implantation," *J. Vac. Sci. Technol.* **B16**, 280–285, Jan./Feb. 1998.

133. W. Vandervorst and T. Clarysse, "On the Determination of Dopant/Carrier Distributions," *J. Vac. Sci. Technol.* **B10**, 302–315, Jan./Feb. 1992.

134. H. Norström, K. Maex, J. Vanhellemont, G. Brijs, W. Vandervorst, and U. Smith, "Simultaneous Formation of Contacts and Diffusion Barriers for VLSI by Rapid Thermal Silicidation of TiW," *Appl. Phys.*, **A51**, 459–466, Dec. 1990.

135. R. Subrahmanyan and M. Duane, "Issues in Two-Dimensional Dopant Profiling," in *Diagnostic Techniques for Semiconductor Materials and Devices 1994* (D.K. Schroder, J.L. Benton, and P. Rai-Choudhury, eds.), Electrochem. Soc., Pennington, NJ, 1994, 65–77.

136. R. Subrahmanyan, "Methods for the Measurement of Two-Dimensional Doping Profiles," *J. Vac. Sci. Technol.* **B10**, 358–368, Jan./Feb. 1992.

137. A.C. Diebold, M.R. Kump, J.J. Kopanski, and D.G. Seiler, "Characterization of Two-Dimensional Dopant Profiles: Status and Review," *J. Vac. Sci. Technol.* **B14**, 196–201, Jan./Feb.1996.

138. M. Barrett, M. Dennis, D. Tiffin, Y. Li, and C.K. Shih, "Two-Dimensional Dopant Profiling of Very Large Scale Integrated Devices Using Selective Etching and Atomic Force Microscopy," *J. Vac. Sci. Technol.* **B14**, 447–451, Jan./Feb.1996.

139. W. Vandervorst, T. Clarysse, P. De Wolf, L. Hellemans, J. Snauwaert, V. Privitera, and V. Raineri, "On the Determination of Two-Dimensional Carrier Distributions," *Nucl. Instrum. Meth.* **B96**, 123–132, March 1995.

140. C.C. Williams, Two-Dimensional Dopant Profiling by Scanning Capacitance Microscopy," *Annu. Rev. Mater. Sci.* **29**, 471–504, 1999.

141. W. Vandervorst, P. Eyben, S. Callewaert, T. Hantschel, N. Duhayon, M. Xu, T. Trenkler and T. Clarysse, "Towards Routine, Quantitative Two-dimensional Carrier Profiling with Scanning Spreading Resistance Microscopy," in *Characterization and Metrology for ULSI Technology*, (D.G. Seiler, A.C. Diebold, T.J. Shaffner, R. McDonald, W.M. Bullis, P.J. Smith, and E.M. Secula, eds.), *Am. Inst. Phys.* **550**, 613–619, 2000.

142. G. Neubauer, A. Erickson, C.C. Williams, J.J. Kopanski, M. Rodgers, and D. Adderton, "Two-Dimensional Scanning Capacitance Microscopy Measurements of Cross-Sectioned Very Large Scale Integration Test Structures," *J. Vac. Sci. Technol.* **B14**, 426–432, Jan./Feb.1996; J.S. McMurray, J. Kim, and C.C. Williams, "Quantitative Measurement of Two-dimensional Dopant Profile by Cross-sectional Scanning Capacitance Microscopy," *J. Vac. Sci. Technol.* **B15**, 1011–1014, July/Aug. 1997.

PROBLEMS

2.1 The $C-V$ curve of a Schottky diode on a p-type Si substrate is shown in Fig. P2.1(a) and (b). The $C-V$ data are also given in tabular form. Determine the $p(x)$ versus x profile for this device; plot as $\log[p(x)]$ in cm^{-3} versus W in μm. $K_s = 11.7$, $A = 10^{-3}$ cm^2.

V (V)	C (F)	V (V)	C (F)	V (V)	C (F)
0	8.39e-11	15.09	6.80e-12	26.29	2.37e-12
0.94	4.63e-11	15.62	6.38e-12	28.03	2.24e-12
2.16	3.20e-11	16.07	6.01e-12	29.86	2.12e-12
3.52	2.44e-11	16.47	5.68e-12	31.79	2.02e-12
4.93	1.97e-11	16.81	5.38e-12	33.82	1.93e-12
6.34	1.66e-11	17.36	4.88e-12	35.94	1.84e-12
7.69	1.43e-11	17.84	4.36e-12	38.16	1.76e-12
8.96	1.26e-11	18.17	3.95e-12	40.48	1.69e-12
10.14	1.12e-11	18.39	3.60e-12	42.89	1.62e-12
11.21	1.01e-11	19.06	3.32e-12	45.40	1.56e-12
12.18	9.22e-12	20.31	3.07e-12	48.01	1.51e-12
13.05	8.47e-12	21.60	2.86e-12	50.71	1.45e-12
13.81	7.83e-12	23.11	2.67e-12		
14.49	7.28e-12	24.65	2.51e-12		

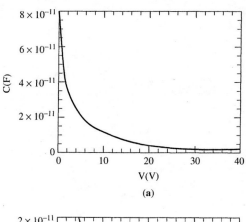

(a)

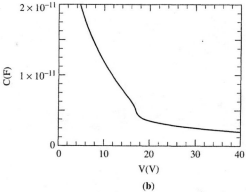

(b)

Fig. P2.1

2.2 The $C-V$ curves and data of the devices in Fig. P2.2 are given. C is the total capacitance. Determine distance d (in cm), doping density N_A (in cm^{-3}), and built-in potential V_{bi} (in V). $K_s = 11.7$, $K_{air} = 1$, $A = 10^{-3}$ cm^2. The semiconductor capacitance is given by

$$C_s = A\sqrt{\frac{qK_s\varepsilon_o N_A}{2(V_{bi} + V)}}.$$

V (V)	C (F)	V (V)	C (F)	V (V)	C (F)	V (V)	C (F)
0	2.276E-11	2.8	1.073E-11	0.000	9.959E-12	4.496	6.681E-12
0.2	2.036E-11	3	1.044E-11	0.430	9.470E-12	4.769	6.569E-12
0.4	1.858E-11	3.2	1.018E-11	0.820	9.067E-12	5.039	6.463E-12
0.6	1.720E-11	3.4	9.933E-12	1.183	8.726E-12	5.307	6.363E-12
0.8	1.609E-11	3.6	9.704E-12	1.527	8.431E-12	5.573	6.269E-12
1	1.517E-11	3.8	9.491E-12	1.857	8.171E-12	5.837	6.179E-12
1.2	1.439E-11	4	9.291E-12	2.175	7.940E-12	6.099	6.094E-12
1.4	1.372E-11	4.4	8.927E-12	2.485	7.732E-12	6.359	6.013E-12
1.6	1.314E-11	4.8	8.602E-12	2.787	7.543E-12	6.618	5.935E-12
1.8	1.262E-11	5.2	8.310E-12	3.083	7.370E-12	6.876	5.861E-12
2	1.217E-11	5.6	8.046E-12	3.374	7.211E-12	7.132	5.790E-12
2.2	1.175E-11	6.0	7.806E-12	3.660	7.064E-12	7.387	5.721E-12
2.4	1.138E-11	6.4	7.586E-12	3.942	6.928E-12	7.640	5.656E-12
2.6	1.104E-11	6.8	7.384E-12	4.221	6.800E-12	4.496	6.681E-12

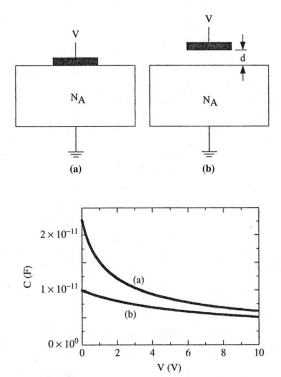

Fig. P2.2

2.3 For a p-type Si MIS capacitor, $C_{inv}/C_{ins} = 0.32$ and $t_{ins} = 30$ nm; "ins" stands for the insulator, which is not SiO_2 in this case.

(a) Determine the doping density for this device using $K_{ins} = 8$, $K_s = 11.7$, $n_i = 10^{10}$ cm^{-3}, $A = 10^{-3}$ cm^2, and $T = 27°C$.

(b) Determine C_{inv}/C_{ins} when $N_A = 10^{16}$ cm^{-3}. Use the approach that leads to Eq. (2.18) in the textbook for this problem.

(c) Use Eq. (2.19) to determine N_A instead of Eq. (2.18).

2.4 The $C-V$ and $1/C^2-V$ curves of a Schottky diode on a *uniformly-doped* substrate, doped to N_A, are shown in Fig. P2.4. Draw the $C-V$ and $1/C^2-V$ curves on the same figures for the case of a p-type layer (doped to N_A) grown on a p-type substrate (doped to N_{A1}) for (a) $N_A > N_{A1}$ and (b) $N_A < N_{A1}$. The voltage required to deplete the p-layer is shown by the vertical dashed line.

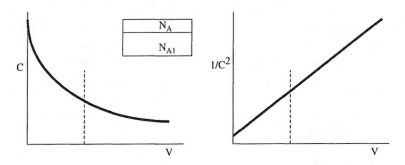

Fig. P2.4

2.5 For a p-type Si MIS capacitor, $C_{inv}/C_{ins} = 0.116$ and $t_{ins} = 100$ nm; "ins" stands for the insulator, which is not SiO_2 in this case.

(a) Determine the doping density N_A (in cm^{-3}) for this device.
Use the approach that leads to Eq. (2.18) in the textbook for this problem.

$$N_A = \frac{4\phi_F}{q K_s \varepsilon_o A^2} \frac{1 - C_{inv}^2}{(1 - C_{inv}/C_{ox})^2} = \frac{4\phi_F}{q K_s \varepsilon_o A^2} \frac{R^2 C_{ox}^2}{(1 - R)^2}. \qquad (2.18)$$

(b) Use Eq. (2.19) to determine N_A instead of Eq. (2.18).

$$\log(N_A) = 30.38759 + 1.68278 \log(C_1) - 0.03177[\log(C_1)]^2. \qquad (2.19)$$

(c) Determine C_{inv}/C_{ins} when $N_A = 10^{16}$ cm^{-3} for $t_{ins} = 100$ nm.
Use: $K_{ins} = 15$, $K_s = 11.7$, $n_i = 10^{10}$ cm^{-3}, $A = 10^{-3}$ cm^2, and $T = 300$ K.

2.6 The capacitance and conductance of semiconductor junction devices are usually measured using the device in Fig. P2.6(a) and its equivalent circuit in Fig. P2.6(b). A capacitance meter assumes the device is represented by the equivalent circuit in Fig. P2.6(c). Such a device can cause measurement problems due to series resistance at the bottom surface, *i.e.*, where the wafer touches the probe station wafer holder,

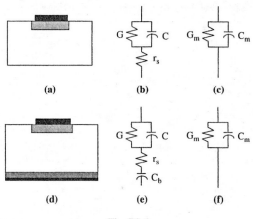

Fig. P2.6

especially if the wafer is not metallized on the back. Such problems can be alleviated by making a capacitive, rather than a resistive contact, as shown in Fig. P2.6(d).

(a) Derive expressions for G_m and C_m in Fig. P2.6(f) in terms of G, C, r_s, and C_b in P2.6(e).

(b) Find the *minimum* back capacitance C_b for this capacitance not to influence the measured capacitance and conductance, *i.e.*, introduce an error of not more than 1%.

(c) What area must be used for C_b if it is an oxide capacitance having an oxide thickness of 100 nm? $K_{ox} = 3.9$. Use $C = 100$ pF, $G = 10^{-6}$ S, $f = 1$ MHz, $r_s = 100$ Ω.

2.7 The C–V_G curve of an MOS capacitor (for positive V_G only) with uniformly doped substrate is shown in Fig. P2.7. Determine the doping density N_A using: (i) Eq. (2.5); (ii) Eq. (2.18); (iii) Eq. (2.19). $A = 5 \times 10^{-4}$ cm^2, $K_s = 11.7$, $K_{ox} = 3.9$, $n_i = 10^{10}$ cm^{-3}, $V_{FB} = 0$.

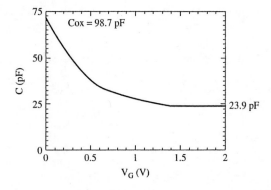

Fig. P2.7

2.8 The capacitance—voltage plot of a Schottky diode on a p-type substrate is shown in P2.8.

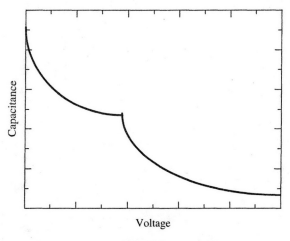

Fig. P2.8

(a) Plot $1/C^2-V$ and N_A-x for this device qualitatively.

(b) Next plot, again qualitatively, the $C-V$ curve and the $1/C^2-V$ curve for another Schottky diode on a p-type substrate with layers 1, 2, and 3 having doping densities $N_{A1} > N_{A2}$, $N_{A2} < N_{A3}$, $N_{A1} < N_{A3}$.

2.9 The $C-V_G$ curve in Problem 2.7 was obtained with a structure of the type shown in Fig. P2.9(a). What would the curve look like for structure in P2.9(b)? *Explain.* The bottom area ≫ top area.

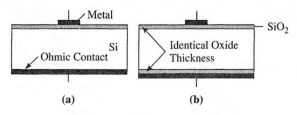

Fig. P2.9

2.10 The threshold voltage V_T of an n-channel MOSFET is given as a function of body or back bias voltage V_{BS}. Determine the doping density N_A and the flatband voltage V_{FB}. $t_{ox} = 25$ nm, $K_{ox} = 3.9$, $K_s = 11.7$, $n_i = 10^{10}$ cm^{-3}, $T = 300$ K.

V_{BS} (V)	0	−2	−4	−6	−8	−10	−12	−14	−16	−18	−20
V_T (V)	0.61	1.17	1.55	1.85	2.11	2.34	2.55	2.75	2.93	3.1	3.26

2.11 The capacitance of a semiconductor device with series resistance is measured as C_m. The data are shown in Fig. P2.11. Determine the true capacitance C and the series resistance r_s of this device.

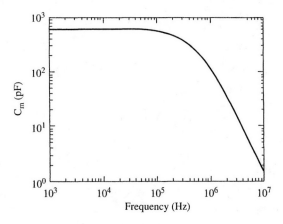

Fig. P2.11

2.12 Someone wants to measure the majority carrier profiles of the structures shown in Fig. P2.12 by $C-V$ profiling. The voltages during the $C-V$ measurements are such that the space-charge region width is confined to the p-region in each case. Comment on the validity of the conventional approach to $C-V$ measurements, *i.e.*, will the correct profile be obtained in each case? Explain why or why not. The space-charge region width is contained within the p-type layer in each case and series resistance is negligible.

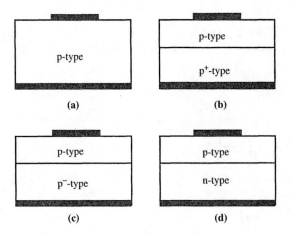

Fig. P2.12

2.13 (a) Calculate and plot C vs. V and $1/C^2$ vs. V for the Schottky barrier diode in Fig. P2.13 from 0 to 50 V for $N_{A1} = 10^{15}$ cm^{-3} and (i) $N_{A2} = 10^{14}$ cm^{-3}, (ii) $N_{A2} = 10^{15}$ cm^{-3}, (iii) $N_{A2} = 10^{16}$ cm^{-3}. Draw all three curves on the same figure.

(b) Calculate V_{BD}, the avalanche-limited breakdown voltage, for each case. Electric field at avalanche breakdown $= 3 \times 10^5$ V/cm, $A = 10^{-3}$ cm^2, $K_s = 11.7$, $V_{bi} = 0.4$ V.

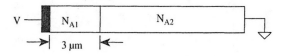

Fig. P2.13

Hint: Starting with Poisson's equation, find a relationship between the space-charge region width W and the applied voltage V using the *depletion approximation*. Then $C = K_s \varepsilon_0 A / W$.

2.14 Calculate and plot C vs. V and $1/C^2$ vs. V for the Schottky barrier diode in Fig. P2.13 with the N_{A1} layer thickness of 1 μm from $V = 0$ to 28 V for $N_{A1}(x) = 2 \times 10^{16} \exp(-kx)$ cm^{-3} and $N_{A2} = 10^{14}$ cm^{-3}. $k = 10^4$ cm^{-1}, $A = 10^{-3}$ cm^2, $K_s = 11.7$, $V_{bi} = 0.5$ V.
Hint: Starting with Poisson's equation, find a relationship between the space-charge region width W and the applied voltage V using the *depletion approximation*. Then $C = K_s \varepsilon_0 A / W$.

2.15 The error ε in the determination of the doping density N_A by the C–V profiling technique is given by

$$\varepsilon = \frac{1.4p}{\Delta C / C}$$

where p is the measurement precision.

(a) Derive and plot $\log(|\varepsilon|)$ versus $\log(W)$, where W is the space-charge region width in microns and ε is in %, for: (a) $\Delta C = 10^{-14}$ F $=$ constant, (b) $\Delta W = 10^{-5}$ cm $=$ constant, (c) $\Delta V = 0.015$ V $=$ constant.

(b) If you had a choice, which of these three approaches would you use for best accuracy?

(c) In your opinion, which one of these three approaches is easiest to implement? Use $p = 0.1\%$, $N_A = 10^{15}$ cm^{-3}, $K_s = 11.7$, $A = 10^{-3}$ cm^2. The following relationships may be useful:

$$C = \frac{K_s \varepsilon_o A}{W}; \quad W^2 = \frac{2 K_s \varepsilon_o V}{q N_A}$$

Use 1 μm $\leq W \leq 10$ μm.

2.16 The capacitance and conductance of an MOS capacitor were measured and are shown in Fig. P2.16. Determine the true capacitance C, the true conductance G and the series resistance r_s.

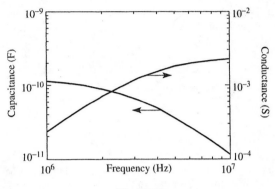

Fig. P2.16

2.17 In the MOSFET threshold voltage doping profiling method, the threshold voltage V_T was measured as function of the substrate bias voltage V_{BS}. The data are:

V_{BS} (V)	V_T (V)
0	2.40
−1	2.84
−2	3.17
−3	3.42
−4	3.64
−5	3.85
−6	4.05
−7	4.22
−8	4.36
−9	4.54
−10	4.70

Determine the doping density profile and the flatband voltage V_{FB}. $t_{ox} = 20$ nm, $K_{ox} = 3.9$, $K_s = 11.7$, $n_i = 10^{10}$ cm^{-3}, $T = 300$ K.

REVIEW QUESTIONS

- How is the *capacitance* measured?
- Why is $1/C^2 - V$ preferred over $C - V$?
- What is important in contactless $C - V$?
- What is measured in most profiling techniques, *i.e.*, doping density or majority carrier density?
- What is the *Debye length*?
- What is measured in the "equilibrium" MOS-C $C - V_G$ method?

- What does series resistance do to capacitance measurements?
- What advantage does the electrochemical profiling technique have?
- How does the threshold voltage technique work?
- What determines the profiling limits?
- What is the Hall effect and how does it work?
- What is *secondary ion mass spectrometry*?
- How does *spreading resistance* profiling work?

3

CONTACT RESISTANCE AND SCHOTTKY BARRIERS

3.1 INTRODUCTION

Since all semiconductor devices have contacts and all contacts have contact resistance, it is important to characterize such contacts. Contacts are generally metal-semiconductor contacts, but they may be semiconductor-semiconductor contacts, where both semiconductors can be single crystal, polycrystalline, or amorphous. In the conceptual discussion of ohmic contacts and contact resistance we will be mainly concerned with metal-semiconductor contacts because they are most common. For the discussion of the measurement techniques the type of contact is unimportant, but the resistance of the contact material is important.

The metal-semiconductor contact, discovered by Braun in 1874, forms the basis of one of the oldest semiconductor devices.[1] The first acceptable theory was developed by Schottky in the 1930s.[2] In his honor metal-semiconductor devices are frequently referred to as *Schottky barrier devices*. Usually this name denotes the use of these devices as rectifiers with distinctly non-linear current-voltage characteristics. A good discussion of the history of metal-semiconductor devices is given by Henisch[3] with a more recent review by Tung.[4]

Ohmic contacts have linear or quasi-linear current-voltage characteristics. It is not necessary, however, that ohmic contacts have linear $I-V$ characteristics. The contacts must be able to supply the necessary device current, and the voltage drop across the contact should be small compared to the voltage drops across the active device regions. An ohmic contact should not degrade the device to any significant extent, and it should not inject minority carriers. Appendix 3.2 lists various metal-semiconductor contacts.

The first comprehensive publication on ohmic contacts was the result of a conference devoted to this topic.[5] The theory of metal-semiconductor contacts with emphasis

Semiconductor Material and Device Characterization, Third Edition, by Dieter K. Schroder
Copyright © 2006 John Wiley & Sons, Inc.

on ohmic contacts was presented by Rideout.[6] Ohmic contacts to III–V devices were reviewed by Braslau[7] and Piotrowska et al.,[8] and ohmic contacts to solar cells were discussed by Schroder and Meier.[9] Yu and Cohen have presented discussions of contact resistance.[10-11] Additional information can be found in the books by Milnes and Feucht,[12] Sharma and Purohit,[13] and Rhoderick.[14] Cohen and Gildenblat give a very good discussion.[15]

3.2 METAL-SEMICONDUCTOR CONTACTS

The *Schottky model* of the metal-semiconductor barrier is shown in Fig. 3.1. The energy bands are shown before contact in the upper part of the figure and after contact in the lower part. We assume intimate contact between the metal and the semiconductor with no interfacial layer. The work function of a solid is defined as the energy difference between the vacuum level and the Fermi level. Work functions for the metal and the semiconductor are shown in Fig. 3.1, with the metal work function Φ_M being less than the semiconductor work function Φ_S in Fig. 3.1(a). The work function is given as the energy Φ_M related to the potential ϕ_M by $\phi_M = \Phi_M/q$.

In Fig. 3.1(b) $\phi_M = \phi_S$, and in Fig. 3.1(c) $\phi_M > \phi_S$. The ideal barrier height after contact for this model is given by[2, 16]

$$\phi_B = \phi_M - \chi \tag{3.1}$$

where χ is the electron affinity of the semiconductor, defined as the potential difference between the bottom of the conduction band and the vacuum level at the semiconductor surface. According to the Schottky theory, the barrier height depends only on the metal work function and on the semiconductor electron affinity and is independent of the semiconductor doping density. This should make it easy to vary the barrier height by merely using metals of the appropriate work function to implement any one of the three barrier types of Fig. 3.1. We have named them *accumulation, neutral,* and *depletion contacts*

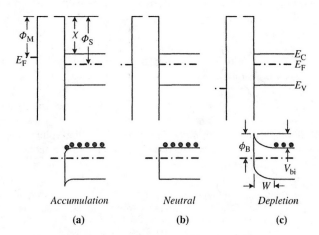

Fig. 3.1 Metal-semiconductor contacts according to the simple Schottky model. The upper and lower parts of the figure show the metal-semiconductor system before and after contact, respectively.

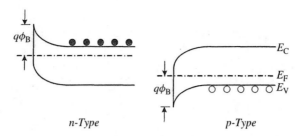

Fig. 3.2   Depletion-type contacts on *n*- and *p*-type substrates.

because the majority carriers are accumulated, unchanged (neutral), or depleted compared to their density in the neutral substrate.

As is evident from Fig. 3.1 an accumulation-type contact is the preferred ohmic contact because electrons in the metal encounter the least barrier to their flow into or out of the semiconductor. In practice it is difficult to alter the barrier height by using metals of varying work functions. It is experimentally observed that the barrier height for the common semiconductors Ge, Si, GaAs, and other III–V materials is relatively independent of the work function of the metal.[17] A *depletion* contact is generally formed on both *n*-type and *p*-type substrates, as shown in Fig. 3.2. For *n*-substrates $\phi_B \approx 2E_g/3$ and for *p*-substrates $\phi_B \approx E_G/3$.[18]

The relative constancy of the barrier height with various work function metals is sometimes attributed to *Fermi level pinning*, where the Fermi level in the semiconductor is pinned at some energy in the band gap to create a depletion-type contact. The details of Schottky barrier formation are not fully understood. It appears, however, that imperfections at the semiconductor surface play an important role during contact formation. Bardeen pointed out the importance of surface states in determining the barrier height.[19] Such surface states may be dangling bonds at the surface or some other types of defects.[17, 20] There is, however, still disagreement between the various proposed mechanisms causing Fermi level pinning.[21–23]

Whatever the mechanisms that cause barrier heights to be relatively independent of the metal work function, it is difficult to engineer an accumulation-type contact. Barrier *height* engineering being impractical, we must look to other means of implementing ohmic contacts. Ohmic contacts are frequently defined as regions of high recombination rates. This implies that highly damaged regions should serve as good ohmic contacts. Such fabrication methods are not practical because damage is usually the last thing one wants in a semiconductor device. Damage-induced ohmic contacts are also not reproducible. This leaves the semiconductor doping density as the only alternative to engineer contacts.[24] As stated earlier, the *barrier height* is relatively independent of the doping density, but the *barrier width* does depend on the doping density. The barrier height does actually depend weakly on doping density through *image force barrier lowering*.

Heavily doped semiconductors have narrow space-charge region (scr) width $W (W \sim N_D^{-1/2})$. For metal-semiconductor contacts with narrow scr widths, electrons can *tunnel* from the metal to the semiconductor and from the semiconductor to the metal. Holes tunnel for *p*-type semiconductors. Some readers may be uncomfortable with the concept of holes tunneling from a metal to a semiconductor. It may be helpful to think of hole tunneling from the metal to the semiconductor as electron tunneling from the semiconductor valence band to the metal.

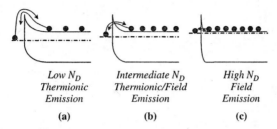

Fig. 3.3 Depletion-type contacts to *n*-type substrates with increasing doping concentrations. The electron flow is schematically indicated by the electrons and their arrows.

The conduction mechanisms for a metal-*n*-type semiconductor are illustrated in Fig. 3.3. For lightly-doped semiconductors the current flows as a result of *thermionic emission* (TE) shown in Fig. 3.3(a) with electrons thermally excited over the barrier.[25] In the intermediate doping range *thermionic-field emission* (*TFE*) dominates with carriers thermally excited to an energy where the barrier is sufficiently narrow for tunneling to take place.[26–27] For high doping densities the barrier is sufficiently narrow at or near the bottom of the conduction band for the electrons to tunnel directly, known as *field emission* (*FE*). The three regimes can be differentiated by considering the characteristic energy E_{00} defined by[26]

$$E_{00} = \frac{qh}{4\pi}\sqrt{\frac{N}{K_s\varepsilon_o m^*_{tun}}} = 1.86 \times 10^{-11}\sqrt{\frac{N(\text{cm}^{-3})}{K_s(m^*_{tun}/m)}}\,[\text{eV}] \qquad (3.2)$$

where N is the doping density, m^*_{tun} is the tunneling effective mass, and m the free electron mass. Equation (3.2) is plotted in Fig. 3.4. A comparison of E_{00} to the thermal energy kT shows thermionic emission to dominate for $kT \gg E_{00}$, for thermionic-field emission $kT \approx E_{00}$ and for field emission $kT \ll E_{00}$. For simplicity we have chosen the demarcation points on Fig. 3.4 as: for TE: $E_{00} \leq 0.5\,kT$, for TFE: $0.5\,kT < E_{00} < 5\,kT$, and for FE: $E_{00} \geq 5\,kT$. For Si with a tunneling effective mass of 0.3 m,[28] this corresponds

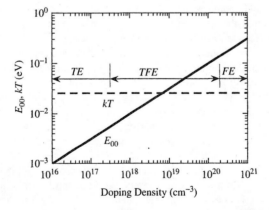

Fig. 3.4 E_{00} and kT as a function of doping density for Si with $m^*_{tun}/m = 0.3$. $T = 300$ K.

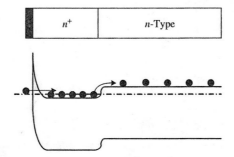

Fig. 3.5   A metal-n^+-n semiconductor contact band diagram.

approximately to TE for $N \leq 3 \times 10^{17}$ cm^{-3}, TFE for $3 \times 10^{17} < N < 2 \times 10^{20}$ cm^{-3}, and FE for $N \geq 2 \times 10^{20}$ cm^{-3}. The tunneling effective mass differs for n-Si and p-Si and also depends on doping density.

The structure of Fig. 3.3(c) is not realized in most real contacts. Generally only the semiconductor directly under the contact is heavily doped; the region farther from the contact being less heavily doped as illustrated in Fig. 3.5. The contact resistance becomes the sum of the metal-semiconductor contact resistance and the n^+n junction resistance. Such a structure has a contact resistance similar to a uniformly doped structure if the metal-semiconductor junction resistance dominates.[29] However, the contact resistance dependence on doping density is expected to be different when the n^+n junction dominates over the metal-semiconductor junction. The inverse dependence of contact resistance on doping density has been attributed to the resistance of the high-low junction.[30–31]

3.3   CONTACT RESISTANCE

Metal-semiconductor contacts fall into two basic categories, illustrated in Fig. 3.6. The current flows either *vertically* or *horizontally* into the contact. Vertical and horizontal or lateral contacts can behave quite differently, because the effective contact area may differ from the true contact area. Let us consider the resistance between points A and B of the sample having metallic conductors lying on an insulator and making ohmic contacts to an n-type layer in a p-type substrate in Fig. 3.7. We divide the total resistance R_T between points A and B into three components: (1) the resistance of the metallic conductor R_m, (2) the contact resistances R_c, and (3) the semiconductor resistance R_{semi}. The total resistance is

$$R_T = 2R_m + 2R_c + R_{semi} \qquad (3.3)$$

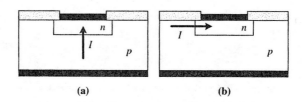

      (a)   **(b)**

Fig. 3.6   (a) "Vertical" and (b) "horizontal" contact.

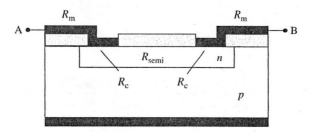

Fig. 3.7 A schematic diagram showing two contacts to a diffused semiconductor layer, with the metal resistance, the contact resistances and the semiconductor resistance indicated.

The semiconductor resistance is determined by the sheet resistance of the n-layer. The contact resistance is less clearly defined. It certainly includes the resistance of the metal-semiconductor contact, sometimes called the *specific interfacial resistivity* ρ_i.[10] But it also includes a portion of the metal immediately above the metal-semiconductor interface, a part of the semiconductor below that interface, current crowding effects, and any interfacial oxide or other layer that may be present between the metal and the semiconductor. How then do we define contact resistance?

The current density J of a metal-semiconductor contact depends on the applied voltage V, the barrier height ϕ_B and the doping density N_D in a manner that varies for each of the three conduction mechanisms in Fig. 3.3. We write that dependence as

$$J = f(V, \phi_B, N_D) \tag{3.4}$$

The contact resistance is characterized by two quantities: the *contact resistance* (ohms) and the *specific contact resistivity*, ρ_c (ohm·cm^2), sometimes referred to as *contact resistivity* or *specific contact resistance*. The specific contact resistivity includes not only the actual interface but the regions immediately above and below the interface.

We define a *specific interfacial resistivity* ρ_i (ohm·cm^2) by

$$\rho_i = \left. \frac{\partial V}{\partial J} \right|_{V=0} \tag{3.5a}$$

As we will see later, the contact area also plays a role in the behavior of the contact. Hence ρ_i is also defined as

$$\rho_i = \left. \frac{\partial V}{\partial J} \right|_{A \to 0} \tag{3.5b}$$

where A is the contact area. This specific interfacial resistivity is a theoretical quantity referring to the metal-semiconductor interface only. It is not actually measurable because of the effects referred to above. The parameter that is determined from measured contact resistance is the *specific contact resistivity*. It is a very useful term for ohmic contacts because it is independent of contact area and is a convenient parameter when comparing contacts of various sizes. We will use ρ_i only when deriving theoretical expressions of metal-semiconductor contacts. Thereafter we use ρ_c when discussing real contacts, their measurements, and measurement interpretations.

The current density of a metal-semiconductor contact, dominated by thermionic emission, is given in its simplest form by[14]

$$J = A^* T^2 e^{-q\phi_B/kT} (e^{qV/kT} - 1) \tag{3.6}$$

where $A^* = 4\pi q k^2 m^*/h^3 = 120(m^*/m)$ A/cm^2·K^2 is Richardson's constant, m is the free electron mass, m^* the effective electron mass, and T the absolute temperature. With Eq. (3.5a) we find the specific interfacial resistivity for *thermionic emission* to be

$$\rho_i(TE) = \rho_1 e^{q\phi_B/kT}; \rho_1 = \frac{k}{qA^*T} \tag{3.7}$$

For *thermionic-field emission* ρ_i is given by[9]

$$\rho_i(TFE) = C_1 \rho_1 e^{q\phi_B/E_0} \tag{3.8}$$

and for *field emission* it is[9]

$$\rho_i(FE) = C_2 \rho_1 e^{q\phi_B/E_{00}} \tag{3.9}$$

C_1 and C_2 are functions of N_D, T, and ϕ_B. E_0 in Eq. (3.8) is related to E_{00} by[26]

$$E_0 = E_{00} \coth (E_{00}/kT) \tag{3.10}$$

Substituting for E_{00} in Eq. (3.9) leads to

$$\rho_i(FE) \sim \exp(C_3/\sqrt{N}) \tag{3.11}$$

where C_3 is a constant and N the doping density under the contact. The actual expression for $\rho_i(FE)$ is more complex.[28] We give merely the very simplest forms here to indicate the dependence of ρ_i on doping density and barrier height. As Eq. (3.11) indicates, $\rho_i(FE)$ is very sensitive to the doping density under the contact. N should be as high as possible for lowest specific interfacial resistivity.

We have given the specific interfacial resistivity by these simple expressions in order not to obscure the main points in this discussion. More complex relations are available for the interested reader.[28, 32-34] The detailed expressions for the various conduction mechanisms are rather complicated and a calculation of the specific interfacial resistivity for each of the three regions is difficult. Various approximations have been proposed and theoretical curves of ρ_i versus N_A or N_D have been generated.[28, 32-34] These curves depend on the effective masses, the barrier height, and various other parameters. The barrier height depends also on the contact metal, and it is therefore impossible to derive "universal" ρ_i versus N_A or ρ_i versus N_D curves. Those that have been derived do not always agree with experimental data. We show in Figs. 3.8 experimental ρ_c versus N_D and N_A data for Si. There is considerable scatter, but a definite trend of lower specific contact resistivity with higher doping densities, predicted by Eq. (3.11), is obvious in the data. Data for GaAs can be found in ref. 37 and 38.

The temperature dependence of the specific contact resistivity for tungsten contacts to n-Si and p-Si, normalized to $T = 305$ K, is shown in Fig. 3.9, showing that there is not a simple ρ_c-T relationship.[39] The temperature behavior of ρ_c is very much dependent on the doping density. For surface doping densities around 10^{20} cm^{-3}, there is almost

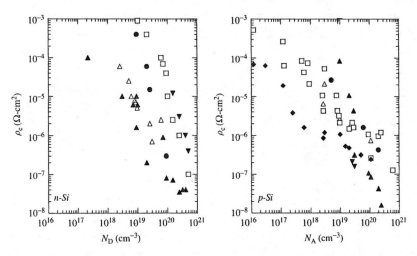

Fig. 3.8 Specific contact resistivity as a function of doping density for Si. The references for n-Si are given in ref. 35 and for p-Si in ref. 36.

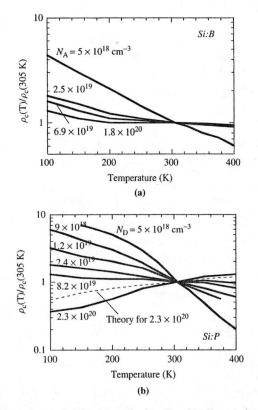

Fig. 3.9 The specific contact resistivity, normalized to $T = 305$ K, as a function of temperature for (a) p-Si and (b) n-Si. The data for $N_D = 2 \times 10^{18}$ cm^{-3} extend from $T = 305$ to 400 K only. The metal is tungsten. Reprinted after ref. 39 by permission of IEEE (© 1986, IEEE).

no temperature dependence whereas for densities above and below that value, there are significant variations of ρ_c with temperature.

3.4   MEASUREMENT TECHNIQUES

Contact resistance measurement techniques fall into four main categories: *two-contact two-terminal, multiple-contact two-terminal, four-terminal,* and *six-terminal methods*. None of these methods is capable of determining the specific interfacial resistivity ρ_i. Instead they determine the specific contact resistivity ρ_c which is not the resistance of the metal-semiconductor interface alone, but it is a practical quantity describing the real contact. It is, therefore, difficult to compare theory with experiment because theory cannot predict ρ_c accurately and experiment cannot determine ρ_i accurately. At times it is even difficult to measure ρ_c unambiguously. We limit ourselves to discussions of measurement techniques. Contact formation and the impact of contact resistance on device behavior can be found in numerous references of which 7,12, 14 and 40 are a few.

3.4.1   Two-Contact Two-Terminal Method

The two-terminal contact resistance measurement method is the earliest method.[41] It is also of questionable accuracy if not properly executed. The simplest implementation is shown in Fig. 3.10. For a homogeneous semiconductor of resistivity ρ and thickness t with two contacts as shown in Fig. 3.10(a), the total resistance $R_T = V/I$, measured by passing a current I through the sample and measuring the voltage V across the two contacts, is

$$R_T = R_c + R_{sp} + R_{cb} + R_p \qquad (3.12a)$$

For Fig. 3.10(b) with both contacts on the top surface

$$R_T = 2R_c + 2R_{sp} + 2R_p \qquad (3.12b)$$

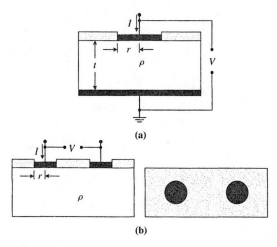

(a)

(b)

Fig. 3.10   (a) A vertical two-terminal contact resistance structure, (b) a lateral two-terminal contact resistance structure.

where R_c is the contact resistance of the top contact, R_{sp} the spreading resistance in the semiconductor directly under the contact, R_{cb} the contact resistance of the bottom contact, and R_p the probe or wire resistance. The bottom contact usually has a large contact area with a concomitant small resistance. Consequently, R_{cb} is often neglected. Similarly, the probe resistance is usually negligible.

The spreading resistance of a flat, non indenting circular top contact of radius r on the surface of a semiconductor of resistivity ρ, thickness t, and a large bottom contact can be approximated by[42]

$$R_{sp} = \frac{\rho}{2\pi r} \arctan{(2t/r)} \tag{3.13}$$

More exact expressions for the spreading resistance have been derived.[43] For $2t \gg r$, Eq. (3.13) can be expressed as

$$R_{sp} = C\frac{\rho}{4r} \tag{3.14}$$

where C is a correction factor that depends on ρ, r, and on the current distribution. For widely separated contacts for the structure in Fig. 3.10(b), on a uniformly-doped, semi-infinite substrate the correction factor $C = 1$. With the current flowing vertically into the top contact as in Fig. 3.10(a), the contact resistance is

$$R_c = \frac{\rho_c}{A_c} = \frac{\rho_c}{\pi r^2} \tag{3.15}$$

For small R_{cb}, Eq. (3.12) shows the contact resistance to be the difference between the total resistance and the spreading resistance. The spreading resistance cannot be measured independently and small errors in R_{sp} can lead to large errors in R_c. The two-terminal method, therefore, works best when $R_{sp} \ll R_c$, approximated by using small-radius contacts.[42, 44–47]

A variation on the two-terminal contact resistance measurement technique is the use of top contacts of varying diameters. Then one measures and plots R_c, calculated from Eq. (3.12) using experimental R_T data, as a function of $1/A_c$ and determines ρ_c from the slope of this plot.[48] Alternately, the total resistance can be plotted against $1/r$ with Eq. (3.12) fitted to this curve.[46] By using various diameters one can see from the shape of the curve whether the data are anomalous.

The two-terminal method is more commonly implemented with the lateral structure of Fig. 3.11. This test structure differs from Fig. 3.10(b) by confining the current to the n-island. The test structure consists of two contacts separated by the spacing d. To confine the current flow, the region on which the contact is located must be isolated from the remainder of the substrate, by either confining the implanted or diffused region (n-type on p-type substrate in Fig. 3.11 or p-on-n) by planar techniques or by etching the region surrounding the island, leaving it as a mesa. The n-type island in this example has width W and ideally the contacts should also be W wide. That is difficult to implement and the contact width Z generally differs from W. The analysis becomes more difficult due to lateral current flow, current crowding at the contacts, and sample geometry.[49] For the geometry of Fig. 3.11, the total resistance is

$$R_T = R_{sh}d/W + R_d + R_w + 2R_c \tag{3.16}$$

where R_{sh} is the sheet resistance of the n-layer, R_d the resistance due to current crowding under the contacts, R_w a contact width correction if $Z < W$, and R_c the contact resistance

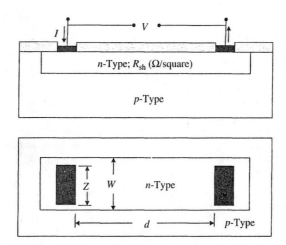

Fig. 3.11   A lateral two-terminal contact resistance structure in cross section and top view.

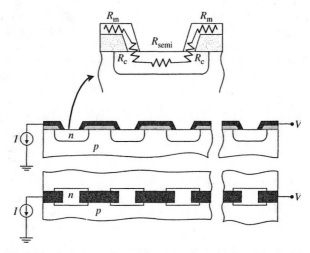

Fig. 3.12   A contact string test structure; cross section and top view.

assumed to be identical for the two contacts. Expressions for these resistances are given in ref. 6.

The *contact chain* or *contact string* in Fig. 3.12 is commonly used for process control, incorporating many contacts (hundreds, thousands, or as many as a million) of the type shown in Fig. 3.11. The total resistance between any two contacts is the sum of the semiconductor resistance, the contact resistance, and the metal resistance. The semiconductor resistance is calculated knowing the sheet resistance and the string geometry. By subtracting the semiconductor resistance from the total resistance one obtains the total contact resistance. The contact resistance for each contact is obtained by dividing by twice the number of contacts. A refined contact string divides the string into sections with intermediate contact pads.[50]

For a contact string consisting of N islands and $2N$ contacts, with contacts separated from each other by spacing d and width W, the total resistance is given by

$$R_T = \frac{N R_{sh} d}{W} + 2 N R_c \qquad (3.17)$$

neglecting the metal resistance. The contact string technique is considered to be a coarse measurement method that is not very useful for detailed evaluations of contact resistance. It is, however, extensively used as a process monitor. If the measured resistance is higher than the norm, it is difficult to know whether all contacts are poor or whether one particular contact is poor unless intermediate probe pads are provided. Frequently the contact string is only accessible at the ends with no intermediate contacts.

Exercise 3.1

Problem: What effect do the np junctions of the contact string have on the measured results?

Solution: The contact string of Fig. 3.12 can be represented by Fig. E3.1. Let us consider the substrate grounded. Suppose $R = R_m + 2R_c + R_{semi} = 50\ \Omega$ and $I = 1$ mA. For 250 islands, we find $V = 12.5$ V. Assume the junctions have a breakdown voltage of 15 V. Clearly, there is no problem in measuring R. What happens if in one process run R_c increases such that $R = 75\ \Omega$. Now $V = IR = 18.8$ V, but the junctions can only withstand 15 V. Since the total voltage cannot exceed 15 V, dictated by the breakdown voltage of the last np junction, an erroneous resistance will be measured. The situation is better if the substrate is not grounded, because now the voltage is divided among the many np junctions. The message here is to be cautious of the layout and measurement connection when making contact string measurements.

3.4.2 Multiple-Contact Two-Terminal Methods

The multiple-contact, two-terminal contact resistance measurement technique, shown in Fig. 3.13, was developed to overcome the deficiencies of the two-contact, two-terminal

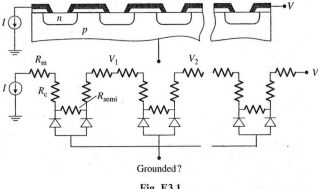

Grounded?

Fig. E3.1

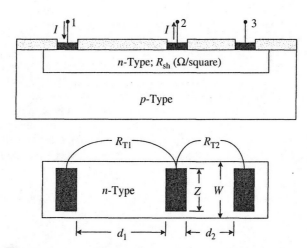

Fig. 3.13 Multiple-contact, two-terminal contact resistance test structure. The contact width and length are Z and L and the diffusion width is W.

method. Three identical contacts are made to the semiconductor with contact spacings d_1 and d_2. Assuming identical contact resistances for each of the three contacts allows the total resistance to be written as

$$R_{Ti} = \frac{R_{sh}d_i}{W} + 2R_c \qquad (3.18)$$

where $i = 1$ or 2. Solving for R_c gives

$$R_c = \frac{(R_{T2}d_1 - R_{T1}d_2)}{2(d_1 - d_2)} \qquad (3.19)$$

This structure does not have the ambiguities of the simpler two-terminal structure, because neither the bulk resistance nor the layer sheet resistance need be known. The assumption of identical contact resistance for all three contacts is somewhat questionable but is reasonable for a sample that is not too large. The contact resistance is obtained by taking the difference of two large numbers. This can present difficulties and is especially troublesome for low resistance contacts. The determination of lengths d_1 and d_2 is a further source of inaccuracy. Occasionally negative contact resistances are obtained by this method.

The structure of Fig. 3.13 only allows the *contact resistance* to be determined. The specific contact resistivity cannot be directly extracted from the two resistance measurements. To find ρ_c requires a more detailed evaluation of the nature of the current flow into and out of the lateral contacts. An early two-dimensional current flow analysis by Kennedy and Murley in diffused semiconductor resistors revealed current crowding at the contacts.[51] The analysis, based on zero contact resistance, showed that only a fraction of the total contact length was active during the transfer of current from the metal to the semiconductor and from the semiconductor to the metal. This fraction was found to be approximately equal to the thickness of the diffused semiconductor sheet.

To take current crowding into account and to be able to extract the specific contact resistivity, a detailed theoretical investigation was undertaken. Murrmann and Widmann used a simple *transmission line model* (TLM) considering both the semiconductor sheet resistance and the contact resistance.[52] They also described a structure to determine the contact resistance using linear and concentric contacts.[53] Berger extended the transmission line method.[54] In contrast to the Kennedy-Murley model, in which the contact resistance is assumed to be zero, in the TLM the contact resistance is non-zero. However, the semiconductor sheet thickness is assumed to be zero in the TLM, with the layer retaining its sheet resistance R_{sh}. This assumption allows one-dimensional current flow only. The "zero sheet thickness" restriction was relaxed by Berger in his extended TLM where he allowed non-zero sheet thickness, but with the current still restricted to one-dimensional flow.[54] The TLM model was later extended to two dimensions by the dual-level transmission line model with the current allowed to flow perpendicularly to the contact interface. A comparison between the simple and the revised TLM shows a maximum contact resistance deviation of 12%.[55]

When current flows from the semiconductor to the metal, it encounters the resistances ρ_c and R_{sh} in Fig. 3.14, choosing the path of least resistance. The potential distribution under the contact is determined by both ρ_c and R_{sh} according to[54]

$$V(x) = \frac{I\sqrt{R_{sh}\rho_c}}{Z} \frac{\cosh[(L-x)/L_T]}{\sinh(L/L_T)} \tag{3.20}$$

where L is the contact length, Z the contact width, and I the current flowing into the contact. Equation (3.20) is plotted in Fig. 3.15 with the potential under the contact normalized to unity at $x = 0$. The voltage is highest near the contact edge $x = 0$ and drops nearly exponentially with distance. The "$1/e$" distance of the voltage curve is defined as the *transfer length* L_T

$$L_T = \sqrt{\rho_c/R_{sh}} \tag{3.21}$$

The transfer length can be thought of as that distance over which most of the current transfers from the semiconductor into the metal or from the metal into the semiconductor. L_T is plotted in Fig. 3.16 against the specific contact resistivity as a function of the sheet resistance. Typical specific contact resistivities are $\rho_c \leq 10^{-6}$ $\Omega \cdot cm^2$ for good contacts. The transfer length is on the order of 1 μm or less for such contacts. Contacts for contact

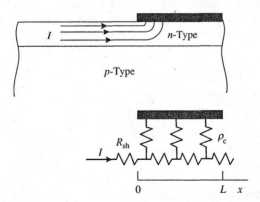

Fig. 3.14 Current transfer from semiconductor to metal represented by the arrows. The semiconductor/metal contact is represented by the ρ_c-R_{sh} equivalent circuit with the current choosing the path of least resistance.

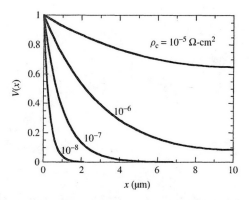

Fig. 3.15 Normalized potential under a contact versus x as function of ρ_c, where $x = 0$ is the contact edge. $L = 10$ μm, $Z = 50$ μm, $R_{sh} = 10$ Ω/square.

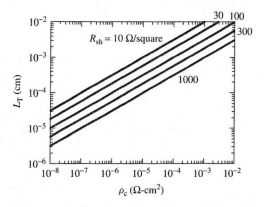

Fig. 3.16 Transfer length as a function of specific contact resistivity and semiconductor sheet resistance.

resistance measurements are often longer than 1 μm. For such contacts, some of the contact is inactive during current transfer.

We will now consider the three contact configurations in Fig. 3.17, with the current flowing from contact 1 to contact 2. In the transmission line method test structure (TLM) in Fig. 3.17(a), also referred to as the *contact front resistance test structure* (CFR), the voltage is measured across the same contacts as the current. In the *contact end resistance test structure* (CER) in Fig. 3.17(b) the voltage is measured between contacts 2 and 3. In the *cross bridge Kelvin resistance test structure* (CBKR) (Fig. 3.17(c)), the voltage is measured at right angles to the current.

With V measured between contacts 1 and 2 at $x = 0$, Eq. (3.20) gives the contact *front* resistance as

$$R_{cf} = \frac{V}{I} = \frac{\sqrt{R_{sh}\rho_c}}{Z} \coth(L/L_T) = \frac{\rho_c}{L_T Z} \coth(L/L_T) \qquad (3.22)$$

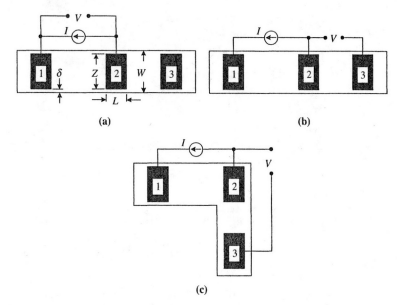

Fig. 3.17   (a) Conventional contact resistance test structure, (b) contact end resistance test structure, and (c) cross bridge Kelvin resistance test structure.

provided $Z = W$. Eq. (3.22) is only an approximation when the sample is wider than Z, because this equation does not consider the current flow around the contacts.

The expression R_{cf} is usually referred to simply as the contact resistance R_c. We will do so here also. Two cases lead to simplifications of Eq. (3.22). For $L \leq 0.5\, L_T$, $\coth(L/L_T) \approx L_T/L$ and

$$R_c \approx \frac{\rho_c}{LZ} \tag{3.23a}$$

and for $L \geq 1.5\, L_T$, $\coth(L/L_T) \approx 1$ and

$$R_c \approx \frac{\rho_c}{L_T Z} \tag{3.23b}$$

The effective contact area is the actual contact area $A_c = LZ$ for the first case. But in the second case the effective contact area is $A_{c,eff} = L_T Z$. In other words, the *effective* contact area can be smaller than the *actual* contact area. This can have important consequences. For example, consider a structure with $R_{sh} = 20\ \Omega/\text{square}$ and $\rho_c = 10^{-7}\ \Omega\cdot\text{cm}^2$. The transfer length $L_T = 0.7\ \mu\text{m}$. For a contact length of $L = 10\ \mu\text{m}$ and width $Z = 50\ \mu\text{m}$, the actual contact area is $LZ = 5 \times 10^{-6}\ \text{cm}^2$. However, the effective contact area is only $L_T Z = 3.5 \times 10^{-7}\ \text{cm}^2$. The current density flowing across the contact is $5 \times 10^{-6}/3.5 \times 10^{-7} = 14$ times higher than if the entire contact were active. This higher current density can cause reliability problems by degrading the contact. The reduced contact area can burn out in extreme cases shifting the effective area along the contact until the entire contact is destroyed.

The effect of contact length on contact resistance is illustrated in Fig. 3.18. It is a plot of the front contact resistance given by Eq. (3.22) multiplied by the contact width Z, for

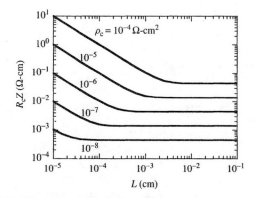

Fig. 3.18 Front contact resistance–contact width product as a function of contact length and specific contact resistivity for $R_{sh} = 20$ Ω/square and $R_{sm} = 0$.

normalization purposes, against the contact length as a function of the specific contact resistivity. Note the initial R_c decrease with contact length. However, $R_c Z$ reaches a minimum at $L \approx L_T$ from which it departs no further no matter how long the contact.

The metal/semiconductor representation of Fig. 3.14 may be too simple for certain contacts. For example, alloyed contacts typically made on GaAs consist of a metal, an alloyed region, and the underlying semiconductor. Similarly contacts formed by depositing a metal on a thin layer of a low band gap material on a higher band gap material fall into this category. This calls for a more complex transmission line model—the trilayer transmission line model. The equations, although similar to the TLM equations, become significantly more complex.[56]

When the voltage is measured between contacts 2 and 3 with the current flowing from 1 to 2, shown in Fig. 3.17(b), the structure is known as the contact end resistor. The voltage is now measured at $x = L$ and Eq. (3.20) leads to the contact *end* resistance

$$R_{ce} = \frac{V}{I} = \frac{\sqrt{R_{sh}\rho_c}}{Z}\frac{1}{\sinh(L/L_T)} = \frac{\rho_c}{L_T Z}\frac{1}{\sinh(L/L_T)} \qquad (3.24)$$

The contact end resistance measurement can be used to determine ρ_c by measuring R_{ce} and using an iteration of Eq. (3.24).[57] For short contacts, R_{ce} is sensitive to contact length variations with the error in determining L limiting the accuracy of the method. For long contacts, R_{ce} becomes very small and the accuracy is limited by instrumentation, seen by looking at the ratio

$$\frac{R_{ce}}{R_{cf}} = \frac{1}{\cosh(L/L_T)} \qquad (3.25)$$

which obviously becomes very small for $L \gg L_T$.

For the cross-bridge Kelvin resistance test structure in Fig. 3.17(c), the voltage contact 3 is located at the side of contact 2. The measured voltage is thus the linear average of the potential over the contact length L. Integrating Eq. (3.20) as

$$V = \frac{1}{L}\int_0^L V(x)\,dx \qquad (3.26)$$

gives the contact resistance as

$$R_c = \frac{V}{I} = \frac{\rho_c}{LZ} \tag{3.27}$$

Equation (3.24) assumes the contact width Z to be identical to the sheet width W. This is rarely realized in practice. Usually $Z < W$. Experiments with $Z = 5$ μm and W ranging from 10 μm to 60 μm showed the contact end resistance to give erroneously high ρ_c. The error increased as ρ_c decreased or as R_{sh} increased.[58] The error arises from the potential difference between the front edge and the rear edge of the contact allowing current to flow around the contact edges. The measured resistance is proportional to the sheet resistance and is insensitive to the contact resistance for large δ. For the simple one-dimensional theory to hold, the test structure should meet the conditions: $L \le L_T$, $Z \gg L$ and $\delta \ll Z$. The one-dimensional analysis is not valid if these conditions are not met. Accurate extraction of ρ_c, however, is possible by fitting numerical simulations to measured data.

The problem of $W \ne Z$ can be avoided with *circular* test structures, consisting of a conducting circular inner region of radius L, a gap of width d, and a conducting outer region.[59] The conducting regions are usually metallic and the gap typically varies form a few microns to tens of microns. For equal sheet resistances under the metal and in the gap, and for the geometry of the circular contact resistance structure in Fig. 3.19(a), the total resistance between the internal and the external contacts is[60]

$$R_T = \frac{R_{sh}}{2\pi} \left[\frac{L_T}{L} \frac{I_0(L/L_T)}{I_1(L/L_T)} + \frac{L_T}{L+d} \frac{K_0(L/L_T)}{K_1(L/L_T)} + \ln\left(1 + \frac{d}{L}\right) \right] \tag{3.28}$$

where I and K denote the modified Bessel functions of the first order. For $L \gg 4L_T$, the Bessel function ratios I_0/I_1 and K_0/K_1 tend to unity and R_T becomes

$$R_T = \frac{R_{sh}}{2\pi} \left[\frac{L_T}{L} + \frac{L_T}{L+d} + \ln\left(1 + \frac{d}{L}\right) \right] \tag{3.29}$$

In the circular transmission line test structure in Fig. 3.19(b), for $L \gg d$, Eq. (3.29) simplifies to

$$R_T = \frac{R_{sh}}{2\pi L} (d + 2L_T)C \tag{3.30}$$

where C is the correction factor[61]

$$C = \frac{L}{d} \ln\left(1 + \frac{d}{L}\right) \tag{3.31}$$

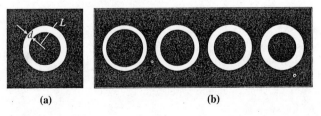

(a)	**(b)**

Fig. 3.19 Circular contact resistance test structure. The dark regions represent metallic regions. Spacing d and radius L are shown in (a).

shown in Fig. 3.20(a). For $d/L \ll 1$, Eq. (3.30) becomes

$$R_T = \frac{R_{sh}}{2\pi L}(d + 2L_T) \qquad (3.32)$$

For practical radii up to about 200 μm and gap spacings of 5–50 μm, the correction factor is necessary to compensate for the difference between the linear transfer length method and the circular TLM layouts to obtain a linear fit to the experimental data. Without the correction factor, the specific contact resistance is underestimated. The total resistance before and after data correction is shown in Fig. 3.20(b) as a function of gap spacing d. Similar to the linear TLM structure, the corrected circular TLM data are linear and yield the contact resistance and the transfer length, from which the specific contact resistivity can be determined.

The circular test structure has one main advantage. It is not necessary to isolate the layer to be measured, because current can only flow from the central contact to the surrounding contact. In the linear TLM test structure, current can flow from contact to

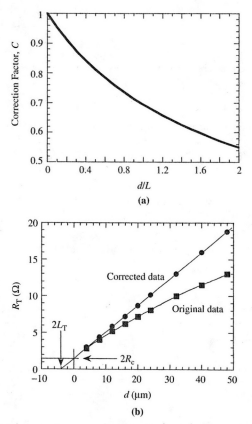

Fig. 3.20   (a) Correction factor C versus d/L ratio for the circular transmission line method test structure, (b) total resistance for the circular TLM test structure before and after data correction. $R_C = 0.75$ ohms, $L_T = 2$ μm, $\rho_c = 4 \times 10^{-6}$ ohm-cm^2, $R_{sh} = 110$ ohms/square. Data courtesy of J.H. Klootwijk and C.E. Timmering, Philips Research Labs.

contact through the region beyond the test structure if it is not isolated. The circular test structure with four metal contacts is very similar to the cross-bridge Kelvin resistor discussed in Section 3.4.3.[62]

Equations (3.22) and (3.24) are derived under the assumption that $\rho_c > 0.2 R_{sh} t^2$, where t is the layer thickness. For $R_{sh} = 20$ ohms/square and $t = 1$ μm, this constraint leads to $\rho_c > 4 \times 10^{-8}$ ohm·cm^2. The TLM method must be modified if that condition is not satisfied, as verified by experiments and by modeling.[63] Most specific contact resistivities are above 4×10^{-8} ohm·cm^2 and the TLM method is valid.

The difficulty of deciding where to measure the voltage in the configuration of Fig. 3.17 has led to a test structure shown in Fig. 3.21(a) and a measurement technique known as the *transfer length method* originally proposed by Shockley.[64] Unfortunately it is also abbreviated as TLM. The TLM test structure is very much like that of Fig. 3.13, but consists of more than three contacts. Two contacts at the ends of the test structure served as entry and exit point for the current in the original *ladder* structure and the voltage was measured between one of the large contacts and each of the successive narrow contacts in Fig. 3.21(a). Later the test structure had unequal spacing between contacts as in Fig. 3.21(b), with the voltage measured between adjacent contacts.

The structure in Fig. 3.21(b) has certain advantages over that of Fig. 3.21(a). When the voltage is measured in the ladder structure between contacts 1 and 4, for example, the current flow may be perturbed by contacts 2 and 3. The effect of contacts 2 and 3 depends on the transfer length L_T and the contact length L. For $L \ll L_T$, the current does not penetrate appreciably into the contact metal and, to first order, contacts 2 and 3 have no effect on the measurement. For $L \gg L_T$, the current does flow into the metal and the contact can be thought of as two contacts, each of length L_T joined by a metallic conductor.[65] The shunting of the current by the metal strips obviously influences the measured voltage or resistance. It is for this reason that the structure in Fig. 3.21(b) is preferred, because there is only bare semiconductor between any two contacts.

For contacts with $L \geq 1.5 L_T$ and for a front contact resistance measurement of the structure in Fig. 3.21(b), the total resistance between any two contacts is

$$R_T = \frac{R_{sh} d}{Z} + 2R_c \approx \frac{R_{sh}}{Z}(d + 2L_T) \qquad (3.33)$$

where we have used the approximation leading from Eq. (3.22) to Eq. (3.23b). Eq. (3.33) is similar to Eq. (3.32) with the contact peripheral length $2\pi L$ replaced by the contact width Z.

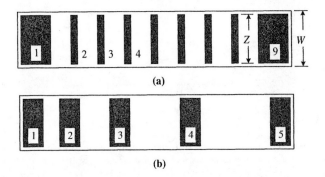

(a)

(b)

Fig. 3.21   Transfer length method test structures.

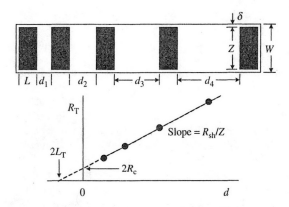

Fig. 3.22 A transfer length method test structure and a plot of total resistance as a function of contact spacing, d. Typical values might be: $L = 50$ μm, $W = 100$ μm, Z-$W = 5$ μm (should be as small as possible), $d \approx 5$ to 50 μm.

The total resistance is measured for various contact spacings and plotted versus d as illustrated in Fig. 3.22. Three parameters can be extracted from such a plot. The slope $\Delta(R_T)/\Delta(d) = R_{sh}/Z$ leads to the *sheet resistance* with the contact width Z independently measured. The intercept at $d = 0$ is $R_T = 2R_c$ giving the *contact resistance*. The intercept at $R_T = 0$ gives $-d = 2L_T$, which leads to the *specific contact resistivity* with R_{sh} known from the slope of the plot. The transfer length method gives a complete characterization of the contact by providing the sheet resistance, the contact resistance, and the specific contact resistivity.

The transfer length method is commonly used, but it has its own problems. The intercept at $R_T = 0$ giving L_T is sometimes not very distinct, leading to incorrect ρ_c values. Perhaps a more serious problem is the uncertainty of the sheet resistance under the contacts. Eq. (3.33) assumes the sheet resistance to be identical under the contacts and between contacts. But the sheet resistance under the contacts may differ from the sheet resistance between contacts due to the effects of contact formation. This would be true for alloyed and silicided contacts where the region under the contact is modified during contact fabrication, leading to the modified expression for the front contact and total resistance,[66]

$$R_{cf} = \frac{\rho_c}{L_{Tk}Z}\coth\,(L/L_{Tk}) \tag{3.34}$$

and

$$R_T = \frac{R_{sh}d}{Z} + 2R_c \approx \frac{R_{sh}d}{Z} + \frac{2R_{sk}L_{Tk}}{Z} = \frac{R_{sh}}{Z}[d + 2(R_{sk}/R_{sh})L_{Tk}] \tag{3.35}$$

where R_{sk} is the sheet resistance *under* the contact and $L_{Tk} = (\rho_c/R_{sk})^{1/2}$. The slope of the R_T versus d plot still gives R_{sh}/Z and the intercept at $d = 0$ gives $2R_c$. However, the intercept at $R_T = 0$ now yields $2L_{Tk}(R_{sk}/R_{sh})$ and it is no longer possible to determine ρ_c since R_{sk} is unknown. Nevertheless, by determining R_{cf} from the transfer length method and R_{ce} from the end resistance method, where

$$R_{ce} = \frac{\sqrt{R_{sk}\rho_c}}{Z\sinh(L/L_{Tk})} = \frac{\rho_c}{ZL_{Tk}\sinh(L/L_{Tk})}; \frac{R_{ce}}{R_{cf}} = \frac{1}{\cosh(L/L_{Tk})} \tag{3.36}$$

one can determine L_{Tk} and ρ_c. In this way it is possible to find the contact resistance *and* the specific contact resistivity in addition to the sheet resistance *between* and *under* the contacts. One can also separate R_{sh} from R_{sk} by etching the semiconductor between the contacts.

Extraction of electrical contact parameters by the TLM method is based on the assumption of constant electrical and geometrical contact parameters across the sample. However, such parameters typically exhibit scatter across a wafer. Statistical modeling has shown that the usual data extraction procedure can lead to errors in the extracted contact parameters even if there is no error in the measured electrical and geometrical parameters.[67] For short contacts ($L < L_T$), ρ_c can be determined accurately regardless of the scatter in other parameters, while R_{sh} and R_{sk} are in error only if ρ_c exhibits scatter over the wafer. For long contacts, the extracted ρ_c and R_{sk} are in error only if R_{sk} or resistance measurements are in error. Best results are obtained for $L \geq 2L_T$. When a wafer exhibits non-uniformities of the electrical parameters of 10–30%, the error in ρ_c and R_{sk} can be as high as 100–1000%. Redundancy through the use of more than one test structure allows the errors to be reduced.

We have so far considered the specific contact resistivity and sheet resistance of the semiconductor, but have neglected the *resistance of the metal*. This generally introduces little error although at times the metal resistance increases with aging and can no longer be neglected. The resistance of silicides is higher than that of pure metals and may not always be negligible. A more serious limitation arises when polysilicon conductors are used instead of metals. Their resistance is significantly higher than that of metals and may need to be considered for proper interpretation of the experimental results. For non-negligible metal resistance, the contact resistance of Eq. (3.22) becomes[68–69]

$$R_{cf} = \frac{\rho_c}{L_{Tm}Z(1+\alpha)^2}\left[(1+\alpha^2)\coth(L/L_{Tm}) + \alpha\left(\frac{2}{\sinh(L/L_{Tm})} + \frac{L}{L_{Tm}}\right)\right] \quad (3.37)$$

where $\alpha = R_{sm}/R_{sk}$, R_{sm} is the metal sheet resistance, and $L_{Tm} = [\rho_c/(R_{sm} + R_{sk})]^{1/2} = L_{Tk}/(1+\alpha)^{1/2}$. Equation (3.37) reduces to Eq. (3.34) for $R_{sm} = 0$ and to Eq. (3.22) for $R_{sk} = R_{sh}$ and $R_{sm} = 0$. The contact front resistance from Eq. (3.37), normalized by multiplying by Z, is plotted in Fig. 3.23 against the contact length as a function of the specific contact resistivity. The main difference between Fig. 3.18 and 3.23 is the

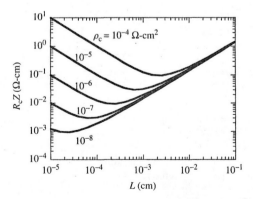

Fig. 3.23 Front contact resistance–contact width product as a function of contact length and specific contact resistivity for $R_{sk} = 20$ Ω/square and $R_{sm} = 50$ Ω/square.

minimum in Fig. 3.23, which is absent when $R_{sm} = 0$. For each combination of ρ_c, R_{sk}, and R_{sm} there is an optimum contact length for minimum contact resistance. For lengths above and below this optimum value, the contact resistance increases. Further discussions of the effects of finite-resistance metal conductors can be found in ref. 70.

We need to consider one more correction. So far we assumed the gap δ in Fig. 3.22 to be zero. The fact that $\delta \neq 0$, can lead to incorrect intercepts of the $R_T - d$ plot. Various corrections have been proposed.[49, 71] We follow the suggestions of ref. 72, where the δ region between the contacts is represented by parallel resistances. As shown in Appendix 3.1, instead of plotting R_T versus d, one plots R' versus d, where

$$R' = 2R_{ce} + \frac{(R_T(\delta \neq 0) - 2R_{ce})R_p}{R_p - R_T(\delta \neq 0) - 2R_{ce}} \tag{3.38}$$

where R_{ce} is the contact end resistance, R_T the measured resistance, and R_p the parallel "strip" resistance. The derivation of Eq. (3.38) and a method to determine R_p are given in Appendix 3.1. Figure 3.24 shows uncorrected and corrected TLM curves for one particular contact area. It clearly shows the different intercepts for the uncorrected lines (solid lines) leading to incorrect contact resistance, transfer length, and specific contact resistivity, but one common intercept for the corrected data (dashed line).

3.4.3 Four-Terminal Contact Resistance Method

The specific contact resistivity measurement techniques discussed so far require the semiconductor bulk resistivity or the semiconductor sheet resistance to be known. However, it is desirable to measure R_c and ρ_c by minimizing or eliminating, if possible, the contribution from bulk or sheet resistance. The measurement technique that comes closest to this goal is the four-terminal Kelvin test structure also known as the cross-bridge Kelvin resistance (CBKR). It appears to have been first used for evaluating metal-semiconductor contacts in 1972[73] but it was only in the early 1980s that it was evaluated seriously.[74–76] In principle, this method allows the specific contact resistivity to be measured without being affected by the underlying semiconductor or the contacting metal conductor.

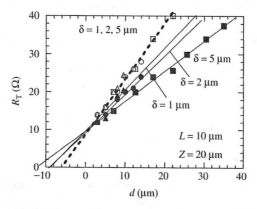

Fig. 3.24 Uncorrected (solid points and lines) and corrected (open points and dashed line) total resistance versus spacing d for Au/Ni/AuGe/n-GaAs contacts annealed at 400°C for 20 s. Reprinted after ref. 72 by permission of IEEE (© 2002, IEEE).

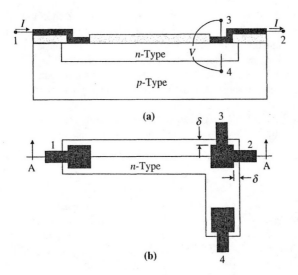

Fig. 3.25 A four-terminal or Kelvin contact resistance test structure. (a) Cross section through section A–A, (b) top view of the structure.

The principle is illustrated in Fig. 3.25. Current is forced between contacts 1 and 2 and the voltage is measured between contacts 3 and 4. There are three voltage drops between pad 1 and pad 2. The first is between pad 1 and the semiconductor n-layer, the second along the semiconductor sheet, and the third between the n-layer and the pad 2/3. A high input impedance voltmeter, for measuring the voltage $V_{34} = V_3 - V_4$, allows very little current flow between pads 3 and 4. Hence, the potential at pad 4 is essentially the same as the potential in the n-region directly under contact 2/3, as illustrated in Fig. 3.25(a) by connection 4 under the contact. V_{34} is solely due to the voltage drop across the contact metal-semiconductor interface. The name "Kelvin Test Structure" refers to the fact that a voltage is measured with little current flow as in four-point probe resistance measurements.
The contact resistance is

$$R_c = \frac{V_{34}}{I} \qquad (3.39)$$

which is simply the ratio of the voltage to the current. The specific contact resistivity is

$$\rho_c = R_c A_c \qquad (3.40)$$

where A_c is the contact area.
Equation (3.40) does not always agree with experimental data. The specific contact resistivity calculated with Eq. (3.40) is an *apparent* specific contact resistivity differing from the true specific contact resistivity by lateral current crowding for contact windows smaller than the diffusion tap, shown as $\delta > 0$ in Fig. 3.25.[77] Contact window to diffused layer misalignment and lateral dopant diffusion account for $\delta > 0$. In the ideal case, $\delta = 0$ as illustrated in Fig. 3.26(a). In an actual contact, some of the current, indicated by the arrows in Fig. 3.26(b), flows *around* the metal contact. In the ideal case with $\delta = 0$, the voltage drop is $V_{34} = I R_c$. For $\delta > 0$, the lateral current flow gives an additional voltage

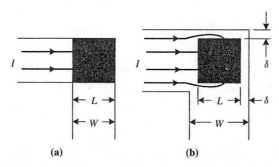

Fig. 3.26   Four-terminal contact resistance test structures. (a) Ideal with only lateral current flow, (b) showing current flowing into and around the contact. The black area is the contact area.

drop that is included in V_{34}, leading to a higher voltage. Therefore, according to Eq. (3.39) R_c is higher and is usually designated R_k. According to Eq. (3.40) ρ_c is also higher if the actual contact area A_c is used. The ρ_c so extracted is known as the *effective* or *apparent* specific contact resistivity. The error introduced by this geometrical factor is highest for low ρ_c and/or high R_{sh} and lowest for high ρ_c and/or low R_{sh}.[78] The vertical voltage drop in the semiconductor normal to the contact plane, usually neglected, leads to an additional correction.[79]

The effect of contact misalignment is shown in Fig. 3.27.[80] Larger δ leads to higher measured resistance. Clearly, for large misalignment, the measured resistance is seriously in error. The true resistance is obtained by extrapolating to $\delta = 0$. The effect of asymmetrical misalignment is illustrated in Fig. 3.28, where the apparent contact resistance is plotted versus misalignments L_1 and L_2. This figure clearly shows the effect of parasitic current paths. In one case R_k increases, in the other it decreases. It is difficult to fabricate test structures with $\delta = 0$. However, a solution is illustrated in Fig. 3.29(a). Here the semiconductor voltage tap consists of individual "strips".[80] The measured voltage for the three taps is shown in Fig. 3.29(b). By extrapolating the data to zero voltage tap spacing, the true resistance is obtained.

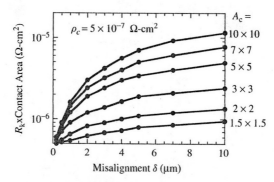

Fig. 3.27   Apparent contact resistance multiplied by the contact area versus misalignment δ. The contact areas are given on the right side of the figure. Under the contact: Arsenic implant, 2×10^{15} cm^{-2}, 50 keV, annealed at 1000°C, 30 s. Contact metal: Ti/TiN/Al/Si/Cu. Adapted from ref. 80.

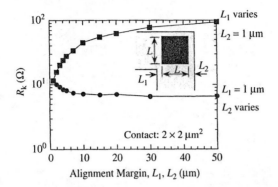

Fig. 3.28 Dependence of contact resistance on misalignment dimensions L_1 and L_2. Under the contact: Arsenic implant, 2×10^{15} cm^{-2}, 50 keV, annealed at 1000°C, 30 s. Contact metal: Ti/TiN/Al/Si/Cu. Adapted from ref. 80.

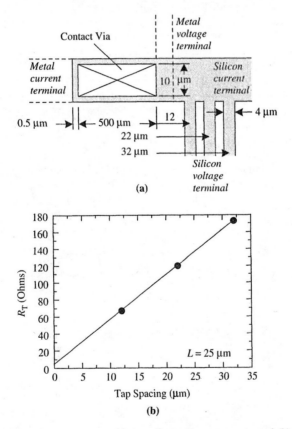

Fig. 3.29 (a) Modified Kelvin contact resistance "tapped" test structure and (b) resistance versus tap spacing. After ref. 80.

A simplified two-dimensional approach gives the contact resistance R_k as[80]

$$R_k = \frac{\rho_c + \sqrt{\rho_c R_{sh}} L_1 \coth(L/L_T) + 0.5 \ R_{sh} L_1^2 + \sqrt{\rho_c R_{sh}} L_2/\sinh(L/L_T)}{(L + L_1 + L_2)W} \qquad (3.41)$$

with the various dimensions shown on Fig. 3.28. Curves calculated with Eq. (3.41) agree qualitatively with the data in Fig. 3.27. Lateral current flow around the contact accounts for the additional resistance. The resistance increase gets worse the lower the specific contact resistivity, further aggravated for higher sheet resistances. Unfortunately, the trend in the technology of today's high-density integrated circuits is toward lower ρ_c and higher R_{sh} due to shallower junctions. Both are in the direction of complicating the interpretation of four-terminal contact resistance test structure measurements. Simple one-dimensional interpretations must be carefully evaluated for their accuracy.

Figure 3.30 shows calculated curves for the *apparent* and the *actual* values of specific contact resistivity for the structure of Fig. 3.31.[79] For the ideal case of $L/W = 1$ or $\delta = 0$ the two are identical indicated by the 45° line for two-dimensional calculations. However, for the more realistic three-dimensional calculations the two are not identical even for $\delta = 0$. As ρ_c decreases the contact resistance voltage decreases and the lateral voltage becomes more important until the contact resistance voltage becomes negligible and $\rho_{c,apparent}$ is independent of the true ρ_c. Universal error corrections curves from *three-dimensional* modeling, including the finite depth of the semiconducting are shown in Fig. 3.31. In these calculations the semiconductor sheet resistance under the contact is assumed identical to the sheet resistance beyond the contacts. R_k in these curves is the contact resistance including parasitic resistances.

Two-dimensional models of the transmission line, the contact end resistance, and the cross-bridge Kelvin resistance structures have been used to calculate and plot the contact resistance normalized by the sheet resistance against the contact length normalized by δ.[81] Deviations from the simple one-dimensional analysis are predicted for all three cases. The TLM has the least sensitivity to δ because it detects the front contact potential, which is only weakly perturbed by peripheral current flow. However, the TLM method relies on extrapolation of experimental data to determine ρ_c. That has a potential error especially if the data points do not lie on a well-defined straight line. Both the CER and the CBKR

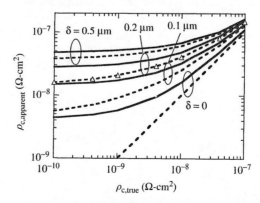

Fig. 3.30 Two-dimensional (dashed) and three-dimensional (solid lines) simulated apparent versus true specific contact resistivity for various tap spacings δ. Reprinted after ref. 79 by permission of IEEE (© 2004, IEEE).

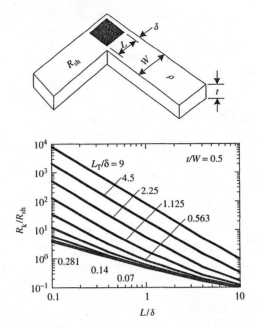

Fig. 3.31   Three-dimensional universal correction curves for CKR structures of R_k/R_{sh} versus L/δ as a function of L_T/δ for tap depth/width ratios of $t/L = 0.5$. Reprinted after ref. 79 by permission of IEEE (© 2004, IEEE).

structures show significant deviations due to peripheral current flow. The contact resistances determined by the CER method are generally low, $R_{ce}(CER) < R_c(CBKR)$, making the measurement more difficult. Contact misalignment introduces further departures from one-dimensional behavior.[82] Self-aligned contacts solve the misalignment problem but not the lateral diffusion problem.[83] Other models of contact resistance calculations are given in refs. 84 and 85.

Contact resistance test structures can also be implemented with a modified MOSFET consisting of three n^+ regions and two gates as illustrated in Fig. 3.32.[86] The "sheet" between contacts 1 and 2 and between contacts 3 and 4 is due to a channel formed by biasing the two MOSFET sections into conduction. This structure is compatible with standard silicide processes. It can be implemented in the CFR, the CER, or in the CKBR configuration.

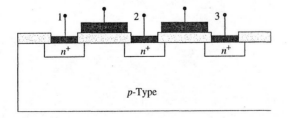

Fig. 3.32   A MOSFET contact resistance test structure.

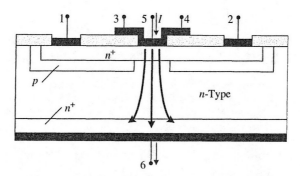

Fig. 3.33   Vertical contact resistance Kelvin test structure.

The *vertical* Kelvin test structure in Fig. 3.33 was developed to overcome the lateral current flow problems of the conventional Kelvin structure.[87] The device requires one additional mask level during its fabrication compared to conventional Kelvin structures. The metal/semiconductor contact is made to a diffused or ion-implanted layer (n^+-layer in Fig. 3.33). Current I confined to the contact area by the oxide window and the isolating np junction, is forced between contact 5 and substrate contact 6. Voltage V_{24} is measured between contacts 2 and 4. V_4 is the voltage of the metal and V_2 is the voltage of the semiconductor layer just below the metal, even though V_2 is measured at some distance from the contact. Just as in a conventional Kelvin structure, there is very little lateral voltage drop along the n^+ layer during the voltage measurement because essentially no current is drawn. The contact resistance and the specific contact resistivity are given by $R_c = V_{24}/I$ and $\rho_c = R_c A_c$.

Lateral effects, so important in all methods that rely on lateral current flow, also play a role in this vertical structure. This comes about not because the current flows laterally to reach a collecting contact, but because of current spreading. The current does not flow strictly vertically. It has a small lateral, spreading component, shown in Fig. 3.33, making the voltage at the sensing contact (contact 2) not exactly equal to the voltage under the metal. The additional spreading resistance causes the measured contact resistance to be higher than the true contact resistance.[88] An additional complication arises when the contact is smaller than the contact opening. The specific contact resistivity is then given approximately as[87]

$$\rho_{c,eff} \approx \rho_c + R_{sh}x_j/2 \tag{3.42}$$

where R_{sh} is the sheet resistance and x_j the junction depth of the upper n^+ layer in Fig. 3.33. Equation (3.42) is valid for $L \geq 10x_j$. The vertical test structure works well the smaller the contact area and the shallower the upper n^+ layer is.

Additional contacts are provided in Fig. 3.33. V_{13} can be used to average the voltage reading with V_{24} to reduce experimental errors. Furthermore, conventional lateral six-terminal measurements can be made to obtain the end resistance R_{ce}, the front resistance R_{cf}, and the sheet resistance R_{sh}. A detailed study of various non-idealities in the vertical test structure has shown the current spreading effect to be small compared to lateral current crowding in horizontal Kelvin test structures.[89] Misalignment between the isolation junction and the metal contact can produce more severe errors, but these can be minimized by averaging the voltage readings on the left and the right arms.

3.4.4   Six-Terminal Contact Resistance Method

The six-terminal contact resistance structure in Fig. 3.34 is related to the four-terminal Kelvin structure with two more contacts for additional measurement options not available with the conventional Kelvin structure.[75] The structure allows the *contact resistance*, the *specific contact resistivity*, the *contact end resistance*, the *contact front resistance*, and the *sheet resistance* under the contact to be determined. For the conventional Kelvin structure contact resistance measurement the current is forced between contacts 1 and 3 in Fig. 3.34 and the voltage is measured between contacts 2 and 4. The analysis is that of Eqs. (3.39) and (3.40) for the one-dimensional case, where $R_c = V_{24}/I$ and $\rho_c = R_c A_c$. All the two-dimensional complications, not reflected in Eqs. (3.39) and (3.40), manifest themselves in the six-terminal structure also.

To measure the contact end resistance $R_{ce} = V_{54}/I$, current is forced between contacts 1 and 3 and the voltage is sensed across contacts 5 and 4. With the contact resistance and the specific contact resistivity determined from the Kelvin part of this structure, the sheet resistance under the contact can be determined from the end resistance using Eq. (3.36) and the contact front resistance, given by R_{cf} in Eqs. (3.22) and (3.36) can be calculated with Eq. (3.36).

3.4.5   Non-Planar Contacts

Thus far we have only concerned ourselves with deviations from simple theory due to two-dimensional current flow. We have assumed the contact itself to be a smooth, intimate contact between the metal and the semiconductor. Real contacts are not this perfect introducing further complications. Contact history in Si integrated circuits is depicted in Fig. 3.35. Initially Al was deposited directly onto Si (Fig. 3.35(a)). For aluminum-silicon contacts, there is a tendency for the silicon to migrate into the aluminum, leaving voids in the silicon.[90] Aluminum can subsequently migrate into these voids creating *spiking*. Under extreme conditions this can lead to junction shorts. Addition of 1 to 3 wt% Si to the Al reduces spiking considerably but creates other problems. For example, it is possible for the Si to precipitate and to grow epitaxially between the original Si surface and the Al film (Fig. 3.35(b)). The epitaxially regrown layer is p^+-type because it contains a high density of aluminum, a p-type dopant in Si, creating a pn junction at the regrown epi/n^+ interface. It has been observed that the propensity for such epitaxial films to form is higher for (100) than for (111)-oriented substrates.[91] This can be a severe problem for small contact areas where the contact resistance for (100)-oriented substrates increases over similar (111) surfaces.[91]

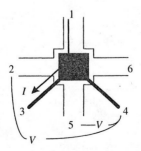

Fig. 3.34   Six-terminal Kelvin structure for the determination of R_c, R_{ce}, R_{cf}, and R_{sk}.

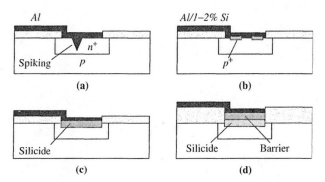

Fig. 3.35 Historic progression of ohmic contacts in Si technology; (a) Al/Si, (b) Al/1-2% Si, (c) Al/silicide/Si, and (d) Al/barrier layer/silicide/Si.

Silicides solved this problem (Fig. 3.35(c)). A silicide is formed by depositing a metal onto Si and heating the sample to form the silicide. Commonly used metals are Ti, Co, and Ni but many other metals form silicides. Silicides penetrate into the Si sample. There is also a chance that Al above the silicide can migrate through the silicide along grain boundaries and form Al/Si contacts. Hence, recent contacts consist of a silicide, a barrier layer (*e.g.*, W plug), and Al or Cu as shown in Fig. 3.35(d). This can give the required low contact resistance and still be chemically stable. Unless the semiconductor is carefully cleaned, there can be interfacial layers between the metal and the semiconductor. These can consist of oxides forming prior to metal deposition. But interfacial layers can also be due to poor substrate cleaning or even due to poor vacuum during metal deposition.[92]

Contacts to GaAs are typically formed by alloying. A Ge-containing alloy is deposited on the device and heated until alloying occurs. The metal-semiconductor interface after contact formation can be very non-planar. It has been suggested that the current in such alloyed contacts flows through Ge-rich islands with the contact resistance largely determined by the spreading resistance under the Ge-rich regions.[93] The effective contact area is likely to be very different from the actual contact area for that model. Very smooth metal-GaAs interfaces can be formed by evaporating Ge, Au, and Cr layers separately and keeping the annealing temperature below the AuGe eutectic temperature.[94] All of these "technological" imperfections make contact resistance measurement interpretation yet more difficult.

3.5 SCHOTTKY BARRIER HEIGHT

The band diagram of a Schottky barrier diode on an *n*-type substrate is shown in Fig. 3.36. The ideal barrier height of ϕ_{B0} is approached only when the diode is strongly forward biased. The actual barrier height ϕ_B is less than ϕ_{B0} due to image force barrier lowering and other factors. V_{bi} is the built-in potential and V_o is the potential of the semiconductor Fermi level with respect to the conduction band. The thermionic current-voltage relationship of a Schottky barrier diode, neglecting series and shunt resistance, is given by

$$I = AA^*T^2 e^{-q\phi_B/kT}(e^{qV/nkT} - 1) = I_{s1}e^{-q\phi_B/kT}(e^{qV/nkT} - 1) = I_s(e^{qV/nkT} - 1)$$

$$(3.43)$$

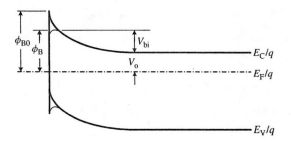

Fig. 3.36   Schottky barrier potential band diagram.

TABLE 3.1   Experimental A^* Values.

Semiconductor	A^* (A/cm$^2 \cdot$ K^2)	Ref.
n-Si	112 ($\pm$6)	95
p-Si	32 ($\pm$2)	95
n-GaAs	4–8	96
n-GaAs	0.41 ($\pm$0.15)	97
p-GaAs	7 ($\pm$1.5)	97
n-InP	10.7	109

where I_s is the saturation current, A the diode area, $A^* = 4\pi q k^2 m^*/h^3 = 120(m^*/m)$ A/cm$^2 \cdot$K^2 Richardson's constant, ϕ_B the effective barrier height, and n the ideality factor. Published values of A^* are given in Table 3.1. Measurements in ref. 97 were made on almost ideal Al/n-GaAs devices with the Al deposited epitaxially by molecular beam epitaxy in ultrahigh vacuum.

The ideality factor n incorporates all those unknown effects that make the device non ideal. A Schottky diode is unlikely to be uniform over its entire area. Barrier height patchiness leads to $n > 1$ and also explains other effects such as n decreasing with temperature and with increasing reverse bias.[99] Equation (3.43) is sometimes expressed as (see Appendix 4.1)

$$I = I_s e^{qV/nkT}(1 - e^{-qV/kT}) \qquad (3.44)$$

Data plotted according to Eq. (3.43) as semilog I versus V are linear only for $V \gg kT/q$ as shown in Fig. 3.37. When plotting $\log[I/(1 - exp(-qV/kT))]$ versus V using Eq. (3.44), the data are linear all the way to $V = 0$, also shown in Fig. 3.37.

3.5.1   Current-Voltage

Among the current-voltage methods, the barrier height is most commonly calculated from the current I_s, determined by extrapolating the semilog I versus V curve to $V = 0$. The barrier height ϕ_B is calculated from I_s in Eq. (3.43) according to

$$\phi_B = \frac{kT}{q} \ln\left(\frac{AA^*T^2}{I_s}\right) \qquad (3.45)$$

The barrier height so determined is ϕ_B for zero bias. The most uncertain of the parameters in Eq. (3.45) is A^*, rendering this method only as accurate as a knowledge of A^*.

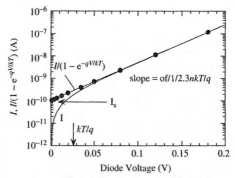

Fig. 3.37 Two ways of plotting current-voltage for a Schottky diode. Reprinted with permission from *Journal of Applied Physics*, **69**, 7142–7145, May 1991. Copyright American Institute of Physics.

Fortunately, A^* appears in the "ln" term and an error of two in A^* gives rise to an error of only $0.7\ kT/q$ in ϕ_B. Nonetheless, errors do occur due to this uncertainty.

An experimental semilog I versus V plot for a Cr/n-Si diode is shown in Fig. 3.38(a). The current deviates from linearity for $V > 0.2$ V due to series resistance (discussed in

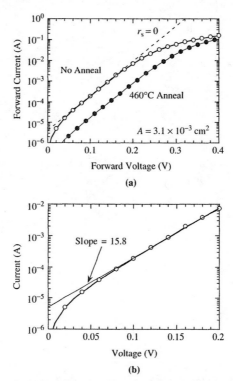

Fig. 3.38 (a) Current-voltage characteristics of a Cr/n-Si diode as deposited and annealed at $460°C$ measured at room temperature, (b) enlarged portion of (a). Courtesy of F. Hossain, Arizona State University.

Sections 4.2 and 4.3). The Schottky barrier diode with area 3.1×10^{-3} cm^2 was fabricated on n-Si.[98] The device contains a p^+ guard ring around the periphery of the Schottky junction area to reduce the edge termination leakage current and it uses chromium (Cr) as the Schottky contact as well as titanium tungsten (TiW) as the diffusion barrier metal and nickel vanadium (NiV)-gold (Au) as the metal overlayer and chromium-nickel-gold as the back ohmic contact. The front and back metal were sputtered and evaporated, respectively. When the device is annealed at $T = 460°$C the barrier height increases and the current decreases. The expanded $I-V$ curve in Fig. 3.38(b) allows the slope to be determined from which $n = 1.05$ and from the $V = 0$ intercept of $I_s = 5 \times 10^{-6}$ A, the barrier height, calculated from Eq. (3.45), is $\phi_B(I-V) = 0.58$ V for $A^* = 110$ A/cm^{2}K^2 for n-Si.

3.5.2   Current—Temperature

For $V \gg kT/q$ Eq. (3.43) can be written as

$$\ln(I/T^2) = \ln(AA^*) - q(\phi_B - V/n)/kT \tag{3.46}$$

A plot of $\ln(I/T^2)$ versus $1/T$ at a constant forward bias voltage $V = V_1$, sometimes called a *Richardson plot*, has a slope of $-q(\phi_B - V_1/n)/k$ and an intercept $\ln(AA^*)$ on the vertical axis. A Richardson plot for the diode of Fig. 3.38 is shown in Fig. 3.39. The slope is usually well defined, but the extraction of A^* from the intercept is prone to error. Generally the $1000/T$ axis covers only a narrow range, 2.6 to about 3.4 in this example. Extrapolating the data from that narrow range to $1/T = 0$ involves extrapolation over a long distance and any uncertainty in the data can produce a large uncertainty in A^*. In Fig. 3.39 the intercept is given by $\log(AA^*)$ from which $A^* = 114$ A/cm^2·K^2.

The barrier height is given by

$$\phi_B = \frac{V_1}{n} - \frac{k}{q}\frac{d[\ln(I/T^2)]}{d(1/T)} = \frac{V_1}{n} - \frac{2.3k}{q}\frac{d[\log(I/T^2)]}{d(1/T)} \tag{3.47}$$

The barrier height is obtained from the slope for a known forward bias voltage, but n must be determined independently. For the data of Fig. 3.39 with $n = 1.05$ determined

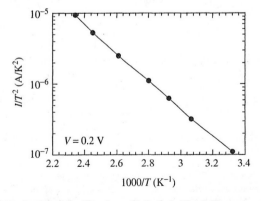

Fig. 3.39 Richardson plot of the "No Anneal" diode in Fig. 3.38 measured at $V = 0.2$ V.

from Fig. 3.38, $V_1 = 0.2$ V, and the slope $d[\log(I/T^2)]/d(1000/T) = -1.97$ we find $\phi_B(I - 1/T) = 0.59$ V, very close to $\phi_B(I-V) = 0.58$ V from the semilog I versus V plot. Sometimes $\ln(I_s/T^2)$ is plotted against $1/T$, with I_s obtained from the intercept of semilog I versus V plots. The current I in Eq. (3.47) should then be replaced by I_s and $V_1 = 0$.

An implicit assumption in the barrier height determination by the Richardson plot method is a temperature-independent barrier height. Should it be temperature dependent, we can write ϕ_B as

$$\phi_B(T) = \phi_B(0) - \xi T \tag{3.48}$$

With this temperature dependence, Eq. (3.46) becomes

$$\ln(I/T^2) = \ln(AA^*) + q\xi/k - q(\phi_B(0) - V/n)/kT \tag{3.49}$$

A Richardson plot now gives the "zero Kelvin" barrier height $\phi_B(0)$, and the intercept is $\ln(AA^*) + q\xi/k$. Now A^* can no longer be determined. Non-linearities are sometimes observed in Richardson plots at low temperatures. These may be due to current mechanisms other than thermionic emission current, usually manifesting themselves as $n > 1.1$. Non-linear Richardson plots are also observed when both the barrier height and the ideality factor are temperature dependent. Accurate extraction of ϕ_B and A^* becomes impossible, but linearity can be restored if $n\ln(I/T^2)$ is plotted against $1/T$.[100]

3.5.3 Capacitance-Voltage

The capacitance per unit area of a Schottky diode is given by[101]

$$\frac{C}{A} = \sqrt{\frac{\pm q K_s \varepsilon_o (N_A - N_D)}{2(\pm V_{bi} \pm V - kT/q)}} \tag{3.50}$$

where the "+" sign applies to p-type ($N_A > N_D$) and the "−" sign to n-type ($N_D > N_A$) substrates and V is the reverse-bias voltage. For n-type substrates $N_D > N_A$, $V_{bi} < 0$, and $V < 0$, whereas for p-type substrates $N_D < N_A$, $V_{bi} > 0$, and $V > 0$. The kT/q in the denominator accounts for the majority carrier tail in the space-charge region which is omitted in the depletion approximation. The built-in potential is related to the barrier height by the relationship

$$\phi_B = V_{bi} + V_o \tag{3.51}$$

as seen in Fig. 3.36. $V_o = (kT/q)\ln(N_c/N_D)$, where N_c is the effective density of states in the conduction band. Plotting $1/(C/A)^2$ versus V gives a curve with the slope $2/[q K_s \varepsilon_o (N_A - N_D)]$, and with the intercept on the V-axis, $V_i = -V_{bi} + kT/q$.

The barrier height is determined from the intercept voltage by

$$\phi_B = -V_i + V_o + kT/q \tag{3.52}$$

The doping density can be determined from the slope as discussed in Chapter 2. $\phi_B(C-V)$ is approximately the flat-band barrier height because it is determined from the $1/C^2-V$ curve for $1/C^2 \rightarrow 0$ or $C \rightarrow \infty$ indicating sufficient forward bias to cause flatband conditions in the semiconductor. A $(C/A)^{-2}$ versus V plot of the diode of Fig. 3.38 is shown in Fig. 3.40. From the slope we find $N_A = 2 \times 10^{16}$ cm^{-3}, and from Eq. (3.52)

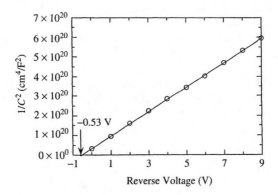

Fig. 3.40 Reverse-bias $1/C^2$ versus voltage of the "No Anneal" diode in Fig. 3.38 measured at room temperature.

the barrier height is $\phi_B(C-V) = 0.74$ V using the intercept voltage $V_i = -0.53$ V and the room temperature $n_i = 10^{10}$ cm^{-3} for Si.

3.5.4 Photocurrent

When a Schottky diode is irradiated with photons of sub band gap energy ($h\nu < E_G$), it is possible to excite carriers from the metal into the semiconductor as shown in Fig. 3.41(a). For $h\nu > \phi_B$, electrons excited from the metal over the barrier into the semiconductor, are

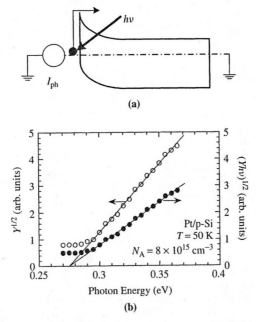

Fig. 3.41 Photoemission yields of a Pt/p-Si Schottky diode. Data adapted from ref. 107.

detected as photocurrent I_{ph}. The light can be incident from the metal or the semiconductor side, since the semiconductor is transparent for these photon energies. The metal must be sufficiently thin for light penetration. The yield Y, defined as the ratio of the photocurrent to the absorbed photon flux, is given by[102]

$$Y = B(h\nu - q\phi_B)^2 \tag{3.53}$$

where B is a constant. $Y^{1/2}$ is plotted versus $h\nu$, and an extrapolation of the linear portion of this curve, sometimes called a *Fowler plot*, to $Y^{1/2} = 0$ gives the barrier height. The yield is also given as[103]

$$Y = C\frac{(h\nu - q\phi_B)^2}{h\nu} \tag{3.54}$$

where C is another constant. Example plots are shown in Fig. 3.41(b). The "toe" below 0.29 eV is due photon-assisted thermionic emission.

A Fowler plot is not always linear as predicted by the theory. When it is non-linear it is difficult to determine ϕ_B. By differentiating Eq. (3.53) the deviation from linearity is much smaller than it is in the conventional Fowler plot, because the extended tail of the Fowler plot in the vicinity of the barrier height is removed by the differentiation.[105] Moreover, the derivative plot is more sensitive to contact non-uniformities and has been used to detect such non-uniformities.[105] The photocurrent technique relies only on photo-excited current flow and is little influenced by tunnel currents, especially if ϕ_B is obtained by extrapolating from $h\nu \gg \phi_B$, where only those electrons well above the barrier height contribute to the photocurrent.

3.5.5 Ballistic Electron Emission Microscopy (BEEM)

Ballistic Electron Emission Microscopy, based on scanning tunneling microscopy is a powerful low-energy tool for non-destructive local characterization of semiconductor heterostructures, such as Schottky diodes and is discussed in more detail in Chapter 9. It can provide information on the homogeneity of the interface electronic structure with extremely high lateral resolution and can yield energy-resolved information on hot-electron transport in the metal film, at the interface, and in the semiconductor.[106]

3.6 COMPARISON OF METHODS

A number of studies have been undertaken to compare barrier heights determined by the current-voltage $(I-V)$, current-temperature $(I-T)$, capacitance-voltage $(C-V)$, and photocurrent (PC) techniques. In one study the barrier height of evaporated Pt films on GaAs substrates was determined as $\phi_B(I-V) = 0.81$ V, $\phi_B(C-V) = 0.98$ V, and $\phi_B(PC) = 0.905$ V.[107] Which is the most reliable value? Any damage at the interface affects the $I-V$ behavior because defects may act as recombination centers or as intermediate states for trap-assisted tunnel currents. Either one of these mechanisms raises n and lowers ϕ_B. $C-V$ measurements are less prone to such defects. However, defects can alter the space-charge region width and hence the intercept voltage. Photocurrent measurements are less sensitive to such defects, and this method is judged to be the most reliable. Nevertheless, Fowler plots are not always linear. The first derivative plot usually does have a straight-line portion, making ϕ_B extraction more reliable.

The sequence $\phi_B(I-V) < \phi_B(PC) < \phi_B(C-V)$ was also observed for a variety of metals deposited on n-GaAs and p-GaAs.[108] Barrier height measurements of Schottky barriers on p-type InP gave $\phi_B(I-T) < \phi_B(C-V)$.[109] The difference was attributed to patchiness of barrier heights across the contact. When two Schottky diodes of different barrier height are connected in parallel, the *lower barrier height* dominates the $I-V$ behavior, but the barrier height with the largest contact area dominates the $C-V$ behavior.[110] In the parallel conduction model, regions with different local barrier heights are assumed to be electrically independent and the total current is simply the sum of the currents flowing through all individual areas. This concept was extended theoretically to mixed-phase contacts of varying dimensions but fixed area ratios, predicting that generally $\phi_B(C-V) > \phi_B(I-V)$.[111] For large contact regions results similar to those in ref. 110 were obtained. For smaller contact regions, however, the low barrier height regions were found to be pinched off by the high barrier height regions.

The barrier height patchiness invoked to explain the differing barrier heights also predicts varying Richardson constants. It is frequently observed that A^* varies with processing conditions such as annealing. It may well be that annealing causes the patchiness to vary and therefore A^* to change. This would rule against using those methods that rely on a knowledge of A^* for ϕ_B determination, favoring $C-V$ and photocurrent measurements over $I-V$ and $I-T$ measurements. For the $C-V$ method it is important that C^{-2} versus V plots be linear and independent of frequency. Photocurrent probes the device from outside the semiconductor, that is, photo emission is from the metal to the semiconductor. The $I-V$ and $C-V$ methods probe the device from the semiconductor side. It is for this reason that the latter two methods are more sensitive to spatial inhomogeneities, insulating layers between the metal and the semiconductor, doping inhomogeneities, surface damage, and tunneling. The PC technique is least influenced by these parameters and is therefore likely to yield the most reliable value of barrier height. For well-behaved contacts with few of these degradation factors, all methods give values that agree reasonably well with one another.

3.7   STRENGTHS AND WEAKNESSES

Two-Terminal Methods: The two-contact, two-terminal contact resistance measurement technique is simple but the least detailed. The contact resistance data are corrupted by either the semiconductor bulk or sheet resistance. The method is only infrequently used today. The two-terminal *contact string* is used mainly as a process monitor. It does not give detailed contact resistance information nor can the specific contact resistivity be reliably extracted. The multiple-contact, two-terminal technique is usually employed in its transfer length method implementation, where the effect of the semiconductor sheet resistance is separated from the contact resistance and both contact resistance as well as specific contact resistivity can be determined. This method allows both front and end contact resistance measurements to be made. Complications in the interpretation of the experimental data arise due to three main effects: (1) the extrapolation of experimental data to obtain intercepts, (2) lateral current flow around the contact, and (3) the sheet resistance under the contact differing from the sheet resistance outside the contact window. Current flows laterally around the contact window whenever the contact window is narrower than the diffusion tap leading to erroneous contact resistances if the experimental data are analyzed by the conventional one-dimensional theory. For the most reliable measurements the test structure should be configured to satisfy the following requirements: $L > L_T$, $Z \gg L$, $\delta = W - Z \ll W$ as defined in Fig. 3.22.

Four-Terminal Method: The four-terminal or Kelvin structure is preferred over the two- and three-terminal structures for several reasons. (1) There is only one metal-semiconductor contact and the contact resistance is measured directly as the ratio of a voltage to a current. R_c can therefore be very small. (2) Neither metal nor semiconductor sheet resistance enter into the R_c determination. Hence there is no practical limit to the value of R_c that can be measured. (3) The contact area can be made small to be consistent with contact areas used in high-density ICs. This makes the method very simple and attractive. However, any lateral current flow obscures the interpretation. Modeling has shown two- and three-dimensional effects to be important, especially for appreciable gaps between the contact window and the diffusion edge.

Six-Terminal Method: The six-terminal method is very similar to the four-terminal technique. It incorporates the Kelvin structure, but additionally allows measurements of the front and end contact resistance as well as the contact sheet resistance. It is only slightly more complex than the four-terminal structure but does not require additional masking operations.

For any of the contact resistance measurement methods it is difficult to determine absolute values of ρ_c. Simple one-dimensional interpretations of the experimental data frequently give incorrect values of specific contact resistivity. Proper interpretation of the experimental data requires more exact modeling. This makes many of the data, determined in the past by simple one-dimensional interpretation, suspect. Nevertheless, ρ_c can be used as a figure of merit but the experimental conditions under which they were obtained should be carefully specified. The *contact resistance* can be measured directly, but the measured resistance may not be the true contact resistance.

Schottky Barrier Height: Strengths and weaknesses of Schottky barrier height measurements are discussed in Section 3.6.

APPENDIX 3.1

Effect of Parasitic Resistance

This discussion follows ref. 72. Equations (3.22) and (3.24) suggest the simple equivalent circuit in Fig. A3.1. When the current I flows as indicated, the resistance between A and ground is R_{cf} and between B and ground it is R_{ce} as required. For the configuration of Fig. A3.2, the equivalent circuit is shown in Fig. A3.3. R_{ce}, the end resistance, is similar to that in Fig. A3.1. The remainder of the contact has the resistance $R_{cf} - R_{ce}$, making the contact resistance R_{cf}. The semiconductor region between the contacts of width Z is characterized by the resistance $R_{sh}d/Z$, where R_{sh} is the sheet resistance, leaving the small overlap regions of length d and width δ, characterized by the parallel resistance R_P. The total resistance between the contacts is then

$$R_T(\delta \neq 0) = 2R_{ce} + [2(R_{cf} - R_{ce}) + R_{sh}d/Z]//R_P/2 \qquad (A3.1)$$

where "//" denotes the parallel resistance combination. For $\delta = 0$

$$R_T(\delta = 0) = 2R_{cf} + R_{sh}d/Z \qquad (A3.2)$$

Multiplying the various terms in Eq. (A3.1) and solving for $2R_{cf} + R_s d/Z$, leads to

$$2R_{cf} + R_{sh}d/Z = 2R_{ce} + \frac{(R_T(\delta \neq 0) - 2R_{ce})R_P}{R_P/2 - R_T(\delta \neq 0) + 2R_{ce}} = R' \qquad (A3.3)$$

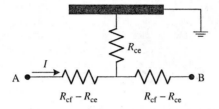

Fig. A3.1   Equivalent circuit of a single contact showing the contact front and end resistances.

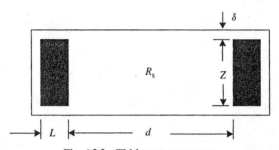

Fig. A3.2   TLM contact structure.

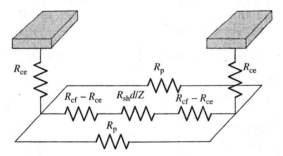

Fig. A3.3   Equivalent circuit of the TLM structure of Fig. A3.2, including the parallel resistances R_p.

$R_T (\delta \neq 0)$ is the measured total resistance between two contacts and R' is the resistance corrected by including the two parallel resistances.

The parallel resistance in Eq. (A3.1) is given by

$$R_p = 2F \, R_{sh} \tag{A3.4}$$

where F is the correction factor

$$F = K(k_0)/K(k_1) \tag{A3.5}$$

and K is the complete elliptic integral

$$K(k) = \int_0^{\pi/2} \frac{d\phi}{\sqrt{1 - (k \sin \phi)^2}} \tag{A3.6}$$

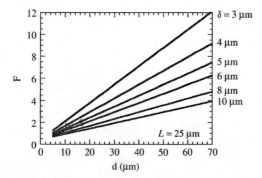

Fig. A3.4   Corrections factor versus d as a function of gap spacing δ for $L = 25$ μm. Reprinted after ref. 72 by permission of IEEE (© 2002, IEEE).

and k_0 and k_1 are given by

$$k_0 = \frac{\tanh(\pi d/4\delta)}{\tanh(\pi(d+4L)/4\delta)}; k_1 = \sqrt{1 - k_0^2} \qquad (A3.7)$$

L is the contact length, d the contact spacing, and δ the gap, all shown on Fig. A3.2.

The correction factor F is plotted in Fig. A3.4 versus contact spacing d as a function of gap spacing δ.

APPENDIX 3.2

Alloys for Contacts to Semiconductors

Material	Alloy	Contact Type
n-Si	Au-Sb	ohmic
p-Si	Au-Ga	ohmic
n-Si	Al	ohmic
p-Si	Al	Schottky
n-GaAs	Au-Ge	ohmic
n-GaAs	Sn	ohmic
p-GaAs	Au-Zn	ohmic
p-GaAs	In	ohmic
n-GaInP	Au-Sn	ohmic
n-InP	Ni/Au-Ge/Ni	ohmic
n-InP	Au-Sn	ohmic
p-InP	Au-Zn	ohmic
n-AlGaAs*	Ni/Au-Ge/Ni	ohmic
p-AlGaAs*	In-Sn	ohmic
GaAs (n or p type)	Ni	Schottky
GaAs (n or p type)	Al	Schottky
GaAs (n or p type)	Au-Ti	Schottky
InP (n or p type)	Au	Schottky
InP (n or p type)	Au-Ti	Schottky

Source: Bio-Rad. Ref. 112.
*with GaAs capping layer

REFERENCES

1. F. Braun "On the Current Transport in Metal Sulfides (in German)," *Annal. Phys. Chem.* **153**, 556–563, 1874.

2. W. Schottky, "Semiconductor Theory of the Blocking Layer (in German)," *Naturwissenschaften* **26**, 843, Dec. 1938; "On the Semiconductor Theory of Blocking and Point Contact Rectifiers (in German)," *Z. Phys.* **113**, 367–414, July 1939; "Simplified and Expanded Theory of Boundary Layer Rectifiers (in German)," *Z. Phys.* **118**, 539–592, Feb. 1942.

3. H.K. Henisch, *Rectifying Semi-Conductor Contacts*, Clarendon Press, Oxford, 1957.

4. R.T. Tung, "Recent Advances in Schottky Barrier Concepts," *Mat. Sci. Eng.* **R35**, 1–138, 2001.

5. B. Schwartz (ed.), *Ohmic Contacts to Semiconductors*, Electrochem. Soc., New York, 1969.

6. V.L. Rideout, "A Review of the Theory and Technology for Ohmic Contacts to Group III–V Compound Semiconductors," *Solid-State Electron.* **18**, 541–550, June 1975.

7. N. Braslau, "Alloyed Ohmic Contacts to GaAs," *J. Vac. Sci. Technol.* **19**, 803–807, Sept./Oct. 1981.

8. A. Piotrowska, A. Guivarch, and G. Pelous, "Ohmic Contacts to III–V Compound Semiconductors: A Review of Fabrication Techniques," *Solid-State Electron.* **26**, 179–197, March 1983.

9. D.K. Schroder and D.L. Meier, "Solar Cell Contact Resistance—A Review," *IEEE Trans. Electron Devices* **ED-31**, 637–647, May 1984.

10. A.Y.C. Yu, "Electron Tunneling and Contact Resistance of Metal-Silicon Contact Barriers," *Solid-State Electron.* **13**, 239–247, Feb. 1970.

11. S.S. Cohen, "Contact Resistance and Methods for Its Determination," *Thin Solid Films* **104**, 361–379, June 1983.

12. A.G. Milnes and D.L. Feucht, *Heterojunction and Metal-Semiconductor Junctions*, Academic Press, New York, 1972.

13. B.L. Sharma and R.K. Purohit, *Semiconductor Heterojunctions*, Pergamon, London, 1974; B.L. Sharma, "Ohmic Contacts to III–V Compound Semiconductors," in *Semiconductors and Semimetals*, (R.K. Willardson and A.C. Beer, eds.), **15**, 1–38, Academic Press, New York, 1981.

14. E.H. Rhoderick and R.H. Williams, *Metal-Semiconductor Contacts*, 2nd ed., Clarendon, Oxford, 1988.

15. S.S. Cohen and G.S. Gildenblat, *Metal-Semiconductor Contacts and Devices*, Academic Press, Orlando, FL, 1986.

16. N.F. Mott, "Note on the Contact Between a Metal and an Insulator or Semiconductor," *Proc. Camb. Phil. Soc.* **34**, 568–572, 1938.

17. W.E Spicer, I. Lindau, P.R. Skeath and C.Y. Su, "The Unified Model for Schottky Barrier Formation and MOS Interface States in 3–5 Compounds," *Appl. Surf. Sci.* **9**, 83–91, Sept. 1981; W. Mönch, "On the Physics of Metal-Semiconductor Interfaces," *Rep. Progr. Phys.* **53**, 221–278, March 1990; L.J. Brillson, "Advances in Understanding Metal-Semiconductor Interfaces by Surface Science Techniques," *J. Phys. Chem. Solids* **44**, 703–733, 1983.

18. C.A. Mead, "Physics of Interfaces," in *Ohmic Contacts to Semiconductors* (B. Schwartz, ed.), Electrochem. Soc., New York, 1969, 3–16.

19. J. Bardeen, "Surface States and Rectification at Metal-Semiconductor Contact," *Phys. Rev.* **71**, 717–727, May 1947.

20. R.H. Williams, "The Schottky Barrier Problem," *Contemp. Phys.* **23**, 329–351, July/Aug. 1982.

21. L.J. Brillson, "Surface Photovoltage Measurements and Fermi Level Pinning: Comment on 'Development and Confirmation of the Unified Model for Schottky Barrier Formation and MOS Interface States on III–V Compounds'," *Thin Solid Films* **89**, L27–L33, March 1982.

22. J. Tersoff, "Recent Models of Schottky Barrier Formation," *J. Vac. Sci. Technol.* **B3**, 1157–1161, July/Aug. 1985.

23. I. Lindau and T. Kendelewicz, "Schottky Barrier Formation on III–V Semiconductor Surfaces: A Critical Evaluation," *CRC Crit. Rev. in Solid State and Mat. Sci.* **13**, 27–55, Jan. 1986.

24. F.A. Kroger, G. Diemer and H.A. Klasens, "Nature of Ohmic Metal-Semiconductor Contacts," *Phys. Rev.* **103**, 279, July 1956.

25. S.M. Sze, *Physics of Semiconductor Devices*, 2nd ed., Wiley, New York, 1981, 255–258.

26. F.A. Padovani and R. Stratton, "Field and Thermionic-Field Emission in Schottky Barriers," *Solid-State Electron.* **9**, 695–707, July 1966; F.A. Padovani, "The Current-Voltage Characteristics of Metal-Semiconductor Contacts," in *Semiconductors and Semimetals* (R.K. Willardson and A.C. Beer, eds.), Academic Press, New York, **7A**, 1971, 75–146.

27. C.R. Crowell and V.L. Rideout, "Normalized Thermionic-Field (TF) Emission in Metal-Semiconductor (Schottky) Barriers," *Solid-State Electron.* **12**, 89–105, Feb. 1969; "Thermionic-Field Resistance Maxima in Metal-Semiconductor (Schottky) Barriers," *Appl. Phys. Lett.* **14**, 85–88, Feb. 1969.

28. K.K. Ng and R. Liu, "On the Calculation of Specific Contact Resistivity on (100) Si," *IEEE Trans. Electron Dev.* **37**, 1535–1537, June 1990.

29. R.S. Popovic, "Metal-N-Type Semiconductor Ohmic Contact with a Shallow N^+ Surface Layer," *Solid-State Electron.* **21**, 1133–1138, Sept. 1978.

30. D.F. Wu, D. Wang and K. Heime, "An Improved Model to Explain Ohmic Contact Resistance of n-GaAs and Other Semiconductors," *Solid-State Electron.* **29**, 489–494, May 1986.

31. G. Brezeanu, C. Cabuz, D. Dascalu and P.A. Dan, "A Computer Method for the Characterization of Surface-Layer Ohmic Contacts," *Solid-State Electron.* **30**, 527–532, May 1987.

32. C.Y. Chang and S.M. Sze, "Carrier Transport Across Metal-Semiconductor Barriers," *Solid-State Electron.* **13**, 727–740, June 1970.

33. C.Y. Chang, Y.K. Fang and S.M. Sze, "Specific Contact Resistance of Metal-Semiconductor Barriers," *Solid-State Electron.* **14**, 541–550, July 1971.

34. W.J. Boudville and T.C. McGill, "Resistance Fluctuations in Ohmic Contacts due to Discreteness of Dopants," *Appl. Phys. Lett.* **48**, 791–793, March 1986.

35. Data for *n*-Si were taken from: 9, 32, 39, 43, 44, 45, 49, 55, 58, 69; S.S. Cohen, P.A. Piacente, G. Gildenblat and D.M. Brown, "Platinum Silicide Ohmic Contacts to Shallow Junctions in Silicon," *J. Appl. Phys.* **53**, 8856–8862, Dec. 1982; S. Swirhun, K.C. Saraswat and R.M. Swanson, "Contact Resistance of LPCVD W/Al and PtSi/Al Metallization," *IEEE Electron Dev. Lett.* **EDL-5**, 209–211, June 1984; S.S. Cohen and G.S. Gildenblat, "Mo/Al Metallization for VLSI Applications," *IEEE Trans. Electron Dev.* **ED-34**, 746–752, April 1987.

36. Data for *p*-Si were taken from: 39, 43, 44, 49, 55, 69; S.S. Cohen, P.A. Piacente, G. Gildenblat and D.M. Brown, "Platinum Silicide Ohmic Contacts to Shallow Junctions in Silicon," *J. Appl. Phys.* **53**, 8856–8862, Dec. 1982; S. Swirhun, K.C. Saraswat and R.M. Swanson, "Contact Resistance of LPCVD W/Al and PtSi/Al Metallization," *IEEE Electron Dev. Lett.* **EDL-5**, 209–211, June 1984; S.S. Cohen and G.S. Gildenblat, "Mo/Al Metallization for VLSI Applications," *IEEE Trans. Electron Dev.* **ED-34**, 746–752, April 1987; G.P. Carver, J.J. Kopanski, D.B. Novotny and R.A. Forman, "Specific Contact Resistivity of Metal-Semiconductor Contacts—A New, Accurate Method Linked to Spreading Resistance," *IEEE Trans. Electron Dev.* **ED-35**, 489–497, April 1988.

37. Data for *n*-GaAs can be found in ref. 6 and references therein.

38. Data for p-GaAs can be found in: 6, C.J. Nuese and J.J. Gannon, "Silver-Manganese Evaporated Ohmic Contacts to *p*-type GaAs," *J. Electrochem. Soc.* **115**, 327–328, March 1968; K.L. Klohn and L. Wandinger, "Variation of Contact Resistance of Metal-GaAs Contacts with Impurity Concentration and Its Device Implications," *J. Electrochem. Soc.* **116**, 507–508, April 1969; H. Matino and M. Tokunaga, "Contact Resistances of Several Metals and Alloys to GaAs," *J. Electrochem. Soc.* **116**, 709–711, May 1969; H.J. Gopen and A.Y.C. Yu, "Ohmic Contacts to Epitaxial GaAs," *Solid-State Electron.* **14**, 515–517, June 1971; O. Ishihara, K.Nishitani, H.Sawano and S.Mitsue, "Ohmic Contacts to P-Type GaAs," *Japan. J. Appl.*

Phys. **15**, 1411–1412, July 1976; C.Y. Su and C. Stolte, "Low Contact Resistance Non Alloyed Ohmic Contacts to Zn Implanted GaAs," *Electron. Lett.* **19**, 891–892, Oct. 1983; R.C. Brooks, C.L. Chen, A. Chu, L.J. Mahoney, J.G. Mavroides, M.J. Manfra and M.C. Finn, "Low-Resistance Ohmic Contacts to *p*-Type GaAs Using Zn/Pd/Au Metallization," *IEEE Electron Dev. Lett.* **EDL-6**, 525–527, Oct. 1985.

39. S.E. Swirhun and R.M. Swanson, "Temperature Dependence of Specific Contact Resistivity," *IEEE Electron Dev. Lett.* **EDL-7**, 155–157, March 1986.

40. D.M. Brown, M. Ghezzo and J.M. Pimbley, "Trends in Advanced Process Technology—Submicrometer CMOS Device Design and Process Requirements," *Proc. IEEE*, **74**, 1678–1702, Dec. 1986.

41. M.V. Sullivan and J.H. Eigler, "Five Metal Hydrides as Alloying Agents on Silicon," *J. Electrochem. Soc.* **103**, 218–220, April 1956.

42. R.H. Cox and H. Strack, "Ohmic Contacts for GaAs Devices," *Solid-State Electron.* **10**, 1213–1218, Dec. 1967.

43. R.D. Brookes and H.G. Mathes, "Spreading Resistance Between Constant Potential Surfaces," *Bell Syst. Tech. J.* **50**, 775–784, March 1971.

44. H. Muta, "Electrical Properties of Platinum-Silicon Contact Annealed in an H_2 Ambient," *Japan. J. Appl. Phys.* **17**, 1089–1098, June 1978.

45. A.K. Sinha, "Electrical Characteristics and Thermal Stability of Platinum Silicide-to-Silicon Ohmic Contacts Metallized with Tungsten," *J. Electrochem. Soc.* **120**, 1767–1771, Dec. 1973.

46. G.Y. Robinson, "Metallurgical and Electrical Properties of Alloyed Ni/Au-Ge Films on *n*-Type GaAs," *Solid-State Electron.* **18**, 331–342, April 1975.

47. G.P. Carver, J.J. Kopanski, D.B. Novotny, and R.A. Forman, "Specific Contact Resistivity of Metal-Semiconductor Contacts—A New, Accurate Method Linked to Spreading Resistance," *IEEE Trans. Electron Dev.* **35**, 489–497, April 1988.

48. A. Shepela, "The Specific Contact Resistance of Pd_2Si Contacts on *n*- and *p*-Si," *Solid-State Electron.* **16**, 477–481, April 1973.

49. C.Y. Ting and C.Y. Chen, "A Study of the Contacts of a Diffused Resistor," *Solid State Electron.* **14**, 433–438, June 1971.

50. J.M. Andrews, "A Lithographic Mask System for MOS Fine-Line Process Development." *Bell Syst. Tech. J.* **62**, 1107–1160, April 1983.

51. D.P. Kennedy and P.C. Murley, "A Two-Dimensional Mathematical Analysis of the Diffused Semiconductor Resistor," *IBM J. Res. Dev.* **12**, 242–250, May 1968.

52. H. Murrmann and D. Widmann, "Current Crowding on Metal Contacts to Planar Devices," *IEEE Trans. Electron Dev.* **ED-16**, 1022–1024, Dec. 1969.

53. H. Murrmann and D. Widmann, "Measurement of the Contact Resistance Between Metal and Diffused Layer in Si Planar Devices (in German)," *Solid-State Electron.* **12**, 879–886, Dec. 1969.

54. H.H. Berger, "Models for Contacts to Planar Devices," *Solid-State Electron.* **15**, 145–158, Feb. 1972; H.H. Berger, "Contact Resistance and Contact Resistivity," *J. Electrochem. Soc.* **119**, 507–514, April 1972.

55. J.M. Pimbley, "Dual-Level Transmission Line Model for Current Flow in Metal-Semiconductor Contacts," *IEEE Trans. Electron Dev.* **ED-33**, 1795–1800, Nov. 1986.

56. G.K. Reeves, P.W. Leech, and H.B. Harrison, "Understanding the Sheet Resistance Parameter of Alloyed Ohmic Contacts Using a Transmission Line Model," *Solid-State Electron.* **38**, 745–751, April 1995; G.K. Reeves and H.B. Harrison, "An Analytical Model for Alloyed Ohmic Contacts Using a Trilayer Transmission Line Model." *IEEE Trans. Electron Dev.* **42**, 1536–1547, Aug. 1995.

57. J.G.J. Chern and W.G. Oldham, "Determining Specific Contact Resistivity from Contact End Resistance Measurements," *IEEE Electron Dev. Lett.* **EDL-5**, 178–180, May 1984.

Comments on this Paper are: J.A. Mazer and L.W. Linholm, "Comments on 'Determining Specific Contact Resistivity from Contact End Resistance Measurements'," *IEEE Electron Dev. Lett.* **EDL-5**, 347–348, Sept. 1984; J. Chern and W.G. Oldham, "Reply to 'Comments on Determining Specific Contact Resistivity from Contact End Resistance Measurements'," *IEEE Electron Dev. Lett.* **EDL-5**, 349, Sept. 1984; M. Finetti, A. Scorzoni and G. Soncini, "A Further Comment on 'Determining Specific Contact Resistivity from Contact End Resistance Measurements'," *IEEE Electron Dev. Lett.* **EDL-6**, 184–185, April 1985.

58. S.E. Swirhun, W.M. Loh, R.M. Swanson and K.C. Saraswat, "Current Crowding Effects and Determination of Specific Contact Resistivity from Contact End Resistance (CER) Measurements," *IEEE Electron Dev. Lett.* **EDL-6**, 639–641, Dec. 1985.

59. G.K. Reeves, "Specific Contact Resistance Using a Circular Transmission Line Model," *Solid-State Electron.* **23**, 487–490, May 1980; A.J. Willis and A.P. Botha, "Investigation of Ring Structures for Metal-Semiconductor Contact Resistance Determination," *Thin Solid Films* **146**, 15–20, Jan. 1987.

60. S.S. Cohen and G.Sh. Gildenblat, *VLSI Electronics*, **13**, Metal-Semiconductor Contacts and Devices, Academic Press, Orlando, FL, 1986, p. 115; G.S. Marlow and M.B. Das, "The Effects of Contact Size and Non-Zero Metal Resistance on the Determination of Specific Contact Resistance," *Solid-State Electron.* **25**, 91–94, Feb. 1982; M. Ahmad and B.M. Arora, "Investigation of AuGeNi Contacts Using Rectangular and Circular Transmission Line Model," *Solid-State Electron.* **35**, 1441–1445, Oct. 1992.

61. J. Klootwijk, Philips Research Labs., private communication.

62. A. Scorzoni, M. Vanzi, and A. Querzè, "The Circular Resistor (CR)—A Novel Structure for the Analysis of VLSI Contacts," *IEEE Trans. Electron Dev.* **37**, 1750–1757, July 1990.

63. E.G. Woelk, H. Kräutle and H. Beneking, "Measurement of Low Resistive Ohmic Contacts on Semiconductors," *IEEE Trans. Electron Dev.* **ED-33**, 19–22, Jan. 1986.

64. W. Shockley in A. Goetzberger and R.M. Scarlett, "Research and Investigation of Inverse Epitaxial UHF Power Transistors," *Rep. No. AFAL-TDR-64-207*, Air Force Avionics Lab., Wright-Patterson Air Force Base, OH, Sept. 1964.

65. L.K. Mak, C.M. Rogers, and D.C. Northrop, "Specific Contact Resistance Measurements on Semiconductors," *J. Phys. E: Sci. Instr.* **22**, 317–321, May 1989.

66. G.K. Reeves and H.B. Harrison "Obtaining the Specific Contact Resistance from Transmission Line Model Measurements," *IEEE Electron Dev. Lett.* **EDL-3**, 111–113, May 1982.

67. L. Gutai, "Statistical Modeling of Transmission Line Model Test Structures—Part I: The Effect of Inhomogeneities on the Extracted Contact Parameters," "Part II: TLM Test Structure with Four or More Terminals: A Novel Method to Characterize Nonideal Planar Contacts in Presence of Inhomogeneities," *IEEE Trans. Electron Dev.* **37**, 2350–2360, 2361–2380, Nov. 1990.

68. D.B. Scott, W.R. Hunter and H. Shichijo, "A Transmission Line Model for Silicided Diffusions: Impact on the Performance of VLSI Circuits," *IEEE Trans. Electron Dev.* **ED-29**, 651–661, April 1982.

69. G.K. Reeves and H.B. Harrison, "Contact Resistance of Polysilicon-Silicon Interconnections," *Electron. Lett.* **18**, 1083–1085, Dec. 1982; G. Reeves and H.B. Harrison, "Determination of Contact Parameters of Interconnecting Layers in VLSI Circuits," *IEEE Trans. Electron Dev.* **ED-33**, 328–334, March 1986.

70. B. Kovacs and I. Mojzes, "Influence of Finite Metal Overlayer Resistance on the Evaluation of Contact Resistivity," *IEEE Trans. Electron Dev.* **ED-33**, 1401–1403, Sept. 1986.

71. I.F. Chang, "Contact Resistance in Diffused Resistors," *J. Electrochem. Soc.* **117**, 368–372, Feb. 1970; A. Scorzoni and U. Lieneweg, "Comparison Between Analytical Methods and Finite-Difference in Transmission-Line Tap Resistors and L-Type Cross-Kelvin Resistors," *IEEE Trans. Electron Dev.* **ED-37**, 1099–1103, June 1990.

72. E.F. Chor and J. Lerdworatawee, "Quasi-Two-Dimensional Transmission Line Model (QTD-TLM) for Planar Ohmic Contact Studies," *IEEE Trans. Electron Dev.* **49**, 105–111, Jan. 2002.

73. K.K. Shih and J.M. Blum, "Contact Resistances of Au-Ge-Ni, Au-Zn and Al to III–V Compounds," *Solid-State Electron.* **15**, 1177–1180, Nov. 1972.

74. S.S. Cohen, G. Gildenblat, M. Ghezzo and D.M. Brown, "Al-0.9%Si/Si Ohmic Contacts to Shallow Junctions," *J. Electrochem. Soc.* **129**, 1335–1338, June 1982.

75. S.J. Proctor and L.W. Linholm, "A Direct Measurement of Interfacial Contact Resistance," *IEEE Electron Dev. Lett.* **EDL-3**, 294–296, Oct. 1982; S.J. Proctor, L.W. Linholm and J.A. Mazer, "Direct Measurements of Interfacial Contact Resistance, End Resistance, and Interfacial Contact Layer Uniformity," *IEEE Trans. Electron Dev.* **ED-30**, 1535–1542, Nov. 1983.

76. J.A. Mazer, L.W. Linholm and A.N. Saxena, "An Improved Test Structure and Kelvin-Measurement Method for the Determination of Integrated Circuit Front Contact Resistance," *J. Electrochem. Soc.* **132**, 440–443, Feb. 1985.

77. A.A. Naem and D.A. Smith, "Accuracy of the Four-Terminal Measurement Techniques for Determining Contact Resistance," *J. Electrochem. Soc.* **133**, 2377–2380, Nov. 1986.

78. M. Finetti, A. Scorzoni and G. Soncini, "Lateral Current Crowding Effects on Contact Resistance Measurements in Four Terminal Resistor Test Patterns," *IEEE Electron Dev. Lett.* **EDL-5**, 524–526, Dec. 1984.

79. A.S. Holland, G.K. Reeves, and P.W. Leech, "Universal Error Corrections for Finite Semiconductor Resistivity in Cross-Kelvin Resistor Test Structures," *IEEE Trans. Electron Dev.* **51**, 914–919, June 2004.

80. M. Ono, A. Nishiyama, and A. Toriumi, "A Simple Approach to Understanding Errors in the Cross-Bridge Kelvin Resistor and a New Pattern for Measurements of Specific Contact Resistivity," *Solid-State Electron.* **46**, 1325–1331, Sept. 2002.

81. W.M. Loh, S.E. Swirhun, T.A. Schreyer, R.M. Swanson and K.C. Saraswat, "Modeling and Measurement of Contact Resistances," *IEEE Trans. Electron Dev.* **ED-34**, 512–524, March 1987.

82. A. Scorzoni, M. Finetti, K. Grahn, I. Suni and P. Cappelletti, "Current Crowding and Misalignment Effects as Sources of Error in Contact Resistivity Measurements—Part I: Computer Simulation of Conventional CER and CKR Structures," *IEEE Trans. Electron. Dev.* **ED-34**, 525–531, March 1987.

83. P. Cappelletti, M. Finetti, A. Scorzoni, I. Suni, N. Cirelli and G.D. Libera, "Current Crowding and Misalignment Effects as Sources of Error in Contact Resistivity Measurements—Part II: Experimental Results and Computer Simulation of Self-Aligned Test Structures," *IEEE Trans. Electron. Dev.* **ED-34**, 532–536, March 1987.

84. U. Lieneweg and D.J. Hannaman, "New Flange Correction Formula Applied to Interfacial Resistance Measurements of Ohmic Contacts to GaAs," *IEEE Electron Dev. Lett.* **EDL-8**, 202–204, May 1987.

85. S.A. Chalmers and B.G. Streetman, "Lateral Diffusion Contributions to Contact Mismatch in Kelvin Resistor Structures," *IEEE Trans. Electron Dev.* **ED-34**, 2023–2024, Sept. 1987.

86. W.T. Lynch and K.K. Ng, "A Tester for the Contact Resistivity of Self-Aligned Silicides," *IEEE Int. Electron Dev. Meet. Digest*, San Francisco, 1988, 352–355.

87. T.F. Lei, L.Y. Leu and C.L. Lee, "Specific Contact Resistivity Measurement by a Vertical Kelvin Test Structure," *IEEE Trans. Electron Dev.* **ED-34**, 1390–1395, June 1987; W.L. Yang, T.F. Lei, and C.L. Lee, "Contact Resistivities of Al and Ti on Si Measured by a Self-Aligned Vertical Kelvin Test Resistor Structure," *Solid-State Electron.* **32**, 997–1001, Nov. 1989.

88. C.L. Lee, W.L. Yang and T.F. Lei, "The Spreading Resistance Error in the Vertical Kelvin Test Resistor Structure for the Specific Contact Resistivity," *IEEE Trans. Electron Dev.* **ED-35**, 521–523, April 1988.

89. L.Y. Leu, C.L. Lee, T.F. Lei, and W.L. Yang, "Numerical Simulation of the Vertical Kelvin Test Structure for Specific Contact Resistivity," *Solid-State Electron.* **33**, 177–188, Feb. 1990.

90. J.G.J. Chern, W.G. Oldham and N. Cheung, "Contact-Electromigration-Induced Leakage Failure in Aluminum-Silicon to Silicon Contacts," *IEEE Trans. Electron Dev.* **ED-32**, 1341–1346, July 1985.

91. H. Onoda, "Dependence of Al-Si/Al Contact Resistance on Substrate Surface Orientation," *IEEE Electron Dev. Lett.* **EDL-9**, 613–615, Nov. 1988.

92. T.J. Faith, R.S. Iven, L.H. Reed, J.J. O'Neill Jr., M.C. Jones and B.B. Levin, "Contact Resistance Monitor for Si ICs," *J. Vac. Sci. Technol.* **B2**, 54–57, Jan./March 1984.

93. N. Braslau, "Alloyed Ohmic Contacts to GaAs," *J. Vac. Sci. Technol.* **19**, 803–807, Sept./Oct. 1981; "Ohmic Contacts to GaAs," *Thin Solid Films* **104**, 391–397, June 1983.

94. J. Willer, D. Ristow, W. Kellner and H. Oppolzer, "Very Stable Ge/Au/Cr/Au Ohmic Contacts to GaAs," *J. Electrochem. Soc.* **135**, 179–181, Jan. 1988.

95. J.M. Andrews and M.P. Lepselter, "Reverse Current-Voltage Characteristics of Metal-Silicide Schottky Diodes," *Solid-State Electron.* **13**, 1011–1023, July 1970.

96. A.K. Srivastava, B.M. Arora, and S. Guha, "Measurement of Richardson Constant of GaAs Schottky Barriers," *Solid-State Electron.* **24**, 185–191, Feb. 1981, and references therein.

97. M. Missous and E.H. Rhoderick, "On the Richardson Constant for Aluminum/Gallium Arsenide Schottky Diodes," *J. Appl. Phys.* **69**, 7142–7145, May 1991.

98. F. Hossain, Arizona State University.

99. R.T. Tung, "Electron Transport of Inhomogeneous Schottky Barriers," *Appl. Phys. Lett.* **58**, 2821–2823, June 1991.

100. A.S. Bhuiyan, A. Martinez and D. Esteve, "A New Richardson Plot for Non-Ideal Schottky Diodes," *Thin Solid Films* **161**, 93–100, July 1988.

101. A.M. Goodman, "Metal-Semiconductor Barrier Height Measurement by the Differential Capacitance Method—One Carrier System," *J. Appl. Phys.* **34**, 329–338, Feb. 1963.

102. R.H. Fowler, "The Analysis of Photoelectric Sensitivity Curves for Clean Metals at Various Temperatures," *Phys. Rev.* **38**, 45–56, July 1931.

103. W. Mönch, *Electronic Properties of Semiconductor Interfaces*, Springer, Berlin, 2004, 63–67.

104. R. Turan, N. Akman, O. Nur, M.Y.A. Yousif, and M. Willander, "Observation of Strain Relaxation in $Si_{1-x}Ge_x$ layers by Optical and Electrical Characterisation of a Schottky Junction," *Appl. Phys.* **A72**, 587–593, May 2001.

105. T. Okumura and K.N. Tu, "Analysis of Parallel Schottky Contacts by Differential Internal Photoemission Spectroscopy," *J. Appl. Phys.* **54**, 922–927, Feb. 1983.

106. L.D. Bell and W.J. Kaiser, "Ballistic Electron Emission Microscopy: A Nanometer-Scale Probe of Interfaces and Carrier Transport," *Annu. Rev. Mat. Sci.* **26**, 189–222, 1996.

107. C. Fontaine, T. Okumura and K.N. Tu, "Interfacial Reaction and Schottky Barrier Between Pt and GaAs," *J. Appl. Phys.* **54**, 1404–1412, March 1983.

108. T. Okumura and K.N. Tu, "Electrical Characterization of Schottky Contacts of Au, Al, Gd and Pt on *n*-Type and *p*-Type GaAs," *J. Appl. Phys.* **61**, 2955–2961, April 1987.

109. Y.P. Song, R.L. Van Meirhaeghe, W.H. Laflère and F. Cardon, "On the Difference in Apparent Barrier Height as Obtained from Capacitance-Voltage and Current-Voltage-Temperature Measurements on Al/p-InP Schottky Barriers," *Solid-State Electron.* **29**, 633–638, June 1986.

110. I. Ohdomari and K.N. Tu, "Parallel Silicide Contacts," *J. Appl. Phys.* **51**, 3735–3739, July 1980.

111. J.L. Freeouf, T.N. Jackson, S.E. Laux and J.M. Woodall, "Size Dependence of "Effective" Barrier Heights of Mixed-Phase Contacts," *J. Vac. Sci. Technol.* **21**, 570–574, July/Aug. 1982.

112. Bio-Rad, *Semiconductor Newsletter*, Winter 1988.

PROBLEMS

3.1   The $I-V$ data of a forward-biased pn junction are given in the following table. Determine the temperature T and the series resistance R_s for this device.

V (V)	I (A)	V (V)	I (A)	V (V)	I (A)
0.0000	0.0000	0.35000	1.0960e-07	0.70000	0.0062910
0.025000	1.2910e-12	0.37500	2.5120e-07	0.72500	0.010050
0.050000	4.2480e-12	0.40000	5.7540e-07	0.75000	0.014290
0.075000	1.1020e-11	0.42500	1.3180e-06	0.77500	0.019610
0.10000	2.6540e-11	0.45000	3.0190e-06	0.80000	0.025430
0.12500	6.2090e-11	0.47500	6.9130e-06	0.82500	0.031850
0.15000	1.4350e-10	0.50000	1.5820e-05	0.85000	0.038330
0.17500	3.3010e-10	0.52500	3.6180e-05	0.87500	0.045040
0.20000	7.5760e-10	0.55000	8.2520e-05	0.90000	0.051940
0.22500	1.7370e-09	0.57500	0.00018720	0.92500	0.058990
0.25000	3.9800e-09	0.60000	0.00041910	0.95000	0.066160
0.27500	9.1190e-09	0.62500	0.00091340	0.97500	0.073440
0.30000	2.0890e-08	0.65000	0.0018820	1.0000	0.080800
0.32500	4.7860e-08	0.67500	0.0035060		

3.2   A portion of a semiconductor test structure is shown in Fig. P3.2. It incorporates a TLM test structure, a one-element contact string and a *circular* Schottky diode. Several measurements were made.

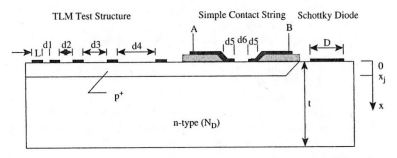

Fig. P3.2

(a) Schottky Diode $I-V$:

V (V)	0.1	0.2	0.3	0.4	0.5
I (A)	5.59×10^{-8}	1.36×10^{-6}	3.04×10^{-5}	6.71×10^{-4}	0.0148

Determine the barrier height ϕ_B (in V) and the ideality factor n.

(b) p^+ Layer:

The p^+ majority carrier profile is approximated by $p(x) \approx N_A(x) = 8 \times 10^{19} \exp(-x/5 \times 10^{-6})$, with x in cm.

Determine the junction depth x_j (in cm), and the sheet resistance R_{sh} (in Ω/square) of the p^+ layer; neglect the contribution of the electrons in the p^+ layer.

(c) <u>TLM Test Structure:</u>

The TLM test structure gave the following values:

d (μm)	$d_1 = 1$	$d_2 = 3$	$d_3 = 7$	$d_4 = 10$
R_T (Ω)	8.2	13.41	23.83	31.65

Determine the sheet resistance R_{sh}, the contact resistance R_c (Ω), and the specific contact resistance ρ_c (Ω-cm^2).

(d) <u>One Element Contact String:</u>

Determine the resistance between points A and B (in Ω). Neglect the metal resistance.

(e) <u>Resistance Through the Wafer:</u>

Suppose two circular contacts of diameter 1 cm are formed on opposite sides of the n-type wafer and that the current flow from top to bottom is confined to this area as it flows through the wafer. Determine the resistance between these two contacts using $\rho_c = (\partial J/\partial V)^{-1}$ evaluated at $V = 0$ assuming current flow is due to thermionic emission. Z (width of the p$^+$*layer*) = 100 μm, $d_5 = $ 50 μm, $d_6 = 500$ μm, $L = 25$ μm, $D = 1$ mm, $A^{**} = 110$ A/cm^2· K, substrate $\rho = 0.1$ Ω-cm (to convert to doping density, use Fig. A1.1), $t = 750$ μm, $T = $ 300 K, $K_s = 11.7$, use $\mu_p = 60$ cm^2/V-s. Neglect the space-charge region width of the p^+n junction in these calculations, *i.e.*, assume it to be zero.

3.3 The I–V and C–V curves of two Schottky diodes were measured. These diodes are fabricated on identical n-type substrates. One diode (device 1) has barrier height ϕ_{B1} and area A and the other consists of a diode with barrier height ϕ_{B1} over half the area and ϕ_{B2} over the other half area. The total area is the same for both devices. The Schottky diode equations are

$$I = AA^*T^2 e^{-q\phi_B/kT}(e^{q(V-Ir_s)/nkT} - 1) = I_o(e^{q(V-Ir_s)/nkT} - 1) \text{ and}$$

$$C = A\sqrt{\frac{K_s \varepsilon_o q N_D}{2(V_{bi} - V)}}$$

I_o is the saturation current. The I–V curve of device 1 is shown in Fig. P3.3.

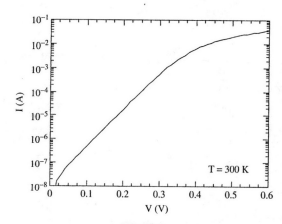

Fig. P3.3

The saturation currents as a function of temperature are:

Device 1:	T (K)	I_{o1} (A)	Device 2:	T (K)	I_{o2} (A)
	300	1.57×10^{-8}		300	3.83×10^{-7}
	350	1.02×10^{-6}		350	1.46×10^{-5}
	400	2.42×10^{-5}		400	2.33×10^{-4}

The room-temperature, zero-biased capacitance is: $C_1(0V) = 4.092 \times 10^{-11} F$; $C_2(0V) = 4.335 \times 10^{-11} F$. $K_s = 11.7$, $\varepsilon_o = 8.854 \times 10^{-14}$ F/cm, $k = 8.617 \times 10^{-5}$ eV/K, $A = 10^{-3}$ cm^2, $N_D = 10^{16}$ cm^{-3}, $n_i = 10^{10}$ cm^{-3}, $E_i = E_G/2 = 0.56$ eV. Determine A^*, n, r_s, ϕ_{B1}, and ϕ_{B2}.

3.4 Consider a Schottky diode whose barrier height is *not constant* over the diode area. Determine the effective barrier height $\phi_{B,eff}$, from

(a) log(I)—V plot
(b) $(A/C)^2$—V plot

where the barrier heights and areas are: $\phi_{B1} = 0.6$ V, $A_1 = 0.2$ A and $\phi_{B2} = 0.7$ V, $A_2 = 0.8$ A, where A is the area given below. Use $A^* = 100$ A/cm$^2 \cdot$ K^2, $A = 10^{-3}$ cm^2, $n = 1$, $T = 300$ K, $K_s = 11.7$, $N_D = 10^{15}$ cm^{-3}, and $N_C = 2.5 \times 10^{19}$ cm^{-3}. The effective barrier height is defined by the equations

$$I = AA^*T^2 e^{-q\phi_{B,eff}/kT}(e^{qV/nkT} - 1) \text{ for the I-V plot, and by}$$

$$V_{bi} = \phi_{B.eff} + V_0 = \phi_{B,eff} + \frac{kT}{q} \ln\left(\frac{N_C}{N_D}\right) \text{ for the } (A/C)^2 \text{—V plot.}$$

Neglect the "kT/q" term in the capacitance equation in the book.

3.5 The transfer length contact resistance test structure is used to measure various electrical parameters. The sheet resistance between contacts R_{sh} is different from the sheet resistance under the contacts R_{sk} in this case.

(a) For negligible metal resistance, the following data were obtained:

d (μm)	3	5	10	20	30	50
V (mV)	43.6	49.6	64.6	94.6	124.6	184.6

$L = 12$ μm, $Z = 100$ μm, $I = 10$ mA. The end resistance for this test structure is $R_e = 3.4 \times 10^{-3} \Omega$. Determine R_{sh}, R_{sk}, R_c, ρ_c, and L_{Tc}.

(b) One day when these measurements were made, it was found that the contact resistance had increased to $R_c = 5.18$ Ω. It is suspected that the *metal resistance* has increased due to a problem with the metal deposition system. All other parameters are unchanged. Determine the metal sheet resistance R_{sm}.

3.6 The $I-V$ curves of a Schottky diode are shown as a function of temperature in Fig. P3.6. The diode has a circular area of 1 mm diameter. The current is given by

$$I = AA^*T^2 e^{-q\phi_B/kT}(e^{qV/nkT} - 1).$$

Determine A^*, n, and ϕ_B.

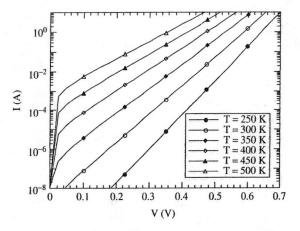

Fig. P3.6

3.7 The I–V curves of a forward-biased pn junction is shown in Fig. P3.7. Determine the temperature T for the "$T =$?" curve and the series resistance r_s for the "$T = 300$ K" curve.

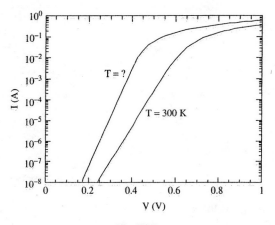

Fig. P3.7

3.8 The pn junction diode $I-V$ equation at high injection levels, neglecting series resistance, is:

$$I = I_{01}(e^{qV/2kT} - 1)$$

With series resistance, but no high injection level effects, the $I-V$ equation is:

$$I = I_{02}(e^{q(V-Ir_s)/kT} - 1)$$

Discuss how one can determine which equation applies for experimental $I-V$ data that fall in that region of the $I-V$ curve where either one of these equations could be valid.

3.9 The doping density versus depth for the n-layer is shown for two cases in Fig. P3.9. Discuss the sheet resistances and the specific contact resistivities for these two cases, *i.e.*, are they the same or not and why or why not.

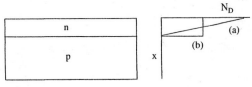

Fig. P3.9

3.10 The contact resistance of contact A is R_{cA} in Fig. P3.10. It is measured between points A–C and between points A–B. For both contacts A and B, we have $L_T < L$, where L_T is the transfer length and L is the contact length. Choose one answer from the following list and explain it briefly.

$\Box R_{cA}(A-C) > R_{cA}(A-B) \quad \Box R_{cA}(A-C) = R_{cA}(A-B) \quad \Box R_{cA}(A-C) < R_{cA}(A-B)$

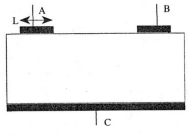

Fig. P3.10

3.11 In the TLM test structure the resistance between adjacent contacts is measured and displayed in Fig. P3.11 as an R_T vs. d plot. What parameters are determined with this test structure? One day something goes wrong during processing and a thin oxide film remains on the n-layer *before* the metal contacts are deposited. Show on the R_T vs. d plot in Fig. P3.11 the data points that would be measured for this case. Contact spacing and size and the n-layer are the same as before. The oxide film is thin enough that current can tunnel through it and can be thought of as a resistive layer.

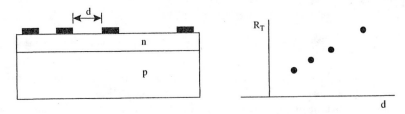

Fig. P3.11

3.12 Two metallic contacts are made on an n-type semiconductor wafer. The $I-V$ curve in Fig. P3.12 is for the case: contact A is a Schottky contact, contact B is an ohmic contact. Draw on the same figure the $I-V$ curve when *both* contacts are Schottky contacts.

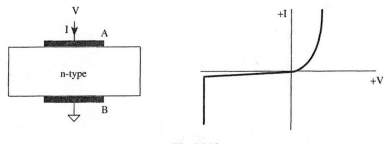

Fig. P3.12

3.13 The $I-V$ curves of two Schottky diodes are shown in Fig. P3.13 for the *same* temperature T. The relevant equation for the current is

$$I = AA^*T^2 \exp(-q\phi_B/kT)(\exp(qV/nkT) - 1)$$

The device parameter that has changed in going from curve (A) to curve (B) is:

$\square A^*$ $\square \phi_B$ $\square n$

Choose one answer and explain.

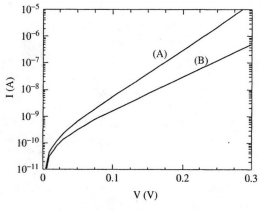

Fig. P3.13

3.14 A Schottky barrier diode is formed on both n- and p-type semiconductor regions as shown in Fig. P3.14.

(a) Draw the $I-V$ curve for this device.

(b) Draw the band diagrams at the surface ($x = 0$) and at $x = x_1$ for $V = 0$. The doping densities and the barrier heights and A* are the same for both semiconductor types and the areas on the two semiconductor types are identical.

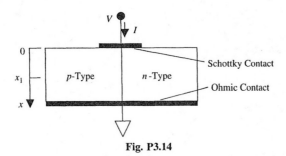

Fig. P3.14

3.15 The $I-V$ and $C-V$ plots of a Schottky diode on a Si substrate are shown in Fig. P3.15. From the $I-V$ curve determine ϕ_B, A*, r_s, T; from the $C-V$ curve determine ϕ_B, N_D. Use $K_s = 11.7$, $k = 8.617 \times 10^{-5}$ eV/K, $\varepsilon_o = 8.854 \times 10^{-14}$ F/cm, diode ideality factor $n = 1$, Area $A = 10^{-3}$ cm², $E_G = 1.12$ eV, $n_i = 9.15 \times 10^{19} (T/300)^2 \exp(-6880/T)$cm⁻³. The saturation current is given at various temperatures as:

T(K)	I_s (A)
250	5.01×10^{-9}
275	7.63×10^{-8}
300	7.49×10^{-7}
325	5.34×10^{-6}
350	2.81×10^{-5}
375	1.21×10^{-4}
400	4.41×10^{-4}

3.16 The resistance of a 100-element ($N = 100$) contact chain R_T is given by

$$R_T = N(2R_c + R_s) \text{ where } R_c = \frac{\rho_c}{L_T Z} \coth\left(\frac{L}{L_T}\right).$$

Two elements of this chain are shown in Fig. P3.16. Determine and plot R_T versus ρ_c for $\rho_c = 10^{-8}$ to $10^{-5}\Omega$-cm² as a log-log plot. Use a sufficient number of points for a smooth curve. $L = 3$ μm, $d = 10$ μm, $Z(n^+$layer width$) = 10$ μm, $R_{sh} = 50$ Ω/square. Neglect the metal resistance.

3.17 The TLM test structure in Fig. P3.17 gave the R_T values in the graph. The doping density N_D in the n^+ layer is uniform.

(a) Determine the sheet resistance R_{sh} (ohms/square), the contact resistance R_c (ohms), the specific contact resistance ρ_c (ohms-cm²), and the doping density N_D (cm⁻³). $Z(n^+$layer width$) = 100$ μm, $L = 25$ μm, $t = 2.5 \times 10^{-4}$ cm, $\mu_n = 50$ cm²/V-s.

(b) Plot a new line for the same parameters as in (i), except $\rho_c = 10^{-7}$ ohms-cm².

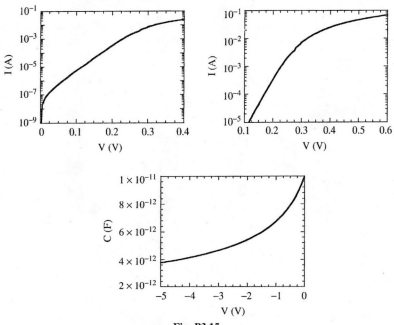

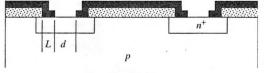

Fig. P3.15

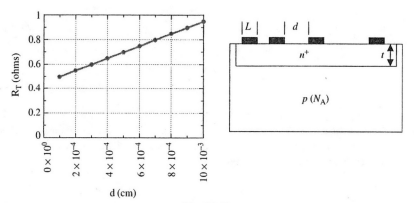

Fig. P3.16

Fig. P3.17

3.18 A transfer length method test structure is shown in P3.18. The n layer is 1 μm thick with resistivity $\rho = 0.001$ ohm-cm. The specific contact resistivity is $\rho_c = 10^{-6}$ ohm-cm^2.

Calculate and plot R_T versus d for $d = 2, 4, 6, 10$ μm. $Z = 20$ μm, $L = 10$ μm.

$$R_T = 2R_c + R_s = \frac{2\rho_c}{L_T Z}\coth\left(\frac{L}{L_T}\right) + \frac{R_{sh}d}{Z}.$$

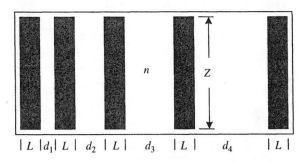

Fig. P3.18

3.19 In the Kelvin contact resistance test structure in Fig. P3.19, it is usually assumed that the voltmeter has very high input resistance and there is negligible voltage drop along the voltage measurement arm. Now suppose the input resistance R_{in} of the voltmeter is finite. For $I = 10^{-3}$ A, $R_{arm} = 100$ Ω, and $R_c = 10$ Ω, determine R_{in} for a 10% error in R_c.

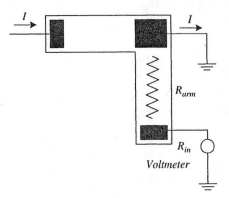

Fig. P3.19

3.20 (a) All contacts in Fig. P3.20 have identical specific contact resistance ρ_c. Is the contact resistance R_c of the three top contacts the same? *Discuss.* $L \gg L_T$, where L_T is the transfer length.

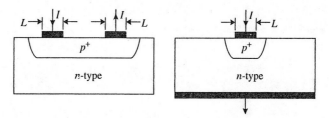

Fig. P3.20

3.21 MOSFETs with different channel lengths, shown in Fig. P3.21, are used to determine the channel length and series resistance R_{SD}. Can such transistors be used to determine the contact resistance R_c and the specific contact resistivity ρ_c? Discuss.

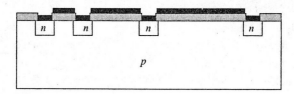

Fig. P3.21

3.22 R_T versus d data points of a *transfer length method* contact resistance measurement are shown in Fig. P3.22 for a uniformly-doped n-type layer on a p-type substrate. The n-type resistivity is ρ, the contact length is L, and the contact width is Z.

(a) Indicate on the figure *three* parameters that can be determined from these data.

(b) Draw the data points when the n-layer thickness is increased; all other parameters remain unchanged.

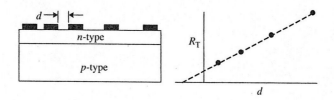

Fig. P3.22

3.23 R_T versus d data points of a *transfer length method* contact resistance measurement are shown in Fig. P3.23 for a uniformly-doped n-type layer on a p-type substrate. The n-type resistivity is ρ, the contact length is L, and the contact width is Z.

(a) Indicate on the figure *three* parameters that can be determined from these data.

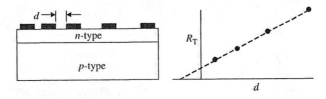

Fig. P3.23

(b) Draw the data points when the *n*-layer resistivity is increased; all other parameters remain unchanged.

REVIEW QUESTIONS

- What is the most important parameter to give low contact resistance?
- What are the three metal-semiconductor conduction mechanisms?
- What is Fermi level pinning?
- What is the specific contact resistivity and what are its units?
- Does the contact chain give detailed contact characterization? Why or why not?
- What is the transfer length method?
- Why is the Kelvin contact test structure best?
- What is the effect of lateral current flow on Kelvin contact resistance measurement?
- How is the barrier height of Schottky diodes determined?
- How can the Richardson constant be measured?

4

SERIES RESISTANCE, CHANNEL LENGTH AND WIDTH, AND THRESHOLD VOLTAGE

4.1 INTRODUCTION

Semiconductor device and circuit performance is generally degraded by series resistance that depends on the series and shunt resistance, on the device, on the current flowing through the device, and on a number of other parameters. The *series resistance* r_s depends on the semiconductor resistivity, on the contact resistance, and sometimes on geometrical factors. Series resistance may be very large before causing device degradation. For example, in a reverse-biased photodiode with a photocurrent in the nano-amperes range, series resistance is a minor consideration. However, series resistances of a few ohms are detrimental for solar cells and power devices. The effect of r_s on capacitance and carrier concentration profiling measurements is discussed in Chapter 2. The aim of the device designer should be a design in which series resistance is negligibly small for that device. However, since r_s cannot be zero, it is important to be able to measure it. The effective channel length and width of a MOSFET are important device parameters because they are required for modeling and they usually differ from the mask-defined and the physical dimensions and the threshold voltage is one of the most important MOSFET parameters. Methods to determine these are discussed.

4.2 PN JUNCTION DIODES

4.2.1 Current-Voltage

The current of a pn junction is often written as a function of the diode voltage V_d as

$$I = I_o(e^{qV_d/nkT} - 1) \qquad (4.1)$$

Semiconductor Material and Device Characterization, Third Edition, by Dieter K. Schroder
Copyright © 2006 John Wiley & Sons, Inc.

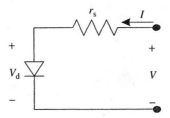

Fig. 4.1   Equivalent circuit of a diode.

where I_o is the saturation current and n the diode ideality factor. The diode voltage V_d is the voltage across the space-charge region and excludes any voltage drops across the p and n quasi-neutral regions. If both I_o and n are constant, then a plot of $\log(I)$ versus V_d yields a straight line for $V_d > nkT/q$.

A semiconductor diode can be represented by the equivalent circuit of Fig. 4.1, consisting of an ideal diode in series with resistance r_s. When current flows through the device, the diode *terminal* voltage V is

$$V = V_d + IR_s \tag{4.2}$$

With series resistance Eq. (4.1) becomes

$$\boxed{I = I_o(e^{q(V-Ir_s)/nkT} - 1)} \tag{4.3}$$

The current in pn junction diodes is due to two components: space-charge region (scr) recombination/generation and quasi-neutral region (qnr) recombination/generation, leading to the $I-V$ relationship

$$I = I_{o,scr}(e^{q(V-Ir_s)/nkT} - 1) + I_{o,qnr}(e^{q(V-Ir_s)/nkT} - 1) \tag{4.4}$$

Equation (4.4) is plotted in Fig. 4.2 for forward bias. There are four distinct regions in the figure. For $Ir_s \ll V \ll nkT/q$, the current depends linearly on voltage ($e^{qV/nkT} - 1 \approx qV/nkT$), giving a non-linear curve on the semilog plot. For $V \gg nkT/q$, the current is dominated by scr recombination at low current and by qnr recombination at higher current. The breakpoint between the two current components occurs at $V = 0.3\ V$ in this example. The $I-V$ curve deviates from linearity at high current due to series resistance r_s.

Extrapolating the two linear regions to $V = 0$ gives $I_{o,scr}$ and $I_{o,qnr}$. The slope is given by

$$m = \frac{d \log I}{dV} \tag{4.5}$$

Knowing the slope and sample temperature allows the ideality factor to be determined from the relationship

$$n = \frac{q}{\ln(10)mkT} = \frac{q}{2.3mkT} \tag{4.6}$$

We will generally use the logarithm to base 10, written as "log", instead of the logarithm to base e, written as "ln", because experimental data are usually plotted on "log", not "ln", scales.

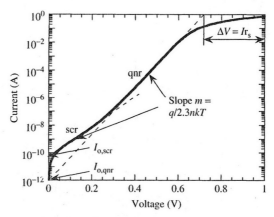

Fig. 4.2 Current versus voltage for a diode with series resistance. Upper dashed line is for $r_s = 0$.

The deviation of the $\log(I)-V$ curve from linearity at high currents is $\Delta V = Ir_s$, allowing r_s to be determined according to

$$r_s = \frac{\Delta V}{I} \tag{4.7}$$

Since the Schottky diode current-voltage behavior is similar to pn junctions, we will use Fig. 3.38 for the r_s extraction. Figure 4.3(a) gives that part of the $I-V$ curve where r_s is negligible and $n = 1.1$ from the slope. Figure 4.3(b) shows the part of the r_s-dominated curve. The deviation from linearity, according to Eq. (4.7), gives $r_s = 0.8\ \Omega$.

The resistance can also be obtained from the diode conductance $g_d = dI/dV$. In the region where r_s is important, qnr recombination dominates and the current

$$I \approx I_{o,qnr}e^{q(V-Ir_s)/nkT} \tag{4.8}$$

gives

$$g_d = \frac{qI(1-r_s g_d)}{nkT} \tag{4.9}$$

We can write Eq. (4.9) as[1]

$$\frac{1}{g_d} = \frac{nkT}{q} + Ir_s \tag{4.10}$$

suggesting a plot of I/g_d versus I. Such a plot has an $I = 0$ intercept of nkT/q and slope r_s, as shown in Fig. 4.4(a).

Equation (4.9) can also be written as

$$\frac{g_d}{I} = \frac{q(1-r_s g_d)}{nkT} \tag{4.11}$$

Plotting g_d/I versus g_d, the $g_d = 0$ intercept is q/nkT, the $g_d/I = 0$ intercept is $1/r_s$ and the slope is qr_s/nkT, as shown in Fig. 4.4(b). Careful measurements have revealed the approach of Eq. (4.11) to give the most reliable results,[2] although Fig. 4.4 shows the scatter in (b) to be more severe than in (a) because both axes require a differentiation of the data. Comparing Figs. 4.3 and 4.4 for r_s extraction brings out an important point. *A slope method is generally more accurate than a single point method to determine an*

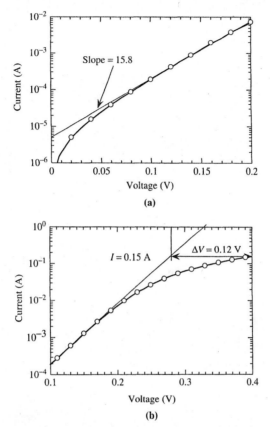

Fig. 4.3   Current versus voltage for the diode in Fig. 3.38. (a) low voltage where r_s can be neglected, (b) high-voltage where r_s dominates.

unknown quantity. Since experimental data exhibit small errors, slope methods allow smoothing of the data, whereas single point measurements incorporate any experimental uncertainties in the parameter determination.

The diode conductance can be measured by superimposing a small ac voltage δV on the dc voltage V and measuring the in-phase component δI with a lock-in amplifier to obtain $g_d = \delta I / \delta V$.[3] Because of the exponential dependence of current on voltage, δV should be kept as low as possible. Alternately, one can differentiate the $I-V$ curve. Again, because of the exponential nature of the curve, dc voltage steps should be less than 1 mV. Using the semilog plot, where $g_d = Id[\ln(I)]/dV$, voltage steps as high as 10 mV are permissible.[2]

4.2.2   Open-Circuit Voltage Decay (OCVD)

Open-circuit voltage decay is a method to determine the minority carrier lifetime of pn junctions as discussed in Chapter 7 and can also be used to determine the diode series resistance, as illustrated in Fig. 4.5. The diode is forward biased. At $t = 0$ switch S is opened, and the open-circuit diode voltage is monitored as a function of time. The *lifetime*

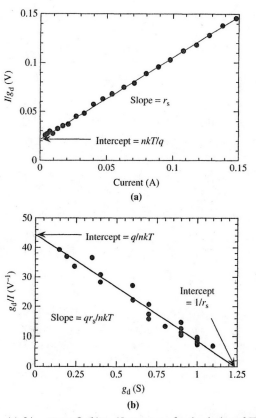

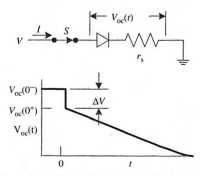

Fig. 4.4 (a) I/g_d versus I, (b) g_d/I versus g_d for the device of Fig. 4.3.

Fig. 4.5 Open-circuit voltage decay of a pn junction showing the voltage discontinuity at $t = 0$.

is determined from the slope of the $V_{oc} - t$ curve. The *series resistance* is obtained from the voltage discontinuity ΔV at $t = 0$.[3]

The voltage drop across the diode just before opening the switch $V_{oc}(0^-)$ consists of the diode voltage V_d and the voltage drop across any device resistances

$$V_{oc}(0^-) = V_d + Ir_s \tag{4.12}$$

When switch S is opened and the current drops to zero, the voltage drops abruptly and $V_{oc}(0^+) = V_d$. With the measured voltage drop given by $\Delta V = V_{oc}(0^-) - V_{oc}(0^+) = Ir_s$ and I measured independently, it is a simple matter to calculate the series resistance $r_s = \Delta V/I$. This absolute measure does not rely on slopes or intercepts and is suitable for low r_s measurements. Diode series resistances as low as 10 to 20 mΩ have been determined this way.

4.2.3   Capacitance-Voltage ($C-V$)

We show the effect of series resistance on capacitance in Chapter 2. For the parallel equivalent circuit configuration, the measured capacitance C_m of a junction device is related to the true capacitance C by

$$C_m = \frac{C}{(1 + r_s G)^2 + (2\pi f r_s C)^2} \tag{4.13}$$

where G is the conductance and f the frequency. For reasonably good junction devices, the condition $r_s G \ll 1$ is generally satisfied, and Eq. (4.13) simplifies to

$$C_m \approx \frac{C}{1 + (2\pi f r_s C)^2} \tag{4.14}$$

Lowering the frequency reduces the second term in the denominator to less than unity and the true capacitance is determined. Then the frequency is raised until the second term dominates, and r_s can be calculated with all other quantities known. This method is only effective when $r_s \gg 1/2\pi f C$. It can also be used when dc current techniques are unable to determine the series resistance, *e.g.*, for an MOS capacitor with no dc current flow.

4.3   SCHOTTKY BARRIER DIODES

4.3.1   Series Resistance

The current-voltage characteristic of a Schottky barrier diode without series resistance is discussed in Section 3.5. The thermionic current-voltage expression of a Schottky barrier diode with series resistance is given by

$$I = I_s(e^{q(V-Ir_s)/nkT} - 1) \tag{4.15}$$

where I_s is the saturation current

$$I_s = AA^*T^2 e^{-q\phi_B/kT} = I_{s1}e^{-q\phi_B/kT} \tag{4.16}$$

where A is the diode area, $A^* = 4\pi q k^2 m^*/h^3 = 120$ (m*/m) A/cm^{2}K^2 is Richardson's constant[4], ϕ_B the effective barrier height, and n the ideality factor. Equation (4.15) is sometimes expressed as (see Appendix 4.1)

$$I = I_s \exp\left(\frac{qV}{nkT}\right)\left(1 - \exp\left(-\frac{qV}{kT}\right)\right) \tag{4.17}$$

valid for $Ir_s \ll V$. Data plotted according to Eq. (4.15) are linear only for $V \gg kT/q$. When plotting $\log[I/(1 - exp(-qV/kT))]$ versus V, using Eq. (4.17), the data are linear to $V = 0$.

The method of extracting r_s, given in Section 4.2.1, can also be used for Schottky diodes. Another method defines the Norde function F as[5]

$$F = \frac{V}{2} - \frac{kT}{q} \ln \left(\frac{I}{I_{s1}} \right) \tag{4.18}$$

With Eqs. (4.15) and (4.16), Eq. (4.18) becomes

$$F = \left(\frac{1}{2} - \frac{1}{n} \right) V + \frac{Ir_s}{n} + \phi_B \tag{4.19}$$

Why is this rather peculiarly defined F function used? When F is plotted against V, it exhibits a minimum which is used to determine r_s and ϕ_B. To see the dependence of F on V, we consider the low and high voltage limits. At low applied voltages, where $Ir_s \ll V$, Eq. (4.19) gives $dF/dV = 1/2 - 1/n \approx -1/2$ for $n \approx 1$. At high voltages, where $Ir_s \gg V$, $dF/dV = 1/2$. Hence, F has a minimum lying between these two limits. The voltage at the minimum is V_{min} and the corresponding current is I_{min}. From $dF/dV = 0$ at the minimum, the series resistance is

$$r_s = \frac{2-n}{I_{min}} \frac{kT}{q} \tag{4.20}$$

The minimum F-value, found by substituting Eq. (4.20) into Eq. (4.19), is

$$F = \left(\frac{1}{2} - \frac{1}{n} \right) V_{min} + \frac{2-n}{n} \frac{kT}{q} + \phi_B \tag{4.21}$$

The series resistance of the Schottky diode is calculated from the ideality factor n and from I_{min}. The ideality factor is obtained from the slope of the $\log(I)$ versus V plot, and I_{min} is the current at $V = V_{min}$. For this method, I_{s1}, and therefore A^*, must be known. This is a disadvantage of this technique, since A^* is not necessarily known. In the absence of an experimentally determined A^*, one must use published values for A^*. That is not always a good assumption since A^* depends on the contact preparation, including the surface cleaning procedure[6] and sample annealing temperature; it even appears to depend on the metal thickness and on the metal deposition method.[7]

The original Norde method of plotting F versus V assumes the ideality factor $n = 1$, and the statistical error is increased by using only a few data points near the minimum of the F versus V curve. A modified Norde increases the accuracy, allowing r_s, n, and ϕ_B to be extracted from an experimental $\log(I)$ versus V plot.[8] Alternately, r_s, n, and ϕ_B can be determined from the $I-V$ curves at two different temperatures.[9]

Barrier height measurements in the absence of series resistance are discussed in Section 3.5. The barrier height is commonly calculated from the saturation current I_s determined by an extrapolation of the $\log(I)$ versus V curve to $V = 0$. Series resistance is not important in this extrapolation because the current I_s is very low. The barrier height ϕ_B is calculated from I_s in Eq. (4.16) according to

$$\phi_B = \frac{kT}{q} \ln \left(\frac{AA^*T^2}{I_s} \right) \tag{4.22}$$

The barrier height so determined is ϕ_B at zero bias. The most uncertain of the parameters in Eq. (4.22) is A^*, rendering this method only as accurate as A^* is known. Fortunately,

A^* appears in the "ln" term. For example, an error of two in A^* gives rise to an error of 0.69 kT/q in ϕ_B.

Several variations of the Norde plot have been proposed to overcome its limitations. In one of these, an H-function is defined as[10]

$$H = V - \frac{nkT}{q} \ln\left(\frac{I}{I_{s1}}\right) = Ir_s + n\phi_B \tag{4.23}$$

A plot of H versus I has a slope of r_s and an H-axis intercept of $n\phi_B$. Like the F plot, the H plot also requires a knowledge of A^*. An approach, *not* requiring a knowledge of A^*, is the modified Norde plot[11]

$$F1 = \frac{qV}{2kT} - \ln\left(\frac{I}{T^2}\right) \tag{4.24}$$

$F1$ is plotted versus V for several different temperatures. Each of these plots exhibits a minimum and each minimum defines an $F1_{min}$, a voltage V_{min}, and a current I_{min}. With Eqs. (4.15), (4.16) and (4.20) and $V \gg kT/q$,

$$2F1_{min} + (2 - n)\ln\left(\frac{I_{min}}{T^2}\right) = 2 - n(\ln(AA^*) + 1) + \frac{qn\phi_B}{kT} \tag{4.25}$$

When the left side of Eq. (4.25) is plotted against q/kT, a straight line results with slope $n\phi_B$ and y-axis intercept $\{2 - n[\ln(AA^*) + 1]\}$. With n independently determined, it is possible to extract both ϕ_B and A^*, provided the area A is known.

The function

$$F(V) = V - V_a \ln(I) \tag{4.26}$$

was also used to determine, I_s, and r_s. One determines the minimum of $F(V)$ for different V_a, where V_a is an independent voltage.[12] Using Eq. (4.26), but with the current I as the dependent variable, and finding the minimum is the basis of yet another method.[13] Methods with different assumed functions but requiring solutions to simultaneous equations have also been proposed.[14] Occasionally it is impossible to extract barrier height and series resistance from $I-V$ measurements using the thermionic emission equation. It may then be necessary to include space-charge region recombination and tunnel currents.[15] When the barrier height is voltage dependent, the extraction of device parameters saturation current, barrier height, diode ideality factor, and series resistance becomes more difficult. One solution to this problem is provided in ref. 16.

4.4 SOLAR CELLS

Solar cells are particularly prone to series resistance, because it reduces the maximum available power. The series resistance should be approximately $r_s < (0.8/X)\Omega$ for 1 cm^2 area cells, where X is the solar concentration.[17] Here $X = 1$ for non-concentrator cells, whereas for concentrator cells X can be several hundred. For $X = 100$, $r_s < 8 \times 10^{-3}$ Ω. Under "one-sun" conditions 10–20% of the maximum power available from a solar cell can be lost due to a series resistance of 1 Ω. Although solar cells are pn junction diodes, their $I-V$ characteristics are often not suitable for the types of measurements of conventional diodes. Since the operation of solar cells in the presence of sunlight

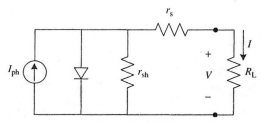

Fig. 4.6   Solar cell equivalent circuit.

may alter the series resistance, r_s should be determined under operating conditions. Shunt resistance is also important for solar cells.

Several methods have been used to determine r_s. They are generally neither simple to implement nor to interpret. A solar cell, represented by the equivalent circuit of Fig. 4.6, consists of a photon or light-induced current generator I_{ph}, a diode, a series resistor r_s, and a shunt resistor r_{sh}. The part of the circuit to the left of the two points is the cell and the part to the right is the load, characterized by the load resistor R_L. Frequently r_s and r_{sh} are assumed to be constant, but they may depend on the cell current. The current I flows through the load resistor and develops a voltage V across it. The current is given by

$$I = I_{ph} - I_o \left(\exp\left(\frac{q(V + Ir_s)}{nkT} \right) - 1 \right) - \frac{V + Ir_s}{r_{sh}} \qquad (4.27)$$

This equation does not take into account that both I_o, and n are not constant over the entire $I-V$ curve. At low voltage, space-charge region (scr) recombination generally dominates, but at higher voltage quasi-neutral region (qnr) recombination is dominant. Equation (4.27) is used for most solar cell analyses in spite of its simplifications, although scr and qnr recombination are occasionally considered separately.

The current-voltage characteristic is measured with conventional $I-V$ techniques or with the quasi-steady-state (Q_{ss}) photoconductance technique, where a flash lamp produces slowly varying illumination and the resulting time dependence of the excess photoconductance of the sample is measured.[18] The Q_{ss} approach can measure the open-circuit voltage of solar cells as a function of the incident light intensity. Monotonically varying illumination produces a voltage versus illumination curve in a fraction of a second. This quasi-steady-state open-circuit voltage method has important advantages over the classic $I_{sc} - V_{oc}$ technique to measure the characteristics of the solar cell free from series resistance effects. An example light intensity-open circuit voltage curve is shown in Fig. 4.7. Care should be exercised when using this technique on high sheet resistance cells, e.g., amorphous solar cells, due to shadows cast by the probe needle, for example.[19] Such shadowing distorts the experimental data.

A current-voltage curve of a solar cell is shown in Fig. 4.8. The open-circuit voltage V_{oc}, the short-circuit current I_{sc}, and the maximum power point V_{max} and I_{max} are also shown. The quantities r_{so} and r_{sho} are the resistances defined by the slopes of the $I-V$ curve at $I = 0$ and at $V = 0$, respectively. The effects of series and shunt resistances are shown on the $I-V$ characteristics in Fig. 4.9 calculated from Eq. (4.27). Series resistances of a few ohms or less degrade the device performance, as do shunt resistances of several hundred ohms. Small $I-V$ degradation has a significant effect on cell efficiency. The maximum power points are shown by the points on Fig. 4.8 and 4.9.

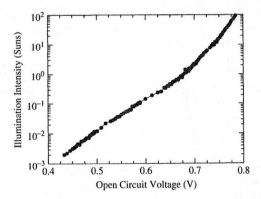

Fig. 4.7   Light intensity versus open circuit voltage. After ref. 18.

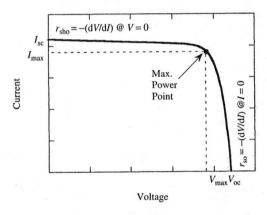

Fig. 4.8   Current-voltage characteristic of a solar cell.

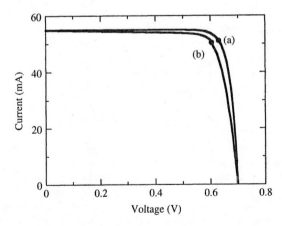

Fig. 4.9   Current-voltage curve of a solar cell with $I_{ph} = 55$ mA, $I_o = 10^{-13}$ A, $n = 1$, $T = 300$ K. The series and shunt resistances are: (a) $r_s = 0$, $r_{sh} = \infty$, (b) $r_s = 0.5$ Ω, $r_{sh} = 500$ Ω.

4.4.1   Series Resistance—Multiple Light Intensities

An early method to determine r_s is based on the measurement of the $I-V$ curves at two different light intensities giving the short-circuit currents I_{sc1} and I_{sc2}, respectively. A current δI below I_{sc}, $I = I_{sc} - \delta I$, is picked on both $I-V$ curves. The currents $I_1 = I_{sc1} - \delta I$ and $I_2 = I_{sc2} - \delta I$ correspond to voltages V_1 and V_2. The series resistance is then[20]

$$r_s = \frac{V_1 - V_2}{I_2 - I_1} = \frac{V_1 - V_2}{I_{sc2} - I_{sc1}} \tag{4.28}$$

By using more than two light intensities more than two points are generated. Drawing a line through all of the points gives the series resistance by the slope of this line, $\Delta I/\Delta V$, as

$$r_s = \frac{\Delta V}{\Delta I} \tag{4.29}$$

The method is illustrated in Fig. 4.10.

The slope method lends itself to r_s determination at any current with no limiting approximations and is generally considered to give good results. It is also independent of I_o, n, and r_{sh}, provided they do not change with the operating point. This is an important consideration. Those techniques that require a knowledge of I_o, n, and r_{sh}, and even I_{ph} in some cases, are at a disadvantage because these parameters may not be accurately known. It is important that the temperature of the cell be constant during the measurements at different light intensities, as temperature variations can alter the series resistance.

Comparison of experimental $I-V$ curves with a theoretical curves ($r_s = 0$) has also been used to determine r_s. The shift of the maximum power point from its theoretical value, $\Delta V_{max} = V_{max}(theory) - V_{max}(exp)$, is given by[22]

$$r_s = \frac{\Delta V_{max}}{I_{max}} \tag{4.30}$$

A weakness of this method is the assumption that parameters like I_o and n are known. If unknown, they must be determined by other means, for they are required to calculate the theoretical $I-V$ curve.

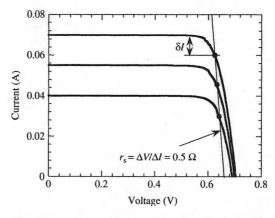

Fig. 4.10   Series resistance determination of a solar cell.

Under short-circuit conditions, where $I = I_{sc}$ and $V = 0$, Eq. (4.27) becomes

$$\ln\left(\frac{I_{ph} - I_{sc}}{I_o}\right) = \frac{qI_{sc}r_s}{nkT} \tag{4.31}$$

A plot of $\ln[(I_{ph} - I_{sc})/I_o]$ versus I_{sc} has a slope of qr_s/nkT.[23] The series resistance is calculated from the slope, provided n and I_{ph} are known.

Another method relies on a *dark* $I-V$ curve, the open-circuit voltage, and the short-circuit current. From Eq. (4.27) with r_{sh} very large, the *dark* voltage is

$$V_{dk} = \frac{nkT}{q} \ln\left(\frac{I_{dk}}{I_o}\right) - I_{dk}r_s \tag{4.32}$$

The open-circuit voltage is given by

$$V_{oc} = \frac{nkT}{q} \ln\left(\frac{I_{ph}}{I_o}\right) \tag{4.33}$$

V_{oc} is independent of r_s since there is no current during an open circuit voltage measurement. Hence, by comparing V_{oc} with V_{dk} at a given current I_{dk}, it is possible to determine r_s at that current. To reduce any error, one should choose that point on the $I_{dk} - V_{dk}$ curve where the diode parameters are the same as those of the open-circuit condition.[24] That corresponds to $I_{dk} = I_{ph}$ and since generally $I_{ph} \approx I_{sc}$,

$$r_s \approx \frac{V_{dk}(I_{sc}) - V_{oc}}{I_{sc}} \tag{4.34}$$

$I_{dk} = I_{sc}$ ensures that the upper limit of the series resistance for a given light intensity is obtained.[24]

4.4.2   Series Resistance—Constant Light Intensity

The series resistance can be determined by the area under the $I-V$ curve,[25] given by the power P_1,

$$P_1 = \int_0^{I_{sc}} V(I)\, dI \tag{4.35}$$

The series resistance, obtained from Eqs. (4.27) and (4.35), is[25]

$$r_s = 2\left(\frac{V_{oc}}{I_{sc}} - \frac{P_1}{I_{sc}^2} - \frac{nkT}{qI_{sc}}\right) \tag{4.36}$$

This method has been used to measure the very low resistances of concentrator solar cells of $r_s = 5$ to $6 \times 10^{-3}\Omega$. Such cells, because they are operated under solar concentrations with high photocurrents, are particularly prone to series resistance degradation.

Series resistances determined by the "area" method have been compared to values determined by the "slope" methods. Such comparisons have shown the "area" method to overestimate r_s at "one-sun" and lower illuminations,[26] because n must be known accurately in Eq. (4.36) and r_{sh} may not be negligible. The results of the two methods at *high* illumination are in reasonably good agreement.

Various analytical techniques have also been used to determine r_s. Some are based on complete curve fitting of the solar equation to experimental $I-V$ curves. Others use several points on the experimental $I-V$ curve to determine the key parameters. In the *five point method* the parameters I_{ph}, I_o, n, r_s, and r_{sh} are calculated from the experimental V_{oc}, I_{sc}, V_m, I_m, r_{so}, and r_{sho} shown in Fig. 4.8.[27] Later simplifications in the equations make the analysis more tractable.[28] A comparison of the five parameters determined by the exact five point, by the approximate five point, and by numerical techniques gave very good agreement for I_{ph}, I_o, and n. The main differences were found for r_s and r_{sh} at low light intensities. In the *three point method*, I_{ph}, I_o, n, r_s, and r_{sh} are determined from the open-circuit voltage, the short circuit current, and the maximum power point. Both five-point and three-point methods give comparable results.[29-30]

Because scr and qnr recombination take place in a solar cell, parameters describing both of these processes should be determined for complete solar cell modeling. Applying small current steps to a solar cell in both the forward and reverse current directions and measuring the resulting voltage, allows I_o(scr), I_o(qnr), n(scr), n(qnr), r_s, and r_{sh} to be determined.[31]

A technique especially suitable for concentrator solar cells with low series resistances is based on high intensity flash illumination.[32] Neglecting the shunt resistance in the circuit in Fig. 4.6, for very high light intensities the output current I approaches but cannot exceed $V_{oc}/(R_L + r_s)$. In order to keep the cell temperature as constant as possible during the measurement, it is best to flash the illumination. Approximating the voltage by $V_{oc} \approx I(R_L + r_s)$ and varying the load resistance at constant light intensity

$$r_s \approx \frac{I_2 R_{L2} - I_1 R_{L1}}{I_1 - I_2} \qquad (4.37)$$

where I_1 and I_2 are the currents for load resistances R_{L1} and R_{L2}. Series resistances as low as 7 to 9 mΩ have been determined with this method for GaAs concentrator solar cells at light intensities approaching 9000 suns with 1 ms light pulses.[32] The value of the load resistance should be on the order of the series resistance.

4.4.3 Shunt Resistance

The shunt resistance r_{sh} can be determined by some of the curve-fitting approaches discussed in the previous section, or it can be determined independently. It is sometimes found from the slope of the reverse-biased current-voltage characteristic before breakdown. Most solar cells, however, exhibit large reverse currents at voltages well below breakdown because solar cells are not designed to operate under high reverse voltages. This makes it difficult to obtain reasonable value for r_{sh} by this method. Furthermore, a solar cell in the dark under reverse bias is a poor representation of a solar cell operating in the light under forward bias.

An alternate method is to rewrite Eq. (4.27) in terms of V_{oc} and I_{sc} as

$$I_{sc}\left(1 + \frac{r_s}{r_{sh}}\right) - \frac{V_{oc}}{r_{sh}} = I_o\left(\exp\left(\frac{qV_{oc}}{nkT}\right) - \exp\left(\frac{qI_{sc}r_s}{nkT}\right)\right) \qquad (4.38)$$

This equation can be simplified for the usual condition of $r_s \ll r_{sh}$. If the measurement is made under low light intensities where $I_{sc}r_s \ll nkT/q$, then Eq. (4.38) becomes

$$I_{sc} - I_o\left(\exp\left(\frac{qV_{oc}}{nkT}\right) - 1\right) = \frac{V_{oc}}{r_{sh}} \qquad (4.39)$$

This approximation is valid for $I_{sc} \leq 3$ mA for series resistances on the order of 0.1 Ω. When measurements of r_{sh} were made under these conditions, r_{sh} was found to be highly sensitive to I_o and n, that may not be known accurately.[33] This problem was alleviated by making the measurements at very low light intensities, allowing the second term on the left side of Eq. (4.39) to be neglected and then

$$I_{sc} \approx \frac{V_{oc}}{r_{sh}} \tag{4.40}$$

The $I_{sc} - V_{oc}$ plot has a linear region of slope $1/r_{sh}$. The curve becomes non-linear at higher light intensities, and the method becomes invalid. Measurements showed that for I_{sc} in the 0 to 200 μA and V_{oc} in the 0 to 50 mV range, the shunt resistances were 65 to 1170 Ω.[33] Example $J_{sc} - V_{oc}$ plots are shown in Fig. 4.11.

4.5   BIPOLAR JUNCTION TRANSISTORS

An integrated-circuit bipolar junction transistor (BJT) with parasitic series resistances is shown in Fig. 4.12. The n^+ emitter and the p-base are formed in an n-collector layer on a p-substrate. The transistor is decoupled from adjacent transistors by oxide isolation regions not shown. The parasitic resistances and their measurement are relevant for our purpose. The emitter resistance R_E is primarily determined by the emitter contact resistance. The base resistance R_B is composed of the intrinsic base resistance R_{Bi}, under the emitter, and the extrinsic base resistance R_{Bx}, from the emitter to the base contact including the base contact resistance. The collector resistance R_C is comprised of two components: R_{C1} and R_{C2}. The resistances are generally functions of the device operating point.

A common method to display the base and collector current is a semilog plot of the logarithm of the current plotted against the emitter-base voltage, shown in Fig. 4.13 and known as a *Gummel plot*.[34] The two currents are expressed as a function of the base-emitter voltage V_{BE} by

$$I_B = I_{B0} \exp\left(\frac{q(V_{BE} - I_B R_B - I_E R_E)}{nkT}\right) \tag{4.41}$$

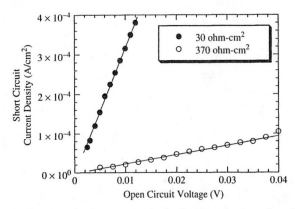

Fig. 4.11   Short circuit current density versus open circuit voltage for two solar cells. Adapted from ref. 21.

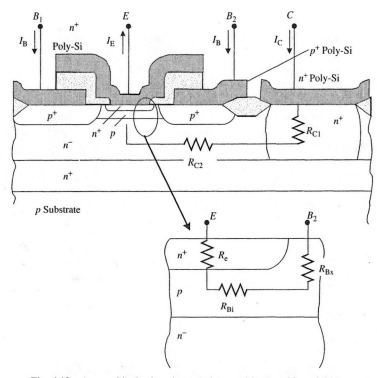

Fig. 4.12   An npn bipolar junction transistor and its parasitic resistances.

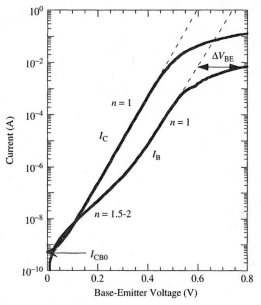

Fig. 4.13   Gummel plots showing the effects of emitter-base space-charge region recombination ($n \approx 1.5$–2), quasi-neutral region recombination ($n \approx 1$), and series resistance.

$$I_C = I_{C0} \exp\left(\frac{q(V_{BE} - I_B R_B - I_E R_E)}{kT}\right) \tag{4.42}$$

I_{B0} depends on whether the dominant recombination mechanism is space-charge region (scr) or quasi-neutral region recombination.

The *collector* current Gummel plot is linear with slope of $q/\ln(10)kT$ over most of its range. It saturates at the collector-base junction leakage current I_{CB0} at low voltages and deviates from linearity at high voltages due to series resistances. For simplicity, additional deviations from linearity at high voltages due to *high-level injection* are not shown.

The base current generally exhibits two linear regions. At low voltages the current is dominated by emitter-base space-charge region recombination with a slope of $q/\ln(10)nkT$, where $n \approx 1.5$ to 2. At intermediate voltages the slope is $q/\ln(10)kT$ just as it is for the collector current due to quasi-neutral region recombination, and at higher voltages the curve deviates from linearity due to series resistances. High-level injection effects are again not shown, for clarity.

The external voltage drop between the base and the emitter terminals V_{BE} is

$$V_{BE} = V'_{BE} + I_B R_B + I_E R_E = V'_{BE} + (R_B + (\beta + 1)R_E)I_B \tag{4.43}$$

and the voltage drop across the parasitic resistances is

$$\Delta V_{BE} = I_B R_B + I_E R_E = (R_B + (\beta + 1)R_E)I_B \tag{4.44}$$

where β is the common emitter current gain, $I_C = \beta I_B$, $I_E = I_C + I_B = (\beta + 1)I_B$, and V_{BE} is the potential drop across the base-emitter junction. Although R_E is generally small, the $(\beta + 1)$ multiplier can make $(\beta + 1)R_E$ appreciable. The emitter and base resistances depress the currents below their ideal values, shown by the curves below the extrapolated dashed lines in Fig. 4.13.

BJT resistance measurement techniques fall into two main categories: dc methods and ac methods. The dc methods are generally fast and easy to implement and can be further subdivided into methods relying on determining the series resistance from $I-V$ curves or from open circuit voltage measurements. The ac techniques require measurement frequencies of typically 50 MHz to several GHz, necessitating a careful consideration of device and measurement circuit parasitics and of the distributed nature of BJT parameters.

4.5.1 Emitter Resistance

The emitter resistance in discrete BJTs is around 1 Ω and for small-area IC transistors it is around 5 to 100 Ω. One method to determine R_E is based on a measure of the collector-emitter voltage V_{CE}[35-36]

$$V_{CE} = \frac{kT}{q} \ln\left(\frac{I_B + I_C(1 - \alpha_R)}{\alpha_R(I_B - I_C(1 - \alpha_R)/\alpha_F)}\right) + R_E(I_B + I_C) + R_C I_C \tag{4.45}$$

neglecting the small reverse saturation current. Here $\alpha_F = \beta_F/(1 + \beta_F)$ and $\alpha_R = \beta_R/(1 + \beta_R)$ are the large-signal forward and reverse common base current gains. With the collector open circuited, $I_C = 0$ and Eq. (4.45) becomes

$$V_{CE} = \frac{kT}{q} \ln\left(\frac{1}{\alpha_R}\right) + R_E I_B = \frac{kT}{q} \ln\left(\frac{1 + \beta_R}{\beta_R}\right) + R_E I_B \tag{4.46}$$

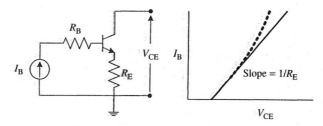

Fig. 4.14 Emitter resistance measurement setup and $I_B - V_{CE}$ plot.

A plot of I_B versus V_{CE} and the measurement setup are shown in Fig. 4.14. The curve is linear with a V_{CE}-axis intercept of $(kT/q)\ln(1/\alpha_R)$ and a slope of $1/R_E$. This behavior is indeed observed for discrete transistors.[36-37] The base current should not be too small for unambiguous measurements. For example, base currents around 10 mA are suitable for $R_E \approx 1 \ \Omega$, and it is important to ensure that zero or very low collector currents are drawn during the measurement. A suitable connection is: BJT base connected to the *collector* terminal, BJT emitter connected to the *emitter* terminal, and BJT collector connected to the *base* terminal of the curve tracer.[38]

Departures of the $I_B - V_{CE}$ curve from linearity occur when α_R is current dependent. This generally happens at low and high currents. Hence an R_E determination may not yield one unique value. The slope of the curve increases at high base currents.[38-39] Intermediate base currents usually give good linearity. Additional complications can arise for integrated circuit transistors where part of the buried layer resistance can add to the emitter resistance due to internally circulating currents even for zero external collector current. The accuracy of this method is also dependent on the sensitivity of the base charge with respect to base current.[40] A method to improve the original open collector measurement, requiring a measurement of forward/reverse current gains and the intrinsic base sheet resistance, allows the $I_B - V_{CE}$ plot to be linearized, making the unambiguous extraction of R_E easier.[41]

A different approach uses two base contacts in Fig. 4.12, by biasing the device in the forward active region with base current supplied through base contact B_1 and no current flowing through contact B_2. The base-emitter voltage V_{BE2} is

$$V_{BE2} = V_{BE0} + R_E I_E \tag{4.47}$$

where V_{BE0} is the base-emitter voltage at the edge closest to B_2.[42] The emitter resistance is

$$R_E = \frac{V_{BE2} - V_{BEeff}}{I_E} \tag{4.48}$$

where V_{BEeff} is determined from the base current expression[42]

$$I_{B1} = \frac{I_{C0}}{\beta}\left(\exp\left(\frac{qV_{BEeff}}{nkT}\right) - 1\right) \tag{4.49}$$

The same method can also be used for base resistance extraction.[42]

Yet another method uses the null in third-order intermodulation as a function of emitter current in a bipolar transistor to find the emitter resistance and the thermal resistance.[43]

4.5.2 Collector Resistance

A problem with collector resistance measurements is the strong dependence of collector resistance on the device operating point. The collector resistance can be determined by the same $I_B - V_{CE}$ method of Section 4.5.1 by interchanging the collector and emitter terminals. With $E \rightarrow C$ and $C \rightarrow E$, the $I_B - V_{CE}$ curve has a V_{CE}-axis intercept of $(kT/q)\ln(1/\alpha_F)$ and a slope of $1/R_C$. Another method uses the parasitic substrate pnp transistor that exists in the structure of Fig. 4.12 and the reverse transistors associated with the npn transistor to determine the internal voltages of the npn BJT, allowing R_C to be determined.[44]

Another method uses the transistor output characteristics. Typical output $I_C - V_{CE}$ curves are shown in Fig. 4.15. The two dashed lines $1/R_{Cnorm}$ and $1/R_{Csat}$ represent the two limiting values of R_C. The $1/R_{Cnorm}$ line is drawn through the knee of each curve, where the output curves tend to horizontal. The collector resistance obtains for the device in its normal, active mode of operation. The $1/R_{Csat}$ line gives the appropriate collector resistance for the transistor in saturation. A good discussion of this measurement technique using a curve tracer is given in Getreu.[38] The collector resistance can also be determined by measuring the substrate current of the parasitic pnp vertical transistor linked with the npn transistor.[45] The pnp device is operated with either the bottom substrate-collector or the top base-collector pn junction forward biased, allowing the separation of the various R_C components.

4.5.3 Base Resistance

The base resistance is difficult to determine accurately because it depends on the device operating point and because its measurement is influenced by the emitter resistance through the term $(\beta + 1)R_E$. The base current flows laterally in BJTs, giving lateral voltage drops in the base, causing V_{BE} to be a function of position. Small V_{BE} variations give rise to large current variations since I_C and I_B depend exponentially on V_{BE}. Most of the emitter current flows at the emitter edge nearest the base contact, referred to as *emitter crowding,* reducing the distance for base current flow with increased emitter current, thereby decreasing R_{Bi} with current.

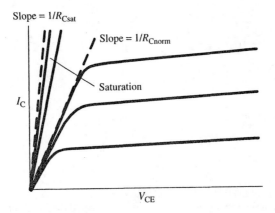

Fig. 4.15 Common emitter output characteristics. The two lines show the limiting values of R_C.

A simple method to determine the total series resistance between emitter and base is shown in Fig. 4.13. The experimental base current deviates from the extrapolated straight line by the voltage drop

$$\Delta V_{BE} = (R_B + (\beta + 1)R_E)I_B \tag{4.50}$$

A plot of $\Delta V_{BE}/I_B$ versus β has a slope of R_E and an intercept on the $\Delta V_{BE}/I_B$-axis of $R_B + R_E$. To vary the current gain β one chooses a device with varying β over some operating range, or use different devices from the same lot. The first method ensures that only one device is measured, but conductivity modulation and other second-order effects may distort the measurement since the current must be changed to vary β. In order to avoid conductivity modulation and other second-order effects, one should make the measurement at a constant emitter current. But, of course, a constant I_E implies constant β. In that case one must use different devices from the same lot whose βs vary over some appropriate range, assuming the resistances to be the same for all devices from that lot.[46]

A variation on this method is based on rewriting Eqs. (4.41) and (4.42) as[39]

$$\frac{nkT}{qI_C} \ln\left(\frac{I_{B1}}{I_B}\right) = R_E + \frac{R_{Bi}}{\beta} + \frac{R_E + R_{Bx}}{\beta} \tag{4.51}$$

where $R_B = R_{Bi} + R_{Bx}$ and $I_{B1} = I_{B0} \exp(qV_{BE}/nkT)$. Then R_{Bi}/β is constant if R_{Bi} is proportional to β.[47] The requirement of $R_{Bi} \sim \beta$ at all I_E is a weakness in this method; it may not always be satisfied. A plot of $(kT/qI_C)\ln(I_{B1}/I_B)$ versus $1/\beta$, for $n = 1$, has a slope of $R_E + R_{Bx}$ and an intercept on the $(kT/qI_C)\ln(I_{B1}/I_B)$-axis of $R_E + R_{Bi}/\beta$, as shown in Fig. 4.16. The intrinsic base resistance must be calculated in this technique. For a rectangular emitter of width W_E and length L_E with a base contact on one side $R_{Bi} = WR_{shi}/3L$, where R_{shi} is the intrinsic base sheet resistance. For a rectangular emitter with two base contacts, $R_{Bi} = WR_{shi}/12L$. For square emitters with contacts on all sides, $R_{Bi} = R_{shi}/32$, and for circular emitters with a base contact all around $R_{Bi} = R_{shi}/8\pi$.[39] The method based on Eq. (4.51) does not take into account lateral voltage drops along the intrinsic base current path. This condition is satisfied for collector currents of less than 10

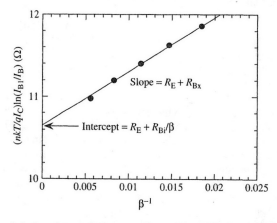

Fig. 4.16 Measured device characteristics according to Eq. (4.51) for a self-aligned, high-speed digital BJT. Reprinted after Ning and Tang[39] by permission of IEEE (© 1984, IEEE).

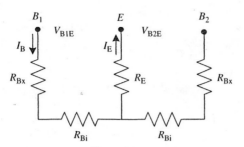

Fig. 4.17   Equivalent emitter-base portion of the "two-base contact" BJT.

to 20 mA for scaled digital BJTs.[39] Current crowding makes the results unreliable, unless such crowding is insignificant, *e.g.*, in very narrow emitter transistors.

The method of Eq. (4.51) must be used with caution for polysilicon emitter contacts when a thin insulating barrier exists between the polysilicon and the single crystal emitter. This can cause the $(kT/qI_C)\ln(I_{B1}/I_B)$ versus $1/\beta$ curve to be non-linear for low $1/\beta$ values. The slope of this plot can even become negative. This behavior cannot be explained by a resistive drop, but is attributable to an interfacial layer between the polysilicon contact and the single crystal emitter.[48]

A quite different approach makes use of the BJT in Fig. 4.12 with two independent base contacts, B_1 and B_2. The BJT emitter-base junction is forward biased using base contact B_1. The voltage is measured between B_1 and the emitter, V_{B1E}, and between B_2 and the emitter, V_{B2E}. For the equivalent circuit of Fig. 4.17, base current flows from B_1 only. For the Kelvin voltage measurement of V_{B2E}, almost no current flows through the right half of the base. The resulting voltages are

$$V_{B1E} = (R_{Bx} + R_{Bi})I_B + R_E I_E; V_{B2E} = R_E I_E \tag{4.52}$$

and

$$\frac{V_{B1E} - V_{B2E}}{I_B} = \frac{\Delta V_{BE}}{I_B} = R_{Bx} + R_{Bi} \tag{4.53}$$

To separate the base resistance into its components, R_B can be written as

$$R_B = R_{Bx} + R_{Bi} = R_{Bx} + \frac{R_{shi}(W_E - 2d)}{3L_E} \tag{4.54}$$

where W_E and L_E are the emitter window width and length, and d describes the deviation between the emitter window and the effective internal base region.[49] The second term on the right side of Eq. (4.54) is discussed earlier with respect to Eq. (4.51). Both R_{Bx} and R_{Bi} can be determined by measuring R_B as a function of W_E for transistors with identical L_E but varying W_E. Such a plot is shown in Fig. 4.18. The sheet resistance R_{shi} is varied by changing the base-emitter bias voltage, due to base conductivity modulation. V_{BE}, however, should not be too high or excessive current crowding will result, but it should be sufficiently high to avoid uncertainties in the potential measurement. The intersection point gives R_{Bx} and 2d. A further refinement of this Kelvin method is given in ref. 50, where more detailed modeling further elucidates the various resistive components.

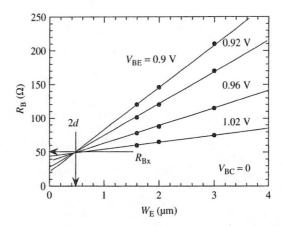

Fig. 4.18 Measured base resistance versus emitter window width as a function of base-emitter voltage. Reprinted after Weng et al.[49] by permission of IEEE (© 1992, IEEE).

Several techniques to measure R_B are based on frequency measurements. In the *input impedance circle method*, the emitter-base input impedance is measured as a function of frequency and is plotted on the complex impedance plane for zero ac collector voltage.[51] The locus of this plot is a semicircle whose real axis intersections at low and high frequencies are

$$R_{in,lf} = R_\pi + R_B + (1+\beta)R_E; R_{in,hf} = R_\pi + R_B \qquad (4.55)$$

Resistance R_π can be calculated from the relationships $R_\pi = \beta/g_m$ with $g_m = qI_C/nkT$. This method allows both R_B and R_E to be determined. The effect of R_π on the measurement of R_B can be reduced by measuring at low temperatures, where R_π is reduced according to the relationships $R_\pi = nkT\beta/qI_C$.[52] The semicircle is sometimes distorted due to parasitic capacitances making the interpretation more difficult. Furthermore, the measurement is very time-consuming and loses accuracy at low collector current when the circle diameter is large. The method is more accurate for $R_B > 40\ \Omega$ and $I \geq 1$ mA.[53]

A variation of this technique is the *phase cancellation method* in which a common base transistor is connected to an impedance bridge, and the input impedance is measured as a function of collector current at a constant frequency of a few MHz. The collector current is varied until the input capacitance becomes zero, and the input impedance is purely resistive at collector current I_{C1}. The input impedance is $Z_i = R_B + R_E$ and the base resistance is given by[51]

$$R_B = \frac{nkT}{qI_{C1}} \qquad (4.56)$$

The phase cancellation method does not lend itself to BJTs with $\beta < 10$ commonly found in lateral pnp transistors, and the base resistance in this method obtains for one value of collector current only. However, the method is fast and relatively unaffected by the emitter resistance, since R_E appears in the input impedance as R_E directly, not as $(\beta+1)R_E$.

In another method the frequency response of $\beta(f)$ and $y_{fb}(f)$, the forward transfer admittance of the BJT in the common base configuration, are measured. The base resistance is[54]

$$R_B = \frac{\beta(0) f_\beta}{y_{fb}(0) f_y} \tag{4.57}$$

where $\beta(0)$ is the low frequency β, $y_{fb}(0)$ the low frequency y_{fb}, f_β the 3 dB frequency of β, and f_y the 3 dB frequency of y_{fb}. The 3 dB frequency is the frequency at which the respective quantity has decreased to 0.7 of its low frequency value. The advantage of this technique is that Eq. (4.57) is relatively unaffected by collector and emitter resistances and that the measurement of y_{fb} is relatively insensitive to stray capacitance. However, it does require measurements of β and y_{fb} over a wide frequency range. In a variation on one of the ac methods, the input impedance of common emitter BJTs is measured at 10 to 50 MHz and R_{Bi}, R_{Bx}, and R_E are extracted from the measurement.[55] The method is suitable for low base-emitter voltages with negligible high current effects. A further variation using a single frequency but varying the emitter-base voltage allows not only the base and emitter resistances but also the base-emitter and the base-collector capacitances to be determined.[56]

The base resistance can also be determined from a pulse measurement similar to the method shown in Fig. 4.5. The base current of a common emitter BJT is pulsed to zero, and the resulting V_{BE} is determined.[57] The base resistance is determined from the sudden drop of the emitter-base voltage $\Delta V_{BE} = R_B I_B$. A cautionary note: extraction of resistances using methods involving kT/q will be in error if self heating causes temperature variations in the device, even with temperature-controlled probe stations.

4.6 MOSFETs

4.6.1 Series Resistance and Channel Length–Current-Voltage

The MOSFET *source/drain series resistance* and the *effective channel length* or *width* are frequently determined with one measurement technique. The resistance between source and drain consists of source resistance, channel resistance, drain resistance, and contact resistances. The source resistance R_S and drain resistance R_D are shown in Fig. 4.19. They are due to the source and drain contact resistance, the sheet resistance of the source and drain, the spreading resistance at the transition from the source diffusion to the channel, and any additional "wire" resistance. The channel resistance is contained in the MOSFET symbol and is not explicitly shown.

Current crowding in the source in the vicinity of the channel gives rise to the spreading resistance R_{sp}. A first-order expression for R_{sp} for a source of constant resistivity is given by

$$R_{sp} = \frac{0.64\rho}{W} \ln\left(\frac{\xi x_j}{x_{ch}}\right) \tag{4.58}$$

where W is the channel width, ρ the source resistivity, x_j the junction depth, x_{ch} the channel thickness, and ξ is a factor that has been given as 0.37,[58] 0.58,[59], 0.75,[60] and 0.9.[61] Its exact value is not that important since it appears in the "ln" term. More realistic expressions for R_{sp} have been derived for junctions with non-uniform dopant profiles.[58]

The effective channel length differs from the mask-defined gate length and even from the physical device gate length due to source and drain junction encroachment under the

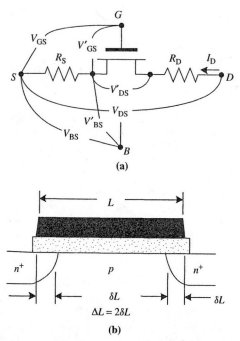

Fig. 4.19 (a) A MOSFET with source and drain resistances, (b) device cross section showing the actual gate length L and $L_{eff} = L - \Delta L$ with $\Delta L = 2\delta L$. The substrate resistance is not shown.

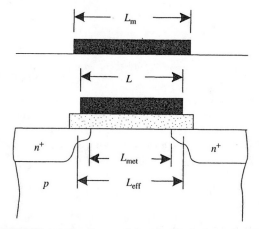

Fig. 4.20 Various MOSFET gate lengths: mask length, physical gate length, metallurgical, and effective channel lengths.

gate, as shown in Fig. 4.20, where L_m is the mask-defined gate length, L the physical gate length, L_{met} the metallurgical channel length (distance between source and drain), and L_{eff} the effective channel length. The effective or electrical channel length is often thought to be the distance between source and drain, *i.e.*, $L_{eff} = L_{met}$. That is not always the case. For highly doped source and drain with steep doping density gradients, the effective length is approximately equal to the physical length between source and drain. However, for *lightly-doped drain* (LDD) structures, the effective length can be larger than the source/drain spacing, because the channel can extend into the lightly-doped source and drain, especially for high gate voltages. L_{eff} can be thought of as that channel length that gives good agreement between theory and experiment when it is substituted into appropriate model equations.

Neglecting the body effect of the ionized bulk charge in the MOSFET space-charge region, the MOSFET current-voltage equation, valid for low drain voltage, is

$$I_D = k(V'_{GS} - V_T - 0.5V'_{DS})V'_{DS} \qquad (4.59)$$

where $k = W_{eff}\mu_{eff}C_{ox}/L_{eff}$, $W_{eff} = W - \Delta W$, $L_{eff} = L - \Delta L$, V_T is the threshold voltage, V'_{GS} and V'_{DS} are defined in Fig. 4.19(a), W is the gate width, L the gate length, C_{ox} the oxide capacitance/unit area, and μ_{eff} the effective mobility. W and L usually refer to the mask dimensions.

With $V_{GS} = V'_{GS} + I_D R_S$ and $V_{DS} = V'_{DS} + I_D(R_S + R_D)$, Eq. (4.59) becomes

$$I_D = k(V_{GS} - V_T - 0.5V_{DS})(V_{DS} - I_D R_{SD}) \qquad (4.60)$$

if $R_S = R_D = R_{SD}/2$, where $R_{SD} = R_S + R_D$. For these measurements the drain voltage is usually low ($V_{DS} \approx 50-100$ mV) ensuring device operation in the *linear* region. For the device in strong inversion, with $(V_{GS} - V_T) \gg 0.5V_{DS}$, Eq. (4.60) becomes

$$I_D = k(V_{GS} - V_T)(V_{DS} - I_D R_{SD}) \qquad (4.61)$$

which can be written as

$$I_D = \frac{W_{eff}\mu_{eff}C_{ox}(V_{GS} - V_T)V_{DS}}{(L - \Delta L) + W_{eff}\mu_{eff}C_{ox}(V_{GS} - V_T)R_{SD}} \qquad (4.62)$$

Equation (4.62) is the basis for most techniques to determine R_{SD}, μ_{eff}, L_{eff}, and W_{eff}. We will discuss the most relevant methods here. The techniques usually require at least two devices of different channel lengths. Comparisons of the various techniques are given by Ng and Brews[62], McAndrew and Layman[63], and Taur.[64] We should make a comment here regarding the threshold voltage V_T which is used in many of the following techniques. As shown later in Section 4.8, one method to determine V_T is the linear extrapolation method. In this technique, $V_T = V_{GSi} - V_{DS}/2$, but the $V_{DS}/2$ term is neglected in Eq. (4.62), leading to some error.

An early method is due to Terada and Muta,[65] and Chern et al.,[66] with $R_m = V_{DS}/I_D$

$$R_m = R_{ch} + R_{SD} = \frac{L - \Delta L}{W_{eff}\mu_{eff}C_{ox}(V_{GS} - V_T)} + R_{SD} \qquad (4.63)$$

where R_{ch} is the channel resistance, *i.e.*, the intrinsic resistance of the MOSFET. Equation (4.63) gives $R_m = R_{SD}$ for $L = \Delta L$. A plot of R_m versus L for devices with differing L and for varying gate voltages in Fig. 4.21, has lines intersecting at one point giving both R_{SD} and ΔL. Which gate lengths should be used in these plots? Should it

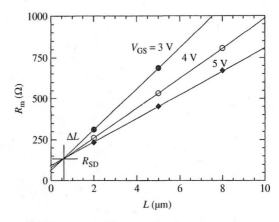

Fig. 4.21 R_m versus L as a function of gate voltage.

be the mask-defined gate lengths or the physical gate lengths? It does not matter. These methods give a ΔL such that L_{eff} will be the correct value, regardless which L is used.

If the R_m versus L lines fail to intersect at a common point, one can carry this technique one step further by writing Eq. (4.63) as

$$R_m = R_{SD} + AL_{eff} = (R_{SD} - A\Delta L) + AL = B + AL \qquad (4.64)$$

The parameters A and B are determined from slope and intercept of R_m versus L plots for different gate voltages. ΔL and R_{SD} are obtained from the slope and intercept, respectively, of a B versus A plot.[67] A and B depend implicitly on $(V_{GS} - V_T)$ and they can be fitted for various gate voltages with a least squares technique. Such a linear regression can be used to extract both R_{SD} and ΔL, with no requirement for a common intersection point.[68] It is, however, assumed that both ΔL and R_{SD} and are only weakly V_{GS} dependent, required for a linear equation. Since neither ΔL nor R_{SD} are fully gate voltage independent, the linear regression will give only approximate results.

The difference between L and L_{eff} is particularly important for short-channel devices. But short-channel devices also have a channel length-dependent threshold voltages, so that each threshold voltage must be determined independently. Furthermore, both the series resistance and the effective channel length may be gate voltage dependent.[69] The effective channel length increases and the series resistance decreases with increasing gate voltage, due to channel broadening in which L_{eff} is modulated by the gate voltage. The effective channel is considered to lie between the transitional points where the current flows from the lateral spread of the source and the drain diffusion to the inversion layer. The end of the channel is where the conductivity of the diffusion resistance is approximately equal to the incremental inversion layer conductivity. Since the inversion layer conductivity increases with gate voltage, it follows that L_{eff} increases and the series resistance decreases with increasing gate voltage.

The dependence of L_{eff} and R_{SD} on gate voltage is particularly acute for LDD devices, containing lightly-doped regions between the source and the channel and between the drain and the channel.[70] The effect of gate voltage-dependent L_{eff} and R_{SD} is one reason for the failure of the R_m versus L lines of Fig. 4.21 to intersect at a common point. As a result no unique value of these two parameters can be obtained. A suggested method to ensure that the lines intersect at one point is to vary V_T in Eq. (4.63) instead of varying V_{GS}.[71] This is most conveniently done by varying the substrate bias V_{BS}, maintaining the gate voltage constant at $V_{GS} \approx 1$ to 2 V. Another approach is to confine the gate voltages to small variations from each other. For example, instead of varying V_{GS} by 1 V as in Fig. 4.21, one might vary V_{GS} by 0.1 V. This brings the several intersection points close to one common point. For LDD devices, R_{SD} and L_{eff} also depend on drain voltage, because the drain space-charge region width varies with V_{DS}.[72] This is usually considered to be a minor effect and is frequently neglected.

The substrate bias technique has yielded unreliable data because substrate bias changes the threshold voltage of MOSFETs of different channel lengths by different amounts. A more serious error is introduced by assuming $dL_{eff}/dV_{BS} = 0$. It has been shown that L_{eff} is reduced by V_{BS} and is no longer clearly defined.[62] An improved method is a combination substrate/gate bias technique.[73] The gate voltage of the longest channel device is held constant while its threshold voltage is changed by substrate bias modulation. When measuring the resistance of shorter-channel devices, the gate voltage is reduced by the amount the threshold voltage has decreased from the long-channel value, ensuring constant gate drive for all devices. Yet another variation on the R_m versus L method uses a "paired gate voltage" approach.[74] Two R_m versus L lines are determined for two gate voltages, one being typically 0.5 V lower than the other. The intersection of these two lines gives a good approximation of R_{SD} and L_{eff}. The gate voltage dependence of R_{SD} and L_{eff} can be found using various V_{GS} pairs. In a variation of the paired gate voltage method, ΔL is determined for a given V_T using one short and one long-channel device. A new ΔL is found for a V_T that differs by about 0.1–0.2 V from the original. This is repeated a number of times and ΔL is plotted against V_T. The intercept on the ΔL axis yields the metallurgical channel length L_{met}.[75]

A different representation of Eq. (4.62) is to define the parameter E as[76]

$$E = R_m(V_{GS} - V_T) = \frac{L - \Delta L}{W_{eff}\mu_{eff}C_{ox}} + R_{SD}(V_{GS} - V_T) \qquad (4.65)$$

There are a number of mobility expressions. One of the simplest and one that is frequently used to interpret channel length and width measurements, is

$$\mu_{eff} = \frac{\mu_o}{1 + \theta(V'_{GS} - V_T)} = \frac{\mu_o}{1 + \theta(V_{GS} - I_D R_S - V_T)} \approx \frac{\mu_o}{1 + \theta(V_{GS} - V_T)} \qquad (4.66)$$

The approximation in Eq. (4.66) is valid for $(V_{GS} - V_T) \gg I_D R_S$. Substituting Eq. (4.66) into Eq. (4.65) gives

$$E = \frac{(L - \Delta L)[1 + \theta(V_{GS} - V_T)]}{W_{eff}\mu_o C_{ox}} + R_{SD}(V_{GS} - V_T) \qquad (4.67)$$

From Eq. (4.65) we find the intercept E_{int} and slope m of E versus $(V_{GS} - V_T)$ plots to be

$$E_{int} = \frac{(L - \Delta L)}{W_{eff}\mu_o C_{ox}}; m = \frac{dE}{dV_{GS}} = \frac{(L - \Delta L)\theta}{W_{eff}\mu_o C_{ox}} + R_{SD} \qquad (4.68)$$

E is plotted against $(V_{GS} - V_T)$ as a function of channel length. The slopes of these plots are $m = (L - \Delta L)\theta/W_{eff}\mu_o C_{ox} + R_{SD}$ and the intercepts on the E-axis are

$E_i = (L - \Delta L)/W_{eff}\mu_o C_{ox}$. E_i varies since devices with varying channel lengths are used. Plots of E_i and m versus L give ΔL and R_{SD} from the intercepts and μ_o and θ from the slopes.

A method related to the method of Eq. (4.65), allowing ΔL, R_{SD}, μ_o and θ to be extracted, is that due to De La Moneda et al., based on writing Eq. (4.63) as[77]

$$R_m = \frac{L - \Delta L}{W_{eff}\mu_o C_{ox}(V_{GS} - V_T)} + \frac{\theta(L - \Delta L)}{W_{eff}\mu_o C_{ox}} + R_{SD} \qquad (4.69)$$

with the effective mobility of Eq. (4.66). First R_m is plotted against $1/(V_{GS} - V_T)$ as shown in Fig. 4.22(a). The slope of this plot is $m = (L - \Delta L)/W_{eff}\mu_o C_{ox}$ and the intercept on the R_m axis is $R_{mi} = [R_{SD} + \theta(L - \Delta L)/W_{eff}\mu_o C_{ox}] = R_{SD} + \theta m$. Next m is plotted against L (Fig. 4.22(b)). This plot has a slope of $1/W_{eff}\mu_o C_{ox}$ and an intercept on the L axis of ΔL, allowing μ_o and ΔL to be determined. Lastly, R_{mi} is plotted against m (Fig. 4.22(c)), giving θ from the slope and R_{SD} from the intercept on the R_{mi} axis.

Two devices suffice for these measurements. The channel lengths of the device pair should be selected to minimize the error in ΔL associated with the extrapolation of the m versus L plot because errors in m are magnified by extrapolation. Errors in ΔL are minimized by choosing channel lengths that differ by about a factor of ten. Furthermore, $(V_{GS} - V_T)$ should be chosen to cover a wide range. One bias point should be for low $(V_{GS} - V_T)$ (about 1 V) where $\mu_o C_{ox}$ is dominant. A second bias point should be for high $(V_{GS} - V_T)$ (about 3–5 V), where θ and R_{SD} dominate. As mentioned earlier, R_{SD} is gate voltage dependent for LDD devices. To find this dependence, one can determine ΔL, plot R_m versus L for various $V_{GS} - V_T$ and determine various R_{SD} at $L = \Delta L$. These R_{SD} can then be plotted as a function of $V_{GS} - V_T$ to illustrate this gate voltage dependence.[78]

A variation of the de la Moneda method is a combination of Eq. (4.60) and (4.66) to give[79]

$$I_D = \frac{k_o(V_{GS} - V_T)(V_{DS} - I_D R_{SD})}{1 + \theta(V_{GS} - V_T)} = k_o(V_{GS} - V_T)(V_{DS} - I_D R') \qquad (4.70)$$

where

$$k_o = \frac{W_{eff}\mu_o C_{ox}}{L_{eff}}; R' = R_{SD} + \frac{\theta}{k_o} \qquad (4.71)$$

Differentiating Eq. (4.70) and using the definition for the transconductance gives

$$g_m = \frac{\partial I_D}{\partial V_{GS}}|_{V_{DS}=constant} = \frac{k_o(V_{DS} - I_D R')}{1 + k_o R'(V_{GS} - V_T)} \qquad (4.72)$$

When combined with Eq. (4.70), we obtain

$$\frac{I_D}{\sqrt{g_m}} = \sqrt{k_o V_{DS}}(V_{GS} - V_T) \qquad (4.73)$$

To determine the various device parameters, we plot $I_D/g_m^{1/2}$ versus V_{GS}. The intercept yields the threshold voltage V_T and the slope gives k_o. The relationship

$$\frac{1}{k_o} = \frac{L - \Delta L}{W_{eff}\mu_o C_{ox}} \qquad (4.74)$$

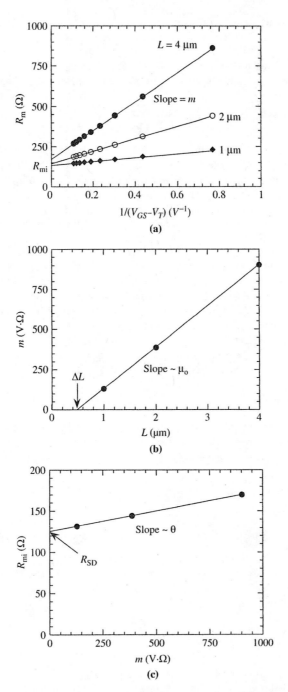

Fig. 4.22   (a) R_m versus $1/(V_{GS} - V_T)$; (b) slope m versus L, and (c) R_{mi} versus m.

suggest a plot of $1/k_o$ versus L. Such a plot has the intercept $L = \Delta L$. R' is obtained from Eq. (4.71). A subsequent plot of R' versus $1/k_o$ yields R_{SD} from the intercept and θ from the slope.

A further variation of Eq. (4.61) for devices with two different channel lengths is the drain current ratio[80]

$$\frac{I_{D1}}{I_{D2}} = \frac{k_1}{k_2}\left(1 - \frac{(I_{D1} - I_{D2})R_{SD}}{V_{DS}}\right) \tag{4.75}$$

for $V_{DS1} \gg I_{D1}R_{SD}$ and $V_{DS2} \gg I_{D2}R_{SD}$ and equal mobilities and equal threshold voltages for the two devices. A plot of I_{D1}/I_{D2} versus $(I_{D1} - I_{D2})$ has a slope of $k_1 R_{SD}/k_2 V_{DS}$ and an intercept on the I_{D1}/I_{D2} axis of k_1/k_2. This method does not work if the conditions $V_{DS1} \gg I_{D1}R_{SD}$ and $V_{DS2} \gg I_{D2}R_{SD}$ are not satisfied. In case these conditions are not satisfied, a modification consists of a plot of $(V_{DS2}/I_{D2} - V_{DS1}/I_{D1})$ versus V_{DS1}/I_{D1},[81] which is linear with an intercept on the V_{DS1}/I_{D1} axis of R_{SD} and a slope $(L_2 - L_1)/(L_1 - \Delta L)$ yielding ΔL.

The transconductance is also used in the *transresistance method*.[82-83] The transconductance g_m and the drain conductance $g_d = \partial I_D/\partial V_{DS}$ are measured in the linear MOSFET region at drain voltages of 25 to 50 mV. The transresistance r is defined by

$$r = \frac{g_m}{g_d^2} \tag{4.76}$$

Two devices are required for the measurement. One is a long-channel device and the other is a short-channel device with known channel length L. The transresistance is determined for each device and a parameter $\Delta\lambda$ is calculated from the two channel lengths and the two transresistances as

$$\Delta\lambda = \frac{Lr_{ref} - L_{ref}r}{r_{ref} - r} \tag{4.77}$$

where $\Delta\lambda$ is plotted against $(V_{GS} - V_T)$ and the extrapolated intercept on the $\Delta\lambda$ axis is ΔL. The series resistance depends on the channel lengths and the drain conductances as

$$R_{SD} = \frac{(L_{ref} - \Delta L)/g_d - (L - \Delta L)/g_{dref}}{L_{ref} - L} \tag{4.78}$$

A comment about techniques that require differentiation: As is well known, differentiation is a noise-producing process, by accentuating small variations in the data. Hence such techniques, *e.g.*, those that require g_d or g_m, tend to be noisier than those not requiring differentiation.

A technique in which the mobility can be any function of gate voltage, and for any R_{SD}, is the *shift and ratio* (S/R) method.[84] It uses one large device and several small devices (varying channel lengths, constant channel width) and starts with Eq. (4.63) rewritten as

$$R_m = R_{SD} + Lf(V_{GS} - V_T) \tag{4.79}$$

where f is a general function of gate overdrive, $V_{GS} - V_T$, common to all devices. Equation (4.79) is differentiated with respect to V_{GS}. The resistance R_{SD} is usually a weak function of gate voltage and its derivative is neglected.

Equation (4.79) becomes

$$S = \frac{dR_m}{dV_{GS}} = L\frac{d[f(V_{GS} - V_T)]}{dV_{GS}} \tag{4.80}$$

S is plotted versus V_{GS} for the large and one small device. To solve for L and V_T, one curve is shifted horizontally by a varying amount δ and the ratio $r = S(V_{GS})/S(V_{GS} - \delta)$ between the two devices is computed as a function of V_{GS}. When S is shifted by a voltage equal to the threshold voltage difference between the two devices, r is nearly constant, which is the key in this measurement. With constant gate overdrive, the mobility is identical or nearly identical, allowing r to be written as

$$r = \frac{S(V_{GS})}{S(V_{GS} - \delta)} = \frac{L_0}{L} \tag{4.81}$$

where L_o and L are the channel lengths of the large and small device, respectively. Plotting the L so obtained versus L_m for several devices gives a line with intercept ΔL on the L_m axis. The method has been successfully used for MOSFETs with channel lengths below 0.2 μm. The best range for V_{GS} is from slightly above V_T to about 1 V above V_T. For LDD devices one should use low gate overdrives to ensure high S allowing dR_{SD}/dV_{GS} to be neglected.[85] Once ΔL is found, R_{SD} can be calculated from Eq. (4.79).

A detailed analysis of various L_{eff} and R_{SD} extraction techniques showed the S/R method to provide the lowest variance and the best accuracy.[85] It is very important, however, to choose a properly optimized gate voltage range in order to satisfy the basic assumption that R_{SD} is V_{GS} independent. It is well known that R_{SD} does depend on V_{GS}, especially near the threshold voltage and in LDD devices. More precise ΔL and V_T extraction is achieved by assuming that $dR_{SD}/dV_{GS} = 0$ only at high gate voltages where ΔL is maximized.[86]

A comprehensive study of the various mechanisms limiting the accuracy of channel length extraction techniques especially for lightly doped drain MOSFETs has shown that low gate overdrives and consistent threshold voltage measurements are very important for reliable channel length extraction.[87]

Other methods of determining the series resistance are based on fitting the current-voltage characteristics using one of several methods. In the least squares method both non-linear and multi-variable least square methods have been used. Two-dimensional device simulators have also been used. A detailed comparison of many of the techniques showed that the various plots, which according to simple theory should be linear, are frequently non-linear.[63] As a result, there are no unique slopes and intercepts rendering the results unreliable. Furthermore, measurement noise can substantially affect intercepts. Experimental noise can sometimes be reduced by using longer integration times during current and voltage measurements. A non-linear optimization procedure gave significantly more accurate and robust results than some of the methods above.[63] A robust method to extract V_T, R_{SD}, ΔL, and ΔW based on optimization using an iterative linear regression procedure has been developed.[88] The parameters are extracted from analytical expressions to a linear set of equations, avoiding differentiations. The method is especially suited to process characterization.

In all methods where series resistance is extracted, it is always R_{SD} that is determined. It is usually assumed that $R_S = R_D$. That may not be always true, especially if a device has been stressed to cause hot electron damage. It is possible to determine the asymmetry between R_S and R_D by measuring the transconductance in the usual MOSFET configuration, i.e., drain is drain and source is source, and in the inverted configuration in which source and drain are interchanged. Combining this measurement with substrate bias and external resistances, allows the asymmetry to be determined.[89]

The conventional current-voltage methods reach their limit when L_{eff} approaches 0.1 μm, because R_{ch} is no longer a linear function of L_{eff} due to short channel effects.

Hence, the key assumption of these methods is no longer satisfied. A method based on an entirely different principle is the *drain-induced barrier lowering* (DIBL) *method.*[90] DIBL, one manifestation of short channel effects, is the threshold voltage reduction with drain voltage, because the drain voltage affects the barrier at the source-substrate junction. In the sub-threshold region, the drain current becomes

$$I_D = I_0 \exp\left(\frac{q(V_{GS} - V_T)}{nkT}\right) \exp\left(\frac{q\lambda V_{DS}}{kT}\right) = I_0 \exp\left(\frac{q(V_{GS} - V_T')}{nkT}\right) \quad (4.82)$$

where λ is the DIBL coefficient and

$$V_T' = V_T - n\lambda V_{DS} \Rightarrow \Delta V_T = V_T' - V_T = -n\lambda V_{DS} \quad (4.83)$$

The effect of DIBL on drain current is shown in Fig. 4.23(a), showing both increased *off* current (I_D at $V_{GS} = 0$) and reduced threshold voltage. The DIBL coefficient is

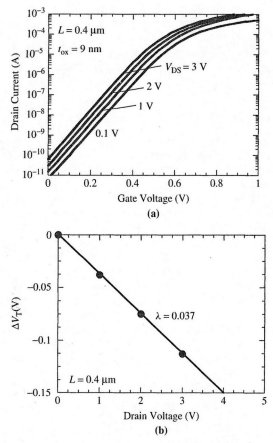

Fig. 4.23 (a) Drain current versus gate voltage as a function of drain voltage illustrating DIBL (b) threshold voltage shift versus drain voltage; the slope gives λ.

determined from the slope of a ΔV_T versus V_{DS} plot, illustrated in Fig. 4.23(b) taking $\Delta V_T = 0$ for $V_{DS} = 0.1$ V.

Drain-induced barrier lowering also depends on the channel length. The shorter the channel the more the drain voltage modulates the source-substrate barrier, suggesting the use of DIBL for effective channel length measurement. The ΔV_T dependence on channel length is[90]

$$\Delta V_T = \alpha + \beta \exp\left(-\frac{L_{eff}}{2L_c}\right) \tag{4.84}$$

where α, β and L_c are constants. The key issue of L_{eff} extraction is to determine these constants. $\alpha = \Delta V_T$ for devices with channel length in the range of 1 $\mu m > L_{eff} > 0.4$ μm. β is determined from the junction built-in and Fermi potentials that depend on the doping density according to

$$\beta = 2\sqrt{(V_{bi} - 2\phi_F)(V_{bi} - 2\phi_F + V_{DS})} \tag{4.85}$$

with β between 0.4 and 0.8 V. The length L_c is determined from

$$L_c = \frac{L_{Ddes1} - L_{Ddes2}}{2[\ln(\Delta V_{T1} - \alpha) - \ln(\Delta V_{T2} - \alpha)]} \tag{4.86}$$

where L_{Ddes} are the design channel lengths of two devices with slightly different channel lengths which should lie between 0.1 and 0.2 μm. The method has been applied for L_{eff} as low as 40 nm.

4.6.2 Channel Length—Capacitance-Voltage

The current-voltage methods of Section 4.6.1 are the most common methods to determine series resistance and effective channel length, largely because of their measurement simplicity. But they do have some limitations, as discussed above. Hence, capacitance techniques are also used to determine L_{eff}. While series resistance cannot be determined by $C-V$ techniques, the measurement is free of ambiguities introduced by series resistance and gate voltage-dependent mobility. We discuss capacitance measurements with reference to the MOSFET in Fig. 4.24.

The capacitance is measured between the gate and the source/drain connected together for devices with varying channel length and wide constant width gates.[91] The substrate is grounded (connected to the shield of the $C-V$ meter cables) to shunt the drain-substrate and source-substrate capacitances from the $C-V$ meter. For $V_G < V_T$, the surface under the gate is accumulated and the capacitance meter reads the two overlap capacitances (Fig. 4.24(a)). For $V_G > V_T$, the surface under the gate is inverted and the capacitance meter reads the two overlap capacitances and the channel capacitance (Fig. 4.24(b)). The effective gate length in this measurement is considered to be the metallurgical channel length L_{met}. C_{ov} and C_{inv} are given by

$$C_{ov} = \frac{K_{ox}\varepsilon_o \Delta L W}{t_{ox}}; C_{inv} = \frac{K_{ox}\varepsilon_o L W}{t_{ox}} \tag{4.87}$$

Rearranging Eq. (4.87) yields L_{met} as

$$L_{met} = L - \Delta L = L\left(1 - \frac{C_{ov}}{C_{inv}}\right) \tag{4.88}$$

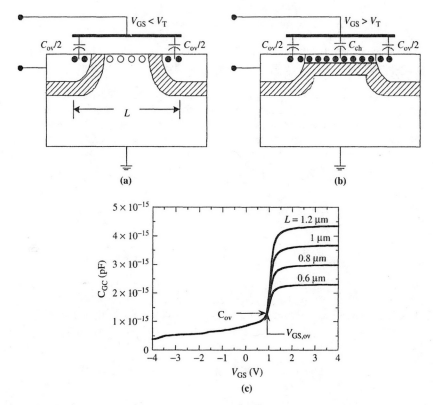

Fig. 4.24 MOSFET for (a) $V_{GS} < V_T$, (b) $V_{GS} > V_T$, and (c) $C_{GC} - V_{GS}$ curves; $W = 10$ μm, $t_{ox} = 10$ nm, $N_A = 1.6 \times 10^{17}$ cm^{-3}.

One can either make a measurement on a single device and use Eq. (4.88) or plot ($C_{inv} - C_{ov}$) versus L, with slope $K_{ox}\varepsilon_o W/t_{ox}$ and intercept ΔL on the L axis. A modified C–V method is given in ref. 92. It has also been applied to DMOSFETs.[93]

At what gate voltage should C_{ov} be measured? Extensive modeling and experimental results place the gate voltage corresponding to C_{ov} at the point where the surface just begins to invert, *i.e.*, $V_{GS} = V_{GS,ov}$ which is near V_T. To determine $V_{GS,ov}$ one measures the capacitance of several devices with different channel lengths. Such curves are shown in Fig. 4.24(c). Then $V_{GS,ov}$ is that gate voltage where the capacitance-gate voltage curves begin to diverge. Figure 4.24(c) shows a single curve in accumulation. Detailed measurements show the curves in accumulation to depend weakly on gate length due to stray capacitances.[94] C_{ov} in the "off" state may differ from that in the "on" state. If it is taken as the capacitance just below the threshold voltage, it contains an unwanted inner fringe term that is absent when the conducting channel is formed. If taken at a negative gate voltage for n-MOSFETs to accumulate the substrate and eliminate the inner fringe component, the overlapped source-drain region can be depleted. Such errors translate into a large error in ΔL for short-channel devices with low intrinsic capacitances.

For small-area MOSFETs the capacitance is very small and the overlap capacitance is still smaller, making for difficult measurements. This problem can be alleviated by connecting many devices in parallel, thereby making the effective area much larger. In one design, 3200 transistors were connected in parallel.[95] A multi-finger gate device for sufficiently high capacitance, may have an offset to the MOSFET device used for I–V characterization due to lithographic proximity effects. For sub-100 nm MOSFETs, the gate oxides are so thin for tunnel currents to be significant, affecting the capacitance measurement.

Once L_{met} is known and if one measures μ_{eff} on a large MOSFET, it is then possible to determine R_{SD} by comparing an ideal with a real device. In this comparison one assumes $L_{met} \approx L_{eff}$. If we use I_D for the drain current of Eq. (4.60) and I_{D0} as the drain current when $R_{SD} = 0$, then by simply taking the ratio $\zeta = I_D / I_{D0}$, R_{SD} is

$$R_{SD} = \frac{(1 - \varsigma)V_{DS}}{I_D} \tag{4.89}$$

In this manner, one can easily generate an R_{SD} versus V_{GS} curve showing the gate voltage dependence of the series resistance.[91]

4.6.3 Channel Width

The methods to determine the channel width W are similar to those for channel length. Several devices with varying gate width and constant gate length are used. An early technique used a plot of the MOSFET drain conductance as a function of W for devices with constant channel length.[96] If source and drain resistances are neglected, then from Eq. (4.60) the drain conductance is

$$g_d = \frac{\partial I_D}{\partial V_{DS}} \bigg|_{V_{GS}=constant} = \frac{(W - \Delta W)\mu_{eff} C_{ox}(V_{GS} - V_T)}{L_{eff}} \tag{4.90}$$

A plot of g_d against W has an intercept on the W-axis of ΔW at $g_d = 0$. This method neglects the source and drain resistances, which is more problematic than it is for channel length measurements. Although it is a reasonably good assumption to take R_S and R_D as constants for devices with varying channel *lengths*, this is no longer true for devices with varying channel *widths*. Both source and drain resistances depend on channel width.

When the drain conductance in Eq. (4.90) is used to extract W_{eff}, it is possible for the intersection point to occur at negative g_d. This can be due to a resistance in parallel with the intrinsic MOSFET due to a leakage path between source and drain at the device periphery. The intersection point yields both W_{eff} and G_P, the parallel conductance.[97]

The drain current can be written as (see Eq. (4.62))

$$I_D = \frac{(W - \Delta W)\mu_{eff} C_{ox}(V_{GS} - V_T)V_{DS}}{L_{eff} + (W - \Delta W)\mu_{eff} C_{ox}(V_{GS} - V_T)R_{SD}} \tag{4.91}$$

Plotting I_D versus W gives $W = \Delta W$ for $I_D = 0$. This has been used to determine W_{eff}.[98]

The measured drain resistance is [see Eq. (4.63)] .

$$R_m = R_{ch} + R_{SD} = \frac{L_{eff}}{(W - \Delta W)\mu_{eff} C_{ox}(V_{GS} - V_T)} + R_{SD} \tag{4.92}$$

The slope of R_m versus $1/(V_{GS} - V_T)$ is $m = L_{eff}/(W - \Delta W)\mu_{eff} C_{ox}$. An mW versus m plot has the slope ΔW.[99] Even if R_{SD} varies with W, it does not vary with L, and differentiating Eq. (4.92) with respect to L gives

$$m = \frac{1}{dR_m/dL} = (W - \Delta W)\mu_{eff}C_{ox}(V_{GS} - V_T) \tag{4.93}$$

Plotting m versus W gives the intercept $W = \Delta W$ at $m = 0$. Both methods require devices of varying gate widths with constant gate length. By varying the gate voltage, it is possible to generate data for W_{eff} as a function of V_{GS}.

A technique using non-linear optimization, similar to that for L_{eff} determination in ref. 63, can also be used for W_{eff} extraction.[100] The drain current is measured for devices with varying widths and constant length and varying lengths and constant width. A non-linear optimization model is fit to the data accounting for the width-dependent V_T, R_{SD}, and W_{eff}. The method is robust, does not assume a linear model, and does not suffer from extrapolation errors in the presence of non-linear or noisy data.

A method that does not rely on current-voltage measurements, not affected by series resistance, is the *capacitance method*. The oxide capacitance of a MOSFET is given by

$$C_{ox} = \frac{K_{ox}\varepsilon_o L_{eff}(W - \Delta W)}{t_{ox}} \tag{4.94}$$

A plot of C_{ox} as a function of W for transistors with identical gate lengths but varying widths gives a straight line with slope $K_{ox}\varepsilon_o L_{eff}/t_{ox}$ and intercept on the W-axis at $W = \Delta W$.[101]

4.7 MESFETs AND MODFETs

A MESFET (metal-semiconductor field-effect transistor) consists of a source, channel, drain, and gate. *Majority* carriers flow from source to drain in response to a drain voltage. The drain current is modulated by a reverse bias on the metal-semiconductor junction gate. With sufficient reverse bias, the space-charge region of the metal-semiconductor contact extends to the insulating substrate and the channel is pinched off. The output current-voltage characteristics resemble those of depletion-mode MOSFETs. However, in contrast to MOSFETs, the MESFET metal-semiconductor junction can be forward biased, leading to high input currents. A MODFET (modulation-doped FET), shown in Fig. 4.25, is similar to a MESFET, with a wide band gap semiconductor interposed between the

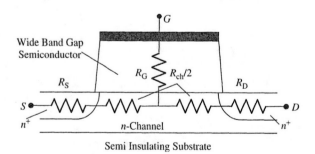

Fig. 4.25 Cross-section of a MODFET showing the various resistances. R_G is the resistance of the wide band gap semiconductor.

n-channel and the gate; in a MESFET the gate is placed directly on the n-channel. We will not distinguish between these two structures.

The ability to forward bias the gate of a MESFET allows additional measurements that are not possible with a MOSFET. With the gate forward biased, the drain-source voltage is

$$V_{DS} = (R_{ch} + R_S + R_D)I_D + (\alpha R_{ch} + R_S)I_G \tag{4.95}$$

where α accounts for the fact that the gate current flows only through a portion of the channel resistance from the gate to the source; $\alpha \approx 0.5$. The gate-source voltage is

$$V_{GS} = \frac{nkT}{q} \ln \left(\frac{I_G}{I_s} \right) + R_S(I_D + I_G) \tag{4.96}$$

where $I_G = I_s \exp(qV_{GS}/nkT)$ is the forward-biased gate Schottky diode current with zero resistance.

I_D versus V_{DS} as a function of I_G has a slope of $1/(R_{ch} + R_S + R_D)$, and V_{DS}/I_G gives $(\alpha R_{ch} + R_S)$ for $I_D = 0$. Furthermore, from the forward-biased $I_G - V_{GS}$ curves as a function of I_D, $\Delta V_{GS}/\Delta I_D = R_S$ for $I_G = constant$, allowing R_S, R_D, and R_{ch} to be determined. When the gate resistance R_G is included, it is determined from the gate current with a voltage between gate and source. However, $\log(I_G)$ is plotted against V_{GD}, not V_{GS}, with the drain open circuited. A deviation of this semilog plot from a straight line is caused by the gate resistance.

Another method relies on a measure of the gate current as a function of the drain-source voltage. The source is grounded and the gate current flows from the gate to the source. The gate current flowing through the source resistance and through a portion of the channel resistance r_{ch} creates a voltage drop. The drain acts as a voltage probe of this voltage drop. The "end" resistance is defined as

$$R_{end} = \frac{\partial V_{DS}}{\partial I_G} \tag{4.97}$$

From Eq. (4.97) the "end" resistance is approximately

$$R_{end} = \alpha R_{ch} + R_S \tag{4.98}$$

In one "end" resistance measurement method, the drain current is zero and the drain contact floats electrically. This gives $\alpha \approx 0.5$. In another version, drain current does flow, but it is constant during the measurement, and the drain does not float. For $I_G \ll I_D$,[102]

$$R_{end} = R_S + \frac{nkT}{qI_D} \tag{4.99}$$

A plot of R_{end} versus $1/I_D$ has a slope nkT/q and an intercept R_S on the R_{end} axis. This plot has a rather limited straight-line portion. Deviation from a straight line at high I_D is the result of the drain current being not much lower than the saturation drain current. At low I_D there is a deviation due to a violation of the $I_G \ll I_D$ requirement, rendering the method of rather limited usefulness. A refinement of this method is given in Chaudhuri and Das.[103]

The transmission line method, discussed in detail in Chapter 3, has also been used for R_S measurement. The technique yields the sheet resistance of the n-channel, from which the source resistance can be calculated, knowing the device dimensions. A disadvantage of this method is the absence of the gate on the TLM structure. Consequently, spreading resistance due to current crowding at the source end cannot be accurately measured.

In another technique, devices with varying channel lengths are used with the devices operated in their linear region.[104] Current-voltage measurements are made with one of the contacts floating. With the gate floating electrically, the various resistances are

$$R_{GS}(fg) = R_S + R_{ch}/2; R_{GD}(fg) = R_D + R_{ch}/2; R_{SD}(fg) = R_S + R_D + R_{ch}$$
(4.100)

A small current is forced from source to drain and the voltage drop between the floating gate and the source is measured with a high-impedance voltmeter to give R_{GS}. Similarly for the other resistances. With the source floating,

$$R_{GS}(fs) = R_G$$
(4.101)

We define

$$R_{ch} = RL_G; R_G = \frac{1}{GL_G}$$
(4.102)

where R represents the channel resistance per unit length of channel and G represents the gate-to-channel conductance/unit length of channel. Substituting Eq. (4.102) into (4.100) and (4.101) gives

$$R_{GS}(fg) = R_S + \frac{R}{2GR_{GS}(fs)}; R_{GD}(fg) = R_D + \frac{R}{2GR_{GS}(fs)};$$

$$R_{SD}(fg) = R_S + R_D + \frac{R}{GR_{GS}(fs)}$$
(4.103)

Plots of $R_{GS}(fg)$, $R_{GD}(fg)$, and $R_{SD}(fg)$ versus $1/R_{GS}(fs)$ are linear with intercepts on the vertical axes of R_S, R_D, and $R_S + R_D$. Examples of such plots are shown in Fig. 4.26. The method can be checked by plotting $1/R_{SD}(fs)$ versus L_m, the mask-defined or drawn channel length. Such a plot should yield a straight line with an intercept

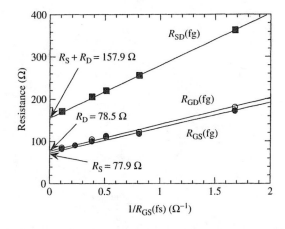

Fig. 4.26 Plots of $R_{GS}(fg)$, $R_{GD}(fg)$, and $R_{SD}(fg)$ versus $1/R_{GS}(fs)$. Reprinted after Azzam et al.[104] by permission of IEEE (© 1990, IEEE).

at $L_m = 0$. Another method uses two drain currents at constant gate current with the gate forward biased. The shift in the $I_G - V_{GS}$ curves corresponding to these two conditions is related to the source resistance.[105] A technique, related to the end contact resistance method, uses the gate electrode instead of the source and drain contacts to measure the source and drain resistances.[106]

4.8   THRESHOLD VOLTAGE

Before discussing threshold voltage measurement techniques, we briefly discuss the concept of threshold voltage. A good overview of threshold voltage measurement techniques is given in ref. 107. The threshold voltage V_T is an important MOSFET parameter required for the channel length/width and series resistance measurements of this chapter. However, V_T is not uniquely defined. Various definitions exist and the reason for this can be found in the $I_D - V_{GS}$ curves of Fig. 4.27. Fig. 4.27(a) shows the $I_D - V_{GS}$ curve of a MOSFET, illustrating the non-linear nature of this curve. Figure 4.27(b) gives an expanded view

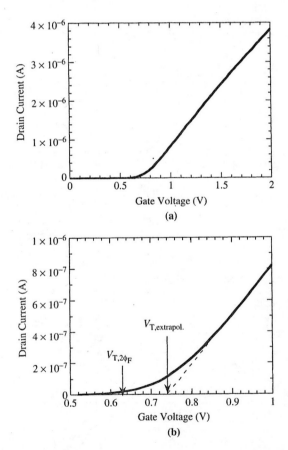

Fig. 4.27 $I_D - V_{GS}$ curve of a MOSFET near the threshold voltage; (b) is an enlarged portion of (a). Modeled using $L_{eff} = 1.5$ μm, $t_{ox} = 25$ nm, $V_{T,start} = 0.7$ V, $V_D = 0.1$ V.

showing the curve near the threshold voltage. There is clearly no unique gate voltage at which drain current begins to flow. A commonly used definition of threshold voltage is that gate voltage for which the surface potential, ϕ_s, in the semiconductor below the gate oxide is given by

$$\phi_s = 2\phi_F = \frac{2kT}{q} \ln\left(\frac{p}{n_i}\right) \approx \frac{2kT}{q} \ln\left(\frac{N_A}{n_i}\right) \tag{4.104}$$

for an n-channel MOSFET. This definition, first proposed in 1953,[108] is based on equating the surface minority carrier density to the majority carrier density in the neutral bulk, *i.e.*, n(surface) $= p$(bulk) and is shown as $V_{T,2\phi F}$ in Fig. 4.27(b). Clearly, it is well below the extrapolated threshold voltage, $V_{T,extrapol}$.

The threshold voltage for large-geometry, n-channel devices on uniformly doped substrates with no short- or narrow-channel effects, when measured from gate to source and the $\phi_s = 2\phi_F$ definition, is

$$V_T = V_{FB} + 2\phi_F + \frac{\sqrt{2qK_s\varepsilon_o N_A(2\phi_F - V_{BS})}}{C_{ox}} \tag{4.105}$$

where V_{BS} is the substrate-source voltage and V_{FB} is the flatband voltage. The threshold voltage for non-uniformly doped, ion-implanted devices depends on the implant dose as well. Additional corrections obtain for short- and narrow-channel devices.

4.8.1 Linear Extrapolation

A common threshold voltage measurement technique is the *linear extrapolation* method with the drain current measured as a function of gate voltage at a low drain voltage of typically 50–100 mV to ensure operation in the linear MOSFET region.[109–111] According to Eq. (4.60) the drain current is zero for $V_{GS} = V_T + 0.5V_{DS}$. But Eq. (4.60) is valid only above threshold. The drain current is not zero below threshold and approaches zero only asymptotically. Hence the I_D versus V_{GS} curve is extrapolated to $I_D = 0$, and the threshold voltage is determined from the extrapolated or intercept gate voltage V_{GSi} by

$$V_T = V_{GSi} - V_{DS}/2 \tag{4.106}$$

Equation (4.106) is strictly only valid for negligible series resistance.[112] Fortunately series resistance is usually negligible at the low drain currents where threshold voltage measurements are made, but it can be appreciable in LDD devices. The linear extrapolation technique can also be used for threshold voltage measurements of depletion-mode or buried channel MOSFETs.[113]

The $I_D - V_{GS}$ curve deviates from a straight line at gate voltages below V_T due to sub-threshold currents and above V_T due to series resistance and mobility degradation effects. It is common practice to find the point of maximum slope on the $I_D - V_{GS}$ curve by a maximum in the transconductance, fit a straight line to the $I_D - V_{GS}$ curve at that point and extrapolate to $I_D = 0$, as illustrated in Fig. 4.28. According to Eq. (4.106), $V_T = 0.9$ V for this device. The linear extrapolation method is sensitive to series resistance and mobility degradation.[87, 112, 114]

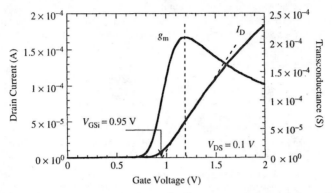

Fig. 4.28 Threshold voltage determination by the linear extrapolation technique. $V_{DS} = 0.1$ V, $t_{ox} = 17$ nm, $W/L = 20$ μm/0.8 μm. Data courtesy of M. Stuhl, Medtronic Corp.

Exercise 4.1

Problem: Does the linearly extrapolated threshold voltage depend on series resistance R_{SD}? Assume μ_{eff} to be independent of V_{GS}.

Solution: First consider the case for $R_{SD} = 0$. As in the linear extrapolation method, the maximum slope of the $I_D - V_{GS}$ curve, the transconductance $g_{m,max}$ is determined. From Fig. E4.1

$$V_{GSi} = V_{GS,max} - \frac{I_{D,max}}{g_{m,max}}, \text{ where}$$

$$I_{D,max} = k(V_{GS,max} - V_T - V_{DS}/2)V_{DS} \text{ and } g_{m,max} = kV_{DS}; k = \frac{W_{eff}\mu_{eff}C_{ox}}{L_{eff}}$$

Substituting $I_{D,max}$ and $g_{m,max}$ into the first equation, and solving for V_T gives

$$V_T = V_{Gsi} - V_{DS}/2, \text{ identical to Eq. (4.106). From Eq. (4.60) with } R_{SD} \neq 0;$$

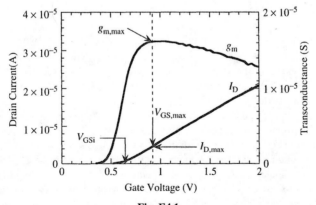

Fig. E4.1

$$I_{D,\max} = k(V_{GS,\max} - V_T - V_{DS}/2)(V_{DS} - I_{D,\max}R_{SD})$$

$$= \frac{k(V_{GS,\max} - V_T - V_{DS}/2)V_{DS}}{1 + kR_{SD}(V_{GS,\max} - V_T - V_{DS}/2)}$$

and

$$g_{m,\max} = \frac{kV_{DS}}{[1 + kR_{SD}(V_{GS,\max} - V_T - V_{DS}/2)]^2}$$

Substituting $I_{D,\max}$ and $g_{m,\max}$ into the V_{Gsi} equation above gives

$$V_{GSi} = V_T + V_{DS}/2 - kR_{SD}(V_{GS,\max} - V_T - V_{DS}/2)^2$$

Solving for the threshold voltage gives

$$V_T = V_{GS,\max} - \frac{V_{DS}}{2} + \frac{1 - \sqrt{1 + 4kR_{SD}(V_{GS,\max} - V_{GSi})}}{2kR_{SD}}$$

Expanding this expression, using $\sqrt{1 + x} \approx 1 + \dfrac{x}{2} - \dfrac{x^2}{8} + \dfrac{3x^3}{48}$ gives

$$V_T \approx V_{GSi} - V_{DS}/2 + kR_{SD}(V_{GS,\max} - V_{GSi})^2 - 2(kR_{SD})^2(V_{GS,\max} - V_{GSi})^3$$

The threshold voltage can also be determined in the MOSFET saturation regime. The drain current in saturation for mobility-dominated MOSFETs is

$$I_{D,sat} = \frac{mW\mu_{eff}C_{ox}}{L}(V_{GS} - V_T)^2 \tag{4.107}$$

where m is a function of doping density; it approaches 0.5 for low doping densities. V_T is determined by plotting $I_D^{1/2}$ versus V_{GS} and extrapolating the curve to zero drain current, illustrated in Fig. 4.29(a).[115-116] Since I_D is dependent on mobility degradation and series resistance, we again extrapolate at the point of maximum slope. Setting $V_{GS} = V_{DS}$ ensures operation in the saturation region.

For short-channel MOSFETs, where the drain current is velocity saturation limited, the saturated drain current is

$$I_D = WC_{ox}(V_{GS} - V_T)v_{sat} \tag{4.108}$$

where v_{sat} is the saturation-limited velocity. The drain current in Eq. (4.108) is linear in $V_{GS} - V_T$ as shown in Fig. 4.29(b). The threshold voltage now is simply the extrapolated gate voltage.

4.8.2 Constant Drain Current

It is obvious from Fig. 4.27 that the drain current at the threshold voltage is higher than zero. This is utilized in the *constant drain current* method where the gate voltage at a specified threshold drain current, I_T, is taken to be the threshold voltage. This measurement is simple with only one voltage measurement necessary and it can be implemented with the circuit of Fig. 4.30(a) or by digital means.[115] It lends itself readily to threshold voltage mapping. The threshold current I_T is forced at the MOSFET source terminal and the op-amp adjusts its output voltage to equal the gate voltage consistent with that I_T.

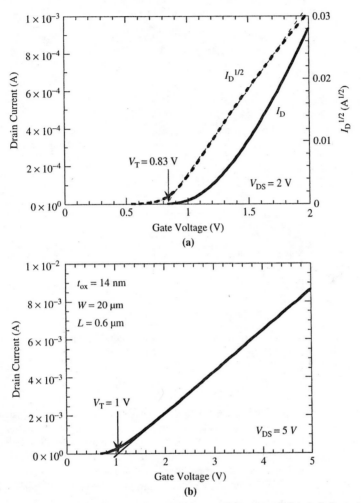

Fig. 4.29 Threshold voltage determination by the saturation extrapolation technique. (a) $V_{DS} = 2$ V, $t_{ox} = 17$ nm, $W/L = 20$ μm/0.8 μm. (b) saturation limited velocity case. Data courtesy of M. Stuhl, Medtronic Corp.

In order to make I_T independent of device geometry, $I_T = I_D/(W_{eff}/L_{eff})$ is sometimes specified at a current around 10 to 50 nA but other values have been used.[114–115] V_T for $I_D = 1$ μA, often used in this type of measurement, is shown in Fig. 4.30(b). Also shown is the "linear extrapolation" V_T. The method has found wide application, provided a consistent drain current is chosen.

4.8.3 Sub-threshold Drain Current

In the *sub-threshold* method the drain current is measured as a function of gate voltage below threshold and plotted as $\log(I_D)$ versus V_{GS}. The sub-threshold current depends

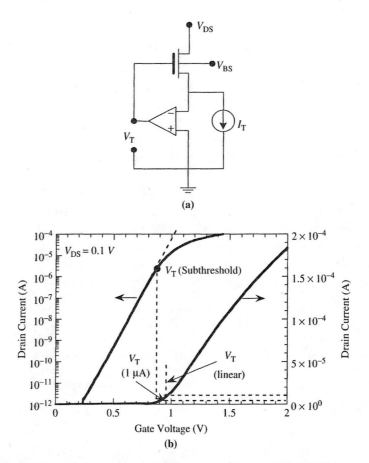

Fig. 4.30 Threshold voltage determination by the sub-threshold and the threshold drain current technique. (a) Measurement circuit, (b) experimental data. $t_{ox} = 17$ nm, $W/L = 20$ μm/0.8 μm. Data courtesy of M. Stuhl, Medtronic Corp.

linearly on gate voltage in such a semilog plot. The gate voltage at which the plot departs from linearity is sometimes taken as the threshold voltage. However, for the data of Fig. 4.30(b) this point yields a threshold voltage of $V_T = 0.87$ V, somewhat lower than that determined by the linear extrapolation method ($V_T = 0.95$ V).

4.8.4 Transconductance

The *transconductance* method uses a linear extrapolation of the $g_m - V_{GS}$ characteristic at its maximum first derivative point.[117] In weak inversion, the transconductance depends exponentially on gate bias, but in strong inversion, if series resistance and mobility degradation are negligible, the transconductance tends to a constant value. In the transition region between weak and strong inversion, the transconductance depends linearly on gate bias. Fig. 4.31 shows an example of this technique with $V_T = 0.83$ V, lower than the previous techniques.

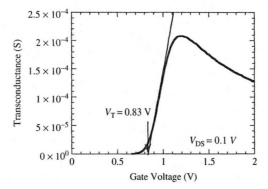

Fig. 4.31 Threshold voltage determination by the transconductance technique. $t_{ox} = 17$ nm, $W/L = 20$ μm/0.8 μm. Data courtesy of M. Stuhl, Medtronic Corp.

4.8.5 Transconductance Derivative

The derivative of the transconductance with gate voltage $\partial g_m / \partial V_{GS}$ is determined at low drain voltage and plotted versus gate voltage in the *transconductance derivative* method. The origin of this method can be understood by considering an ideal MOSFET, where $I_D = 0$ for $V_{GS} < V_T$ and $I_D \sim V_{GS}$ for $V_{GS} > V_T$. Hence the first derivative dI_D/dV_{GS} is a step function and the second derivative $d^2 I_D = dV_{GS}^2$ will tend to infinity at $V_{GS} = V_T$. In a real device the second derivative is not infinite, but exhibits a maximum. An example plot is shown in Fig. 4.32 for the device of Fig. 4.28. The threshold voltage is about the same as for the method in Figs. 4.28. The method is not affected by series resistance and mobility degradation.[112]

4.8.6 Drain Current Ratio

The *drain current ratio* method was developed to avoid the dependence of the extracted V_T on mobility degradation and parasitic series resistance.[114] The drain current, given in Eq. (4.62), is reproduced here

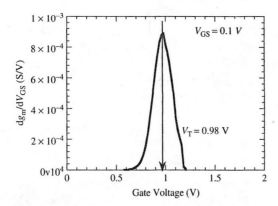

Fig. 4.32 Threshold voltage determination by the transconductance change technique. $t_{ox} = 17$ nm, $W/L = 20$ μm/0.8 μm. Data courtesy of M. Stuhl, Medtronic Corp.

$$I_D = \frac{W_{eff}\mu_{eff}C_{ox}(V_{GS} - V_T)V_{DS}}{(L - \Delta L) + W_{eff}\mu_{eff}C_{ox}(V_{GS} - V_T)R_{SD}} \tag{4.109}$$

Using

$$\mu_{eff} = \frac{\mu_o}{1 + \theta(V_{GS} - V_T)} \tag{4.110}$$

allows Eq. (4.109) to be written as

$$I_D = \frac{WC_{ox}}{L}\frac{\mu_o}{1 + \theta_{eff}(V_{GS} - V_T)}(V_{GS} - V_T)V_{DS} \tag{4.111}$$

where

$$\theta_{eff} = \theta + (W/L)\mu_o C_{ox}R_{SD} \tag{4.112}$$

The transconductance is given by

$$g_m = \frac{\partial I_D}{\partial V_{GS}} = \frac{WC_{ox}}{L}\frac{\mu_o}{[1 + \theta_{eff}(V_{GS} - V_T)]^2}V_{DS}$$

The $I_D/g_m^{1/2}$ ratio

$$\frac{I_D}{\sqrt{g_m}} = \sqrt{\frac{WC_{ox}\mu_o}{L}V_{DS}(V_{GS} - V_T)} \tag{4.113}$$

is a linear function of gate voltage, whose intercept on the gate-voltage axis is the threshold voltage. This method is valid provided the gate voltage is confined to small variations near V_T and the assumptions $V_{DS}/2 \ll (V_{GS} - V_T)$ and $\partial R_{SD}/\partial V_{GS} \approx 0$ are satisfied. The plot is shown in Fig. 4.33 giving $V_T = 0.97$ V. The low-field mobility μ_o can be determined from the slope of the $I_D - g_m^{1/2}$ versus $V_{GS} - V_T$ plot and the mobility degradation factor is

$$\theta_{eff} = \frac{I_D - g_m(V_{GS} - V_T)}{g_m(V_{GS} - V_T)^2} \tag{4.114}$$

from which θ can de determined provided R_{SD} is known.

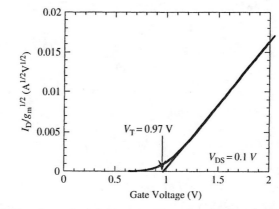

Fig. 4.33 Threshold voltage determination by the drain current/transconductance technique. $t_{ox} = 17$ nm, $W/L = 20$ μm/0.8 μm. Data courtesy of M. Stuhl, Medtronic Corp.

A comparison of several methods was carried out as a function of channel length.[118] The results are shown in Fig. 4.34. It is clear from this plot, as it is from the data in this section, that the threshold voltage can vary widely depending on how it is measured. In all threshold voltage measurements it is important to state the sample measurement temperature since V_T does depend on temperature. A typical V_T temperature coefficient is -2 mV/°C, but it can be higher.[119]

4.9 PSEUDO MOSFET

The *pseudo MOSFET* is a simple test structure to characterize the Si layer of silicon-on-insulator (SOI) wafers without having to fabricate test devices.[120] The original implementation is illustrated in Fig. 4.35(a), with the bulk Si substrate the "gate", the buried oxide (BOX) the "gate" oxide, and the Si film the transistor "body". Mechanical probes on the film surface form the *source* and *drain*. Biasing the gate drives the Si at the bottom interface into inversion, depletion, or accumulation, allowing both *electron* and *hole* conduction to be characterized. Drain current-gate voltage and drain current-time measurements yield the effective electron *and* hole mobilities, threshold voltage, dopant type, dopant density, interface and oxide charge densities, series resistance, and layer defects. To reduce the effect of BOX leakage due to BOX defects, it is advantageous to etch the Si layer into islands.

A more recent implementation is the mercury probe HgFET in Fig. 4.35(b), with Hg the source S, the concentric drain D, and the concentric guard ring GR.[121] While changing the probe configuration from Fig. 4.35(a) to Fig. 4.35(b) may appear to be trivial, this change is actually quite profound. In the two-probe configuration, the probe contact resistance and contact area depend on the probe pressure that may be difficult to control. The Hg probe configuration has well-defined source and drain areas, as well as a guard ring to suppress surface leakage currents. However, the HgFET relies on Hg-Si interfaces, *i.e.*, Schottky barrier source and drain. It turns out that the Hg-Si interface is

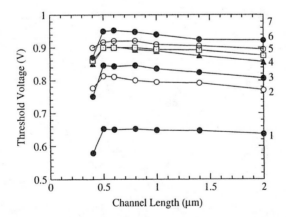

Fig. 4.34 Threshold voltage versus channel length determined by various methods: 1: constant current for $I_D = 1$ nA/(W/L), 2: transconductance, 3: saturation drain current extrapolation, 4: V_{GS} where $d^2 \log I_D/dV_{GS}{}^2$ is a minimum, 5: drain current linear extrapolation, 6: transconductance derivative, 7: linear extrapolation corrected for mobility. From ref. 118.

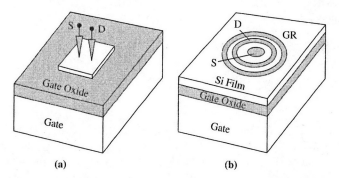

Fig. 4.35 Pseudo MOSFETs (a) probe and (b) Hg contact configurations.

very sensitive to surface treatment and this interface is extremely important during HgFET measurements. A common method to control the Hg-Si barrier, is to rinse the Si sample in dilute HF (*e.g.*, 1 HF:20 H_2O). This gives a low *electron* barrier height.[122] With time, as the surface conditions change, the electron barrier height increases and the hole barrier height decreases.[123]

4.10 STRENGTHS AND WEAKNESSES

This chapter covers such a variety of characterization techniques that it is difficult to summarize the strengths and weaknesses of each method here. Instead, we have chosen to mention the strengths and weaknesses throughout the chapter.

APPENDIX 4.1

Schottky Diode Current-Voltage Equation

The current-voltage equation of a Schottky diode with series resistance is

$$I = AA^*T^2 e^{-q\phi_B/kT} (e^{q(V - Ir_s)/nkT} - 1) \tag{A4.1}$$

It has been suggested that Eq. (A4.1) is incorrect because it predicts the non-ideality, included through the parameter n, to affect only the current flow from the semiconductor to the metal but not from the metal to the semiconductor,[124] as is obvious from Eq. (A4.1). For high forward bias only the first term in the "exp" bracket is important and it contains the factor n. For reverse bias the second term is important and it does not contain n.

To overcome this problem, consider the voltage dependence of the barrier height. The barrier height ϕ_B depends on voltage due to image force barrier lowering, due to voltage drops across any interfacial layers between the metal and the semiconductor, and other possible effects. Assuming the barrier height depends linearly on voltage according to

$$\phi_B(V) = \phi_{B0} + \gamma(V - Ir_s) \tag{A4.2}$$

where $\gamma > 0$ because the barrier height increases with increased forward bias, Eq. (A4.1) becomes

$$I = AA^*T^2 e^{-q\phi_{B0}/kT} e^{-q\gamma(V - Ir_s)/kT} (e^{q(V - Ir_s)/kT} - 1) \tag{A4.3}$$

Defining the diode ideality factor n by

$$\frac{1}{n} = 1 - \gamma = 1 - \frac{\partial \phi_B}{\partial V} \tag{A4.4}$$

allows Eq. (A4.3) to be written as

$$I = AA^* T^2 e^{-q\phi_{B0}/kT} e^{q(V-Ir_s)/nkT} (1 - e^{-q(V-Ir_s)/kT}) \tag{A4.5}$$

To determine n, it is common practice to use that range of the $\log(I)-V$ plot where series resistance is negligible ($V \ll Ir_s$). Under those restrictions Eq. (A4.5) becomes

$$I = AA^* T^2 e^{-q\phi_{B0}/kT} e^{qV/nkT} (1 - e^{-qV/kT}) \tag{A4.6}$$

Instead of plotting $\log(I)$ versus V, Eq. (A4.6) predicts that $\log[I/(1 - exp(-qV/kT))]$ versus V should be plotted. Such a plot exhibits a straight line all the way to $V = 0$, giving a wider range of the curve from which n is determined.[125] The ideality factor is near unity for well-behaved Schottky diodes. However, it can deviate from unity as a result of current flow due to mechanisms other than thermionic emission, *e.g.*, thermionic-field emission current, interface damage, and interfacial layers all tend to raise n above unity.

REFERENCES

1. J.S. Escher, H.M. Berg, G.L. Lewis, C.D. Moyer, T.U. Robertson and H.A. Wey, "Junction-Current-Confinement Planar Light-Emitting Diodes and Optical Coupling into Large-Core Diameter Fibers Using Lenses," *IEEE Trans. Electron Dev.* **ED-29**, 1463–1469, Sept. 1982.

2. J.H. Werner, "Schottky Barrier and pn-Junction I/V Plots—Small Signal Evaluation," *Appl. Phys.* **A47**, 291–300, Nov. 1988.

3. K. Schuster and E. Spenke, "The Voltage Step at the Switching of Alloyed pin Rectifiers," *Solid-State Electron.* **8**, 881–882, Nov. 1965.

4. S.M. Sze, *Physics of Semiconductor Devices*, 2nd ed., Wiley, New York, 1981, 256–263.

5. H. Norde, "A Modified Forward I–V Plot for Schottky Diodes with High Series Resistance," *J. Appl. Phys.* **50**, 5052–5053, July 1979.

6. N.T. Tam and T. Chot, "Experimental Richardson Constant of Metal-Semiconductor Schottky Barrier Contacts," *Phys. Stat. Sol.* **93a**, K91–K95, Jan. 1986.

7. N. Toyama, "Variation in the Effective Richardson Constant of a Metal-Silicon Contact Due to Metal Film Thickness," *J. Appl. Phys.* **63**, 2720–2724, April 1988.

8. C.D. Lien, F.C.T. So and M.A. Nicolet, "An Improved Forward I–V Method for Non-Ideal Schottky Diodes with High Series Resistance," *IEEE Trans. Electron Dev.* **ED-31**, 1502–1503, Oct. 1984.

9. K. Sato and Y. Yasumura, "Study of the Forward I–V Plot for Schottky Diodes with High Series Resistance," *J. Appl. Phys.* **58**, 3655–3657, Nov. 1985.

10. S.K. Cheung and N.W. Cheung, "Extraction of Schottky Diode Parameters from Forward Current-Voltage Characteristics," *Appl. Phys. Lett.* **49**, 85–87, July 1986.

11. T. Chot, "A Modified Forward I-U Plot for Schottky Diodes with High Series Resistance," *Phys. Stat. Sol.* **66a**, K43–K45, July 1981.

12. R.M. Cibils and R.H. Buitrago, "Forward I–V Plot for Nonideal Schottky Diodes With High Series Resistance," *J. Appl. Phys.* **58**, 1075–1077, July 1985.

13. T.C. Lee, S. Fung, C.D. Beling, and H.L. Au, "A Systematic Approach to the Measurement of Ideality Factor, Series Resistance, and Barrier Height for Schottky Diodes," *J. Appl. Phys.* **72**, 4739–4742, Nov. 1992.

14. K.E. Bohlin, "Generalized Norde Plot Including Determination of the Ideality Factor," *J. Appl. Phys.* **60**, 1223–1224, Aug. 1986; J.C. Manifacier, N. Brortyp, R. Ardebili, and J.P. Charles, "Schottky Diode: Comments Concerning Diode Parameter Determination from the Forward I–V Plot," *J. Appl. Phys.* **64**, 2502–2504, Sept. 1988.

15. D. Donoval, M. Barus, and M. Zdimal, "Analysis of I–V Measurements on PtSi-Si Schottky Structures in a Wide Temperature Range," *Solid-State Electron.* **34**, 1365–1373, Dec. 1991.

16. V. Mikhelashvili, G. Eisenstein, and R. Uzdin, "Extraction of Schottky Diode Parameters with a Bias Dependent Barrier Height," *Solid-State Electron.* **45**, 143–148, Jan. 2001.

17. D.K. Schroder and D.L. Meier, "Solar Cell Contact Resistance—A Review," *IEEE Trans. Electron Dev.* **ED-31**, 637–647, May 1984.

18. M.J. Kerr, A. Cuevas, and R.A. Sinton, "Generalized Analysis of Quasi-Steady-State and Transient Decay Open Circuit Voltage Measurements," *J. Appl. Phys.* **91**, 399–404, Jan. 2002.

19. N.P. Harder, A.B. Sproul, T. Brammer, and A.G. Aberle, "Effects of Sheet resistance and Contact Shading on the Characterization of Solar Cells by Open-Circuit Voltage Measurements," *J. Appl. Phys.* **94**, 2473–2479, Aug. 2003.

20. M. Wolf and H. Rauschenbach, "Series Resistance Effects on Solar Cell Measurements," *Adv. Energy Conv.* **3**, 455–479, Apr./June 1963.

21. D.H. Neuhaus, N.-P. Harder, S. Oelting, R. Bardos, A.B. Sproul, P. Widenborg, and A.G. Aberle, "Dependence of the Recombination in Thin-Film Si Solar Cells Grown by Ion-Assisted Deposition on the Crystallographic Orientation of the Substrate," *Solar Energy Mat. and Solar Cells*, **74**, 225–232, Oct. 2002.

22. G.M. Smirnov and J.E. Mahan, "Distributed Series Resistance in Photovoltaic Devices; Intensity and Loading Effects," *Solid-State Electron.* **23**, 1055–1058, Oct. 1980.

23. S.K. Agarwal, R. Muralidharan, A. Agarwala, V.K. Tewary and S.C. Jain, "A New Method for the Measurement of Series Resistance of Solar Cells," *J. Phys. D.* **14**, 1643–1646, Sept. 1981.

24. K. Rajkanan and J. Shewchun, "A Better Approach to the Evaluation of the Series Resistance of Solar Cells," *Solid-State Electron.* **22**, 193–197, Feb. 1979.

25. G.L. Araujo and E. Sanchez, "A New Method for Experimental Determination of the Series Resistance of a Solar Cell," *IEEE Trans. Electron Dev.* **ED-29**, 1511–1513, Oct. 1982.

26. J.C.H. Phang, D.S.H. Chan and Y.K. Wong, "Comments on the Experimental Determination of Series Resistance in Solar Cells," *IEEE Trans. Electron Dev.* **ED-31**, 717–718, May 1984.

27. K.L. Kennerud, "Analysis of Performance Degradation in CdS Solar Cells," *IEEE Trans. Aerosp. Electr. Syst.* **AES-5**, 912–917, Nov. 1969.

28. D.S.H. Chan, J.R. Phillips and J.C.H. Phang, "A Comparative Study of Extraction Methods for Solar Cell Model Parameters," *Solid-State Electron.* **29**, 329–337, March 1986.

29. J.P. Charles, M. Abdelkrim, Y.H. Muoy and P. Mialhe, "A Practical Method for Analysis of the I–V Characteristics of Solar Cells," *Solar Cells* **4**, 169–178, Sept. 1981.

30. P. Mialhe, A. Khoury and J.P. Charles, "A Review of Techniques to Determine the Series Resistance of Solar Cells," *Phys. Stat. Sol.* **83a**, 403–409, May 1984.

31. D. Fuchs and H. Sigmund, "Analysis of the Current-Voltage Characteristic of Solar Cells," *Solid-State Electron.* **29**, 791–795, Aug. 1986.

32. J.E. Cape, J.R. Oliver and R.J. Chaffin, "A Simplified Flashlamp Technique for Solar Cell Series Resistance Measurements," *Solar Cells* **3**, 215–219, May 1981.

33. D.S. Chan and J.C.H. Phang, "A Method for the Direct Measurement of Solar Cell Shunt Resistance," *IEEE Trans. Electron Dev.* **ED-31**, 381–383, March 1984.

34. H.K. Gummel, "Measurement of the Number of Impurities in the Base Layer of a Transistor," *Proc. IRE*, **49**, 834, April 1961.

35. J.J. Ebers and J.L. Moll, "Large Signal Behavior of Junction Transistors," *Proc. IRE*, **42**, 1761–1772, Dec. 1954.

36. W. Filensky and H. Beneking, "New Technique for Determination of Static Emitter and Collector Series Resistances of Bipolar Transistors," *Electron. Lett.* **17**, 503–504, July 1981.

37. L.J. Giacoletto, "Measurement of Emitter and Collector Series Resistances," *IEEE Trans. Electron Dev.* **ED-19**, 692–693, May 1972.

38. I. Getreu, *Modeling the Bipolar Transistor*, Tektronix, Beaverton, OR, 1976. This book provides a very good discussion of BJT characterization methods, especially for using curve tracers.

39. T.H. Ning and D.D. Tang, "Method for Determining the Emitter and Base Series Resistances of Bipolar Transistors," *IEEE Trans. Electron Dev.* **ED-31**, 409–412, April 1984.

40. J. Choma, Jr., "Error Minimization in the Measurement of Bipolar Collector and Emitter Resistances," *IEEE J. Solid-State Circ.* **SC-11**, 318–322, April 1976.

41. K. Morizuka, O. Hidaka, and H. Mochizuki, "Precise Extraction of Emitter Resistance from an Improved Floating Collector Measurement," *IEEE Trans. Electron Dev.* **42**, 266–273, Feb. 1995.

42. M. Linder, F. Ingvarson K.O. Jeppson, J.V. Grahn S-L Zhang, and M. Östling, "Extraction of Emitter and Base Series Resistances of Bipolar Transistors from a Single DC Measurement," *IEEE Trans. Semicond. Manufact.* **13**, 119–126, May 2000.

43. J.B. Scott, "New Method to Measure Emitter Resistance of Heterojunction Bipolar Transistors," *IEEE Trans. Electron Dev.* **50**, 1970–1973, Sept. 2003.

44. W.D. Mack and M. Horowitz, "Measurement of Series Collector Resistance in Bipolar Transistors," *IEEE J. Solid-State Circ.* **SC-17**, 767–773, Aug. 1982.

45. J.S. Park, A. Neugroschel, V. de la Torre, and P.J. Zdebel, "Measurement of Collector and Emitter Resistances in Bipolar Transistors," *IEEE Trans. Electron Dev.* **38**, 365–372, Feb. 1991.

46. J. Logan, "Characterization and Modeling for Statistical Design," *Bell Syst. Tech. J.* **50**, 1105–1147, April 1971.

47. D.D. Tang, "Heavy Doping Effects in pnp Bipolar Transistors," *IEEE Trans. Electron Dev.* **ED-27**, 563–570, March 1980.

48. B. Ricco, J.M.C. Stork and M. Arienzo, "Characterization of Non-Ohmic Behavior of Emitter Contacts of Bipolar Transistors," *IEEE Electron Dev. Lett.* **EDL-5**, 221–223, July 1984.

49. J. Weng, J. Holz, and T.F. Meister, "New Method to Determine the Base Resistance of Bipolar Transistors," *IEEE Electron Dev. Lett.* **13**, 158–160, March 1992.

50. R.C. Taft and J.C. Plummer, "An Eight-Terminal Kelvin-Tapped Bipolar Transistor for Extracting Parasitic Series Resistance," *IEEE Trans. Electron Dev.* **38**, 2139–2154, Sept. 1991.

51. W.M.C. Sansen and R.G. Meyer, "Characterization and Measurement of the Base and Emitter Resistances of Bipolar Transistors," *IEEE J. Solid-State Circ.* **SC-7**, 492–498, Dec. 1972.

52. T.E. Wade, A. van der Ziel, E.R. Chenette and G. Roig, "Base Resistance Measurements on Bipolar Junction Transistors Via Low Temperature Bridge Techniques," *Solid-State Electron.* **19**, 385–388, May 1976.

53. R.T. Unwin and K.F. Knott, "Comparison of Methods Used for Determining Base Spreading Resistance," *Proc. IEE Pt.I* **127**, 53–61, April 1980.

54. G.C.M. Meijer and H.J.A. de Ronde, "Measurement of the Base Resistance of Bipolar Transistors," *Electron. Lett.* **11**, 249–250, June 1975.

55. A. Neugroschel, "Measurement of the Low-Current Base and Emitter Resistances of Bipolar Transistors," *IEEE Trans. Electron Dev.* **ED-34**, 817–822, April 1987; "Corrections to "Measurement of the Low-Current Base and Emitter Resistances of Bipolar Transistors"," *IEEE Trans. Electron Dev.* **ED-34**, 2568–2569, Dec. 1987.

56. J.S. Park and A. Neugroschel, "Parameter Extraction for Bipolar Transistors," *IEEE Trans. Electron Dev.* **ED-36**, 88–95, Jan. 1989.

57. P. Spiegel, "Transistor Base Resistance and Its Effect on High Speed Switching," *Solid State Design* 15–18, Dec. 1965.

58. K.K. Ng and W.T. Lynch, "Analysis of the Gate-Voltage-Dependent Series Resistance of MOSFET's," *IEEE Trans. Electron Dev.* **ED-33**, 965–972, July 1986.

59. K.K. Ng, R.J. Bayruns and S.C. Fang, "The Spreading Resistance of MOSFET's," *IEEE Electron Dev. Lett.* **EDL-6**, 195–198, April 1985.

60. G. Baccarani and G.A. Sai-Halasz, "Spreading Resistance in Submicron MOSFET's," *IEEE Electron Dev. Lett.* **EDL-4**, 27–29, Feb. 1983.

61. J.M. Pimbley, "Two-Dimensional Current Flow in the MOSFET Source-Drain," *IEEE Trans. Electron Dev.* **ED-33**, 986–996, July 1986.

62. K.K. Ng and J.R. Brews, "Measuring the Effective Channel Length of MOSFETs," *IEEE Circ. Dev.* **6**, 33–38, Nov. 1990.

63. C.C. McAndrew and P.A. Layman, "MOSFET Effective Channel Length, Threshold Voltage, and Series Resistance Determination by Robust Optimization," *IEEE Trans. Electron Dev.* **39**, 2298–2311, Oct. 1992.

64. Y. Taur, "MOSFET Channel Length: Extraction and Interpretation," *IEEE Trans. Electron Dev.* **47**, 160–170, Jan. 2000.

65. K. Terada and H. Muta, "A New Method to Determine Effective MOSFET Channel Length," *Japan. J. Appl. Phys.* **18**, 953–959, May 1979.

66. J.G.J. Chern, P. Chang, R.F. Motta and N. Godinho, "A New Method to Determine MOSFET Channel Length," *IEEE Electron Dev. Lett.* **EDL-1**, 170–173, Sept. 1980.

67. D.J. Mountain, "Application of Electrical Effective Channel Length and External Resistance Measurement Techniques to a Submicrometer CMOS Process," *IEEE Trans. Electron Dev.* **ED-36**, 2499–2505, Nov. 1989.

68. S.E. Laux, "Accuracy of an Effective Channel Length/External Resistance Extraction Algorithm for MOSFET's," *IEEE Trans. Electron Dev.* **ED-31**, 1245–1251, Sept. 1984.

69. K.L. Peng, S.Y. Oh, M.A. Afromowitz and J.L. Moll, "Basic Parameter Measurement and Channel Broadening Effect in the Submicrometer MOSFET," *IEEE Electron Dev. Lett.* **EDL-5**, 473–475, Nov. 1984.

70. S. Ogura, P.J. Tsang, W.W. Walker, D.L. Critchlow and J.F. Shepard, "Design and Characteristics of the Lightly Doped Drain-Source (LDD) Insulated Gate Field-Effect Transistor," *IEEE J. Solid-State Circ.* **SC-15**, 424–432, Aug. 1980.

71. B.J. Sheu, C. Hu, P. Ko and F.C. Hsu, "Source-and-Drain Series Resistance of LDD MOSFET's," *IEEE Electron Dev. Lett.* **EDL-5**, 365–367, Sept. 1984; C. Duvvury, D.A.G. Baglee and M.P. Duane, "Comments on 'Source-and-Drain Series Resistance of LDD MOSFET's'," *IEEE Electron Dev. Lett.* **EDL-5**, 533–534, Dec. 1984; B.J. Sheu, C. Hu, P. Ko and F.C. Hsu, "Reply to 'Comments on "Source-and-Drain Series Resistance of LDD MOSFET's"'," *IEEE Electron Dev. Lett.* **EDL-5**, 535, Dec. 1984.

72. S.L. Chen and J. Gong, "Influence of Drain Bias Voltage on Determining the Effective Channel Length and Series Resistance of Drain-Engineered MOSFETs Below Saturation," *Solid-State Electron.* **35**, 643–649, May 1992.

73. M.R. Wordeman, J.Y.C. Sun and S.E. Laux, "Geometry Effects in MOSFET Channel Length Extraction Algorithms," *IEEE Electron Dev. Lett.* **EDL-6**, 186–188, April 1985.

74. G.J. Hu, C. Chang, and Y.T. Chia, "Gate-Voltage Dependent Effective Channel Length and Series Resistance of LDD MOSFET's," *IEEE Trans. Electron Dev.* **ED-34**, 2469–2475, Dec. 1987.

75. S. Hong and K. Lee, "Extraction of Metallurgical Channel Length in LDD MOSFET's," *IEEE Trans. Electron Dev.* **42**, 1461–1466, Aug. 1995.

76. P.I. Suciu and R.L. Johnston, "Experimental Derivation of the Source and Drain Resistance of MOS Transistors," *IEEE Trans. Electron Dev.* **ED-27**, 1846–1848, Sept. 1980.

77. F.H. De La Moneda, H.N. Kotecha and M. Shatzkes, "Measurement of MOSFET Constants," *IEEE Electron Dev. Lett.* **EDL-3**, 10–12, Jan. 1982.

78. S.S. Chung and J.S. Lee, "A New Approach to Determine the Drain-and-Source Series Resistance of LDD MOSFET's," *IEEE Trans. Electron Dev.* **40**, 1709–1711, Sept. 1993.

79. M. Sasaki, H. Ito, and T. Horiuchi, "A New Method to Determine Effective Channel Length, Series Resistance, and Threshold Voltage," *Proc. IEEE Int. Conf. Microelectr. Test Struct.* 1996, 139–144.

80. K.L. Peng and M.A. Afromowitz, "An Improved Method to Determine MOSFET Channel Length," *IEEE Electron Dev. Lett.* **EDL-3**, 360–362, Dec. 1982.

81. J.D. Whitfield, "A Modification on 'An Improved Method to Determine MOSFET Channel Length'," *IEEE Electron Dev. Lett.* **EDL-6**, 109–110, March 1985.

82. S. Jain, "A New Method for Measurement of MOSFET Channel Length," *Japan. J. Appl. Phys.* **27**, L1559–L1561, Aug. 1988; "Generalized Transconductance and Transresistance Methods for MOSFET Characterization," *Solid-State Electron.* **32**, 77–86, Jan. 1989.

83. S. Jain, "Equivalence and Accuracy of MOSFET Channel Length Measurement Techniques," *Japan. J. Appl. Phys.* **28**, 160–166, Feb. 1989.

84. Y. Taur, D.S. Zicherman, D.R. Lombardi, P.R. Restle, C.H. Hsu, H.I. Hanafi, M.R. Wordeman, B. Davari, and G.G. Shahidi, "A New "Shift and Ratio" Method for MOSFET Channel-Length Extraction," *IEEE Electron Dev. Lett.* **13**, 267–269, May 1992.

85. S. Biesemans, M. Hendriks, S. Kubicek, and K.D. Meyer, "Practical Accuracy Analysis of Some Existing Effective Channel Length and Series Resistance Extraction Method for MOSFET's," *IEEE Trans. Electron Dev.* **45**, 1310–1316, June 1998.

86. G. Niu, S.J. Mathew, J.D. Cressler, and S. Subbanna, "A Novel Channel Resistance Ratio Method for Effective Channel Length and Series Resistance Extraction in MOSFETs," *Solid-State Electron.* **44**, 1187–1189, July 2000.

87. J.Y.-C. Sun, M.R. Wordeman and S.E. Laux, "On the Accuracy of Channel Length Characterization of LDD MOSFET's," *IEEE Trans. Electron Dev.* **ED-33**, 1556–1562, Oct. 1986.

88. P.R. Karlsson and K.O. Jeppson, "An Efficient Method for Determining Threshold Voltage, Series Resistance and Effective Geometry of MOS Transistors," *IEEE Trans. Semic. Manufact.* **9**, 215–222, May 1996.

89. A. Raychoudhuri, M.J. Deen, M.I.H. King, and J. Kolk, "Finding the Asymmetric Parasitic Source and Drain Resistances from the ac Conductances of a Single MOS Transistor," *Solid-State Electron.* **39**, 900–913, June 1996.

90. Q. Ye and S. Biesemans, "L_{eff} Extraction for Sub-100 nm MOSFET Devices," *Solid-State Electron.* **48**, 163–166, Jan. 2004.

91. S.W. Lee, "A Capacitance-Based Method for Experimental Determination of Metallurgical Channel Length of Submicron LDD MOSFET's," *IEEE Trans. Electron Dev.* **41**, 403–412, March 1994; J.C. Guo, S.S. Chung, and C.H. Hsu, "A New Approach to Determine the Effective Channel Length and the Drain-and-Source Series Resistance of Miniaturized MOSFET's," *IEEE Trans. Electron Dev.* **41**, 1811–1818, Oct. 1994.

92. H.S. Huang, J.S. Shiu, S.J. Lin, J.W. Chou, R. Lee, C. Chen, and G. Hong, "A Modified Capacitance-Voltage Method Used for L_{eff} Extraction and Process Monitoring in Advanced 0.15 μm Complementary Metal-Oxide-Semiconductor Technology and Beyond," *Japan. J. Appl. Phys.* **40**, 1222–1226, March 2001; H.S. Huang, S.J. Lin, Y.J. Chen, I.K. Chen, R. Lee, J.W. Chou, and G. Hong, "A Capacitance Ratio Method Used for L_{eff} Extraction of an Advanced Metal-Oxide-Semiconductor Device With Halo Implant," *Japan. J. Appl. Phys.* **40**, 3992–3995, June 2001.

93. R. Valtonen, J. Olsson, and P. De Wolf, "Channel Length Extraction for DMOS Transistors Using Capacitance-Voltage Measurements," *IEEE Trans. Electron Dev.* **48**, 1454–1459, July 2001.

94. C.H. Wang, "Identification and Measurement of Scaling-Dependent Parasitic Capacitances of Small-Geometry MOSFET's," *IEEE Trans. Electron Dev.* **43**, 965–972, June 1996.

95. P. Vitanov, U. Schwabe and I. Eisele, "Electrical Characterization of Feature Sizes and Parasitic Capacitances Using a Single Test Structure," *IEEE Trans. Electron Dev.* **ED-31**, 96–100, Jan. 1984.

96. Y.R. Ma and K.L. Wang, "A New Method to Electrically Determine Effective MOSFET Channel Width," *IEEE Trans. Electron Dev.* **ED-29**, 1825–1827, Dec. 1982.

97. M.J. Deen and Z.P. Zuo, "Edge Effects in Narrow-Width MOSFET's," *IEEE Trans. Electron Dev.* **38**, 1815–1819, Aug. 1991.

98. Y.T. Chia and G.J. Hu, "A Method to Extract Gate-Bias-Dependent MOSFET's Effective Channel Width," *IEEE Trans. Electron Dev.* **38**, 424–437, Feb. 1991.

99. N.D. Arora, L.A. Bair, and L.M. Richardson, "A New Method to Determine the MOSFET Effective Channel Width," *IEEE Trans. Electron Dev.* **37**, 811–814, March 1990.

100. C.C. McAndrew, P.A. Layman, and R.A. Ashton, "MOSFET Effective Channel Width Determination by Nonlinear Optimization," *Solid-State Electron.* **36**, 1717–1723, Dec. 1993.

101. B.J. Sheu and P.K. Ko, "A Simple Method to Determine Channel Widths for Conventional and LDD MOSFET's," *IEEE Electron Dev. Lett.* **EDL-5**, 485–486, Nov. 1984.

102. K. Lee, M.S. Shur, A.J. Valois, G.Y. Robinson, X.C. Zhu and A. van der Ziel, "A New Technique for Characterization of the "End" Resistance in Modulation-Doped FET's," *IEEE Trans. Electron Dev.* **ED-31**, 1394–1398, Oct. 1984.

103. S. Chaudhuri and M.B. Das, "On the Determination of Source and Drain Series Resistances of MESFET's," *IEEE Electron Dev. Lett.* **EDL-5**, 244–246, July 1984.

104. W.A. Azzam and J.A. Del Alamo, "An All-Electrical Floating-Gate Transmission Line Model Technique for Measuring Source Resistance in Heterostructure Field-Effect Transistors," *IEEE Trans. Electron Dev.* **37**, 2105–2107, Sept. 1990.

105. L. Yang and S.I. Long, "New Method to Measure the Source and Drain Resistance of the GaAs MESFET," *IEEE Electron Dev. Lett.* **EDL-7**, 75–77, Feb. 1986.

106. R.P. Holmstrom, W.L. Bloss and J.Y. Chi, "A Gate Probe Method of Determining Parasitic Resistance in MESFET's," *IEEE Electron Dev. Lett.* **EDL-7**, 410–412, July 1986.

107. A. Ortiz-Conde, F.J. Garcia Sanchez, J.J. Liou, A. Cerdeira, M. Estrada, and Y. Yue, "A Review of Recent MOSFET Threshold Voltage Extraction Methods," *Microelectr. Rel.* **42**, 583–596, April-May 2002.

108. W.L. Brown, "n-Type Surface Conductivity on p-Type Germanium," *Phys. Rev.* **91**, 518–537, Aug. 1953.

109. S.C. Sun and J.D. Plummer, "Electron Mobility in Inversion and Accumulation Layers on Thermally Oxidized Silicon Surfaces," *IEEE Trans. Electron Dev.* **ED-27**, 1497–1508, Aug. 1980.

110. R.V. Booth, M.H. White, H.S. Wong and T.J. Krutsick, "The Effect of Channel Implants on MOS Transistor Characterization," *IEEE Trans. Electron Dev.* **ED-34**, 2501–2509, Dec. 1987.

111. ASTM Standard F617M-95, "Standard Method for Measuring MOSFET Linear Threshold Voltage," *1996 Annual Book of ASTM Standards*, Am. Soc. Test. Mat., Conshohocken, PA, 1996.

112. H.S. Wong, M.H. White, T.J. Krutsick and R.V. Booth, "Modeling of Transconductance Degradation and Extraction of Threshold Voltage in Thin Oxide MOSFET's," *Solid-State Electron.* **30**, 953–968, Sept. 1987.

113. S.W. Tarasewicz and C.A.T. Salama, "Threshold Voltage Characteristics of Ion-Implanted Depletion MOSFETs," *Solid-State Electron.* **31**, 1441–1446, Sept. 1988.

114. G. Ghibaudo, "New Method for the Extraction of MOSFET Parameters," *Electron. Lett.* **24**, 543–545, April 1988; S. Jain, "Measurement of Threshold Voltage and Channel Length of Submicron MOSFETs," *Proc. IEE Pt.I* **135**, 162–164, Dec. 1988.

115. H.G. Lee, S.Y. Oh and G. Fuller, "A Simple and Accurate Method to Measure the Threshold Voltage of an Enhancement-Mode MOSFET," *IEEE Trans. Electron Dev.* **ED-29**, 346–348, Feb. 1982.

116. ASTM Standard F1096, "Standard Method for Measuring MOSFET Saturated Threshold Voltage," *1996 Annual Book of ASTM Standards*, Am. Soc. Test. Mat., Conshohocken, PA, 1996.

117. M. Tsuno, M. Suga, M. Tanaka, K. Shibahara, M. Miura-Mattausch, and M. Hirose, "Physically-Based Threshold Voltage Determination for MOSFET's of All Gate Lengths," *IEEE Trans. Electron Dev.* **46**, 1429–1434, July 1999.

118. K. Terada, K. Nishiyama, and K-I, Hatanaka, "Comparison of MOSFET-Threshold-Voltage Extraction Methods," *Solid-State Electron.* **45**, 35–40, Jan. 2001.

119. F.M. Klaassen and W. Hes, "On the Temperature Coefficient of the MOSFET Threshold Voltage," *Solid-State Electron.* **29**, 787–789, Aug. 1986.

120. S. Cristoloveanu, D. Munteanu, and M. Liu, "A Review of the Pseudo-MOS Transistor in SOI Wafers: Operation, Parameter Extraction, and Applications," *IEEE Trans. Electron Dev.*, **47**, 1018–1027, May 2000.

121. H.J. Hovel, "Si Film Electrical Characterization in SOI Substrates by the HgFET Technique," *Solid-State Electron.* **47**, 1311–1333, Aug. 2003.

122. Y.J. Liu and H.Z. Yu, "Effect of Organic Contamination on the Electrical Degradation of Hydrogen terminated Silicon upon Exposure to Air under Ambient Conditions," *J. Electrochem. Soc.* **150**, G861–G865, Dec. 2003.

123. J.Y. Choi, S. Ahmed, T. Dimitrova, J.T.C. Chen, and D.K. Schroder, "The Role of the Mercury-Si Schottky Barrier Height in Pseudo-MOSFETs," *IEEE Trans. Electron Dev.* **51** 1164–1168, July 2004.

124. E.H. Rhoderick, "Metal-Semiconductor Contacts," *Proc. IEE* Pt.I **129**, 1–14, Feb. 1982; E.H. Rhoderick and R.H. Williams, *Metal-Semiconductor Contacts*, 2nd ed., Clarendon Press, Oxford, 1988.

125. J.D. Waldrop, "Schottky-Barrier Height of Ideal Metal Contacts to GaAs," *Appl. Phys. Lett.* **44**, 1002–1004, March 1984.

PROBLEMS

4.1   The $I-V$ data of a forward-biased pn junction are shown. Determine the temperature T and the series resistance r_s for this device.

V (V)	I (A)	V (V)	I (A)	V (V)	I (A)
0.00	0.0000	0.350	1.096e-07	0.700	0.006291
0.0250	1.291e-12	0.375	2.512e-07	0.725	0.01005
0.0500	4.248e-12	0.400	5.754e-07	0.750	0.01429
0.0750	1.102e-11	0.425	1.318e-06	0.775	0.01961
0.100	2.654e-11	0.450	3.019e-06	0.800	0.02543
0.125	6.209e-11	0.475	6.913e-06	0.825	0.03185
0.150	1.435e-10	0.500	1.582e-05	0.850	0.03833
0.175	3.301e-10	0.525	3.618e-05	0.875	0.04504
0.200	7.576e-10	0.550	8.252e-05	0.900	0.05194
0.225	1.737e-09	0.575	0.0001872	0.925	0.05899
0.250	3.980e-09	0.600	0.0004191	0.950	0.06616
0.275	9.119e-09	0.625	0.0009134	0.975	0.07344
0.300	2.089e-08	0.650	0.001882	1.00	0.08080
0.325	4.786e-08	0.675	0.003506		

4.2 The $I-V$ curves of a forward-biased pn junction are shown in Fig. P4.2. Determine the temperature T for the "$T = ?$" curve and the series resistance r_s for the "$T = 300$ K" curve.

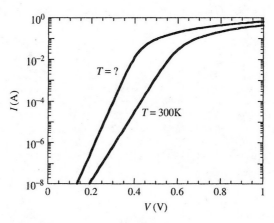

Fig. P4.2

4.3 The current voltage relationship for a pn junction is

$$I = I_{o,scr}\left(\exp\left(\frac{q(V - Ir_s)}{nkT}\right) - 1\right) + I_{o,qnr}\left(\exp\left(\frac{q(V - Ir_s)}{nkT}\right) - 1\right).$$

From the $I-V$ curve in Fig. P4.3 or data determine $I_{o,scr}$, $I_{o,qnr}$, n in the scr, n in the qnr, and r_s. $T = 300$ K. Determine r_s and n also from I/g_d versus I and g_d/I versus g_d plots.

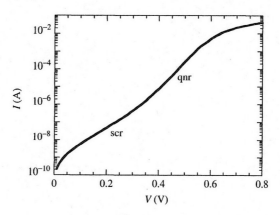

Fig. P4.3

$V(V)$	I (A)	V	I	V	I	V	I
0.0	0.0	0.20	4.916e-08	0.40	7.533e-06	0.6	0.005193
0.01	2.141e-10	0.21	6.046e-08	0.41	1.049e-05	0.61	0.006188
0.02	4.738e-10	0.22	7.445e-08	0.42	1.472e-05	0.62	0.007770

$V(V)$	I (A)	V	I	V	I	V	I
0.03	7.890e-10	0.23	9.183e-08	0.43	2.079e-05	0.63	0.009066
0.04	1.172e-09	0.24	1.135e-07	0.44	2.952e-05	0.64	0.01073
0.05	1.637e-09	0.25	1.408e-07	0.45	4.211e-05	0.65	0.01224
0.06	2.201e-09	0.26	1.751e-07	0.46	6.029e-05	0.66	0.01402
0.07	2.887e-09	0.27	2.187e-07	0.47	8.657e-05	0.67	0.01569
0.08	3.721e-09	0.28	2.745e-07	0.48	0.0001245	0.68	0.01757
0.09	4.734e-09	0.29	3.464e-07	0.49	0.0001792	0.69	0.01937
0.10	5.966e-09	0.30	4.398e-07	0.50	0.0002575	0.70	0.02112
0.11	7.466e-09	0.31	5.623e-07	0.51	0.0003691	0.71	0.02321
0.12	9.291e-09	0.32	7.243e-07	0.52	0.0005260	0.72	0.02506
0.13	1.151e-08	0.33	9.406e-07	0.53	0.0007432	0.73	0.02720
0.14	1.422e-08	0.34	1.232e-06	0.54	0.001037	0.74	0.02912
0.15	1.753e-08	0.35	1.629e-06	0.55	0.001421	0.75	0.03130
0.16	2.157e-08	0.36	2.173e-06	0.56	0.001904	0.76	0.03327
0.17	2.651e-08	0.37	2.925e-06	0.57	0.002479	0.77	0.03549
0.18	3.257e-08	0.38	3.975e-06	0.58	0.003122	0.78	0.03765
0.19	4.001e-08	0.39	5.449e-06	0.59	0.003792	0.79	0.03976

4.4 The current–voltage curves of a Schottky diode are shown in Fig. P4.4 for various temperatures. Determine n, ϕ_B, A^*, and r_s. $A = 10^{-3}$ cm^2.

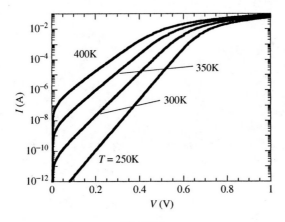

Fig. P4.4

4.5 A solar cell obeys the "light" and "dark" equations

$$I = I_L - I_o \left(\exp\left(\frac{q(V + Ir_s)}{nkT} \right) - 1 \right) \; ; \; I_{dk} = I_o \left(\exp\left(\frac{q(V - Ir_s)}{nkT} \right) - 1 \right).$$

From the curves in Fig. P4.5 determine: I_o, n and r_s. $T = 290$ K. To determine r_s use three methods: (i) the "light" curves only; (ii) the "dark" curve only; (iii) both curves.

4.6 Consider a resistor R placed externally in either the base lead or the emitter lead in the bipolar junction transistor in Fig. P4.6. Which placement has the largest effect on the collector I_C?

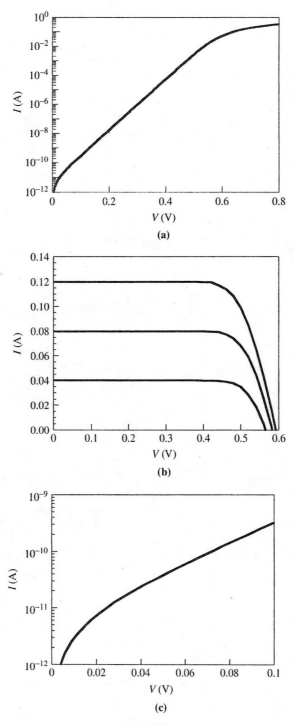

Fig. P4.5

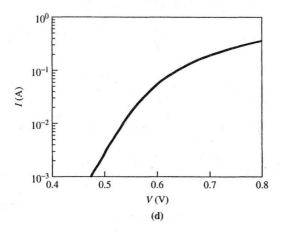

Fig. P4.5 *(continued)*

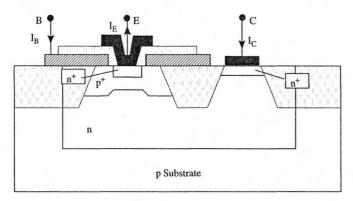

Fig. P4.6

4.7 L_{eff} and $R_{SD} = R_S + R_D$ of a MOSFET can be obtained from a plot of the measured drain resistance R_m vs. L. Consider two R_m versus L curves of an LDD (lightly-doped drain) MOSFET for V_{GS1} and V_{GS2}, where $V_{GS2} = V_{GS1} + \Delta V_1$. Draw the two lines for V_{GS1} and V_{GS2} on an R_m versus L plot. On the same figure, draw the line for $V_{GS3} = V_{GS1} + \Delta_2$, where $\Delta V_2 < \Delta V_1$. Remember, in LDD devices, both L_{eff} and R_{SD} are gate voltage dependent. Give reasons for your answer.

4.8 Consider the two *n*-channel MOSFETs in Fig. P4.8. $N_{A2} > N_{A1}$. Discuss whether the *threshold voltages* and the *drain currents for a given drain and gate voltage*

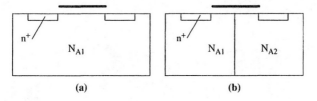

Fig. P4.8

are the same for these devices. Justify your answers. Assume the source and substrate to be grounded.

4.9 Consider the four *n*-channel MOSFETs in Fig. P4.9. $N_{A2} > N_{A1}$. Discuss whether the *threshold voltages* and the *drain currents for a given drain and gate voltage* are the same for these devices. Justify your answers. Assume the source and substrate to be grounded.

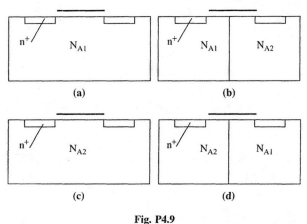

Fig. P4.9

4.10 Consider two MOSFETs of the type shown in Fig. P4.10.

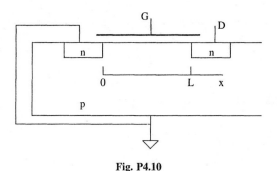

Fig. P4.10

(a) Uniform gate oxide thickness $t_{ox} = t_{ox1}$.

(b) Graded gate oxide thickness between source and drain, according to

$$t_{ox}(x) = (t_{ox1} - t_{ox2})(1 - x/L) + t_{ox2}; t_{ox2} < t_{ox1}.$$

Are the threshold voltages for these two structures identical? Are the drain currents, measured at low drain voltage, identical for these two structures? Give reasons for your answers. $V_{FB} = 0$.

4.11 The measured resistance of a MOSFET is shown in Fig. P4.11 for various gate lengths as a function of gate voltage. Choose one answer.

$$\Box V_{GI} > V_{G2}, \Box V_{GI} = V_{G2}, \Box V_{GI} < V_{G2}.$$

What is determined by point A? Draw on the same figure the lines for the same gate voltages when $R_m = 0$ and $\Delta L = 0$. All other parameters are unchanged.

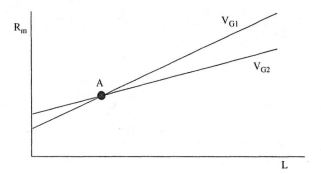

Fig. P4.11

4.12 $R_m = V_{DS}/I_D$ is shown in Fig. P4.12 for the MOSFET on the left for gate voltages V_{G1} and V_{G2}. Draw the V_{G2} line for the LDD structure on the right. V_{G1} is that gate voltage at which a channel is formed between the two n-regions without changing the conductivity of these regions.

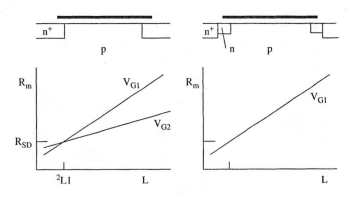

Fig. P4.12

4.13 The current-voltage relationship of a MOSFET in the presence of series resistance is (source and substrate are grounded):

$$I_D \approx \frac{W_{eff} C_{ox}}{L_{eff}} \frac{\mu_o}{[1 + \theta(V_{GS} - V_T)]} (V_{GS} - V_T - 0.5 V_{DS}) V'_{DS},$$

where $V'_{DS} = V_{DS} - I_D(R_S + R_D)$, $W_{eff} = W - \Delta W$, and $L_{eff} = L - \Delta L$. Using the $I_D - V_{GS}$ data determine V_T, μ_o, θ, ΔL, and $R_{SD} = R_S + R_D$; assume $\Delta W = 0$. $t_{ox} = 10$ nm, $W = 50$ μm, $V_D = 50$ mV. The drain current for various channel lengths and various gate voltages is listed in the following table:

| | | I_D (A) | | |
V_{GS} (V)	$L = 20$ μm	12 μm	7 μm	1 μm
0.725	4.935e-07	8.326e-07	1.460e-06	1.517e-05
1.025	6.176e-06	1.026e-05	1.749e-05	0.0001132
1.325	1.145e-05	1.876e-05	3.119e-05	0.0001527
1.625	1.636e-05	2.645e-05	4.304e-05	0.0001740
1.925	2.094e-05	3.345e-05	5.339e-05	0.0001873
2.225	2.523e-05	3.985e-05	6.250e-05	0.0001964
2.525	2.924e-05	4.572e-05	7.058e-05	0.0002031
2.825	3.301e-05	5.113e-05	7.781e-05	0.0002081
3.125	3.656e-05	5.612e-05	8.430e-05	0.0002121
3.425	3.991e-05	6.075e-05	9.017e-05	0.0002153
3.725	4.307e-05	6.504e-05	9.550e-05	0.0002179
4.025	4.606e-05	6.905e-05	0.0001004	0.0002202
4.325	4.889e-05	7.278e-05	0.0001048	0.0002220
4.625	5.157e-05	7.628e-05	0.0001089	0.0002237
4.925	5.412e-05	7.957e-05	0.0001127	0.0002251
5.225	5.655e-05	8.265e-05	0.0001162	0.0002263

4.14 Draw $I_D - V_{DS}$ for $V_{GS} = V_{GS1} > V_T$ and $I_D - V_{GS}$ for low V_{DS}, with region [1]: (i) p^+, (ii) n^+, as shown in Fig. P4.14. Draw both curves on the same figure in each case. What device characteristics are determined from $I_D - V_{DS}$ curves? What device characteristics are determined from $I_D - V_{GS}$ curves?

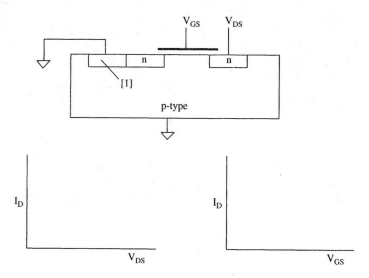

Fig. P4.14

4.15 The $I_D - V_{GS}$ and $I_D - V_{DS}$ plots of two MOSFETs with different gate lengths are shown in Fig. P4.15. Determine V_T, R_{SD} and ΔL for each device. Determine the effective mobility for the $L = 2$ μm device at $V_{GS} = 2$ V, using

$$\mu_{eff} = \frac{g_d L_{eff}}{W C_{ox}(V_{GS} - V_T)}.$$

MOSFET1: $t_{ox} = 5$ nm, $L = 0.5$ μm, $W = 10$ μm, $K_{ox} = 3.9$.
MOSFET2: $t_{ox} = 10$ nm, $L = 2$ μm, $W = 10$ μm, $K_{ox} = 3.9$.

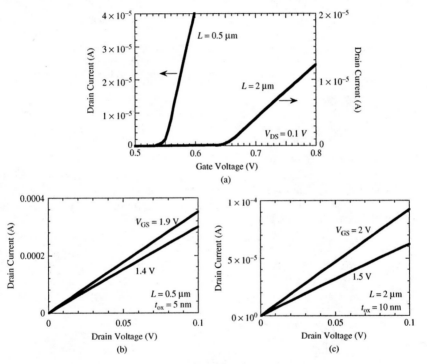

Fig. P4.15

4.16 The $I_D - V_{GS}$ and $I_D - V_{DS}$ plots of two MOSFETs with different gate lengths are shown ion Fig. P4.16. Determine V_T, R_{SD} and ΔL. Determine the effective mobility for the $L = 2$ μm device at $V_{GS} = 2$ V, using

$$\mu_{eff} = \frac{g_d L_{eff}}{W C_{ox}(V_{GS} - V_T)}.$$

MOSFET1: $t_{ox} = 5$ nm, $L = 0.25$ μm, $W = 5$ μm, $K_{ox} = 3.9$.
MOSFET2: $t_{ox} = 5$ nm, $L = 2$ μm, $W = 5$ μm, $K_{ox} = 3.9$.

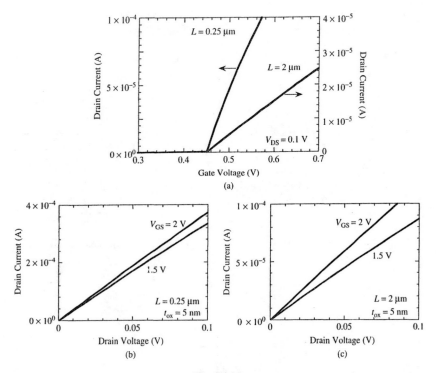

Fig. P4.16

4.17 $R_m = V_{DS}/I_D$ versus $1/(V_{GS} - V_T)$ curves are measured on MOSFETs with various gate lengths and shown in Fig. P4.17. Determine ΔL(in μm), R_{SD}, μ_o, and θ. $W = 10$ μm, $t_{ox} = 5$ nm, $K_{ox} = 3.9$, $V_T = 0.4$ V.

Curve fitting gives: $y = 198.7 + 50x$; $y = 200.6 + 112x$; $y = 203 + 173x$; $y = 207.3 + 263x$.

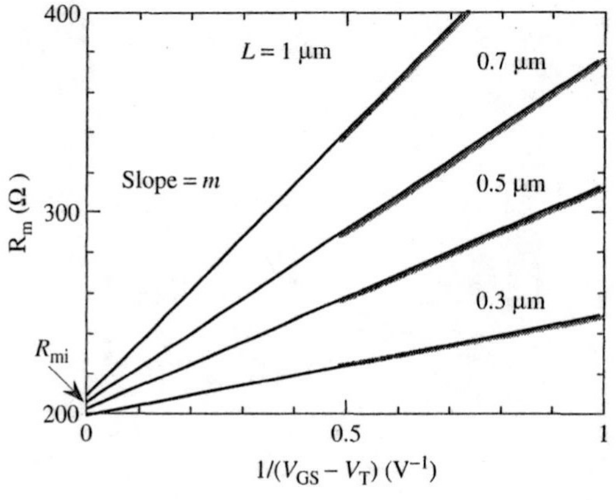

Fig. P4.17

4.18 On Fig. P4.18, show the *physical gate length* and the *metallurgical channel length*. Can the *effective channel length* be larger than L_1? Discuss.

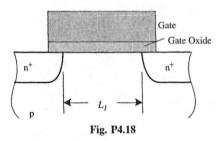

Fig. P4.18

4.19 Two R_m versus L lines for a MOSFET are shown in Fig. P4.19. $R_m = V_{DS}/I_D$. Determine the source and drain resistance R_{SD} and $\Delta L = L - L_{eff}$. Then, on the same figure, draw the two lines when the MOSFET oxide thickness t_{ox} is *decreased*.

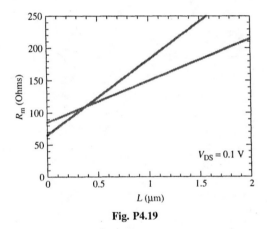

Fig. P4.19

4.20 The total resistance R_m defined as V_{DS}/I_D is shown in Fig. P4.20 for MOSFETs with different gate lengths.

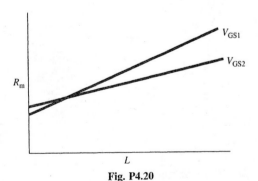

Fig. P4.20

Choose one answer: □$V_{GS1} > V_{GS2}$ □$V_{GS1} = V_{GS2}$ □$V_{GS1} < V_{GS2}$

What parameters can be determined from this plot? Draw the two lines for the same gate voltages V_{GS1} and V_{GS2} when the oxide thickness is reduced. Assume the threshold voltage remains unchanged.

4.21 The total resistance R_m defined as V_{DS}/I_D is shown in Fig. P4.21 for MOSFETs with different gate lengths.

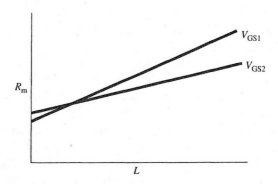

Fig. P4.21

Choose one answer: □$V_{GS1} > V_{GS2}$ □$V_{GS1} = V_{GS2}$ □$V_{GS1} < V_{GS2}$

What parameters can be determined from this plot? Draw the two lines for the same gate voltages V_{GS1} and V_{GS2} when the source and drain contact resistances are increased.

4.22 The R_m versus L plot of MOSFET (a) is shown in Fig. P 4.22.

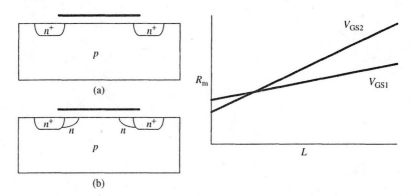

Fig. P4.22

(a) What is L and R_m at the point of intersection?

(b) □$V_{GS1} > V_{GS2}$ □$V_{GS1} = V_{GS2}$ □$V_{GS1} < V_{GS2}$

(c) Draw on the R_m versus L plot the two lines for the LDD MOSFET (b) for the same gate voltages. The gate overlap over the n^+ source and drain in (a) is the same as the overlap over the n regions in (b). For MOSFET (b): At the lower gate voltage, a channel exists between the two n-regions; at the higher gate voltage, the n-regions are accumulated by the gate voltage.

REVIEW QUESTIONS

- Why is the $I-V$ curve a straight line on a semilog plot?
- Why does a Si diode log $I-V$ curve have two slopes?
- How does series resistance affect the diode current?
- How is the barrier height of Schottky diodes determined?
- Why can the Schottky diode barrier heights be different when determined from $I-V$ or $C-V$ data?
- Why are series and shunt resistance important in solar cells?
- How are emitter and base resistances in BJT determined?
- Name three device/material parameters that influence the threshold voltage?
- Why does the effective channel length differ from the physical gate length?
- Which effective channel length methods are useful for short-channel MOSFETs?
- What is an advantage of the capacitance-voltage technique over current-voltage techniques for effective channel length determination?
- How is the threshold voltage measured?

5

DEFECTS

5.1 INTRODUCTION

All semiconductors contain defects. They may be foreign atoms (impurities) or crystalline defects. Impurities are intentionally introduced as dopant atoms (shallow-level impurities), recombination centers (deep-level impurities) to reduce the device lifetime, or deep-level impurities to increase the substrate resistivity. Impurities are also unintentionally incorporated during crystal growth and device processing. Various types of defects are shown schematically in Fig. 5.1. The open circles represent the host atoms (*e.g.*, silicon). The defects are: (1) foreign interstitial (*e.g.*, oxygen in silicon), (2) foreign substitutional (*e.g.*, dopant atom), (3) vacancy, (4) self interstitial, (5) stacking fault, (6) edge dislocation, and (7) precipitate. The corncob illustrates a vacancy and an interstitial and the saguaro cactus a stacking fault and edge dislocation. Today's silicon is grown very pure with metallic densities on the order of 10^{10} cm^{-3} or less. Processing tends to introduce higher densities, but many of these impurities are gettered during subsequent processing with densities of typically $10^{10} - 10^{12}$ cm^{-3} after processing.

Metallic impurities affect various device parameters. We show in Fig. 5.2 some regions where metals cause problems. A major concern is metallic contamination at the semiconductor/oxide interface because it degrades the gate oxide integrity. Metals also degrade devices if located at high stress points and in junction space-charge regions. The effect of iron and copper contamination in silicon is illustrated in Fig. 5.3. Fig. 5.3(a) shows the % failure versus oxide breakdown electric field as a function of iron contamination in Si wafers. Fig. 5.3(b) shows a similar plot for copper contamination. Typically metal contamination leads to more severe oxide breakdown degradation for thicker oxides, but as these figure show there is degradation even for 3 nm oxides. Thinner oxides show

Semiconductor Material and Device Characterization, Third Edition, by Dieter K. Schroder
Copyright © 2006 John Wiley & Sons, Inc.

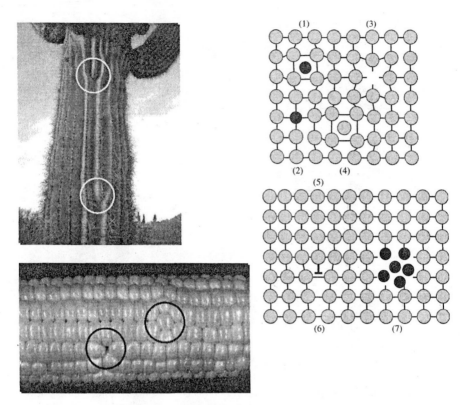

Fig. 5.1 Schematic representation of defects in semiconductors described in the text.

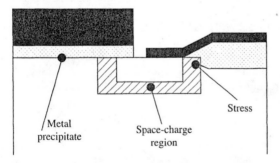

Metal precipitate

Space-charge region

Stress

Fig. 5.2 MOSFET regions sensitive to metal contamination.

less degradation due to the higher leakage currents through such thin oxides even in the absence of metal contamination.

The characterization of shallow-level or dopant impurities is discussed in Chapters 2, 10, and 11. Shallow-level impurity densities are best measured electrically, but their energy levels are best determined optically. In this chapter we discuss predominantly the measurement of deep-level impurities whose densities and energy

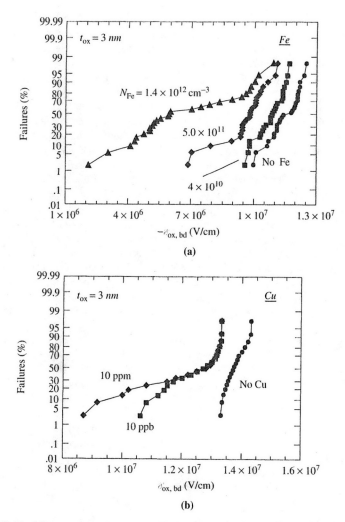

Fig. 5.3 Oxide failure percentage versus oxide breakdown electric field as a function of metal contamination for (a) Fe-contaminated Si and (b) Cu-contaminated Si; the wafers were dipped in a 10 ppb or 10 ppm CuSO₄ solution and annealed at 400°C. Data after ref. 1.

levels are best measured electrically. Milnes gives a good review of impurities in semiconductors.[2-3] Jaros treats the theoretical aspects of deep-level impurities.[4]

5.2   GENERATION-RECOMBINATION STATISTICS

5.2.1   A Pictorial View

The band diagram of a perfect single crystal semiconductor consists of a valence band and a conduction band separated by the band gap, with no energy levels within the band gap. When the periodicity of the single crystal is perturbed by foreign atoms or crystal defects, discrete energy levels are introduced into the band gap, shown by the E_T lines in Fig. 5.4. Each line represents one such defect. Such defects are commonly called generation-recombination (G-R) centers or traps. G-R centers lie deep in the band gap and are known as deep energy level impurities, or simply *deep-level impurities*. They act as recombination centers when there are excess carriers in the semiconductor and as generation centers when the carrier density is below its equilibrium value as in the reverse-biased space-charge region (scr) of pn junctions or MOS-capacitors, for example.

For single crystal semiconductors like silicon, germanium, and gallium arsenide, deep level impurities are usually metallic impurities, but they can be crystal imperfections, such as dislocations, stacking faults, precipitates, vacancies, or interstitials. Usually they are undesirable, but occasionally they are deliberately introduced to alter a device characteristic, *e.g.*, the switching time of bipolar devices. In some semiconductors like GaAs and InP, deep-level impurities raise the substrate resistivity, creating semi-insulating substrates. For amorphous semiconductors, defects are mainly due to structural imperfections.

Let us consider the deep-level impurity in Fig. 5.4 with an energy E_T and density N_T impurities/cm^3. The energy E_T is an effective energy discussed in Appendix 5.1. The semiconductor has n electrons/cm^3 in the conduction band and p holes/cm^3 in the valence band introduced by shallow-level dopants, not shown on the figure. To follow the various capture and emission processes, let the center first capture an electron from the conduction band (Fig. 5.4(a)), characterized by the capture coefficient c_n. After electron capture one of two events takes place. The center can either emit the electron back to the conduction band, called electron emission e_n (Fig. 5.4(b)), or it can capture a hole from the valence band, shown in Fig. 5.4(c) as c_p. After either of these events, the G-R center is occupied by a hole and again has two choices. Either it emits the hole back to the valence band e_p in Fig. 5.4(d) or captures an electron (Fig. 5.4(a)). These are the only four possible events between the conduction band, the impurity energy level, and the valence band. Process (d) is sometimes viewed as electron emission from the valence band to the impurity shown by the dashed arrow. We will, however, use the hole emission process in (d) because it lends itself more readily to mathematical analysis.

A *recombination* event is Fig. 5.4(a) followed by (c) and a *generation* event is (b) followed by (d). The impurity is a G-R center and *both* the conduction and valence bands participate in recombination and generation. These mechanisms are the topic of

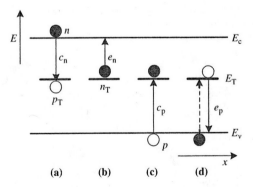

Fig. 5.4 Electron energy band diagram for a semiconductor with deep-level impurities. The capture and emission processes are described in the text.

Chapter 7. A third event that is neither recombination nor generation, is the *trapping* event (a) followed by (b) or (c) followed by (d). In either case a carrier is captured and subsequently emitted back to the band from which it came. Only one of the two bands and the center participate and the impurity is a trap. Impurities are frequently referred to as *traps*, regardless of whether they act as recombination, generation, or trapping centers. The subscript "T" in the following equations stands for trap.

Whether an impurity acts as a trap or a G-R center depends on E_T, the location of the Fermi level in the band gap, the temperature, and the capture cross-sections of the impurity. Generally those impurities with energies near the middle of the band gap behave as G-R centers, whereas those near the band edges act as traps. Generally the electron emission rate for centers in the upper half of the band gap is higher than the hole emission rate. Similarly the hole emission rate is generally higher than the electron emission rate for centers in the lower half of the band gap. For most centers one emission rate dominates, and the other can frequently be neglected.

5.2.2   A Mathematical Description

A G-R center can exist in one of two states. When occupied by an electron, it is in the n_T state and when occupied by a hole, it is in the p_T state (both shown in Fig. 5.4). If the G-R center is a donor, n_T is neutral and p_T is positively charged. For an acceptor, n_T is negatively charged and p_T is neutral. The density of G-R centers occupied by electrons n_T and holes p_T must equal the total density N_T or $N_T = n_T + p_T$. In other words, a center is either occupied by an electron or a hole. When electrons and holes recombine or are generated, the electron density in the conduction band n, the hole density in the valence band p, and the charge state of the center n_T or p_T are all functions of time. For that reason we will first address the question, "what is the time rate of change of n, p, and n_T?" We develop the appropriate equations for electrons. The equations for holes are analogous, and their derivation follows similar paths. A good discussion of the equations and their derivations is given by Sah et al.[5]

The electron density in the conduction band is diminished by electron capture (process (a) in Fig. 5.4) and increased by electron emission (process (b) in Fig. 5.4) and the electron time rate of change due to G-R mechanisms is[6-7]

$$\frac{dn}{dt}\Big|_{G-R} = (b) - (a) = e_n n_T - c_n n p_T \qquad (5.1)$$

The subscript "G-R" signifies that we are only considering emission and capture processes through G-R centers. We are not considering radiative or Auger processes. However, later in the chapter we address briefly optical emission as a mechanism to excite carriers into or out of G-R centers. Electron emission depends on the density of G-R centers occupied by electrons and the emission rate through the relation $(b) = e_n n_T$. This relationship does not contain n because it is not necessary for there to be electrons in the conduction band during the emission process. But the G-R centers must be occupied by electrons, for if there are no electrons on the centers, none can be emitted.

The capture process is slightly more complicated because it depends on n, p_T and the capture coefficient c_n through the relation $(a) = c_n n p_T$. The electron density n is important because, to capture electrons, there must be electrons in the conduction band. For holes we find the parallel expression

$$\frac{dp}{dt}\Big|_{G-R} = (d) - (c) = e_p p_T - c_p p n_T \qquad (5.2)$$

The emission rate e_n represents the electrons emitted per second from electron-occupied G-R centers. The capture rate $c_n n$ represents the density of electrons captured per second from the conduction band. The units are: e_n in $1/s$ and c_n in cm^3/s. You may wonder how there can be more than one electron emitted from a G-R center. After an electron has been emitted, the center finds itself in the p_T state and subsequently emits a hole, returning it to the n_T state. Then the cycle repeats.

Where do the electrons and holes come from for this cycle to continue? Surely they cannot come from the center itself. It may be helpful to view hole emission from the G-R center as electron emission from the valence band to the G-R center, indicated by the dashed line in Fig. 5.4(d). In this picture the electron-hole emission process is nothing more than an electron being excited from the valence band to the conduction band with an intermediate stop at the E_T level. However, it is easier to deal with the equations if we consider hole and electron emission as shown by the solid lines in Fig. 5.4.

The capture coefficient c_n is defined by

$$c_n = \sigma_n v_{th} \tag{5.3}$$

where v_{th} is the electron thermal velocity and σ_n is the electron capture cross-section of the G-R center. A physical explanation of c_n can be gleaned from Eq. (5.3). We know that electrons move randomly at their thermal velocity and that G-R centers remain immobile in the lattice. Nevertheless, it is helpful to change the frame of reference by letting the electrons be immobile and the G-R centers move at velocity v_{th}. The centers then sweep out a volume per unit time of $\sigma_n v_{th}$. Those electrons that find themselves in that volume have a very high probability of being captured. Capture cross-sections vary widely depending on whether the center is neutral, negatively, or positively charged. A center with a negative or repulsive charge has a smaller electron capture cross section than one that is neutral or attractively charged. Neutral capture cross-sections are on the order of 10^{-15} cm^2—roughly the physical size of the atom.

Whenever an electron or hole is captured or emitted, the center occupancy changes, and that rate of change is, from Eqs. (5.1) and (5.2), given by

$$\frac{dn_T}{dt}\Big|_{G-R} = \frac{dp}{dt} - \frac{dn}{dt} = (c_n n + e_p)(N_T - n_T) - (c_p p + e_n)n_T \tag{5.4}$$

This equation is non-linear, with n and p being time-dependent variables. If the equation can be linearized, it can be solved easily. Two cases allow this simplification. (1) In a reverse-biased space-charge region both n and p are small and can, to first order, be neglected. (2) In the quasi-neutral regions n and p are reasonably constant. Solving Eq. (5.4) for condition (2) gives $n_T(t)$ as

$$n_T(t) = n_T(0)\exp\left(-\frac{t}{\tau}\right) + \frac{(e_p + c_n n)N_T}{e_n + c_n n + e_p + c_p p}\left(1 - \exp\left(-\frac{t}{\tau}\right)\right) \tag{5.5}$$

where $n_T(0)$ is the density of G-R centers occupied by electrons at $t = 0$ and $\tau = 1/(e_n + c_n n + e_p + c_p p)$. The *steady-state density* as $t \to \infty$ is

$$n_T = \frac{e_p + c_n n}{e_n + c_n n + e_p + c_p p}N_T \tag{5.6}$$

This equation shows the steady-state occupancy of n_T to be determined by the electron and hole densities as well as by the emission and capture rates. Equations (5.5) and (5.6) are the basis for most deep-level impurity measurements.

Equation (5.5) is difficult to solve because neither capture nor emission rates may be known. Furthermore, n and p vary with time and generally also with distance in a device. Certain experimental simplifications are usually made to allow data interpretation. We will show the results of those simplifications here and the experimental implementations later.

For an n-type substrate where, to first order, p can be neglected, Equation (5.5) becomes

$$n_T(t) = n_T(0) \exp\left(-\frac{t}{\tau_1}\right) + \frac{(e_p + c_n n)N_T}{e_n + c_n n + e_p}\left(1 - \exp\left(-\frac{t}{\tau_1}\right)\right) \tag{5.7}$$

with $\tau_1 = 1/(e_n + c_n n + e_p)$. There are two cases of particular interest for the Schottky diode on an n-substrate in Fig. 5.5. The diode is at zero bias in Fig. 5.5(a). With n mobile electrons, capture dominates emission, and the steady-state G-R center density from Eq. (5.7) is $n_T \approx N_T$. When the diode is pulsed from zero to reverse bias as shown in Fig. 5.5(b), with most G-R centers initially occupied by electrons for $t \le 0$, electrons are emitted from the G-R centers for $t > 0$. Emission dominates during this reverse-bias phase because the emitted electrons are swept out of the reverse-biased space-charge region very quickly, thereby reducing the chance of being recaptured. The electron sweep-out or transit time is $t_t \approx W/v_n$. For $v_n \approx 10^7$ cm/s and W being a few microns, t_t is a few tens of picoseconds. This time is significantly shorter than typical capture times. However, near the edge of the scr the mobile electron density tails off into the scr from the quasi-neutral region even under reverse bias. This implies that the $c_n n$ term in Eq. (5.7) is

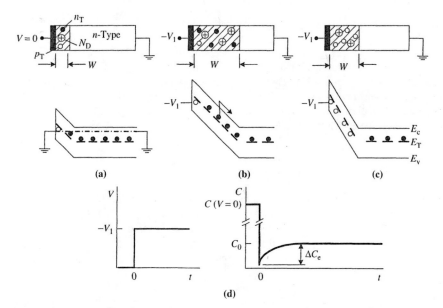

Fig. 5.5 A Schottky diode for (a) zero bias, (b) reverse bias at $t = 0$, (c) reverse bias as $t \to \infty$. The applied voltage and resultant capacitance transient are shown in (d).

not negligible in that part of the scr and electron emission competes with electron capture. With n not spatially homogeneous, τ is not constant, and the time dependence of $n(t)$ can be non-exponential.

Let us consider traps in the upper half of the band gap with $e_n \gg e_p$, allowing e_p to be neglected in Eq. (5.7). During the initial *emission* period, the time dependence of n_T simplifies to

$$n_T(t) = n_T(0) \exp\left(-\frac{t}{\tau_e}\right) \approx N_T \exp\left(-\frac{t}{\tau_e}\right) \tag{5.8}$$

with $\tau_e = 1/e_n$. Following electron emission from traps, holes remain and are subsequently emitted followed by electron emission, and so on. The steady-state trap density n_T in the reverse-biased scr is

$$n_T = \frac{e_p}{e_n + e_p} N_T \tag{5.9}$$

Some traps will be in the n_T and some will be in the p_T state. When the diode is pulsed from reverse bias to zero bias, electrons rush in to be captured by traps in the p_T state. The time dependence of n_T during the *capture* period is

$$n_T(t) = N_T - (N_T - n_T(0)) \exp\left(-\frac{t}{\tau_c}\right) \tag{5.10}$$

where $\tau_c = 1/c_n n$ and $n_T(0)$ is the initial steady-state density given by Eq. (5.9).

Similar equations to those in this section also hold for interface trapped charge. The relevant electron and hole densities are those at the surface, the traps are interface traps, and the capture and emission coefficients are those of the interface traps. The concepts, however, remain unchanged.

5.3 CAPACITANCE MEASUREMENTS

The equations in Section 5.2.2 describe the traps in terms of their densities and their emission and capture coefficients. With impurities being charged or neutral, and with electrons or holes emitted or captured, any measurement that detects charged species can be used for their characterization, *i.e.*, *capacitance*, *current*, or *charge* measurements. We will first discuss capacitance measurements and later address the other two. The capacitance of the Schottky diode of Fig. 5.5 is

$$C = A\sqrt{\frac{q K_s \varepsilon_o}{2}} \sqrt{\frac{N_{scr}}{V_{bi} - V}} \tag{5.11}$$

where N_{scr} is the ionized impurity density in the space-charge region. The ionized shallow-level donors (dopant atoms) in the scr are positively charged and $N_{scr} = N_D{}^+ - n_T^-$ for deep-level *acceptor* impurities that are negatively charged when occupied by electrons. When occupied by holes the deep level acceptors are neutral and $N_{scr} = N_D{}^+$. For shallow-level donors and deep-level donors occupied by electrons, $N_{scr} = N_D{}^+$. For deep-level donors occupied by holes, $N_{scr} = N_D{}^+ + p_T{}^+$.

The time-dependent capacitance reflects the time dependence of $n_T(t)$ or $p_T(t)$. Two chief methods are utilized to determine deep-level impurities. In the first, the steady-state capacitance is measured at $t = 0$ and at $t = \infty$. In the second, the time-varying capacitance is monitored.

5.3.1   Steady-State Measurements

We saw in Chapter 2 that plots of $1/C^2$ versus V yield the doping density. It is possible to determine N_T from such plots. For shallow-level donors and deep-level acceptors $1/C^2$ is given as

$$\frac{1}{C^2} = \frac{1}{K^2} \frac{V_{bi} - V}{N_D - n_T(t)} \tag{5.12}$$

For the reverse-biased diode of Fig. 5.5, $n_T(t)$ is negatively charged when occupied by electrons. With time, as electrons are emitted and the traps become neutral, $(N_D - n_T(t))$ increases and $1/C^2$ decreases. In steady-state measurements the reverse-biased capacitance at $t = 0$ is compared with the reverse-biased capacitance as $t \to \infty$. If we define a slope $S(t) = -dV/d(1/C^2)$, then

$$S(\infty) - S(0) = K^2[n_T(0) - n_T(\infty)] \tag{5.13}$$

For $n_T(0) \approx N_T$ and $n_T(\infty) \approx 0$, applicable for $e_n \gg e'_p$, the difference of the two slopes gives the deep-level impurity density. This method was used during early impurity measurements.[8] A slightly more detailed analysis takes account of those traps with energy levels below the Fermi level.[9] They do not emit and capture electrons as those levels above the Fermi level, perturbing the charge distributions somewhat, but is usually a minor effect.

5.3.2   Transient Measurements

Figure 5.5 shows the space-charge region width W to change when electrons are emitted from traps. In transient measurements it is this time-varying W that is detected as a time-varying capacitance. From Eq. (5.11)

$$C = A\sqrt{\frac{qK_s\varepsilon_o N_D}{2(V_{bi} - V)}} \sqrt{1 - \frac{n_T(t)}{N_D}} = C_0\sqrt{1 - \frac{n_T(t)}{N_D}} \tag{5.14}$$

where C_0 is the capacitance of a device with no deep-level impurities at reverse bias $-V$. It is, of course, possible to measure C and analyze the data as C^2 to avoid taking the square root. We address that method at the end of this section. However, for the most common use of transient capacitance measurements, the deep-level impurities form only a small fraction of the scr impurity density, *i.e.*, $N_T \ll N_D$. In other words, one is looking for trace amounts of impurities. Using a first-order expansion of Eq. (5.14) gives

$$C \approx C_0\left(1 - \frac{n_T(t)}{2N_D}\right) \tag{5.15}$$

Emission—Majority Carriers: Carrier emission is most commonly measured. The junction device is initially zero biased, allowing impurities to capture majority carriers (Fig. 5.5(a)). The capacitance is the zero-biased value $C(V = 0)$. Following a reverse bias pulse, majority carriers are emitted as a function of time (Fig. 5.5(b)). Equation (5.8) is the appropriate equation. When substituted into Eq. (5.15), we find

$$C = C_0\left[1 - \left(\frac{n_T(0)}{2N_D}\right)\exp\left(-\frac{t}{\tau_e}\right)\right] \tag{5.16}$$

Equation (5.16) is shown in Fig. 5.5(d) for $t > 0$. The scr is widest and the capacitance is lowest immediately after the device is reverse biased. As majority carriers are emitted from the traps (Fig. 5.5(b)), W decreases and C increases until steady state is attained (Fig. 5.5(c)). In Fig. 5.5(c) holes remain on the traps. What happens, of course, is that after electrons are emitted, holes will be emitted, then electrons, and so forth. This is the leakage current of reverse-biased diodes. Here we are only concerned with the initial electron emission to characterize the traps.

The same time dependence of the capacitance is observed for deep-level donor impurities in n-type substrates. In that case the impurities are neutral, when initially occupied by electrons, and the scr impurity density at $t = 0^+$ is N_D. As electrons are emitted, the traps become positively charged, and the final charge is $q[N_D + p_T(\infty)]$. Both charge and capacitance increase with time. The capacitance increases with time regardless of whether the deep-level impurities are donors or acceptors. Using the same arguments, it is straightforward to show that this is also true for p-type substrates with either donor or acceptor traps. *The capacitance increases with time for majority carrier emission whether the substrate is n- or p-type and whether the impurities are donors or acceptors.*

From the decay time constant of the C-t curve one derives τ_e and from the reverse-biased capacitance change, one obtains $n_T(0)$. Defining $\Delta C_e = C(t = \infty) - C(t = 0)$ we have

$$\Delta C_e = \frac{n_T(0)}{2N_D} C_0 \tag{5.17}$$

Plotting the capacitance difference

$$C(\infty) - C(t) = \frac{n_T(0)}{2N_D} C_0 \exp\left(-\frac{t}{\tau_e}\right) \tag{5.18}$$

as $\ln[C(\infty) - C(t)]$ versus t, gives a curve with slope $-1/\tau_e$ and intercept on the ln-axis of $\ln[n_T(0)C_0/2N_D]$. The emission time constant contains parameters describing the trap. To bring these out, we have to return to the capture and emission coefficients.

The capture and emission coefficients are related to each other through Eqs. (5.1) and (5.2). In equilibrium we invoke the *principle of detailed balance*, which states that under equilibrium conditions each fundamental process and its inverse must balance independent of any other process that may be occurring inside the material.[10-11] This requires fundamental process (a) in Fig. 5.4 to self-balance with its inverse process (b). Consequently $dn/dt = 0$ under *equilibrium conditions* and

$$e_{no}n_{To} = c_{no}n_o p_{To} = c_{no}n_o(N_T - n_{To}) \tag{5.19}$$

where the subscript "o" stands for equilibrium. n_o and n_{To} are defined as[10]

$$n_o = n_i \exp((E_F - E_i)/kT); n_{To} = \frac{N_T}{1 + \exp((E_T - E_F)/kT)} \tag{5.20}$$

Combining Eqs. (5.19) and (5.20) gives

$$e_{no} = c_{no}n_i \exp((E_T - E_i)/kT) = c_{no}n_1 \tag{5.21}$$

The derivation for holes gives an expression similar to Eq. (5.21).

Then a crucial assumption is made: *the emission and capture coefficients remain equal to their equilibrium values under non-equilibrium conditions.* This gives

$$e_n = c_n n_1; e_p = c_p p_1 \tag{5.22}$$

where

$$n_1 = n_i \exp((E_T - E_i)/kT); p_1 = n_i \exp(-(E_T - E_i)/kT) \tag{5.23}$$

The validity of the equilibrium assumption under non-equilibrium conditions is open to question. For small deviations from equilibrium, it may be assumed that the emission and capture coefficients do not deviate significantly from their equilibrium values.[12] Certainly it is a poor approximation in the reverse-biased junction scr where high electric fields exist, but that is precisely where most capacitance transient measurements are made. Capture cross-sections determined from emission measurements generally do not give true cross-section values, as discussed in Appendix 5.1. The equilibrium assumption is nevertheless a common assumption, and any measured results are subject to this uncertainty.

We show the electric field effect in Fig. 5.6. An electron energy diagram at zero electric field is shown by (1). An energy $E_c - E_T$ is required for electron emission from the trap to the conduction band. An applied electric field causes the bands to be slanted, as shown by (2), and the emission energy is reduced by the energy δE. Poole-Frenkel emission over the lowered barrier is shown as (a).[13] Even less energy is required for phonon-assisted tunneling, shown as (b), in which the electron is excited by phonons for only part of the energy barrier and then tunnels through the remaining barrier. As an example, the electric field dependence of the emission coefficient for the gold acceptor level in silicon is negligible for electric fields up to 10^4 V/cm, but for fields around 10^5 V/cm the emission coefficient increases by about a factor of two and continues to increase with higher fields.[14]

With $e_n = 1/\tau_e$ and $c_n = \sigma_n v_{th}$, the emission time constant is

$$\tau_e = \frac{\exp((E_i - E_T)/kT)}{\sigma_n v_{th} n_i} = \frac{\exp((E_c - E_T)/kT)}{\sigma_n v_{th} N_c} \tag{5.24}$$

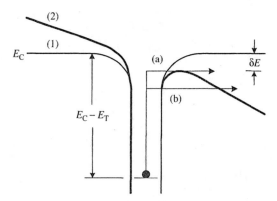

Fig. 5.6 Electron energy diagram in equilibrium (1) and in the presence of an electric field (2) showing field-enhanced electron emission: (a) Poole-Frenkel emission, (b) phonon-assisted tunneling.

TABLE 5.1   Coefficients $\gamma_{n,p}$ for Si and GaAs.

Semiconductor	$\gamma_{n,p}$ (cm^{-2}s^{-1}K^{-2})
n-Si	1.07×10^{21}
p-Si	1.78×10^{21}
n-GaAs	2.3×10^{20}
p-GaAs	1.7×10^{21}

A similar expression for holes is

$$\tau_e = \frac{\exp((E_T - E_i)/kT)}{\sigma_p v_{th} n_i} = \frac{\exp((E_T - E_v)/kT)}{\sigma_p v_{th} N_c} \tag{5.25}$$

where N_c and N_v are the effective conduction and valence band densities of state and the thermal velocities v_{th} differ slightly for electrons and holes. The emission time constant τ_e depends on the energy E_T and the capture cross-section σ_n. The emission time constants in Eqs. (5.24) and (5.25) are somewhat simplified. The energy differences $\Delta E_c = (E_c - E_T)$ and $\Delta E_v = (E_T - E_v)$ are actually Gibbs free energies ΔG, that differ from ΔE, discussed in Appendix 5.1.

The electron thermal velocity is

$$v_{th} = \sqrt{\frac{3kT}{m_n}} \tag{5.26}$$

and the effective density of states in the conduction band is

$$N_c = 2 \left(\frac{2\pi m_n kT}{h^2} \right)^{3/2} \tag{5.27}$$

allowing the emission time constant to be written as

$$\tau_e T^2 = \frac{\exp((E_c - E_T)/kT)}{\gamma_n \sigma_n} \tag{5.28}$$

with $\gamma_n = (v_{th}/T^{1/2})(N_c/T^{3/2}) = 3.25 \times 10^{21}(m_n/m_o)$ cm^{-2}s^{-1}K^{-2}, where m_n is the electron density-of-states effective mass.[15-16] The γ values for Si and GaAs[17] are given in Table 5.1. Modified GaAs values $\gamma_n = 1.9 \times 10^{20}$ cm^{-2}s^{-1}K^{-2} and $\gamma_p = 1.8 \times 10^{21}$ cm^{-2}s^{-1}K^{-2} have been proposed, based on a critical evaluation of GaAs parameters.[18]

Exercise 5.1

Problem: What are typical emission times for impurities with energy levels in the semiconductor band gap?

Solution: The emission time constant τ_e, given by Eq. (5.24), is plotted in Fig. E5.1, illustrating the large range of τ_e for a change in energy level $\Delta E = E_c - E_T$.

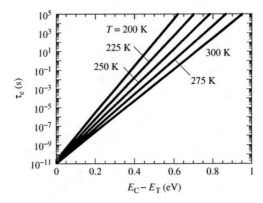

Fig. E5.1 Emission time constants for $\gamma_n = 1.07 \times 10^{21}$ cm^{-2}s^{-1}K^{-2} and $\sigma_n = 10^{-15}$ cm^2.

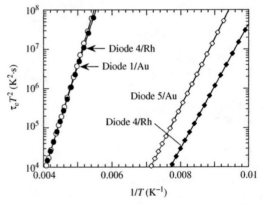

Fig. 5.7 $\tau_e T^2$ versus $1/T$ plots for Si diodes containing Au and Rh. Reprinted with permission after Pals. Ref. 19.

A plot of $\ln(\tau_e T^2)$ versus $1/T$, has a slope of $(E_c - E_T)/k$ and an intercept on the $\ln(\tau_e T^2)$ axis of $\ln[1/(\gamma_n \sigma_n)]$, leading to σ_n. Although this method of determining the capture cross-section is fairly common, the values so obtained should be viewed with caution. The cross-sections are affected by the electric fields in the scr as well as by other effects discussed in Appendix 5.1. An example plot for Au and Rh in Si is shown in Fig. 5.7, whose E_T and σ are shown in Table 5.2.

The energy levels and the capture cross-sections in Table 5.2 are determined from the intercept of the $\ln(\tau_e T^2)$ versus $1/T$ lines and by another method—the filling pulse method that is described in the sub-section "Capture—Majority Carriers". Note the large discrepancy between the two methods, with the intercept method giving values at least ten times larger. There are various reasons for this large discrepancy. Electric field enhanced emission tends to give larger cross-sections. As discussed in Appendix 5.1, the term $(\gamma_n \sigma_n)$ contains possible degeneracy factors and entropy terms, rendering the extrapolated cross-sections questionable.

TABLE 5.2   Energy Levels and Capture Cross Sections for the Diodes of Fig. 5.7.

Diode	$E_c - E_T$ (eV)	$E_c - E_T$ (eV)	$\sigma_{n,p}$ (intercept) (cm^2)	$\sigma_{n,p}$ (filling pulse) (cm^2)
$1 - p^+n$	0.56		2.8×10^{-14}	1.3×10^{-16}
$4 - p^+n$	0.315		1.6×10^{-13}	3.6×10^{-15}
$4 - p^+n$	0.534		7.5×10^{-15}	4×10^{-15}
$5 - n^+p$		0.346	1.5×10^{-13}	1.6×10^{-15}

The time constant τ_e can also be determined by combining Eqs. (5.12), (5.13), and (5.8) as

$$S(\infty) - S(t) = K^2 n_T(t) = K^2 n_T(0) \exp(-t/\tau_e) \tag{5.29}$$

and plotting $\ln[S(\infty) - S(t)]$ versus t. This was one of the earliest approaches.[9] However, the slope $-dV/d(1/C^2)$ is more complex to measure with automatic equipment than just C, and the method of Eq. (5.29) is rarely used today. Yet, Eq. (5.29) does not entail a small-signal expansion and is *not* subject to the limitation $N_T \ll N_D$.

Transient C-t data no longer follow a simple exponential time dependence when the emission rate is electric field dependent, when there are multiple exponentials due to several trapping levels with similar emission rates, and when the trap density is not negligibly small compared to the shallow-level dopant density. The analysis becomes more complicated for the last case, and we do not derive the relevant equations. This problem has been treated elsewhere.[20-23]

Emission—Minority Carriers: The preceding section considered the capacitance response to majority carrier capture and emission when a Schottky diode is pulsed between zero and reverse bias. Similar results obtain when a pn junction is pulsed between zero and reverse bias. With the pn junction there is an additional option. Under forward bias, minority carriers are injected. Let us consider a p^+n junction and neglect the p^+ region in this discussion. During the forward-bias phase, holes are injected into the n-substrate and capture dominates emission. The steady-state G-R center occupancy is from Eq. (5.6):

$$n_T = \frac{c_n n}{c_n n + c_p p} N_T \tag{5.30}$$

which depends on both capture coefficients and both carrier densities. The occupancy is difficult to predict, but the traps are no longer solely occupied by electrons as they are for the zero bias case; a certain fraction is occupied by holes. Schottky diodes do not inject minority carriers efficiently, and pn junctions should be used for electrical minority carrier injection. It is possible to inject minority carriers from high-barrier-height Schottky diodes with minority carrier storage at the inverted surface.[24-25]

For the sake of our discussion here, we assume $c_p \gg c_n$ and $p \approx n$. Then most traps are occupied by holes and for the deep-level acceptor impurities we have considered so far, the centers are neutral with $n_T \approx 0$ and $N_{scr} \approx N_D$ at $t = 0$ after the junction has been forward biased. When pulsed to reverse bias, minority holes are emitted from the traps, their charge changes from neutral to negative, and $N_{scr} \approx (N_D - n_T)$ for $t \to \infty$. The total ionized scr density *decreases*, the scr width increases, and the capacitance *decreases*

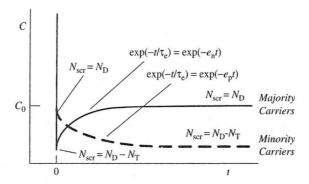

Fig. 5.8 The capacitance-time transients following majority carrier emission and minority carrier emission.

with time. This is shown in Fig. 5.8 and is opposite to majority carrier behavior. For simplicity, we assume in Fig. 5.8 all deep-level impurities to be filled with electrons (majority carrier emission) or holes (minority carrier emission) at $t = 0$. The capacitance transient is still described by an expression of the type in Eq. (5.16), with the emission time constant now $\tau_e = 1/e_p$.

Traps in the upper half of the band gap are generally detected with majority carrier pulses; those in the lower half of the band gap are observed with minority carrier pulses for n-type substrates. Traps with energies around the middle of the band gap can respond to either majority or minority carrier excitation. Minority carriers can also be injected optically as discussed later.

Capture—Majority Carriers: Consider the Schottky diode of Fig. 5.5(c). It has been reverse biased sufficiently long that all majority carriers have been emitted and the traps are in the p_T state. When the diode is pulsed from reverse bias (5.5(c)) to zero bias (5.5(a)), electrons rush into the scr to be captured by unoccupied traps. The density of traps able to capture majority carriers, for negligible emission, is given by

$$n_T(t) = N_T - [N_T - n_T(0)]\exp(-t_f/\tau_c) \tag{5.31}$$

where t_f is the capture or "filling" time. If there is sufficient time, i.e., $t_f \gg \tau_c$, essentially all traps capture electrons and $n_T(t_f \to \infty) \approx N_T$. If the time available for electron capture is short, only a fraction of the traps will be occupied by electrons when the diode returns to reverse bias. In the limit of very short times, i.e., $t_f \ll \tau_c$, very few electrons are captured and $n_T(t_f \to 0) \approx 0$.

When the device is reverse biased, $n_T(0)$ in Eq. (5.16) is given by Eq. (5.31), with the initial density during the emission phase equal to the final density of the capture phase. The reverse-bias capacitance at $t = 0$ then depends on the filling pulse width, shown by substituting Eq. (5.31) into (5.16) to give

$$C(t) = C_0 \left(1 - \frac{N_T - [N_T - n_T(0)]\exp(-t_f/\tau_c)}{2N_D} \exp\left(-\frac{t - t_f}{\tau_e}\right) \right) \tag{5.32}$$

Equation (5.32) is shown in Fig. 5.9(a).

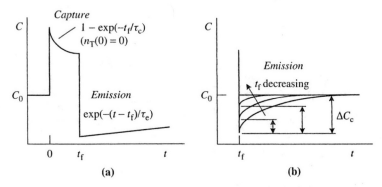

Fig. 5.9 (a) *C-t* response showing the capture and initial part of the emission process, (b) the emission *C-t* response as a function of capture pulse width.

The capture time τ_c can be determined by varying t_f, the filling pulse width. The capture time is usually much shorter than the emission time. We show the *C-t* curves during emission as a function of t_f in Fig. 5.9(b). The capacitance at $t = t_f^+$ is dependent on the capture time and is given by

$$C(t_f^+) = C_0 \left(1 - \frac{N_T - [N_T - n_T(0)]\exp(-t_f/\tau_c)}{2N_D} \right) \tag{5.33}$$

Equation (5.33) can be written as

$$\Delta C_C = C(t_f) - C(t_f = \infty) = \frac{N_T - n_T(0)}{2N_D} C_0 \exp\left(-\frac{t_f}{\tau_c}\right) \tag{5.34}$$

with ΔC_c shown on Fig. 5.9(b). Then t_f can be extracted from Eq. (5.34) by writing it as

$$\ln(\Delta C_C) = \ln\left(\frac{N_T - n_T(0)}{2N_D} C_0\right) - \frac{t_f}{\tau_c} \tag{5.35}$$

A plot of $\ln(\Delta C_c)$ versus t_f has a slope of $-1/\tau_c = -\sigma_n v_{th} n$ and an intercept on the $\ln(\Delta C_c)$ axis of $\ln\{[N_T - n_T(0)]C_0/2N_D\}$, obtained by varying the capture pulse width during the capacitance transient measurement. In this manner the capture cross-section is determined from *capture*, not emission. Since capture times are much shorter than emission times, the instrumentation is more demanding. Modifications to capacitance meters to accommodate the necessary narrow pulses are given in ref. 26. Sometimes one obtains non-linear $\ln(\Delta C_c)$ versus t_f plots due to slow capture from carrier tails extending into the scr. Models to derive σ_n from these curves are frequently too imprecise or involve complicated curve fitting routines, but are required for non-linear experimental data.[27]

A variation on this method is not to measure the capacitance as a function of time, but instead to keep the capacitance constant during the measurement through a feedback circuit and measure the voltage required to keep the capacitance constant.[28-29] The data analysis is similar and a plot of the voltage change ΔV required to keep the capacitance constant shows the expected semi-logarithmic behavior.

Equation (5.31) gives the capture time as $\tau_c = (\sigma_n v_{th} n)^{-1}$. The actual trap filling process is more complicated because not all traps empty during the emission process. Those traps with energy levels below the Fermi level will tend to remain occupied by electrons during the emission transient[29] and do not capture electrons during the filling pulse. This should be taken into account during the data analysis.

Capture—Minority Carriers: There are several methods to determine the capture properties of minority carriers. One method is very similar to that of the previous section, except that during the filling pulse the diode is forward biased. Various pulse widths are used to determine the capture properties.[26,30-31] Neglecting carrier emission, the capture time constant during the filling pulse is given by Eq. (5.5) as

$$\tau_c = \frac{1}{c_n n + c_p p} \tag{5.36}$$

and the trap occupancy will be that of Eq. (5.30). It depends not only on n and p, but also on c_n and c_p. The injected minority carrier density is varied by changing the injection level, and both c_n and c_p can be determined.[26] The narrow pulse widths (nanoseconds or lower) necessary to fill the centers partially are a decided disadvantage. A more fundamental limit is the turn-on time of junction diodes, because they do not turn on instantly following a sharp pulse. The minority carrier density builds up in a time related to the minority carrier lifetime. For the narrow pulses required for the capture measurements, it is very likely that the minority carrier density does not reach its steady-state value.

In an alternate method, the traps are populated with minority carriers not with constant-amplitude, varying-width bias pulses, but with constant-width, varying-amplitude pulses. The diode is forward biased with a long pulse, around 1 ms, and then reverse biased. The reverse-bias capacitance transient is observed. The minority carrier density is related to the injection current.[26] One must pay attention that minority carrier recombination with majority carriers is not significant.

It is also possible to inject minority carriers optically in pn junctions or Schottky diodes. We mention the method only briefly here and discuss it in more detail in Section 5.6.3. Consider a reverse-biased pn junction or Schottky barrier diode. A light pulse with photon energy $h\nu > E_G$ is flashed on the device, creating electron-hole pairs in the scr and in the quasi-neutral region. The minority carriers from the quasi-neutral region diffuse to the reverse-biased space-charge region to be captured by traps. With the light turned off, those captured minority carriers are emitted and detected as *C-t* or *I-t* transients. From the transient one determines E_T, σ_p, and N_T.[32]

5.4 CURRENT MEASUREMENTS

The carriers emitted from traps can be detected as a *capacitance*, a *charge*, or a *current*.[5,33-34] We saw earlier that the capacitance is given by Eq. (5.16). As the temperature changes, only the time constant changes; the initial capacitance step remains constant. For transient current measurements, the integral of the *I-t* curve represents the total charge emitted by the traps. For high temperatures, the time constant is short, but the initial current is high. For low temperatures, the time constant increases and the current decreases, but the area under the *I-t* curve remains constant. This makes current measurements difficult at low temperatures. By combining *C-t* measurements at the lower temperatures with I-t measurements at the higher temperatures, it is possible to obtain time constant data over ten orders of magnitude.[33]

Current measurements are more complicated because the current consists of emission current I_e, displacement current I_d, and junction leakage current I_l. The emission current is

$$I_e = qA \int_0^W \frac{dn}{dt} dx \qquad (5.37)$$

The displacement current is[5]

$$I_d = qA \int_0^W \frac{dn_T}{dt} \frac{x}{W} dx \qquad (5.38)$$

The lower limit of the integral in Eqs. (5.37) and (5.38) should have been the zero-biased scr width. However, for simplicity we have set the lower limit to zero. With $dn/dt \approx e_n n_T$ (Eq. (5.1)), $dn_T/dt \approx -e_n n_T$ (Eq. (5.4)), and electron emission dominating for the reverse-biased diode of Fig. 5.4, we find

$$I(t) = \frac{qAW(t)e_n n_T(t)}{2} + I_l = \frac{qAW_0 n_T(t)}{2\tau_e \sqrt{1 - n_T(t)/N_D}} + I_l \qquad (5.39)$$

using

$$W(t) = \sqrt{\frac{2K_s\varepsilon_o(V_{bi} - V)}{q(N_D - n_T(t))}} = \sqrt{\frac{2K_s\varepsilon_o(V_{bi} - V)}{qN_D(1 - n_T(t)/N_D)}} = \frac{W_0}{\sqrt{1 - n_T(t)/N_D}} \qquad (5.40)$$

For $n_T \ll N_D$ and using Eq. (5.8), the current becomes

$$I(t) = \frac{qAW_0}{2\tau_e} \frac{n_T(0)\exp(-t/\tau_e)}{1 - (n_T(0)/2N_D)\exp(-t/\tau_e)} + I_l \qquad (5.41)$$

The interpretation of current measurements is more complex than capacitance measurements because the I-t curve does not have a simple dependence on τ_e, i.e., τ_e appears in the numerator *and* the denominator of Eq. (5.41). If the second term in the denominator is small compared to unity for $n_T(0) \ll 2N_D$ and may be neglected, the current exhibits an exponential time dependence. The addition of the leakage current generally presents no problems since it is constant unless it is sufficiently high to mask the current transient. The instrumentation must be able to handle the large current transients during the pulse. The amplifier should be non-saturable, or the large circuit transients must be eliminated from the current transient of interest. A circuit with these properties is described in ref. 26.

Current transients do *not* allow a distinction between *majority* and *minority* carrier emission. Another feature of current measurements is a shift of the peak to higher temperatures relative to capacitance for the same rate window because the current is inversely proportional to the emission time constant (see Eq. (5.41)) while the capacitance is not. This property causes the current to increase very rapidly with temperature, effectively skewing the line shape toward higher temperatures.

Current measurements are preferred when it is difficult to make capacitance measurements. For example, the low capacitance of small-geometry MOSFETs or MESFETs is difficult to measure and the capacitance change is even smaller. In that case it is possible to detect the presence of deep-level impurities by pulsing the gate voltage and monitoring

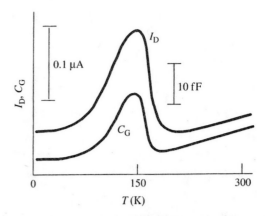

Fig. 5.10 Drain current I_D and gate capacitance C_G transients of a 100 μm × 150 μm gate MES-FET. Reprinted with permission after Hawkins and Peaker. Ref. 38.

the drain current as a function of time, known as *conductance* or *current DLTS*. Consider a MOSFET biased to some drain voltage and pulsed from accumulation to inversion, that is, from "off" to "on". Traps have captured majority carriers during the "off" state. A space-charge region is created when the device is turned "on" and drain current flows. As carriers are emitted from traps, the scr width and the threshold voltage change, causing a time-dependent drain current.[35] In *constant-resistance DLTS*, the MOSFET conductance is applied as an input signal to a feedback circuit, providing the voltage to compensate for the charge loss from traps during emission.[36] The mobility or transconductance need not be known. This technique is similar to the constant capacitance DLTS as it compensates for the emission of trapped carriers by adjusting the applied bias.

Current measurements work best in devices in which the channel can be totally depleted. In a MESFET, for example, the gate is pulsed from zero to reverse bias, creating a deep space-charge region. Electron or hole emission from traps changes the scr width and is measured as a drain current change that can be detected with the gate voltage held constant, or the gate voltage change can be detected with the current held constant through a feedback circuit.[37] Examples of MESFET drain current and capacitance data are shown in Fig. 5.10.[38] For these measurements it was necessary to use gate areas of 100 μm × 150 μm to obtain sufficiently large capacitances to be measurable.

Drain current measurements are relatively simple to implement, but they are more difficult to interpret than capacitance measurements for trap density extraction because the current is a change in drain current brought about by a changing scr width. Interpretation of the data requires a knowledge of the mobility.[39] This difficulty is circumvented by holding the drain current constant, changing the gate voltage, and converting gate voltage changes to current changes through the device transconductance.[38]

5.5 CHARGE MEASUREMENTS

Carriers emitted from traps can be detected directly as a *charge* with the circuit of Fig. 5.11. Switch S is closed to discharge the feedback capacitor C_F. At $t = 0$ the diode is reverse biased, S is opened, and from Eq. (5.41), with the second term in the denominator

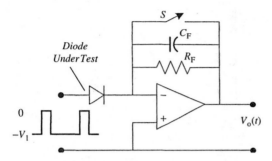

Fig. 5.11   Circuit for charge transient measurements.

neglected, the current through the diode for $t \geq 0$ is

$$I(t) = \frac{q A W_0}{2 \tau_e} n_T(0) \exp(-t/\tau_e) + I_1 \tag{5.42}$$

With the input current into the op-amp approximately zero, the diode current must flow through the $R_F C_F$ feedback circuit, giving the output voltage

$$V_0(t) = \frac{q A W_0 R_F n_T(0)}{2(t_F - \tau_e)} \left(\exp\left(-\frac{t}{t_F}\right) - \exp\left(-\frac{t}{\tau_e}\right) \right) + I_1 R_F \left(1 - \exp\left(-\frac{t}{t_F}\right) \right) \tag{5.43}$$

where $t_F = R_F C_F$. Choosing the feedback network such that $t_F \gg \tau_e$ reduces Eq. (5.43) to

$$V_0(t) \approx \frac{q A W_0 n_T(0)}{2 C_F} \left(1 - \exp\left(-\frac{t}{\tau_e}\right) \right) + \frac{I_1 t}{C_F} \tag{5.44}$$

Charge transient measurements have been implemented with the relatively simple circuit shown in Fig. 5.11.[40] The integrator replaces the high-speed capacitance meter in C-t measurements or the high-gain current amplifier in I-t measurements. The output voltage depends only on the total charge released during the measurement and is independent of τ_e. Charge measurements can also be used for MOS capacitor characterization.[41]

5.6   DEEP-LEVEL TRANSIENT SPECTROSCOPY (DLTS)

5.6.1   Conventional DLTS

The early C-t and I-t measurements and methods were developed by Sah and his students.[5,33] The initial implementation was time-consuming and tedious because the measurements were single-shot measurements. The power of emission and capture transient analysis was only fully realized when automated data acquisition techniques were adopted. The first of these was Lang's dual-gated integrator or double boxcar approach named *deep-level transient spectroscopy* (DLTS).[42–43]

Lang introduced the *rate window* concept to deep level impurity characterization. If the C-t curve from a transient capacitance experiment is processed so that a selected decay rate produces a maximum output, then a signal whose decay time changes monotonically with time reaches a peak when the rate passes through the rate window of a boxcar averager or the frequency of a lock-in amplifier. When observing a repetitive C-t transient through such a rate window while varying the decay time constant by varying the sample temperature, a peak appears in the capacitance versus temperature plot. Such a plot is a *DLTS spectrum*.[44−45] The technique, which is merely a method to extract a maximum in a decaying waveform, applies to capacitance, current, and charge transients.

We explain DLTS using capacitance transients. Assume the C-t transient follows the exponential time dependence

$$C(t) = C_0 \left[1 - \frac{n_T(0)}{2N_D} \exp\left(-\frac{t}{\tau_e}\right) \right] \tag{5.45}$$

with τ_e depending on temperature as

$$\tau_e = \frac{\exp((E_c - E_T)/kT)}{\gamma_n \sigma_n T^2} \tag{5.46}$$

The time constant τ_e decreases with increasing temperature, illustrated by the C-t curves in Fig. 5.12(a).

The capacitance decay waveform is typically corrupted with noise, and the heart of DLTS is the extraction of the signal from the noise in an automated manner. The technique is a correlation technique, which is a signal-processing method with the input signal multiplied by a reference signal, the weighting function $w(t)$, and the product filtered

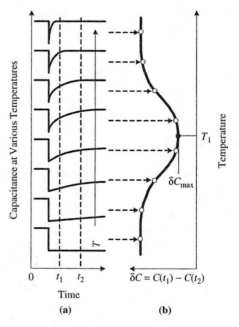

Fig. 5.12 Implementation of the rate window concept with a double boxcar integrator. The output is the average difference of the capacitance amplitudes at sampling times t_1 and t_2. Reprinted with permission after Miller et al.[44]

(averaged) by a linear filter. The properties of such a correlator depend strongly on the weighting function and on the filtering method. The filter can be an integrator or a low-pass filter. The correlator output is

$$\delta C = \frac{1}{T} \int_0^T f(t) w(t) \, dt = \frac{C_0}{T} \int_0^T \left(1 - \frac{n_T(0)}{2N_D} \exp\left(-\frac{t}{\tau_e}\right)\right) w(t) \, dt \qquad (5.47)$$

where T is the period and we use Eq. (5.45) for $f(t)$.

Boxcar DLTS: Suppose that the C-t waveforms in Fig. 5.12(a) are sampled at times $t = t_1$ and $t = t_2$ and that the capacitance at t_2 is subtracted from the capacitance at t_1, i.e., $\delta C = C(t_1) - C(t_2)$. Such a difference signal is a standard output feature of a double boxcar instrument. The temperature is slowly scanned while the device is repetitively pulsed between zero and reverse bias. There is no difference between the capacitance at the two sampling times for very slow or for very fast transients, corresponding to low and high temperatures. A difference signal is generated when the time constant is on the order of the gate separation $t_2 - t_1$, and the capacitance difference passes through a maximum as a function of temperature, as shown in Fig. 5.12(b). This is the DLTS peak. The capacitance difference, or DLTS signal, is obtained from Eq. (5.47), using the weighting function $w(t) = \delta(t - t_1) - \delta(t - t_2)$, as

$$\delta C = C(t_1) - C(t_2) = \frac{n_T(0)}{2N_D} C_0 \left(\exp\left(-\frac{t_2}{\tau_e}\right) - \exp\left(-\frac{t_1}{\tau_e}\right)\right) \qquad (5.48)$$

where $T = t_1 - t_2$ in Eq. (5.47).

δC in Fig. 5.12(b) exhibits a maximum δC_{max} at temperature T_1. Differentiating Eq. (5.48) with respect to τ_e and setting the result equal to zero gives $\tau_{e,max}$ at δC_{max} as

$$\tau_{e,max} = \frac{t_2 - t_1}{\ln(t_2/t_1)} \qquad (5.49)$$

Equation (5.49) is independent of the magnitude of the capacitance and the signal baseline need not be known. By generating a series of C-t curves at different temperatures for a given gate setting t_1 and t_2, one value of τ_e corresponding to a particular temperature is generated, giving one datum point on a $\ln(\tau_e T^2)$ versus $1/T$ plot. The measurement sequence is then repeated for another t_1 and t_2 gate setting for another point. In this manner, a series of points are obtained to generate an Arrhenius plot. δC-t plots for t_2/t_1 fixed, t_1 and t_2 varied are shown in Fig. 5.13. The effect of other t_1, t_2 variations on δC-t plots is discussed in Exercise 5.2.

Example DLTS spectra of iron-doped Si are shown in Fig. 5.14.[46] As discussed in Chapter 7, iron forms Fe-B pairs in boron-doped p-type Si with a DLTS peak at around $T = 50$ K. When the sample is heated at 180–200°C for a few minutes, the Fe-B pairs dissociate into interstitial iron and substitutional boron and the DLTS peak for the interstitial Fe occurs around $T = 250$ K. After a few days the interstitial iron again forms Fe-B pairs and the "$T = 50$ K" peak returns as shown in Fig. 5.14. Example DLTS spectra of Au-doped Si samples are shown in Fig. 5.15 showing both majority and minority carrier peaks.[47] The opposite polarity peaks correspond to the schematic diagrams in Fig. 5.8. The majority carrier peaks are measured with DLTS pulsed between zero and reverse bias. The minority carrier peaks are determined by optical minority carrier injection, where above band gap light, incident on the semitransparent Schottky diode, creates

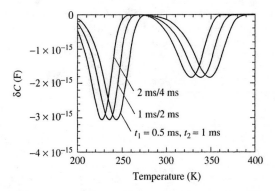

Fig. 5.13 DLTS spectra for t_2/t_1 fixed, t_1 and t_2 varied. $E_c - E_{T1} = 0.37$ eV, $\sigma_{n1} = 10^{-15}$ cm^2, $N_{T1} = 5 \times 10^{12}$ cm^{-3}, $E_c - E_{T2} = 0.6$ eV, $\sigma_{n2} = 5 \times 10^{-15}$ cm^2, $N_{T2} = 2 \times 10^{12}$ cm^{-3}, $C_0 = 4.9 \times 10^{-12} F$, $N_D = 10^{15}$ cm^{-3}.

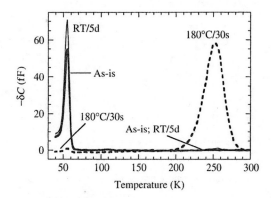

Fig. 5.14 DLTS spectra for iron-contaminated Si wafer; "As-is", after 180°C/30 s dissociation anneal, and room temperature storage for 5 days. Data after ref. 46.

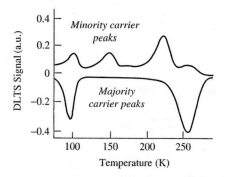

Fig. 5.15 Majority and minority carrier DLTS peaks for a Au-doped Si sample. Adapted from ref. 47.

electron-hole pairs. The sampling or gate width should be relatively wide, because the signal/noise ratio is proportional to the square root of the gate width.[45] Equation (5.49) then needs to be modified by changing t_1 to $(t_1 + \Delta t)$ and t_2 to $(t_2 + \Delta t)$ where Δt is the gate width.[48]

Exercise 5.2

Problem: What is the effect of varying the sampling times t_1 and t_2?

Solution: The sampling times can be varied by: (1) t_1 fixed, t_2 varied (Fig. E5.2(a)); (2) t_2 fixed, t_1 varied (Fig. E5.2(b)); (3) t_2/t_1 fixed, t_1 and t_2 varied (Fig. 5.14). Method (3) is best because the peaks shift with temperature with no curve shape change, making peak location easier. Additionally $\ln(t_2/t_1)$ remains constant. For methods (1) and (2) the peaks change both in size and in shape. Alternatively, one can vary t_2-t_1 at a constant temperature with t_2/t_1 constant. Then one would change the temperature and repeat to generate an Arrhenius plot from a single temperature scan.

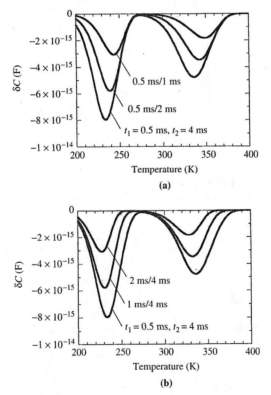

Fig. E5.2   DLTS spectra for (a) t_1 fixed, t_2 varied, (b) t_2 fixed, t_1 varied. $E_c - E_{T1} = 0.37$ eV, $\sigma_{n1} = 10^{-15}$ cm^2, $N_{T1} = 5 \times 10^{12}$ cm^{-3}, $E_c - E_{T2} = 0.6$ eV, $\sigma_{n2} = 5 \times 10^{-15}$ cm^2, $N_{T2} = 2 \times 10^{12}$ cm^{-3}, $C_0 = 4.9 \times 10^{-12} F$, $N_D = 10^{15}$ cm^{-3}.

The DLTS signal does not give the capacitance step ΔC_e of Fig. 5.5 ($\delta C_{max} < \Delta C_e$), and the impurity density cannot be determined from the DLTS signal using Eq. (5.17). The impurity density, derived from the maximum capacitance δC_{max} of the $\delta C\text{-}T$ curves, is given by

$$N_T = \frac{\delta C_{max}}{C_0} \frac{2N_D \exp\{[r/(r-1)]\ln(r)\}}{1-r} = \frac{\delta C_{max}}{C_0} \frac{2r^{r/(r-1)}}{1-r} N_D \qquad (5.50)$$

where $r = t_2/t_1$. Equation (5.50) is derived from Eqs. (5.48) and (5.49) with $\delta C_{max} = \delta C$, assuming $n_T(0) = N_T$. For $r = 2$, a common ratio, $N_T = -8N_D\delta C_{max}/C_0$, and for $r = 10$, $N_T = -2.87N_D\delta C_{max}/C_o$. The minus sign accounts for the fact that $\delta C < 0$ for majority carrier traps.

Well-maintained DLTS systems can detect $\delta C_{max}/C_0 \approx 10^{-5}$ to 10^{-4}, allowing trap densities on the order of $(10^{-5}$ to $10^{-4})N_D$ to be determined. High-sensitivity bridges allow measurements as low as $\delta C_{max}/C_0 \approx 10^{-6}$.[49] Capacitance meters often have response times of 1 to 10 ms and should be modified to allow faster transients to be measured. In addition, difficulties arise from overloads during device pulsing. Overload recovery delays are avoided by installing a fast relay that grounds the input of the amplifier during the pulse, deactivating the internal overload detection circuitry.[50]

Several refinements of the basic boxcar DLTS technique have been implemented. In the *Double-Correlation DLTS* (*D*-DLTS) method, pulses of two different amplitudes are used instead of the one-amplitude pulse of the basic technique. However, *D*-DLTS retains the conventional DLTS rate window concepts as shown in Fig. 5.16.[51] The weighting function gives the signal

$$[C'(t_1) - C(t_1)] - [C'(t_2) - C(t_2)] = \Delta C(t_1) - \Delta C(t_2) \qquad (5.51)$$

In the first correlation the transient capacitances after the two pulses are related to form the differences $\Delta C(t_1)$ and $\Delta C(t_2)$ at corresponding delay times after each pulse shown in Fig. 5.16. In a second step, the correlation $[\Delta C(t_1) - \Delta C(t_2)]$ is performed as in conventional DLTS to resolve the time constant spectrum during the temperature scan. The

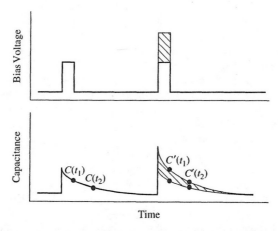

Fig. 5.16 Bias pulses and capacitance transients for double correlation DLTS. Reprinted with permission after Lefèvre and Schulz.[51]

measurement requires either a four-channel boxcar integrator or an external modification to a two-channel boxcar integrator.[52]

This added complexity sets an observation window within the space-charge region, allowing the impurities within this spatial window to be detected. By setting the window well within the scr, away from the quasi-neutral region scr edge, all traps are well above the Fermi level, and the capacitance transient is due to emission only. Traps near the Fermi level are excluded from the measurement and all traps within the window experience approximately the same electric field. Trap density profiles are obtained by varying the observation window or by changing the pulse amplitudes or the dc reverse bias.

Constant Capacitance DLTS: In *Constant Capacitance DLTS* (CC-DLTS) the capacitance is held constant during the carrier emission measurement by dynamically varying the applied voltage during the transient through a feedback path.[26,53–54] Miller pioneered the feedback method and applied it originally to carrier density profiling.[55] Just as the transient capacitance contains the trap information in the constant voltage method, so the time-varying voltage contains the trap information in the constant capacitance method. The approximate capacitance transient expression in Eq. (5.15) is valid for $N_T \ll N_D$. For $N_T > 0.1 N_D$ large changes occur in W and the C-t signal becomes non-exponential. Equation (5.14), which does not have this limitation, gives

$$V = -\frac{q K_s \varepsilon_o A}{2 C^2} \left(N_D - n_T(0) \exp\left(-\frac{t}{\tau_e}\right) \right) + V_{bi} \qquad (5.52)$$

valid for arbitrary N_T because the scr width is held constant and the resulting voltage change is directly proportional to the change in scr charge.

Equation (5.52) shows the V-t response to be exponential in time. Sometimes a non-exponential portion to the V-t curve occurs near $t = 0$, *e.g.*, by carrier capture even during the emission phase of the measurement. The majority carrier density does not drop abruptly to zero at the scr edge but tails into the scr, and electron emission competes with electron capture in that tail region. Electron capture dominates at the scr edge and most of the traps remain filled with electrons, leading to a non-exponential V-t curve.[56]

One of the limitations of CC-DLTS is the slower circuit response due to the feedback circuits. An early implementation was limited to transients with time constants on the order of a second,[57] that was reduced to about 10 ms for the same meter by using double feedback amplifiers.[58] The response time was later further reduced and the sensitivity increased.[59] However, feedback circuitry generally degrades the sensitivity of CC-DLTS compared to *Constant Voltage DLTS* (CV-DLTS). CC-DLTS is well suited for trap density depth profiling.[60] It has also been used for interface trapped charge measurements due to its high-energy resolution, and it permits more accurate DLTS measurements of defect profiles for high trap densities. Further refinements are possible by combining *D*-DLTS with CC-DLTS.[61]

Lock-in Amplifier DLTS: *Lock-in amplifier DLTS* is attractive because lock-in amplifiers are more standard lab instruments than boxcar integrators,[62] and they have a better signal/noise ratio than boxcar DLTS.[63] Lock-in amplifiers use a square wave weighting function whose period is set by the frequency of the lock-in amplifier. A DLTS peak is observed when this frequency bears the proper relationship to the emission time constant. A lock-in amplifier can be thought of as a one-component Fourier analyzer to analyze a repetitive signal. The weighting function resembles that of a boxcar integrator but is wider, increasing the signal/noise ratio but also posing an overload problem.

The device junction capacitance is very high during the forward-biased phase and tends to overload the relatively slow (response time ~ 1 ms unless modified) capacitance meter. A lock-in amplifier is very sensitive to the meter transient and overloads easily since its square wave weighting function has unit amplitude at all times. The boxcar does not have this problem because the first sampling window is delayed past the initial transient. The lock-in amplifier sensitivity to overloads can be reduced by preceding the weighting function by a narrow-band filter. This leads to an approximate sinusoidal weighting function. A better solution is to gate off the first 1 to 2 ms of the capacitance meter output, eliminating the overloading problems.[48,64] The analysis of the lock-in amplifier signal must include this gate-off time. The gate-off time also affects the base line which may become non-zero after the signal is suppressed part of the time.[65] The phase setting also affects the signal.[66] Details of three basic modes of lock-in DLTS operation and the relevant precautions to observe are discussed in ref. 48. Choosing a gate-off time that is always the same fraction of the repetition rate avoids problems of erroneous DLTS peaks.[67]

The details of a lock-in amplifier-based DLTS system are given by Rohatgi et al.[64] For the weighting function $w(t) = 0$ for $0 \leq t < t_d$, $w(t) = 1$ for $t_d < t < T/2$, $w(t) = -1$ for $T/2 < t < (T - t_d)$, and $w(t) = 0$ for $(T - t_d) < t < T$, the output from the lock-in amplifier is[63]

$$\delta C = -\frac{GC_0 n_T(0)}{N_D} \frac{\tau_e}{T} \exp\left(-\frac{t_d}{\tau_e}\right) \left[1 - \exp\left(-\frac{T - 2t_d}{2\tau_e}\right)\right]^2 \tag{5.53}$$

where G is the lock-in amplifier and capacitance meter gain, T is the pulse period, and the delay time t_d is the interval between the end of the bias pulse and the end of the holding interval. Equation (5.53) exhibits a maximum, similar to that of Eq. (5.48). Differentiating Eq. (5.53) with respect to τ_e and setting the result equal to zero allows $\tau_{e,\text{max}}$ to be determined from the transcendental equation

$$1 + \frac{t_d}{\tau_{e,\text{max}}} = \left(1 + \frac{T - t_d}{\tau_{e,\text{max}}}\right) \exp\left(-\frac{T - 2t_d}{2\tau_{e,\text{max}}}\right) \tag{5.54}$$

For a typical delay time of $t_d = 0.1T$, $\tau_{e,\text{max}} = 0.44T$. A $\ln(\tau_e T^2)$ versus $1/T$ plot is generated as described in the previous section once pairs of τ_e and T are known. The trap density, derived from Eqs. (5.53) and (5.54) for $\delta C = \delta C_{\text{max}}$ under the assumption that $n_T(0) = N_T$ and $t_d = 0.1T$, is given by

$$N_T = \frac{8\delta C_{\text{max}}}{C_0} \frac{N_D}{G}. \tag{5.55}$$

Instead of holding the lock-in frequency constant and varying the sample temperature, it is also possible to keep the temperature constant and vary the frequency.[68]

Correlation DLTS: Correlation DLTS is based on optimum filter theory, which states that the optimum weighting function of an unknown signal corrupted by white noise has the form of the noise-free signal itself. This can be implemented in DLTS by multiplying the exponential capacitance or current waveforms by a repetitive decaying exponential generated with an RC function generator and integrating the product.[63]

Correlation DLTS has a higher signal/noise ratio than either boxcar or lock-in DLTS.[69] Since the small capacitance transient rides on a dc background, it is not sufficient to use a simple exponential because the weighting function and baseline restoration are required.[70] The method has not found much application, but it has been used to study impurities in high-purity germanium.[71]

Isothermal DLTS: In the isothermal DLTS method, the sample temperature is held constant and the sampling time is varied.[72] The technique is also based on Eq. (5.45), repeated here

$$C(t) = C_0 \left[1 - \left(\frac{n_T(0)}{2N_D} \right) \exp\left(-\frac{t}{\tau_e} \right) \right] \tag{5.56}$$

Differentiating this expression and multiplying by time t, gives

$$t \frac{dC(t)}{dt} = -\frac{t}{\tau_e} \frac{n_T(0)}{2N_D} C_0 \exp\left(-\frac{t}{\tau_e} \right) \tag{5.57}$$

The function $t \, dC(t)/dt$ plotted versus t has a maximum value $(n_T(0)C_0/2N_D)(1/e)$ at $t = \tau_e$. Generating a series of $t \, dC(t)/dt$ versus t plots at several *constant* temperatures allows an Arrhenius plot of $\ln(\tau_e T^2)$ versus $1/T$, similar to a conventional DLTS plot. The chief difference is the constancy of the temperature during the measurement, easing the requirements on the temperature control/measurement. Instead, the measurement difficulty shifts to the time domain, where $C(t)$ measurements have to be made over a wide time range, requiring fast capacitance meters. Differentiating may introduce additional "noise" into the data. A plot of $t \, dC(t)/dt$ versus t for the same data as Fig. 5.13, is shown in Fig. 5.17. Note the close correspondence between the temperature dependence and the time dependence of the capacitance signal.

Computer DLTS: Computer DLTS refers to DLTS systems in which the capacitance waveform is digitized and stored electronically for further data management.[73] One temperature sweep of the sample is sufficient since the entire C-t curve is obtained at each of a number of different temperatures. It is readily established whether the signal is exponential; this is not possible with the boxcar or lock-in methods since those methods only

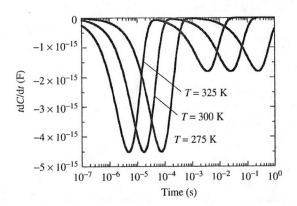

Fig. 5.17   DLTS spectra for T fixed, t varied. $E_c - E_{T1} = 0.37$ eV, $\sigma_{n1} = 10^{-15}$ cm^2, $N_{T1} = 5 \times 10^{12}$ cm^{-3}, $E_c - E_{T2} = 0.6$ eV, $\sigma_{n2} = 5 \times 10^{-15}$ cm^2, $N_{T2} = 2 \times 10^{12}$ cm^{-3}, $C_0 = 4.9 \times 10^{-12} F$, $N_D = 10^{15}$ cm^{-3}.

give maxima at selected temperatures but lose the waveform itself. Various signal processing functions can be applied to the C-t data: fast Fourier transforms, the method of moments to analyze simple and multiple exponential decays,[74-76] Laplace transform,[77] spectroscopic line fitting,[78] the covariance method of linear predictive modeling,[79] linear regression,[80] and an algorithm allowing the separation of closely spaced peaks.[81] One implementation uses a pseudo logarithmic sample storage scheme allowing 11 different sampling rates and 3–5 decades of time constants to be taken, that can separate closely spaced deep levels, where conventional DLTS fails.[82]

Laplace DLTS: There are two broad DLTS categories: analog and digital signal processing. Analog signal processing is done in real time as the sample temperature is ramped, choosing only one or two decay components at a time with filters producing an output proportional to the signal within a particular time constant range, by multiplying the capacitance meter output signal by a time-dependent weighting function. Digital schemes digitize the analog transient output of the capacitance meter and averaging many digitized transients to reduce the noise level. The time constant resolution of conventional DLTS is too poor for studying fine structure in the emission process due to the filter rather than thermal broadening. Even a perfect defect produces a broad line on the DLTS spectrum due to instrumental effects. Any emission time constant variation results in additional peak broadening. Some improvement in resolution is possible by changing the filter characteristic.[77]

A common approach to the quantitative description of non-exponential behavior in the capacitance transients is to assume that they are characterized by a spectrum of emission rates

$$f(t) = \int_0^\infty F(s)e^{-st}\, ds \qquad (5.58)$$

where $f(t)$ is the recorded transient and $F(s)$ is the spectral density function.[77] For simplicity, this spectrum is sometimes represented by a Gaussian distribution overlaying the logarithmic emission rate scale. In this way it is possible to describe the non-exponential transient in terms of broadening of the emission activation energy.

A mathematical representation of the capacitance transients given by Eq. (5.58) is the Laplace transform of the true spectral function $F(s)$. To find a real spectrum of the emission rates in the transient it is necessary to use an algorithm that effectively performs an inverse Laplace transform for the function $f(t)$, yielding a spectrum of delta-like peaks for multi-, mono-exponential transients, or a broad spectrum with no fine structure for continuous distribution. It is not necessary to make any *a priori* assumptions about the functional shape of the spectrum, except that all decays are exponential in the same direction.

Laplace DLTS (L-DLTS) gives an intensity output as a function of emission rate. The area under each peak is directly related to the initial trap concentration. The measurement is carried out at a fixed temperature, and several thousand capacitance transients are captured and averaged. L-DLTS can provide an order of magnitude higher energy resolution than conventional DLTS techniques, provided a good signal-to-noise ratio exists. In practice this limits the application to cases where the defect density is 5×10^{-4} to 5×10^{-2} of the shallow donor or acceptor density. Given these limitations, L-DLTS enables a range of measurements which are not practical in other systems. It is very important to reduce all noise contributions. For example, it is very important to use very stable power supplies and pulse generators.

An obvious application of L-DLTS is to separate states with very similar emission rates. The poor resolution of conventional DLTS has resulted in considerable confusion over

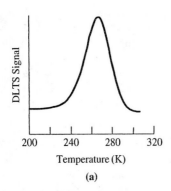 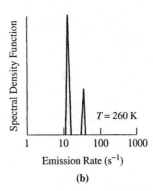

(a)   **(b)**

Fig. 5.18 (a) DLTS and (b) Laplace DLTS spectra of hydrogenated silicon containing gold. The DLTS peak is attributed to electron emission from the gold acceptor and gold-hydrogen levels. The Laplace spectrum clearly separates the gold-acceptor level and the gold-hydrogen. Adapted from Deixler et al.[83]

the "identity" of particular DLTS fingerprints. Using conventional DLTS, it is sometimes possible to separate states with very similar emission rates, provided they have different activation energies, by conducting the DLTS experiment over a very wide range of rate windows. An example is shown in Fig. 5.18. Figure 5.18(a) gives a conventional DLTS peak of gold in Si. This sample was hydrogen annealed and there should be a hydrogen-gold peak, which is not obvious, however. The *L*-DLTS spectrum, which is a plot of spectral density function versus emission rate, in Fig. 5.18(b) clearly shows two distinct peaks.[83] Knowing the emission rate allows the energy level to be determined.

Laplace DLTS has been used for Pt-doped Si, EL2 in GaAs, and DX defects in AlGaAs, GaSb, GaAsP, and δ-doped GaAs.[84] In each case the standard DLTS gave featureless peaks while the Laplace DLTS spectra revealed the fine structure in the thermal emission process.

5.6.2 Interface Trapped Charge DLTS

The instrumentation for *interface trapped charge DLTS* is identical to that for bulk deep-level DLTS. However, the data interpretation is different because interface traps are continuously distributed in energy through the band gap, whereas bulk traps have discrete energy levels. We illustrate the interface trapped charge majority carrier DLTS concept for the MOS capacitor (MOS-*C*) in Fig. 5.19(a). For a positive gate voltage electrons are captured and most interface traps are occupied by majority electrons for *n*-substrates (Fig. 5.19(b)). A negative gate voltage drives the device into deep depletion, and electrons are emitted from interface traps (Fig. 5.19(c)). The emitted electrons give rise to a capacitance, current, or charge transient. Although electrons are emitted over a broad energy spectrum, emission from interface traps in the upper half of the band gap dominates. DLTS is very sensitive, allowing interface trap density determination in the mid 10^9 cm$^{-2}eV^{-1}$ range.

Interface trap characterization by DLTS was first implemented with MOSFETs.[85] MOS-FETs, being three-terminal devices, have an advantage over MOS capacitors (MOS-Cs). By reverse biasing the source/drain and pulsing the gate, majority electrons are captured and emitted without interference from minority holes that are collected by the source-drain.

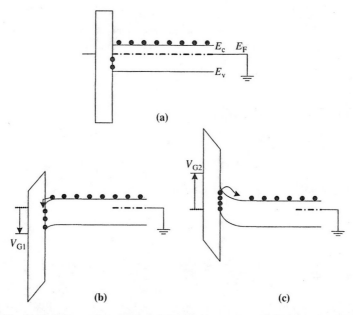

Fig. 5.19   (a) Majority carrier capture and (b) majority carrier emission from interface traps.

This allows interface trap majority carrier characterization in the upper half of the band gap. With the source-drain forward biased, an inversion layer forms, allowing interface traps to be filled with minority holes. Minority carrier characterization is then possible and the lower half of the band gap can be explored. This is not possible with MOS-Cs because there is no minority carrier source. When an inversion layer does form through thermal ehp generation, especially at higher temperatures and at high ehp generation rates, it can interfere with majority carrier trap DLTS measurements.

MOS capacitors are, nevertheless, used for interface trap characterization.[53,86-87] Unlike the conductance technique discussed in Chapter 6, DLTS measurements are independent of surface potential fluctuations. The derivation of the capacitance expression is more complex for MOS-Cs than it is for diodes. We quote the main results whose derivations can be found in Johnson[54] and Yamasaki et al.[87] For $q^2 D_{it} = C_{it} \ll C_{ox}$ and $\delta C = C_{hf}(t_1) - C_{hf}(t_2) \ll C_{hf}$

$$\dot{\delta C} = \frac{C_{hf}^3}{K_s \varepsilon_o N_D C_{ox}} \int_{-\infty}^{\infty} D_{it}(e^{-t_2/\tau_e} - e^{-t_1/\tau_e}) \, dE_{it} \tag{5.59}$$

where

$$\tau_e = \frac{e^{(E_c - E_{it})/kT}}{\gamma_n \sigma_n T^2} \tag{5.60}$$

E_{it} is the energy of the interface traps. The maximum emission time is $\tau_{e,\max} = (t_2 - t_1)/\ln(t_2/t_1)$ from Eq. (5.49). In conjunction with Eq. (5.60) where $\tau_{e,\max}$ corresponds to $E_{it,\max}$, we find, when the electron capture cross-section is not a strong function of energy,

$$E_{it,\max} = E_c - kT \ln \left(\frac{\gamma_n \sigma_n T^2 (t_2 - t_1)}{\ln(t_2/t_1)} \right) \tag{5.61}$$

where $E_{it,max}$ is sharply peaked. If D_{it} varies slowly in the energy range of several kT around $E_{it,max}$, it can be considered reasonably constant and can be taken outside the integral of Eq. (5.59). The remaining integral becomes

$$\int_{-\infty}^{\infty} (e^{-t_2/\tau_e} - e^{-t_1/\tau_e}) \, dE_{it} = -kT \ln(t_2/t_1) \tag{5.62}$$

allowing Eq. (5.59) to be written as

$$\delta C \approx \frac{C_{hf}^3}{K_s \varepsilon_o N_D C_{ox}} kT D_{it} \ln(t_2/t_1) \tag{5.63}$$

From Eq. (5.63) the interface trap density is

$$D_{it} = -\frac{K_s \varepsilon_o N_D C_{ox}}{kT C_{hf}^3 \ln(t_2/t_1)} \delta C \tag{5.64}$$

determined from electrons emitted from interface traps in time $(t_2 - t_1)$ in the energy interval $\Delta E = kT \ln(t_2/t_1)$ at energy $E_{it,max}$. A plot of D_{it} versus E_{it} is constructed by varying t_1 and t_2. For each t_1, t_2 combination, an E_{it} is obtained from Eq. (5.60) and a D_{it} from Eq. (5.64). If the sample contains bulk as well as interface traps, it is possible to differentiate bulk traps from interface traps by the shape and the peak temperature of the DLTS plot.[87]

For the *constant capacitance* DLTS technique an equation analogous to Eq. (5.64) is[54]

$$D_{it} = \frac{C_{ox}}{qkT A \ln(t_2/t_1)} \Delta V_G \tag{5.65}$$

where A is the device area and ΔV_G is the gate voltage change required to keep the capacitance constant. Equation (5.65) is easier to use than (5.64) because neither the high-frequency capacitance nor the doping density need be known. Figure 5.20 shows the interface trap distribution for n-Si, with D_{it} measured by the quasi-static and the

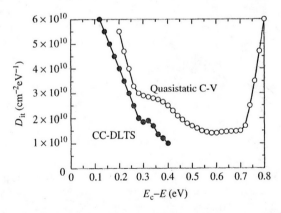

Fig. 5.20 Interface trapped charge density for n-Si measured by the CC-DLTS and quasi-static methods. Reprinted with permission after Johnson et al. Ref. 88.

CC-DLTS technique.[88] The discrepancy between the two curves may be due to the assumption of constant capture cross-sections in the DLTS analysis.

MOS capacitors can also be measured by the current DLTS method. Using the small pulse method,[89] in which pulses of tens of millivolts are used, both interface trap density and capture cross-sections can be measured.[90] Small filling pulses are applied as the quiescent bias is scanned at constant temperature and constant rate window. As the Fermi level scans the band gap, a DLTS peak is observed when τ_e in a small energy region around the Fermi level matches the rate window. Varying the rate window or the temperature gives the interface trap distribution.

5.6.3   Optical and Scanning DLTS

Optical DLTS comes in various implementations. Light can be used (1) to determine optical properties of traps, such as optical capture cross-sections, (2) to create electron-hole pairs for minority carrier injection, and (3) to create ehps in semi-insulating materials, where electrical injection is difficult. Light does two basic things: it imparts energy to a trapped carrier, causing its emission from a trap to the conduction or to the valence band, and it changes n and/or p by creating ehps, thereby changing the capture properties of the center. An electron beam in a scanning electron microscope also creates ehps and can be used for DLTS measurements.

Optical Emission: For conventional majority carrier emission, a Schottky diode on an n-type substrate is zero biased and traps are filled with electrons at low temperatures. Instead of raising the temperature and detecting the capacitance or current transient due to thermal emission, the sample is held at a sufficiently low temperature for negligible thermal emission. Light is shone on the sample provided with a transparent or semitransparent contact. For $h\nu < (E_c - E_T)$ there is no band gap optical absorption. For $h\nu > E_c - E_T$ photons excite electrons from the traps into the conduction band. Equation (5.8) holds, but the emission rate e_n becomes $e_n + e_n^o$, where e_n^o is the *optical emission rate* $e_n^o = \sigma_n^o\Phi$, with σ_n^o the *optical* capture cross-section and Φ the photon flux density. The trap density is obtained from the capacitance step just as it is during thermal emission measurements. The light is used in these experiments to determine optical trap properties, such as the optical cross-section, using either capacitance or current transients.[30,91-93]

It is possible to determine the multiplicity of charge states by varying the energy of the incident light. For a center with two donor levels, for example, one increases the light energy to excite electrons from the upper level into the conduction band, detected by a capacitance change. Increasing the energy further leaves the capacitance unchanged, provided all electrons have been excited out of that level, until the energy is sufficient to excite electrons from the second level into the conduction band, giving a second capacitance rise. This has been used to determine the double-donor nature of sulfur in silicon.[94]

In the two-wavelength method, a steady-state, above band gap background light creates a steady-state population of holes on traps below the Fermi level and of electrons on traps above the Fermi level. A variable-energy probe light excites carriers from the traps into either of the bands while the junction is pulsed electrically,[95] or ehps are generated optically by above band-gap light.[96] Both electrons and holes can be captured by traps in the scr. When the light is turned off, the carriers are thermally emitted. In this method, the light merely generates ehps; the transient is due to thermal emission. Other optical techniques were mentioned earlier when we discussed the use of light to generate ehps for the measurement of the minority carrier capture cross-sections.[26,32]

Photoinduced Current Transient Spectroscopy: The optical techniques of the previous section supplement electrical measurements. Although the measurements can generally be done electrically, the optical input makes the measurement easier (minority carrier generation) or gives additional information (optical cross-section). But purely electrical measurements are difficult in high-resistivity or semi-insulating substrates, *e.g.*, GaAs and InP. Optical inputs can then be a decided advantage and in some cases are the only way to obtain information of deep level impurities.

In the *photoinduced current transient spectroscopy* (PITS or PICTS) method the current is measured as a function of time. The sample is provided with a top semitransparent ohmic contact. Capacitance cannot be measured because the substrate resistance is too high. During the PITS measurement light is pulsed on the sample, and the photocurrent rises to a steady-state value. The light pulse can have above band-gap or below band-gap energy.[97] The photocurrent transient at the end of the light pulse consists of a rapid drop followed by a slower decay. The initial rapid drop is due to ehp recombination and the slow decay is due to carrier emission. The slow current transient can be analyzed by DLTS rate window methods.[98] It is sometimes possible to determine whether the level is an electron or a hole trap by measuring the peak height as the bias polarity is changed. However, this identification is not as simple as it is for capacitance transients.

For electron traps and sufficient light intensity to saturate the photocurrent, the transient current is[99]

$$\delta I = \frac{C N_T}{\tau_e} \exp(-t/\tau_e) \tag{5.66}$$

where C is a constant [see Eq. (5.42)]. When plotted against temperature, δI exhibits a maximum for $t = \tau_e$ as determined by differentiating Eq. (5.66) with respect to temperature,

$$\frac{d(\delta I)}{dT} = \frac{K N_T}{\tau_e^3}(t - \tau_e) \exp(-t/\tau_e)\frac{d\tau_e}{dT} \tag{5.67}$$

and setting Eq. (5.67) equal to zero.

PITS is not well suited for trap density determination, and the reliability of information extracted from the data for trap identification falls off as the trap energy approaches the intrinsic Fermi level.[108] Additional complications occur when carriers emitted from traps recombine. The recombination lifetime for semi-insulating materials is usually quite low. In addition, emitted carriers can be retrapped. All of these effects make the method difficult to use.[100] Unfortunately, there are few techniques other than PITS to characterize such materials.

Scanning DLTS: Scanning DLTS (*S*-DLTS) uses a scanning electron microscope electron beam as the excitation source. The high spatial resolution—in the micron range—is its main advantage, but also one of its disadvantages because such a small sampling area produces very small DLTS signals. For conventional DLTS the diode diameter is typically in the 0.5 to 1 mm range, and the entire area is active during the measurement. For *S*-DLTS the diode diameter is similar, giving rise to a large steady-state capacitance. But the emission-active area, defined by the electron beam diameter, can be much smaller and gives very small capacitance changes. The original *S*-DLTS used current DLTS because it can be more sensitive than capacitance DLTS.[101] Equation (5.41) shows the current to be inversely proportional to the emission time constant. As T increases, τ_e decreases, and hence I increases. Later developments of an extremely sensitive capacitance meter with 10^{-6} pF sensitivity, consisting of a resonance-tuned LC bridge at 28 MHz with permanent slow automatic zero balance to ensure operation in a tuned state at all times, allowed capacitance DLTS measurements.[102] Quantitative measurements are difficult to implement in S-DLTS,[103] but one can map a distribution of a particular impurity by scanning the device area, choosing an appropriate temperature and rate window. A few hundred impurity atoms per scanning point have been detected.[104]

5.6.4 Precautions

Leakage Current: Several measurement precautions have already been mentioned throughout this chapter. Here we point out a few more. Devices sometimes exhibit high reverse-bias leakage currents. During DLTS measurements of leaky MOS capacitors, the DLTS peak amplitude decreases much more strongly with slower rate windows than expected. This was attributed to competition between carrier capture due to leakage current and thermal emission. The thermal emission rate then becomes an apparent rate given by

$$e_{n,app} = e_n + c_n n \tag{5.68}$$

We can write the leakage current density as

$$J_{leak} = qnv = \frac{qnvc_n}{c_n} = \frac{qnvc_n}{\sigma_n v_{th}} \approx \frac{qnc_n}{\sigma_n} \tag{5.69}$$

assuming $v \approx v_{th}$. Substituting Eq. (5.69) into Eq. (5.68) gives

$$e_{n,app} = e_n + \frac{J_{leak}\sigma_n}{q} \tag{5.70}$$

If we assume the leakage current to be of the form[105]

$$J_{leak} = q A^* T^2 e^{-E_A/kT} \tag{5.71}$$

then Eq. (5.28) becomes

$$\tau_e T^2 = \frac{\exp((E_c - E_T)/kT)}{\sigma_n \gamma_n (1 - (A^*/\gamma_n) \exp((E_c - E_T - E_A)/kT))} \tag{5.72}$$

If Eq. (5.72) applies, errors in the trap energy and capture cross-section extracted from an Arrhenius plot will result.[105] For leaky diodes, an experimental system with two diodes, having similar $C-V$ and $I-V$ characteristics, is driven 180° out of phase.[106]

Series Resistance: Another device anomaly that can affect the DLTS response is the device series resistance and parallel conductance. A pn or Schottky diode consisting of junction capacitance C, junction conductance G, and series resistance r_s in Fig. 5.21(a). Capacitance meters assume the device to be represented by either the parallel equivalent circuit in Fig. 5.21(b) or the series equivalent circuit in Fig. 5.21(c). C_P and C_S can be written as

$$C_P = \frac{C}{(1 + r_s G)^2 + (\omega r_s C)^2} \approx \frac{C}{1 + (\omega r_s C)^2} ; C_S = C\left(1 + \left(\frac{G}{\omega C}\right)^2\right) \tag{5.73}$$

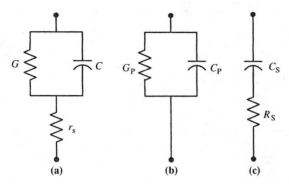

Fig. 5.21 (a) Actual circuit, (b) parallel equivalent circuit, and (c) series equivalent circuit for a pn or Schottky diode.

where $\omega = 2\pi f$ and the "$r_s G$" term in the denominator was neglected in the approximate expression.

A DLTS measurement records the change in capacitance given by

$$\Delta C_P = \frac{\Delta C}{1 + (\omega r_s C)^2}\left(1 - \frac{2(\omega r_s C)^2}{1 + (\omega r_s C)^2}\right); \Delta C_S = \Delta C\left(1 - \left(\frac{G}{\omega C}\right)^2\right) \qquad (5.74)$$

where ΔC_P depends on r_s and ΔC_S depends on G. For $r_s = 0$ and $G = 0$, $\Delta C_P = \Delta C_S = \Delta C$. However, as r_s increases, ΔC_P decreases. ΔC_P and the DLTS signal can become zero and even reverse sign and majority carrier traps can be mistaken for minority carrier traps.[78,107] Similarly, as G increases, ΔC_S decreases and can also become negative.

If series resistance is anticipated to be a problem, one can insert additional external resistance into the circuit and check for sign reversal.[108] If sign reversal is not observed, there is a good chance that it has already taken place without any additional external resistance, and the measured data must be carefully evaluated. Occasionally an additional capacitance is introduced by an oxide layer at the back of the sample, which can also lead to DLTS signal reversal.[109] Series resistance is not a particular problem for current DLTS because it is essentially a dc measurement, not requiring the high probe frequency of capacitance DLTS.

Instrumentation Considerations: The temperature of the sample has to be precisely controlled and measured for precise energy level extraction. Temperature control and measurement to 0.1 K is desirable. That is not always easy to do, since the thermocouple or diode used for temperature measurements is usually located in a heat sink block away from the sample under test. The capacitance meter should be sufficiently fast to be able to follow the smallest transient of interest. For some instruments it is necessary to block the large capacitance during the filling pulse to prevent instrument overload. A good discussion of instrument considerations is given in ref. 43.

Incomplete Trap Filling: We have assumed that all traps fill with majority carriers during the capture time and emit majority carriers during the emission time. That is only an assumption as illustrated with the band diagram in Fig. 5.22.[110] For the zero-biased device in Fig. 5.22(a), traps within W_1 do not fill because they are above the Fermi level;

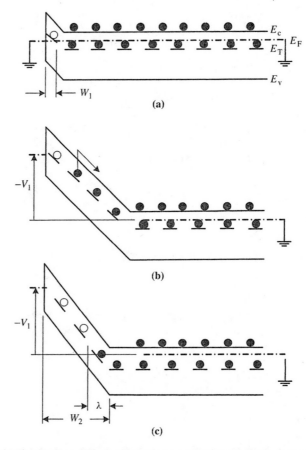

Fig. 5.22 Band diagram for a Schottky diode on an n-substrate. (a) Diode at zero bias during the filling phase, (b) immediately after the reverse bias pulse, (c) steady-state reverse bias.

those traps to the right of W_1, but near W_1, fill more slowly than those further to the right because the electron density tails off. Consequently, for narrow filling pulses, not all traps to the right of W_1 become occupied by electrons. When the bias switches to reverse bias, Fig. 5.22(b), electrons are emitted. However, those traps within λ do not emit electrons because they are below the Fermi level (Fig. 5.22(c)), where W_2 is the final scr width and λ is given by[45]

$$\lambda = \sqrt{\frac{2K_s\varepsilon_o(E_F - E_T)}{q^2 N_D}} \tag{5.75}$$

Only those traps within $(W\text{-}W_1\text{-}\lambda)$ participate during the DLTS measurement.[111] W_1 is almost always neglected; frequently λ is neglected too. When λ is not neglected, the capacitance step ΔC_e of Eq. (5.17) becomes[45,112]

$$\Delta C_e = \frac{n_T(0)}{2N_D} C_0 f(W) \tag{5.76}$$

where

$$f(W) = 1 - \frac{(2\lambda/W(V))(1 - C(V)/C(0)}{1 - [C(V)/C(0)]^2} \tag{5.77}$$

$C(0)$ and $C(V)$ are the capacitances at voltages zero and V, respectively. If the edge region can be neglected, $f(W)$ becomes unity. However, with $f(W) < 1$, neglecting the edge region can introduce appreciable error.[113]

Blackbody Radiation: The usual assumption is that the device is in the dark during DLTS measurements. This is true if the device is encapsulated with the case at the measurement temperature. If, however, the device is in wafer form and it "sees" a part of the dewar at a temperature higher than the measurement temperature, *e.g.*, room temperature, it is possible for photons in the blackbody radiation spectrum to cause optical emission to add to thermal emission and give erroneous activation energies. If this is a concern, it is experienced at low temperatures and at low scanning rates.[114]

5.7 THERMALLY STIMULATED CAPACITANCE AND CURRENT

Thermally stimulated capacitance (TSCAP) and *current* (TSC) measurements were popular before DLTS. The techniques were originally used for insulators and later adapted to lower resistivity semiconductors when it was recognized that the reverse-biased scr is a region of high resistance.[115] During the measurement the device is cooled and the traps are filled with majority carriers at zero bias or traps can be filled with minority carriers by optical injection or by forward biasing a pn junction. Then the device is reverse biased, heated at a constant rate, and the steady-state capacitance or current is measured as a function of temperature. Capacitance steps or current peaks are observed as traps emit their carriers, shown in Fig. 5.23.

The temperature of the TSC peak or the midpoint of the TSCAP step T_m is related to the activation energy $\Delta E = E_c - E_T$ or $\Delta E = E_T - E_v$ by[116]

$$\Delta E = kT_m \ln\left(\frac{\gamma_n\sigma_n kT_m^4}{\beta(\Delta E + 2kT_m)}\right) \tag{5.78}$$

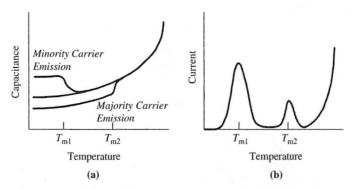

Fig. 5.23 Schematic of (a) TSCAP and (b) TSC for a sample with a majority carrier trap of density N_T and a shallower minority carrier trap of density $2N_T$. The current increase at higher temperatures is due to thermally generated current. Reprinted with permission after Lang. Ref. 45.

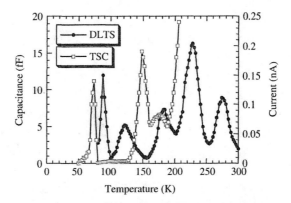

Fig. 5.24 DLTS and TSC data for high resistivity silicon. Reprinted from ref. 117 with kind permission from Elsevier Science-NL, Burgerhartsraat 2S, 1055 KV Amsterdam, The Netherlands.

For p-type samples the subscript n should be replaced by p. The trap density is obtained from the area under the TSC curve or from the step height of the TSCAP curve.

The equipment is simpler than that for DLTS, but the information obtained from TSC and TSCAP is more limited and more difficult to interpret. The thermally stimulated techniques allow a quick sweep of the sample to survey the entire range of traps in a sample and work well for $N_T \geq 0.1 N_D$ and $\Delta E \geq 0.3$ eV. The TSC peaks depend on the heating rate, but the TSCAP steps do not. TSC is influenced by leakage currents. TSCAP allows discrimination between minority and majority carrier traps by the *sign* of the capacitance change as indicated in Fig. 5.23(a); TSC does not. Thermally stimulated measurements have been largely replaced with DLTS. However, in high-resistivity materials, where it is difficult to make DLTS measurements, TSC can be used. An example is shown in Fig. 5.24 where both DLTS and TSC were used to determine the energy levels in high resistivity Si.[117] The defect energy levels extracted from the data agree quite well between the two methods.

5.8 POSITRON ANNIHILATION SPECTROSCOPY (PAS)

Positron annihilation spectroscopy (PAS) is the spectroscopy of *gamma (γ) rays* emerging from the annihilation of positrons and electrons. It can be used to examine defects in semiconductors without any special test structures, is independent of the sample conductivity, and is non-destructive.[118] Before discussing PAS, we will briefly describe positrons, since they are rarely mentioned in semiconductor books. A *positron* is similar to an electron. Its mass is the same as that of an electron and its charge is the same magnitude but of opposite sign to that of an electron. The positron was predicted by Dirac in 1928 and was observed experimentally in 1932 by Anderson during cosmic ray cloud chamber experiments. Positrons diffusing through matter may be captured at certain trapping sites and the character and the density of these lattice defects can be investigated.

An excellent discussion of PAS is given by Krause-Rehberg and Leipner.[118] The energy and momentum conservation during the annihilation of electrons with positrons can be used to study solids because the annihilation parameters are sensitive to lattice imperfections. The positron may be trapped in crystal defects, based on the formation of an

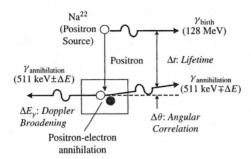

Fig. 5.25 Schematic illustration of positron annihilation showing positron creation, positron-electron annihilation, γ ray emission and the three main experimental techniques for PAS.

attractive potential at open-volume defects, such as vacancies, vacancy agglomerates, and dislocations. When a positron is trapped in an open-volume defect, the annihilation parameters are changed in a characteristic way. Its lifetime increases due to the lower electron density. Momentum conservation leads to a small angular spread of the collinear γ-quanta or a Doppler shift of the annihilation energy. Most positron lifetimes for the important semiconductors and lifetimes for various vacancy-type defects have been experimentally determined. Neutral and negative vacancy-type defects, as well as negative ions, are the dominant positron traps in semiconductors. Temperature-dependent lifetime measurements may distinguish between both defect types.

Positrons are most commonly produced during nuclear decay, when a proton of proton rich nuclei decays into a neutron with the emission of a positron and a neutrino. For example $_{11}Na^{22} \rightarrow _{10}Ne^{22} + positron + neutrino$. The Na^{22} isotope has a half life of 2.6 years and emits a 1.27 MeV γ ray within 10 ps of emitting a positron. This γ ray is used in lifetime spectroscopy measurements. Radioactive decay positrons possess a wide energy range. To produce a monochromatic positron beam for PAS from such a broad spectrum, the positrons pass through a moderator, e.g., W, Ni, and Mo. The positron energy is typically $kT \approx 25$ meV after moderation.

A positron is a stable particle by itself, but when it is combined with an electron, the two annihilate each other with the mass of the positron-electron pair converted into energy, i.e., gamma rays, as illustrated in Fig. 5.25. The released energy is twice the electron rest mass energy $2mc^2 = 2 \times 8.19 \times 10^{-14}$ $J = 2 \times 5.11 \times 10^5$ eV, where m is the electron rest mass and c the speed of light. The most probable decay is by the emission of two γ rays, moving in opposite directions. The energies, emission directions, and time of emission of these γ rays provide information about the behavior of positron-electron pairs and thus about the material where they annihilate. Energy and momentum conservation requires each γ ray to have one half the energy of the positron-electron system, i.e., 511 keV. The probability of annihilation depends on the density of available electrons.

When annihilation occurs, the gamma rays have an energy and directional distribution which depends on the electron motion before annihilation. The angle between the two γ rays differs slightly from 180°, with the angular deviation $\Delta\theta$ depending on the component of electron momentum perpendicular to the emission direction, p_{perp}. The energy of each γ ray, E_γ, depends on the component of electron momentum parallel to the emission direction, p_{par}

$$\Delta\theta = \frac{p_{perp}}{mc}; E_\gamma = mc^2 + \frac{p_{par}c}{2}; \Delta E_\gamma = E_\gamma - mc^2 = \frac{p_{par}c}{2} \qquad (5.79)$$

The terms $\Delta\theta$ and ΔE_γ provide information about the electron momentum components in a material. It is chiefly the electron momenta that determine $\Delta\theta$ and ΔE_γ, since positrons have low energy before annihilation. Additional information about the state of the electron before annihilation can be obtained by measuring the positron lifetime Δt. The annihilation positron lifetime is in the low ns range, but is affected by processes that alter the local density of electrons, making the lifetime one measure of crystal perfection. The positron lifetime is inversely proportional to the electron density of the material sampled by the positron, making it a unique probe of open volume lattice defects. The lifetime is the time between the creation of the positron and the creation of the gamma rays. For pure Si it is 219 ps, for monovacancies in Si about 266 ps, and for divacancies in Si about 320 ps.[119] Most defects produce two effects related to positron annihilation. Defects producing local region of negative charge, attract positrons and defects alter the *electron density* and *momentum* distribution near the trapped positron. This leads to changes in Δt, $\Delta\theta$, and ΔE_γ.

The positron *lifetime* is measured with two fast γ ray detectors and a timing circuit. Many positron sources, including Na, emit gamma rays (γ_{birth} in Fig. 5.25) within a few picoseconds of the positron emission. Detection of this γ ray signals the positron injection into the material under test. $\Delta\theta$ is measured with a positron angular correlation spectrometer. In angular correlation of annihilation radiation, one measures the *angle* between the directions of photons in two γ annihilations. Momentum conservation during annihilation of a positron-electron pair requires the γ rays to move in opposite directions if the pair is at rest. If the pair has a finite momentum, it causes a deviation of the angle between the gamma rays from 180°. The measurement consists of counting pairs of annihilating γ rays emitted at angles that differ slightly from 180° as illustrated in Fig. 5.25. Typical values for $\Delta\theta$ are on the order of 0.01 rad. An example of the positron lifetime after electron irradiation is shown in Fig. 5.26, where 2 MeV electron irradiation produced vacancies which were annealed and the lifetime is a measure of the vacancy density. The initial vacancy density was estimated to be 3×10^{17} cm^{-3}.[118]

The motion of the annihilating positron-electron pair causes *a Doppler shift* in the energy of the 511 keV γ rays. The energy E_γ is measured with a positron Doppler

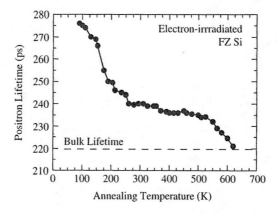

Fig. 5.26 Positron lifetime versus annealing temperature for float-zone Si. The sample was electron irradiated at 2 MeV, $T = 4$ K, and 10^{18} cm^{-2} dose. The bulk lifetime refers to a vacancy-free sample. Adapted from Krause-Rehberg and Leipner.[118]

broadening line-shape spectrometer. The shape of the 511 keV gamma ray line is broadened due to the electron momenta and is usually characterized by the "S parameter". The S parameter is defined as the number of counts in the central region of the 511 keV peak, containing about half of the total area, divided by the total number of counts in the peak. Lifetime and Doppler broadening experiments are more commonly used than angular correlation. The latter requires more complex equipment.

PAS exploits the high sensitivity of positrons to regions of lower-than-average electron density such as vacancies, vacancy clusters, voids, and other defects in semiconductors, e.g., dislocations, grain boundaries, and interfaces. Any process that produces vacancies is suitable for PAS, e.g., ion implantation, where small vacancy clusters, too small for electron microscope detection, can be detected. PAS has also been used to study radiation damage and the SiO_2-Si interface.[119] Doppler broadening reflects the momentum state of the electron annihilated by the positron. Positrons trapped at vacancies have a higher probability of annihilation with electrons having low momentum and consequently the S parameter increases with the presence of vacancies or vacancy-type defects. Measuring the S parameter as a function of annealing allows ion implanted samples to be characterized in terms of vacancy creation during implantation and their subsequent destruction during the implant damage anneal.[120] To study depth-dependent defects, positron beams with 0.1–30 keV energies were implanted. However, there is a depth resolution limit, because the positron implantation profile is broadened with increasing positron energy and its full width half maximum is comparable to the mean implanted depth. The trap sites of the implanted positrons also depend on their thermal diffusion following implantation. Enhanced depth resolution was achieved by repeated chemical etching and positron measurement.[121] The defect profile and annealing behavior in B and P ion implantations to Si showed that defects were induced beyond the implanted ion profile. Positron emission has also been applied to microscopy where positrons are used instead of electrons in a scanning electron microscope.[122]

5.9 STRENGTHS AND WEAKNESSES

DLTS is the most common deep-level characterization technique today, having replaced thermally stimulated current and capacitance. It lends itself to a number of different implementations and equipment is commercially available. Although DLTS is spectroscopic in nature, giving trap energies, it is frequently not easy to assign a specific impurity to a particular DLTS spectrum. Identification of impurities is not always straightforward.

Capacitance Transient Spectroscopy: Its strength lies in the ease of measurement. Most systems use commercial capacitance meters or bridges and add signal-processing functions (lock-in amplifiers, boxcar integrators, or computers). One can distinguish between majority and minority carrier traps, and its sensitivity is independent of the emission time constant. Its major weakness is the inability to characterize high resistivity substrates. The fact that its sensitivity is independent of a time constant can be a disadvantage because the sensitivity cannot be changed. Laplace DLTS produces very high resolution plots allowing trap with close lying energy levels to be distinguished.

Current Transient Spectroscopy: Its strength lies in the ability to characterize conducting as well as semi-insulating substrates. The fact that the current depends inversely on the emission time constant allows the sensitivity of the method to be changed by changing the time constant. This has led to its use in scanning DLTS. Its weakness is its dependence on diode quality, where leakage current can interfere with the measurement.

Optical DLTS: Its strength lies in the ability to create minority carriers without the need for pn junctions. This allows materials in which it is difficult to make pn junctions to be characterized. O-DLTS is useful to determine impurity optical cross-sections. Its major weakness lies in the requirement for light. The low temperature dewar must have transparent windows, and monochromators or pulsed light sources must be available.

Positron Annihilation Spectroscopy: Its strength lies in the contactless, non-destructive characterization of defects in solids. It allows depth-dependent defect characterization. Its weaknesses are that it is chiefly sensitive to void-like defects such as vacancies and requires elaborate equipment that is not readily available to most researchers.

APPENDIX 5.1

Activation Energy and Capture Cross-Section

The relationship between the emission rate and the capture cross-section is often written as

$$e_n = \sigma_n v_{th} N_c \exp((E_c - E_T)/kT) \tag{A5.1}$$

This relationship is frequently used to determine E_T and σ_n. However, when the capture cross-section is determined from the intercept of a $\ln(\tau_e T^2)$ versus $1/T$ plot, considerable error can result.

From thermodynamics we find the following definitions:[123]

$$G = H - TS; H = E + pV \tag{A5.2}$$

where G is the Gibbs free energy, H the enthalpy, E the internal energy, T the temperature, S the entropy, p the pressure and V the volume. The energy to excite an electron thermally from a trap into the conduction band is ΔG_n.[124] Equation (A5.1) then becomes

$$e_n = \sigma_n v_{th} N_c \exp(-\Delta G_n/kT) \tag{A5.3}$$

From Eq. (A5.2), $\Delta G_n = \Delta H_n - T \Delta S_n$ for constant T. When substituted into Eq. (A5.3), the emission rate is

$$e_n = \sigma_n X_n v_{th} N_c \exp(-\Delta H_n/kT) \tag{A5.4}$$

where $X_n = \exp(\Delta S_n/k)$ is an "entropy factor", that accounts for the entropy change accompanying electron emission from a trap to the conduction band. The entropy change can be expressed as $\Delta S_n = \Delta S_{ne} + \Delta S_{na}$, where ΔS_{ne} is the change due to electronic degeneracy and ΔS_{na} is due to atomic vibrational changes.

The electronic contribution may be expressed in terms of two degeneracy factors: g_0 is the degeneracy of the trap unoccupied by an electron, and g_1 is the degeneracy of the trap occupied by one electron, giving

$$X_n = (g_0/g_1) \exp(\Delta S_{na}/k) \tag{A5.5}$$

The degeneracy factors are not well known for deep-level impurities. Using values from shallow levels and with $\Delta S_{na} \approx a$ few k, X_n can easily be $10-100$.

Equation (A5.4) states that the energy determined from a $\ln(\tau_e T^2)$ or $\ln(T^2/e_n)$ versus $1/T$ plot is an enthalpy, and the prefactor can be written as $\sigma_{n,eff} v_{th} N_C$, with $\sigma_{n,eff} = \sigma_n X_n$. In other words, the effective capture cross-section differs from the true capture cross-section by X_n. If that distinction is not made, then obviously the extracted cross-section can be seriously in error. Effective cross-sections larger by factors of 50 or more from true cross-sections are not uncommon.[15] Examples are shown in Table 5.2.

Additional complications occur when σ_n is temperature dependent. Some cross-sections follow the relationship

$$\sigma_n = \sigma_\infty \exp(-E_b/kT) \tag{A5.6}$$

where σ_∞ is the cross-section as $T \to \infty$ and E_b is the cross-section activation energy. Equation (A5.4) becomes

$$e_n = \sigma_n X_n v_{th} N_c \exp\left(-\frac{\Delta H_n + E_b}{kT}\right) \tag{A5.7}$$

Under these conditions the Arrhenius plot gives neither the trap energy level nor its extrapolated cross-section correctly. If in addition the capture cross-section is electric-field dependent, further inaccuracies arise. A good discussion of energy levels, enthalpies, entropies, capture cross-sections, *etc.*, can be found in the work of Lang et al.[15] Further thermodynamic derivations can be found in the work by Thurmond and Van Vechten.[125-126]

A non-thermodynamic approach defines the energy $\Delta E_T = E_c - E_T$ as being temperature dependent according to $\Delta E_T = \Delta E_{T0} - \alpha T$. The degeneracy ratio in Eq. (A5.5) is written as g_n.[127] Equation (A5.1) becomes

$$e_n = \sigma_n X_n v_{th} N_c \exp(-\Delta E_{T0}/kT) \tag{A5.8}$$

where now $X_n = g_n \exp(\alpha/k)$. We find the energy as that as $T \to 0K$ and the cross-section is again $\sigma_n X_n$, although now X_n is defined differently.

APPENDIX 5.2

Time Constant Extraction

The capacitance of a Schottky barrier or p^+n junction containing impurities is from Eq. (5.11)

$$C = K\sqrt{\frac{N_D - N_T \exp(-t/\tau_e)}{V_{bi} - V}} \tag{A5.9}$$

where $n_T(0) = N_T$, if we confine ourselves to emission transients for simplicity.

How is τ_e determined? One method to extract τ_e is to take $dV/d(1/C^2)$ from Eq. (A5.9) as[8]

$$\frac{dV}{d(1/C^2)}\Big|_{t=\infty} - \frac{dV}{d(1/C^2)}\Big|_t = K^2 N_T \exp(-t/\tau_e) \tag{A5.10}$$

and to plot the ln(left side of Eq. (A5.10)) versus t. The slope of this plot gives τ_e, and the intercept at $t = 0$ is $\ln(K^2 N_T)$. This method places no limitation on the magnitude of N_T with respect to N_D.

Another method defines $f(t) = C(t)^2 - C_0{}^2 = [-K^2 N_T/(V_{bi} - V)]\exp(-t/\tau_e)$, where C_0 is the capacitance in Eq. (A5.9) for $N_T = 0$. The measurement is made at constant temperature. Differentiating $f(t)$ and multiplying by t gives

$$t\frac{df}{dt} = \frac{K^2 N_T}{V_{bi} - V}\frac{t}{\tau_e}\exp(-t/\tau_e) \tag{A5.11}$$

When plotted against t, $t\,df/dt$ has a maximum of $K^2 N_T/[e(V_{bi} - V)]$ at $t = \tau$.[72] Hence determining the maximum in the curve gives the time constant.

For $N_T \ll N_D$, we can write [see Eq. (5.16)]

$$C = C_0\left[1 - \left(\frac{n_T(0)}{2N_D}\right)\exp(-t/\tau_e)\right] = C_0\left[1 - \left(\frac{N_T}{2N_D}\right)\exp(-t/\tau_e)\right] \tag{A5.12}$$

Equation (A5.12) has been used in a number of implementations to extract τ_e. In the two-point method, the C-t exponential time-varying curve is sampled $t = t_1$ and $t = t_2$.[42] From Eq. (5.49)

$$\tau_{e,\max} = \frac{t_2 - t_1}{\ln(t_2/t_1)} \tag{A5.13}$$

In the three-point method, three points are measured on the C-t curve at a *constant* temperature, $C = C_1$ at $t = t_1$, $C = C_2$ at $t = t_2$, and $C = C_3$ at $t = t_3$.[128] From Eq. (A5.12)

$$\frac{C_1 - C_2}{C_2 - C_3} = \frac{\exp(\Delta t/\tau_e) - 1}{1 - \exp(\Delta t/\tau_e)} \tag{A5.14}$$

where $\Delta t = t_2 - t_1 = t_3 - t_2$. A solution of Eq. (A5.14) for τ_e is

$$\tau_e = \frac{\Delta t}{\ln[(C_1 - C_2)/(C_2 - C_3)]} \tag{A5.15}$$

A good choice for Δt is $\tau_e/2$, but of course τ_e is not known *a priori*, although a first-order value for it can be obtained from the "1/e point" on the capacitance decay curve.

Another technique is based on a very different approach. Consider the function $y_1 = y(t) = A\exp(-t/\tau) + B$, i.e., an exponentially decaying function superimposed on a dc background. We define a second function $y_2 = y(t + \Delta t) = A\exp[-(t + \Delta t)/\tau] + B$. The second function is obtained from the first by simply adding a constant increment Δt to the time t. A plot of y_2 versus y_1 is a straight line with slope $m = \exp(-\Delta t/\tau)$ and intercept on the y_2 axis of $B(1 - m)$.[129] Then τ is calculated from the slope and Δt and B are found from the intercept and the slope. Δt should be smaller than τ, but not much smaller, e.g., $\Delta t \approx 0.1$ to 0.5τ.

An excellent discussion of decay time extraction is given by Istratov and Vyvenko.[130] For a single energy level impurity with a single exponential decay, the transient is characterized by

$$f(t) = A\exp(-\lambda t) + B \tag{A5.16}$$

where A is the decay amplitude, B is a constant (the baseline offset), and λ is the decay rate, decay constant or rate constant, which is the inverse of the decay time constant τ

$(\tau = 1/\lambda)$. If the decay consists of a sum of n exponentials of the form Eq. (A5.16), then

$$f(t) = \sum_{i=1}^{n} A_i \exp(-\lambda_i t) \qquad (A5.17)$$

neglecting the baseline offset B. This behavior is expected from more than one energy level. The goal of any multi-exponential analysis is to determine the number of exponential components n, their amplitudes A_i, and decay rates λ_i. When the decay is due to a continuous distribution of emission rates given by a spectral function $g(\lambda)$ rather than by a sum of discrete exponential transients

$$f(t) = \int_0^{\infty} g(\lambda) \exp(-\lambda t)\, d\lambda \qquad (A5.18)$$

where $g(\lambda)$ is the spectral function. Such behavior is exhibited by interface traps with a continuous distribution of energy in the band gap at the SiO_2/Si interface, for example.

The major goal of exponential analysis is to distinguish exponential components with close time constants in the experimentally measured decay. To achieve high resolution in exponential analysis, it is very important to record the transient until it decays completely. Since the ratio of amplitudes of two exponentials with close decay rates: $\exp(-\lambda_1 t)$ and $\exp(-\lambda_2 t)$ increases with the time as $\exp[((\lambda_2 - \lambda_1)t]$, these exponentials always can, at least theoretically, be distinguished if the decay is monitored for a sufficiently long time. Since the exponential is a decaying function of time, the transient should be monitored as long as the signal amplitude exceeds the noise level. For a signal-to-noise ratio, $S/R = 100$, the measurement time should be at least 4.6τ, for $S/R = 1000$ about 6.9τ, and for $S/R = 10^4$ at least 9.2τ.[130] This is frequently ignored in experiments and numerical simulations.

Consider the example in Fig. A5.1. Twenty-four data points were fitted by a double exponential $f_2(t) = 2.202 \exp(-4.45t) + 0.305 \exp(-1.58t)$ and by a triple exponential $f_3(t) = 0.0951 \exp(-t) + 0.8607 \exp(-3t) + 1.5576 \exp(-5t)$ in Fig. A5.1(a). Lanczos showed that a sum of two exponentials could be reproduced to within two decimal places by a sum of three exponentials with entirely different time constants and amplitudes.[131] However, a discrepancy is observed when the data are extended to longer times as shown in Fig. A5.1(b). However, the difference between the two curves does not exceed 0.001 of the decay amplitude, and can be detected only if the S/R exceeds 1000.

APPENDIX 5.3

Si and GaAs Data

Arrhenius plots for Si and GaAs are shown in Figs. A5.2 and A5.3. In Fig. A5.2, $(300/T)^2 e_n$ and $(300/T)^2 e_p$ are plotted instead of τ_{nT}^2 and τ_{pT}^2, giving negative slopes. The deep level impurity metals are shown wherever possible, and the numbers listed below the elements are their energy levels calculated from the slopes. The superscripts are the references given in the review paper by Chen and Milnes.[3]

Table A5.1 lists typical trace contamination in Si most commonly produced during device processing or after 1-MeV electron beam irradiation.[132] The impurities were determined from transient capacitance spectroscopy. DLTS spectra have been correlated with metallic impurities, growth-related defects, oxidation, heat treatments, electron and proton

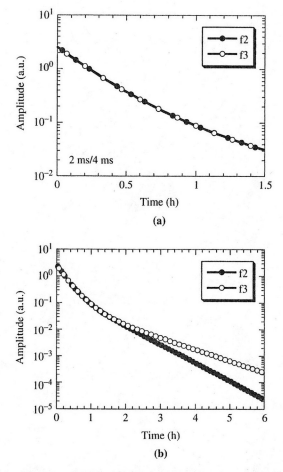

Fig. A5.1   (a) Data points were fitted by a double exponential $f_2(t) = 2.202 \exp(-4.45t) + 0.305 \exp(-1.58t)$ and by a triple exponential $f_3(t) = 0.0951 \exp(-t) + 0.8607 \exp(-3t) + 1.5576 \exp(-5t)$. The difference between $f_2(t)$ and $f_3(t)$ is less than the line width. (b) The curves separated after 2 h, but the absolute value of the separation is less than only 0.001 of the decay amplitude. Adapted from ref. 131.

irradiation, dislocation-related states, electronically stimulated defects, and laser anneal. Established temperature regimes of defect and impurity reactions are indicated.

An unknown DLTS peak can be compared with the data in Table A5.1 by two methods.[132] First, an Arrhenius plot of $\tau_e T^2$ versus $1/T$ can be constructed using the point given by the temperature of the known peak (T) at a time constant of 1.8 ms (τ) and the slope given by the activation energy (E_T) in the table. Alternatively, the temperature at which a signal from a listed defect should occur using any time constant of the analyzing instrument can be determined by iteration. A simple computer program sets the ratio R,

$$R = \frac{\tau_1 T_1^2 \exp(-E_T/kT_1)}{\tau_2 T_2^2 \exp(-E_T/kT_2)} \tag{A5.16}$$

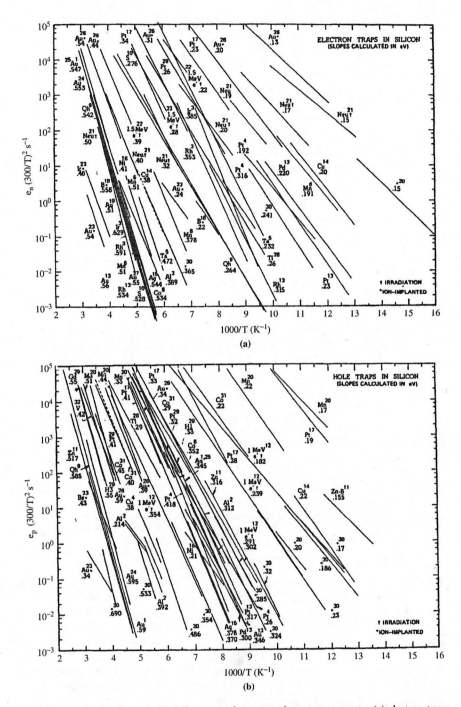

Fig. A5.2 Arrhenius plots obtained from capacitance transient measurements: (a) electron traps, (b) hole traps in Si. The vertical axis is $(300/T^2)e_{n,p}$ instead of $\tau_{n,p}T^2$. Reprinted, with permission, from the *Annual Review of Material Science*, Vol. 10, © 1980 by Annual Reviews Inc.

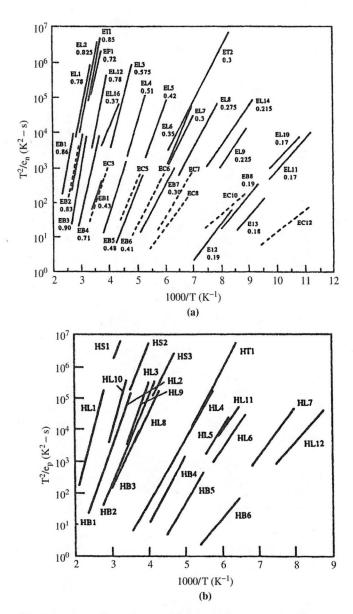

Fig. A5.3 Arrhenius plots obtained from capacitance transient measurements: (a) electron traps, (b) hole traps in GaAs. The vertical axis is $T^2/e_{n,p}$ instead of $\tau_{n,p}T^2$. Reprinted, with permission after Martin et al.[17] and Mitonneau et al.[17] © Institution of Electrical Engineers.

where subscript 1 refers to the value of Table A5.1 and subscript 2 refers to the value for the particular measurement. For $\tau_1 > \tau_2$ the temperature T_2 is increased; for $\tau_1 < \tau_2$ the temperature T_2 is decreased until $R = 1$.

TABLE A5.1   Capacitance Transient Spectral Features for Silicon.

Defect	T (K) 1.8 ms	E_T (eV)	σ_{maj} (cm^2)	Anneal	Comments[a]
Ag	286	E (0.51)	10^{-16}		Q, *, FZ
	184	H (0.38)	—		Q, *, FZ
Au	288	E (0.53)	2×10^{-16}		Q, *, FZ
	173	H (0.35)	$> 10^{-15}$		Q, *, FZ
Cu	112	H (0.22)	$> 6 \times 10^{-14}$	Out 150°C	Q, *, FZ
	242	H (0.41)	8×10^{-14}		Q, *, FZ
Fe	181	E (0.35)	6×10^{-15}		Q, *, FZ
(Fe-B)	59	H (0.10)	$> 4 \times 10^{-15}$	Out > 150°C	Q, *, FZ
(Fe_i)	267	H (0.46)		In > 150°C, out > 200°C	Q, *, FZ
	208	E (0.21)	—		S, FZ
	299	E (0.46)	—		S, FZ
	184	H (0.23)	—		S, FZ
	170	E (0.35)	—		Q, CG
	168	H (0.30)	5×10^{-15}		Q, CG
	237	H (0.43)	—		Q, CG
	220	H (0.47)	—		Q, CG
Mn	68	E (0.11)	—		Q, FZ
	216	E (0.41)	10^{-15}		Q, FZ
	81	H (0.13)	$> 2 \times 10^{-15}$		Q, FZ
Ni	257	E (0.43)	5×10^{-16}		Q, *, FZ
	88	E (0.14)	10^{-16}	Out 150°C	Q, *, FZ
Pt	114	E (0.22)	$> 4 \times 10^{-15}$		Q, *, FZ
	174	E (0.30)	$\sim 10^{-15}$		Q, *, FZ
	87	H (0.22)			Q, *, FZ
O-Donor	Below Freezeout	E (0.07)	$\sim 10^{-15}$	In 400°C, out 600°C	*, CG
			—		
	58	E (0.15)	—	In 400°C, out 600°C	*, CG
Heat Treatment	59,60	E (0.15)		In 900°C	*, CG
	112	E (0.22)	$> 3 \times 10^{-15}$	In 900°C	*, CG
	228	E (0.47)	2×10^{-16}	In 900°C	*, CG
Laser Donor	115	E (0.19)	7×10^{-16}	Out 550°C	Q, FZ, CG
	200	E (0.33–0.36)	5×10^{-16}	Out 650°C	Q, FZ, CG
	211	H (0.36)	5×10^{-19}		Q, *, FZ, CG
Vacancy-O	98	E (0.18)	5×10^{-16}	In -43°C, out 350°C	1 MeV, CG
Vacancy-Vacancy	139	E (0.23)		Out 300°C	1 MeV, CG, FZ
	245	E (0.41)	—	Out 300°C	1 MeV, CG, FZ
	123	H (0.21)	10^{-14}	Out 300°C	1 MeV, CG, FZ
P-Vacancy	237	E (0.44)	2×10^{-16}	Out 150°C	1 MeV, CG, FZ
$C_s - C_i$	204	H (0.36)	4×10^{-15}	In 43°C	1 MeV, CG, FZ
Dislocation	225	E (0.38)	2×10^{-16}		FZ
	206	H (0.35)	$> 10^{-16}$		FZ
Point Defect Debris	288	E (0.63–0.68)	8×10^{-17}	Out 800°C	FZ, cross slip
			1.4×10^{-15}		
			$> 5 \times 10^{-17}$		

Source: Ref. 132. (a) Symbols: Q = quenched material, * = diffused junction, S = slow cool, FZ = float zone growth, CG = crucible growth, and 1 MeV = electron bombardment.

REFERENCES

1. B.D. Choi and D.K. Schroder, "Degradation of Ultrathin Oxides by Iron Contamination," *Appl. Phys. Lett.* **79**, 2645–2647, Oct. 2001; Y.H. Lin, Y.C. Chen, K.T. Chan, F.M. Pan, I.J. Hsieh, and A. Chin, "The Strong Degradation of 30 Å Gate Oxide Integrity Contaminated by Copper," *J. Electrochem. Soc.* **148**, F73–F76, April 2001.

2. A.G. Milnes, *Deep Impurities in Semiconductors*, Wiley-Interscience, New York, 1973.

3. J.W. Chen and A.G. Milnes, "Energy Levels in Silicon," in *Annual Review of Material Science* (R.A. Huggins, R.H. Bube and D.A. Vermilyea, eds.), Annual Reviews, Palo Alto, CA, **10**, 157–228, 1980; A.G. Milnes, "Impurity and Defect Levels (Experimental) in Gallium Arsenide," in *Advances in Electronics and Electron Physics* (P.W. Hawkes, ed.), Academic Press, Orlando, FL, **61**, 63–160, 1983.

4. M. Jaros, *Deep Levels in Semiconductors*, A. Hilger, Bristol, 1982.

5. C.T. Sah, L. Forbes, L.L. Rosier and A.F. Tasch Jr., "Thermal and Optical Emission and Capture Rates and Cross Sections of Electrons and Holes at Imperfection Centers in Semiconductors from Photo and Dark Junction Current and Capacitance Experiments," *Solid-State Electron.* **13**, 759–788, June 1970.

6. R.N. Hall, "Electron-Hole Recombination in Germanium," *Phys. Rev.* **87**, 387, July 1952.

7. W. Shockley and W.T. Read, "Statistics of the Recombinations of Holes and Electrons," *Phys. Rev.* **87**, 835–842, Sept. 1952.

8. R. Williams, "Determination of Deep Centers in Conducting Gallium Arsenide," *J. Appl. Phys.* **37**, 3411–3416, Aug. 1966; R.R. Senechal and J. Basinski, "Capacitance of Junctions on Gold-Doped Silicon," *J. Appl. Phys.* **39**, 3723–3731, July 1968; "Capacitance Measurements on Au-GaAs Schottky Barriers," *J. Appl. Phys.* **39**, 4581–4589, Sept. 1968.

9. M. Bleicher and E. Lange, "Schottky-Barrier Capacitance Measurements for Deep Level Impurity Determination," *Solid-State Electron.* **16**, 375–380, March 1973.

10. R.F. Pierret, *Advanced Semiconductor Fundamentals*, Addison-Wesley, Reading, MA, 1987, 146–152.

11. W. Shockley, "Electrons, Holes, and Traps," *Proc. IRE* **46**, 973–990, June 1958.

12. C.T. Sah, "The Equivalent Circuit Model in Solid-State Electronics—Part I: The Single Energy Level Defect Centers," *Proc. IEEE* **55**, 654–671, May 1967; "The Equivalent Circuit Model in Solid-State Electronics—Part II: The Multiple Energy Level Impurity Centers," *Proc. IEEE* **55**, 672–684, May 1967.

13. P.A. Martin, B.G. Streetman and K. Hess, "Electric Field Enhanced Emission from Non-Coulombic Traps in Semiconductors," *J. Appl. Phys.* **52**, 7409–7415, Dec. 1981.

14. A.F. Tasch, Jr. and C.T. Sah, "Recombination-Generation and Optical Properties of Gold Acceptor in Silicon," *Phys. Rev.* **B1**, 800–809, Jan. 1970.

15. D.V. Lang, H.G. Grimmeiss, E. Meijer and M. Jaros, "Complex Nature of Gold-Related Deep Levels in Silicon," *Phys. Rev.* **B22**, 3917–3934, Oct. 1980.

16. H.D. Barber, "Effective Mass and Intrinsic Concentration in Silicon," *Solid-State Electron.* **10**, 1039–1051, Nov. 1967.

17. G.M. Martin, A. Mitonneau and A Mircea, "Electron Traps in Bulk and Epitaxial GaAs Crystals," *Electron. Lett.* **13**, 191–193, March 1977; A. Mitonneau, G.M. Martin, and A Mircea, "Hole Traps in Bulk and Epitaxial GaAs Crystals," *Electron. Lett.* **13**, 666–668, Oct. 1977.

18. W.B. Leigh, J.S. Blakemore and R.Y. Koyama, "Interfacial Effects Related to Backgating in Ion-Implanted GaAs MESFET's," *IEEE Trans. Electron Dev.* **ED-32**, 1835–1841, Sept. 1985.

19. J.A. Pals, "Properties of Au, Pt, Pd and Rh Levels in Silicon Measured with a Constant Capacitance Technique," *Solid-State Electron.* **17**, 1139–1145, Nov. 1974.

20. H. Okushi and Y. Tokumaru, "A Modulated DLTS Method for Large Signal Analysis (C^2–DLTS)," *Japan. J. Appl. Phys.* **20**, L45–L47, Jan. 1981.

21. W.E. Phillips and J.R. Lowney, "Analysis of Nonexponential Transient Capacitance in Silicon Diodes Heavily Doped with Platinum," *J. Appl. Phys.* **54**, 2786–2791, May 1983.

22. A.C. Wang and C.T. Sah, "Determination of Trapped Charge Emission Rates from Nonexponential Capacitance Transients Due to High Trap Densities in Semiconductors," *J. Appl. Phys.* **55**, 565–570, Jan. 1984.

23. D. Stiévenard, M. Lannoo and J.C. Bourgoin, "Transient Capacitance Spectroscopy in Heavily Compensated Semiconductors," *Solid-State Electron.* **28**, 485–492, May 1985.

24. F.D. Auret and M. Nel, "Detection of Minority-Carrier Defects by Deep Level Transient Spectroscopy Using Schottky Barrier Diodes," *J. Appl. Phys.* **61**, 2546–2549, April 1987.

25. L. Stolt and K. Bohlin, "Deep-Level Transient Spectroscopy Measurements Using High Schottky Barriers," *Solid-State Electron.* **28**, 1215–1221, Dec. 1985.

26. C.H. Henry, H. Kukimoto, G.L. Miller and F.R. Merritt, "Photocapacitance Studies of the Oxygen Donor in GaP. II. Capture Cross Sections," *Phys. Rev.* **B7**, 2499–2507, March 1973; A.C. Wang and C.T. Sah, "New Method for Complete Electrical Characterization of Recombination Properties of Traps in Semiconductors," *J. Appl. Phys.* **57**, 4645–4656, May 1985.

27. J.A. Borsuk and R.M. Swanson, "Capture-Cross-Section Determination by Transient-Current Trap-Filling Experiments," *J. Appl. Phys.* **52**, 6704–6712, Nov. 1981.

28. S.D. Brotherton and J. Bicknell, "The Electron Capture Cross Section and Energy Level of the Gold Acceptor Center in Silicon," *J. Appl. Phys.* **49**, 667–671, Feb. 1978.

29. A. Zylbersztejn, "Trap Depth and Electron Capture Cross Section Determination by Trap Refilling Experiments in Schottky Diodes," *Appl. Phys. Lett.* **33**, 200–202, July 1978.

30. H. Kukimoto, C.H. Henry and F.R. Merritt, "Photocapacitance Studies of the Oxygen Donor in GaP. I. Optical Cross Sections, Energy Levels, and Concentration," *Phys. Rev.* **B7**, 2486–2499, March 1973.

31. S.D. Brotherton and J. Bicknell, "Measurement of Minority Carrier Capture Cross Sections and Application to Gold and Platinum in Silicon," *J. Appl. Phys.* **53**, 1543–1553, March 1982.

32. B. Hamilton, A.R. Peaker and D.R. Wight, "Deep-State-Controlled Minority-Carrier Lifetime in n-Type Gallium Phosphide," *J. Appl. Phys.* **50**, 6373–6385, Oct. 1979; R. Brunwin, B. Hamilton, P. Jordan and A.R. Peaker, "Detection of Minority-Carrier Traps Using Transient Spectroscopy," *Electron. Lett.* **15**, 349–350, June 1979.

33. C.T. Sah, "Bulk and Interface Imperfections in Semiconductors," *Solid-State Electron.* **19**, 975–990, Dec. 1976.

34. J.A. Borsuk and R.M. Swanson, "Current Transient Spectroscopy: A High-Sensitivity DLTS System," *IEEE Trans. Electron Dev.* **ED-27**, 2217–2225, Dec. 1980.

35. P.K. McLarty, D.E. Ioannou and H.L. Hughes, "Deep States in Silicon-on-Insulator Substrates Prepared by Oxygen Implantation Using Current Deep Level Transient Spectroscopy," *Appl. Phys. Lett.* **53**, 871–873, Sept. 1988.

36. P.V. Kolev and M.J. Deen, "Constant-Resistance Deep-Level Transient Spectroscopy in Submicron Metal-Oxide-Semiconductor Field-Effect Transistors," *J. Appl. Phys.* **83**, 820–825, Jan. 1998.

37. M.G. Collet, "An Experimental Method to Analyse Trapping Centres in Silicon at Very Low Concentrations," *Solid-State Electron.* **18**, 1077–1083, Dec. 1975.

38. I.D. Hawkins and A.R. Peaker, "Capacitance and Conductance Deep Level Transient Spectroscopy in Field-Effect Transistors," *Appl. Phys. Lett.* **48**, 227–229, Jan. 1986.

39. J.M. Golio, R.J. Trew, G.N. Maracas and H. Lefèvre, "A Modeling Technique for Characterizing Ion-Implanted Material Using C-V and DLTS Data," *Solid-State Electron.* **27**, 367–373, April 1984.

40. J.W. Farmer, C.D. Lamp and J.M. Meese, "Charge Transient Spectroscopy," *Appl. Phys. Lett.* **41**, 1063–1065, Dec. 1982.

41. K.I. Kirov and K.B. Radev, "A Simple Charge-Based DLTS Technique," *Phys. Stat. Sol.* **63a**, 711–716, Feb. 1981.

42. D.V. Lang, "Deep-Level Transient Spectroscopy: A New Method to Characterize Traps in Semiconductors," *J. Appl. Phys.* **45**, 3023–3032, July 1974; D.V. Lang, "Fast Capacitance Transient Apparatus: Application to ZnO and O Centers in GaP p-n Junctions," *J. Appl. Phys.* **45**, 3014–3022, July 1974.

43. ASTM Standard F 978-90, "Standard Test Method for Characterizing Semiconductor Deep Levels by Transient Capacitance," *1996 Annual Book of ASTM Standards*, Am. Soc. Test., Conshohocken, PA, 1996.

44. G.L. Miller, D.V. Lang and L.C. Kimerling, "Capacitance Transient Spectroscopy," in *Annual Review Material Science* (R.A. Huggins, R.H. Bube and R.W. Roberts, eds.), Annual Reviews, Palo Alto, CA, **7**, 377–448, 1977.

45. D.V. Lang, "Space-Charge Spectroscopy in Semiconductors," in *Topics in Applied Physics*, **37**, *Thermally Stimulated Relaxation in Solids* (P. Bräunlich, ed.), Springer, Berlin, 1979, 93–133.

46. B.D. Choi, D.K. Schroder, S. Koveshnikov, and S. Mahajan, "Latent Iron in Silicon," *Japan. J. Appl. Phys.* **40**, L915–L917, Sept. 2001.

47. M.A. Gad and J.H. Evans-Freeman, "High Resolution Minority Carrier Transient Spectroscopy of Si/SiGe/Si Quantum Wells," *J. Appl. Phys.* **92**, 5252–5258, Nov. 2002.

48. D.S. Day, M.Y. Tsai, B.G. Streetman and D.V. Lang, "Deep-Level-Transient Spectroscopy: System Effects and Data Analysis," *J. Appl. Phys.* **50**, 5093–5098, Aug. 1979.

49. S. Misrachi, A.R. Peaker and B. Hamilton, "A High Sensitivity Bridge for the Measurement of Deep States in Semiconductors," *J. Phys. E: Sci. Instrum.* **13**, 1055–1061, Oct. 1980.

50. T.I. Chappell and C.M. Ransom, "Modifications to the Boonton 72BD Capacitance Meter for Deep-Level Transient Spectroscopy Applications," *Rev. Sci. Instrum.* **55**, 200–203, Feb. 1984.

51. H. Lefèvre and M. Schulz, "Double Correlation Technique (DDLTS) for the Analysis of Deep Level Profiles in Semiconductors," *Appl. Phys.* **12**, 45–53, Jan. 1977.

52. K. Kosai, "External Generation of Gate Delays in a Boxcar Integrator—Application to Deep Level Transient Spectroscopy," *Rev. Sci. Instrum.* **53**, 210–213, Feb. 1982.

53. G. Goto, S. Yanagisawa, O. Wada and H. Takanashi, "Determination of Deep-Level Energy and Density Profiles in Inhomogeneous Semiconductors," *Appl. Phys. Lett.* **23**, 150–151, Aug. 1973.

54. N.M. Johnson, "Measurement of Semiconductor-Insulator Interface States by Constant-Capacitance, Deep-Level Transient Spectroscopy," *J. Vac. Sci. Technol.* **21**, 303–314, July/Aug. 1982.

55. G.L. Miller, "A Feedback Method for Investigating Carrier Distributions in Semiconductors," *IEEE Trans. Electron Dev.* **ED-19**, 1103–1108, Oct. 1972.

56. J.M. Noras, "Thermal Filling Effects on Constant Capacitance Transient Spectroscopy," *Phys. Stat. Sol.* **69a**, K209–K213, Feb. 1982.

57. M.F. Li and C.T. Sah, "New Techniques of Capacitance-Voltage Measurements of Semiconductor Junctions," *Solid-State Electron.* **25**, 95–99, Feb. 1982.

58. R.Y. DeJule, M.A. Haase, D.S. Ruby and G.E. Stillman, "Constant Capacitance DLTS Circuit for Measuring High Purity Semiconductors," *Solid-State Electron.* **28**, 639–641, June 1985.

59. P. Kolev, "An Improved Feedback Circuit for Constant-Capacitance Voltage Transient Measurements," *Solid-State Electron.* **35**, 387–389, March 1992.

60. M.F. Li and C.T. Sah, "A New Method for the Determination of Dopant and Trap Concentration Profiles in Semiconductors," *IEEE Trans. Electron Dev.* **ED-29**, 306–315, Feb. 1982.

61. N.M. Johnson, D.J. Bartelink, R.B. Gold, and J.F. Gibbons, "Constant-Capacitance DLTS Measurement of Defect-Density Profiles in Semiconductors," *J. Appl. Phys.* **50**, 4828–4833, July 1979.

62. L.C. Kimerling, "New Developments in Defect Studies in Semiconductors," *IEEE Trans. Nucl. Sci.* **NS-23**, 1497–1505, Dec. 1976.

63. G.L. Miller, J.V. Ramirez and D.A.H. Robinson, "A Correlation Method for Semiconductor Transient Signal Measurements," *J. Appl. Phys.* **46**, 2638–2644, June 1975.

64. A. Rohatgi, J.R. Davis, R.H. Hopkins and P.G. McMullin, "A Study of Grown-In Impurities in Silicon by Deep-Level Transient Spectroscopy," *Solid-State Electron.* **26**, 1039–1051, Nov. 1983.

65. G. Couturier, A. Thabti and A.S. Barrière, "The Baseline Problem in DLTS Technique," *Rev. Phys. Appliqué* **24**, 243–249, Feb. 1989.

66. J.T. Schott, H.M. DeAngelis and P.J. Drevinsky, "Capacitance Transient Spectra of Processing- and Radiation-Induced Defects in Silicon Solar Cells," *J. Electron. Mat.* **9**, 419–434, March 1980.

67. G. Ferenczi and J. Kiss, "Principles of the Optimum Lock-In Averaging in DLTS Measurement," *Acta Phys. Acad. Sci. Hung.* **50**, 285–290, 1981.

68. P.M. Henry, J.M. Meese, J.W. Farmer and C.D. Lamp, "Frequency-Scanned Deep-Level Transient Spectroscopy," *J. Appl. Phys.* **57**, 628–630, Jan. 1985.

69. K. Dmowski and Z. Pióro, "Noise Properties of Analog Correlators with Exponentially Weighted Average," *Rev. Sci. Instrum.* **58**, 2185–2191, Nov. 1987.

70. M.S. Hodgart, "Optimum Correlation Method for Measurement of Noisy Transients in Solid-State Physics Experiments," *Electron. Lett.* **14**, 388–390, June 1978; C.R. Crowell and S. Alipanahi, "Transient Distortion and nth Order Filtering in Deep Level Transient Spectroscopy ($D^n LTS$)," *Solid-State Electron.* **24**, 25–36, Jan. 1981.

71. E.E. Haller, P.P. Li, G.S. Hubbard and W.L. Hansen, "Deep Level Transient Spectroscopy of High Purity Germanium Diodes/Detectors," *IEEE Trans. Nucl. Sci.* **NS-26**, 265–270, Feb. 1979.

72. H. Okushi and Y. Tokumaru, "Isothermal Capacitance Transient Spectroscopy for Determination of Deep Level Parameters," *Japan. J. Appl. Phys.* **19**, L335–L338, June 1980.

73. K. Hölzlein, G. Pensl, M. Schulz and P. Stolz, "Fast Computer-Controlled Deep Level Transient Spectroscopy System for Versatile Applications in Semiconductors" *Rev. Sci. Instrum.* **57**, 1373–1377, July 1986.

74. P.D. Kirchner, W.J. Schaff, G.N. Maracas, L.F. Eastman, T.I. Chappell and C.M. Ransom, "The Analysis of Exponential and Nonexponential Transients in Deep-Level Transient Spectroscopy," *J. Appl. Phys.* **52**, 6462–6470, Nov. 1981.

75. K. Ikeda and H. Takaoka, "Deep Level Fourier Spectroscopy for Determination of Deep Level Parameters," *Japan. J. Appl. Phys.* **21**, 462–466, March 1982.

76. S. Weiss and R. Kassing, "Deep Level Transient Fourier Spectroscopy (DLTFS)—A Technique for the Analysis of Deep Level Properties," *Solid-State Electron.* **31**, 1733–1742, Dec. 1988.

77. L. Dobaczewski, A.R. Peaker, and K. Bonde Nielsen, "Laplace-transform Deep-level Spectroscopy: The Technique and Its Applications to the Study of Point Defects in Semiconductors," *J. Appl. Phys.* **96**, 4689–4728, Nov. 2004.

78. J.E. Stannard, H.M. Day, M.L. Bark and S.H. Lee, "Spectroscopic Line Fitting to DLTS Data," *Solid-State Electron.* **24**, 1009–1013, Nov. 1981.

79. F.R. Shapiro, S.D. Senturia and D. Adler, "The Use of Linear Predictive Modeling for the Analysis of Transients from Experiments on Semiconductor Defects," *J. Appl. Phys.* **55**, 3453–3459, May 1984.

80. M. Henini, B. Tuck and C.J. Paull, "A Microcomputer-Based Deep Level Transient Spectroscopy (DLTS) System," *J. Phys. E: Sci. Instrum.* **18**, 926–929, Nov. 1985.

81. R. Langfeld, "A New Method of Analysis of DLTS-Spectra," *Appl. Phys.* **A44**, 107–110, Oct. 1987.

82. W.A. Doolittle and A. Rohatgi, "A Novel Computer Based Pseudo-Logarithmic Capacitance/Conductance DLTS System Specifically Designed for Transient Analysis," *Rev. Sci. Instrum.* **63**, 5733–5741, Dec. 1992.

83. P. Deixler, J. Terry, I.D. Hawkins, J.H. Evans-Freeman, A.R. Peaker, L. Rubaldo, D.K. Maude, J.-C. Portal, L. Dobaczewski, K. Bonde Nielsen, A. Nylandsted Larsen, and A. Mesli, "Laplace-transform Deep-level Transient Spectroscopy Studies of the G4 Gold–hydrogen Complex in Silicon," *Appl. Phys. Lett.* **73**, 3126–3128, Nov. 1998.

84. L. Dobaczewski, P. Kaczor, M. Missous, A.R. Peaker, and Z.R. Zytkiewicz "Evidence for Substitutional-Interstitial Defect Motions Leading to DX Behavior by Donors in $Al_x Ga_{1-x} As$," *Phys. Rev. Lett.* **68**, 2508–2511, April 1992; L. Dobaczewski, P. Kaczor, I.D. Hawkins, and A.R. Peaker, "Laplace Transform Deep-level Transient Spectroscopic Studies of Defects in Semiconductors," *J. Appl. Phys.* **76**, 194–198, July 1994.

85. K.L. Wang and A.O. Evwaraye, "Determination of Interface and Bulk-Trap States of IGFET's Using Deep- Level Transient Spectroscopy," *J. Appl. Phys.* **47**, 4574–4577, Oct. 1976; K.L. Wang, "Determination of Processing-Related Interface States and Their Correlation with Device Properties," in *Semiconductor Silicon 1977* (H.R. Huff and E. Sirtl, eds.), Electrochem. Soc., Princeton, NJ, 404–413; K.L. Wang, "MOS Interface-State Density Measurements Using Transient Capacitance Spectroscopy," *IEEE Trans. Electron Dev.* **ED-27**, 2231–2239, Dec. 1980.

86. M. Schulz and N.M. Johnson, "Transient Capacitance Measurements of Hole Emission from Interface States in MOS Structures," *Appl. Phys. Lett.* **31**, 622–625, Nov. 1977; T.J. Tredwell and C.R. Viswanathan, "Determination of Interface-State Parameters in a MOS Capacitor by DLTS," *Solid-State Electron.* **23**, 1171–1178, Nov. 1980.

87. K. Yamasaki, M. Yoshida and T. Sugano, "Deep Level Transient Spectroscopy of Bulk Traps and Interface States in Si MOS Diodes," *Japan. J. Appl. Phys.* **18**, 113–122, Jan. 1979.

88. N.M. Johnson, D.J. Bartelink and M. Schulz, "Transient Capacitance Measurements of Electronic States at the Si-SiO$_2$ Interface," in *The Physics of SiO$_2$ and Its Interfaces* (S.T. Pantelides, ed.), Electrochem. Soc., Pergamon Press, New York, 1978, pp. 421–427.

89. T. Katsube, K. Kakimoto and T. Ikoma, "Temperature and Energy Dependences of Capture Cross Sections at Surface States in Si Metal-Oxide-Semiconductor Diodes Measured by Deep Level Transient Spectroscopy," *J. Appl. Phys.* **52**, 3504–3508, May 1981.

90. W.D. Eades and R.M. Swanson, "Improvements in the Determination of Interface State Density Using Deep Level Transient Spectroscopy," *J. Appl. Phys.* **56**, 1744–1751, Sept. 1984; W.D. Eades and R.M. Swanson, "Determination of the Capture Cross Section and Degeneracy Factor of Si-SiO$_2$ Interface States," *Appl. Phys. Lett.* **44**, 988–990, May 1984.

91. B. Monemar and H.G. Grimmeiss, "Optical Characterization of Deep Energy Levels in Semiconductors," *Progr. Cryst. Growth Charact.* **5**, 47–88, Jan. 1982; H.G. Grimmeiss, "Deep Level Impurities in Semiconductors," in *Annual Review Material Science* (R.A. Huggins, R.H. Bube and R.W. Roberts, eds.), Annual Reviews, Palo Alto, CA, **7**, 341–376, 1977.

92. A. Chantre, G. Vincent and D. Bois, "Deep-Level Optical Spectroscopy in GaAs," *Phys. Rev.* **B23**, 5335–5339, May 1981.

93. P.M. Mooney, "Photo-Deep Level Transient Spectroscopy: A Technique to Study Deep Levels in Heavily Compensated Semiconductors," *J. Appl. Phys.* **54**, 208–213, Jan. 1983.

94. C.T. Sah, L.L. Rosier and L. Forbes, "Direct Observation of the Multiplicity of Impurity Charge States in Semiconductors from Low-Temperature High-Frequency Photo-Capacitance," *Appl. Phys. Lett.* **15**, 316–318, Nov. 1969.

95. A.M. White, P.J. Dean and P. Porteous, "Photocapacitance Effects of Deep Traps in Epitaxial GaAs," *J. Appl. Phys.* **47**, 3230–3239, July 1976.

96. S. Dhar, P.K. Bhattacharya, F.Y. Juang, W.P. Hong and R.A. Sadler, "Dependence of Deep-Level Parameters in Ion-Implanted GaAs MESFET's on Material Preparation," *IEEE Trans. Electron Dev.* **ED-33**, 111–118, Jan. 1986.

97. R.E. Kremer, M.C. Arikan, J.C. Abele and J.S. Blakemore, "Transient Photoconductivity Measurements in Semi-Insulating GaAs. I. An Analog Approach," *J. Appl. Phys.* **62**, 2424–2431, Sept. 1987 and references therein; J.C. Abele, R.E. Kremer and J.S. Blakemore, "Transient Photoconductivity Measurements in Semi-Insulating GaAs. II. A Digital Approach," *J. Appl. Phys.* **62**, 2432–2438, Sept. 1987.

98. D.C. Look, "The Electrical and Photoelectronic Properties of Semi-Insulating GaAs," in *Semiconductors and Semimetals* (R.K. Willardson and A.C. Beer, eds.) Academic Press, New York, **19**, 75–170, 1983.

99. M.R. Burd and R. Braunstein, "Deep Levels in Semi-Insulating Liquid Encapsulated Czochralski-Grown GaAs," *J. Phys. Chem. Sol.* **49**, 731–735, 1988.

100. J.C. Balland, J.P. Zielinger, C. Noguet and M. Tapiero, "Investigation of Deep Levels in High-Resistivity Bulk Materials by Photo-Induced Current Transient Spectroscopy: I. Review and Analysis of Some Basic Problems," *J. Phys. D.: Appl. Phys.* **19**, 57–70, Jan. 1986; J.C. Balland, J.P. Zielinger, M. Tapiero, J.G. Gross and C. Noguet, "Investigation of Deep Levels in High-Resistivity Bulk Materials by Photo-Induced Current Transient Spectroscopy: II. Evaluation of Various Signal Processing Methods," *J. Phys. D.: Appl. Phys.* **19**, 71–87, Jan. 1986.

101. P.M. Petroff and D.V. Lang, "A New Spectroscopic Technique for Imaging the Spatial Distribution of Non-radiative Defects in a Scanning Transmission Electron Microscope," *Appl. Phys. Lett.* **31**, 60–62, July 1977.

102. O. Breitenstein, "A Capacitance Meter of High Absolute Sensitivity Suitable for Scanning DLTS Application," *Phys. Stat. Sol.* **71a**, 159–167, May 1982.

103. K. Wada, K. Ikuta, J. Osaka and N. Inoue, "Analysis of Scanning Deep Level Transient Spectroscopy," *Appl. Phys. Lett.* **51**, 1617–1619, Nov. 1987.

104. J. Heydenreich and O. Breitenstein, "Characterization of Defects in Semiconductors by Combined Application of SEM (EBIC) and SDLTS," *J. Microsc.* **141**, 129–142, Feb. 1986.

105. K. Dmowski, B. Lepley, E. Losson, and M. El Bouabdellati, "A Method to Correct for Leakage Current Effects in Deep Level Transient Spectroscopy Measurements on Schottky Diodes," *J. Appl. Phys.* **74**, 3936–3943, Sept. 1993.

106. D.S. Day, M.J. Helix, K. Hess and B.G. Streetman, "Deep Level Transient Spectroscopy for Diodes with Large Leakage Currents," *Rev. Sci. Instrum.* **50**, 1571–1573, Dec. 1979.

107. E. Simoen, K. De Backker, and C. Claeys, "Deep-Level Transient Spectroscopy of Detector Grade High Resistivity, Float Zone Silicon," *J. Electron. Mat.* **21**, 533–541, May 1992.

108. A. Broniatowski, A. Blosse, P.C. Srivastava and J.C. Bourgoin, "Transient Capacitance Measurements on Resistive Samples," *J. Appl. Phys.* **54**, 2907–2910, June 1983.

109. T. Thurzo and F. Dubecky, "On the Role of the Back Contact in DLTS Experiments with Schottky Diodes," *Phys. Stat. Sol.* **89a**, 693–698, June 1985.

110. J.H. Zhao, J.C Lee, Z.Q. Fang, T.E. Schlesinger and A.G. Milnes, "The Effects of the Nonabrupt Depletion Edge on Deep-Trap Profiles Determined by Deep-Level Transient Spectroscopy," *J. Appl. Phys.* **61**, 5303–5307, June 1987; errata *ibid.* p. 5489.

111. S.D. Brotherton, "The Width of the Non-Steady State Transition Region in Deep Level Impurity Measurements," *Solid-State Electron.* **26**, 987–990, Oct. 1983.

112. D. Stievenard and D. Vuillaume, "Profiling of Defects Using Deep Level Transient Spectroscopy," *J. Appl. Phys.* **60**, 973–979, Aug. 1986.

113. D.C. Look, Z.Q. Fang, and J.R. Sizelove, "Convenient Determination of Concentration and Energy Level in Deep-Level Transient Spectroscopy," *J. Appl. Phys.* **77**, 1407–1410, Feb. 1995; "Depletion Approximation in Semiconductor Trap Filling Analysis: Application to EL2

in GaAs," *Solid-State Electron.* **39**, 1398–1400, Sept. 1996; D.C. Look and J.R. Sizelove, "Depletion Width and Capacitance Transient Formulas for Deep Traps of High Concentration," *J. Appl. Phys.* **78**, 2848–2850, Aug. 1995.

114. K.B. Nielsen and E. Andersen, "Significance of Blackbody Radiation in Deep-Level Transient Spectroscopy," *J. Appl. Phys.* **79**, 9385–9387, June 1996.

115. L.R. Weisberg and H. Schade, "A Technique for Trap Determination in Low-Resistivity Semiconductors," *J. Appl. Phys.* **39**, 5149–5151, Oct. 1968.

116. M.G. Buehler and W.E. Phillips, "A Study of the Gold Acceptor in a Silicon p^+n Junction and an n-Type MOS Capacitor by Thermally Stimulated Current and Capacitance Measurements," *Solid-State Electron.* **19**, 777–788, Sept. 1976.

117. C. Dehn, H. Feick, P. Heydarpoor, G. Lindström, M. Moll, C. Schütze, and T. Schulz, "Neutron Induced Defects in Silicon Detectors Characterized by DLTS and TSC," *Nucl. Instrum. Meth.* **A377**, 258–274, Aug. 1996.

118. C. Szeles and K.G. Lynn, "Positron-Annihilation Spectroscopy," in *Encycl. Appl. Phys.* **14**, 607–632, 1996; R. Krause-Rehberg and H.S. Leipner, *Positron Annihilation in Semiconductors*, Springer, Berlin, 1999.

119. P. Asoka-Kumar, K.G. Lynn, and D.O. Welch, "Characterization of Defects in Si and SiO_2-Si Using Positrons," *J. Appl. Phys.* **76**, 4935–4982, Nov. 1994.

120. M. Fujinami, A. Tsuge, and K. Tanaka, "Characterization of Defects in Self-Ion Implanted Si Using Positron Annihilation Spectroscopy and Rutherford Backscattering Spectroscopy," *J. Appl. Phys.* **79**, 9017–9021, June 1996.

121. M. Fujinami, T. Miyagoe, T. Sawada, and T. Akahane, "Improved Depth Profiling With Slow Positrons of Ion implantation-induced Damage in Silicon," *J. Appl. Phys.* **94**, 4382–4388, Oct. 2003.

122. A. Rich and J. Van House, "Positron Microscopy," in *Encycl. Phys. Sci. Technol.*, Academic Press, **13**, 365–372, 1992; G.R. Brandes, K.F. Canter, T.N. Horsky, P.H. Lippel, and A.P. Mills, Jr., "Scanning Positron Microscopy," *Rev. Sci. Instrum.* **59**, 228–232, Feb. 1988.

123. F. Reif, *Fundamentals of Statistical and Thermal Physics*, McGraw-Hill, New York, 1965, 161–166.

124. O. Engström and A. Alm, "Thermodynamical Analysis of Optimal Recombination Centers in Thyristors," *Solid-State Electron.* **21**, 1571–1576, Nov./Dec. 1978; "Energy Concepts of Insulator-Semiconductor Interface Traps," *J. Appl. Phys.* **54**, 5240–5244, Sept. 1983.

125. C.D. Thurmond, "The Standard Thermodynamic Functions for the Formation of Electrons and Holes in Ge, Si, GaAs, and GaP," *J. Electrochem. Soc.* **122**, 1133–1141, Aug. 1975.

126. J.A. Van Vechten and C.D. Thurmond, "Entropy of Ionization and Temperature Variation of Ionization Levels of Defects in Semiconductors," *Phys. Rev.* **B14**, 3539–3550, Oct. 1976.

127. A. Mircea, A. Mitonneau and J. Vannimenus, "Temperature Dependence of Ionization Energies of Deep Bound States in Semiconductors," *J. Physique* **38**, L41–L43, Jan. 1972.

128. F. Hasegawa, "A New Method (the Three-Point Method) of Determining Transient Time Constants and Its Application to DLTS," *Japan. J. Appl. Phys.* **24**, 1356–1358, Oct. 1985; J.M. Steele, "Hasegawa's Three Point Method for Determining Transient Time Constant," *Japan. J. Appl. Phys.* **25**, 1136–1137, July 1986.

129. P.C. Mangelsdorf, Jr., "Convenient Plot for Exponential Functions with Unknown Asymptotes," *J. Appl. Phys.* **30**, 442–443, March 1959.

130. A.A. Istratov and O.F. Vyvenko, "Exponential Analysis in Physical Phenomena," *Rev. Sci. Instrum.* **70**, 1233–1257, Feb. 1999.

131. C. Lanczos, *Applied Analysis*, Prentice-Hall, Englewood Cliffs, NJ, 1959, 272 ff–.

132. J.L. Benton and L.C. Kimerling, "Capacitance Transient Spectroscopy of Trace Contamination in Silicon," *J. Electrochem. Soc.* **129**, 2098–2102, Sept. 1982.

133. V. Pandian and V. Kumar, "Single-gate Deep Level Transient Spectroscopy Technique," *J. Appl. Phys.* **67**, 560–563, Jan. 1990.

134. E. Losson and B. Lepley, "New Method of Deep Level Transient Spectroscopy Analysis: A Five Emission Rate Method", *Mat. Sci. Eng.* **B20**, 214–220, June 1993.

PROBLEMS

5.1 Using

$$\tau_e T^2 = \frac{\exp((E_c - E_T)/kT)}{\gamma_n \sigma_n}; \delta C = \frac{C_0 n_T(0)}{2N_D}\left(\exp\left(-\frac{t_2}{\tau_e}\right) - \exp\left(-\frac{t_1}{\tau_e}\right)\right)$$

$$= \Delta C_0 \left(\exp\left(-\frac{t_2}{\tau_e}\right) - \exp\left(-\frac{t_1}{\tau_e}\right)\right)$$

(a) Show that when δC is plotted versus temperature, the peak DLTS value, $\delta C_{\max}$, occurs for

$$\tau_e = \frac{t_2 - t_1}{\ln(t_2/t_1)} = \frac{t_1(r-1)}{\ln(r)} \quad \text{where } r = t_2/t_1.$$

(b) Show that $\delta C_{\max} = \Delta C_0((1-r)/r^{r/(r-1)})$. *Hint*: Define $x = \exp(-t_1/\tau_e)$.

5.2 Using the equations in Problem 5.1,

(a) Show that when $t_2 \gg t_1$,

$$\ln\left[\frac{\ln(\Delta C_o/\delta C)}{T^2}\right] \approx \ln(\gamma_n \sigma_n t_1) - \frac{\Delta E}{kT} \quad \text{where } \Delta E = E_C - E_T.$$

(b) Show that a plot of $\ln\left[\dfrac{\ln(\Delta C_o/\delta C)}{T^2}\right]$ versus $1/T$ allows ΔE and σ_n to be extracted.

(c) Plot δC versus T for $\Delta C_o = 10^{-13} F$, $\Delta E = 0.4$ eV, $\sigma_n = 10^{-15}$ cm^2, and $\gamma_n = 1.07 \times 10^{21}$ cm^{-2}s^{-1}K^{-2}, for $t_1 = 1$ ms and $r = 2, 5, 10, 100,$ and 500. Plot all five curves on the same figure.

(d) Plot $\ln\left[\dfrac{\ln(\Delta C_o/\delta C)}{T^2}\right]$ versus $1/T$ for the *high temperature* branch of the δC-T curve of (iii) for $r = 500$ and extract ΔE and σ_n. This technique is discussed in Ref. 133.

5.3 In the boxcar DLTS approach, the peak of the δC-T curve is used to determine τ_e and the relevant temperature T for points on an Arrhenius plot. This gives only one point per temperature scan. More data points lead to better Arrhenius plots. One way to obtain more data points is to use more points of a given δC-T curve than just the peak value. For example, one can use points at $\delta C_{\max}$, $0.75\delta C_{\max}$, and $0.5\delta C_{\max}$, as shown in the Fig. P5.3. We know that $\tau_e(\delta C_{\max}) = (t_2 - t_1)/\ln(t_2/t_1) = t_1(r-1)/\ln(r)$ (see Eq. (5.49)). Determine the two values each for:

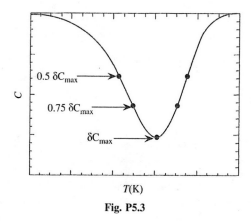

Fig. P5.3

(a) $\tau_{e,0.5} = \tau_e(0.5\delta C_{max})$

(b) $\tau_{e,0.75} = \tau_e(0.75\delta C_{max})$, all four in terms of t_1 for $r = 2$. This technique is discussed in Ref. 134.

5.4 The deep-level transient spectroscopy (DLTS) curve in Fig. P5.4 was obtained by the boxcar method on a Schottky barrier diode on an n-type Si substrate for $t_1 = 0.5$ ms, $t_2 = 1$ ms.

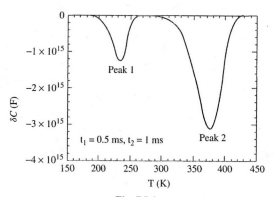

Fig. P5.4

Other curves gave:

t_1 (ms)	t_2 (ms)	$T_{1\,max}(K)$	$\delta C_{1\,max}$ (F)	$T_{2\,max}(K)$	$\delta C_{2\,max}$ (F)
0.5	1	234	-1.25×10^{-15}	376	-3.125×10^{-15}
1	2	227	-1.25×10^{-15}	364	-3.125×10^{-15}
2	4	220	-1.25×10^{-15}	352	-3.125×10^{-15}
4	8	213	-1.25×10^{-15}	341	-3.125×10^{-15}
8	16	207	-1.25×10^{-15}	331	-3.125×10^{-15}

Determine $\Delta E = E_c - E_T$, N_T and the intercept σ_n for both peaks. $C_0 = 5 \times 10^{-12} F$, $N_D = 10^{15}$ cm^{-3}, $\gamma_n = 1.07 \times 10^{21}$ cm^{-2}s^{-1}K^{-2}.

5.5 The capacitance transients for peak 2 in Problem 5.4, were measured for filling pulse widths $t_f = 5$ ns and $t_f = \infty$, for $t_1 = 1$ ms and $t_2 = 2$ ms. Other curves gave:

t_f (ns)	0.5	1	2	3	5
δC (F)	5.9×10^{-17}	1.15×10^{-16}	2.19×10^{-16}	3.31×10^{-16}	4.77×10^{-16}
	7	10	20	∞	
	6.13×10^{-16}	7.72×10^{-16}	1.07×10^{-15}	1.25×10^{-15}	

Determine τ_c, c_n, σ_n and N_T from these data. Use $v_{th} = 10^7 (T/300)^{0.5}$, $n \approx N_D$.

5.6 Consider a Schottky diode at zero bias with deep-level impurities N_T. Light is incident on this device generating electron-hole pairs uniformly. All deep-level impurities are filled with holes while the light is "on" as shown in Fig. P5.6(a). Then at $t = 0$, the light is turned off and a reverse-bias voltage $-V_1$ is applied simultaneously.

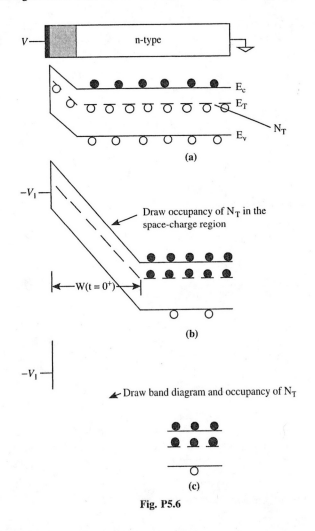

Fig. P5.6

(a) On the band diagram, Fig. P5.6(b), draw the occupancy of N_T at $t = 0^+$, *i.e.*, immediately after the light is turned off.

(b) Draw the band diagram and the occupancy of N_T as $t \to \infty$ in Fig. P5.6(c). In both cases concern yourself only with the space-charge region. Don't worry about the quasi-neutral region. The deep-level impurities are *acceptors*, $N_T < N_D$.

5.7 The deep-level transient spectroscopy data in Fig. P5.7(a) were obtained by the box-car method on a Schottky barrier diode on an *n*-type Si substrate. $\gamma_n = 1.07 \times 10^{21}$ cm^{-2}s^{-1}K^{-2}, $N_D = 10^{15}$ cm^{-3}, $C_o = 1$ pF. In this device it is known that the emission rate can be represented by

$$e_n = \sigma_n v_{th} N_c \exp(-\Delta E / kT), \quad \text{where } \sigma_n = \sigma_{no} \exp(-E_b / kT)$$

The data for σ_n versus T are given in Fig. P5.7(b).
Determine $\Delta E = E_c - E_T$, N_T, σ_{no}, and E_b. (See Appendix 5.1)

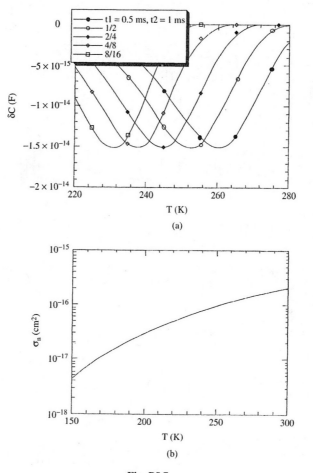

Fig. P5.7

5.8 Plot δC versus T ($150K \leq T \leq 300K$), similar to Fig. P5.7(a), using the boxcar DLTS Eq. (5.48) for an n-Si sample with two energy levels in the band gap using the values:

$$\gamma_n = 1.07 \times 10^{21} \text{ cm}^{-2}\text{s}^{-1}\text{K}^{-2},$$

$$N_D = 10^{15} \text{ cm}^{-3}, C_o = 1 \text{ pF}, \Delta E_1 = 0.25 \text{ eV}, \Delta E_2 = 0.4 \text{ eV},$$

$$\sigma_{n1} = 10^{-16} \text{ cm}^2, \sigma_{n2} = 10^{-15} \text{ cm}^2,$$

$$N_{T1} = 5 \times 10^{12} \text{ cm}^{-3}, N_{T2} = 8 \times 10^{12} \text{ cm}^{-3}, t_1 = 1 \text{ ms}, t_2 = 2 \text{ ms}.$$

5.9 The deep-level transient spectroscopy data in Fig. P5.9 were obtained by the box-car method on a Schottky barrier diode on a p-type Si substrate. The diode area is 0.02 cm^2 and the diode bias voltage was varied from zero to reverse bias voltage of $5V$ during the measurement. $K_s = 11.7, \gamma_p = 1.78 \times 10^{21} \text{ cm}^{-2}\text{s}^{-1}\text{K}^{-2}, N_A = 10^{15}$ $\text{cm}^{-3}, V_{bi} = 0.87 \text{ V}$. Determine $E_T - E_v$, N_T, and the intercept σ_p for each of the impurities.

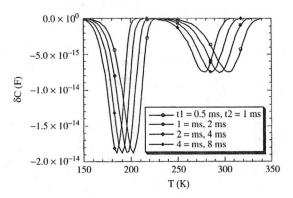

Fig. P5.9

5.10 Determine and plot δC versus T for a Schottky diode on an n-type Si substrate containing two types of impurities. Use the following parameters: $\gamma_n = 1.07 \times 10^{21}$ $\text{cm}^{-2}\text{s}^{-1}\text{K}^{-2}, N_D = 5 \times 10^{15} \text{ cm}^{-3}, C_o = 104 \text{ pF}$.

Impurity 1 : $E_c - E_{T1} = 0.3 \text{ eV}, N_{T1} = 10^{12} \text{ cm}^{-3}, \sigma_{n1} = 10^{-15} \text{ cm}^2$;

Impurity 2 : $E_c - E_{T2} = 0.5 \text{ eV}, N_{T2} = 5 \times 10^{11} \text{ cm}^{-3}, \sigma_{n2} = 5 \times 10^{-16} \text{ cm}^2$.

Use the boxcar equations with $t_1 = 1 \text{ ms}, t_2 = 2 \text{ ms}$ and the temperature range $150 \leq T \leq 350 \text{ K}$.

5.11 A deep-level *acceptor* impurity is diffused uniformly into an n-type Si wafer. The wafer was originally doped with arsenic to $N_D = 10^{15} \text{ cm}^{-3}$.
Calculate and plot the resistivity versus deep-level impurity density ($10^{14} \leq N_T \leq 10^{17} \text{ cm}^{-3}$) on a log-log plot for $E_T = 0.46, 0.56$, and 0.66 eV. Plot all three curves on one figure to compare.

You have to first solve for E_F using the equations below. Knowing E_F you can then find n and p and then determine ρ.

Charge neutrality requires

$$p + n_D^+ - n - n_T^- = 0$$

where

$$p = n_i \exp((E_i - E_F)/kT); n = n_i \exp((E_F - E_i)/kT)$$

$$n_D^+ = \frac{N_D}{1 + \exp((E_F - E_D)/kT)}; n_T^- = \frac{N_T}{1 + \exp((E_T - E_F)/kT)}$$

$$\rho = \frac{1}{q(\mu_n n + \mu_p p)}$$

Use: $N_D = 10^{15}$ cm^{-3}, $n_i = 10^{10}$ cm^{-3}, $T = 300$ K, $\mu_n = 1400$ cm$^2/V$-s, $\mu_p = 450$ cm$^2/V$-s, $E_G = 1.12$ eV, $E_i = 0.56$ eV, $E_D = E_c - 0.045$ eV. It is easiest to use $E_v = 0$ as a reference energy.

5.12 The deep-level transient spectroscopy data in Fig. P5.12 were obtained on a Schottky barrier diode on an n-type Si substrate containing two impurities. $\gamma_n = 1.07 \times 10^{21}$ cm^{-2}s^{-1}K^{-2}, $N_D = 10^{15}$ cm^{-3}, $C_0 = 1$ pF. Determine $\Delta E = E_c - E_T$, σ_n, and N_T for each deep-level impurity.

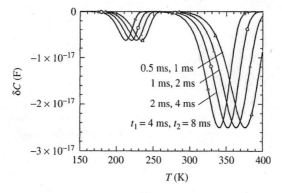

Fig. P5.12

5.13 Consider interface trapped charge or interface state density D_{it} at the SiO$_2$/Si interface of an MOS device. The device is heavily inverted and all interface states are filled with electrons. Determine the density of interface states still filled with electrons, $N_{it} = D_{it}\Delta E$, 100 µs *after* the surface is driven into depletion and electrons are emitted from interface states during the 100 µs. $D_{it} = 5 \times 10^{10}$ cm^{-2}eV^{-1}, $T = 300$ K, $\sigma_n = 10^{-15}$ cm^2, $v_{th} = 10^7$ cm/s, $N_c = 2.5 \times 10^{19}$ cm^{-3}, $k = 8.617 \times 10^{-5}$ eV/K, E_G(Si) $= 1.12$ eV. The electron emission time constant from interface states at energy E_{it} is given by

$$\tau_e = \frac{\exp[(E_c - E_{it})/kT]}{\sigma_n v_{th} N_c}$$

5.14 The Arrhenius plot of a deep-level impurity in Si is shown in Fig. P5.14. Determine $E_c - E_T$ and σ_n.

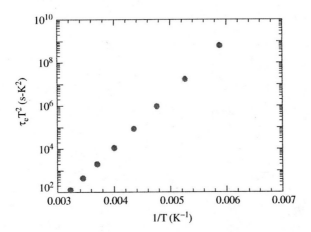

Fig. P5.14

Use $\gamma_n = 1.07 \times 10^{21}$ cm^{-2}s^{-1}K^{-1}, $k = 8.617 \times 10^{-5}$ eV/K.

5.15 Calculate and plot $C(t)/C_o$ given by $\dfrac{C(t)}{C_o} = 1 - \dfrac{N_T}{2N_D} \exp(-t/\tau_e)$ for $N_D = 10^{15}$ cm^{-3}, $N_T = 5 \times 10^{12}$ cm^{-3}, $\sigma_n = 10^{-15}$ cm^2, $E_c - E_T = 0.35$ eV, $\gamma_n = 1.07 \times 10^{21}$ cm^{-2}s^{-1}K^{-1}, for $T = 200$ K, 225 K and 250 K over the time interval: $0 < t < 0.002$ s.

5.16 A Schottky diode on an n-type substrate, containing a deep-level impurity, is *zero* biased for some time. Next, the device is *reverse* biased at $t = 0$. The charge density in the reverse-biased space-charge region at $t = 0^+$, immediately after applying the reverse-bias pulse, is $\rho = qN_D$. The deep-level impurity is: ☐ donor ☐ acceptor. Give your reason.

5.17 Identify the two deep-level impurities in Fig. P 5.17.

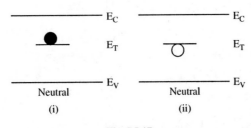

Fig. P5.17

Deep-level impurity (i) is a: ☐ donor ☐ acceptor. Give your reason.
Deep-level impurity (ii) is a: ☐ donor ☐ acceptor. Give your reason.

5.18 There are two defects in the transmission electron micrograph in Fig. P 5.18.

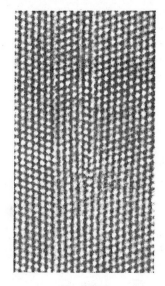

Fig. P5.18

Identify them and state whether they are *point, line, plane,* or *volume* defects.

5.19 A DLTS plot of δC versus T is shown in Fig. P5.19 for a certain impurity with energy level $E_T = E_{T1}$ and density $N_T = N_{T1}$. On the same figure draw the curve for an impurity with $E_T = E_{T2} > E_{T1}$ and $N_T = N_{T2} < N_{T1}$. t_2/t_1 is unchanged.

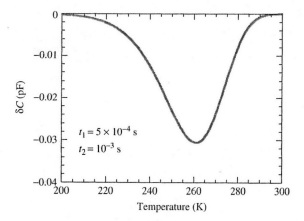

Fig. P5.19

5.20 A DLTS plot of δC versus T is shown in Fig. P5.19. On the same figure draw the curve when both t_1 and t_2 are *increased*, but t_2/t_1 is unchanged.

5.21 Consider an n-type semiconductor doped with N_D donor atoms/cm^3 with energy level E_D shown in "Before" in Fig. P5.21. All donors are ionized. Next, a deep-level impurity at energy level E_T is introduced into the n-type semiconductor wafer, shown in "After".

The deep-level impurity is a: ☐ donor ☐ acceptor. Give your reason.

The wafer resistivity: ☐ increases ☐ decreases ☐ remains unchanged. Give your reason.

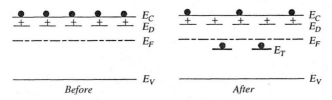

Before   After

Fig. P5.21

5.22 The DLTS spectrum of impurity 1 in an n-type semiconductor is shown in Fig. P5.22. It has an energy level E_{T1}, density N_{T1}, and capture cross section σ_{n1}.

(a) Show the effect of *decreasing* σ_{n1} on the spectrum in Fig. P5.22(a) and on the $\ln(\tau_e T^2) - 1/T$ plot in Fig. P5.22(b).

(b) On Fig.5.22(c), draw the DLTS spectrum for impurity 2 with energy level E_{T2}, where $E_c - E_{T2} < E_c - E_{T1}$, $N_{T2} < N_{T1}$, and $\sigma_{n2} = \sigma_{n1}$ in this case.

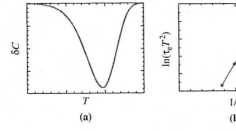
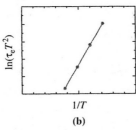
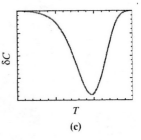

(a)   (b)   (c)

Fig. P5.22

REVIEW QUESTIONS

- Name some common defects in Si wafers.
- What do metallic impurities do in Si devices?
- Name some defect sources.
- What are point defects? Name three point defects.
- Name a line defect, an area defect, and a volume defect.
- How do *oxidation-induced stacking faults* originate?
- Why is emission generally slower than capture?
- What determines the capacitance transient?

- Where does the energy for thermal emission come from?
- Why do *minority* and *majority* carrier emission have opposite behavior?
- What is *deep-level transient spectroscopy* (DLTS)?
- What parameters can be determined with DLTS?
- What advantage does *Laplace DLTS* have?
- What is *positron annihilation spectroscopy* and for what defect measurement is it most useful?

6

OXIDE AND INTERFACE TRAPPED CHARGES, OXIDE THICKNESS

6.1 INTRODUCTION

The discussions in this chapter are applicable to all insulator-semiconductor systems. However, the examples are generally directed at the SiO_2-Si system. The most important aspect of device scaling for this chapter is the thinner oxide with each successive technology node. Thin oxides with their respective higher leakage currents, have a pronounced effect on many of the methods in this chapter. Capacitance-voltage and oxide thickness measurements must be more carefully interpreted for thin, leaky oxides.

Oxide Charges:[1]   There are four general types of charges associated with the SiO_2-Si system shown on Fig. 6.1. They are *fixed oxide charge, mobile oxide charge, oxide trapped charge* and *interface trapped charge*. This nomenclature was standardized in 1978. The abbreviations of the various charges are given below. In each case, Q is the net effective charge per unit area at the SiO_2-Si interface (C/cm^2), N is the net effective number of charges per unit area at the SiO_2-Si interface (number/cm^2), and D_{it} is given in units of number/$cm^2 \cdot eV$. $N = |Q|/q$, where Q can be positive or negative, but N is always positive.

(1) Interface Trapped Charge (Q_{it}, N_{it}, D_{it}):   These are positive or negative charges, due to structural defects, oxidation-induced defects, metal impurities, or other defects caused by radiation or similar bond breaking processes (*e.g.*, hot electrons). The interface trapped charge is located at the Si–SiO_2 interface. Unlike fixed charge or trapped charge, interface trapped charge is in electrical communication with the underlying silicon. Interface traps can be charged or discharged, depending on the surface potential. Most of the interface trapped charge can be neutralized by low-temperature ($\sim$450°C) hydrogen or

Semiconductor Material and Device Characterization, Third Edition, by Dieter K. Schroder
Copyright © 2006 John Wiley & Sons, Inc.

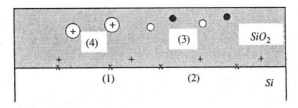

Fig. 6.1   Charges and their location for thermally oxidized silicon. Reprinted after Deal by permission of IEEE (© 1980, IEEE).

forming gas (hydrogen/nitrogen mixture) anneals. This charge type has been called *surface states, fast states, interface states* and so on. It has been designated by N_{ss}, N_{st} and other symbols in the past.

   (2) Fixed Oxide Charge (Q_f, N_f):   This is a positive charge near the Si–SiO$_2$ interface. The charge density, whose origin is related to the oxidation process, depends on the oxidation ambient and temperature, cooling conditions, and on silicon orientation. Since the fixed oxide charge cannot be determined unambiguously in the presence of moderate densities of interface trapped charge, it is usually measured after a low-temperature (450°C) hydrogen or forming gas anneal which minimizes interface trapped charge. The fixed oxide charge is not in electrical communication with the underlying silicon. Q_f depends on the final oxidation temperature. The higher the oxidation temperature, the lower is Q_f. However, if it is not permissible to oxidize at high temperatures, it is possible to lower Q_f by annealing the oxidized wafer in a nitrogen or argon ambient after oxidation. This has resulted in the well-known "Deal triangle" in Fig. 6.2, which shows the reversible relationship between Q_f and oxidation and annealing.[2] An oxidized sample may be prepared at any temperature and then subjected to dry oxygen at any other temperature, with the resulting value of Q_f being associated with the final temperature and any Q_f value resulting from a previous oxidation can be reduced to a constant value. Fixed charge was often designated as Q_{ss} in the past.

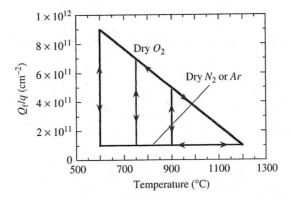

Fig. 6.2   "Deal triangle" showing the reversibility of heat treatment effects on Q_f. Reprinted after Deal et al.[2] with permission of the publisher, the Electrochemical Society, Inc.

(3) Oxide Trapped Charge (Q_{ot}, N_{ot}): This positive or negative charge may be due to holes or electrons trapped in the oxide. Trapping may result from ionizing radiation, avalanche injection, Fowler-Nordheim tunneling, or other mechanisms. Unlike fixed charge, oxide trapped charge is sometimes annealed by low-temperature (<500°C) treatments, although neutral traps may remain.

(4) Mobile Oxide Charge (Q_m, N_m): This is caused primarily by ionic impurities such as Na^+, Li^+, K^+, and possibly H^+. Negative ions and heavy metals may contribute to this charge.

6.2 FIXED, OXIDE TRAPPED, AND MOBILE OXIDE CHARGE

6.2.1 Capacitance-Voltage Curves

The various charges can be determined by the capacitance-voltage $(C-V)$ of metal-oxide-semiconductor capacitors (MOS-C). Before discussing measurement methods, we derive the capacitance-voltage relationships and describe the $C-V$ curves. The energy band diagram of an MOS capacitor on a p-type substrate is shown in Fig. 6.3. The intrinsic energy level E_i or potential ϕ in the neutral part of the device is taken as the zero reference potential. The surface potential ϕ_s is measured from this reference level. The capacitance is defined as

$$C = \frac{dQ}{dV} \tag{6.1}$$

It is the change of charge due to a change of voltage and is most commonly given in units of farad/unit area. During capacitance measurements, a small-signal ac voltage is applied to the device. The resulting charge variation gives rise to the capacitance. Looking at an MOS-C from the gate, $C = dQ_G/dV_G$, where Q_G and V_G are the gate charge and the gate voltage. Since the total charge in the device must be zero, $Q_G = -(Q_S + Q_{it})$ assuming no oxide charge. The gate voltage is partially dropped across the oxide and partially across the semiconductor. This gives $V_G = V_{FB} + V_{ox} + \phi_s$, where V_{FB} is the

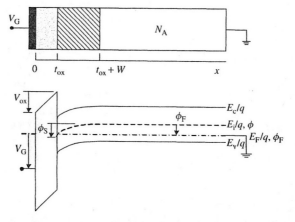

Fig. 6.3 Cross-section and potential band diagram of an MOS capacitor.

flatband voltage, V_{ox} the oxide voltage, and ϕ_s the surface potential, allowing Eq. (6.1) to be rewritten as

$$C = -\frac{dQ_S + dQ_{it}}{dV_{ox} + d\phi_s} \tag{6.2}$$

The semiconductor charge density Q_S, consists of hole charge density Q_p, space-charge region bulk charge density Q_b, and electron charge density Q_n. With $Q_S = Q_p + Q_b + Q_n$, Eq. (6.2) becomes

$$C = -\cfrac{1}{\cfrac{dV_{ox}}{dQ_S + dQ_{it}} + \cfrac{d\phi_s}{dQ_p + dQ_b + dQ_n + dQ_{it}}} \tag{6.3}$$

Utilizing the general capacitance definition of Eq. (6.1), Eq. (6.3) becomes

$$C = \cfrac{1}{\cfrac{1}{C_{ox}} + \cfrac{1}{C_p + C_b + C_n + C_{it}}} = \frac{C_{ox}(C_p + C_b + C_n + C_{it})}{C_{ox} + C_p + C_b + C_n + C_{it}} \tag{6.4}$$

The positive accumulation charge Q_p dominates for negative gate voltages for p-substrate devices. For positive V_G, the semiconductor charges are negative. The minus sign in Eq. (6.3) cancels in either case.

Equation (6.4) is represented by the equivalent circuit in Fig. 6.4(a). For negative gate voltages, the surface is heavily accumulated and Q_p dominates. C_p is very high approaching a short circuit. Hence, the four capacitances are shorted as shown by the heavy line in Fig. 6.4(b) and the overall capacitance is C_{ox}. For small positive gate voltages, the surface is depleted and the space-charge region charge density, $Q_b = -qN_A W$, dominates. Trapped interface charge capacitance also contributes. The total capacitance is the combination of C_{ox} in series with C_b in parallel with C_{it} as shown in Fig. 6.4(c). In weak inversion C_n begins to appear. For strong inversion, C_n dominates because Q_n is very high. If Q_n is able to follow the applied ac voltage, the low-frequency equivalent circuit (Fig. 6.4(d)) becomes the oxide capacitance again. When the inversion charge is unable to follow the ac voltage, the circuit in Fig. 6.4(e) applies in inversion, with $C_b = K_s \varepsilon_o / W_{inv}$ with W_{inv} the inversion space-charge region width discussed in Chapter 2.

The inversion capacitance dominates only if the inversion charge is able to follow the frequency of the applied ac voltage, also called the ac probe frequency. With the MOS-C biased in inversion, the ac voltage drives the device periodically above and below the dc bias point. During the phase when the device is driven to a slightly higher gate voltage, an increased gate charge requires an increased semiconductor charge (inversion charge or space-charge region (scr) charge). For the inversion charge to increase, electron-hole pairs (ehp) must be thermally generated in the scr. The scr generation current density, given by $J_{scr} = qn_i W / \tau_g$ and discussed in more detail in Chapter 7, dominates at room temperature in silicon. The current flowing through the oxide is the displacement current density $J_d = CdV_G/dt$. In order for the inversion charge to respond, the scr current must be able to supply the required displacement current or $J_d \leq J_{scr}$. This leads to

$$\frac{dV_G}{dt} \leq \frac{qn_i W}{\tau_g C_{ox}} \tag{6.5}$$

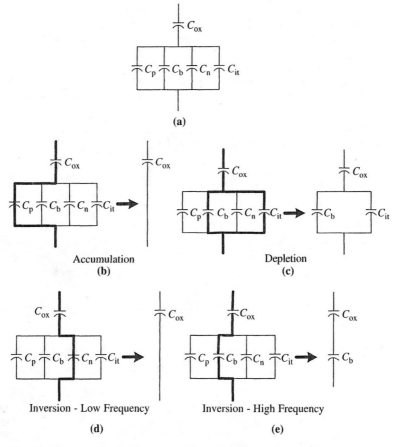

Fig. 6.4 Capacitances of an MOS capacitor for various bias conditions as discussed in the text.

with C approximated by C_{ox}. For Si at $T = 300$ K with $n_i = 10^{10}$ cm^{-3}

$$\frac{dV_G}{dt} \le \frac{0.046 W t_{ox}}{\tau_g} V/s \tag{6.6}$$

with W in μm, t_{ox} in nm, and τ_g in μs. When the MOSFET gate capacitance is measured, the low-frequency C–V_G characteristic is typically obtained when the source and drain are grounded, because the S/D can supply carriers to the channel easily even at high frequencies without thermal generation.

Generation lifetimes lie in the 10 μs to 10 ms range. For $t_{ox} = 5$ nm, $W = 1$ μm, and $\tau_g = 10$ μs, $dV_G/dt = 0.023$ V/s—not a severe constraint. However, for $\tau_g = 1$ ms, $dV_G/dt = 0.23$ mV/s—a very severe constraint. This constraint can be somewhat relaxed by measuring at elevated temperatures because n_i increases. By raising the temperature from 300 K to 350 K, n_i increases from 10^{10} cm^{-3} to 3.6×10^{11} cm^{-3} relaxing the ramp rate by a factor of 36, i.e., from 0.23 to 8.3 mV/s. Defining an *effective frequency* as $f_{eff} = (dV_G/dt)/v$, where v is the ac voltage, we find $f_{eff} \approx 1.5$ Hz for the former and

0.015 Hz for the latter using $v = 15$ mV. These first-order numbers show that extremely low frequencies are required to obtain low-frequency $C-V$ curves at room temperature. Increased generation rates at higher temperatures allow higher frequencies. Since typical $C-V$ measurement frequencies lie in the $10^4 - 10^6$ Hz range, it is obvious that high-frequency curves are usually observed.

The *low-frequency* semiconductor capacitance $C_{S,lf}$ is given by

$$C_{S,lf} = \hat{U}_S \frac{K_s \varepsilon_0}{2L_{Di}} \frac{[e^{U_F}(1 - e^{-U_S}) + e^{-U_F}(e^{U_S} - 1)]}{F(U_S, U_F)} \tag{6.7}$$

where the dimensionless semiconductor surface electric field $F(U_S, U_F)$ is defined by

$$F(U_S, U_F) = \sqrt{e^{U_F}(e^{-U_S} + U_S - 1) + e^{-U_F}(e^{U_S} - U_S - 1)} \tag{6.8}$$

The Us are normalized potentials, defined by $U_S = q\phi_s/kT$ and $U_F = q\phi_F/kT$, where the surface potential ϕ_s and the Fermi potential $\phi_F = (kT/q)\ln(N_A/n_i)$ are defined in Fig. 6.3. The symbol $\hat{U}_S$ stands for the sign of the surface potential and is given by

$$\hat{U}_S = \frac{|U_S|}{U_S} \tag{6.9}$$

where $\hat{U}_s = 1$ for $U_s > 0$ and $\hat{U}_s = -1$ for $U_s < 0$. The intrinsic Debye length L_{Di} is

$$L_{Di} = \sqrt{\frac{K_s \varepsilon_0 kT}{2q^2 n_i}} \tag{6.10}$$

The *high-frequency* $C-V$ curve results when the minority carriers in the inversion charge are unable to follow the ac voltage. The majority carriers at the scr edge are able to follow the ac signal thereby exposing more or less ionized dopant atoms. The dc voltage sweep rate, given by Eq. (6.5), must be sufficiently low to generate the necessary inversion charge. The high-frequency semiconductor capacitance in inversion is[3]

$$C_{S,hf} = \hat{U}_S \frac{K_s \varepsilon_0}{2L_{Di}} \frac{[e^{U_F}(1 - e^{-U_S}) + e^{-U_F}(e^{U_S} - 1)/(1 + \delta)]}{F(U_s, U_F)} \tag{6.11}$$

with δ given by

$$\delta = \frac{(e^{U_S} - U_S - 1)/F(U_S, U_F)}{\int_0^{U_S} \dfrac{e^{U_F}(1 - e^{-U})(e^U - U - 1)}{2[F(U, U_F)]^3} \, dU} \tag{6.12}$$

An approximate expression, accurate to 0.1–0.2% in strong inversion, is[4]

$$C_{S,hf} = \sqrt{\frac{q^2 K_s \varepsilon_0 N_A}{2kT\{2|U_F| - 1 + \ln[1.15(|U_F| - 1)]\}}} \tag{6.13}$$

When the dc bias voltage is changed rapidly with insufficient time for inversion charge generation, the *deep-depletion* curve results. Its high- or low-frequency semiconductor capacitance is

$$C_{S,dd} = \frac{C_{ox}}{\sqrt{[1 + 2(V_G - V_{FB})/V_0]} - 1} \qquad (6.14)$$

where $V_0 = qK_s\varepsilon_0 N_A/C_{ox}^2$.

The total capacitance is given by

$$C = \frac{C_{ox}C_S}{C_{ox} + C_S} \qquad (6.15)$$

The gate voltage is related to the oxide voltage, the surface potential, and the flatband voltage V_{FB} through the relationship

$$V_G = V_{FB} + \phi_s + V_{ox} = V_{FB} + \phi_s + \hat{U}_S \frac{kT K_s t_{ox} F(U_S, U_F)}{q K_{ox} L_{Di}} \qquad (6.16)$$

Ideal low-frequency (lf), high-frequency (hf) and deep depletion (dd) $C-V$ curves are shown in Fig. 6.5 for $Q_{it} = 0$ and $V_{FB} = 0$. They coincide in accumulation and depletion but deviate in inversion, because the inversion charge is unable to follow the applied ac voltage for the hf case and does not exist for the dd case.

Which of these three curves is obtained during a $C-V$ measurement depends on the measurement conditions. Consider an MOS-C on a p-substrate with the dc gate voltage swept from negative to positive voltages. Superimposed on the dc voltage is an ac voltage of typically 10–15 mV amplitude. All three curves are identical in accumulation and depletion. The curves deviate from one another when the device enters inversion. If the dc voltage is swept sufficiently slowly to allow the inversion charge to form and if the ac voltage is of a sufficiently low frequency for the inversion charge to be able to respond to the ac probe frequency, then the low-frequency curve is obtained. If the dc voltage is

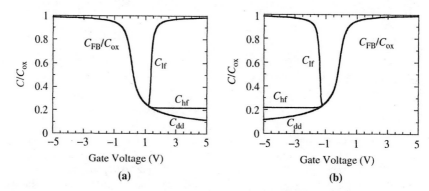

Fig. 6.5 Low-frequency (lf), high-frequency (hf), and deep-depletion (dd) normalized SiO$_2$-Si capacitance-voltage curves of an MOS-C; (a) p-substrate $N_A = 10^{17}$ cm^{-3}, (b) n-substrate $N_D = 10^{17}$ cm^{-3}, $t_{ox} = 10$ nm, $T = 300$ K.

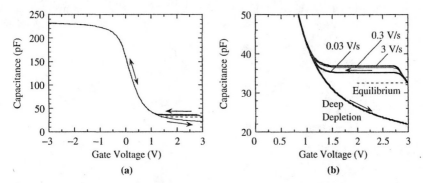

Fig. 6.6 Effect of sweep direction and sweep rate on the hf MOS-C capacitance on p-substrate, (a) entire $C–V_G$ curve, (b) enlarged portion of (a) showing the dc sweep direction; $f = 1$ MHz. Data courtesy of Y.B. Park, Arizona State University.

swept sufficiently slowly to allow the inversion charge to form but the ac probe frequency is too high for the inversion charge to be able to respond, then the high-frequency curve is obtained. The deep-depletion curve obtains for either high- or low-frequency if the dc sweep rate is too high and no inversion charge can form during the sweep.

The most commonly measured curve is the high-frequency curve. However, the true hf curve is not always easy to obtain. Consider the $C–V_G$ curve in Fig. 6.6. The true or equilibrium curve is shown by the dashed line. If the bias is swept from $-V_G$ to $+V_G$ there is a tendency for the $C–V$ curve to go into partial deep depletion and the resulting curve will be *below* the true curve, especially for high generation lifetime material. We showed the limitation on the ac frequency in Eq. (6.5). This limitation also holds for the dc bias sweep rate; the sweep rate for high lifetime material must be extremely low.

When the bias is swept from $+V_G$ to $-V_G$, inversion charge is injected into the substrate. The inversion layer/substrate junction becomes forward biased and the resulting capacitance will be *above* the true curve. The true curve is, in general, only obtained by setting the bias voltage and waiting for the device to come to equilibrium, then repeating this procedure to generate the $C–V$ curve point-by-point. If the point-by-point procedure is inconvenient, then the $+V_G \rightarrow -V_G$ sweep direction is preferred for p-substrates since the deviation of the capacitance from its true value is generally less than it is for the $-V_G \rightarrow +V_G$ sweep.

Exercise 6.1

Problem: What happens to C_{hf} when the measurement temperature is raised?

Solution: According to Eq. (6.5) the minority carriers respond to higher sweep rates when n_i increases and they respond to higher probe frequencies as T increases, *i.e.*, low frequency behavior should be observed at high probe frequencies. This is illustrated in Fig. E6.1. The data points are experimental data and the solid lines are calculated lf curves. At room temperature the hf curve is measured and there is large discrepancy between the measured and calculated lf curves. As temperature increases, some of the inversion layer carriers are able to respond and the hf curve begins to show lf characteristics. Finally

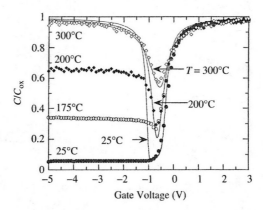

Fig. E6.1 Measured hf (points) and calculated lf (lines) curves of an MOS-C. $N_D = 2.6 \times 10^{14}$ cm^{-3}, $t_{ox} = 30$ nm, $f = 10$ kHz. Data courtesy of S.Y. Lee, Arizona State University.

at $T = 300°C$, the hf curve coincides with the lf curve. Hence C_{hf} and C_{lf} measured at $T = 300°C$ are identical in this example. The temperature at which this happens, also depends on parameters other than n_i, e.g., τ_g, W, and C_{ox}.

6.2.2 Flatband Voltage

The flatband voltage is determined by the metal-semiconductor work function difference ϕ_{MS} and the various oxide charges through the relation

$$V_{FB} = \phi_{MS} - \frac{Q_f}{C_{ox}} - \frac{Q_{it}(\phi_s)}{C_{ox}} - \frac{1}{C_{ox}} \int_0^{t_{ox}} \frac{x}{t_{ox}} \rho_m(x) \, dx - \frac{1}{C_{ox}} \int_0^{t_{ox}} \frac{x}{t_{ox}} \rho_{ot}(x) \, dx$$

(6.17)

where $\rho(x) =$ oxide charge per unit volume. The fixed charge Q_f is located very near the Si–SiO$_2$ interface and is considered to be at that interface. Q_{it} is designated as $Q_{it}(\phi_s)$, because the occupancy of the interface trapped charge depends on the surface potential. Mobile and oxide trapped charges may be distributed throughout the oxide. The x-axis is defined in Fig. 6.3. The effect on flatband voltage is greatest, when the charge is located at the oxide-semiconductor substrate interface, because then it images all of its charge in the semiconductor. When the charge is located at the gate-oxide interface, it images all of its charge in the gate and has no effect on the flatband voltage. For a given charge density, the flatband voltage is reduced as the oxide capacitance increases, *i.e.*, for thinner oxides. Hence, oxide charges usually contribute little to flatband or threshold voltage shifts for thin-oxide MOS devices.

The flatband voltage of Eq. (6.17) is for a uniformly doped substrate, with the gate voltage referenced to the grounded back contact. For an epitaxial layer of doping density N_{epi} on a substrate of doping density N_{sub}, the built-in potential at the epi-substrate junction modifies the flatband voltage to[5]

$$V_{FB}(epi) = V_{FB}(bulk) \pm \frac{kT}{2q} \ln \left(\frac{N_{sub}}{N_{epi}} \right)$$

(6.18)

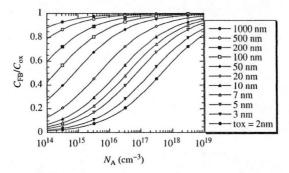

Fig. 6.7 C_{FB}/C_{ox} versus N_A as a function of t_{ox} for the SiO$_2$-Si system at $T = 300$ K.

The plus sign in Eq. (6.18) is for p-type and the minus sign for n-type material, assuming the substrate and the epitaxial layer doping densities are of the same type, either both acceptors or both donors.

To determine the various charges, one compares theoretical and experimental capacitance-voltage curves. The experimental curves are usually shifted with respect to the theoretical curves as a result of the charges and the work function difference of Eq. (6.17). The voltage shift can be measured at any capacitance, however, it is frequently measured at the *flatband capacitance* C_{FB} and is designated the *flatband voltage* V_{FB}. For ideal curves, V_{FB} is zero. The flatband capacitance is given by Eq. (6.15) with $C_S = K_s\varepsilon_o/L_D$, where $L_D = [kT K_s\varepsilon_o/q^2(p+n)]^{1/2} \approx [kT K_s\varepsilon_o/q^2 N_A]^{1/2}$ is the Debye length defined in Eq. (2.11). For Si with SiO$_2$ as the insulator, C_{FB} normalized by C_{ox}, is given as

$$\frac{C_{FB}}{C_{ox}} = \left(1 + \frac{136\sqrt{T/300}}{t_{ox}\sqrt{N_A \text{ or } N_D}}\right)^{-1} \tag{6.19}$$

with t_{ox} in cm and N_A (N_D) in cm^{-3}. In Fig. 6.7, C_{FB}/C_{ox} is plotted versus N_A as a function of oxide thickness.

The flatband capacitance can be easily calculated when the doping density is uniform and when the wafer is sufficiently thick. The calculation becomes more difficult when the doping is non-uniform and numerical techniques may have to be employed.[6] For thin silicon layers, *e.g.*, silicon-on-insulator, the active semiconductor layer may be so thin that it cannot accommodate the space-charge region of the MOS-C. Then special precautions must be used to determine C_{FB}. Graphical and analytical methods have been used.[7] The analytical methods rely on a measure of the capacitance, which is 90% or 95% of the oxide capacitance. The voltage for this capacitance is then related to the flatband voltage.[8]

Exercise 6.2

Problem: Determine the flatband voltage of an MOS capacitor.

Solution: The flatband voltage must be accurately known to determine C_{FB}. Calculating C_{FB}, as described, allows V_{FB} to be determined, provided all the parameters in Eq. (6.17)

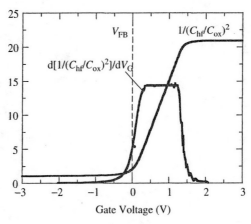

Fig. E6.2

are well known. That may not always be the case. One way to determine V_{FB} experimentally is to plot $(1/C_{hf})^2$ or $1/(C_{hf}/C_{ox})^2$ versus V_G as shown in Fig. E6.2. This curve corresponds to the data in Fig. 6.5(a). The lower knee of this curve occurs at $V_G = V_{FB}$. Such a transition is sometimes difficult to determine. Differentiating this curve and finding the maximum slope of the left flank of this differentiated curve occurs at V_{FB}. Differentiating this differentiated curve a second time results in a sharply peaked curve whose peak coincides with V_{FB}. The second differentiation usually introduces a great deal of noise, but smoothing the data helps. This method is discussed in R.J. Hillard, J.M. Heddleson, D.A. Zier, P. Rai-Choudhury, and D.K. Schroder, "Direct and Rapid Method for Determining Flatband Voltage from Non-equilibrium Capacitance Voltage Data," in *Diagnostic Techniques for Semiconductor Materials and Devices* (J.L. Benton, G.N. Maracas, and P. Rai-Choudhury, eds.), Electrochem. Soc., Pennington, NJ, 1992, 261–274.

Finite Gate Doping Density.   We have so far neglected the effect the gate may have on the $C-V_G$ curve, other than the metal-semiconductor work function difference. Polycrystalline Si is a common gate material, with doping densities around $10^{19}-10^{20}$ cm^{-3}. What is the effect of this? Consider the MOS-C in Fig. 6.8, consisting of a p-type substrate and an n^+ polysilicon gate. For negative gate voltage, substrate and gate are accumulated and we can treat the gate as a metal. However, for positive gate voltage, not only is the substrate depleted and eventually inverted, but the gate can also be depleted and perhaps inverted. Instead of C_{ox} in series with C_S, there is now an additional gate capacitance C_{gate}, reducing overall capacitance. The measurement of gate doping density by a $C-V$ technique is discussed in Chapter 2.

The effect of gate depletion is illustrated on the $C/C_{ox}-V_G$ curves in Fig. 6.9. Note the additional capacitance drop for $+V_G$. This drop increases as N_D in the gate decreases. Such polysilicon gate depletion changes the threshold voltage of MOSFETs, reduces the drain current, and increases the gate resistance. All of these effects reduce circuit speed. On the other hand, gate and source/drain overlap capacitances are also reduced, which tends to increase circuit speed. A recent study has shown the overall effect to be negative, *i.e.*, circuit speed is reduced.[9]

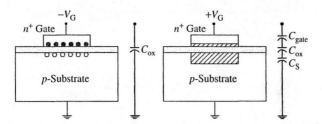

Fig. 6.8   Schematic illustration of an MOS-C with finite gate doping density, showing gate depletion for positive gate voltage.

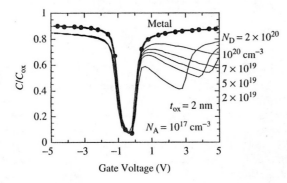

Fig. 6.9   Low-frequency capacitance-voltage curves for a metal gate and various n^+ poly-Si gate doping densities. Simulation courtesy of $D.$ Vasileska, Arizona State University.

Exercise 6.3

Problem: How are the $C-V$ curves of MOS devices affected by quantization and Fermi-Dirac statistics?

Solution: Equations describing the $C-V$ curves above are frequently derived using simplified assumptions. One modification to these assumptions is the depletion of the poly-Si gate. Other modifications, significant for sub 10 nm oxide thicknesses, include Fermi-Dirac $(F-D)$ instead of Maxwell-Boltzmann statistics and inversion layer quantization. Both of these effects must be considered for devices in strong accumulation or inversion. In this degenerate condition, the free carriers occupy discrete energy states in the conduction band reducing the substrate capacitance. Simulations and experiments confirm these effects. Simulated results are shown in Fig. E6.3, where $t_{ox,phys}$ is the physical oxide thickness. These curves include $F-D$, quantization, and gate depletion effects. The substrate is inverted and the gate accumulated ($C_{gate} = C_{inv}$) at $+V_G$ and for $-V_G$ the substrate is accumulated and the gate inverted ($C_{gate} = C_{acc}$). C_{inv} is calculated at $V_G = V_{FB} - 4$ V and C_{acc} is calculated at $V_G = V_{FB} + 3$ V. This figure shows the gate capacitance to be less than the oxide capacitance by at least 10% for $t_{ox} < 10$ nm. Hence extracting oxide thicknesses from $C-V$ measurements will yield incorrect t_{ox} if the data are not properly analyzed. These effects are discussed in K.S. Krisch, J.D. Bude, and L. Manchanda,

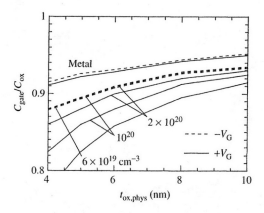

Fig. E6.3 Simulated C_{gate}/C_{ox} ratio versus $t_{ox,phys}$ for metal and n^+ poly-Si/p-Si structure ($N_D = 10^{17}$ cm^{-3}). Oxide leakage current is neglected. Simulation courtesy of D. Vasileska, Arizona State University.

"Gate Capacitance Attenuation in MOS Devices With Thin Gate Dielectrics," *IEEE Electron Dev. Lett.* **17**, 521–524, Nov. 1996; D. Vasileska, D.K. Schroder, and D.K. Ferry, "Scaled Silicon MOSFET's: Degradation of the Total Gate Capacitance," *IEEE Trans Electron Dev.* **44**, 584–587, April 1997.

6.2.3 Capacitance Measurements

High Frequency: High-frequency $C-V$ curves are typically measured at 10 kHz–1 MHz. The basic capacitance measuring circuit in Fig. 6.10 consists of the device to be measured and an output resistor R. The MOS device is represented by the parallel G/C circuit, with G the conductance of the scr and C its capacitance. An ac current i flows through the device and the resistor, giving the output voltage as

$$v_o = iR = \frac{R}{Z}v_i = \frac{R}{R + (G + j\omega C)^{-1}}v_i = \frac{RG(1 + RG) + (\omega RC)^2 + j\omega RC}{(1 + RG)^2 + (\omega RC)^2}v_i$$
$$(6.20)$$

For $RG \ll 1$ and $(\omega RC)^2 \ll RG$, Eq. (6.20) reduces to

$$v_o \approx (RG + j\omega RC)v_i \qquad (6.21)$$

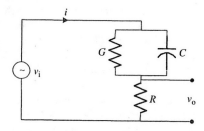

Fig. 6.10 Simplified capacitance measuring circuit.

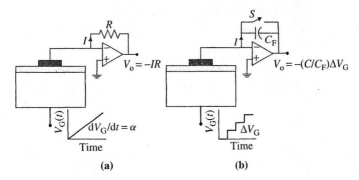

Fig. 6.11 Block diagram of circuits to measure the current and charge of an MOS capacitor.

The output voltage has two components: the in-phase RG and the out-of-phase $j\omega RC$, with $v_o = RGv_i$ for the $0°$ phase and ωRCv_i for the $90°$ phase components. Using a phase sensitive detector, one can determine the conductance G or the capacitance C, knowing R and $\omega = 2\pi f$.

Low Frequency: Current-Voltage: The low-frequency capacitance of an MOS-C is usually *not* obtained by measuring the capacitance, but rather by measuring a current or a charge, because capacitance measurements at low frequencies are very noisy. In the quasi-static or linear ramp voltage method, the current is measured in response to a slowly varying voltage ramp in Fig. 6.11(a).[10] The op-amp circuit with a resistive feedback connected to the MOS-C gate is an ammeter. The resulting displacement current is given by

$$I = \frac{dQ_G}{dt} = \frac{dQ_G}{dV_G}\frac{dV_G}{dt} = C\frac{dV_G}{dt} \tag{6.22}$$

For a linear voltage ramp, dV_G/dt is constant, I is proportional to C, and the low-frequency C–V curve is obtained, if dV_G/dt is sufficiently low.

Exercise 6.4

Problem: What is the effect of gate leakage current on the lf C–V curve?

Solution: It is important that the gate leakage current be as low as possible, because gate current adds or subtracts from the displacement current. This leads to an erroneous capacitance, because the current is no longer proportional to the capacitance in that case. The gate capacitor becomes very lossy due to high leakage, and the gate capacitance rolls up or down in the inversion and accumulation regions of the C–V curve and it is no longer possible to extract C_{ox} directly. The roll-off varies with the gate leakage current, so that for two gate dielectrics with the same thickness and different leakage currents, different C_{ox} and t_{ox} are obtained. Example C–V curves are shown in Fig. E6.4. A good discussion of these problems can be found in C. Scharrer and Y. Zhao, "High Frequency Capacitance Measurements Monitor EOT (Equivalent Oxide Thickness) of Thin Gate Dielectrics," *Solid State Technol.* **47**, Febr. 2004.

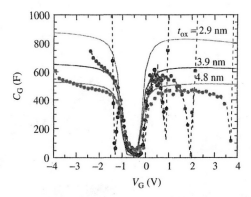

Fig. E6.4 Quasi-static curves for no oxide leakage (lines), oxide leakage current (points).

Low-Frequency: Charge-Voltage: In the quasi-static $I-V$ method in Fig. 6.11(a), leakage currents are included in the $I-V$ plot. Moreover, the ammeter in conjunction with the capacitor is a differentiator and tends to exaggerate noise spikes or non-linearities in the voltage ramp. The $Q-V$ quasi-static method alleviates some of the limitations of the $I-V$ quasi-static method. Initially the MOS-C was placed in the feedback loop of an op-amp and it was charged with a constant current,[11] and later modified.[12] Analog and digital versions[13] have been proposed and a commercial version is shown schematically in Fig. 6.11(b).[14] This circuit is an integrator, reducing the effects of spurious signals. The MOS-C is connected with its gate to the op-amp and its substrate to the voltage source in Fig. 6.11 to minimize stray capacitance and noise.

This technique, also called the feedback charge method, uses a voltage step input ΔV to the virtual ground op-amp. The capacitance is determined by measuring the transfer of charge in response to this voltage increment. The feedback capacitor C_F is initially discharged by closing the low-leakage current switch S. When the measurement starts, S is opened and ΔV_G causes charge ΔQ to flow onto capacitor C_F, giving the output voltage

$$\Delta V_o = -\frac{\Delta Q}{C_F} \qquad (6.23)$$

With $\Delta Q = C\Delta V_G$

$$\Delta V_o = -\frac{C}{C_F}\Delta V_G \qquad (6.24)$$

with the output voltage proportional to the MOS-C capacitance. Gain is introduced into the measurement for $C > C_F$ by choosing the capacitance ratio C/C_F appropriately. Incrementing ΔV_G generates a C_{lf} versus V_G curve. Additionally, when Q changes, a current Q/t flows. This current should only flow during the transient time period until the device reaches equilibrium. Hence, Q/t is a measure of whether equilibrium has been established and is used to determine the time increments at which ΔV_G should be changed to measure the equilibrium low-frequency $C-V$ curve.[14] The method is well suited for MOS measurements since it has high noise immunity, because sizable voltages rather than low currents are measured, and since voltage steps rather than precise linear voltage ramps are used.

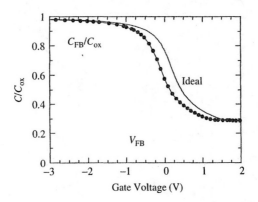

Fig. 6.12   Ideal (*line*) and experimental (*points*) MOS-C curves. $N_A = 5 \times 10^{16}$ cm^{-3}, $t_{ox} = 20$ nm, $T = 300$ K, $C_{FB}/C_{ox} = 0.77$.

6.2.4   Fixed Charge

The *fixed charge* is determined by comparing the flatband voltage shift of an experimental $C-V$ curve with a theoretical curve and measure the voltage shift, as shown in Fig. 6.12. C_{FB} is calculated from Eq. (6.19) or taken from Fig. 6.7, provided the oxide thickness and the doping density are known or determined as in Exercise 6.2. To determine Q_f, one should eliminate or at least reduce the effects of all other oxide charges and reduce the interface trapped charge to as low a value as possible. Q_{it} is reduced by annealing in a hydrogen ambient at temperatures around 400–450°C. Pure hydrogen is rarely used due to its explosive nature. *Forming gas*, a hydrogen-nitrogen mixture (~5–10% H$_2$), is commonly used. When the SiO$_2$ is covered by Si$_3$N$_4$, Q_{it} annealing is more difficult due to the imperviousness of the nitride.[15]

Q_f is related to the flatband voltage by the equation

$$Q_f = (\phi_{MS} - V_{FB})C_{ox} \qquad (6.25)$$

where ϕ_{MS} must be known in order to determine Q_f. Equation (6.25) assumes that interface traps play a negligible role in fixed charge density measurements. Methods to determine ϕ_{MS} are given in Section 6.2.5. The normalized flatband capacitance is 0.77 and $V_{FB} = -0.3$ V for the example in Fig. 6.12. Since ϕ_{MS} is required to determine Q_f from $C-V$ flatband voltage shifts, there is as much uncertainty in the fixed charge as there is in ϕ_{MS}. For example, the uncertainty in $N_f = Q_f/q$, according to Eq. (6.25), is related to the uncertainty in ϕ_{MS} for SiO$_2$ with $K_{ox} = 3.9$ by

$$\Delta N_f = \frac{K_{ox}\varepsilon_o}{qt_{ox}}\Delta\phi_{MS} = \frac{2.16 \times 10^{13}}{t_{ox}(nm)}\Delta\phi_{MS}(V) \text{ cm}^{-2} \qquad (6.26)$$

For an uncertainty in the metal-semiconductor work function difference of $\Delta\phi_{MS} = 0.05$ V, $\Delta N_f = 5.4 \times 10^{11}$ cm^{-2} for $t_{ox} = 2$ nm. This kind of uncertainty is higher than typical fixed charge densities, showing the importance of knowing ϕ_{MS} accurately.

A second method to determine Q_f dispenses with a knowledge of ϕ_{MS}. Rewriting Eq. (6.25) as

$$V_{FB} = \phi_{MS} - \frac{Q_f}{C_{ox}} = \phi_{MS} - \frac{Q_f t_{ox}}{K_{ox}\varepsilon_o} \tag{6.27}$$

suggests a plot of V_{FB} versus t_{ox} with slope $Q_f/K_{ox}\varepsilon_0$ and intercept ϕ_{MS}. This method, described in more detail in the next section, requires MOS capacitors with differing t_{ox}. However, it is more accurate because it is independent of ϕ_{MS}. Since the published literature shows variations of ϕ_{MS} by as much as 0.5 V, it is obviously important to determine ϕ_{MS} for a given process and not rely on published values.

6.2.5 Gate-Semiconductor Work Function Difference

The metal-semiconductor work function difference ϕ_{MS} is indicated in Fig. 6.13 for a flatband metal-oxide-semiconductor potential band diagram with zero oxide charges. $V_G = V_{FB}$ assures that the bands in the semiconductor and in the oxide are flat. For zero oxide or interface charge, $V_{FB} = \phi_{MS}$ from Eq. (6.17). Note that all quantities are given in potentials in Fig. 6.13, not in energies. ϕ_M and ϕ_M' are the metal and effective metal work function, ϕ_S is the semiconductor work function, χ and χ' are the electron and effective electron affinity. All other symbols have their usual meanings. From Fig. 6.13,

$$\phi_{MS} = \phi_M - \phi_S = \phi_M' - (\chi' + (E_c - E_F)/q) \tag{6.28}$$

Here ϕ_M', χ', and $(E_c - E_F)/q$ are constants for a given gate material, semiconductor, and temperature. For p- and n-substrates, Eq. (6.28) becomes

$$\phi_{MS} = K - \phi_F = K - \frac{kT}{q}\ln\left(\frac{N_A}{n_i}\right); \phi_{MS} = K + \phi_F = K + \frac{kT}{q}\ln\left(\frac{N_D}{n_i}\right) \tag{6.29}$$

where $K = \phi_M' - \chi' - (E_c - E_i)/q$ and $(E_c - E_F)/q = (E_c - E_i)/q + \phi_F = (E_c - E_i)/q + (kT/q)\ln(N_A/n_i)$. ϕ_{MS} depends not only on the semiconductor and the gate material, but also on the substrate doping type and density.

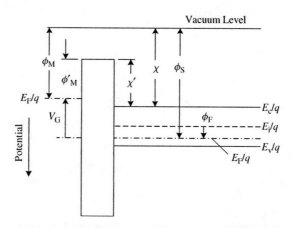

Fig. 6.13 Potential band diagram of a metal-oxide-semiconductor system at flatband.

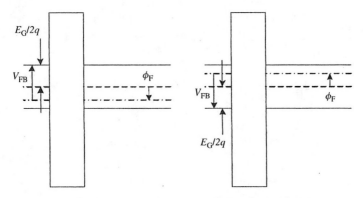

Fig. 6.14 Potential band diagram of (a) n^+ poly-Si/p substrate, and (b) p^+ poly-Si/n substrate at flatband.

Figure 6.14 shows the band diagram for an n^+ poly-Si-p substrate and for a p^+ poly-Si-n substrate MOS-C. Since both gate and substrate have the same electron affinity, we find

$$\phi_{MS} = \phi_F(gate) - \phi_F(substrate) \tag{6.30}$$

The Fermi level for n^+ poly-Si gates coincides approximately with the conduction band and with the valence band for p^+ poly-Si gates, giving $\phi_{MS}(n^+gate) \approx -E_G/2q - (kT/q)\ln(N_A/n_i)$ and ϕ_{MS} (p^+ gate) $\approx E_G/2q + (kT/q)\ln(N_D/n_i)$. For n^+ gates on n-substrates, $\phi_{MS}(n^+gate) \approx -E_G/2q + (kT/q)\ln(N_D/n_i)$, where N_A and N_D are the substrate doping densities.

Early ϕ_{MS} determinations used photoemission measurements.[16] With a voltage applied between a semitransparent gate and the substrate, no current flows in the absence of light because of the insulating nature of the oxide. Photons of sufficient energy strike the gate and excite electrons from the gate into the oxide. Some of these electrons drift through the oxide to be collected as photocurrent. Electrons are excited from the semiconductor into the oxide and flow to the gate for positive gate voltages and the barrier height of the *semiconductor/oxide* interface is determined. For negative gate voltages, electrons are excited from the gate into the oxide and flow to the semiconductor leading to the *gate/oxide* barrier.

Photoemission measurements determine ϕ_{MS} only indirectly. A more direct measure utilizes Eq. (6.27), repeated here

$$V_{FB} = \phi_{MS} - \frac{Q_f}{C_{ox}} = \phi_{MS} - \frac{Q_f t_{ox}}{K_{ox}\varepsilon_o} \tag{6.31}$$

A plot of V_{FB} versus oxide thickness has a *slope* of $-Q_f/K_{ox}\varepsilon_0$ and an *intercept* on the V_{FB} axis of ϕ_{MS}.[17] This method is more direct, as it measures the capacitance of MOS capacitors. Furthermore, since the flatband voltage is measured, it ensures zero electric field at the semiconductor surface eliminating Schottky barrier lowering corrections. The oxide thickness can be varied by oxidizing the wafer to a given thickness, measuring V_{FB}, etching a portion of the oxide, remeasuring V_{FB} and so on. This method ensures that the same spot on the oxide is measured each time. Oxide etching does not affect the fixed charge, since Q_f is located very near the SiO$_2$-Si interface. Sometimes the oxide is etched in strips to different thicknesses, or oxides can be grown to different thicknesses on different wafers and MOS capacitors formed, assuming Q_f to be the same for all samples.

Plots of $V_{FB} - t_{ox}$ are shown in Fig. 6.15.[18] The MOS capacitors with SiO_2 gate dielectric were fabricated on p-type Si substrates. 40–200 nm thick poly-Si was deposited on the gate dielectric followed by 80–200 nm hafnium. Silicidation was done by furnace annealing at 420°C or rapid thermal annealing at temperatures from 600°C to 750°C for 1 min. and the samples were annealed in forming gas at 420°C for 30 min.

ϕ_{MS} depends on oxidation temperature, wafer orientation, interface trap density, and on the low temperature D_{it} anneal.[19] The work function of poly-Si gate devices should depend on the doping density of the gate. One report shows a ϕ_{MS} maximum at phosphorus and arsenic densities of 5×10^{19} cm^{-3}, with the work function difference decreasing above and below this density.[20] The dependence of ϕ_{MS} on doping density is shown in Fig. 6.16 for the SiO_2/Si system with poly-Si gates.

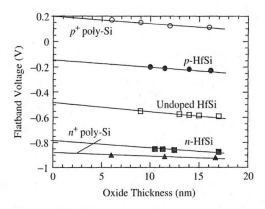

Fig. 6.15 Flatband voltage versus oxide thickness; p-type Si substrates. 40–200 nm thick poly-Si plus 80–200 nm hafnium silicided at 420°C or rapid thermal annealed at 600°C to 750°C for 1 min. Annealed in forming gas at 420°C for 30 min. Adapted from ref. 18.

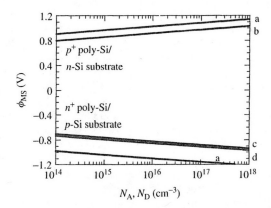

Fig. 6.16 ϕ_{MS} as a function of doping density for poly-Si/SiO$_2$/Si MOS devices. The numbers refer to references. The references are: a,[21] b,[22] c,[23] and d[20].

6.2.6 Oxide Trapped Charge

Charge can become trapped in the oxide during device *operation* even if not introduced during device fabrication. Electrons and/or holes can be injected from the substrate or from the gate. Energetic radiation also produces electron-hole pairs in the oxide and some of these electrons and/or holes are subsequently trapped in the oxide. The *flatband voltage shift* ΔV_{FB} due to oxide trapped charge Q_{ot} is obtained from

$$\Delta V_{FB} = V_{FB}(Q_{ot}) - V_{FB}(Q_{ot} = 0) \tag{6.32}$$

assuming all other charges remain unchanged during the oxide trapped charge introduction. Contrary to Q_f, the oxide trapped charge is usually not located at the oxide/semiconductor interface, but is distributed through the oxide. The distribution of Q_{ot} must be known for proper interpretation of $C-V$ curves. Trapped charge distributions are measured most commonly by the *etch-off* and the *photo $I-V$* methods.

In the etch-off method, thin layers of the oxide are etched. The $C-V$ curve is measured after each etch and the oxide charge profile is determined from these $C-V$ curves. The photo $I-V$ method is non-destructive and more accurate than the etch-off method. It is based on the optical injection of electrons from the gate or from the substrate into the oxide. Electron injection depends on the distance of the energy barrier from the injecting surface and on the barrier height. Both barrier distance and barrier height are affected by oxide charge and gate bias. Photo $I-V$ curves yields both the barrier distance and the barrier height. A good discussion of the method can be found in ref. 24 and references therein. Occasionally the technique is useful to monitor the flatband voltage continuously.[25]

A determination of the charge distribution in the oxide is tedious and therefore not routinely done. In the absence of such information, the flatband voltage shift due to charge injection is generally interpreted by assuming the charge is at the oxide-semiconductor interface using the expression

$$\boxed{Q_{ox} = -C_{ox}\Delta V_{FB}} \tag{6.33}$$

6.2.7 Mobile Charge

Mobile charge in SiO_2 is due primarily to the ionic impurities Na^+, Li^+, K^+, and perhaps H^+. Sodium is the dominant contaminant. Lithium has been traced to oil in vacuum pumps and potassium can be introduced during chemical-mechanical polishing. The practical application of MOSFETs was delayed due to mobile oxide charges in the early 1960s. MOSFETs were found to be very unstable for positive gate bias but relatively stable for negative gate voltages. Sodium was the first impurity to be related to this gate bias instability.[26] By intentionally contaminating MOS-Cs and measuring the $C-V$ shift after bias-temperature stress, it was shown that alkali cations could easily drift through thermal SiO_2 films. Chemical analysis of etched-back oxides by neutron activation analysis and flame photometry was used to determine the Na profile.[27] The drift has been measured with the isothermal transient ionic current method, the thermally stimulated ionic current method, and the triangular voltage sweep method.[28]

The mobility some oxide contaminants is given by the expression[29]

$$\mu = \mu_o \exp(-E_A/kT) \tag{6.34}$$

where for Na: $\mu_0 = 3.5 \times 10^{-4}$ cm^2/V·s (within a factor of 10) and $E_A = 0.44 \pm 0.09$ eV; for Li: $\mu_0 = 4.5 \times 10^{-4}$ cm^2/V·s (within a factor of 10) and $E_A = 0.47 \pm 0.08$ eV, for K: $\mu_0 = 2.5 \times 10^{-3}$ cm^2/V·s (within a factor of 8) and $E_A = 1.04 \pm 0.1$ eV, and for Cu, $\mu_0 = 4.8 \times 10^{-7}$ cm^2/V·s and $E_A = 0.93 \pm 0.2$ eV.[29] The oxide electric field is given by V_G/t_{ox}, neglecting the small voltage drop across the semiconductor and gate. The drift velocity of mobile ions through the oxide is $v_d = \mu V_G/t_{ox}$ and the transit time t_t is

$$t_t = \frac{t_{ox}}{v_d} = \frac{t_{ox}^2}{\mu V_G} = \frac{t_{ox}^2}{\mu_o V_G} \exp(E_A/kT) \qquad (6.35)$$

Equation (6.35) is plotted in Fig. 6.17 for the three alkali ions and for Cu. For this plot the oxide electric field is 10^6 V/cm, a common oxide electric field for such measurements, and the oxide thickness is 100 nm. For thinner or thicker oxides, the transit time change according to Eq. (6.35). Na and Li drift very rapidly through the oxide. Typical measurement temperatures lie in the 200 to 300°C range and only a few milliseconds suffice for the charge to transit the oxide. Mobile charge densities in the $5 \times 10^9 – 10^{10}$ cm^{-2} range are generally acceptable in integrated circuits.

Bias-Temperature Stress: The *bias-temperature stress* (BTS) method is one of two techniques to determine the mobile charge. However, in contrast to room-temperature $C–V$ measurements for Q_f determination, for mobile charge measurements the temperature must be sufficiently high for the charge to be mobile. Typically the device is heated to 150 to 250°C, and a gate bias to produce an oxide electric field of around 10^6 V/cm is applied for 5–10 min. for the charge to drift to one oxide interface. The device is then cooled to room temperature *under bias* and a $C–V$ curve is measured. The procedure is then repeated with the opposite bias polarity. The mobile charge is determined from the flatband voltage shift, according to the equation

$$\boxed{Q_m = -C_{ox}\Delta V_{FB}} \qquad (6.36)$$

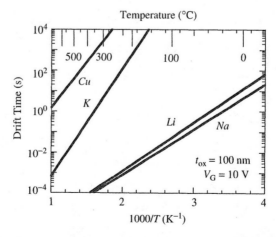

Fig. 6.17   Drift time for Na, Li, K, and Cu for an oxide electric field of 10^6 V/cm and $t_{ox} = 100$ nm.

The reproducibility of BTS measurements becomes questionable as mobile ion densities approach 10^9 cm^{-2}. For example, the flatband voltage shift in a 10 nm thick oxide due to the drift of a 10^9 cm^{-2} mobile ion density is 0.5 mV. Changing the gate area does not help since one measures voltage shifts, not capacitance.

There is sometimes a question of whether a measured flatband voltage shift is due to oxide trapped charge or due to mobile charge. A simple check to discriminate between the two is the following: Consider an MOS-C on a p-type substrate whose $C-V$ curve is initially measured with moderate gate voltage excursions giving $C-V$ curve (a) in Fig. 6.18. We assume that as a result of the modest gate voltage excursion charge is neither injected into the oxide nor does mobile charge move. Next, a BTS test is done with positive gate voltage. Keeping the oxide electric field around 1 MV/cm causes mobile charge to drift, but the electric field is insufficient for appreciate charge injection. If the $C-V$ curve after the BTS is curve (b) in Fig. 6.18, then the drift is due to positive mobile charge. For higher gate voltages at room temperature, there is a good chance that electrons and/or holes can be injected into the oxide and mobile charge may also drift, making that measurement less definitive.

Triangular Voltage Sweep: In the *triangular voltage sweep* (TVS) method the current is measured instead of the capacitance.[30] The MOS-C is held at an elevated, constant temperature of 200 to 300°C and the low-frequency $C-V$ curve is measured. C_{lf} is usually *not* obtained by measuring the capacitance, but rather by measuring a current or charge, as discussed in Section 6.2.3. TVS is based on measuring the charge flow through the oxide at an elevated temperature in response to an applied time-varying voltage. The charge flow is detected either as a current or as a charge. For a mobile ion density of 10^9 cm^{-2}, the resulting current is $I = 34$ pA for a sweep rate of 0.01 V/s and gate area of 0.01 cm^2. The charge in a charge sensing measurement is $Q = 1.6$ pC. Both of these are within typical measurement capability.

The current is determined by applying a slowly varying voltage ramp, as shown in Fig. 6.11(a), and measuring the current. If the ramp rate is sufficiently low, the measured current is the sum of displacement and conduction current due to the mobile charge. The current I is defined by

$$I = \frac{dQ_G}{dt} \tag{6.37}$$

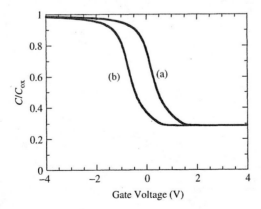

Fig. 6.18 $C-V_G$ curves illustrating the effects of mobile charge motion.

With $Q_G = -(Q_s + Q_{it} + Q_f + Q_{ot} + Q_m)$, the current can be written as[24]

$$I = C_{lf}\left(\alpha - \frac{dV_{FB}}{dt}\right) \tag{6.38}$$

where $\alpha = dV_G/dt$ is the gate voltage ramp rate. Integrating both sides from $-V_{G1}$ to $+V_{G2}$ gives

$$\int_{-V_{G1}}^{V_{G2}} (I/C_{lf} - \alpha)\, dV_G = -\alpha\{V_{FB}[t(V_{G2})] - V_{FB}[t(-V_{G1})]\} \tag{6.39}$$

Let us assume that at $-V_{G1}$ all mobile charges are located at the gate-oxide interface $(x = 0)$ and at V_{G2} all mobile charges are located at the semiconductor-oxide interface $(x = t_{ox})$. Then considering mobile charge only we find from Eq. (6.17),

$$-\alpha\{V_{FB}[t(V_{G2})] - V_{FB}[t(-V_{G1})]\} = \alpha\frac{Q_m}{C_{ox}} \tag{6.40}$$

and Eq. (6.39) becomes

$$\boxed{\int_{-V_{G1}}^{V_{G2}} (I/C_{lf} - \alpha)C_{ox}\, dV_G = \alpha Q_m} \tag{6.41}$$

As shown in Exercise 6.1, the hf and lf C–V curves coincide at high temperatures and the mobile charge is obtained by measuring the hf and lf curves and taking the area between the two curves, as illustrated in Fig. 6.19.[31] The integral of Eq. (6.41) represents the area between the lf and the hf curves in Fig. 6.19. One may ask why the lf curve exhibits the mobile charge hump, when C_{lf} and C_{hf} coincide. The reason is that during the lf *current* measurement, not only does the inversion charge respond to the probe frequency, but the mobile charge also drifts. For high temperature and high frequency *capacitance* measurements, only the inversion charge is detected.

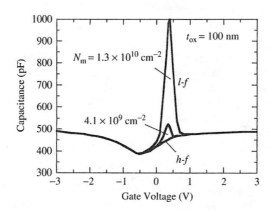

Fig. 6.19 C_{lf} and C_{hf} measured at $T = 250°C$. The mobile charge density is determined form the area between the two curves.

Sometimes two peaks are observed in $I - V_G$ curves at different gate voltages. These have been attributed to mobile ions with different mobilities. For an appropriate temperature and sweep rate, high-mobility ions (*e.g.*, Na^+) drift at lower oxide electric fields than low-mobility ions (*e.g.*, K^+). Hence, the Na peak occurs at lower gate voltages than the K peak. Such discrimination between different types of mobile impurities is not possible with the bias-temperature method. This also explains why sometimes the total number of impurities determined by the BTS and the TVS methods differ. In the BTS method one usually waits long enough for all the mobile charge to drift through the oxide. If in the TVS method the temperature is too low or the gate ramp rate is too high, it is possible that only one type of charge is detected. For example, it is conceivable that high-mobility Na drifts but low-mobility K does not. The TVS method also lends itself to mobile charge determination in *interlevel dielectrics*, not just gate oxides, since a current or charge is measured instead of a capacitance.

Other Methods: The electrical characterization methods are dominant because they are easily implemented and are very sensitive. The BTS method has a sensitivity of about 10^{10} cm^{-2} and the TVS method can detect densities as low as about 10^9 cm^{-2}. However, electrical methods cannot detect neutral impurities nor the sodium content in chemicals, furnace tubes etc. Analytical methods that have been employed for sodium detection include radiotracer,[32] neutron activation analysis,[33] flame photometry,[34] and secondary ion mass spectrometry (SIMS). For SIMS it is important to take surface charging by the positive or negative ion beam into account, because it can alter the ionic distribution and give erroneous distribution curves.[35]

6.3 INTERFACE TRAPPED CHARGE

Interface trapped charge, also known as interface traps or states, are attributed to dangling bonds at the semiconductor/insulator interface. Their density is most commonly reduced by forming gas anneal. A good overview of the nature of interface trapped charge and methods for its characterization can be found in refs. 24, 36, 37.

6.3.1 Low Frequency (Quasi-static) Methods

The low-frequency or *quasi-static method* is a common interface trapped charge measurement method. It provides information only on the interface trapped charge density, but not on their capture cross-sections. In this chapter we use the terms "interface trapped charge" and "interface traps" interchangeably. Before discussing characterization techniques, it is useful to discuss the nature of interface traps. One model attributes donor-like behavior to D_{it} below E_i and acceptor-like behavior to D_{it} above E_i as shown in Fig. 6.20(a). Although this model is not universally accepted, there is experimental evidence for it.[38] Donor interface traps below E_F are occupied by electrons and hence neutral. Those with energies $E_F < E < E_i$ are unoccupied donors and hence positively charged. Those above E_i are unoccupied acceptors and hence neutral. As a result, at flatband, D_{it} contributes a positive net charge. For positive gate voltage (Fig. 6.20(b)) some of the acceptor states lie below E_F and there is a net negative charge while for negative gate voltage (Fig. 6.20(c)) there is a more net positive charge. Hence, according to Eq. (6.17) the $C - V$ curves shift to the left for negative gate voltage and to the right for positive gate voltage.

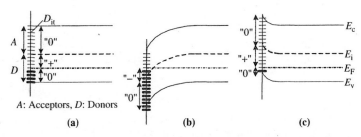

Fig. 6.20 Semiconductor band diagrams illustrating the effects of interface traps; (a) $V_G = 0$, (b) $V_G > 0$, (c) $V_G < 0$. Electron-occupied interface traps are indicated by the small horizontal *heavy* lines and unoccupied traps by the *light* lines.

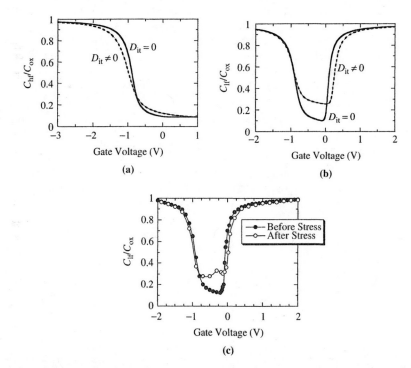

Fig. 6.21 Effect of D_{it} on MOS-C capacitance-voltage curves. (a) Theoretical high-frequency, (b) theoretical low-frequency and (c) experimental low-frequency curves. Gate voltage stress generated interface traps.

The effect of interface traps on both hf and lf $C-V$ curves is illustrated in Fig. 6.21. If interface traps cannot follow the ac probe frequency, they do not contribute a capacitance and the equivalent circuits are those of Fig. 6.4 with $C_{it} = 0$. However, interface traps can follow the slowly varying dc bias. As the gate voltage is swept from accumulation to inversion, the gate charge is $Q_G = -(Q_S + Q_{it})$ assuming no oxide charges. In contrast to the ideal case, where $Q_{it} = 0$, now both semiconductor and interface traps must be charged. The relationship of surface potential to gate voltage differs from Eq. (6.16) and the hf $C-V$ curve *stretches out* as shown in Fig. 6.21(a). This stretch-out is not the result of interface traps contributing excess capacitance, but rather it is the result of the $C-V$ curve stretch-out along the gate voltage axis. Interface traps do respond to the probe frequency at low measurement frequencies, and the curve distorts because the interface traps contribute interface trap capacitance C_{it} and the curve stretches out along the voltage axis, shown in Fig. 6.21(b). For $\phi_s = \phi_F$, the upper half band gap donor-type and lower half band gap acceptor-type interface traps cancel one another, leading to the coincidence of the ideal and distorted $C-V$ curves. Experimental curves are shown in Fig. 6.21(c) before and after oxide stress induced by gate current through the oxide.

The basic theory of the quasi-static method was developed by Berglund.[38] The method compares a low-frequency $C-V$ curve with one free of interface traps. The latter can be a theoretical curve, but is usually an hf $C-V$ curve determined at a frequency where interface traps are assumed not to respond. "Low frequency" means that interface traps *and* minority carrier inversion charges must be able to respond to the measurement ac probe frequency. The constraints for minority carrier response are discussed in Section 6.2.1. The interface trap response has similar limitations. Fortunately, the limitations are usually less severe than for minority carrier response and frequencies low enough for inversion layer response are generally low enough for interface trap response.

The lf capacitance is given by Eq. (6.4) in depletion-inversion as

$$C_{lf} = \left(\frac{1}{C_{ox}} + \frac{1}{C_S + C_{it}} \right)^{-1} \tag{6.42}$$

where we have replaced $C_b + C_n$ by C_S, the lf semiconductor capacitance. C_{it} is related to the interface trap density D_{it} by $D_{it} = C_{it}/q^2$, giving

$$\boxed{D_{it} = \frac{1}{q^2} \left(\frac{C_{ox}C_{lf}}{C_{ox} - C_{lf}} - C_S \right)} \tag{6.43}$$

Equation (6.43) is suitable for interface trap density determination over the entire band gap.

Exercise 6.5

Problem: Why is $C_{it} = q^2 D_{it}$ used here when most text use $C_{it} = q D_{it}$?

Solution: $C_{it} = q D_{it}$ is quoted in well respected texts, *e.g.*, Nicollian and Brews on p. 195.[24] But...if we substitute units, something is not right. With D_{it} in cm^{-2} eV^{-1} (the usual units) and q in Coul the units for C_{it} are $\dfrac{Coul}{cm^2 eV} = \dfrac{Coul}{cm^2 Coul - Volt} = \dfrac{F}{cm^2 Coul}$ using eV $= Coul - Volt$; $Volt = \dfrac{Coul}{F}$. This suggests that the correct definition should be $C_{it} = q^2 D_{it}$. We must keep in mind, however, that in the expression $E(eV) = qV$, $q = 1$ not $1.6 \times q 10^{-19}$! Hence $C_{it} = q^2 D_{it} = 1 \times 1.6 \times 10^{-19} D_{it}$. If D_{it} is given in cm^{-2} J^{-1}, then $C_{it} = (1.6 \times 10^{-19})^2 D_{it}$. This was pointed out to me by Kwok Ng and can be found in his book K.K. Ng, *Complete Guide to Semiconductor Devices*, 2nd Ed., Wiley-Interscience, New York, 2002, p. 183.

C_{lf} and C_S must be known to determine D_{it}. C_{lf} is measured as a function of gate voltage and C_S is calculated from Eq. (6.7). In Eq. (6.7), the capacitance is calculated as a function of surface potential ϕ_s but in Eq. (6.43) C_{lf} is measured as a function of gate voltage. Hence, we need a relationship between ϕ_s and V_G. Berglund proposed[39]

$$\phi_s = \int_{V_{G1}}^{V_{G2}} (1 - C_{lf}/C_{ox}) \, dV_G + \Delta \tag{6.44}$$

where Δ is an integration constant given by the surface potential at $V_G = V_{G1}$. The integrand is obtained by integrating the measured C_{lf}/C_{ox} versus V_G curve with V_{G1} and V_{G2} arbitrarily chosen, since the integration constant Δ is unknown. Integration from $V_G = V_{FB}$ makes $\Delta = 0$, because band bending is zero at flatband. Integration from V_{FB} to accumulation and from V_{FB} to inversion gives the surface potential over most of the band gap range. If the integration is carried out from strong accumulation to strong inversion, the integral should give $[\phi_s(V_{G2}) - \phi_s(V_{G1})] = E_G/q$. A value higher than E_G/q indicates gross non-uniformities in the oxide or at the oxide-semiconductor interface, making the analysis invalid. Various approaches to determine the surface potential based on lf and hf $C-V$ curves have been proposed.[40] Kuhn proposed fitting the experimental and theoretical C_{lf} versus ϕ_s curves in accumulation and strong inversion.[10] Plotting $(1/C_s)^2$ against ϕ_s gives a line with slope N_A and intercept Δ if N_A is uniform. If it is non-uniform, then no unique value of Δ is obtained. These methods are generally based on measuring charge using an operational amplifier with a capacitor in the feedback loop. In one circuit, D_{it} is determined and plotted directly as a function of ϕ_s.[41]

The determination of D_{it} from Eq. (6.43) and (6.44) is quite time consuming and a simplified approach was proposed by Castagné and Vapaille.[42] It eliminated the uncertainty associated with the calculation of C_S in Eq. (6.43) and replaced it with a measured C_S. From the hf $C-V$ curve, we find from Eq. (6.15),

$$C_S = \frac{C_{ox} C_{hf}}{C_{ox} - C_{hf}}. \tag{6.45}$$

Substituting Eq. (6.45) into (6.43) gives D_{it} in terms of the *measured* lf and hf $C-V$ curves as

$$\boxed{D_{it} = \frac{C_{ox}}{q^2} \left(\frac{C_{lf}/C_{ox}}{1 - C_{lf}/C_{ox}} - \frac{C_{hf}/C_{ox}}{1 - C_{hf}/C_{ox}} \right)} \tag{6.46}$$

Equation (6.46) gives D_{it} over only a limited range of the band gap, typically from the onset of inversion, but not strong inversion, to a surface potential towards the majority carrier band edge where the ac measurement frequency equals the inverse of the interface trap emission time constant. This corresponds to an energy about 0.2 eV from the majority carrier band edge. The higher the frequency the closer to the band edge can be probed. Typical hf and lf curves are shown in Fig. 6.22.

Data for $D_{it} - \phi_s$ typically have a U-shaped distribution with a minimum near midgap and sharp increases toward either band edge, as shown in Fig. 6.26. It is very important when using the technique based on Eq. (6.43), that the integration constant Δ be well known. Small errors in Δ have a large effect on D_{it} near the band edges.[43] Errors can also be introduced by surface potential fluctuations due to inhomogeneities in oxide charge and/or substrate doping density.[44] Errors in D_{it} extraction are also introduced by neglecting quantum-mechanical effects in the inversion capacitance.[45] The conventional

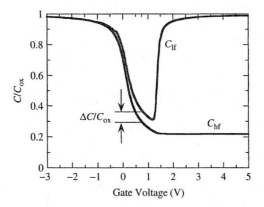

Fig. 6.22 High- and low-frequency C–V_G curves showing the offset $\Delta C/C_{ox}$ due to interface traps.

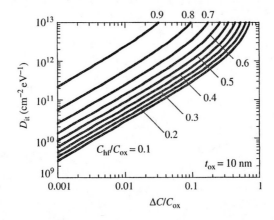

Fig. 6.23 Interface trapped charge density from the hf curve and the offset $\Delta C/C_{ox}$.

quasi-static technique underestimates the interface state density if the quantum-mechanical effect is significant, which becomes more critical as the doping density is increased.

It is not always necessary to determine D_{it} as a function of surface potential. For example, for process monitoring it is frequently sufficient to determine D_{it} at one point on the C–V curve. A convenient choice is the minimum C_{lf} where the technique is most sensitive. This point corresponds to a surface potential in the light inversion region near midgap, ($\phi_F < \phi_s < 2\phi_F$). To extract D_{it}, Eq. (6.46) is plotted in Fig. 6.23 for SiO$_2$ with $t_{ox} = 10$ nm. To use the figure, measure C_{lf}/C_{ox} and C_{hf}/C_{ox}, then determine $\Delta C/C_{ox} = C_{lf}/C_{ox} - C_{hf}/C_{ox}$ and find D_{it} from the graph ($\Delta C/C_{ox}$ is defined in Fig. 6.22).[46] For oxide thicknesses other than 10 nm, multiply D_{it} from Fig. 6.22 by $10/t_{ox}$ with t_{ox} in nm. Other graphical techniques have also been proposed.[47]

For high-frequency curves, the measurement frequency must be sufficiently high that interface traps do not respond. The usual 1 MHz frequency may suffice, but for devices with high D_{it} there will be some response due to interface traps. If possible, one should use higher frequencies, but care must be used to ascertain that series resistance effects do not become important. It is easier to measure C_{lf} when sweeping from inversion to accumulation, because minority carriers need not be generated thermally since they already exist in the inversion layer. Series resistance and stray light can also influence the curve.[48] A detailed accounting of the errors in extracting D_{it} is given by Nicollian and Brews.[24] The lower limit of D_{it} that can be determined with the quasi-static technique lies around 10^{10} cm^{-2} eV^{-1}. However, as oxide thickness decreases, the lf curve contains an appreciable oxide leakage current component, rendering quasi-static results questionable.

The charge voltage method is well suited for MOS measurements and can also determine the additive constant Δ of Eq. (6.44) by comparing experimental and theoretical ϕ_s versus W curves, where W is the space-charge region width obtained from the experimental hf $C-V$ curve.

6.3.2 Conductance

The *conductance method*, proposed by Nicollian and Goetzberger in 1967, is one of the most sensitive methods to determine D_{it}.[49] Interface trap densities of 10^9 cm^{-2} eV^{-1} and lower can be measured. It is also the most complete method, because it yields D_{it} in the depletion and weak inversion portion of the band gap, the capture cross-sections for majority carriers, and information about surface potential fluctuations. The technique is based on measuring the equivalent parallel conductance G_P of an MOS-C as a function of bias voltage and frequency. The conductance, representing the loss mechanism due to interface trap capture and emission of carriers, is a measure of the interface trap density.

The simplified equivalent circuit of an MOS-C appropriate for the conductance method is shown in Fig. 6.24(a). It consists of the oxide capacitance C_{ox}, the semiconductor capacitance C_S, and the interface trap capacitance C_{it}. The capture-emission of carriers by D_{it} is a lossy process, represented by the resistance R_{it}. It is convenient to replace the circuit of Fig. 6.24(a) by that in Fig. 6.24(b), where C_P and G_P are given by

$$C_P = C_S + \frac{C_{it}}{1 + (\omega \tau_{it})^2} \tag{6.47}$$

$$\frac{G_P}{\omega} = \frac{q \omega \tau_{it} D_{it}}{1 + (\omega \tau_{it})^2} \tag{6.48}$$

where $C_{it} = q^2 D_{it}$, $\omega = 2\pi f$ (f = measurement frequency) and $\tau_{it} = R_{it} C_{it}$, the interface trap time constant, given by $\tau_{it} = [v_{th}\sigma_p N_A \exp(-q\phi_s/kT]^{-1}$. Dividing G_P by ω makes Eq. (6.48) symmetrical in $\omega \tau_{it}$. Equations (6.47) and (6.48) are for interface traps

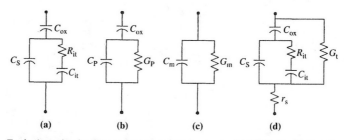

Fig. 6.24 Equivalent circuits for conductance measurements; (a) MOS-C with interface trap time constant $\tau_{it} = R_{it} C_{it}$, (b) simplified circuit of (a), (c) measured circuit, (d) including series r_s resistance and tunnel conductance G_t.

with a single energy level in the band gap. Interface traps at the SiO_2-Si interface, how-ever, are continuously distributed in energy throughout the Si band gap. Capture and emission occurs primarily by traps located within a few kT/q above and below the Fermi level, leading to a time constant dispersion and giving the normalized conductance as[49]

$$\frac{G_P}{\omega} = \frac{qD_{it}}{2\omega\tau_{it}}ln[1 + (\omega\tau_{it})^2] \tag{6.49}$$

Equations (6.48) and (6.49) show that the conductance is easier to interpret than the capacitance, because Eq. (6.48) does not require C_S. The conductance is measured as a function of frequency and plotted as G_P/ω versus ω. G_P/ω has a maximum at $\omega = 1/\tau_{it}$ and at that maximum $D_{it} = 2G_P/q\omega$. For Eq. (6.49) we find $\omega \approx 2/\tau_{it}$ and $D_{it} = 2.5G_P/q\omega$ at the maximum. Hence we determine D_{it} from the maximum G_P/ω and determine τ_{it} from ω at the peak conductance location on the ω-axis. G_P/ω versus f plots, calculated according to Eqs. (6.48) and (6.49), are shown in Fig. 6.25. The calculated curves are based on D_{it} values from a detailed interface extraction routine from the experimental data also shown on the figure. Note the much broader experimental peak.

Experimental G_P/ω versus ω curves are generally broader than predicted by Eq. (6.49), attributed to interface trap time constant dispersion caused by surface potential fluctuations due to non-uniformities in oxide charge and interface traps as well as doping density. Surface potential fluctuations are more pronounced in p-Si than in n-Si.[50] Surface potential fluctuations complicate the analysis of the experimental data. When such fluctuations are taken into account, Eq. (6.49) becomes

$$\frac{G_P}{\omega} = \frac{q}{2}\int_{-\infty}^{\infty}\frac{D_{it}}{\omega\tau_{it}}\ln[1 + (\omega\tau_{it})^2]P(U_s)\,dU_s \tag{6.50}$$

where $P(U_s)$ is a probability distribution of the surface potential fluctuation given by

$$P(U_s) = \frac{1}{\sqrt{2\pi\sigma^2}}\exp\left(-\frac{(U_s - \overline{U}_s)^2}{2\sigma^2}\right) \tag{6.51}$$

with $\overline{U}_S$ and σ the normalized mean surface potential and standard deviation, respectively.

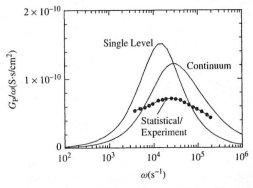

Fig. 6.25 G_p/ω versus ω for a single level [Eq. (6.48)], a continuum [Eq. (6.49)], and experimental data.[37] For all curves: $D_{it} = 1.9 \times 10^9$ cm^{-2}eV^{-1}, $\tau_{it} = 7 \times 10^{-5}$ s.

The line through the data points in Fig. 6.25 is calculated from Eq. (6.50). Note the good agreement between theory and experiment when ϕ_s fluctuations are considered. An approximate expression giving the interface trap density in terms of the measured maximum conductance is[49]

$$D_{it} \approx \frac{2.5}{q} \left(\frac{G_P}{\omega} \right)_{max}$$

(6.52)

Capacitance meters generally assume the device to consist of the parallel $C_m - G_m$ combination in Fig. 6.24(c). A circuit comparison of Fig. 6.24(b) to 6.24(c) gives G_P/ω in terms of the measured capacitance C_m, the oxide capacitance, and the measured conductance G_m as

$$\frac{G_P}{\omega} = \frac{\omega G_m C_{ox}^2}{G_m^2 + \omega^2 (C_{ox} - C_m)^2}$$

(6.53)

assuming negligible series resistance. The conductance measurement must be carried out over a wide frequency range. A comparison of interface traps determined by the quasi-static and the conductance techniques is shown in Fig. 6.26. Note the broad energy range over which the quasi-static method yields D_{it} and the good agreement over the narrower range where the conductance method is valid. The portion of the band gap probed by conductance measurements is typically from flatband to weak inversion. The measurement frequency should be accurately determined and the signal amplitude should be kept at around 50 mV or less to prevent harmonics of the signal frequency giving rise to spurious conductances. The conductance depends only on the device area for a given D_{it}. However, a capacitor with thin oxide has a high capacitance relative to the conductance, especially for low D_{it} and the resolution of the capacitance meter is dominated by the out-of-phase capacitive current component. Reducing C_{ox} by increasing the oxide thickness helps this measurement problem.

For *thin oxides*, there may be appreciable oxide leakage current. In addition, the device has series resistance which has so far been neglected. In the more complete circuit in Fig. 6.24(d), G_t represents the tunnel conductance and r_s the series resistance.

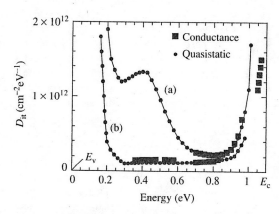

Fig. 6.26 Interface trapped charge density versus energy from the quasi-static and conductance methods. (a) (111) *n*-Si, (b) (100) *n*-Si. After ref. 50 and 51.

Equation (6.53) now becomes[52]

$$\frac{G_P}{\omega} = \frac{\omega(G_c - G_t)C_{ox}^2}{G_c^2 + \omega^2(C_{ox} - C_c)^2} \tag{6.54}$$

where

$$C_c = \frac{C_m}{(1 - r_s G_m)^2 + (\omega r_s C_m)^2} \tag{6.55}$$

$$G_c = \frac{\omega^2 r_s C_m C_c - G_m}{r_s G_m - 1} \tag{6.56}$$

C_m and G_m are the measured capacitance and conductance. The series resistance is determined by biasing the device into accumulation according to[24]

$$r_s = \frac{G_{ma}}{G_{ma}^2 + \omega^2 C_{ma}^2} \tag{6.57}$$

where G_{ma} and C_{ma} are the measured conductance and capacitance in accumulation. The tunnel conductance is determined from Eq. (6.56) as $\omega \to 0$.[52] Equation (6.54) reverts to Eq. (6.53) when $r_s = G_t = 0$.

Several models have been assumed to explain the experimental conductances.[53] In general it is necessary to use one of these models to extract D_{it} and σ_p with confidence. Schemes have been proposed for analyzing data by taking pairs of values of G_p/ω having a predetermined relationship of either frequency[54] or magnitude.[55] For example, G_p/ω curves can be determined at two frequencies and the appropriate parameters are found from universal curves. Brews uses a single G_p/ω curve and determines the points where the curve has fallen to a fraction of its peak value and then utilizes universal curves to determine D_{it} and σ_p.[55] Noras presents an algorithm to extract the relevant parameters.[55] In yet another simplification, a single hf $C-V$ and $G-V$ curve suffices to determine D_{it}.[56]

Instead of changing the frequency and holding the temperature constant, it is also possible to change the temperature and hold the frequency constant.[57] This has the advantage of not requiring measurements over a wide frequency range and one can chose a frequency for which series resistance is negligible. Elevated temperature measurements enhance the sensitivity near mid-gap allowing the detection of trap energy levels and capture cross-sections.[58] It also is possible to use MOSFETs instead of MOS-Cs and measure the transconductance instead of the conductance but still use the concepts of the conductance method.[59] This permits interface trap density determination on devices with the small gate areas associated with MOSFETs without the need for special MOS-C test structures.

6.3.3 High Frequency Methods

Terman Method: The room-temperature, high-frequency capacitance method developed by Terman was one of the first methods for determining the interface trap density.[60] The method relies on a hf $C-V$ measurement at a frequency sufficiently high that interface traps are assumed not to respond. They should, therefore, not contribute any capacitance.

How can one measure interface traps if they do not respond to the applied ac signal? Although interface traps do not respond to the ac probe frequency, they *do* respond to the slowly varying dc gate voltage and cause the hf $C-V$ curve to stretch out along the gate voltage axis as interface trap occupancy changes with gate bias illustrated in Fig. 6.21(a). In other words, for an MOS-C in depletion or inversion additional charge placed on the gate induces additional semiconductor charge $Q_G = -(Q_b + Q_n + Q_{it})$. With

$$V_G = V_{FB} + \phi_s + V_{ox} = V_{FB} + \phi_s + Q_G/C_{ox} \tag{6.58}$$

it is obvious that for a given surface potential ϕ_s, V_G varies when interface traps are present, leading to the $C-V$ "stretch-out" in Fig. 6.21. The stretch-out produces a *non parallel* shift of the $C-V$ curve. Interface traps distributed uniformly through the semiconductor band gap produce a fairly smoothly varying but distorted $C-V$ curve. Interface traps with distinct structure, for example peaked distributions, produce more abrupt distortions in the $C-V$ curve.

The relevant equivalent circuit of the hf MOS-C is that in Fig. 6.4(c) with $C_{it} = 0$, that is $C_{hf} = C_{ox}C_S/(C_{ox} + C_S)$ where $C_S = C_b + C_n$. C_{hf} is the same as that of a device without interface traps provided C_S is the same. The variation of C_S with surface potential is known for an ideal device. Knowing ϕ_s for a given C_{hf} in a device without Q_{it} allows us to construct a ϕ_s versus V_G curve of the actual capacitor as follows: From the ideal MOS-C $C-V$ curve, find ϕ_s for a given C_{hf}. Then find V_G on the experimental curve for the same C_{hf}, giving one point of a ϕ_s versus V_G curve. Repeat for other points until a satisfactory $\phi_s - V_G$ curve is constructed. This $\phi_s - V_G$ curve contains the relevant interface trap information. The experimental ϕ_s versus V_G curve is a stretched-out version of the theoretical curve and the interface trap density is determined from this curve by[24]

$$D_{it} = \frac{C_{ox}}{q^2}\left(\frac{dV_G}{d\phi_s} - 1\right) - \frac{C_S}{q^2} = \frac{C_{ox}}{q^2}\frac{d\Delta V_G}{d\phi_s} \tag{6.59}$$

where $\Delta V_G = V_G - V_G(ideal)$ is the voltage shift of the experimental from the ideal curve, and V_G the experimental gate voltage.

The method is generally considered to be useful for measuring interface trap densities of 10^{10} cm^{-2} eV^{-1} and above,[61] and has been widely critiqued. Its limitations were originally pointed out to be due to inaccurate capacitance measurements and insufficiently high frequencies.[62] A later, theoretical study concluded that D_{it} in the 10^9 cm^{-2} eV^{-1} range can be determined provided the capacitance is measured to a precision of 0.001 to 0.002 pF.[63]

For thinner oxides, the voltage shift associated with the interface traps also decreases. An assumption of the Terman method is that the measured C_{hf} curve does not contain appreciable interface state capacitance. Simulations have shown that the difference between the true high-frequency $C-V$ curve and the 1 MHz curve is on the same order of the difference between the "no D_{it}" curve and the 1 MHz curve, because the interface state capacitance is small, but non-negligible, compared to the voltage stretch-out for thin dielectrics.[64] For thicker dielectrics, the interface state capacitance is the same, but the voltage stretch-out increases. Both interface trap capacitance and voltage stretch-out scale with D_{it} making this method questionable for thin oxides.

To compare experimental with theoretical curves, one needs to know the doping density exactly. Any dopant pile up or out-diffusion introduces errors. Surface potential fluctuations can cause fictitious interface trap peaks near the band edges. The assumption that interface traps do not follow the ac probe frequency may not be satisfied for surface potentials near flatband and towards accumulation unless exceptionally high frequencies are used. Lastly, differentiation of the ϕ_s versus V_G curve can cause errors. Large discrepancies were found for D_{it} determined by the Terman technique compared with deep level transient spectroscopy (DLTS).[65]

Gray-Brown and Jenq Method: In the *Gray-Brown method*, the high-frequency capacitance is measured as a function of temperature.[66] Reducing the temperature causes the Fermi level to shift towards the majority carrier band edge and the interface trap time constant τ_{it} increases at lower temperatures. Hence interface traps near the band edges should not respond to typical ac probe frequencies at low temperatures whereas at room temperature they do respond. This method should extend the range of interface traps measurements to D_{it} near the majority carrier band edge.

The hf $C-V$ curves are measured from room temperature to typically $T = 77$ K. The interface trap density is obtained from the flatband voltages at those temperatures. Just as the interface trap occupancy changes with gate voltage in the Terman method, so it changes with temperature in this method. It is this change that is analyzed and D_{it} is extracted from the experimental data. The original measurements were made at 150 kHz and gave characteristic peaks of interface traps near the band edges. Theoretical calculations later indicated that these peaks were an artifact by using too low ac probe frequencies.[67] Frequencies near 200 MHz should be used to maintain high-frequency conditions near the band edges. It is useful as a fast, qualitative indicator of interface traps. In particular, an hf $C-V$ measurement at 77 K shows a "ledge" in the curve.[66, 68] This ledge voltage is related to the interface trap density over part of the band gap.

A method related to the Gray-Brown technique is the *Jenq technique*.[69] The MOS device is biased into accumulation at room temperature. Then it is cooled to $T = 77$ K and swept from accumulation to deep depletion, driven into inversion by illumination or short circuiting the source-drain of a MOSFET, and then swept from inversion to accumulation. The hysteresis between the two curves is proportional to the average interface trap density over typically the central 0.7–0.8 eV of the band gap. A comparison of average D_{it} determined by this technique and by charge pumping shows excellent agreement over the $3 \times 10^{10} \leq D_{it} \leq 10^{12}$ cm^{-2} eV^{-1} range.[70]

6.3.4 Charge Pumping

In the *charge pumping* method, originally proposed in 1969,[71] a MOSFET is used as the test structure, making it suitable for interface trap measurements on small-geometry MOSFETs instead of large-diameter MOS capacitors. We explain the technique with reference to Fig. 6.27. The MOSFET source and drain are tied together and slightly reverse biased with voltage V_R. The time varying gate voltage is of sufficient amplitude for the surface under the gate to be driven into inversion and accumulation. The pulse train can be square, triangular, trapezoidal, sinusoidal, or trilevel. The charge pumping current is measured at the substrate, at the source/drain tied together, or at the source and drain separately.

Let us begin by considering the MOSFET in inversion shown in Fig. 6.27(a). The corresponding semiconductor band diagram—from the Si surface into the substrate—is shown in Fig. 6.27(c). For clarity we show only the semiconductor substrate on this energy band diagram. The interface traps, continuously distributed through the band gap, are represented by the small horizontal lines at the semiconductor surface with the filled circles representing electrons occupying interface traps. When the gate voltage changes from positive to negative potential, the surface changes from inversion to accumulation

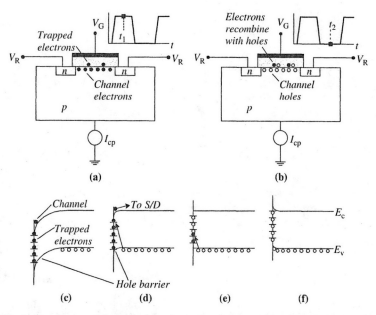

Fig. 6.27 Device cross-sections and energy bands for charge pumping measurements. The figures are explained in the text.

and ends up as in Fig. 6.27(b) and (f). However, the important processes take place during the transition from inversion to accumulation and from accumulation to inversion.

When the gate pulse falls from its high to its low value during its finite transition time, most electrons in the inversion layer drift to source and drain and electrons on those interface traps near the conduction band are thermally emitted into the conduction band (Fig. 6.27(d)) and also drift to source and drain. Those electrons on interface traps deeper within the band gap do not have sufficient time to be emitted and will remain on interface traps. Once the hole barrier is reduced (Fig. 6.27(e)), holes flow to the surface where some are captured by those interface traps still occupied by electrons. Holes are indicated by the open circles on the band diagrams. Finally, most traps are filled with holes as shown in Fig. 6.27(f). Then, when the gate returns to its positive voltage, the inverse process begins and electrons flow into the interface to be captured. Eight holes flow into the device in Fig. 6.27(b). Two are captured by interface traps. When the device is driven into inversion, six holes leave. Hence, eight holes in, six out result in a net charge pumping current, I_{cp}, that is proportional to D_{it}.

The time constant for electron emission from interface traps is

$$\tau_e = \frac{\exp(E_c - E_1)/kT}{\sigma_n v_{th} N_c} \qquad (6.60)$$

where E_1 is the interface trap energy measured from the bottom of the conduction band. The concepts of electron and hole capture, emission, time constants, and so on are discussed in Chapter 5. For a square wave of frequency f, the time available for electron emission is half the period $\tau_e = 1/2f$. The energy interval over which electrons are emitted is, from Eq. (6.60),

$$E_c - E_1 = kT \ln(\sigma_n v_{th} N_c/2f) \qquad (6.61)$$

For example, $E_c - E_1 = 0.28$ eV for $\sigma_n = 10^{-16}$ cm^2, $v_{th} = 10^7$ cm/s, $N_c = 10^{19}$ cm^{-3}, $T = 300$ K and $f = 100$ kHz. Hence, electrons from E_c to $E_c - 0.28$ eV are emitted while those below $E_c - 0.28$ eV are not emitted and therefore recombine with holes, when holes come rushing in. The hole capture time constant is

$$\tau_c = \frac{1}{\sigma_p v_{th} p_s} \qquad (6.62)$$

where $p_s = hole$ density/cm^3 at the surface. τ_c is very small for any appreciable hole density. In other words, emission, not capture, is the rate limiting process.

During the reverse cycle when the surface changes from accumulation to inversion, the opposite process occurs. Holes within an energy interval

$$E_2 - E_v = kT \ln(\sigma_p v_{th} N_v/2f) \qquad (6.63)$$

are emitted into the valence band and the remainder recombine with electrons flowing in from source and drain. E_2 is the interface trap energy measured from the top of the valence band. Those electrons on interface traps within the energy interval $\Delta E = E_G - (E_c - E_1) - (E_2 - E_v)$

$$\Delta E \approx E_G - kT[ln(\sigma_n v_{th} N_c/2f) + ln(\sigma_p v_{th} N_v/2f)] \qquad (6.64)$$

recombine. A detailed discussion of these concepts is given in ref. 72.

Q_n/q electrons/cm^2 flow into the inversion layer from the source and drain but only $(Q_n/q - D_{it}\Delta E)$ electrons/cm^2 flow back into the source-drain. $D_{it}\Delta E$ electrons/cm^2 recombine with holes. For each electron-hole pair recombination event, an electron and a hole must be supplied. Hence $D_{it}\Delta E$ holes/cm^2 also recombine. In other words, more holes flow into the semiconductor than leave, giving rise to the charge pumping current I_{cp} in Fig. 6.27. $D_{it}\Delta E$ holes being supplied at rate of f Hz to a MOSFET with gate area A_G gives the charge pumping current $I_{cp} = qA_G f D_{it}\Delta E$. In our example $\Delta E \approx 1.12 - 0.56 = 0.56$ eV. Substituting numerical values for a 10 $\mu m \times$ 10 μm gate area, a 100 kHz pump frequency, an interface trap density $D_{it} = 10^{10}$ cm^{-2} eV^{-1}, and $\Delta E = 0.56$ eV gives $I_{cp} \approx 10^{-10}$ A. As predicted, I_{cp} has been found to be proportional to both gate area and pump frequency.

The gate voltage waveform can be of various shapes. Early work used square waves. Later trapezoidal[73] and sinusoidal[74] waveforms were used. The waveforms can be constant base voltage in accumulation and pulsing with varying voltage amplitude ΔV into inversion as illustrated in Fig. 6.28(a), or varying the base voltage from inversion to accumulation keeping ΔV constant as in Fig. 6.28(b). The current saturates for the former, while for the latter it reaches a maximum and then decreases. The letters "a" to "e" on Fig. 6.28 correspond to the points on the current waveforms.

The plot of charge pumping current versus gate voltage in Fig. 6.28(a) depends somewhat on source-drain voltage V_R in Fig. 6.27. The non-saturating characteristic sometimes observed for $V_R = 0$ has been attributed to the recombination of those channel electrons

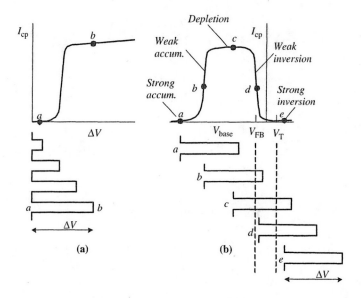

Fig. 6.28 Bilevel charge pumping waveforms.

unable to drift back to source and drain. This current is the "geometrical component" of I_{cp}, with the total charge pumping current given by[73]

$$I_{cp} = A_G f [q D_{it} \Delta E + \alpha C_{ox}(V_{GS} - V_T)] \tag{6.65}$$

where α is the fraction of the inversion charge that recombines with holes before drifting back to the source-drain and A_G is the gate area. The geometrical component is negligible for MOSFETs with short gate lengths or for gate pulse trains with moderate rise and fall times, giving the channel electrons sufficient time to drift back to source and drain.

The basic charge pumping technique gives an average value of D_{it} over the energy interval ΔE. It does not give an energy distribution of the interface traps. Various refinements have been proposed to obtain energy-dependent interface trap distributions. Elliot varied the pulse base level from inversion to accumulation keeping the amplitude of the gate pulse constant.[75] Groeseneken[73] varied the rise and fall times of the gate pulses while Wachnik[75] used small pulses with small rise and fall times to determine the energy distribution of D_{it}. For a trapezoidal waveform, the recombined charge per cycle, $Q_{cp} = I_{cp}/f$, is given by[73]

$$Q_{cp} = 2qkT\overline{D}_{it} A_G \ln \left(v_{th} n_i \sqrt{\sigma_n \sigma_p} \sqrt{\zeta 1 - \zeta} \frac{|V_{FB} - V_T|}{|\Delta V_{GS}|f} \right) \tag{6.66}$$

where $\overline{D}_{it}$ is the average interface trap density, ΔV_{GS} the gate pulse peak-peak amplitude, and ζ the gate pulse duty cycle. The slope of a Q_{cp} versus log(f) plot gives D_{it} and the intercept on the log(f) axis yields $(\sigma_n \sigma_p)^{1/2}$. By using a voltage controlled oscillator, one can sweep the frequency continuously and plot Q_{cp} versus log(f) to extract $\overline{D}_{it}$ and $(\sigma_n \sigma_p)^{1/2}$.[76] A plot of Q_{cp} as a function of log(f) in Fig. 6.29 shows the expected linear

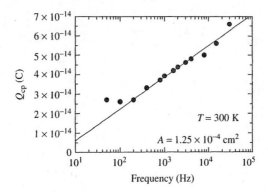

Fig. 6.29 MOSFET Q_{cp} versus frequency; $\overline{D}_{it} = 7 \times 10^9$ cm^{-2} eV^{-1}. Data adapted from. ref. 77.

dependence. The departure from linearity is due to traps not at the SiO$_2$-Si interface, but within the oxide, discussed later in this section.

The interface trap distribution through the band gap and capture cross-sections can be determined with a *trilevel* waveform with an intermediate voltage level V_{step},[78] illustrated in Fig. 6.30, switching the device from inversion to an intermediate state near midgap, and then to accumulation instead of from inversion to accumulation directly. At point (a), the device is in strong inversion with interface traps filled with electrons. As the waveform changes to (b) electrons begin to be emitted from interface traps, starting with the traps nearest the conduction band. The gate voltage remains constant to point (c). For $t_{step} \gg \tau_e$, where τ_e is the emission time constant of interface traps being probed, all traps above E_T have emitted their electrons and only those below E_T are available for recombination when holes come in to recombine with the electrons at point (d) on the waveform. This gives a charge pumping current that saturates as t_{step} increases. For $t_{step} < \tau_e$, fewer electrons have time to be emitted and more are available for hole recombination giving a correspondingly higher charge pumping current.

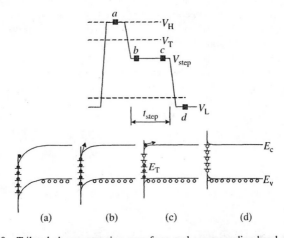

Fig. 6.30 Trilevel charge pumping waveform and corresponding band diagrams.

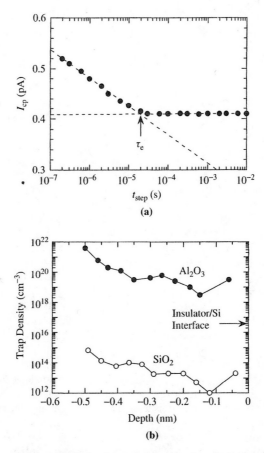

Fig. 6.31 (a) I_{cp} as a function of t_{step} showing τ_e at the point where I_{cp} begins to saturate. Reprinted after Saks et al. (Ref. 79) by permission of IEEE (© 1990, IEEE); (b) insulator trap density versus insulator depth from the insulator/Si interface for Al_2O_3 and SiO_2. Data adapted from. ref. 80.

A typical I_{cp} versus t_{step} plot in Fig. 6.31(a) shows the I_{cp} saturation and the $t_{step} = \tau_e$ breakpoint. From the emission time τ_e one can determine the capture cross-section according to the expression

$$\tau_e = \frac{\exp(E_c - E_T)/kT}{\sigma_n v_{th} N_c} \tag{6.67}$$

For a discussion of Eq. (6.67) see Chapter 5. By varying V_{step} one can probe interface traps through the band gap. Of course, the surface potential must be related to V_{step} by one of the techniques discussed in Section 6.3.1. The interface trap density is determined from the slope of the I_{cp} versus t_{step} curve according to the expression[79]

$$D_{it} = -\frac{1}{qkT A_G f} \frac{dI_{cp}}{d \ln t_{step}} \tag{6.68}$$

The trilevel charge pumping current can be expressed as[80]

$$I_{cp} = q A_G f D_{it} \left[E_T - kT \ln \left(1 - \left(1 - \exp \left(\frac{E_T - E_c}{kT} \right) \right) \exp \left(-\frac{t_{step}}{\tau_e} \right) \right) \right] \quad (6.69)$$

Equation (6.69) simplifies for low and high t_{step}

$$I_{cp}(t_{step} \to 0) \approx q A_G f D_{it} E_G; \; I_{cp}(t_{step} \to \infty) \approx q A_G f D_{it} E_T \quad (6.70)$$

demonstrating that various portions of the band gap can be probed with the trilevel charge pumping approach. Furthermore, by reducing the pulse frequency, one can probe traps *within the insulator*. In this case, electrons tunnel into and out of those traps from the channel with the tunneling time depending exponentially on the trap distance from the interface.[80] Example trap distributions are shown in Fig. 6.31(b) illustrating the higher trap density in Al_2O_3 compared to SiO_2.

Charge pumping can also determine the spatial variation of interface traps *along* the MOSFET channel by varying the drain and/or source bias leading to "A_G" variations caused by the drain-source space-charge region extending into the channel region.[81] Another method is the variation of voltage pulse amplitudes, thereby probing regions of the channel with varying threshold and flatband voltage.[81-82] Charge pumping has also been used to determine the oxide trap density close to the SiO_2-Si interface.[83] The charge recombined per cycle, $Q_{cp} = I_{cp}/f$, should be independent of frequency. However, Q_{cp} increases as the waveform frequency is reduced from typical frequencies of $10^4 - 10^6$ Hz to 10–100 Hz. At low frequencies there is sufficient time for electrons to tunnel to traps located in the oxide and to recombine there. Such traps are sometimes referred to as *border traps*.[84] Charge pumping can also be implemented by varying the temperature and keeping the gate waveform frequency constant.[85] For silicon-on-insulator MOSFETs, there are two SiO_2/Si interfaces and charge pumping currents depend on the state of the back interface. It is highest with the bottom interface in depletion.[86] Interface trap densities determined by various measurement techniques are shown in Fig. 6.32.

The charge pumping current is assumed to be due electron-hole pair recombination at interface traps with I_{cp} given by Eq. (6.65). For thin oxides, there is an additional gate

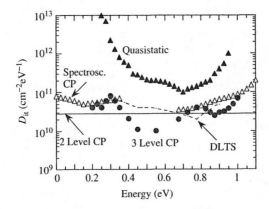

Fig. 6.32 Interface trap density as a function of energy through the band gap for various measurement techniques. Data after ref. 88.

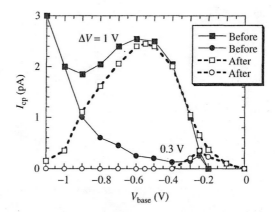

Fig. 6.33 Charge pumping current versus base voltage for two voltage pulse heights before and after gate leakage current correction. $t_{ox} = 1.8$ nm, $f = 1$ kHz. Adapted from ref. 87.

current that adds to the charge pumping current. $J_{cp} = 4 \times 10^{-3}$ A/cm² for $f = 1$ MHz, $D_{it} = 5 \times 10^{10}$ cm⁻² eV⁻¹, and $\Delta E = 0.5$ eV. The gate oxide leakage current can easily exceed this value. The charge pumping to gate oxide leakage current density ratio is

$$\frac{J_{cp}}{J_G} = \frac{qfD_{it}\Delta E}{J_G} \approx \frac{4 \times 10^{-3}}{J_G} \tag{6.71}$$

Fig. 6.33 shows the effect of gate oxide leakage current on I_{cp}.[87] At sufficiently low frequencies, the gate leakage current dominates and can be subtracted from the total current.

6.3.5 MOSFET Sub-threshold Current

The drain current of a MOSFET operated at gate voltages below threshold (sub-threshold) is[89]

$$I_D = I_{D1} \exp\left(\frac{q(V_{GS} - V_T)}{nkT}\right)\left(1 - \exp\left(-\frac{qV_{DS}}{kT}\right)\right) \tag{6.72}$$

where I_{D1} depends on temperature, device dimensions and substrate doping density; n, given by $n = 1 + (C_b + C_{it})/C_{ox}$, accounts for the charge placed on the gate that does not result in inversion layer charge. Some gate charge is imaged as space-charge region charge and some as interface trap charge. Ideally $n = 1$, but $n > 1$ as the doping density increases ($C_b \sim N_A^{1/2}$) and as the interface trap density increases ($C_{it} \sim D_{it}$).

The usual sub-threshold plot is $\log(I_D)$ versus V_{GS} for $V_{DS} \gg kT/q$. Such a plot has a slope of $q/[\ln(10)nkT]$. The slope is usually expressed as the sub-threshold swing S, which is that gate voltage necessary to change the drain current by one decade, and is given by

$$S = \frac{1}{Slope} = \frac{\ln(10)nkT}{q} \approx \frac{60nT}{300} \text{ mV/decade} \tag{6.73}$$

with T in Kelvin.

The interface trap density, obtained from a plot of $\log(I_D)$ versus V_G is

$$D_{it} = \frac{C_{ox}}{q^2}\left(\frac{qS}{\ln(10)kT} - 1\right) - \frac{C_b}{q^2} \tag{6.74}$$

requiring an accurate knowledge of C_{ox} and C_b. The slope also depends on surface potential fluctuations. This is the reason that this method is usually used as a comparative technique in which the sub-threshold swing is measured, then the device is degraded and remeasured. The change in D_{it} is given

$$\Delta D_{it} = \frac{C_{ox}}{\ln(10)qkT}(S_{after} - S_{before}) \tag{6.75}$$

The assumption in Eq. (6.75) is that the interface trap creation is uniform along the MOSFET channel. This is generally not the case when the MOSFET is stressed with gate and drain voltages and ΔD_{it} gives an average value.

Sub-threshold MOSFET curves are shown in Fig. 6.34 before and after stress, causing a threshold voltage shift and a slope change. For the SiO_2-Si interface, interface traps in the upper half of the band gap are acceptors and those in the lower half are donors with the demarcation between the two occurring at about half the band gap. Hence when the surface potential coincides with the Fermi level, as shown in Fig. 6.35(a) by $\phi_s = \phi_F$ at the surface, interface traps in the upper half are empty of electrons and neutral, and those in the lower half are occupied by electrons, hence also neutral, and the traps do not contribute to a gate voltage shift. We define a voltage V_{so} as

$$V_{so} = V_T - V_{mg} \tag{6.76}$$

where V_{mg} is the midgap gate voltage, which is typically the gate voltage at $I_D \approx$ 0.1–1 pA. Increasing the gate voltage from V_{mg} to V_T fills interface traps in the upper half of the band gap with electrons (Fig. 6.35(b)). The sub-threshold curve shifts, causing V_{so} to change from V_{so1} to V_{so2}. From this shift the interface trap density change ΔN_{it} is[90]

$$\Delta V_{it} = V_{so2} - V_{so1} \text{ and } \Delta N_{it} = \Delta D_{it}\Delta E = \frac{\Delta V_{it}C_{ox}}{q} \tag{6.77}$$

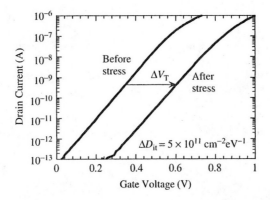

Fig. 6.34 MOSFET sub-threshold characteristics before and after MOSFET stress. The change in slope results in a stress-generated $\Delta D_{it} = 5 \times 10^{11}$ cm^{-2} eV^{-1}.

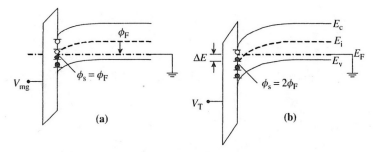

Fig. 6.35   Band diagrams for midgap and threshold voltages.

where ΔN_{it} is the increased interface trap density within the energy interval ΔE shown in Fig. 6.35(b). ΔE usually covers the range from midgap to strong inversion. Since at midgap the interface traps do not contribute any voltage shift, a shift at V_{mg} must then be due to oxide trapped charge according to

$$\Delta V_{ot} = V_{mg2} - V_{mg1} \text{ and } \Delta N_{ot} = \frac{\Delta V_{ot} C_{ox}}{q} \tag{6.78}$$

6.3.6   DC-IV

The *DC-IV* method is a dc current technique.[91] We explain it with reference to the MOSFET in Fig. 6.36(a). With the source S forward biased, electrons are injected into the p-well. Some electrons diffuse to the drain to be collected and measured as drain current I_D. Some electrons recombine with holes in the p-well bulk (not shown) and some recombine with holes at the surface below the gate. Only the surface-recombining electrons are influenced by the gate voltage. The holes lost by recombination are replaced by holes from the body contact leading to body current I_B. In contrast to a regular MOSFET with the source usually grounded, here the source is forward biased. In some DC-IV publications the source is referred to as the emitter, the drain as the collector, the

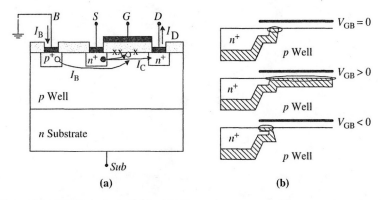

Fig. 6.36   (a) MOSFET configuration for DC-IV measurements and (b) cross-sections showing the space-charge regions and the encircled surface generation regions.

body as the base, and the currents as the collector and base currents and the n-substrate has been used as the electron injector/source.

The electron-hole pair surface recombination rate depends on the surface condition. With the surface in strong inversion or accumulation, the recombination rate is low. The rate is highest with the surface in depletion.[92] The body current is given by

$$\Delta I_B = q A_G n_i s_r \exp(q V_{BS}/2kT) \qquad (6.79)$$

where s_r is the surface recombination velocity given by

$$s_r = (\pi/2)\sigma_o v_{th} \Delta N_{it} \qquad (6.80)$$

with σ_o the capture cross-section (assuming $\sigma_n = \sigma_p = \sigma_o$).

Although the MOSFET in Fig. 6.36 resembles a bipolar junction transistor, it has the additional feature that the region between source (S) and drain (D) can be varied with the gate voltage. When the gate voltage exceeds the flatband voltage, a channel forms between S and D and the drain current will increase significantly. For $V_{GB} = V_T$, the $I_D - V_{GB}$ curve saturates. If charge is injected into the oxide, leading to a V_T shift, the drain current will also shift. It is this shift that can be used to determine oxide charge. We should point out that the interface trap density determined with the sub-threshold slope method samples the band gap between midgap and strong inversion, while the DC-IV body current samples the band gap between sub-threshold and weak accumulation, *i.e.*, surface depletion. By varying the gate voltage, different regions of the device are depleted (Fig. 6.36(b)) and those regions can be characterized, allowing spatial D_{it} profiling. Experimental DC-IV data are shown in Fig. 6.37 for a MOSFET before and after gate current stress.[93] A clear peak is observed at maximum surface recombination around $V_{GB} = 0$. In this example the method was used to determine interface trap generation caused by gate oxide current stress and plasma charging damage. A comparison of interface traps determined by charge pumping and DC-IV, gave very similar results.[81] Both techniques allow lateral trap profiling.

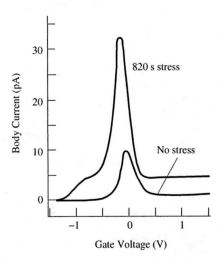

Fig. 6.37   DC-IV measured body currents. (a) control wafer, (b) stressed with -12 mA/cm^2 gate current density. $V_{BS} = 0.3$ V, $W/L = 20/0.4$ μm, $t_{ox} = 5$ nm. Data adapted from ref. 93.

6.3.7   Other Methods

A sensitive method to determine D_{it} is *deep-level transient spectroscopy*, covered in Chapter 5. The charge transfer loss in charge-coupled devices (CCD) is also a sensitive indicator of interface trap densities,[94] but is not practical if a CCD has to be specially fabricated as the test structure. In the *surface charge analyzer* method, the oxide in an MOS-C is replaced by a mylar sheet and the gate is replaced by an optically transparent, electrically conducting layer.[95] By exposing the sample to above band gap light, that creates ehp in the semiconductor through the transparent gate, the ac surface photovoltage is given by[95]

$$\delta V_{SPV} = \frac{q(1-R)\Phi W}{4fK_s\varepsilon_o} \tag{6.81}$$

where Φ is the incident photon flux density, W the space-charge region width, and f the modulated light frequency. W is determined from a measurement of δV_{SPV}. With the mylar sheet about 10 μm thick, the measured series mylar-oxide capacitance is dominated by C_{mylar} and the total charge is

$$Q = Q_S + Q_{ox} + Q_{it} = -CV_G \approx -C_{mylar}V_G \tag{6.82}$$

Knowing W allows Q_S to be determined. Q_{ox} and Q_{it} are then determined by the usual MOS-C analyses. Changing the bias voltage drives the Si surface into inversion, depletion, or accumulation. Since the electrode is separated from the sample by the 10 μm thick mylar film, its small probe capacitance is dominant and leakage current is suppressed. The interface trap density and energy are given by[96]

$$D_{it}(E) = \frac{K_s\varepsilon_o}{q^2W}\left(\frac{1}{qN_A}\frac{dQ}{dW} - 1\right) \tag{6.83}$$

$$E = E_F - E_i + q\phi_s = kT\ln\left(\frac{N_A}{n_i}\right) - \frac{qN_AW^2}{2K_s\varepsilon_o} \tag{6.84}$$

Since the space-charge region width W is measured instead of the capacitance, this technique is independent of oxide thickness, in contrast to some of the earlier methods that depend sensitively on t_{ox} and their interpretation becomes difficult for thin oxides with high oxide leakage currents. Furthermore, there is no need for quantum mechanical and gate depletion corrections. It is, however, influenced by the substrate doping density and N_A should not be higher than about 10^{17} cm^{-3}.

The technique can be used as an *in-line* method to obtain surface charge information, *e.g.*, follow various cleaning cycles. In one comparison between the SCA and conventional MOS-C methods, the SCA method fared very well, especially due to its shorter measurement cycle, since devices need not be fabricated.[95] It has also been used to determine D_{it} for SiO$_2$, HfO$_2$, and Si$_3$N$_4$ for equivalent oxide thicknesses of 1–3 nm.[96]

Crystallographic structural information on interface traps can be obtained from *electron spin resonance* (ESR) measurements,[97] but the method is relatively insensitive and $D_{it} \geq 10^{11}$ cm^{-2} eV^{-1} is required. ESR was instrumental in identifying dangling bonds at the SiO$_2$/Si interface as interface traps.[98] Figure 6.38 shows the two major Si oriented surfaces and the associated dangling bonds, designated P_b, P_{b0}, and P_{b1} centers.

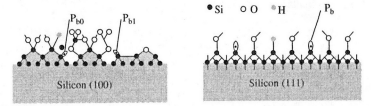

Fig. 6.38   Silicon surface for (100) and (111) orientation showing the P_{b0}, P_{b1}, and P_b centers.

6.4   OXIDE THICKNESS

The *oxide thickness* is an important parameter for the interpretation of many of the techniques discussed in this chapter. Electrical, optical and physical methods are used in its determination, including $C-V$, $I-V$, ellipsometry, transmission electron microscopy (TEM), X-ray photoelectron spectroscopy (XPS), medium energy ion scattering spectrometry (MEIS), nuclear reaction analysis (NRA), Rutherford backscattering (RBS), elastic backscattering spectrometry (EBS), secondary ion mass spectrometry (SIMS), grazing incidence X-ray reflectometry (GIXRR), and neutron reflectometry. We discuss the $C-V$ method here and mention other methods briefly. Some of them are detailed in later chapters. A recent joint study by numerous techniques (MEIS, NRA, RBS, EBS, XPS, SIMS, ellipsometry, GIXRR, neutron reflectometry and TEM) compared oxide thicknesses of 10 carefully prepared samples covering oxide thicknesses of 1.5 to 8 nm.[99] There are three thickness offsets: water and carbonaceous contamination equivalent to ~1 nm and adsorbed oxygen mainly from water at an equivalent thickness of 0.5 nm.

The existence of an interfacial layer between silicon dioxide and silicon is accepted by a majority of the technical community. There is approximately 1 monolayer (ML) of an interfacial layer at the SiO₂/Si interface.[100] There is evidence for up to ~1 ML of additional sub-stoichiometric oxide located within the first 0.5 to 1 nm of the interface. Each characterization method probes slightly different aspects of the interface. X-ray reflectivity and X-ray photoelectron spectroscopy support the presence of stress as do infrared IR measurements. X-ray photoelectron spectroscopy shows the presence of at least a monolayer film of incompletely oxidized silicon. Infrared spectroscopy further supports the presence of sub-stoichiometry at the interface. Thus ellipsometry observes a slab of mixed dielectric constant. Stress within the oxide layer itself, *i.e.*, above the interface plane, is supported by X-ray reflectivity and X-ray photoelectron spectroscopy. Ellipsometry determines thickness based on optical models that include an interfacial layer. The long wavelength of ellipsometry and the need to sample a large area results in an averaged sampling of interfacial optical properties.

6.4.1   Capacitance-Voltage

It would seem that capacitance-voltage data lend themselves to oxide thickness determination with the MOS device in strong accumulation. Complications arise for thin oxides that render conventional methods questionable. These complications include Fermi-Dirac rather than Boltzmann statistics, quantization of carriers in the accumulation layer, poly-Si gate depletion, and oxide leakage current. The capacitance of the depleted gate and of the accumulation layer, being in series with the oxide capacitance, lead to thicker effective oxides than simple theory would predict.[101]

In the Maserjian, the McNutt and Sah, and the Kar methods, the following assumptions are made: the interface trap capacitance is negligible in accumulation at 100 kHz-1 MHz, the differential interface trap charge density, between flatband and accumulation is negligible, the oxide charge density is negligible, and quantization effects are neglected. The relevant equations are for the McNutt-Sah method[102]

$$\left| \frac{dC_{hf,acc}}{dV} \right|^{1/2} = \sqrt{\frac{q}{2kTC_{ox}}} (C_{ox} - C_{hf,acc}) \tag{6.85}$$

where $C_{hf,acc}$ is the high-frequency accumulation capacitance. A plot of $(dC_{hf,acc}/dV)^{1/2}$ versus $C_{hf,acc}$ yields C_{ox} as the intercept on the $C_{hf,acc}$ axis and from the slope. For the Maserjian method[103]

$$\frac{1}{C_{hf,acc}} = \frac{1}{C_{ox}} + \left(\frac{2}{b^2} \right)^{1/3} \sqrt{\frac{1}{C_{hf,acc}}} \left| \frac{dC_{hf,acc}}{dV} \right|^{1/6} \tag{6.86}$$

where b is a constant. One plots $C_{hf,acc}^{-1/2}(dC_{hf,acc}/dV)^{1/6}$ versus $1/C_{hf,acc}$. If a linear fit is obtained, then its intercept on the $1/C_{hf,acc}$ axis yields $1/C_{ox}$. With quantization effects, the equation becomes[104]

$$\frac{1}{C_{hf,acc}} = \frac{1}{C_{ox}} + s \left| \frac{d(1/C_{hf,acc}^2)}{dV} \right|^{1/4} \tag{6.87}$$

where s is a constant. Equation (6.87) has a simpler form than Eq. (6.86). In this case, one plots $1/C_{hf,acc}$ versus $(d(1/C_{hf,acc}^2)/dV)^{1/4}$. For a linear fit, its intercept on the $1/C_{hf,acc}$ axis yields $1/C_{ox}$. For the Kar method[105]

$$\frac{1}{C_{hf,acc}} = \frac{1}{C_{ox}} + \left(\frac{1}{2\beta} \left| \frac{d(1/C_{hf,acc}^2)}{dV} \right| \right)^{1/2}, \tag{6.88}$$

where β is a constant. Here, one plots $1/C_{hf,acc}$ versus $(d(1/C_{hf,acc}^2)/dV)^{1/2}$. For a linear fit, its intercept on the $1/C_{hf,acc}$ axis yields $1/C_{ox}$. This method has been successfully used for 1–8 nm thick high-K dielectrics.

A variation of the Maserjian method is based on the following equations.[106] The capacitance with the device in accumulation is

$$\frac{1}{C} = \frac{1}{C_{ox}} + \frac{1}{C_S}; C_{ox} = \frac{K_{ox}\varepsilon_o A}{t_{ox}}; C_S = \frac{dQ_{acc}}{d\phi_s} \tag{6.89}$$

with

$$Q_{acc} = K \exp\left(\frac{q\phi_s}{2kT} \right) \text{ giving } C_S = \frac{qQ_{acc}}{2kT}. \tag{6.90}$$

Using

$$V_G = V_{FB} + \phi_s - \frac{Q_{acc}}{C_{ox}} \rightarrow V_G - V_{FB} - \phi_s = -\frac{2kT}{q}\frac{C_S}{C_{ox}}. \tag{6.91}$$

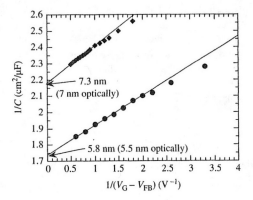

Fig. 6.39 $1/C$ versus $1/(V_G - V_{FB})$ for two oxide thicknesses. Reprinted after Vincent et al. (Ref. 106) by permission of IEEE (© 1997, IEEE).

Combining Eqs. (6.89) and (6.91) gives

$$\frac{1}{C} = \frac{1}{C_{ox}} - \frac{2kT}{qC_{ox}} \frac{1}{V_G - V_{FB} - \phi_s} \approx \frac{1}{C_{ox}} - \frac{2kT}{qC_{ox}} \frac{1}{V_G - V_{FB}}. \tag{6.92}$$

The approximation in Eq. (6.92) holds for $(V_G - V_{FB}) \gg \phi_s$, valid in strong accumulation.

Equation (6.92) suggests a plot of $1/C$ versus $1/(V_G - V_{FB})$, as illustrated in Fig. 6.39. The $1/C$ axis intercept is $1/C_{ox}$. Although poly-Si gate depletion affects the second term of Eq. (6.92), it does not alter the intercept and can be neglected. A more accurate approach without the Eq. (6.92) approximation is given in ref. 107. The oxide thickness can also be determined from a plot of gate corona charge versus gate voltage of an MOS capacitor discussed in Chapter 9.

One can also vary the frequency of the applied signal. Measuring the circuits in Figs. E6.5(a) and (b) at two different frequencies, allows the various components in Fig. 6.4(a) to be determined[108]

$$C = \frac{f_1^2 C_{P1}^2 (1 + D_1^2) - f_2^2 C_{P2}^2 (1 + D_2^2)}{f_1^2 - f_2^2}; D = \frac{G_P}{\omega C_P} = \frac{G_t(1 + r_s G_t)}{\omega C} + \omega r_s C \tag{6.93}$$

where D_1 and C_{P1} refer to measured values at frequency f_1 and D_2 and C_{P2} at f_2.

$$G_t = \sqrt{\omega^2 C_P C (1 + D^2) - (\omega C)^2} \tag{6.94}$$

$$r_s = \frac{D}{\omega C_P (1 + D^2)} - \frac{G_t}{G_t^2 + (\omega C)^2} \tag{6.95}$$

A detailed analysis of the two-frequency method has shown that D should be less than 1.1.[109] For thin oxides, the device area must be reduced for $D < 1.1$ but the device must remain sufficiently large not to be limited by the capacitance meter's lower measurement limit. Reductions of G_t and r_s lead to higher D, implemented by reducing the device area because $G_t \sim$ area and $r_s \sim 1/area^{1/2}$ due to spreading resistance. The minimum radial frequency, determined from Eq. (6.93)

$$\omega_{min} = \frac{G_t}{C} \sqrt{1 + \frac{1}{r_s G_t}}$$ (6.96)

leads to the minimum dissipation factor

$$D_{min} = 2\sqrt{r_s G_t (1 + r_s G_t)}$$ (6.97)

Figure 6.40 shows the dependence of measurement error on device area and oxide thickness. For a $f = 1$ MHz oxides to about 1.5 nm can be measured. The frequency in Fig. 6.40 refers to the higher of the two frequencies.

Treating the MOSFET as a transmission line leads to the capacitance[110]

$$C \approx C_m \frac{1 + \cosh(K)}{1 + \sinh(K)/K}$$ 6.98

where $K = (r_s' G_t' L^2)^{1/2}$ and C_m is the measured capacitance, L the gate length and

$$r_s' = \frac{W}{L} \sqrt{\frac{Z_{dc}}{Y_{dc}} \frac{4}{4 - Z_{dc}Y_{dc}}} \cosh^{-1}\left(\frac{2}{2 - Z_{dc}Y_{dc}}\right) (\Omega/square)$$ (6.99)

$$G_t' = \frac{1}{WL} \frac{\cosh^{-1}\left(\dfrac{2}{2 - Z_{dc}Y_{dc}}\right)}{\sqrt{\dfrac{Z_{dc}}{Y_{dc}} \dfrac{4}{4 - Z_{dc}Y_{dc}}}} (S/cm^2)$$ (6.100)

where W is the gate width. The measurement is a dc measurement with the MOSFET source and substrate (or CMOS well) grounded. The gate voltage is swept over an appropriate voltage range and the dc gate admittance Y_{dc} is determined from the slope of the $I_G - V_{GS}$ curve. At each gate voltage, the drain voltage is swept from -15 mV to

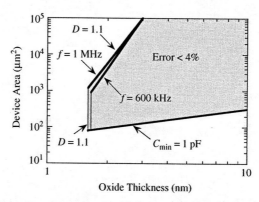

Fig. 6.40 Measurement error dependence on device area and oxide thickness. The two-frequency measured capacitance is in error less than 4% in the shaded region. At higher frequencies the $D = 1.1$ border shifts to thinner oxides. Adapted from ref. 109.

+15 mV and the slope of the $I_D - V_{DS}$ yields the dc drain impedance Z_{dc}. Both r_s' and G_t' are strongly gate voltage dependent and need to be accurately measured. Corrections are required for longer gates, because the increased channel resistance leads to reduced capacitance. Similarly, thinner oxides lead to higher gate current and increased channel voltage drop and require corrections. The method has proven successful for oxides as thin as 0.9 nm.

Exercise 6.6

Problem: What is the effect of gate leakage current and series resistance on $C-V$ behavior?

Solution: In accumulation with no interface traps, the equivalent circuit from Fig. 6.24 becomes Fig. E6.5(a). Following Chapter 2, we convert it to the parallel and series equivalent circuits in Figs. E6.5(b) and (c) where

$$C_P = \frac{C}{(1 + r_s G)^2 + (\omega r_s C)^2}; C_S = C\left(1 + \left(\frac{G}{\omega C}\right)^2\right).$$

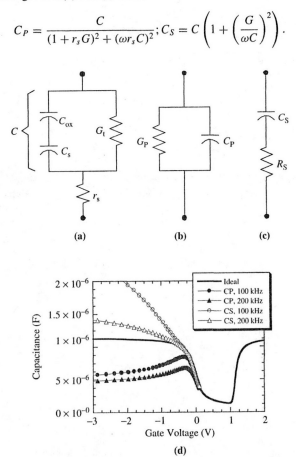

Fig. E6.5 (a) MOS-C equivalent circuit with tunnel conductance and series resistance, (b) parallel, (c) series equivalent circuits, and (d) calculated $C-V_G$ curves.

To understand the basic concepts, we have used a simple constant series resistance $r_s = 0.5$ Ω, $t_{ox} = 3$ nm, $N_A = 10^{17}$ cm^{-3}, and $G_t = \exp(1/V_G)$ for $V_G < 0$. The resulting C_P and C_S as well as the ideal ($r_s = G_t = 0$) capacitances are shown in Fig. E6.5(d). C_P decreases and C_S increases as a result of G_t, making oxide thickness extraction more difficult. Of course, the actual dependence of G_t on gate voltage differs from this simple model, but it illustrates the main concept. This kind of behavior has been experimentally verified, e.g., D.P. Norton, "Capacitance-Voltage Measurements on Ultrathin Gate Dielectrics," *Solid-State Electron.* **47**, 801–805, May 2003.

6.4.2   Current-Voltage

Oxide current-voltage characteristics are discussed in Chapter 12. Here we briefly give the relevant equations and how they relate to oxide thickness. The current flowing through an insulator is either Fowler-Nordheim (FN) or direct tunnel current. The FN current density is

$$J_{FN} = A\mathcal{E}_{ox}^2 \exp\left(-\frac{B}{\mathcal{E}_{ox}}\right) \tag{6.101}$$

where $\mathcal{E}_{ox}$ is the oxide electric field and A and B are constants. The direct tunnel current density is

$$J_{dir} = \frac{AV_G}{t_{ox}^2}\frac{kT}{q}C \exp\left(-\frac{B(1 - (1 - qV_{ox}/\Phi_B)^{1.5})}{\mathcal{E}_{ox}}\right) \tag{6.102}$$

where Φ_B is the semiconductor-insulator barrier height and V_{ox} the oxide voltage. Both currents are very sensitive to oxide thickness. Tunneling currents also contain a small oscillatory component. These oscillations arise due to the quantum interference of electrons and show a strong dependence on oxide thickness, suggesting that these oscillations can be used for oxide thickness determination.[111]

6.4.3   Other Methods

Ellipsometry, discussed in Chapter 10, is suitable for oxides into the 1–2 nm regime. Variable angle, spectroscopic ellipsometry is especially suited for oxide thickness measurements.

   Transmission Electron Microscopy, discussed in Chapter 11, is very precise and usable to very thin oxides, but sample preparation is tedious.

   X-ray Photoelectron Spectroscopy and other beam techniques are discussed in Chapter 11.

6.5   STRENGTHS AND WEAKNESSES

Mobile Oxide Charge: The strength of the *bias temperature stress* method is its simplicity requiring merely the measurement of a $C-V$ curve, albeit at elevated temperatures. Its weakness is that the total mobile charge density is measured. Separation of various species is not possible. Furthermore, occasionally the $C-V$ curve becomes distorted due to interface trapped charge and the flatband voltage is difficult to determine.

   The main strengths of the *triangular voltage sweep* method are its ability to differentiate between different mobile charge species, its higher sensitivity, and the fact that the method is fast because the sample does not need to be heated and cooled; it needs only to be heated. Since a current or charge is measured, this method lends itself to determination of mobile charge in interlevel dielectrics, which is not possible with capacitance methods. Its weakness is the increasing oxide leakage current for thin oxides.

Interface Trapped Charge: For MOS capacitors the choice for the most practical methods lies between the *conductance* and the *quasi-static* methods. These are the two most widely used techniques. The strength of the conductance method lies in its high sensitivity and its ability to give the majority carrier capture cross sections. Its major weakness is the limited surface potential range over which D_{it} is obtained and the required effort to extract D_{it}, although simplified methods have been proposed.

The main strengths of the quasi-static method (both the $I-V$ and the $Q-V$) are the relative ease of measurement and the large surface potential range over which D_{it} is obtained. A weakness for the $I-V$ version is the current measurement requirement. The currents are usually low because the sweep rates must be low to ensure quasi-equilibrium. The $Q-V$ version alleviates some of these problems. For both techniques, increased gate oxide leakage currents are problematic for thin oxides, making the methods difficult or impossible.

For MOSFETs the choice is *charge pumping, sub-threshold current,* and *DC-IV* methods. The chief strengths are the direct measurement of the current, which is proportional to D_{it} and the fact that measurements can be made on regular MOSFETs with no need for special test structures. Charge pumping has been used to determine a single interface trap.[112] It can also determine the insulator trap density. Its main weaknesses are that unless special measurement variations and interpretations are used, one gets a single value for an average interface trap density - not the energy distribution of D_{it} and the measurement is sensitive to gate leakage current. The *sub-threshold method* is simpler to implement than charge pumping but is difficult to interpret for interface trap measurement. It is more useful when determining the change of interface trap density following hot electron stressing or energetic radiation exposure. *DC-IV* yields results similar to CP, but the measured current is related to the surface recombination velocity and the capture cross-section needs to be known to extract the interface trap density.

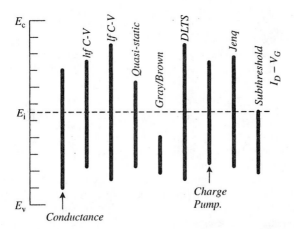

Fig. 6.41   Ranges of energy in the band gap of a *p*-type Si substrate over which interface trap charges are determined by various characterization techniques.

The various energies over which interface trap charges can be determined are shown in Fig. 6.41. A good discussion of various interface trap charge measurement techniques with their strengths and weaknesses is given in ref. 113.

Oxide Thickness: Among the electrical techniques, MOS $C-V$ measurements are most common. However, thin oxide leakage currents make the measurement interpretation more difficult. Occasionally, $I-V$ data are used for thickness extraction. Ellipsometry is routinely used for oxide thickness measurements, being sensitive to very thin oxides. However, the optical parameters of the layer must be known and for thin oxides the insulator may be inhomogeneous. Among the physical characterization techniques, XPS is suitable for very thin oxides. An excellent overview of SiO_2 and nitrided oxide including fabrication and characterization issues is given by Greene et al.[114]

APPENDIX 6.1

Capacitance Measurement Techniques

Most capacitance measurements are made with capacitance bridges or capacitance meters. In the vector voltage-current method of Fig. A6.1, ac signal v_i is applied to the device under test (DUT) and the device impedance Z is calculated from the ratio of v_i to the sample current i_i. A high-gain operational amplifier with feedback resistor R_F operates as a current-to-voltage converter. With the input to the op-amp at virtual ground, the negative terminal is essentially at ground potential, because the high input impedance allows no input current to the op-amp, $i_i \sim i_o$. With $i_i = v_i/Z$ and $i_o = -v_o/R_F$, the device impedance can be derived from v_o and v_i as

$$Z = -\frac{R_F v_i}{v_o} \tag{A6.1}$$

where the device impedance of the parallel $G-C$ circuit in Fig. A6.1 is given by

$$Z = \frac{G}{G^2 + (\omega C)^2} - \frac{j\omega C}{G^2 + (\omega C)^2} \tag{A6.2}$$

It consists of a conductance, the first term, and a susceptance, the second term. The voltages v_o and v_i are fed to a phase detector and the conductance and susceptance of

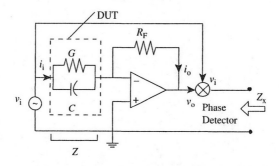

Fig. A6.1   Schematic circuit diagram of a capacitance-conductance meter.

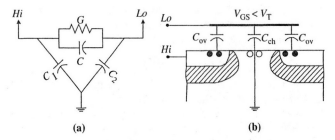

Fig. A6.2 Three-terminal capacitance measurement connections: (a) the measurement principle, (b) a MOSFET.

the sample are obtained by using the $0°$ and $90°$ phase angles of v_o referenced to v_i. The zero degree phase angle gives the conductance G while the $90°$ phase angle gives the susceptance or the capacitance C.

Although this method uses a simple circuit configuration and has relatively high accuracy, it is difficult to design a feedback resistor amplifier with i_o in exact proportion to i_i at high frequencies. An auto-balance circuit incorporating a null detector and a modulator overcomes this problem.[115] More detailed discussions of capacitance measurement circuits, probe stations, and other capacitance measurement hints can be found in the book by Nicollian and Brews.[24]

Some capacitance meters are *three-terminal* while others are *five-terminal* instruments. One of the terminals in either instrument is ground while the others connect to the device under test. The five-terminal instrument operates much like a four-point probe with the outer two terminals supplying the current and the inner two terminals measuring the potential. The ground terminal on these instruments gives additional flexibility by eliminating stray capacitances. Two examples with the ground terminal in a capacitance meter are shown in Fig. A6.2. Consider a three-terminal device with conductance G and capacitance C, which also has stray capacitances C_1 and C_2 shown in Fig. A6.2(a). By connecting the DUT to the capacitance meter (Hi-Lo) and the two stray capacitances to ground, C_1 and C_2 are eliminated from the measurement by shunting them to ground. The MOSFET of Fig. A6.2(b) is arranged to determine the gate-source and gate-drain overlap capacitances C_{ov}, by shunting the oxide capacitance in the channel region, C_{ch}, to ground. To determine C_{ch}, one connects the gate and substrate to the capacitance meter and shunts the source and drain to ground. The internal structure of the device, *e.g.*, substrate resistance or CMOS well resistance, play a role in capacitance measurements of the type in Fig. A6.2(b) especially for small capacitances.[116]

APPENDIX 6.2

Effect of Chuck Capacitance and Leakage Current

When device capacitance is measured at the wafer level, with the wafer resting on a chuck, precautions must be observed for the measurement setup not to influence the results. Consider the experimental arrangement in Fig. A6.3(a). The "Hi" terminal of

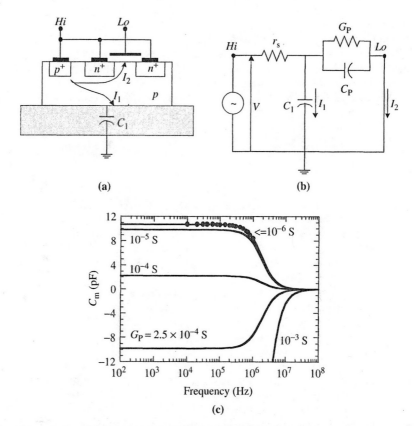

Fig. A6.3 (a) Cross-section of a MOSFET showing the effect of chuck capacitance, (b) equivalent circuit, and (c) theoretical and experimental measured capacitances. $r_s = 124\ \Omega$, $C_1 = 680$ pF, $C_P = 10.7$ pF. Lines: theory, points: experimental data from ref. 118.

the capacitance meter should be connected to the substrate/source/drain and the "Lo" terminal to the gate.[117] The capacitance is measured by applying a time varying voltage and the resulting current is proportional to the capacitance. However, the current has two paths: through the device capacitance and through the parasitic chuck capacitance. The equivalent circuit in Fig. A6.3(b), consists of the device capacitance C_P, the leakage conductance G_P, e.g., due to tunneling, series resistance r_s, and parasitic capacitance C_1. The capacitance meter assumes the circuits consists of a parallel C_m, G_m circuit, given by

$$C_m = \frac{C_1(C_P/C_1 - r_s G_P)}{(1 + r_s G_P)^2 + (\omega r_s C_1(1 + C_P/C_1))^2} \tag{A6.3a}$$

$$G_m = \frac{G_P + r_s G_P^2 + \omega^2 r_s C_P C_1(1 + C_P/C_1)}{(1 + r_s G_P)^2 + (\omega r_s C_1(1 + C_P/C_1))^2} \tag{A6.3b}$$

For negligibly small C_1, Eq. (A6.3) simplifies to Eq. (2.32).

Equation (A6.3a) is plotted in Fig. A6.3(c) for various values of G_P. Note the drop off at the higher frequencies due to the high chuck capacitance, which is also observed experimentally as indicated by the points.[118] C_m becomes negative for $C_P/C_1 < r_s G_P$. This is observed during MOS capacitance measurements for high gate voltages and thin oxides where the oxide becomes very leaky.[119] One solution to the capacitance droop at the higher frequencies, is to nullify the chuck capacitance by connecting the top chuck layer to the "Hi" terminal and the middle layer of a triaxial chuck to the guard terminal of the capacitance meter with the wafer resting on the chuck.[118]

REFERENCES

1. B.E. Deal, "Standardized Terminology for Oxide Charges Associated with Thermally Oxidized Silicon," *IEEE Trans. Electron Dev.* **ED-27**, 606–608, March 1980.

2. B.E. Deal, M. Sklar, A.S. Grove and E.H. Snow, "Characteristics of the Surface-State Charge (Q_{ss}) of Thermally Oxidized Silicon," *J. Electrochem. Soc.* **114**, 266–274, March 1967.

3. J.R. Brews, "An Improved High-Frequency MOS Capacitance Formula," *J. Appl. Phys.* **45**, 1276–1279, March 1974.

4. A. Berman and D.R. Kerr, "Inversion Charge Redistribution Model of the High-Frequency MOS Capacitance," *Solid-State Electron.* **17**, 735–742, July 1974.

5. W.E. Beadle, J.C.C. Tsai and R.D. Plummer, *Quick Reference Manual for Silicon Integrated Circuit Technology*, Wiley-Interscience, New York, 1985, 14–28.

6. H. El-Sissi and R.S.C. Cobbold, "Numerical Calculation of the Ideal C/V Characteristics of Nonuniformly Doped MOS Capacitors," *Electron. Lett.* **9**, 594–596, Dec. 1973.

7. J. Hynecek, "Graphical Method for Determining the Flatband Voltage for Silicon on Sapphire," *Solid-State Electron.* **18**, 119–120, Feb. 1975; K. Lehovec and S.T. Lin, "Analysis of C-V Data in the Accumulation Regime of MIS Structures," *Solid-State Electron.* **19**, 993–996, Dec. 1976.

8. F.P. Heiman, "Thin-Film Silicon-on-Sapphire Deep Depletion MOS Transistors," *IEEE Trans. Electron Dev.* **ED-13**, 855–862, Dec. 1966; K. Iniewski and A. Jakubowski, "New Method of Determination of the Flat-Band Voltage in SOI MOS Structures, *Solid-State Electron.* **29**, 947–950, Sept. 1986.

9. W.W. Lin and C.L. Liang, "Separation of dc and ac Competing Effects of Poly Silicon Gate Depletion in Deep Submicron CMOS Circuit Performance," *Solid-State Electron.* **39**, 1391–1393, Sept. 1996.

10. R. Castagné, "Determination of the Slow Density of an MOS Capacitor Using a Linearly Varying Voltage," (in French) *C.R. Acad. Sc. Paris* **267**, 866–869, Oct. 1968; M. Kuhn, "A Quasi-Static Technique for MOS C-V and Surface State Measurements," *Solid-State Electron.* **13**, 873–885, June 1970; W.K. Kappallo and J.P. Walsh, "A Current Voltage Technique for Obtaining Low-Frequency C-V Characteristics of MOS Capacitors," *Appl. Phys. Lett.* **17**, 384–386, Nov. 1970.

11. J. Koomen, "The Measurement of Interface State Charge in the MOS System," *Solid-State Electron.* **14**, 571–580, July 1971; K. Ziegler and E. Klausmann, "Static Technique for Precise Measurements of Surface Potential and Interface State Density in MOS Structures," *Appl. Phys. Lett.* **26**, 400–402, Apr. 1975.

12. J.R. Brews and E.H. Nicollian, "Improved MOS Capacitor Measurements Using the Q-C Method," *Solid-State Electron.* **27**, 963–975, Nov. 1984.

13. E.H. Nicollian and J.R. Brews, "Instrumentation and Analog Implementation of the Q-C Method for MOS Measurements," *Solid-State Electron.* **27**, 953–962, Nov. 1984; D.M. Boulin, J.R. Brews and E.H. Nicollian, "Digital Implementation of the Q-C Method for MOS Measurements," *Solid-State Electron.* **27**, 977–988, Nov. 1984.

14. T.J. Mego, "Improved Feedback Charge Method for Quasistatic CV Measurements in Semi-conductors," *Rev. Sci. Instrum.* **57**, 2798–2805, Nov. 1986.

15. P.L. Castro and B.E. Deal, "Low-Temperature Reduction of Fast Surface States Associated with Thermally Oxidized Silicon," *J. Electrochem. Soc.* **118**, 280–286, Feb. 1971.

16. R. Williams, "Photoemission of Electrons from Silicon into Silicon Dioxide," *Phys. Rev.* **140**, A569–A575, Oct. 1965; R. Williams, "Properties of the Silicon-SiO_2 Interface," *J. Vac. Sci. Technol.* **14**, 1106–1111, Sept./Oct. 1977.

17. W.M. Werner, "The Work Function Difference of the MOS-System with Aluminium Field Plates and Polycrystalline Silicon Field Plates," *Solid-State Electron.* **17**, 769–775, Aug. 1974.

18. C.S. Park, B.J. Cho, and D.L. Kwong, "Thermally Stable Fully Silicided Hf-Silicide Metal-Gate Electrode," *IEEE Electron Dev. Lett.* **25**, 372–374, June 2004.

19. R.R. Razouk and B.E. Deal, "Hydrogen Anneal Effects on Metal-Semiconductor Work Function Difference," *J. Electrochem. Soc.* **129**, 806–810, April 1982; A.I. Akinwande and J.D. Plummer, "Process Dependence of the Metal Semiconductor Work Function Difference," *J. Electrochem. Soc.* **134**, 2297–2303, Sept. 1987.

20. N. Lifshitz, "Dependence of the Work-Function Difference Between the Polysilicon Gate and Silicon Substrate on the Doping Level in Polysilicon," *IEEE Trans. Electron Dev.* **ED-32**, 617–621, March 1985.

21. W.M. Werner, "The Work Function Difference of the MOS System with Aluminum Field Plates and Polycrystalline Silicon Field Plates," *Solid-State Electron.* **17**, 769–775, Aug. 1974.

22. T.W. Hickmott and R.D. Isaac, "Barrier Heights and Polycrystalline Silicon-SiO_2 Interface," *J. Appl. Phys.* **52**, 3464–3475, May 1981.

23. D.B. Kao, K.C. Saraswat and J.P. McVittie, "Annealing of Oxide Fixed Charges in Scaled Polysilicon Gate MOS Structures," *IEEE Trans. Electron Dev.* **ED-32**, 918–925, May 1985.

24. E.H. Nicollian and J.R. Brews, *MOS Physics and Technology*, Wiley, New York, 1982.

25. S.P. Li, M. Ryan and E.T. Bates, "Rapid and Precise Measurement of Flatband Voltage," *Rev. Sci. Instrum.* **47**, 632–634, May 1976.

26. E.H. Snow, A.S. Grove, B.E. Deal and C.T. Sah, "Ion Transport Phenomena in Insulating Films," *J. Appl. Phys.* **36**, 1664–1673, May 1965.

27. W.A. Pliskin and R.A. Gdula, "Passivation and Insulation," in *Handbook on Semiconductors*, Vol. 3 (S.P. Keller, ed.), North Holland, Amsterdam, 1980 and references therein.

28. N.J. Chou, "Application of Triangular Voltage Sweep Method to Mobile Charge Studies in MOS Structures," *J. Electrochem. Soc.* **118**, 601–609, April 1971; G. Derbenwick, "Mobile Ions in SiO_2: Potassium," *J. Appl. Phys.* **48**, 1127–1130, March 1977; J.P. Stagg, "Drift Mobilities of Na^+ and K^+ Ions in SiO_2 Films," *Appl. Phys. Lett.* **31**, 532–533, Oct. 1977; M.W. Hillen, G. Greeuw and J.F. Verwey, "On the Mobility of Potassium Ions in SiO_2," *J. Appl. Phys.* **50**, 4834–4837, July 1979; M. Kuhn and D.J. Silversmith, "Ionic Contamination and Transport of Mobile Ions in MOS Structures," *J. Electrochem. Soc.* **118**, 966–970, June 1971.

29. G. Greeuw and J.F. Verwey, "The Mobility of Na^+, Li^+ and K^+ Ions in Thermally Grown SiO_2 Films," *J. Appl. Phys.* **56**, 2218–2224, Oct. 1984; Y. Shacham-Diamand, A. Dedhia, D. Hoffstetter, and W.G. Oldham, "Copper Transport in Thermal SiO_2," *J. Electrochem. Soc.* **140**, 2427–2432, Aug. 1993.

30. M. Kuhn and D.J. Silversmith, "Ionic Contamination and Transport of Mobile Ions in MOS Structures," *J. Electrochem. Soc.* **118**, 966–970, June 1971; M.W. Hillen and J.F. Verwey, "Mobile Ions in SiO_2 Layers on Si," in *Instabilities in Silicon Devices: Silicon Passivation and Related Instabilities* (G. Barbottin and A. Vapaille, eds.), Elsevier, Amsterdam, 1986, 403–439.

31. L. Stauffer, T. Wiley, T. Tiwald, R. Hance, P. Rai-Choudhury, and D.K. Schroder, "Mobile Ion Monitoring by Triangular Voltage Sweep," *Solid-State Technol.* **38**, S3–S8, August 1995.

32. T.M. Buck, F.G. Allen, J.V. Dalton and J.D. Struthers, "Studies of Sodium in SiO_2 Films by Neutron Activation and Radiotracer Techniques," *J. Electrochem. Soc.* **114**, 862–866, Aug. 1967.

33. E. Yon, W.H. Ko and A.B. Kuper, "Sodium Distribution in Thermal Oxide on Silicon by Radiochemical and MOS Analysis," *IEEE Trans. Electron Dev.* **ED-13**, 276–280, Feb. 1966.

34. B. Yurash and B.E. Deal, "A Method for Determining Sodium Content of Semiconductor Processing Materials," *J. Electrochem. Soc.* **115**, 1191–1196, Nov. 1968.

35. H.L. Hughes, R.D. Baxter and B. Phillips, "Dependence of MOS Device Radiation-Sensitivity on Oxide Impurities," *IEEE Trans. Nucl. Sci.* **NS-19**, 256–263, Dec. 1972.

36. A. Goetzberger, E. Klausmann and M.J. Schulz, "Interface States on Semiconductor/Insulator Interfaces," *CRC Crit. Rev. Solid State Sci.* **6**, 1–43, Jan. 1976.

37. G. DeClerck, "Characterization of Surface States at the Si–SiO$_2$ Interface," in *Nondestructive Evaluation of Semiconductor Materials and Devices* (J.N. Zemel, ed.), Plenum Press, New York, 1979, 105–148.

38. P.V. Gray and D.M. Brown, "Density of SiO$_2$-Si Interface States," *Appl. Phys. Lett.* **8**, 31–33, Jan. 1966; D.M. Fleetwood, "Long-term Annealing Study of Midgap Interface-trap Charge Neutrality," *Appl. Phys. Lett.* **60**, 2883–2885, June 1992.

39. C.N. Berglund, "Surface States at Steam-Grown Silicon-Silicon Dioxide Interfaces," *IEEE Trans. Electron Dev.* **ED-13**, 701–705, Oct. 1966.

40. T.C. Lin and D.R. Young, "New Methods for Using the Q-V Technique to Evaluate Si–SiO$_2$ Interface States," *J. Appl. Phys.* **71**, 3889–3893, April 1992; J.M. Moragues, E. Ciantar, R. Jérisian, B. Sagnes, and J. Qualid, "Surface Potential Determination in Metal-Oxide-Semiconductor Capacitors," *J. Appl. Phys.* **76**, 5278–5287, Nov. 1994.

41. S. Nishimatsu and M. Ashikawa, "A Simple Method for Measuring the Interface State Density," *Rev. Sci. Instrum.* **45**, 1109–1112, Sept. 1984.

42. R. Castagné and A. Vapaille, "Description of the SiO$_2$-Si Interface Properties by Means of Very Low Frequency MOS Capacitance Measurements," *Surf. Sci.* **28**, 157–193, Nov. 1971.

43. G. Declerck, R. Van Overstraeten and G. Broux, "Measurement of Low Densities of Surface States at the Si–SiO$_2$ Interface," *Solid-State Electron.* **16**, 1451–1460, Dec. 1973.

44. R. Castagné and A. Vapaille, "Apparent Interface State Density Introduced by the Spatial Fluctuations of Surface Potential in an MOS Structure," *Electron. Lett.* **6**, 691–694, Oct. 1970.

45. Y. Omura and Y. Nakajima, "Quantum Mechanical Influence and Estimated Errors on Interface-state Density Evaluation by Quasi-static C-V Measurement," *Solid-State Electron.* **44**, 1511–1514, Aug. 2000.

46. S. Wagner and C.N. Berglund, "A Simplified Graphical Evaluation of High-Frequency and Quasistatic Capacitance-Voltage Curves," *Rev. Sci. Instrum.* **43**, 1775–1777, Dec. 1972.

47. R. Van Overstraeten, G. Declerck and G. Broux, "Graphical Technique to Determine the Density of Surface States at the Si–SiO$_2$ Interface of MOS Devices Using the Quasistatic C-V Method," *J. Electrochem. Soc.* **120**, 1785–1787, Dec. 1973.

48. A.D. Lopez, "Using the Quasistatic Method for MOS Measurements," *Rev. Sci. Instrum.* **44**, 200–204, Feb. 1972.

49. E.H. Nicollian and A. Goetzberger, "The Si–SiO$_2$ Interface—Electrical Properties as Determined by the Metal-Insulator-Silicon Conductance Technique," *Bell Syst. Tech. J.* **46**, 1055–1133, July/Aug. 1967.

50. M. Schulz, "Interface States at the SiO$_2$-Si Interface," *Surf. Sci.* **132**, 422–455, Sept. 1983.

51. A.K. Aggarwal and M.H. White, "On the Nonequilibrium Statistics and Small Signal Admittance of Si–SiO$_2$ Interface Traps in the Deep-Depleted Gated-Diode Structure," *J. Appl. Phys.* **55**, 3682–3694, May 1984.

52. E.M. Vogel, W.K. Henson, C.A. Richter, and J.S. Suehle, "Limitations of Conductance to the Measurement of the Interface State Density of MOS Capacitors with Tunneling Gate Dielectrics," *IEEE Trans. Electron Dev.* **47**, 601–608, March 2000; T.P. Ma and R.C. Barker, "Surface-State Spectra from Thick-oxide MOS Tunnel Junctions," *Solid-State Electron.* **17**, 913–929, Sept. 1974.

53. E.H. Nicollian, A. Goetzberger and A.D. Lopez, "Expedient Method of Obtaining Interface State Properties from MIS Conductance Measurements," *Solid-State Electron.* **12**, 937–944, Dec. 1969; W. Fahrner and A. Goetzberger, "Energy Dependence of Electrical Properties of Interface States in Si–SiO₂ Interfaces," *Appl. Phys. Lett.* **17**, 16–18, July 1970; H. Deuling, E. Klausmann and A. Goetzberger, "Interface States in Si–SiO₂ Interfaces," *Solid-State Electron.* **15**, 559–571, May 1972; J.R. Brews, "Admittance of an MOS Device with Interface Charge Inhomogeneities," *J. Appl. Phys.* **43**, 3451–3455, Aug. 1972; P.A. Muls, G.J. DeClerck and R.J. Van Overstraeten "Influence of Interface Charge Inhomogeneities on the Measurement of Surface State Densities in Si–SiO₂ Interfaces by Means of the MOS ac Conductance Technique," *Solid-State Electron.* **20**, 911–922, Nov. 1977 and references therein.

54. J.J. Simonne, "A Method to Extract Interface State Parameters from the MIS Parallel Conductance Technique," *Solid-State Electron.* **16**, 121–124, Jan. 1973.

55. J.R. Brews, "Rapid Interface Parameterization Using a Single MOS Conductance Curve," *Solid-State Electron.* **26**, 711–716, Aug. 1983; J.M. Noras, "Extraction of Interface State Attributes from MOS Conductance Measurements," *Solid-State Electron.* **30**, 433–437, April 1987, "Parameter Estimation in MOS Conductance Studies," *Solid-State Electron.* **31**, 981–987, May 1988.

56. W.A. Hill and C.C. Coleman, "A Single-Frequency Approximation for Interface-State Density Determination," *Solid-State Electron.* **23**, 987–993, Sept. 1980.

57. A. De Dios, E. Castán, L. Bailón, J. Barbolla, M. Lozano, and E. Lora-Tamayo, "Interface State Density Measurement in MOS Structures by Analysis of the Thermally Stimulated Conductance," *Solid-State Electron.* **33**, 987–992, Aug. 1990.

58. E. Duval and E. Lheurette, "Characterisation of Charge Trapping at the Si–SiO₂ (100) Interface Using High-temperature Conductance Spectroscopy," *Microelectron. Eng.* **65**, 103–112, Jan. 2003.

59. H. Haddara and G. Ghibaudo, "Analytical Modeling of Transfer Admittance in Small MOS-FETs and Application to Interface State Characterisation," *Solid-State Electron* **31**, 1077–1082, June 1988.

60. L.M. Terman, "An Investigation of Surface States at a Silicon/Silicon Oxide Interface Employing Metal-Oxide-Silicon Diodes," *Solid-State Electron.* **5**, 285–299, Sept./Oct. 1962.

61. C.C.H. Hsu and C.T. Sah, "Generation-Annealing of Oxide and Interface Traps at 150 and 298 K in Oxidized Silicon Stressed by Fowler-Nordheim Electron Tunneling," *Solid-State Electron.* **31**, 1003–1007, June 1988.

62. K.H. Zaininger and G. Warfield, "Limitations of the MOS Capacitance Method for the Determination of Semiconductor Surface Properties," *IEEE Trans. Electron Dev.* **ED-12**, 179–193, April 1965.

63. C.T. Sah, A.B. Tole and R.F. Pierret, "Error Analysis of Surface State Density Determination Using the MOS Capacitance Method," *Solid-State Electron.* **12**, 689–709, Sept. 1969.

64. E.M. Vogel and G.A. Brown, "Challenges of Electrical Measurements of Advanced Gate Dielectrics in Metal-Oxide-Semiconductor Devices," in *Characterization and Metrology for VLSI Technology: 2003 Int. Conf.* (D.G. Seiler, A.C. Diebold, T.J. Shaffner, R. McDonald, S. Zollner, R.P. Khosla, and E.M. Secula, eds.), Am. Inst. Phys., 771–781, 2003.

65. E Rosenecher and D. Bois, "Comparison of Interface State Density in MIS Structure Deduced from DLTS and Terman Measurements," *Electron. Lett.* **18**, 545–546, June 1982.

66. P.V. Gray and D.M. Brown, "Density of SiO₂-Si Interface States," *Appl. Phys. Lett.* **8**, 31–33, Jan. 1966; D.M. Brown and P.V. Gray, "Si–SiO₂ Fast Interface State Measurements," *J. Electrochem. Soc.* **115**, 760–767, July 1968; P.V. Gray, "The Silicon-Silicon Dioxide System," *Proc. IEEE* **57**, 1543–1551, Sept. 1969.

67. M.R. Boudry, "Theoretical Origins of N_{ss} Peaks Observed in Gray-Brown MOS Studies," *Appl. Phys. Lett.* **22**, 530–531, May 1973.

68. D.K. Schroder and J. Guldberg "Interpretation of Surface and Bulk Effects Using the Pulsed MIS Capacitor," *Solid-State Electron.* **14**, 1285–1297, Dec. 1971.

69. C.S. Jenq, "High-Field Generation of Interface States and Electron Traps in MOS Capacitors," Ph.D. Dissertation, Princeton University, 1978; A. Mir and D. Vuillaume, "Positive Charge and Interface State Creation at the Si–SiO₂ Interface During Low-Fluence and High-Fluence Electron Injections," *Appl. Phys. Lett.* **62**, 1125–1127, March 1993.

70. N. Saks, "Comparison of Interface Trap Densities Measured by the Jenq and Charge Pumping Techniques," *J. Appl. Phys.* **74**, 3303–3306, Sept. 1993.

71. J.S. Brugler and P.G.A. Jespers, "Charge Pumping in MOS Devices," *IEEE Trans. Electron Dev.* **ED-16**, 297–302, March 1969.

72. D. Bauza, "Rigorous Analysis of Two-Level Charge Pumping: Application to the Extraction of Interface Trap Concentration Versus Energy Profiles in Metal–Oxide–Semiconductor Transistors," *J. Appl. Phys.* **94**, 3229–3248, Sept. 2003.

73. G. Groeseneken, H.E. Maes, N. Beltrán and R.F. De Keersmaecker, "A Reliable Approach to Charge-Pumping Measurements in MOS Transistors," *IEEE Trans. Electron Dev.* **ED-31**, 42–53, Jan. 1984; P. Heremans, J. Witters, G. Groeseneken and H.E. Maes, "Analysis of the Charge Pumping Technique and Its Application for the Evaluation of MOSFET Degradation," *IEEE Trans. Electron Dev.* **36**, 1318–1335, July 1989.

74. J.L. Autran and C. Chabrerie, " Use of the Charge Pumping Technique with a Sinusoidal Gate Waveform," *Solid-State Electron.* **39**, 1394–1395, Sept. 1996.

75. A.B.M. Elliot, "The Use of Charge Pumping Currents to Measure Surface State Densities in MOS Transistors," *Solid-State Electron.* **19**, 241–247, March 1976; R.A. Wachnik and J.R. Lowney, "A Model for the Charge-Pumping Current Based on Small Rectangular Voltage Pulses," *Solid-State Electron.* **29**, 447–460, April 1986; "The Use of Charge Pumping to Characterize Generation by Interface Traps," *IEEE Trans. Electron.Dev.* **ED-33**, 1054–1061, July 1986.

76. W.L. Chen, A. Balasinski, and T.P. Ma, "A Charge Pumping Method for Rapid Determination of Interface-Trap Parameters in Metal-Oxide-Semiconductor Devices," *Rev. Sci. Instrum.* **63**, 3188–3190, May 1992.

77. M. Katashiro, K. Matsumoto, and R. Ohta, "Analysis and Application of Hydrogen Supplying Process in Metal-Oxide-Semiconductor Structures," *J. Electrochem. Soc.* **143**, 3771–3777, Nov. 1996.

78. W.L. Tseng, "A New Charge Pumping Method of Measuring Si–SiO₂ Interface States," *J. Appl. Phys.* **62**, 591–599, July 1987; F. Hofmann and W.H. Krautschneider, "A Simple Technique for Determining the Interface-Trap Distribution of Submicron Metal-Oxide-Semiconductor Transistors by the Charge Pumping Method," *J. Appl. Phys.* **65**, 1358–1360, Feb. 1989.

79. N.S. Saks and M.G. Ancona, "Determination of Interface Trap Capture Cross Sections Using Three-Level Charge Pumping," *IEEE Electron Dev. Lett.* **11**, 339–341, Aug. 1990; R.R. Siergiej, M.H. White, and N.S. Saks, "Theory and Measurement of Quantization Effects on Si–SiO₂ Interface Trap Modeling," *Solid-State Electron.* **35**, 843–854, June 1992.

80. S. Jakschik, A. Avellan, U. Schroeder, and J.W. Bartha, "Influence of Al₂O₃ Dielectrics on the Trap-Depth Profiles in MOS Devices Investigated by the Charge-Pumping Method," *IEEE Trans. Electron Dev.* **51**, 2252–2255, Dec. 2004.

81. A. Melik-Martirosian and T.P. Ma, "Lateral Profiling of Interface Traps and Oxide Charge in MOSFET Devices: Charge Pumping Versus *DCIV*," *IEEE Trans. Electron Dev.* **48**, 2303–2309, Oct. 2001; C. Bergonzoni and G.D. Libera, "Physical Characterization of Hot-Electron-Induced MOSFET Degradation Through an Improved Approach to the Charge-Pumping Technique," *IEEE Trans. Electron Dev.* **39**, 1895–1901, Aug. 1992.

82. M. Tsuchiaki, H. Hara, T. Morimoto, and H. Iwai, "A New Charge Pumping Method for Determining the Spatial Distribution of Hot-Carrier-Induced Fixed Charge in p-MOSFET's," *IEEE Trans. Electron Dev.* **40**, 1768–1779, Oct. 1993.

83. R.E. Paulsen and M.H. White, "Theory and Application of Charge Pumping for the Characterization of Si–SiO₂ Interface and Near-Interface Oxide Traps," *IEEE Trans. Electron Dev.* **41**, 1213–1216, July 1994.

84. D.M. Fleetwood, "Fast and Slow Border Traps in MOS Devices" *IEEE Trans Nucl. Sci.* **43**, 779–786, June 1996.

85. G. Van den bosch, G.V. Groeseneken, P. Heremans, and H.E. Maes, "Spectroscopic Charge Pumping: A New Procedure for Measuring Interface Trap Distributions on MOS Transistors," *IEEE Trans. Electron Dev.* **38**, 1820–1831, Aug. 1991.

86. Y. Li and T.P. Ma, "A Front-Gate Charge-Pumping Method for Probing Both Interfaces in SOI Devices," *IEEE Trans. Electron Dev.* **45**, 1329–1335, June 1998.

87. D. Bauza, "Extraction of Si–SiO$_2$ Interface Trap Densities in MOS Structures with Ultrathin Oxides," *IEEE Electron Dev. Lett.* **23**, 658–660, Nov. 2002; D. Bauza, "Electrical Properties of Si–SiO$_2$ Interface Traps and Evolution with Oxide Thickness in MOSFET's with Oxides from 2.3 to 1.2 nm Thick," *Solid-State Electron.* **47**, 1677–1683, Oct. 2003; P. Masson, J-L Autran, and J. Brini, "On the Tunneling Component of Charge Pumping Current in Ultrathin Gate Oxide MOSFET's," *IEEE Electron Dev. Lett.* **20**, 92–94, Feb. 1999.

88. J.L. Autran, F. Seigneur, C. Plossu, and B. Balland, "Characterization of Si–SiO$_2$ Interface States: Comparison Between Different Charge Pumping and Capacitance Techniques," *J. Appl. Phys.* **74**, 3932–3935, Sept. 1993.

89. P.A. Muls, G.J. DeClerck and R.J. van Overstraeten, "Characterization of the MOSFET Operating in Weak Inversion," in *Adv. in Electron. and Electron Phys.* **47**, 197–266, 1978.

90. P.J. McWhorter and P.S. Winokur, "Simple Technique for Separating the Effects of Interface Traps and Trapped-Oxide Charge in Metal-Oxide-Semiconductor Transistors," *Appl. Phys. Lett.* **48**, 133–135, Jan. 1986.

91. A. Neugroschel, C.T. Sah, M. Han, M.S. Carroll. T. Nishida, J.T. Kavalieros, and Y. Lu, "Direct-Current Measurements of Oxide and Interface Traps on Oxidized Silicon," *IEEE Trans. Electron Dev.* **42**, 1657–1662, Sept. 1995.

92. D.J. Fitzgerald and A.S. Grove, "Surface Recombination in Semiconductors," *Surf. Sci.* **9**, 347–369, Feb. 1968.

93. H. Guan, Y. Zhang, B.B. Jie, Y.D. He, M-F. Li, Z. Dong, J. Xie, J.L.F. Wang, A.C. Yen, G.T.T. Sheng, and W. Li, "Nondestructive DCIV Method to Evaluate Plasma Charging in Ultrathin Gate Oxides," *IEEE Electron Dev. Lett.* **20**, 238–240, May 1999.

94. R.J. Kriegler, T.F. Devenyi, K.D. Chik and J. Shappir, "Determination of Surface-State Parameters from Transfer-Loss Measurements in CCDs," *J. Appl. Phys.* **50**, 398–401, Jan. 1979.

95. E. Kamieniecki, "Surface Photovoltage Measured Capacitance: Application to Semiconductor/Electrolyte System," *J. Appl. Phys.* **54**, 6481–6487, Nov. 1983; V. Murali, A.T. Wu, A.K. Chatterjee, and D.B. Fraser, "A Novel Technique for *In-Line* Monitoring of Micro-Contamination and Process Induced Damage," *IEEE Trans. Semicond. Manufact.* **5**, 214–222, Aug. 1992; L.A. Lipkin, "Real-Time Monitoring with a Surface Charge Analyzer," *J. Electrochem. Soc.* **140**, 2328–2332, Aug. 1993.

96. H. Takeuchi and T.J. King, "Surface Charge Analysis of Ultrathin HfO$_2$, SiO$_2$, and Si$_3$N$_4$," *J. Electrochem. Soc.* **151**, H44–H48, Feb. 2004.

97. E.H. Poindexter and P.J. Caplan, "Characterization of Si/SiO$_2$ Interface Defects by Electron Spin Resonance," *Progr. Surf. Sci.* **14**, 201–294, 1983.

98. E.H. Poindexter, "MOS Interface States: Overview and Physicochemical Perspective," *Semicond. Sci. Technol.* **4**, 961–969, Dec. 1989.

99. M.P. Seah, S.J. Spencer, F. Bensebaa, I. Vickridge, H. Danzebrink, M. Krumrey, T. Gross, W. Oesterle, E. Wendler, B. Rheinländer, Y. Azuma, I. Kojima, N. Suzuki, M. Suzuki, S. Tanuma, D.W. Moon, H.J. Lee, Hyun Mo Cho, H.Y. Chen, A.T.S. Wee, T. Osipowicz, J.S. Pan, W.A. Jordaan, R. Hauert, U. Klotz, C. van der Marel, M. Verheijen, Y. Tamminga, C. Jeynes, P. Bailey, S. Biswas, U. Falke, N.V. Nguyen, D. Chandler-Horowitz, J.R. Ehrstein, D. Muller, and J.A. Dura, "Critical Review of the Current Status of Thickness Measurements for Ultrathin SiO$_2$ on Si, Part V: Results of a CCQM Pilot Study," *Surf. Interface Anal.* **36**, 1269–1303, Sept. 2004.

100. A.C. Diebold, D. Venables, Y. Chabal, D. Muller, M. Weldonc, and E. Garfunkel, "Characterization and Production Metrology of Thin Transistor Gate Oxide Films," *Mat. Sci. in Semicond. Proc.* **2**, 103–147, July 1999.

101. K.S. Krisch, J.D. Bude, and L. Manchanda, "Gate Capacitance Attenuation in MOS Devices With Thin Gate Dielectrics," *IEEE Electron Dev. Lett.* **17**, 521–524, Nov. 1996; D. Vasileska, D.K. Schroder, and D.K. Ferry, "Scaled Silicon MOSFET's: Degradation of the Total Gate Capacitance," *IEEE Trans Electron Dev.* **44**, 584–587, April 1997.

102. M.J. McNutt and C.T. Sah, "Determination of the MOS Capacitance," *J. Appl. Phys.* **46**, 3909–3913, Sept. 1975.

103. J. Maserjian, G. Peterson, and C. Svensson, "Saturation Capacitance of Thin Oxide MOS Structures and the Effective Surface State Density of Silicon," *Solid-State Electron.* **17**, 335–339, April 1974.

104. J. Maserjian, in *The Physics and Chemistry of SiO_2 and the Si/SiO_2 Interface*, (C.R. Helms and B.E. Deal, eds.), Plenum Press, New York, 1988.

105. S. Kar, "Determination of the Gate Dielectric Capacitance of Ultrathin High-k Layers," *J. Electrochem. Soc.* **151**, G476–G481, July 2004.

106. E. Vincent, G. Ghibaudo, G. Morin, and C. Papadas, "On the Oxide Thickness Extraction in Deep-Submicron Technologies," *Proc. 1997 IEEE Int. Conf. on Microelectron. Test Struct.* 105–110, 1997.

107. G. Ghibaudo, S. Bruyère, T. Devoivre, B. DeSalvo, and E. Vincent, "Improved Method for the Oxide Thickness Extraction in MOS Structures with Ultrathin Gate Dielectrics," *IEEE Trans. Semicond. Manufact.* **13**, 152–158, May 2000.

108. J.F. Lønnum and J.S. Johannessen, "Dual-Frequency Modified C/V Technique," *Electron. Lett.* **22**, 456–457, April 1986; K.J. Yang and C. Hu, "MOS Capacitance Measurements for High-Leakage Thin Dielectrics," *IEEE Trans. Electron Dev.* **46**, 1500–1501, July 1999.

109. A. Nara, N. Yasuda, H. Satake, and A. Toriumi, "Applicability Limits of the Two-Frequency Capacitance Measurement Technique for the Thickness Extraction of Ultrathin Gate Oxide," *IEEE Trans. Semicond. Manufact.* **15**, 209–213, May 2002.

110. D.W. Barlage, J.T. O'Keeffe, J.T. Kavalieros, M.N. Nguyen, and R.S. Chau, "Inversion MOS Capacitance Extraction for High-Leakage Dielectrics Using a Transmission Line Equivalent Circuit," *IEEE Electron Dev. Lett.* **21**, 406–408, Sept. 2000.

111. S. Zafar, Q. Liu, and E.A. Irene, "Study of Tunneling Current Oscillation Dependence on SiO_2 Thickness and Si Roughness at the Si/SiO_2 Interface," *J. Vac. Sci. Technol.* **A13**, 47–53, Jan./Feb. 1995; K.J. Hebert and E.A. Irene, "Fowler-Nordheim Current Oscillations at Metal/Oxide/Si Interfaces," *J. Appl. Phys.* **82**, 291–296, July 1997; L. Mao, C. Tan, and M. Xu, "Thickness Measurements for Ultrathin-Film Insulator Metal-Oxide-Semiconductor Structures Using Fowler-Nordheim Tunneling Current Oscillations," *J. Appl. Phys.* **88**, 6560–6563, Dec. 2000.

112. L. Militaru and A. Souifi, "Study of a Single Dangling Bond at the SiO_2/Si Interface in Deep Submicron Metal-Oxide-Semiconductor Transistors," *Appl. Phys. Lett.* **83**, 2456–2458, Sept. 2003.

113. S.C. Witczak, J.S. Suehle, and M. Gaitan, "An Experimental Comparison of Measurement Techniques to Extract $Si-SiO_2$ Interface Trap Density," *Solid-State Electron.* **35**, 345–355, March 1992.

114. M.L. Green, E.P. Gusev, R. Degraeve, and E.L. Garfunkel, "Ultrathin (<4 nm) SiO_2 and Si–O–N Gate Dielectric Layers for Silicon Microelectronics: Understanding the Processing, Structure, and Physical and Electrical Limits," *J. Appl. Phys.* **90**, 2057–2121, Sept. 2001.

115. Service Manual for HP 4275-A Multi Frequency LCR Meter, Hewlett-Packard, 1983, p. 8–4.

116. W.W. Lin and P.C. Chan, "On the Measurement of Parasitic Capacitances of Devices with More Than Two External Terminals Using an LCR Meter," *IEEE Trans. Electron Dev.* **38**, 2573–2575, Nov. 1991.

117. Accurate Capacitance Characterization at the Wafer Level, Agilent Technol. Application Note 4070–2, 2000.

118. P.A. Kraus, K.A. Ahmed, and J.S. Williamson, Jr., "Elimination of Chuck-Related Parasitics in MOSFET Gate Capacitance Measurements," *IEEE Trans. Electron Dev.* **51**, 1350–1352, Aug. 2004.

119. Y. Okawa, H. Norimatsu, H. Suto, and M. Takayanagi, "The Negative Capacitance Effect on the C-V Measurement of Ultra Thin Gate Dielectrics Induced by the Stray Capacitance of the Measurement System," *IEEE Proc. Int. Conf. Microelectronic Test Struct.* 197–202, 2003.

PROBLEMS

6.1 Consider an MOS capacitor with a p^+ poly-Si gate ($E_F = E_v$) and a p-type substrate with $N_A = 10^{16}$ cm^{-3}. $t_{ox} = 15$ nm, $n_i = 10^{10}$ cm^{-3}, $T = 300$ K, $K_s = 11.7$, $K_{ox} = 3.9$, E_G(poly-Si) $= E_G$(Si $= 1.12$ eV).

  (a) Determine the flatband voltage V_{FB} and the normalized flatband capacitance C_{FB}/C_{ox}.

  (b) Determine V_{FB} when the p^+ poly-Si gate is replaced with an n^+ poly-Xx gate ($E_F = E_c$), where Xx is a semiconductor with electron affinity $\chi(Xx) = \chi(Si)$, but with band gap $E_G(Xx) = E_G(Si)/2$. $Q_f = Q_{it} = Q_m = Q_{ot} = 0$.

6.2 The flatband voltage V_{FB} data are given in the following table as a function of oxide thickness t_{ox} for an MOS capacitor. This device has a fixed charge density Q_f (C/cm^2) and a *uniform* oxide trapped charge density ρ_{ot} (C/cm^3). The flatband voltage is given by

$$V_{FB} = \phi_{MS} - \frac{Q_f}{C_{ox}} - \frac{1}{C_{ox}} \int_0^{t_{ox}} (x/t_{ox})\rho_{ot}(x)\,dx$$

Determine the work function difference ϕ_{MS}, the fixed charge density $N_f = Q_f/q$ (cm^{-2}), the oxide trapped charge density ρ_{ot}/q (cm^{-3}) and N_{ot} (cm^{-2}). Determine N_{ot} for $t_{ox} = 10^{-5}$ cm. $K_{ox} = 3.9$, $Q_{it} = Q_m = 0$. *Note:* You have to think of the effect of a *uniform* ρ_{ot} on V_{FB}.

t_{ox} (cm)	V_{FB} (V)	t_{ox} (cm)	V_{FB} (V)
10^{-6}	0.265	6×10^{-6}	-0.256
2×10^{-6}	0.207	7×10^{-6}	-0.429
3×10^{-6}	0.126	8×10^{-6}	-0.626
4×10^{-6}	0.0219	9×10^{-6}	-0.846
5×10^{-6}	-0.105	10^{-5}	-1.09

6.3 Consider the low-frequency C_{lf}/C_{ox} versus V_G curve in Fig. P6.3. It is for an MOS capacitor with a p-type substrate ($N_A = 10^{15}$ cm^{-3}), a metal gate, and $V_{FB} = 0$. Draw the C_{lf}/C_{ox} versus V_G curve for this device on the same figure with the metal gate replaced by an n-type poly-Si gate doped to $N_D = N_A$ (substrate). $T = 300$ K, $n_i = 10^{10}$ cm^{-3}.

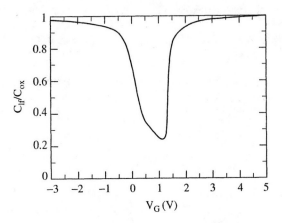

Fig. P6.3

6.4   Consider the low-frequency C_{lf}/C_{ox} versus V_G curve in Fig. P6.4. It is for an MOS capacitor with a p-type substrate ($N_A = N_{A1}$), a metal gate, and $V_{FB} = 0$. Draw the C_{lf}/C_{ox} versus V_G curve for this device if the metal gate is replaced with a p-type poly-Si gate doped to $N_A = N_{A1}$.

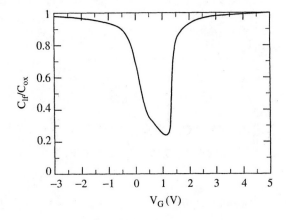

Fig. P6.4

6.5   Consider an MOS capacitor with $t_{ox} = 40$ nm and $V_{FB} = 0$. Now consider a similar device except the oxide is contaminated with mobile ions. These are very peculiar mobile ions. The upper half of the oxide (the side nearest the gate) contains a uniform density of *positively* charged ions with $\rho_{m1} = 0.04$ C/cm^3. The lower half of the oxide (the side nearest the substrate) contains a uniform density of *negatively* charged ions with $\rho_{m2} = -0.06$ C/cm^3. Determine V_{FB} for this case. The device undergoes a bias-temperature stress at elevated temperature with *positive* gate voltage and all charges move. Determine V_{FB} for this case.

6.6 The $C_{hf}/C_{ox} - V_G$ curve of an ideal MOS-C is shown in Fig. P6.6(a). Draw on the same figure the $C_{hf}/C_{ox} - V_G$ curve for an MOS-C with identical dimensions in which the oxide of half of the gate area contains positive charge and the other half does not (Fig. 6.6(b)). The flatband voltage of the contaminated half of the device is $V_{FB} = -2$ V.

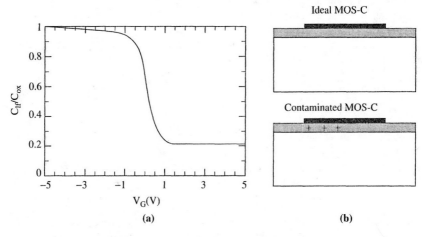

(a)   (b)

Fig. P6.6

6.7 (a) Draw the $C_{lf}/C_{ox} - V_G$ curve *qualitatively* for an ideal MOS-C ($V_{FB} = 0$) when the semiconductor is *intrinsic* ($N_A = N_D = 0$). Use $t_{ox} = 10$ nm.

   (b) Does the $C_{lf}/C_{ox} - V_G$ curve change if t_{ox} increases from 10 nm to, say, 100 nm? Discuss. Assume that series resistance is not a problem.

6.8 The high-frequency $C - V_G$ curve of an MOS capacitor is shown in Fig. P6.8 $C_{FB}/C_{ox} = 0.6$.

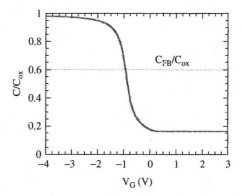

Fig. P6.8

Determine the fixed charge density N_f in units of cm^{-2}. Then by some magical process the fixed charge is removed from half the area of this device, but remains in the other half. The device has area A. For A/2 the fixed charge is the same as the original, for the other A/2 it is zero. Draw the new $C-V_G$ curve. $t_{ox} = 20$ nm, $K_{ox} = 3.9$, $T = 300$ K, $\phi_{MS} = 0$, there are no other oxide charges.

6.9 An MOS capacitor consists of a polycrystalline Si gate, a thick thermally grown oxide, and a p-Si substrate. Flatband voltage measurements as a function of oxide thickness give:

V_{FB} (V)	-1.98	-1.76	-1.59	-1.42	-1.20	-1.05
t_{ox} (μm)	0.3	0.25	0.2	0.15	0.1	0.05

(a) Determine the fixed oxide charge density N_f in units of cm^{-2} and the work function difference ϕ_{MS} in units of V. Assume the fixed charge is all located in the oxide at the SiO_2/Si interface.

(b) Is the gate n^+ or p^+ poly-Si? Why?

(c) Next consider a *positive* mobile charge *uniformly* distributed through the oxide of this device with a volume density of $N_m = 10^{16}$ cm^{-3}. This oxide has the same N_f as in (i). Determine the flatband voltage for $t_{ox} = 0.1$ μm. $K_{ox} = 3.9$.

6.10 The $C - V_G$ curve of an MOS capacitor is measured as curve (A) in Fig. P 6.10. This device has mobile charge uniformly distributed throughout the oxide. Next, a gate voltage is applied and all of the charge drifts to one side of the oxide, giving curve (B). $T = 300$ K, $K_{ox} = 3.9$, $K_s = 11.7$.

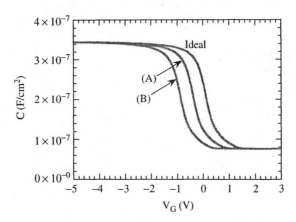

Fig. P6.10

(a) Determine the oxide thickness (in nm) and the doping concentration (in cm^{-3}) (from the flatband capacitance).

(b) Choose one answer for each of the *three* choices and *justify* your answers.

(i) The applied voltage during the mobile ion drift experiment is: ☐ positive ☐ negative

(ii) The mobile ion charge is: □ positive □ negative

(iii) The mobile ions drift to the: □ oxide/gate interface □ oxide/substrate interface

6.11 The sub-threshold $I_D - V_{GS}$ curves of a MOSFET are shown in Fig. P6.11 above *before* and *after* stressing the device. Determine the interface trap density change ΔD_{it} (in cm^{-2} eV^{-1}) induced by the stress. $T = 300$ K, $K_{ox} = 3.9$, $t_{ox} = 10$ nm.

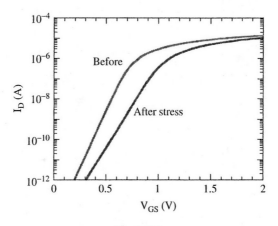

Fig. P6.11

6.12 During charge pumping measurements, electrons and holes are captured by interface states leading to electron-hole pair recombination and electron/hole emission. The charge pumping current is given by

$$I_{cp} = q A f D_{it} \Delta E$$

where ΔE is the energy interval over which electrons/holes are *not* emitted to E_c or E_v. Determine and plot ΔE versus log(f) and log(I_{cp}) versus log(f) for $T = 250, 300, 350$ K over the frequency range $10^4 \le f \le 10^6$ Hz. Use $A = 10^{-6}$ cm^2, $D_{it} = 5 \times 10^{10}$ cm^{-2} eV^{-1}, $\sigma_n = \sigma_p = 10^{-15}$ cm^2, $v_{th} = 10^7 (T/300)^{1/2}$ cm/s, $N_c = 2.5 \times 10^{19} (T/300)^{1.5}$ cm^{-3}, $E_G = 1.12$ eV.

6.13 The electron and hole emission time constants from interface traps are given by

$$\tau_{e,n} = \frac{\exp[(E_c - E_{it})/kT]}{\sigma_n v_{th} N_c} ; \tau_{e,p} = \frac{\exp[(E_{it} - E_v)/kT]}{\sigma_p v_{th} N_v}$$

In the charge pumping method, the interface trap density N_{it} around the central portion of the band gap (ΔE) of a MOSFET is determined ($N_{it} = D_{it} \Delta E$), depending on how many electrons and holes drift back to the source/drain and substrate and how many remain on interface traps to recombine. During the charge pumping measurement, a square wave of frequency f is applied to the gate. Consider two measurements with two different frequencies, $f = f_1$ and $f = f_2$, where $f_1 < f_2$. For which frequency, f_1 or f_2, is a larger portion of the interface traps in the band gap determined? Discuss your answer. Use equations and/or band diagrams if appropriate.

6.14 Draw the band diagram of the MOS capacitor in Fig. P 6.14 biased at $V_G = -0.75$ V, *i.e.*, at the flatband voltage point. This device has a metal gate and $Q_m = Q_f = Q_{ot} = Q_{it} = 0$.

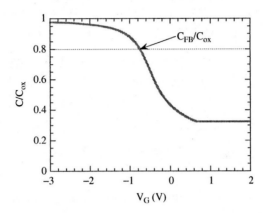

Fig. P6.14

6.15 The $I_D - V_{GS}$ curves of two MOSFETs are shown in Fig. P 6.15. Curve (a) is for an ideal device with $V_{FB} = 0$ and curve (b) is for a device with uniform gate oxide charge. Determine the charge density ρ_{ox} (C/cm^3). $C_{ox} = 10^{-8}$ F/cm^2, $t_{ox} = 10$ nm, $\phi_{MS} = 0$, $Q_f = 0$, $D_{it} = 0$.

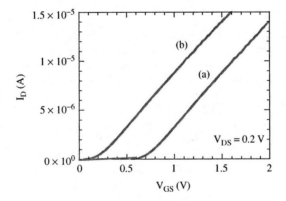

Fig. P6.15

6.16 V_{FB} versus t_{ox} of an MOS capacitor, is shown in Fig. P6.16. Draw and justify the V_{FB} versus t_{ox} plot for an MOS capacitor *qualitatively* for the *same* ϕ_{MS} and Q_f but in addition having a *uniform positive* oxide charge density ρ_{ox} (C/cm^3) throughout the oxide.

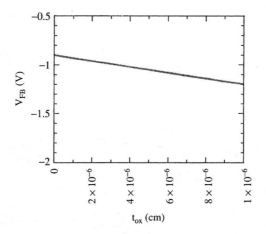

Fig. P6.16

REVIEW QUESTIONS

- Name the four main charges in thermal oxides.
- How is the low-frequency capacitance measured?
- Why do the lf and hf $C-V$ curves differ in inversion?
- What is the flatband voltage and flatband capacitance?
- What is the effect of gate depletion on $C-V$ curves?
- How does *bias-temperature stress* differ from *triangular voltage sweep*?
- Describe charge pumping.
- How is the interface trapped charge measured?
- How does the conductance method work?
- How does the sub-threshold slope yield the interface trap density?
- How does the DC-IV method work?
- Briefly describe two oxide thickness measurement techniques.

7

CARRIER LIFETIMES

7.1 INTRODUCTION

The theory of electron-hole pair (ehp) recombination through recombination centers (also called traps) was put forth in 1952 in the well-known papers by Hall[1] and Shockley and Read[2]. Hall later expanded on his original brief letter.[3] Even though lifetimes and diffusion lengths are routinely measured in the IC industry their measurement and measurement interpretation are frequently misunderstood. Lifetime is one of few parameters giving information about the low defect densities in semiconductors. No other technique can detect defect densities as low as $10^9 - 10^{11}$ cm^{-3} in a simple, contactless room temperature measurement. In principle, there is no lower limit to the defect density determined by lifetime measurements. It is for these reasons that the IC community, largely concerned with unipolar MOS devices in which lifetime plays a minor role, has adopted lifetime measurements as a "process cleanliness monitor." Here, we discuss lifetimes, their dependence on material and device parameters like energy level, injection level, and surfaces, and how lifetimes are measured.

Different measurement methods can give widely differing lifetimes for the same material or device. In most cases, the reasons for these discrepancies are fundamental and are not due to a deficiency of the measurement. The difficulty with defining a lifetime is that we are describing a property of a carrier within the semiconductor rather than the property of the semiconductor itself. Although we usually quote a single numerical value, we are measuring some weighted average of the behavior of carriers influenced by surfaces, interfaces, energy barriers, and the density of carriers besides the properties of the semiconductor material and its temperature.

Lifetimes fall into two primary categories: *recombination lifetimes* and *generation lifetimes*.[4] The concept of recombination lifetime τ_r holds when excess carriers decay

---

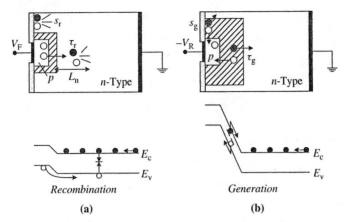

Fig. 7.1 (a) Forward-biased and (b) reverse-biased junction, illustrating the various recombination and generation mechanisms.

as a result of recombination. Generation lifetime τ_g applies when there is a paucity of carriers, as in the space-charge region (scr) of a reverse-biased device and the device tries to attain equilibrium. During recombination an electron-hole pair ceases to exist on average after a time τ_r, illustrated in Fig. 7.1(a). The generation lifetime, by analogy, is the time that it takes on average to generate an ehp, illustrated in Fig. 7.1(b). Thus generation lifetime is a misnomer, since the creation of an ehp is measured and generation time would be more appropriate. Nevertheless, the term "generation lifetime" is commonly accepted.

When these recombination and generation events occur in the bulk, they are characterized by τ_r and τ_g. When they occur at the surface, they are characterized by the *surface recombination velocity* s_r and the *surface generation velocity* s_g, also illustrated in Fig. 7.1. Both bulk and surface recombination or generation occur simultaneously and their separation is sometimes quite difficult. The measured lifetimes are always *effective* lifetimes consisting of bulk and surface components.

Before discussing lifetime measurement techniques, it is instructive to consider τ_r and τ_g in more detail. Those readers not interested in these details can skip these sections and go directly to the measurement methods. The excess ehps may have been generated by photons or particles of energy higher than the band gap or by forward biasing a pn junction. There are more carriers after the stimulus than before, and the excess carriers return to equilibrium by recombination. A detailed derivation of the relevant equations is given in Appendix 7.1.

7.2 RECOMBINATION LIFETIME/SURFACE RECOMBINATION VELOCITY

The bulk recombination rate R depends non-linearly on the departure of the carrier densities from their equilibrium values. We consider a *p*-type semiconductor throughout this chapter and are chiefly concerned with the behavior of the *minority electrons*. Confining ourselves to linear, quadratic, and third order terms, R can be written as

$$R = A(n - n_o) + B(pn - p_o n_o) + C_p(p^2 n - p_o^2 n_o) + C_n(pn^2 - p_o n_o^2) \qquad (7.1)$$

where $n = n_o + \Delta n$, $p = p_o + \Delta p$, n_o, p_o are the equilibrium and Δn, Δp the excess carrier densities. In the absence of trapping, $\Delta n = \Delta p$, allowing Eq. (7.1) to be simplified to

$$R \approx A\Delta n + B(p_o + \Delta n)\Delta n + C_p(p_o^2 + 2p_o\Delta n + \Delta n^2)\Delta n$$
$$+ C_n(n_o^2 + 2n_o\Delta n + \Delta n^2)\Delta n \qquad (7.2)$$

where some terms containing n_o have been dropped because $n_o \ll p_o$ in a p-type material.

The recombination lifetime is defined as

$$\tau_r = \frac{\Delta n}{R} \qquad (7.3)$$

giving

$$\tau_r = \frac{1}{A + B(p_o + \Delta n) + C_p(P_o^2 + 2p_o\Delta n + \Delta n^2) + C_n(n_o^2 + 2n_o\Delta n + \Delta n^2)} \qquad (7.4)$$

Three main recombination mechanisms determine the recombination lifetime: *Shockley-Read-Hall* (SRH) or *multiphonon* recombination characterized by τ_{SRH}, *radiative* recombination characterized by τ_{rad} and *Auger* recombination characterized by τ_{Auger}. The three recombination mechanisms are illustrated in Fig. 7.2. The recombination lifetime τ_r is determined according to the relationship

$$\tau_r = \frac{1}{\tau_{SRH}^{-1} + \tau_{rad}^{-1} + \tau_{Auger}^{-1}} \qquad (7.5)$$

During SRH recombination, electron-hole pairs recombine through deep-level impurities or traps, characterized by the density N_T, energy level E_T, and capture cross-sections σ_n and σ_p for electrons and holes, respectively. The energy liberated during the recombination event is dissipated by lattice vibrations or phonons, illustrated in Fig. 7.2(a). The SRH lifetime is given by[2]

$$\tau_{SRH} = \frac{\tau_p(n_o + n_1 + \Delta n) + \tau_n(p_o + p_1 + \Delta p)}{p_o + n_o + \Delta n} \qquad (7.6)$$

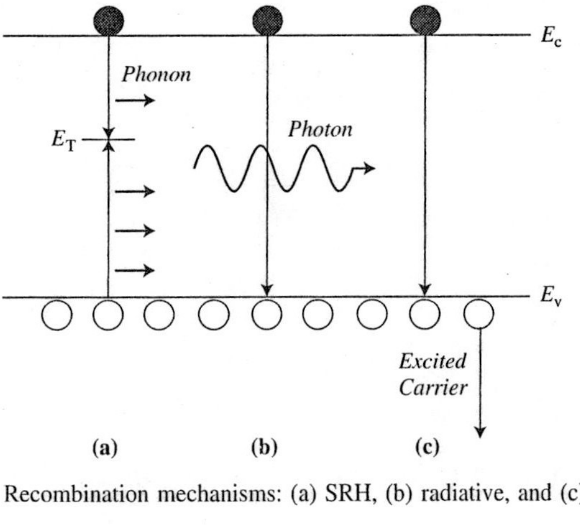

Fig. 7.2 Recombination mechanisms: (a) SRH, (b) radiative, and (c) Auger.

where n_1, p_1, τ_n, and τ_p are defined as

$$n_1 = n_i \exp\left(\frac{E_T - E_i}{kT}\right); p_1 = n_i \exp\left(-\frac{E_T - E_i}{kT}\right) \tag{7.7}$$

$$\tau_p = \frac{1}{\sigma_p v_{th} N_T}; \tau_n = \frac{1}{\sigma_n v_{th} N_T} \tag{7.8}$$

During radiative recombination ehps recombine directly from band to band with the energy carried away by photons in Fig. 7.2(b). The radiative lifetime is[5]

$$\boxed{\tau_{rad} = \frac{1}{B(p_o + n_o + \Delta n)}} \tag{7.9}$$

B is the radiative recombination coefficient. The radiative lifetime is inversely proportional to the carrier density because in band-to-band recombination both electrons and holes must be present simultaneously.

During Auger recombination, illustrated in Fig. 7.2(c), the recombination energy is absorbed by a third carrier and the Auger lifetime is inversely proportional to the carrier density squared. The Auger lifetime is given by

$$\boxed{\begin{aligned}\tau_{Auger} &= \frac{1}{C_p(p_o^2 + 2p_o\Delta n + \Delta n^2) + C_n(n_o^2 + 2n_o\Delta n + \Delta n^2)} \\ &\approx \frac{1}{C_p(p_o^2 + 2p_o\Delta n + \Delta n^2)}\end{aligned}} \tag{7.10}$$

where C_p is the Auger recombination coefficient for a holes and C_n for electrons. Values for radiative and Auger coefficients are given in Table 7.1.

Equations (7.6) to (7.10) simplify for both low-level and high-level injection. Low-level injection holds when the *excess minority* carrier density is low compared to the *equilibrium majority* carrier density, $\Delta n \ll p_o$. Similarly, high-level injection holds when

TABLE 7.1 Recombination Coefficients.

Semiconductor	Temperature (K)	Radiative Recombination Coefficient, B (cm^3/s)	Auger Recombination Coefficient, C (cm^6/s)
Si	300	4.73×10^{-15} [10]	$C_n = 2.8 \times 10^{-31}, C_p = 10^{-31}$ [11 D/S]
Si	300	—	$C_n + C_p = 2\text{–}35 \times 10^{-31}$ [11 B/G]
Si	77	8.01×10^{-14} [10]	—
Ge	300	5.2×10^{-14} [5]	$C_n = 8 \times 10^{-32}, C_p = 2.8 \times 10^{-31}$
GaAs	300	1.7×10^{-10} [8 S/R]	$C_n = 1.6 \times 10^{-29}, C_p = 4.6 \times 10^{-31}$ [6]
GaAs	300	1.3×10^{-10} [8 't Hooft]	$C_n = 5 \times 10^{-30}, C_p = 2 \times 10^{-30}$ [8 S/R]
GaP	300	5.4×10^{-14} [5]	—
InP	300	$1.6\text{–}2 \times 10^{-11}$ [7]	$C_n = 3.7 \times 10^{-31}, C_p = 8.7 \times 10^{-30}$ [6]
InSb	300	4.6×10^{-11} [5]	—
InGaAsP	300	4×10^{-10} [8]	$C_n + C_p = 8 \times 10^{-29}$ [9]

$\Delta n \gg p_o$. The injection level is important during lifetime measurements. The appropriate expressions for *low-level* (ll) and for *high-level* (hl) injection become

$$\tau_{SRH}(ll) \approx \frac{n_1}{p_o}\tau_p + \left(1 + \frac{p_1}{p_o}\right)\tau_n \approx \tau_n; \tau_{SRH}(hl) \approx \tau_p + \tau_n \qquad (7.11)$$

where the second approximation in the $\tau_{SRH}(ll)$ expression holds when $n_1 \ll p_o$ and $p_1 \ll p_o$. A more detailed discussion of injection level is given by Schroder.[12]

$$\tau_{rad}(ll) = \frac{1}{Bp_o}; \tau_{rad}(hl) = \frac{1}{B\Delta n} \qquad (7.12)$$

$$\tau_{Auger}(ll) = \frac{1}{C_p p_o^2}; \tau_{Auger}(hl) = \frac{1}{(C_p + C_n)\Delta n^2} \qquad (7.13)$$

The Si recombination lifetimes according to Eq. (7.5) are plotted in Fig. 7.3. At high carrier densities, the lifetime is controlled by Auger recombination and at low densities by SRH recombination. Auger recombination has the characteristic $1/n^2$ dependence. The high carrier densities may be due to high doping densities or high excess carrier densities. Whereas SRH recombination is controlled by the cleanliness of the material, Auger recombination is an intrinsic property of the semiconductor. Radiative recombination plays almost no role in Si except for very high lifetime substrates (see τ_{rad} in Fig. 7.3), but is important in direct band gap semiconductors like GaAs. The data for *n*-Si in Fig. 7.3 can be reasonably well fitted with $C_n = 2 \times 10^{-31}$ cm^6/s. However, the fit is not perfect and detailed Auger considerations suggest different Auger coefficients.[13]

The *bulk SRH recombination rate* is given by[2]

$$R = \frac{\sigma_n \sigma_p v_{th} N_T (pn - n_i^2)}{\sigma_n(n + n_1) + \sigma_p(p + p_1)} = \frac{(pn - n_i^2)}{\tau_p(n + n_1) + \tau_n(p + P_1)} \qquad (7.14)$$

leading to the SRH lifetime expression (7.6). The *surface SRH recombination rate* is

$$R_s = \frac{\sigma_{ns}\sigma_{ps}v_{th}N_{it}(p_s n_s - n_i^2)}{\sigma_{ns}(n_s + n_{1s}) + \sigma_{ps}(p_s + p_{1s})} = \frac{s_n s_p (p_s n_s - n_i^2)}{s_n(n_s + n_{1s}) + s_p(p_s + p_{1s})} \qquad (7.15)$$

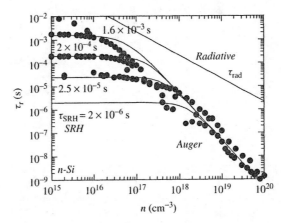

Fig. 7.3 Recombination lifetime versus majority carrier density for *n*-Si with $C_n = 2 \times 10^{-31}$ cm^6/s and $B = 4.73 \times 10^{-15}$ cm^3/s. More detailed Auger considerations suggest $C_n = 1.8 \times 10^{-24} n^{1.65}$.[13] Data from ref. 11 and 13.

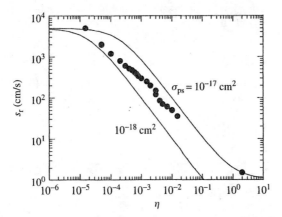

Fig. 7.4 s_r versus injection level η as a function of σ_{ps} for $N_{it} = 10^{10}$ cm^{-2}, $p_{os} = 10^{16}$ cm^{-3}, $E_{Ts} = 0.4$ eV, $\sigma_{ns} = 5 \times 10^{-14}$ cm^2. Data from ref. 15.

where

$$s_n = \sigma_{ns} v_{th} N_{it}; s_p = \sigma_{ps} v_{th} N_{it} \tag{7.16}$$

The subscript "s" refers to the appropriate quantity at the surface; p_s and n_s are the hole and electron densities (cm^{-3}) at the surface. The interface trap density N_{it} (cm^{-2}) is assumed constant in Eq. (7.15). If not constant, the interface trap density D_{it} (cm^{-2} eV^{-1}) must be integrated over energy with N_{it} in these equations given by $N_{it} \approx kT D_{it}$.[14]

The surface recombination velocity s_r is

$$s_r = \frac{R_s}{\Delta n_s} \tag{7.17}$$

From Eq. (7.15)

$$s_r = \frac{s_n s_p (p_{os} + n_{os} + \Delta n_s)}{s_n (n_{os} + n_{1s} + \Delta n_s) + s_p (p_{os} + p_{1s} + \Delta p_s)} \tag{7.18}$$

The surface recombination velocity for low-level and high-level injection becomes

$$s_r(ll) = \frac{s_n s_p}{s_n(n_{1s}/p_{os}) + s_p(1 + p_{1s}/p_{os})} \approx s_n; s_r(hl) = \frac{s_n s_p}{s_n + s_p} \tag{7.19}$$

s_r depends strongly on injection level for the SiO$_2$/Si interface as shown in Fig. 7.4.

7.3 GENERATION LIFETIME/SURFACE GENERATION VELOCITY

Each of the recombination processes of Fig. 7.2 has a generation counterpart. The inverse of multiphonon recombination is thermal ehp generation in Fig. 7.1(b). The inverse of radiative and Auger recombination are optical and impact ionization generation. Optical generation is negligible for a device in the dark and with negligible blackbody radiation from its surroundings. Impact ionization is usually considered to be negligible for devices biased sufficiently below their breakdown voltage. However, impact ionization at low ionization rates can occur at low voltages, and care must be taken to eliminate this generation mechanism during τ_g measurements.

From the SRH recombination rate expression in Eq. (7.14), it is obvious that generation dominates for $pn < n_i^2$. Furthermore the smaller the pn product, the higher is the generation rate. R becomes negative and is then designated as the *bulk generation rate G*

$$G = -R = \frac{n_i^2}{\tau_p n_1 + \tau_n p_1} = \frac{n_i}{\tau_g} \tag{7.20}$$

for $pn \approx 0$ with

$$\tau_g = \tau_p \exp\left(\frac{E_T - E_i}{kT}\right) + \tau_n \exp\left(-\frac{E_T - E_i}{kT}\right) \tag{7.21}$$

The condition pn $\rightarrow$ 0 is approximated in the scr of a reverse-biased junction.

The quantity τ_g, defined in Eq. (7.21), is the *generation lifetime*[16] that depends inversely on the impurity density and on the capture cross-section for electrons and holes, just as recombination does. It also depends exponentially on the energy level E_T. The generation lifetime can be quite high if E_T does not coincide with E_i. Generally, τ_g is higher than τ_r, at least for Si devices, where detailed comparisons have been made and $\tau_g \approx (50-100)\tau_r$.[12, 16]

When $p_s n_s < n_i^2$ at the surface, we find from Eq. (7.15), the *surface generation rate*

$$G_s = -R_s = \frac{s_n s_p n_i^2}{s_n n_{1s} + s_p p_{1s}} = n_i s_g \tag{7.22}$$

where s_g is the *surface generation velocity*, sometimes designated as s_o (see note in Grove[17]), given by

$$s_g = \frac{s_n s_p}{s_n \exp((E_{it} - E_i)/kT) + s_p \exp(-(E_{it} - E_i)/kT)} \tag{7.23}$$

For $E_{it} \neq E_i$, we find $s_r > s_g$ from Eqs. (7.18) and (7.23).

7.4 RECOMBINATION LIFETIME—OPTICAL MEASUREMENTS

Before discussing lifetime characterization techniques, we will briefly give the relevant equations for the common optical methods. More details are given Appendix 7.1. Consider a *p*-type semiconductor with light incident on the sample. The light may be steady state or transient. The continuity equation for uniform ehp generation and zero surface recombination is[18]

$$\frac{\partial \Delta n(t)}{\partial t} = G - R = G - \frac{\Delta n(t)}{\tau_{eff}} \tag{7.24}$$

where $\Delta n(t)$ is the time dependent excess minority carrier density, G the ehp generation rate, and τ_{eff} the effective lifetime. Solving for τ_{eff} gives

$$\tau_{eff}(\Delta n) = \frac{\Delta n(t)}{G(t) - d\Delta n(t)/dt} \tag{7.25}$$

In the *transient photoconductance decay* (PCD) method, with $G(t) \ll d\Delta n(t)/dt$

$$\tau_{eff}(\Delta n) = -\frac{\Delta n(t)}{d\Delta n(t)/dt} \tag{7.26}$$

In the *steady-state* method, with $G(t) \gg d\Delta n(t)/dt$

$$\tau_{eff}(\Delta n) = \frac{\Delta n}{G} \tag{7.27}$$

and in the *quasi-steady-state photoconductance* (QSSPC) method, Eq. (7.25) obtains. Both Δn and G need to be known in the steady-state and QSSPC methods to determine the effective lifetime.

The excess carrier density decay for low level injection is given by $\Delta n(t) = \Delta n(0)\exp(-t/\tau_{eff})$ where τ_{eff} is

$$\frac{1}{\tau_{eff}} = \frac{1}{\tau_B} + D\beta^2 \tag{7.28}$$

with β found from the relationship

$$\tan\left(\frac{\beta d}{2}\right) = \frac{s_r}{\beta D} \tag{7.29}$$

where τ_B is the bulk recombination lifetime, D the minority carrier diffusion constant under low injection level and the ambipolar diffusion constant under high injection level, s_r the surface recombination velocity, and d the sample thickness. Equation (7.28) holds for any optical absorption depth provided the excess carrier density has ample time to distribute uniformly, *i.e.*, $d \ll (Dt)^{1/2}$. The effective lifetime of Eq. (7.28) is plotted in Fig. 7.5 versus d as a function of s_r, showing the dependence on d and s_r. For thin samples, τ_{eff} no longer bears any resemblance to τ_B, the bulk lifetime, and is dominated by surface recombination. The surface recombination velocity must be known to determine τ_B unambiguously unless the sample is sufficiently thick. Although the surface recombination velocity of a sample is generally not known, by providing the sample with high s_r, by

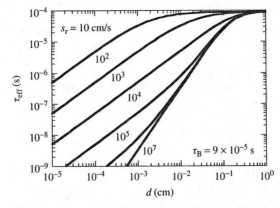

Fig. 7.5 Effective lifetime versus wafer thickness as a function of surface recombination velocity. $D = 30 \text{ cm}^2/\text{s}$.

sandblasting for example, it is possible to determine τ_B directly. However, the sample must be extraordinarily thick. Equation (7.28) can be written as

$$\frac{1}{\tau_{eff}} = \frac{1}{\tau_B} + \frac{1}{\tau_S} \tag{7.30}$$

where τ_S is the surface lifetime.

Two limiting cases are of particular interest: $s_r \to 0$ gives $\tan(\beta d/2) \approx \beta d/2$ and $s_r \to \infty$ gives $\tan(\beta d/2) \approx \infty$ or $\beta d/2 \approx \pi/2$, making the surface lifetime

$$\tau_S(s_r \to 0) = \frac{d}{2s_r}; \tau_S(s_r \to \infty) = \frac{d^2}{\pi^2 D} \tag{7.31}$$

For $s_r \to 0$, a plot of $1/\tau_{eff}$ versus $1/d$ has a slope of $2s_r$ and an intercept of $1/\tau_B$, allowing both s_r and τ_B to be determined. For $s_r \to \infty$, a plot of $1/\tau_{eff}$ versus $1/d^2$ has a slope of $\pi^2 D$ and an intercept of $1/\tau_B$. Both examples are illustrated in Fig. 7.6. The approximation $\tau_S = d/2s_r$ holds for $s_r < D/4d$.

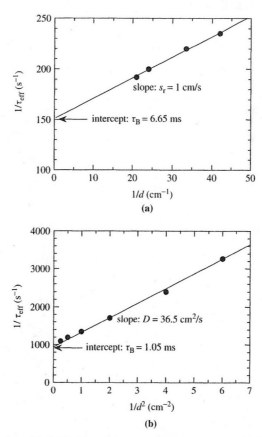

Fig. 7.6 Determination of bulk lifetime, surface recombination velocity, and diffusion coefficient from lifetime measurements. Data from ref. 19.

Equations (7.28)–(7.31) hold for samples with one dimension much smaller than the other two dimensions, for example, a wafer. For samples with none of the three dimensions very large, Eq. (7.30) becomes for $s_r \rightarrow \infty$

$$\frac{1}{\tau_{eff}} = \frac{1}{\tau_B} + \pi^2 D \left(\frac{1}{a^2} + \frac{1}{b^2} + \frac{1}{c^2} \right) \tag{7.32}$$

where a, b, and c are the sample dimensions. It is recommended that the sample surfaces have high surface recombination velocities, by sandblasting the sample surfaces, for example.[20] The recommended dimensions and the maximum bulk lifetimes that can be determined through Eq. (7.32) for Si samples are given in Table 7.2.

The time dependence of the carrier decay after cessation of an optical pulse is a complicated function, as discussed in Appendix 7.1.[21-22] We show in Fig. 7.7 calculated excess carrier decay curves with the time dependence

$$\Delta n(t) = \Delta n(0) \exp \left(-\frac{t}{\tau_{eff}} \right) \tag{7.33}$$

According to Eq. (7.30) the effective lifetime is

$$\frac{1}{\tau_{eff}} = \frac{1}{\tau_B} + \frac{1}{\tau_S} = \frac{1}{\tau_B} + D\beta^2 \tag{7.34}$$

TABLE 7.2   Recommended Dimensions for PCD Samples and Maximum Bulk Lifetimes for Si.

Sample Length (cm)	Sample Width × Height (cm × cm)	Maximum τ_B (µs) n-Si	Maximum τ_B (µs) p-Si
1.5	0.25 × 0.25	240	90
2.5	0.5 × 0.5	950	350
2.5	1 × 1	3600	1340

Source: ASTM Standard F28. Ref. 20.

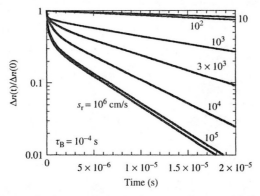

Fig. 7.7   Calculated normalized excess carrier density versus time as a function of surface recombination velocity. $d = 400$ µm, $\alpha = 292$ cm^{-1}.

where β is determined from Eq. (7.28), which has a series of solutions for $\beta d/2$ in the ranges 0 to $\pi/2$, π to $3\pi/2$, 2π to $5\pi/2$, and so on. For each combination of s_r, d, and D, we find a series of β values, giving a series of τ_S. One way to solve Eq. (7.29) is to write it as

$$\frac{\beta_m d}{2} - (m-1)\pi = \arctan\left(\frac{s_r}{\beta_m D}\right) \tag{7.35}$$

where $m = 1, 2, 3, \ldots$ and solve iteratively for β_m. The higher order terms decay much more rapidly than the first term. Hence, the semi-log curves are non-linear for short times and then become linear for longer times. From Eq. (7.33), the slope of this plot is

$$Slope = \frac{d\ln(\Delta n(t))}{dt} = \frac{\ln(10)d\log(\Delta n(t))}{dt} = -\frac{1}{\tau_{eff}} \tag{7.36}$$

Taking the slope in the linear portion of the plot gives τ_{eff}. To be safe, one should wait for the transient to decay to about half of its maximum value before measuring the time constant.

7.4.1 Photoconductance Decay (PCD)

The *photoconductance decay* lifetime characterization technique was proposed in 1955[23] and has become one of the most common lifetime measurement techniques. As the name implies, ehps are created by optical excitation, and their decay is monitored as a function of time following the cessation of the excitation. Other excitation means such as high-energy electrons and gamma rays can also be used. The samples may either be contacted with the current being monitored or the measurement can be contactless.

In PCD, the conductivity σ

$$\sigma = q(\mu_n n + \mu_p p) \tag{7.37}$$

is monitored as a function of time. $n = n_o + \Delta n$, $p = p_o + \Delta p$ and we assume both equilibrium and excess carriers to have identical mobilities. This is true under low-level injection when Δn and Δp are small compared to the equilibrium majority carrier density, but not for high optical excitation, because carrier-carrier scattering reduces the mobilities.

In some PCD methods the time-dependent excess carrier density is measured directly; in others indirectly. For insignificant trapping, $\Delta n = \Delta p$, and the excess carrier density is related to the conductivity by

$$\Delta n = \frac{\Delta\sigma}{q(\mu_n + \mu_p)} \tag{7.38}$$

A measure of $\Delta\sigma$ is a measure of Δn, provided the mobilities are constant during the measurement.

A schematic measurement circuit for PC decay is shown in Fig. 7.8. We follow Ryvkin for the derivation of the appropriate equations.[24] For a sample with dark resistance r_{dk} and steady-state photoresistance r_{ph}, the output voltage change between the dark and the illuminated sample is

$$\Delta V = (i_{ph} - i_{dk})R \tag{7.39}$$

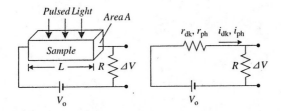

Fig. 7.8   Schematic diagram for contact photoconductance decay measurements.

where i_{ph}, i_{dk} are the photocurrent and the dark current. With

$$\Delta g = g_{ph} - g_{dk} = \frac{1}{r_{ph}} - \frac{1}{r_{dk}} \tag{7.40}$$

Equation (7.39) becomes

$$\Delta V = \frac{r_{dk}^2 R \Delta g V_o}{(R + r_{dk})(R + r_{dk} + R r_{dk} \Delta g)} \tag{7.41}$$

where $\Delta g = \Delta \sigma A / L$. According to Eq. (7.41), there is no simple relationship between the time dependence of the measured voltage and the time dependence of the excess carrier density.

There are two main versions of the technique in Fig. 7.8: the *constant voltage* method and the *constant current* method. The load resistor R is chosen to be small compared to the sample resistance in the *constant voltage* method, and Eq. (7.41) becomes

$$\Delta V \approx \frac{R \Delta g V_o}{1 + R \Delta g} \approx R \Delta g V_o \left(1 - \frac{\Delta V}{V_o}\right) \tag{7.42}$$

For low-level excitation ($\Delta g R \ll 1$ or $\Delta V \ll V_o$) $\Delta V \sim \Delta g \sim \Delta n$; the voltage decay is proportional to the excess carrier density. For the *constant current* case, R is very large, and

$$\Delta V \approx \frac{(r_{dk}^2 / R) \Delta g V_o}{1 + r_{dk} \Delta g} \approx r_{dk} \Delta g V_o \left(\frac{r_{dk}}{R} - \frac{\Delta V}{V_o}\right) \tag{7.43}$$

For $r_{dk} \Delta g \ll 1$ or $\Delta V / V_o \ll r_{dk/R}$, $\Delta V \sim \Delta g \sim \Delta n$ again.

For the measurements in Fig. 7.8, the contacts should not inject minority carriers and the illumination should be restricted to the non-contacted part of the sample to avoid contact effects or minority carrier sweep-out. The electric field in the sample should be held to a value $\mathscr{E} = 0.3/(\mu \tau_r)^{1/2}$, where μ is the minority carrier mobility.[20] The excitation light should penetrate the sample. A $\lambda = 1.06$ μm laser is suitable for Si. One can also pass the light through a filter made of the semiconductor to be measured to remove the higher energy light. The carrier decay can also be monitored without sample contacts, allowing for a fast, non-destructive measure of $\Delta n(t)$, using the rf bridge circuit of Fig. 7.9(a)[25–26] or the microwave circuit of Fig. 7.9(b) in the reflected or transmitted microwave mode.[27]

Low surface recombination velocities can be achieved by treating the surface in one of several ways. Oxidized Si surfaces have been reported with $s_r \approx 20$ cm/s.[28] Immersing

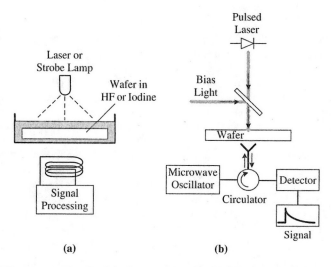

Fig. 7.9 PCD measurement schematic for contactless (a) rf bridge and (b) microwave reflectance measurements.

a bare Si sample in one of several solutions can reduce s_r even below this value. For example, immersion in HF has given $s_r = 0.25$ cm/s for high level injection.[29] Immersing the sample in iodine in methanol has given $s_r \approx 4$ cm/s.[22] Low temperature silicon nitride deposited in a remote plasma CVD system has yielded $s_r \approx 4–5$ cm/s.[30] The contactless PCD technique has been extended to lifetime measurements on GaAs by using a Q-switched Nd:YAG laser as the light source.[31] By using inorganic sulfides as passivating layers, surface recombination velocities as low as 1000 cm/s were obtained on GaAs samples.

In the *microwave reflection method* of Fig. 7.9(b),[32-33] the photoconductivity is monitored by microwave reflection or transmission. Microwaves at ~10 GHz frequency are directed onto the wafer through a circulator to separate the reflected from the incident microwave signal. The microwaves are reflected from the wafer, detected, amplified, and displayed. In the small perturbation range, the relative change in reflected microwave power $\Delta P/P$ is proportional to the incremental wafer conductivity $\Delta\sigma$[33]

$$\frac{\Delta P}{P} = C\Delta\sigma \tag{7.44}$$

where C is a constant. The microwaves penetrate a skin depth into the sample. Typical skin depths in Si at 10 GHz are 350 μm for $\rho = 0.5$ ohm-cm to 2200 μm for $\rho = 10$ ohm-cm. Skin depth is discussed in Section 1.5.1. Consequently, a good part of the wafer thickness is sampled by the microwaves and the microwave reflected signal is characteristic of the bulk carrier density. The lower limit of τ_r that can be determined depends on the wafer resistivity. Lifetimes as low as 100 ns have been measured.

If a resonant microwave cavity is used, it is important that the signal decay is indeed that of the photoconductor and not that of the measurement apparatus. When the cavity is off resonance the system response is very fast, while an on-resonance cavity results in a large increase in the system fall time.[34]

7.4.2   Quasi-Steady-State Photoconductance (QSSPC)

In the QSSPC method the sample is illuminated with a flash lamp with a decay time constant of several ms and an illumination area of several cm^2.[35] Due to the slow decay time, the sample is under quasi steady-state conditions during the measurement as the light intensity varies from its maximum to zero. The steady-state condition is maintained as long as the flash lamp time constant is longer than the effective carrier lifetime. The time-varying photoconductance is detected by inductive coupling. The excess carrier density is calculated from the photoconductance signal. The generation rate, required in Eq. (7.25), is determined from the light intensity measured with a calibrated detector. Semiconductors absorb only a fraction of the incident photons, depending on the reflectivity of the front and back surfaces, possible faceting of those surfaces, and the thickness of the wafer. The value of the absorption fraction for a polished, bare silicon wafer is $f \approx 0.6$. If the wafer has an optimized antireflection coating, $f \approx 0.9$, while a textured wafer with antireflection coating can approach $f \approx 1$.[36] The generation rate per unit volume G can then be evaluated from the incident photon flux and the wafer thickness, according to

$$G = \frac{f\,\Phi}{d} \tag{7.45}$$

where Φ is the photon flux density and d the sample thickness.

Assuming the flash lamp light decay is exponential in time, the generation rate is

$$G(t) = 0 \ \text{ for } t \le 0; G_o \exp(-t/\tau_{flash}) \ \text{ for } t > 0 \tag{7.46}$$

and the solution of Eq. (7.25) is[18]

$$\Delta n(t) = \frac{\tau_{eff}}{1 - \tau_{eff}/\tau_{flash}} G_o \left(\exp\left(-\frac{t}{\tau_{flash}}\right) - \exp\left(-\frac{t}{\tau_{eff}}\right) \right) \tag{7.47}$$

For $\tau_{eff} < \tau_{flash}$, the sample is in quasi steady-state during the measurement. Hence, the flash lamp decay time must be sufficiently long for the QSSCP measurement to be valid. An example QSSCP plot is shown in Fig. 7.10, illustrating the increasing SRH lifetime with injection level followed by lifetime decrease due to Auger recombination.

7.4.3   Short-Circuit Current/Open-Circuit Voltage Decay (SCCD/OCVD)

The recombination lifetime can be determined by monitoring the pn junction voltage, current, and short circuit current decay after optical generation of excess carriers.[38-40] The combination *open-circuit voltage decay/short-circuit current decay* method was developed for characterizing the lifetime, diffusion length, and surface recombination velocity of solar cells in which the base width is typically on the order of or less than the minority carrier diffusion length, making the determination of these parameters difficult. In contrast to most other methods in which only a single parameter is measured, two measurements - the short-circuit current and the open-circuit voltage - are necessary to determine τ_r and s_r.

The theory is based on a solution of the minority carrier differential equation [Eq. A7.13] subject to the boundary conditions[40]

$$\frac{1}{\Delta n(x, t)} \frac{\partial \Delta n(x, t)}{\partial x} = -\frac{s_r}{D_n} \ \text{ for } x = d \tag{7.48a}$$

$$\Delta n(0, t) = 0 \tag{7.48b}$$

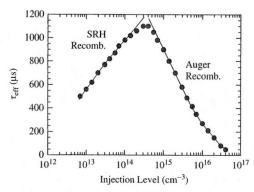

Fig. 7.10 Effective recombination lifetime versus injection carrier density obtained with the QSSPC technique. Adapted from ref. 37.

for the short-circuit current, and

$$\frac{\partial \Delta n(x, t)}{\partial x} = 0 \ \ for \ x = 0 \tag{7.49}$$

for the open-circuit voltage method.

So far we have only concerned ourselves with substrate minority carrier recombination in n^+p junctions. There is, of course, also minority carrier recombination in the scr and in the heavily doped n^+ emitter. The minority carriers are swept out of the scr by the electric field in times on the order of 10^{-11} s under short-circuit conditions. The emitter lifetime is generally much lower than the base lifetime, and emitter contributions play a role only during the early phase of the current decay.[41] Emitter recombination causes carriers from the base to be injected into the emitter where they recombine at a faster rate. However, the voltage decay is determined by the base recombination parameters for long times.[42] If the asymptotic decay rate is measured after the initial transient, then a decay time, representative of base recombination, is observed.[41]

The current decay is found to be exponential with time, with the time constant determined by the time dependence of the excess carrier density. The voltage decay can be significantly influenced by the junction RC time constant, which can be very large for large-area junction devices. This effect is reduced by measuring the small-signal voltage decay with a steady-state bias light to reduce R.[43] One might expect the current and voltage decays to be identical for devices with the base much thicker than the minority carrier diffusion length because s_r is no longer important. This is indeed the case. Both have the asymptotic time dependence

$$I_{sc}, V_{oc} \sim \frac{\exp(-t/\tau_B)}{\sqrt{t}} \tag{7.50}$$

This method is one of few allowing *both* the lifetime and surface recombination velocity at the back surface to be determined, by measuring the current and voltage decays of the same device. Being a transient technique, it is subject to higher-order decay time constants and possible trapping. These potential sources of error are considerably reduced by measuring the time constants asymptotically toward the end of the decay and using a bias light.

7.4.4   Photoluminescence Decay (PLD)

Photoluminescence decay is another method of monitoring the time dependence of excess carriers. Excess carriers are generated by a short pulse of incident photons with energy $hv > E_G$. The excess carrier density is monitored by detecting the time dependence of the light emitted by the recombining electron-hole pairs. The PL signal is higher for efficient light-emitting direct band gap semiconductors, *e.g.*, GaAs or InP, than for indirect band gap semiconductors, *e.g.*, Si or Ge, for which photoluminescence is quite inefficient. Instead of optical excitation, electron-beam excitation can also been used in *transient cathodoluminescence*.

The excess carrier density and time decay expressions are those discussed in Section 7.4.1. We expect PL decay to follow those considerations, except that the PL intensity is given by

$$\Phi_{PL}(t) = K \int_0^d \Delta n(x, t) \, dx \qquad (7.51)$$

where K is a constant accounting for the solid angle over which the light is emitted and for the reflectivity for the radiation emitted from the sample and d is the sample thickness.

A complication arises if self-absorption takes place, where some of the photons generated by the recombination radiation are absorbed by the semiconductor. Once absorbed they can create ehps. The lifetime expression becomes[44]

$$\frac{1}{\tau_{PL}} = \frac{1}{\tau_{non-rad}} + \frac{1}{\tau_S} + \frac{1}{\gamma \tau_{rad}} \qquad (7.52)$$

where $\tau_{non-rad}$, τ_{rad} and τ_s are the non-radiative, the radiative, and the surface lifetimes; γ is the photon recycling factor. Self-absorption is not important for indirect band-gap semiconductors since the optical absorption coefficient is low for near band-gap photons, but it can be important for direct band-gap semiconductors. A discussion of PL lifetime determination is given in ref. 45. PL decay has been used to map the lifetime in Si power devices by scanning the excitation beam across the device.[46]

7.4.5   Surface Photovoltage (SPV)

The steady-state *surface photovoltage* method determines the *minority carrier diffusion length* using optical excitation. The diffusion length is related to the recombination lifetime through the relation $L_n = (D\tau_r)^{1/2}$. SPV is an attractive technique, because (1) it is non-destructive and contactless, (2) sample preparation is simple (no contacts, junctions, or high temperature processing required), (3) it is a steady-state method relatively immune to the slow trapping and detrapping effects that can influence transient measurements, and (4) the equipment is commercially available.

The SPV technique was first described in 1957[47] to determine diffusion lengths in Si[48-49] and GaAs.[49] The sample is assumed to be homogeneous and of thickness d in Fig. 7.11. One surface is chemically treated to induce a surface space-charge region (scr) of width W. The scr is the result of surface charges, not due to a bias voltage. The surface with the induced scr is uniformly illuminated by chopped monochromatic light of energy higher than the band gap, with the back surface kept in the dark. The light is chopped to enhance the signal/noise ratio using lock-in techniques. The wavelength is varied during the measurement. Some of the optically generated minority carriers diffuse toward the illuminated surface to be collected by the scr, establishing a surface potential or surface

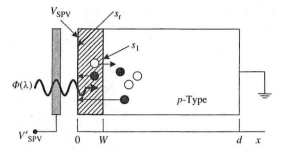

Fig. 7.11 Sample cross-section for SPV measurements. The optically transparent, electrically conducting contact to the left of the sample allows light to reach the sample and the voltage to be measured.

photovoltage voltage V_{SPV} relative to the grounded back surface. V_{SPV} is proportional to the excess minority carrier density $\Delta n(W)$ at the edge of the scr. The precise relationship between $\Delta n(W)$ and V_{SPV} need not be known, but it must be a monotonic function. Light reaching the back surface produces an undesirable SPV signal that can be detected by its large amplitude, by a reversal in signal polarity over the SPV wavelength range, or by a signal decrease with increasing illumination at the longer wavelengths.

The excess carrier density through the wafer for low-level injection is given by Eq. (A7.4). In principle, it is possible to extract the diffusion length L_n from that expression for arbitrary W, d, and α. In practice, several constraints are imposed on the system to simplify data extraction. The undepleted wafer should be much thicker than the diffusion length and the scr width should be small compared to L_n. The absorption coefficient should be sufficiently low for $\alpha W \ll 1$, but sufficiently high for $\alpha(d - W) \gg 1$. The light diameter should be large compared to the sample thickness, allowing a one-dimensional analysis and low-level injection should prevail. The assumptions

$$d - W \geq 4L_n; W \ll L_n; \alpha W \ll 1; \alpha(d - W) \gg 1; \Delta n \ll p_o \qquad (7.53)$$

allow Eq. (A7.4) to be reduced to

$$\Delta n(W) \approx \frac{(1 - R)\Phi}{(s_1 + D_n/L_n)} \frac{\alpha L_n}{(1 + \alpha L_n)} \qquad (7.54)$$

The excess carrier density at $x = W$ is related to the surface photovoltage by

$$\Delta n(W) = n_{po}\left(\exp\left(\frac{qV_{SPV}}{kT}\right) - 1\right) \approx n_{po}\frac{qV_{SPV}}{kT} \text{ for } V_{SPV} \ll \frac{kT}{q} \qquad (7.55)$$

giving

$$V_{SPV} = \frac{(kT/q)(1 - R)\Phi L_n}{n_{po}(s_1 + D_n/L_n)(L_n + 1/\alpha)} \qquad (7.56)$$

V_{SPV} is proportional to Δn for $V_{SPV} < 0.5kT/q$. Typical surface photovoltages are in the low millivolt range, ensuring a linear relationship. s_1 is the surface recombination velocity at $x = W$, not at the surface, where s_r is the surface recombination velocity, as illustrated in Fig. 7.11.

During SPV measurements, D_n and L_n are assumed to be constant. Furthermore over a restricted wavelength range the reflectivity R can also be considered constant. The surface recombination velocity s_1 is usually unknown. However, if $\Delta n(W)$ is held constant during the measurement, the surface potential is also constant, and s_1 can be considered reasonably constant. This leaves α and Φ as the only variables. There are two SPV implementations: (1) *constant surface photovoltage* and (2) *constant photon flux density.* In method (1), $V_{SPV} = constant$ implies $\Delta n(W)$ is constant. A series of different wavelengths is selected during the measurement with each wavelength providing a different α. The photon flux density Φ is adjusted for each wavelength to hold V_{SPV} constant, allowing Eq. (7.56) to be written as

$$\Phi = \frac{n_{po}(s_1 + D_n/L_n)(L_n + 1/\alpha)}{(kT/q)(1 - R)L_n}V_{SPV} = C_1\left(L_n + \frac{1}{\alpha}\right) \qquad (7.57)$$

where C_1 is a constant.

Then Φ is plotted against $1/\alpha$ for constant V_{SPV}. The result is a line whose extrapolated intercept on the negative $1/\alpha$ axis ($\Phi = 0$) is the minority carrier diffusion length L_n, shown in Fig. 7.12(a). The slope of such a plot is C_1 which contains the surface

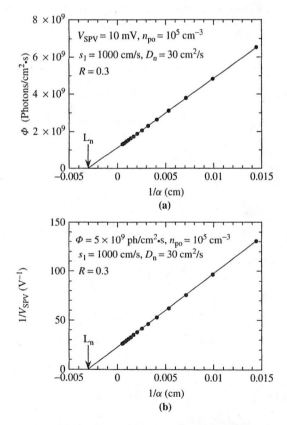

Fig. 7.12   (a) Constant voltage, (b) constant photon flux density SPV plots for Si samples.

recombination velocity s_1. While it is difficult to extract s_1 from all the other parameters contained in C_1, it is possible to observe changes in s_1 by comparing SPV plots before and after a process that changes surface recombination.

For the constant photon flux density implementation, we write Eq. (7.56) as

$$\frac{1}{V_{SPV}} = \frac{n_{po}(s_1 + D_n/L_n)(L_n + 1/\alpha)}{(kT/q)(1 - R)\Phi L_n} = C_2\left(L_n + \frac{1}{\alpha}\right) \qquad (7.58)$$

where C_2 is a constant. A plot of $1/V_{SPV}$ versus $1/\alpha$ gives L_n as illustrated in Fig. 7.12(b). V_{SPV} changes during the measurement, hence surface recombination may vary during the measurement.

The Φ versus $1/\alpha$ plot is a straight line for well-behaved samples. A detailed theoretical study has shown the constant surface photovoltage method to give correct results even considering recombination in the space-charge region.[50] A detailed theoretical and experimental comparison of the PCD and SPV methods has shown the lifetimes determined by these techniques to be identical, provided one considers effects such as surface recombination, sample thickness, and so on.[51]

Exercise 7.1

Problem: How can *iron in Si* be detected with lifetime/diffusion length measurements?

Solution: Since $\tau_{SRH} \sim 1/N_T$, it should be possible to determine N_T by measuring τ_{SRH}. Further, since $L_n \sim \tau_{SRH}^{1/2}$, one should be able to determine N_T from minority carrier diffusion length measurements also. Some impurities in Si have unique characteristics, *e.g.*, iron forms pairs with boron in *p*-type Si. For a Fe-contaminated, *B*-doped Si wafer at room temperature, the iron forms *Fe-B* pairs. Upon heating at 200°C for a few minutes or illuminating the device (>0.1 W/cm^2 light intensity), the *Fe-B* pairs dissociate into interstitial iron (Fe_i) and substitutional *B*. The recombination properties of Fe_i differ from those of *Fe-B*, as shown by the effective diffusion lengths in Fig. E7.1(a). By measuring the diffusion length or lifetime before ($L_{n,i}$, $\tau_{eff,i}$) and after ($L_{n,f}$, $\tau_{eff,f}$) Fe-B pair dissociation, N_{Fe} is

$$N_{Fe} = 1.06 \times 10^{16}\left(\frac{1}{L_{n,f}^2} - \frac{1}{L_{n,i}^2}\right) = C\left(\frac{1}{\tau_{eff,f}} - \frac{1}{\tau_{eff,i}}\right) \text{ [cm}^{-3}]$$

with diffusion lengths in μm and lifetimes in μs. The diffusion lengths as a function of Fe density for a range of N_{Fe} are shown in Fig. E7.1(b). The prefactor, usually assumed as 1.06×10^{16} μm^2/cm^3, varies from 2.5×10^{16} μm^2/cm^3 at $N_B = 10^{13}$ cm^{-3} to 7.5×10^{15} μm^2/cm^3 at $N_B = 10^{17}$ cm^{-3} (D. H. Macdonald, L. J. Geerligs, and A. Azzizi, "Iron Detection in Crystalline Silicon by Carrier Lifetime Measurements for Arbitrary Injection and Doping," *J. Appl. Phys.* **95**, 1021–1028, Feb. 2004).

The measurement has some restrictions. The diffusion lengths must be measured under low-injection conditions. The most reliable technique for this is SPV, since it operates in true low injection. PCD and QSSPC suffer from reduced sensitivity at low-injection and are also affected by minority carrier trapping at low injection causing a majority carrier excess, which distorts the photoconductance that is due to minority and majority carriers. Voltage-based techniques such as SPV are not affected by trapping because

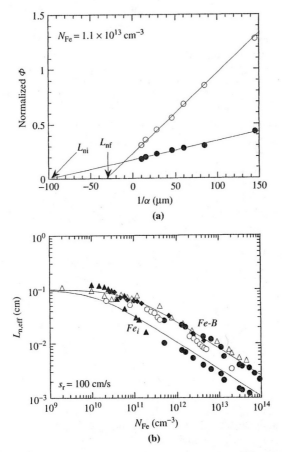

Fig. E7.1 (a) Surface photovoltage plot for iron-contaminated Si sample, (b) effective minority carrier diffusion length versus iron density showing the "before" and "after" FeB pair breaking data.

they detect only minority carriers. As a result of these considerations, the widely used photoconductance-based lifetime techniques generally operate at mid to high injection levels. However, even for low-injection SPV measurements, if the doping density is outside the $1–3 \times 10^{15}$ cm^{-3} range, the prefactor is not constant, due to the properties of Fe_i and $Fe\text{-}B$: the energy level of the $Fe\text{-}B$ center is relatively shallow, and its impact on the low-injection lifetime depends on the doping density. On the other hand Fe_i, being a deep center, yields a doping density-independent low-injection lifetime. Since the prefactor C is determined by the *difference* of the inverse lifetimes, it also varies with the doping density. The factor C varies sensitively with injection level from $C = 3 \times 10^{13}$ μs/cm^3 to -3×10^{13} μs/cm^3. It becomes negative for $\Delta n > 2 \times 10^{14}$ cm^{-3} (see McDonald et al. above), *i.e.*, the lifetime decreases after dissociation for low injection but increases for high injection! The $Fe\text{-}B$ pairing time constant, after dissociation, is given by

$$\tau_{pairing} = \frac{4.3 \times 10^5 T}{N_A} \exp\left(\frac{0.68}{kT}\right)$$

A good discussion can be found in G. Zoth and W. Bergholz, "A Fast, Preparation-Free Method to Detect Iron in Silicon," *J. Appl. Phys.* **67**, 6764–6771, June 1990 and Macdonald et al., above. The references for the experimental data are: O.J. Antilla and M.V. Tilli, "Metal Contamination Removal on Silicon Wafers Using Dilute Acidic Solutions," *J. Electrochem. Soc.* **139**, 1751–1756, June 1992; Y. Kitagawara, T. Yoshida, T. Hamaguchi, and T. Takenaka, "Evaluation of Oxygen-Related Carrier Recombination Centers in High-Purity Czochralski-Grown Si Crystals by the Bulk Lifetime Measurements," *J. Electrochem. Soc.* **142**, 3505–3509, Oct. 1995; M. Miyazaki, S. Miyazaki, T. Kitamura, T. Aoki, Y. Nakashima, M. Hourai, and T. Shigematsu, "Influence of Fe Contamination in Czochralski-Grown Silicon Single Crystals on LSI-Yield Related Crystal Quality Characteristics," *Japan. J. Appl. Phys.* **34**, 409–413, Feb. 1995; A.L.P. Rotondaro, T.Q. Hurd, A. Kaniava, J. Vanhellemont, E. Simoen, M.M. Heyns, and C. Claeys, "Impact of Cu and Fe Contamination on the Minority Carrier Lifetime of Silicon Substrates," *J. Electrochem. Soc.* **143**, 3014–3019, Sept. 1996.

Chromium in silicon forms $Cr\text{-}B$ pairs. When these pairs dissociate, the lifetime increases (K. Mishra, "Identification of Cr in p-type Silicon Using the Minority Carrier Lifetime Measurement by the Surface Photovoltage Method," *Appl. Phys. Lett.* **68**, 3281–3283, June 1996).

The condition $W \ll L_n$ is generally satisfied for single-crystal Si samples, but that may not be true for other semiconductors. For example, the diffusion length in GaAs is often only a few microns. In amorphous Si it is even shorter. In such a situation the intercept is given by[52]

$$\frac{1}{\alpha} = -L_n \left(1 + \frac{(W/L_n)^2}{2(1 + W/L_n)} \right) \tag{7.59}$$

Equation (7.59) reduces to (7.57) for $W \ll L_n$. For $W \gg L_n$ the $1/\alpha$ intercept is $-W/2$, independent of the diffusion length. The scr width can be reduced with steady-state light on the device when $W \gg L_n$.

In Eqs. (7.57) and (7.58) the photon flux density is plotted against the inverse absorption coefficient. It is not the absorption coefficient, however, but the wavelength that is varied during the measurement. An accurate wavelength-absorption coefficient relationship is therefore very important for SPV measurements. Any error in that relationship leads to incorrect diffusion lengths. Various equations have been proposed. A fit to recent $\alpha - \lambda$ data for silicon is given by[53]

$$\alpha = \left(\frac{83.15}{\lambda} - 74.87 \right)^2 \quad [\text{cm}^{-1}] \tag{7.60}$$

with the wavelength λ in μm, valid for the 0.7 to 1.1 μm wavelength range typically used for Si.

An expression that gives reasonable agreement with experimental GaAs absorption data[54] is

$$\alpha = \left(\frac{286.5}{\lambda} - 237.13 \right)^2 \quad [\text{cm}^{-1}] \tag{7.61}$$

for the 0.75 to 0.87 μm wavelength range. For InP[55]

$$\alpha = \left(\frac{252.1}{\lambda} - 163.2 \right)^2 \quad [\text{cm}^{-1}] \tag{7.62}$$

is a reasonable approximation for the 0.8 to 0.9 μm wavelength range.

The reflectance R in Eq. (7.54) is usually considered to be constant. However, there is a weak wavelength dependence for Si, given by[56]

$$R = 0.3214 + \frac{0.03565}{\lambda} - \frac{0.03149}{\lambda^2} \qquad (7.63)$$

for $0.7 < \lambda < 1.05$ μm with λ in μm.

SPV measurements have become very common in the semiconductor industry, largely due to the availability of commercial equipment. Diffusion lengths are routinely measured, because they are a good measure of process cleanliness. Monitoring of furnace tube cleanliness, detection of metallic contamination of incoming chemicals, and control of photoresist ashing are but a few examples for SPV applications.[57]

A crucial component of SPV is the surface treatment to create the surface scr. The ASTM method recommends boiling n-Si in water for one hour.[56] For p-Si a one-minute etch in 20 ml concentrated HF + 80 ml H_2O is recommended. This method works best when care is taken in earlier preparation steps not to produce a stain film by withdrawing the sample from an HF-containing etch directly into air. Otherwise, a low or unstable SPV is likely to result. The stain can be avoided by quenching the HF-containing etch thoroughly with deionized water before withdrawing the sample into air. Another surface treatment for Si samples is a standard Si clean/etch,[58] removing any residual SiO_2 in buffered HF and treating n-Si in an aqueous solution of $KMnO_4$. For p-Si, the $KMnO_4$ step is omitted.

Schottky and pn junction diodes are also suitable for SPV measurements. In both cases, one makes contact to the device directly, without the need for capacitive contacts, yielding a higher surface photovoltage. The metal must be partially transparent for Schottky diodes.[59] It may be necessary to observe certain precautions.[60] Aluminum, 10–20 nm thick, is sufficiently transparent to be suitable. It also possible to use liquid contacts.[61]

The size of the optical beam has an influence on the measured diffusion length. For a beam diameter less than about $30L_n$, the diffusion length is reported to be larger than the true value.[62] As with all diffusion length measuring techniques, the true diffusion length can only be determined for samples thicker than $4L_n$. Effective diffusion lengths are determined for thinner samples,[63] but for samples thinner than L_n, it is difficult to extract the correct diffusion length.[64] Further complications arise if the sample consists of regions of different diffusion lengths as found in Si wafers that have undergone a denuding and oxygen precipitation cycle. The extraction of the diffusion length then becomes quite complicated.[65] The SPV technique has also been implemented with ac photosignals.[66] A photon beam is scanned across the sample and the resulting ac photovoltage is detected with a capacitive probe and displayed on a TV monitor.

Exercise 7.2

Problem: Is it possible to determine L_n when $d < L_n$?

Solution: The Φ term was calculated and plotted versus $1/\alpha$ as a function of L_n using Eq. (A7.4) for Si. Fig. E7.2(a) shows a good linear fit to the calculated data for $d \approx 4L_n$ as expected, but beyond that there is poor linearity and the simple analysis of Eq. (7.57) does not work. For $x = 0$, $d \ll L_n$, and $\alpha d \gg 1$ Eq. (A7.4) becomes

$$\Delta n(0) = \frac{(1 - R)\Phi\alpha\tau}{(\alpha^2 L_n^2 - 1)} \frac{s_{r2}\alpha d + \alpha D - Dd/L_n^2 - s_{r2}}{s_{r1}s_{r2}d/D + Dd/L_n^2 + s_{r1} + s_{r2}} \approx \frac{(1 - R)\Phi}{(1 - \alpha^{-2}L_n^{-2})} \frac{d - 1/\alpha}{s_{r1}d + D}$$

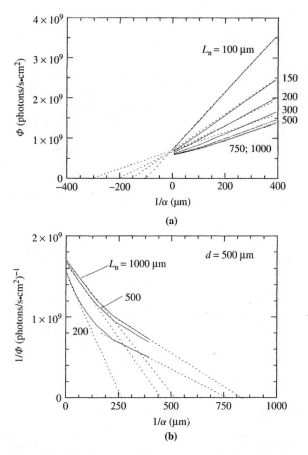

Fig. E7.2   Constant voltage SPV plots (a) exact equation, (b) approximate equation. $s_{r1} = 10^4$ cm/s, $s_{r2} = 10^4$ cm/s, $D_n = 30$ cm^2/s, $V_{SPV} = 10$ mV, $R = 0.3$, $n_{po} = 10^5$ cm^{-3}, $d = 500$ μm.

where the approximation holds for high s_{r2}. A plot of $1/\Phi$ versus $1/\alpha$, according to

$$\frac{1}{\Phi} \approx \frac{(1-R)}{\Delta n(0)(1-\alpha^{-2}L_n^{-2})} \frac{d - 1/\alpha}{s_{r1} d + D}$$

has a $1/\alpha$ intercept that is neither the sample thickness d nor L_n. It is obvious from these figures that the diffusion length cannot be reliably determined when L_n exceeds the sample thickness.

7.4.6   Steady-State Short-Circuit Current (SSSCC)

The *steady-state short-circuit current* method is related to the SPV method. The sample must contain a collecting junction such as a pn junction or Schottky diode, and the short-circuit current is measured as a function of wavelength. Using the same assumptions as

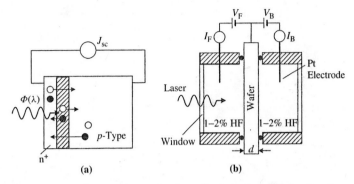

Fig. 7.13 Measurement schematic for (a) the short-circuit current diffusion length measurement method and (b) for the ELYMAT double surface method.

those of the SPV method [Eq. (7.53)], the short-circuit current density of the n^+p junction of Fig. 7.13(a) is, according to Eq. (A 7.9), given by

$$J_{sc} \approx q(1 - R)\Phi \left(\frac{L_n}{L_n + 1/\alpha} + \frac{L_p}{L_p + 1/\alpha} \right) \tag{7.64}$$

The diffusion length is generally low for heavily doped layers, allowing the second term to be neglected for an n^+p junction, and the short-circuit current density becomes

$$J_{sc} \approx q(1 - R)\Phi \left(\frac{L_n}{L_n + 1/\alpha} \right) \tag{7.65}$$

Neglecting ehp generation in the n^+ layer and in the space-charge region is permissible if these regions are narrow and if α is not too high.

Equation (7.65) has been used in two ways to extract the diffusion length. In one technique the current is held constant by adjusting the photon flux density as the wavelength is changed.[67] Equation (7.65) then becomes

$$\Phi = C_1(L_n + 1/\alpha) \tag{7.66}$$

where $C_1 = J_{sc}/q(1 - R)L_n$. L_n is the intercept on the negative $1/\alpha$ axis when Φ is plotted against $1/\alpha$. In a second technique, Eq. (7.65) is written as[68]

$$\frac{1}{\alpha} = (X - 1)L_n \tag{7.67}$$

where $X = q(1 - R)\Phi/J_{sc}$. Here $1/\alpha$ is plotted against $(X - 1)$ and the diffusion length is given by the slope of this plot. A check on the data is provided by the extrapolated lines passing through the origin. Both methods neglect carrier collection from the n^+ region and the scr.

The short-circuit current methods are in principle similar to SPV, but they require a junction to collect the minority carriers. In practice, it is easier to measure a current than an open-circuit voltage. However, junction formation may alter the diffusion length. Two implementations not requiring permanent junctions make use of mercury contacts or liquid semiconductor contacts. In the mercury contact method, two Hg probes are pressed against one side of the sample and modulated light is incident on the other side. From an analysis of the frequency-dependent photocurrent, one can extract the lifetime and diffusion coefficient.[69] A different implementation, the *electrolytical metal tracer* (ELYMAT), shown schematically in Fig. 7.13(b), uses electrolyte-semiconductor junctions at the front and at the back surface to map the photo response which is related to the diffusion length.[70] The wafer is immersed in an electrolyte (normally, but not necessarily, 1%–2% HF in H_2O). The electrolyte serves the dual function of photocurrent collection as well as wafer surface passivation.

The front and back induced photocurrents I_F and I_B are measured in response to laser beam excitation of the sample. For light with short penetration depth, moderate wafer diffusion lengths, and negligible back surface recombination, the front current I_F is measured. For penetrating light and low front surface recombination, the back current I_B is measured. The currents are given by[70]

$$I_B \approx \frac{I_{max}(1 + s_{rf}/D_n\alpha)}{\cosh(d/L_n) + (s_{rf}L_n/D_n)\sinh(d/L_n)} \approx \frac{I_{max}}{\cosh(d/L_n)} \qquad (7.68a)$$

$$I_F \approx I_{max}; I_{max} \approx qA\Phi(1 - R)(1 - \exp(-\alpha d)) \qquad (7.68b)$$

where A is the area, s_{rf} the surface recombination velocity at the front surface, and Φ the photon flux density. The approximation in the first equation holds for low s_{rf}, as observed in an HF solution. Measuring I_F and I_B allows L_n to be determined. Using laser excitation with two different wavelengths and voltage bias, allows the surface recombination velocity and the minority carrier diffusion length as well as depth-dependent diffusion length to be determined. Scanning the laser produces diffusion length maps rapidly with no mechanical motion. Although submersing the sample in a dilute HF solution ensures low surface recombination, at times it is useful to control surface effects further, *e.g.*, using an oxidized Si wafer with a solution that does not etch SiO_2, *e.g.*, CH_3COOH. Then biasing the solution with respect to the sample, the Si surface can be accumulated, depleted or inverted. Such "electrostatic passivation" further reduces surface recombination.[71]

7.4.7 Free Carrier Absorption

The *free carrier absorption* lifetime method is a non-contacting technique, relying on optical ehp generation and optical detection using two different wavelengths. As illustrated in Fig. 7.14, a pump beam using photons with energy $h\nu > E_G$ creates ehps. The readout is based on the dependence of the free carrier absorption of photons with $h\nu < E_G$ on the density of free carriers. The probe beam transmitted photon flux density Φ_t is given by

$$\Phi_t = \frac{(1 - R)^2\Phi_i \exp(-\alpha_{fc}d)}{1 - R^2\exp(-2\alpha_{fc}d)} \qquad (7.69)$$

where α_{fc} is the free carrier absorption coefficient, d the sample thickness, and R the reflectivity. For n-type semiconductors the absorption coefficient is[72]

$$\alpha_{fc} = K_n\lambda^2 n \qquad (7.70)$$

where K_n is a materials constant and λ the wavelength of the probe beam. For n-Si, $K_n \approx 10^{-18}$ $cm^2/\mu m^2$, and for p-Si, $K_p \approx (2-2.7) \times 10^{-18}$ $cm^2/\mu m^2$.[72–73] A small correction to K_n in Eq. (7.70) has been suggested.[74]

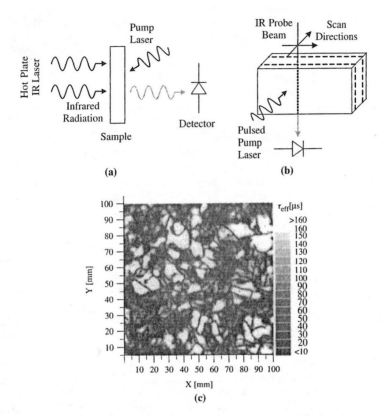

Fig. 7.14 (a) Schematic free carrier absorption arrangement, (b) schematic for lifetime mapping, (c) free carrier absorption lifetime map after Isenberg et al.[79]

The method can be used in both steady-state and transient modes. A probe beam, for example, a CO_2 laser ($\lambda = 10.6$ μm), HeNe ($\lambda = 3.39$ μm) or black body radiation, is incident on the sample in the steady-state embodiment. The transmitted beam is detected by an infrared detector. The pump beam is chopped at a few hundred Hz for synchronous detection by a lock-in amplifier. In the transient method the pump beam is pulsed, and the time-dependent carrier density is detected through the transmitted probe beam. A further implementation of the technique is the phase shift method.[75] Excess carriers are generated by sine-wave modulated light. A phase shift occurs between the generation and infrared transmission through the sample. This phase shift leads to the lifetime.

The change in the transmitted probe beam as a result of a chopped or pulsed pump beam is

$$\Delta\Phi_t \approx -\frac{(1-R)\Phi_i \Delta\alpha_{fc}d}{1+R} \tag{7.71}$$

using $\exp(-2\alpha_{fc}d) \approx \exp(-\alpha_{fc}d) \approx 1$ in Eq. (7.71) with $\alpha_{fc}d \ll 1$. The change in the absorption coefficient is

$$\Delta\alpha_{fc} = K_n\lambda^2\Delta n = \frac{K_n\lambda^2}{d}\int_0^d \Delta n(x)\,dx \tag{7.72}$$

In turn, Δn is related to the minority carrier lifetime and the surface recombination velocity through Eq. (A7.4). In addition Δn contains the sample reflectivity, the pump beam absorption coefficient, and the photon flux density. The fractional change in transmitted photon flux density, under certain simplifying assumptions, is[76]

$$\frac{\Delta\Phi_t}{\Phi_t} \approx \frac{(1-R)K_n\lambda^2\Phi_i\tau_n(1+s_{r1}/\alpha_{fc}D_n)}{1+s_{r1}L_n/D_n} \tag{7.73}$$

It is obvious that lifetime extraction is not simple, even if the assumptions leading to Eq. (7.73) are satisfied since a number of sample parameters must be known. However, the measurement requires neither high-speed light sources nor detectors because it is a steady-state measurement and is therefore suitable for short lifetime determination.

The transient version data interpretation is simpler since the transient carrier decay contains the recombination information. A 3.39 μm HeNe probe beam and a pulsed 1.06 μm Nd:YAG pump beam (150 ns pulse width) were used in one implementation.[77] The lifetime so determined agreed well with the lifetimes measured by open-circuit voltage decay and by photoconductance decay. As illustrated in Fig. 7.14(b), the probe and pump beams can be perpendicular to one another and by scanning the probe beam, it is possible to map the lifetime through the wafer thickness, for example.[78]

An interesting free carrier lifetime characterization approach uses infrared (IR) radiation from a black body transmitted through the sample and detected by an infrared light detecting charge-coupled device as the detector (mercury-cadmium-telluride or AlGaAs/GaAs).[79] The black body source can be as simple as a hot plate. A laser with $h\nu > E_G$ creates ehps in the sample. By taking the difference of the IR radiation through the sample with and without the laser, one measures the free carrier absorption due to the excess carriers. Taking two-dimensional images of the IR radiation over the entire wafer, allows for rapid measurements. The system is calibrated with a set of Si wafers of varying doping densities. The transmissivity of these wafers successively placed between the camera and the black body is measured. The signal differences are then due to the differences in free-carrier absorption of the samples. One needs to apply a correction to account for the fact that in the calibration procedure of p-type wafers only the IR absorption of holes is measured while in an actual measurement laser-generated electron-hole pair generation must be considered.

Knowing the laser generation rate $G' = (1-R)\Phi$ (cm^{-2} s^{-1}) and the sample thickness d, the effective lifetime is

$$\tau_{eff} = \frac{d\Delta n}{G'} \tag{7.74}$$

A two-dimensional lifetime map obtained in 50 s with this technique is shown in Fig. 7.14(c). No scanning is required, since both black body and excitation laser are broad area sources, covering the entire sample. The black body emits over a wide wavelength range with a peak wavelength at

$$\lambda_{peak} \approx \frac{3000}{T} \; \mu m \tag{7.75}$$

A hot plate at $T = 350$ K, has its peak wavelength at $\lambda_{peak} \approx 8.6$ μm—a suitable wavelength for free carrier absorption measurements. Just as carriers absorb IR radiation, they also emit IR radiation. According to Kirchhoff's law they emit the same power as they absorb to remain at a given temperature. Hence, the sample itself will emit IR radiation and can be used to determine the lifetime. The sample is still excited with a laser and the difference signal is acquired as in the transmission system. Both emission and absorption have been used for lifetime measurements.[80]

7.4.8 Electron Beam Induced Current (EBIC)

Electron beam induced current is used to measure minority carrier diffusion length, minority carrier lifetime, and defect distribution. In contrast to photons that typically create one ehp pair upon absorption, an absorbed electron of energy E creates

$$N_{ehp} = \frac{E}{E_{ehp}}\left(1 - \frac{\gamma E_{bs}}{E}\right) \tag{7.76}$$

electron-hole pairs.[81] E_{bs} is the mean energy of the backscattered electrons, γ the backscattering coefficient and E_{ehp} the average energy required to create one ehp ($E_{ehp} \approx 3.2E_G$; for Si $E_{ehp} = 3.64 \pm 0.03$ eV).[82] The backscattering term $\gamma E_{bs}/E$ is approximately equal to 0.1 for Si and 0.2–0.25 for GaAs over the 2 to 60 keV electron energy range. The electron penetration depth or range R_e is given by[83]

$$R_e = \frac{2.41 \times 10^{-11}}{\rho} E^{1.75} \quad [cm] \tag{7.77}$$

where ρ is the semiconductor density (g/cm^3) and E the incident energy (eV). For Si and GaAs, $R_e(Si) = 1.04 \times 10^{-11} E^{1.75}$ cm and $R_e(GaAs) = 4.53 \times 10^{-12} E^{1.75}$ cm.

It is instructive to calculate the ehp density generated by an electron beam of energy E and beam current I_b. The generation volume tends to be pear shaped, as shown in Chapter 11, for atomic numbers $Z < 15$. For $15 < Z < 40$ it approaches a sphere, and for $Z > 40$ it becomes hemispherical. We approximate it as a sphere of volume $(4/3)\pi (R_e/2)^3$, for simplicity. Combining Eqs. (7.76) and (7.77) gives the generation rate

$$G = \frac{N_{ehp} I_b}{(4/3)\pi q (R_e/2)^3} = \frac{8.5 \times 10^{50} \rho^3 I_b}{E_{ehp} E^{4.25}} \quad [cm^{-3} \, s^{-1}] \tag{7.78}$$

neglecting the backscattered term in Eq. (7.76). For Si with a beam current of 10^{-10} A, $E_{ehp} = 3.64$ eV and $E = 10^4$ eV the generation rate is $G = 3 \times 10^{24}$ ehp/cm$^3 \cdot$ s.

The interaction of an electron beam with the semiconductor sample can take place for a variety of geometries. One of these is shown in Fig. 7.15(a). The electron beam induced current I_{EBIC} collected by the junction changes by moving the beam in the x-direction. Changes in the z-direction are produced by changing the beam energy. The e-beam creates ehps at a distance d from the edge of the scr. Some of the minority carriers diffuse to the junction to be collected, and I_{EBIC} decreases with increasing d due to bulk and surface recombination.

I_{EBIC} can be expressed as[84]

$$I_{EBIC} = \frac{qG' R_e L_n^n}{(2\pi)^{1/2} d^n} = Cd^{-n} \exp\left(-\frac{d}{L_n}\right) \tag{7.79}$$

where $G' = I_b N_{ehp}/q$, provided $s_r \gg D_n/L_n$, $L_n \ll d$, $R_e \ll d$, $R_e L_n \ll d^2$, and low-level injection prevails. The exponent n depends on surface recombination. For $s_r \to 0$, $n = 1/2$ and for $s_r \to \infty$ $n = 3/2$. A plot of $\ln(I_{EBIC} d^n)$ versus d should give a straight

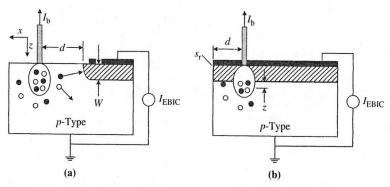

Fig. 7.15   (a) Conventional EBIC implementation, (b) depth modulation by electron beam energy.

line of slope $-1/L_n$. Since s_r is generally not known, n is also unknown. One method to determine n is to plot $\ln(I_{EBIC}d^n)$ versus d and vary n until a straight line results.[85]

For the configuration in Fig. 7.15(b), I_{EBIC} is[86]

$$I_{EBIC} = I_1 \left(\exp\left(-\frac{z}{L_n}\right) - \frac{2s_r F}{\pi} \right)$$ (7.80)

where I_1 is a constant and F depends on s_r and on the ehp generation point. The second term in Eq. (7.80) vanishes for $d = L_n$ and L_n is found by recording I_{ph} versus z.[87] Surface recombination plays an important role in EBIC measurements.[88]

Instead of determining the diffusion length from the steady-state photocurrent as a function of lateral motion or beam penetration, one can use a stationary pulsed beam and extract the minority carrier lifetime from the transient analysis. An approximate expression for I_{EBIC} for high s_r, is[88]

$$I_{EBIC}(t) = K_1 \left(\frac{\tau_n}{t}\right)^2 \exp\left(\frac{d}{L_n}\left(1 - \frac{\tau_n}{4t}\right) - \frac{t}{\tau_n}\right)$$ (7.81)

valid for $d \gg L_n$ for Fig. 7.15(b). Theory predicts that I_{EBIC} does not decay immediately after the injection has ceased. Instead, there is a delay that is more pronounced the further the beam is from the junction. For optical excitation, the technique is known as *optical beam induced current* (OBIC), with considerations very similar to EBIC except for a different generation expression.[89-90]

Most EBIC measurements are made as illustrated in Fig. 7.15 and the method is fairly straightforward for long diffusion lengths. For short diffusion length measurements, the sample can be beveled to enhance the depth.[91] Surface recombination effects are reduced if the beam penetration is increased. This can be directly tested by plotting $\ln(I_{EBIC})$ versus d for various beam energies. The plot should approach a straight line for higher energies.

7.5   RECOMBINATION LIFETIME—ELECTRICAL MEASUREMENTS

7.5.1   Diode Current-Voltage

The pn junction diode forward current depends on recombination of excess carriers and is the sum of space-charge region, quasi-neutral region (qnr), and surface recombination

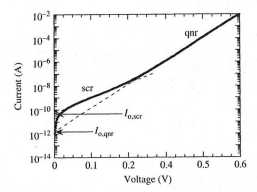

Fig. 7.16   pn junction $I-V$ curve showing space-charge and quasi neutral region currents.

currents. In reverse bias, it is generation in the various regions that is measured. In most analyses, surface recombination is neglected and the current density is

$$J = J_{0,scr}\left(\exp\left(\frac{qV}{nkT}\right) - 1\right) + J_{0,qnr}\left(\exp\left(\frac{qV}{kT}\right) - 1\right)$$

$$J_{0,scr} = \frac{qn_i W}{\tau_{scr}}; \quad J_{0,qnr} = qn_i^2 F\left(\frac{D_n}{N_A L_n} + \frac{D_p}{N_D L_p}\right) \tag{7.82}$$

where F is a correction factor that depends on the sample geometry, *e.g.*, denuded zones on defective substrates, epitaxial layers on heavily or lightly doped substrates, silicon-on-insulator (SOI), etc. It is, in general, a complicated function of the active layer thickness, the diffusion lengths and doping densities in the layer and the substrates, and possible interface recombination velocity at the layer-substrate interface.

Equation (7.82) is plotted in Fig. 7.16. Extrapolating the qnr line to $V = 0$ yields $I_{0,qnr}$. For a p^+n junction $N_A \gg N_D$ and although τ_n in a heavily-doped region is much lower than τ_p in the lightly-doped substrate, it is often permissible to neglect the first term in the $J_{0,qnr}$ term because N_A is very high, giving

$$J_{0,qnr} \approx qn_i^2 F\frac{D_p}{N_D L_p} \tag{7.83}$$

Knowing n_i, F, D_p, and N_D allows L_p to be determined.

Consider the device cross-sections in Fig. 7.17. Neglecting electron injection into the p^+ region for simplicity, the current in the forward-biased p^+n junction in Fig. 7.17(a) with $d < L_n$ depends on hole recombination in the scr (1), in the qnr (2), and at the surface (3). The correction factor is given by[65]

$$F = \frac{(s_r L_p/D_p)\cosh(d/L_p) + \sinh(d/L_p)}{\cosh(d/L_p) + (s_r L_p/D_p)\sinh(d/L_p)} \tag{7.84}$$

Figure 7.17(b) shows an n-substrate consisting of a denuded zone (1) of width d (L_{p1}, N_A) on a precipitated substrate (2) (L_{p2}, N_A). An epitaxial layer (1) of thickness d (L_{p1}, N_{A1}) on a substrate (2) (L_{p2}, N_{A2}) is shown in (c) and (d) shows an SOI wafer. Correction factors have been derived for these cases.[92] For the epitaxial device in Fig. 7.17(c)

$$F \approx \frac{(1 + N_{A2}/N_{A1})\exp(D_p/L_p) + (1 - N_{A2}/N_{A1})\exp(-D_p/L_p)}{(1 + N_{A2}/N_{A1})\exp(D_p/L_p) - (1 - N_{A2}/N_{A1})\exp(-D_p/L_p)} \tag{7.85}$$

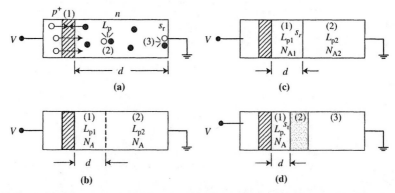

Fig. 7.17 pn junction cross sections (a) recombination mechanisms in the n-substrate for $d < L_p$, (b) denuded zone on precipitated substrate, (c) epitaxial layer on substrate, and (d) SOI wafer.

Although correction factors can, in principle, be applied, it is fraught with difficulties, because the lifetimes in the epi layer and the substrate and the interface recombination velocity at the epi-substrate interface are rarely known. Some of the pitfalls in the interpretation of such measurements are given in ref. 93. A recent study concluded that the most effective techniques to characterize epitaxial layers are generation lifetime techniques.[94]

Instead of extrapolating the forward-biased current-voltage characteristics, one can also use the reverse-bias current-voltage curve.[95] Under reverse bias, where $V < 0$, Eq. (7.82) becomes

$$J_r = -\frac{qn_iW}{\tau_g} - qn_i^2 F\left(\frac{D_n}{N_A L_n} + \frac{D_p}{N_D L_p}\right) \approx -\frac{qn_iW}{\tau_g} - qn_i^2 F\frac{D_p}{N_D L_p} \qquad (7.86)$$

Plotting J_r versus W gives a curve with slope related to the generation lifetime τ_g and intercept giving $J_{0,qnr}$, illustrated in Fig. 7.18. The scr width is determined from reverse-biased capacitance-voltage data, but the true capacitance must be measured,[96] especially for small area diodes where perimeter, corner, and parasitic capacitances are important.

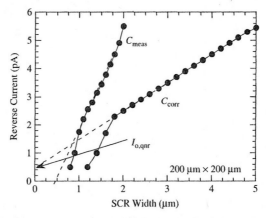

Fig. 7.18 Reverse leakage current versus scr width for measured and corrected capacitance. Adapted from ref. 96.

The actual diode leakage current consists of areal, peripheral, corner, and parasitic currents according to[97]

$$I_r = AJ_A + PJ_P + N_C J_C + I_{par}$$

where A is the diode area, P the diode perimeter (J_P in units of A/cm), N_C the number of corners (J_C in units of A/corner) and I_{par} is a parasitic current.

7.5.2 Reverse Recovery (RR)

The diode *reverse-recovery* method was one of the first electrical lifetime characterization techniques.[98-100] A measurement schematic and current-time and voltage-time responses are shown in Fig. 7.19. In Fig. 7.19(b) the current is suddenly switched from forward to reverse current by changing switch position S, whereas in 7.19(c) the current is gradually changed, typical of power devices in which currents cannot be switched very abruptly.

For a description of the method, let us consider Figs. 7.19(a) and (b). A forward current I_f flows through the diode for $t < 0$ and the diode voltage is V_f. Excess carriers are injected into the quasi-neutral regions, leading to low device resistance. At $t = 0$ the current is switched from I_f to I_r, with $I_r \approx (V_r - V_f)/R$. The small diode resistance is neglected because the diode remains forward biased during the initial time of I_r flow. Currents can be switched very quickly in minority carrier devices because only a change

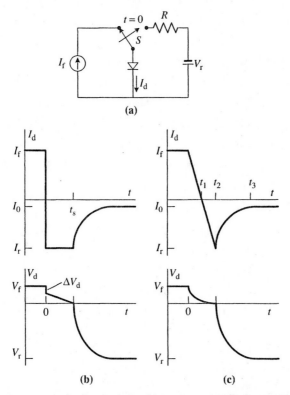

Fig. 7.19 Reverse recovery circuit schematic, (b) current and voltage waveforms for abruptly switched current, and (c) current and voltage waveforms for ramped currents.

in the *slope* of the minority carrier density gradient at the edge of the scr is required. The diode voltage, in contrast, is proportional to the log(excess carrier *density*) at the scr edge. The voltage hardly changes during this period and the diode remains forward biased although the current has reversed direction. The voltage step ΔV_d is due to the ohmic voltage drop in the device.[101]

The excess carrier density decreases during the reverse current phase as some carriers are swept out of the device by the reverse current and some carriers recombine. The excess minority carrier densities at the edges of the scr are approximately zero at $t = t_s$, and the diode becomes zero biased. For $t > t_s$, the voltage approaches the reverse-bias voltage V_r and the current approaches the leakage current I_0.

The $I_d - t$ curve is conveniently divided into the constant-current storage phase, $0 \leq t \leq t_s$, and the recovery phase, $t > t_s$. The storage time t_s is related to the lifetime by[99]

$$erf\sqrt{\frac{t_s}{\tau_r}} = \frac{1}{1 + I_r/I_f} \tag{7.87}$$

with "erf", the error function, defined and approximated by

$$erf(x) = \frac{2}{\sqrt{\pi}} \int_0^x e^{-z^2}\,dz \approx 1 - \left(\frac{0.34802}{1 + 0.4704x} - \frac{0.095879}{(1 + 0.4704x)^2} + \frac{0.74785}{(1 + 0.4704x)^3}\right)$$
$$\times \exp(-x^2) \tag{7.88}$$

An approximate charge storage analysis that considers the charge Q_s remaining at $t = t_s$ gives[102]

$$t_s = \tau_r\left[\ln\left(1 + \frac{I_f}{I_r}\right) - \ln\left(1 + \frac{Q_s}{I_f \tau_r}\right)\right] \tag{7.89}$$

$Q_s/I_r\tau_r$ can be considered a constant for many cases.

A plot of t_s versus $\ln(1 + I_f/I_r)$ is shown in Fig. 7.20. The lifetime is found from the slope and the intercept is $(1 + Q_s/I_r\tau_r)$. The slope is constant only if the second term in Eq. (7.89) is constant. Various approximations have been derived for Q_s, and it is found to be approximately constant provided $I_r \ll I_f$.[103] The effect of recombination in the heavily doped emitter can be virtually eliminated by keeping $I_r \ll I_f$.[104] The plot

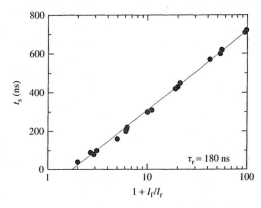

Fig. 7.20 Storage time versus $(1 + I_f/I_r)$. Reprinted after Kuno[102] by permission of IEEE (© 1964, IEEE).

of Fig. 7.20 becomes highly curved if these conditions are not met and a unique lifetime can no longer be extracted. For Fig. 7.19(c) the lifetime is related to t_1, t_2, and t_3 by[105]

$$\tau_r \approx \sqrt{(t_2 - t_1)(t_3 - t_1)} \tag{7.90}$$

where t_3 is defined as the time for $I_d = 0.1 I_r$. The junction displacement current $I_j = C_j \, dV_j/dt$ is neglected in all of these expressions because it constitutes only a small fraction of the total current.[100]

What is τ_r in Eqs. (7.87) and (7.90)? To first order it would seem to be the base lifetime in pn junctions. For short-base diodes, it is an effective lifetime representing both bulk and surface recombination.[106] A problem in forward-biased pn junctions is the existence of excess carriers in *both* quasi-neutral regions and in the scr. The emitter is generally much more heavily doped than the base and the emitter lifetime is much lower than the base lifetime. Hence, one would expect emitter recombination to have a significant influence on the RR transient. This is particularly troublesome for high injection conditions, leading to appreciably reduced lifetimes.[107–108] However, the emitter can alter the measured lifetime from its true base value even at low and moderate injection levels.

7.5.3 Open-Circuit Voltage Decay (OCVD)

The *open-circuit voltage decay* method measurement principle is shown in Fig. 7.21(a).[109–110] The diode is forward biased and at $t = 0$ switch S is opened and the voltage, decaying due to recombination of excess carriers, is detected as in Fig. 7.21(b). The voltage step $\Delta V_d = I_f r_s$ is due to the ohmic voltage drop in the diode when the current ceases and can be used to determine the device series resistance as discussed in Chapter 4.[101] OCVD is similar to the optically excited, open-circuit voltage decay method in Section 7.4.3. In contrast to RR, in the OCVD method the excess carriers all recombine; none are swept out of the device by a reverse current since the current is zero.

The excess minority carrier density Δn_p in the quasi-neutral region at the edge of the scr in a p-substrate, is related to the time-varying junction voltage $V_j(t)$ by

$$\Delta n_p(t) = n_{po} \left(\exp\left(\frac{q V_j(t)}{kT} \right) - 1 \right) \tag{7.91}$$

where n_{po} is the equilibrium minority carrier density. The junction voltage is

$$V_j(t) = \frac{kT}{q} \ln\left(\frac{\Delta n_p(t)}{n_{po}} + 1 \right) \tag{7.92}$$

A measure of the voltage time dependence is a measure of the excess carrier time dependence.

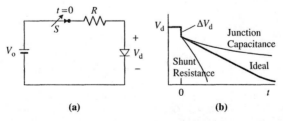

Fig. 7.21 Open circuit voltage decay (a) circuit schematic and (b) voltage waveform.

The diode voltage is $V_d = V_j + V_b$, where V_b is the base voltage, neglecting the voltage across the emitter. How can there be a base voltage when there is no current flow during the decay? The base or *Dember voltage* is the result of unequal electron and hole mobilities and is given by[111]

$$V_b(t) = \frac{kT}{q} \frac{b-1}{b+1} \ln\left(1 + \frac{(b+1)\Delta n_p(t)}{n_{po} + b p_{po}}\right) \tag{7.93}$$

with $b = \mu_n/\mu_p$. The Dember voltage is negligible for low injection levels, and we will not consider it further, but may not be negligible for high injection levels. We assume that $V_d(t) \approx V_j(t)$, given by Eq. (7.92), and will simply use $V(t)$ for the time-varying device voltage.

For $d \gg L_n$ and low-level injection[110]

$$V(t) = V(0) + \frac{kT}{q} \ln\left(erfc\sqrt{\frac{t}{\tau_r}}\right) \tag{7.94}$$

where $V(0)$ is the diode voltage before opening the switch and $erfc(x) = 1 - erf(x)$ is the complementary error function. Equation (7.94), plotted in Fig. 7.22, obtains for $V(t) \gg kT/q$. The curve has an initial rapid decay followed by a linear region with constant slope. The slope is

$$\frac{dV(t)}{dt} = -\frac{(kT/q)\exp(-t/\tau_r)}{\sqrt{\pi t \tau_r}\, erfc\sqrt{1/\tau_r}} \approx -\frac{kT/q}{\tau_r(1 - \tau_r/2t)} \tag{7.95}$$

where the approximation holds for $t \geq 4\tau_r$. Equation (7.95) can be further simplified by neglecting the second term in the bracket. For $t \geq 4\tau_r$ the lifetime is determined from the slope according to

$$\tau_r = -\frac{kT/q}{dV(t)/dt} \tag{7.96}$$

As Fig. 7.22 shows, the curve becomes linear for $t > 4\tau_r$.

A word of caution regarding Eq. (7.96). The assumption in the derivation leading to this equation is that recombination is dominated by quasi-neutral region recombination

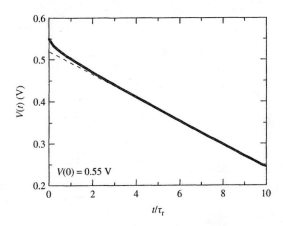

Fig. 7.22 Open circuit voltage decay waveform according to Eq. (7.94).

with the simple exponential voltage dependence $\exp(qV/kT)$. For scr recombination the dependence becomes $\exp(qV/nkT)$, where the diode ideality factor n lies typically between 1 and 2. Equation (7.96) should contain n as a prefactor. Of course, as the diode voltage drops from $V(0) \approx 0.7$ V or so to zero, n is likely to vary from 1 to a value closer to 2, and since one usually does not know what n is, it is generally taken as unity.

Due to the low emitter lifetime, excess emitter carriers recombine more rapidly than excess base carriers causing carriers from the base to be injected into the emitter during the voltage decay, and reducing the voltage decay time. Fortunately this effect becomes negligible for $t \geq 2.5\tau_b$, where τ_b is the base lifetime, and the $V(t) - t$ curve becomes linear with slope $(kT/q\tau_b)$ regardless of emitter recombination or band-gap narrowing.[112] Under high level injection, the lifetime is given by[113]

$$\tau_r = -\frac{2kT/q}{dV(t)/dt} \tag{7.97}$$

subject to the restrictions: the excess carrier density in the base is uniform and the base excess carrier density is higher than the base doping density. The 2 accounts for high injection effects. The high injection level $V - t$ curve frequently exhibits two distinct slopes.

Unusual $V - t$ responses, shown in Fig. 7.21(b), are sometimes observed for non-negligible diode capacitance or low junction shunt resistance. Capacitance tends to extend the $V - t$ curve, giving the curve a smaller slope leading to too high lifetimes.[114] Space-charge region recombination and shunt resistance cause the $V - t$ curve to drop faster than observed for quasi-neutral bulk recombination only. A variation of the OCVD method that has been found to be useful for devices exhibiting such decay curves is to switch an external resistor and capacitor into the measurement circuit and differentiate the curve to extract the lifetime.[115] Another possible anomaly is a peak in the $V - t$ curve near $t = 0$ due to emitter recombination.[116]

A variation of OCVD is the *small-signal OCVD* method in which the diode is biased to a steady-state voltage, by illuminating the device and imposing a small electrical pulse on the "optical" bias.[43, 117] With the pulse "on," additional carriers are injected, and with it "off," these additional carriers recombine. This method is used to measure τ_r under bias conditions and also to reduce capacitance and shunt resistance effects.

A comparison of the RR and the OCVD techniques favored OCVD for its ease and accuracy.[107] In OCVD the lifetime can be extracted from that part of the $V - t$ curve where base recombination dominates, whereas in RR storage time measurements there is some averaging over a voltage range that includes at the lower current the scr recombination current. During OCVD the experimental considerations are relaxed since the carriers decay by recombination only.

7.5.4 Pulsed MOS Capacitor

The principle of the *pulsed MOS capacitor* (MOS-C) *recombination* lifetime measurement technique is divided into two methods. In the first of these for an MOS-C biased into strong inversion in Fig. 7.23(a) and point A in 7.23(d), the inversion charge density is

$$Q_{n1} = (V_{G1} - V_T)C_{ox} \tag{7.98}$$

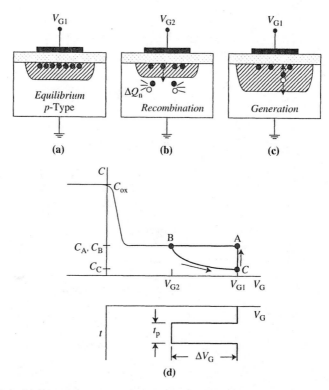

Fig. 7.23 Pulsed MOS capacitor recombination lifetime measurement. The device behavior at various voltages is shown in (a), (b), and (c) and the $C-V_G$ and $V_G - t$ curves in (d).

A voltage pulse of amplitude $-\Delta V_G$ and pulse width t_p superimposed on V_{G1} reduces the gate voltage during the pulse period to $V_{G2} = V_{G1} - \Delta V_G$ shown in Fig. 7.23(b) and by point B in 7.23(d). The inversion charge is

$$Q_{n2} = (V_{G2} - V_T)C_{ox} < Q_{n1} \qquad (7.99)$$

The charge difference $\Delta Q_n = (Q_{n1} - Q_{n2})$ is injected into the substrate, as indicated in Fig. 7.21(b).

What happens to ΔQ_n? Minority carriers in an inversion layer do not recombine with majority carriers because they are separated by the electric field of the scr. However, those minority carriers injected into the substrate are surrounded by holes and can recombine.

Let us now consider two extremes. First, for a wide pulse ($t_p > \tau_r$) the injected minority carriers have sufficient time to recombine. When the gate voltage returns to V_{G1} only Q_{n2} is available, and the MOS-C is driven into partial deep depletion, shown in Fig. 7.23(c) and by point C in 7.23(d). Thermal ehp generation subsequently returns the device to equilibrium, point A. Second, for a narrow pulse ($t_p \ll \tau_r$), the device goes through similar stages as in the first case except the injected minority carriers have insufficient time to recombine because the pulse width is less than the recombination lifetime and the capacitance sequence in Fig. 7.23(d) is $C_A \to C_B \to C_A$. For intermediate pulse widths, the capacitance lies between C_C and C_A.

The capacitance at the end of the injection pulse is a measure of how many minority carriers have recombined during the pulse period. For a simple exponential decay of the minority carriers[118]

$$\Delta Q_n(t) = \Delta Q_n(0) \exp\left(-\frac{t}{\tau_r}\right) = K\left(\frac{1}{C_A^2} - \frac{1}{C_C^2}\right) \tag{7.100}$$

where K is a constant. To determine the lifetime, the pulse width is varied and the capacitance C_C is measured for each pulse width and $\ln(1/C_A{}^2 - 1/C_C{}^2)$ is plotted against t_p; τ_r is obtained from the slope of this plot. A more detailed theory shows the exponential time decay of the carriers in Eq. (7.100) to be too simplistic because minority carriers recombine not only in the quasi-neutral substrate but also in the scr and at the surface.[119] This pulsed MOS-C recombination lifetime measurement method has not found wide acceptance because most capacitance meters are unable to pass the required narrow pulses undistorted. It is easier to modify the experimental arrangement by coupling the device to the capacitance meter through a pulse transformer at its input terminals.[120]

A variation of the pulsed MOS-C technique, based on charge pumping, has been proposed for MOSFETs.[121] When a MOSFET is pulsed from inversion into accumulation, most of the inversion charge leaves the channel through the source and the drain. However, a small fraction of the charge is unable to reach either source or drain and recombines with majority carriers. This fraction, proportional to the pulse frequency, is detected as a substrate current. As the frequency increases to the point where the time between successive pulses is on the order of τ_r, the substrate current-pulse frequency relationship becomes non-linear, and τ_r can be extracted from the current.

The second pulsed MOS-C method is based on an entirely different principle - the measurement of the relaxation time of an MOS-C when pulsed into deep depletion. We assume that prior to applying the depleting gate voltage, the device is in equilibrium and illustrate the technique in Fig. 7.24, where the MOS-C capacitance is driven from A to B by a depleting voltage step. Thermal generation returns the device to equilibrium, shown by the path B to C, in Fig. 7.24(a). The return to equilibrium on the $C - t$ diagram is typically as shown in Fig. 7.24(b). The recovery time t_f is determined by the thermal ehp generation in the semiconductor and at the oxide-semiconductor interface.

The thermal generation rates in Fig. 7.25 are (1) bulk scr generation characterized by the generation lifetime τ_g, (2) lateral surface scr generation characterized by the surface generation velocity s_g, (3) surface scr generation under the gate characterized by the

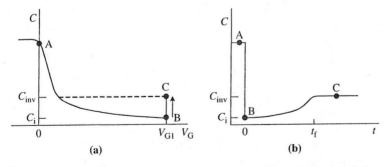

Fig. 7.24   The $C-V_G$ and $C - t$ behavior of an MOS-C pulsed into deep depletion.

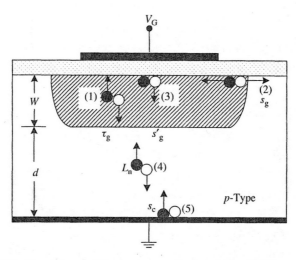

Fig. 7.25   Thermal generation components of a deep-depleted MOS capacitor.

surface generation velocity s'_g, (4) quasi-neutral bulk generation characterized by the minority carrier diffusion length L_n, and (5) back surface generation characterized by the generation velocity s_c. Components 1 and 2 depend on the scr width, discussed in Section 7.6.2, and 3–5 are independent of the scr width.

The capacitance depends on the gate voltage and on the inversion charge Q_n as[122]

$$C(t) = \frac{C_{ox}}{\sqrt{1 + 2(V'_G(t) + Q_n(t)/C_{ox})/V_0}} \qquad (7.101)$$

where $V'_G(t) = V_G(t) - V_{FB}$ and $V_0 = qK_s\varepsilon_0 N_A/C_{ox}^2$. Solving Eq. (7.101) for V'_G and differentiating with respect to t gives

$$\frac{dV_G}{dt} = -\frac{1}{C_{ox}}\frac{dQ_n}{dt} - \frac{qK_s\varepsilon_0 N_A}{C^3}\frac{dC}{dt} \qquad (7.102)$$

with $dV_{FB}/dt = 0$. In Eq. (7.102) we have dropped the time dependence designation "(t)" for simplicity.

Equation (7.102) is an important equation relating the gate voltage rate of change with time to inversion charge and capacitance rate of change with time. For the *pulsed* capacitor, V_G is constant, $dV_G/dt = 0$, and Eq. (7.102) solved for dQ_n/dt becomes

$$\frac{dQ_n}{dt} = -\frac{qK_s\varepsilon_o C_{ox} N_A}{C^3}\frac{dC}{dt} \qquad (7.103)$$

where dQ_n/dt represents the thermal generation rates in Fig. 7.25

$$\frac{dQ_n}{dt} = G_1 + G_2 + G_3 + G_4 + G_5 = -\frac{qn_i W}{\tau_g} - \frac{qn_i s_g A_S}{A_G} - qn_i s'_g - \frac{qn_i^2 D_n}{N_A L'_n} \quad (7.104)$$

where $A_S = 2\pi r W$ is the area of the lateral scr (assuming the lateral scr width to be identical to the vertical scr width W) and $A_G = \pi r^2$ is the gate area. L_n' is an effective diffusion length that couples bulk and back surface generation and is given by[122]

$$L_n' = L_n \frac{\cosh(d/L_n) + (s_c L_n/D_n) \sinh(d/L_n)}{(s_c L_n/D_n) \cosh(d/L_n) + \sinh(d/L_n)} \tag{7.105}$$

The surface generation velocity s_c depends on the type of back contact. For a p-semiconductor-metal contact, the surface generation velocity is very high. A p-p$^+$ semiconductor-metal has low s_c because the low-high p-p$^+$ contact is a minority carrier barrier.[123]

The first two terms in Eq. (7.104) are scr width-dependent generation rates. Here we consider the last two scr width-independent rates. Those with an n_i dependence all exhibit identical temperature dependence. However, G_4 has a n_i^2 dependence, causing it to increase faster with temperature. This is the basis for the recombination lifetime measurement. G_4 dominates for temperatures above about 75°C.

When $dQ_n/dt = -qn_i^2 D_n/N_A L_n'$ is substituted into Eq. (7.103)[122]

$$C = \frac{C_i}{\sqrt{1 - t/t_1}} \tag{7.106}$$

where $t_1 = (K_s/K_{ox})(C_{ox}/C_i)^2(N_A/n_i)^2(t_{ox}/2)(L_n'/D_n)$ and C_i is defined in Fig. 7.24.

The measurement consists of a C-t plot. When quasi-neutral region generation dominates, $1 - (C_i/C)^2$ plotted against t has a slope of $1/t_1$. The diffusion length is determined from t_1. To ensure that qnr generation dominates, the $[1 - (C_i/C)^2]$ versus t curve should be linear. If it is not, then the measurement temperature is probably too low.

7.5.5 Other Techniques

Short-Circuit Current Decay: For reverse-recovery, the diode current is switched from forward to reverse; for open-circuit voltage it is switched from forward to zero current. In the *short-circuit current decay method* the current is switched from forward current to short-circuit current or zero voltage. Emitter minority carriers play a relatively minor role in this measurement because they recombine very quickly when the diode is short-circuited.[124-125]

Conductivity Modulation: The *conductivity modulation* technique was developed to measure the recombination lifetime in epitaxial layers with thicknesses less than the minority carrier diffusion length. The measured lifetime is neither affected by substrate recombination nor by recombination in heavily doped regions. The structure consists of alternate n$^+$ and p$^+$ stripes diffused or implanted into a *p*-epi layer on a p$^+$ substrate. All p$^+$ stripes are connected to each other, and all n$^+$ stripes are connected to each other forming a lateral n$^+$ pp$^+$ diode. The spacing between stripes is smaller than the minority carrier diffusion length. The lateral diode is forward biased. A small ac voltage is superimposed on the dc bias, and the ac current due to recombination in the epitaxial layer is measured and related to the recombination lifetime.[126] By applying a dc voltage to the p$^+$ substrate, it is possible to profile the lifetime through the epitaxial layer.

7.6   GENERATION LIFETIME—ELECTRICAL MEASUREMENTS

7.6.1   Gate-Controlled Diode

The generation lifetime τ_g is determined by junction leakage current and MOS capacitor storage time measurements. One device to characterize the generation parameters is the three-terminal *gate-controlled diode*, consisting of a p substrate, an n^+ region (D), a circular gate (G) surrounding the n^+ region, and a circular guard ring (GR) surrounding the gate in Fig. 7.26. The gate is sometimes located in the center surrounded by a circular n^+ region. The gate should overlap the n^+ region slightly to prevent potential barriers. The guard ring should be close to the gate and bias the semiconductor into accumulation to isolate the gate-controlled diode from the rest of the wafer. Devices can also be decoupled by doping the semiconductor between the devices more heavily.

We give here the necessary background for generation lifetime and surface generation velocity measurements. Figure 7.26 shows three generation regions: (1) the diode scr (J), (2) the gate-induced scr (GIJ), and (3) the depleted surface (S) under the gate. Each region contributes a current with the total current $I_J + I_{GIJ} + I_S$. Let us first consider the semiconductor under the gate with the diode short-circuited to the substrate ($V_D = 0$). The surface is accumulated for $V_G < V_{FB}$, at flatband for $V_G = V_{FB}$, depleted for $V_{FB} < V_G < V_T$ and inverted for $V_G > V_T$. Accumulation and flatband conditions remain unchanged when the diode is reverse biased, but depletion and inversion conditions change. For diode voltage $V_D \neq 0$, depletion holds for $0 < \phi_s < V_D + 2\phi_F$ and inversion for $\phi_s \geq V_D + 2\phi_F$, with the surface potential ϕ_s related to the gate voltage by Eq. (6.16) and $\phi_F = (kT/q)\ln(N_A/n_i)$.

The diode is biased at a constant voltage V_{D1}, and the gate voltage is varied. The surface under the gate is accumulated for negative gate voltage $-V_{G1}$, illustrated in Fig. 7.27(a). The measured current is the diode scr generated current I_J shown in Fig. 7.27(a) and by point A in (d). The current increase for more negative gate voltages has been attributed to weak breakdown of the gate-induced n^+-p^+ junction at the surface. At $V_G = V_{FB}$ the semiconductor is at flatband, the diode scr width is the same at the surface as in the bulk.

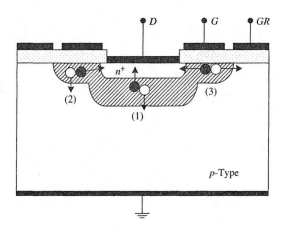

Fig. 7.26   The gate-controlled diode. D is the n^+ p diode, G the gate, and GR the guard ring, illustrating the various generation mechanisms and locations.

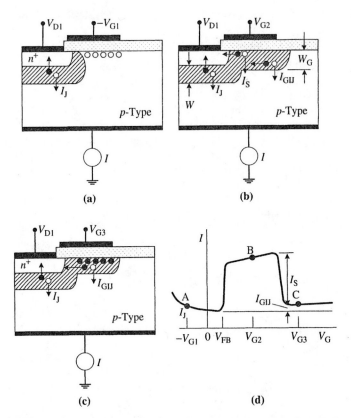

Fig. 7.27 Gate-controlled diode in (a) accumulation, (b) depletion, (c) inversion; (d) shows the current-voltage characteristic with points A, B, and C corresponding to (a), (b), and (c).

The surface under the gate depletes for $V_G > V_{FB}$, and the current increases rapidly, due to the surface generation current I_S and the gate-induced scr current I_{GIJ} in Fig. 7.27(b). Higher gate voltages lead to a more gradual current increase as the scr under the gate widens. Gate voltage V_{G2} (point B in Fig. 7.27(d)) is characteristic of this part of the current-voltage curve, the surface potential lies in the range $0 < \phi_s < V_{D1} + 2\phi_F$, and the scr width under the gate is given by

$$W_{G,dep} = \frac{K_s t_{ox}}{K_{ox}} \left(\sqrt{1 + \frac{2(V_G - V_{FB})}{V_0}} - 1 \right) \tag{7.107}$$

assuming no inversion charge. The surface inverts for surface potential $\phi_s \geq V_{D1} + 2\phi_F$, and the gate scr width pins to

$$W_{G,inv} = \sqrt{\frac{2K_s \varepsilon_o (V_{D1} + 2\phi_F)}{q N_A}} \tag{7.108}$$

The junction scr width is

$$W_J = \sqrt{\frac{2K_s\varepsilon_o(V_{D1} + V_{bi})}{qN_A}} \qquad (7.109)$$

with $V_{bi} = (kT/q)\ln(N_AN_D/n_i^2)$ the built-in potential.

Surface generation drops precipitously when the surface inverts and I_S effectively disappears, as shown by the inverted surface in Fig. 7.27(c). Further gate voltage increases beyond the inversion voltage give no further current changes. This is also evident from Eq. (7.15), which shows high surface generation for a depleted surface when p_s and n_s are low and low surface generation when either p_s or n_s is high. Thermal generation is reduced for heavily inverted surfaces, because most interface traps are occupied by electrons. Assuming zero surface generation, the current is due to the junction current I_J and the field-induced junction current I_{FIJ}, shown in Fig. 7.27(c) and by C in Fig. 7.27(d). Experimental current-voltage curves are shown in Fig. 7.28. In these curves the current does increase for $+V_G$ due to diode non-idealities not discussed here.

There is a sufficiently high reverse bias for the mobile carrier density to be negligible in the scr and at the depleted surface. The bulk and surface generation rates are given by Eqs. (7.20) and (7.22). The bulk scr generation current is $qG \times$ volume and the surface component is $qG_S \times$ area, where volume and area are the thermal generation volume and area. The total current is $I = I_J + I_{GIJ} + I_S$ with

$$I_J = \frac{qn_iW_JA_J}{\tau_{g,J}}; I_{GIJ} = \frac{qn_iW_GA_G}{\tau_{g,G}}; I_S = qn_is_gA_G \qquad (7.110)$$

where $\tau_{g,J}$ and $\tau_{g,G}$ are the generation lifetimes in the diode and gate region, respectively.

To extract the generation parameters one must measure or calculate the various scr widths. The widths can be experimentally determined from capacitance measurements, but it is usually more convenient to calculate them using Eqs. (7.107) to (7.109). To determine the surface generation velocity, one usually makes $I-V_G$ measurements at low diode voltages ($V_D \approx 0.5-1V$), thereby increasing the importance of surface current relative to bulk current. It is also possible, of course, to determine the generation lifetime

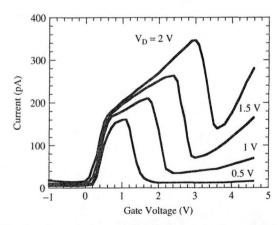

Fig. 7.28 Experimental gate-controlled diode current-voltage characteristics.

under the gate separately from the generation lifetime under the n^+ diffusion and to profile both as a function of depth.[127]

The theory discussed so far was originally proposed by Grove and Fitzgerald.[128] It is based on several simplifying assumptions. It assumes the current to be due to scr-generated current only. This is a reasonable assumption for Si devices at room temperature, but the quasi-neutral current component may not be negligible for high-lifetime devices. The ratio of the bulk scr current to the bulk quasi-neutral region current, that is, component 1 to component 4 in Eq. (7.104), becomes

$$\frac{I_{scr}}{I_{qnr}} = \frac{N_A W L_n}{n_i D_n \tau_g} = \frac{N_A W \sqrt{\tau_r}}{n_i \sqrt{D_n} \tau_g} \approx 36 \frac{\sqrt{\tau_r}}{\tau_g} \tag{7.111}$$

for $N_A/n_i = 10^6$, $W = 2$ μm, $D_n = 30$ cm²/s and $L_n = (D_n \tau_r)^{1/2}$. For $\tau_g = \tau_r = 1$ μs we find the ratio to be 36,000, and scr current clearly dominates and for $\tau_g = 1$ ms and $\tau_r = 100$ μs, it is 360. The ratio approaches unity for temperatures above room temperature. When the ratio in Eq. (7.111) approaches unity, quasi-neutral region current becomes important and Eq. (7.110) no longer holds.

The assumption of total depletion of the surface before inversion has been shown not to be the case.[129] The lateral surface current inverts the surface weakly for all but a very small fraction of the channel for gate biases far below the gate voltage required to invert the surface strongly. Active devices are frequently surrounded by implantation-doped/thick-oxide channel stops. These channel-stop sidewalls contribute additional current, and the gate-controlled diode is an effective test structure to measure this current.[130]

7.6.2 Pulsed MOS Capacitor

The pulsed MOS capacitor lifetime measuring technique is commonly used to determine τ_g. Many papers have been written on the basic method, first proposed by Zerbst in 1966,[131] and on subsequent variations. A review of the various methods can be found in Kang and Schroder.[132] We give the most relevant concepts and equations here for three popular versions of this method, leaving the details to the published literature.

Zerbst Plot: The MOS capacitor is pulsed into deep depletion, and the capacitance-time curve is measured, as shown in Fig. 7.29. An experimental room-temperature $C - t$ curve is shown in Fig. 7.29(a). The capacitance relaxation is determined by thermal ehp generation, which can be written as

$$\frac{dQ_n}{dt} = -\frac{qn_i(W - W_{inv})}{\tau_g} - \frac{qn_i s_g A_S}{A_G} - qn_i s_{eff} = -\frac{qn_i(W - W_{inv})}{\tau_{g,eff}} - qn_i s_{eff} \tag{7.112}$$

where $W_{inv} = (4 K_s \varepsilon_o \phi_F / q N_A)^{1/2}$ and $\tau_{g,eff}$ is defined by considering only the scr generation rates

$$\frac{dQ_{n,scr}}{dt} = -\frac{qn_i(W - W_{inv})}{\tau_g} - \frac{qn_i s_g A_S}{A_G} = -qn_i \left(\frac{W - W_{inv}}{\tau_g} + \frac{2\pi r s_g(W - W_{inv})}{\pi r^2} \right)$$
$$= -\frac{qn_i(W - W_{inv})}{\tau_g} \left(1 + \frac{2 s_g \tau_g}{r} \right) = -\frac{qn_i(W - W_{inv})}{\tau_{g,eff}} \tag{7.113}$$

The effective scr width $(W - W_{inv})$ approximates the actual generation width and ensures that at the end of the $C - t$ transient the scr generation becomes zero. The term $qn_i s_{eff}$

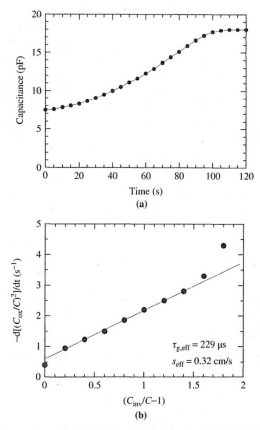

Fig. 7.29 (a) $C - t$ response and (b) Zerbst plot. Reprinted with permission after Kang and Schroder.[132]

accounts for the scr width-independent generation rates (surface generation under the gate and in the quasi-neutral region) with

$$s_{eff} = s'_g + \frac{n_i D_n}{N_A L'_n} \tag{7.114}$$

The scr width is related to the capacitance C through

$$W = K_s \varepsilon_o \frac{C_{ox} - C}{C_{ox} C} \tag{7.115}$$

Combining Eqs. (7.103), (7.112) and (7.115) gives

$$-\frac{d}{dt}\left(\frac{C_{ox}}{C}\right)^2 = \frac{2n_i}{\tau_{g,eff} N_A} \frac{C_{ox}}{C_{inv}} \left(\frac{C_{inv}}{C} - 1\right) + \frac{2K_{ox} n_i s_{eff}}{K_s t_{ox} N_A} \tag{7.116}$$

using the identity $(2/C^3)\, dC/dt = -[d(1/C)^2/dt]$.

Equation (7.116) is the basis of the well-known *Zerbst plot*, $-d(C_{ox}/C)^2/dt$ versus $(C_f/C - 1)$, shown in Fig. 7.29(b). The curved portion near the origin is when the device approaches equilibrium and the curvature at the other end of the straight line has been attributed to field-enhanced emission from interface and/or bulk traps.[133] The slope of the straight line is $2n_i C_{ox}/N_A C_f \tau_{g,eff}$ and its extrapolated intercept on the vertical axis is $2n_i K_{ox} s_{eff}/K_s t_{ox} N_A$. The slope is a measure of the scr generation parameters τ_g and s_g, whereas the intercept is related to the scr width-independent generation parameters s'_g, L_n and s_c. s_{eff} obtained from the intercept should not be interpreted as the surface generation velocity as is sometimes done. It includes not only the quasi-neutral bulk generation rates, but a more detailed analysis of the $C - t$ response shows that the inherent inaccuracy of the $(W - W_{inv})$ approximation for the generation width can lead to a non-zero intercept even if $s_{eff} = 0$.[134]

It is instructive to examine the two axes of the *Zerbst plot* for a better insight into the physical meaning of such a plot. For the identity that leads to Eq. (7.116), we find from Eq. (7.103) and (7.116)

$$-\frac{d}{dt}\left(\frac{C_{ox}}{C}\right)^2 \sim \frac{dQ_n}{dt}; \frac{C_{inv}}{C} - 1 \sim W - W_{inv} \tag{7.117}$$

The Zerbst plot vertical axis is proportional to the total ehp carrier generation rate or to the generation current and the horizontal axis is proportional to the scr generation width. So this rather complicated plot is nothing more than a plot of generation current versus scr width.

The measured $C - t$ transient times are usually quite long with times of tens of seconds to minutes being common. The relaxation time t_f is related to $\tau_{g,eff}$ by[135-136]

$$t_f \approx \frac{10N_A}{n_i}\tau_{g,eff} \tag{7.118}$$

This equation brings out a very important feature of the pulsed MOS-C technique, which is the magnification factor N_A/n_i built into the measurement. Values of $\tau_{g,eff}$ range over many orders of magnitude, but representative values for high quality silicon devices lie in the range of 10^{-4} to 10^{-2} s. Equation (7.118) predicts the actual $C - t$ transient time to be 10 to 10^4 s. These long times point out a virtue of this measurement technique. To measure lifetimes in the microsecond range, it is only necessary to measure capacitance recovery times on the order of seconds.

The time magnification factor in Eq. (7.118) is also a disadvantage. Such long measurement times preclude mapping of large number of devices. Several approaches have been proposed to reduce the measurement time. Equations (7.112) and (7.114) show $s_{eff} \sim n_i^2$. As the temperature is raised, this scr width-independent term becomes more important and the relaxation time is considerably reduced. The Zerbst plot is shifted vertically retaining the slope determined by $\tau_{g,eff}$.[136] As pointed out in Section 7.5.4, the temperature should not be so high that quasi-neutral generation dominates, for then it is impossible to extract $\tau_{g,eff}$. A t_f reduction is also attained by illuminating the sample.[137]

The measurement time can also be reduced by driving the MOS-C into deep depletion by a voltage pulse and then into inversion by a light pulse. Subsequently, a series of small pulses of opposite polarity and varying amplitudes are superimposed on the depleting voltage, driving the device into weaker inversion and then into depletion. C and dC/dt are determined after each pulse to construct a Zerbst plot. The total measurement time can be reduced by as much as a factor of 10.[138] In yet another simplification the scr width is calculated from the $C - t$ response, and $\ln(W)$ is plotted against time.[139] Such a plot is nearly linear. The line can be extrapolated to t_f without recording the entire curve by using only the initial portion of the $C - t$ response.

One caution about MOS-C generation lifetime measurement is the possibility of gate oxide currents for thin oxides.[140] After pulsing the device into deep depletion, the inversion layer builds up as a function of time until equilibrium is attained. If part of this inversion layer leaks through the oxide during the measurement, obviously the measurement time is extended, leading to an incorrect $\tau_{g,eff}$. Solutions to this problem are using lower gate voltages for which gate oxide leakage current is negligible or using a constant charge approach, as discussed for the corona-oxide-semiconductor method later in this section. Another issue for thin oxides is that tunneling electrons or holes when they enter the semiconductor from the gate have sufficient energy to generate additional ehps by impact ionization.[141] A contactless capacitance measurement technique uses a metal probe held slightly less than one micrometer above the sample. $C–V$ and $C - t$ measurements have been implemented without the need of a permanent contact on the sample.[142] Pulsed capacitor measurements can also be implemented in silicon-on-insulator (SOI) samples, where the SOI MOSFET drain current is measured, illustrated in Fig. 7.30. The device is biased above threshold ($V_G > V_T$) with some back gate bias V_{GB1}. The back gate bias is then pulsed to V_{GB2} and the resulting drain current transient is measured. The analysis is similar to $C - t$ analyses and $\tau_{g,eff}$ can be extracted.[143]

It is also possible to extract the SOI recombination lifetime. With the SOI MOSFET back gate grounded, the front gate is switched from depletion or accumulation to strong inversion. The minority carriers to form the inversion channel are rapidly supplied by the source/drain regions. The positive front gate pulse leads to a scr region extension, but the majority carriers expelled from this region cannot be removed instantaneously and are stored in the neutral body, inducing a temporary increase in the body potential, reducing the threshold voltage and increasing the drain current. Equilibrium is reached through carrier *recombination* to remove the excess majority carriers.[144]

Current-Capacitance: The *Zerbst* technique requires differentiation of the experimental data and a knowledge of N_A. The *current-capacitance* technique requires neither, but one must measure the current and the capacitance of a pulsed MOS-C. The current is

$$I = A_G \left(\frac{dQ_n}{dt} + qN_A \frac{dW}{dt} \right) \tag{7.119}$$

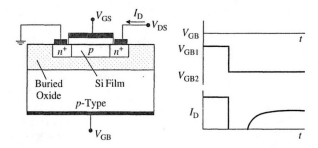

Fig. 7.30   Generation lifetime measurement schematic for SOI devices.

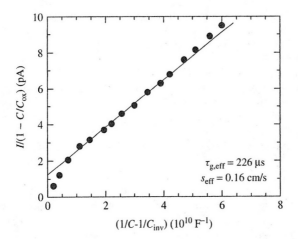

Fig. 7.31 Current versus inverse capacitance plot for the device whose Zerbst plot is shown in Fig. 7.29. Reprinted with permission after Kang and Schroder.[132]

where the first term is the generation and the second term the displacement current. The current-capacitance relationship is[133]

$$\frac{I}{1 - C/C_{ox}} = \frac{q K_s \varepsilon_o A_G^2 n_i}{\tau_{g,eff}} \left(\frac{1}{C} - \frac{1}{C_{inv}} \right) + q A_G n_i s_{eff} \tag{7.120}$$

From the $C - t$ and the $I - t$ curve one plots $I/(1 - C/C_{ox})$ versus $(1/C - 1/C_f)$. The slope of this curve gives $\tau_{g,eff}$ and the intercept gives s_{eff}, as shown in Fig. 7.31.

For a generation lifetime profile, Eq. (7.120) can be written as

$$\tau_{g,eff} = \frac{q K_s \varepsilon_o A_G^2 n_i}{C_{ox}} \frac{d(C_{ox}/C)/dt}{d[I/(1 - C/C_{ox})]/dt} \tag{7.121}$$

By measuring current and capacitance simultaneously and differentiating the data, it is possible to plot a profile of $\tau_{g,eff}$ directly without knowing the doping profile.

In a modified current-capacitance method, leading to *much reduced measurement times*, the scr generation current density[145]

$$J_{scr} = \frac{C_{ox}}{C_{ox} - C} J = \frac{q n_i (W - W_{inv})}{\tau_{g,eff}} + q n_i s_{eff} \tag{7.122}$$

is combined with the scr width

$$W = K_S \varepsilon_o A_g \left(\frac{1}{C} - \frac{1}{C_{ox}} \right) \tag{7.123}$$

The current and the high-frequency capacitance are measured simultaneously immediately after pulsing into deep depletion. The pulse duration is just long enough to measure the capacitance and the current. From these data the scr width W and the scr current density J_{scr} are determined. The pulse height is continuously increased, probing deeper into the sample. The measurement time is determined solely by the acquisition time of the capacitance meter and the ammeter multiplied by the number of data points. J_{scr} is plotted versus W and the slope of this curve yields the effective generation lifetime according to

$$\tau_{g,eff} = \frac{q n_i (W - W_{inv})}{d J_{scr}/d W} \tag{7.124}$$

Such a plot is similar to Fig. 7.31. The doping concentration need not be known.

Linear Sweep: In the *linear sweep* technique a linearly varying voltage is applied to the gate of an MOS-C of a polarity to drive the device into depletion. We showed in Chapter 6 that for sufficiently slow sweep rates, the equilibrium $C-V_G$ curve is traced out. We also know that when the sweep rate is high, the pulsed MOS-C deep-depletion curve is obtained. For intermediate sweep rates, an intermediate trace is swept out, shown in Fig. 7.32, lying between the deep-depletion and equilibrium curves.

The interesting point about this curve is its saturation characteristic.[146] Assume the voltage sweeps from point A in Fig. 7.32 to the right. For voltages more positive than V_B, the scr widens beyond W_{inv}, with the capacitance driven below C_{inv}. Electron-hole pair generation attempts to re-establish equilibrium, but the gate voltage continues to drive the device into deep depletion, further increasing W. This in turn enhances the generation rate that is proportional to W. At the voltage V_{sat} the attempt by the linearly varying gate voltage to drive the device into deeper depletion is exactly balanced by the generation rate holding it at that capacitance. The capacitance-voltage curve saturates at C_{sat}.

For a constant sweep rate, $dV_G/dt = R$, Eq. (7.102) becomes

$$\frac{d Q_n}{dt} = -\frac{q K_s \varepsilon_o C_{ox} N_A}{C^3}\frac{dC}{dt} - C_{ox} R \tag{7.125}$$

Using the generation rate expression Eq. (7.112), leads to

$$-\frac{d}{dt}\left(\frac{C_{ox}}{C}\right)^2 = \frac{2}{V_0}\left(\frac{q K_s \varepsilon_o n_i (C_{inv}/C - 1)}{C_{inv} C_{ox} \tau_{g,eff}} + \frac{q n_i s_{eff}}{C_{ox}} - R\right) \tag{7.126}$$

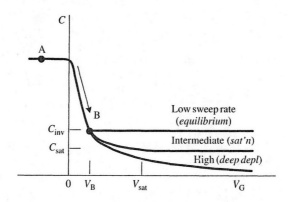

Fig. 7.32 Inversion, saturation, and deep-depletion MOS-C curves.

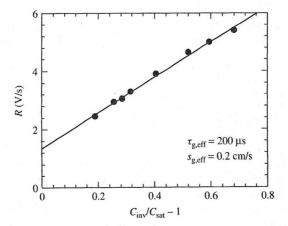

Fig. 7.33 Linear ramp plot for the device whose Zerbst plot is shown in Fig. 7.29. Reprinted with permission after Kang and Schroder.[132]

When the device enters saturation, C_{sat} changes neither with voltage nor with time and the left side of Eq. (7.126) becomes zero and

$$R = \frac{q K_s \varepsilon_o n_i (C_{inv}/C_{sat} - 1)}{C_{inv} C_{ox} \tau_{g,eff}} + \frac{q n_i s_{eff}}{C_{ox}} \tag{7.127}$$

Equation (7.127) gives the relationship between the linear sweep rate R and the generation parameters $\tau_{g,eff}$ and s_{eff}. In the experiment, a series of C–V_G curves at different linear sweep rates are plotted. The C_{sat} values are taken from these curves, and a plot of R versus $(C_{inv}/C_{sat} - 1)$ yields a straight line of slope $q K_s \varepsilon_o n_i / C_{inv} C_{ox} \tau_{g,eff}$ and intercept $q n_i s_{eff}/C_{ox}$. Similar to the *Zerbst plot*, $\tau_{g,eff}$ is obtained from the slope and s_{eff} from the intercept.

Experimental data for the linear sweep method are shown in Fig. 7.33 for the device whose *Zerbst plot* is shown in Fig. 7.29. Note the good agreement between the experimentally determined values for $\tau_{g,eff}$ and s_{eff}. The linear sweep technique does not require the acquisition of an entire $C-t$ curve, nor the differentiation of the experimental data. It does, however, require multiple saturating C–V_G curves. For those devices with high lifetimes and consequent long $C-t$ transients, it is found that very low sweep rates are required, with resultant long data acquisition times. The use of a feedback circuit, with the capacitance preset to a certain value and the linear sweep rate adjusting itself through feedback to maintain this preset value, reduces the data acquisition time.[147] Computer automation of the linear sweep technique has also been developed.[148] The technique has also been adapted for lifetime measurements in silicon-on-insulator materials.[149]

Corona-Oxide-Semiconductor: The *corona-oxide-semiconductor* (COS) technique is illustrated in Fig. 7.34 (corona charge characterization is discussed in Chapter 9).[150] Positive or negative charge from a corona source is deposited on the semiconductor sample surface. In Fig. 7.34, negative charge is deposited on an oxidized p-Si wafer to bias the substrate into accumulation followed by a smaller area positive charge driving the sample

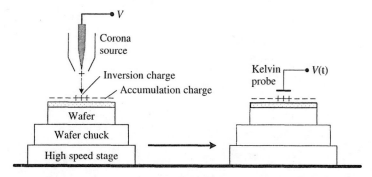

Fig. 7.34 Corona-pulsed deep-depletion measurement apparatus.

into deep depletion. The relaxation to equilibrium is monitored by measuring the Kelvin probe voltage as a function of time. The ability to deposit both negative and positive charge has the distinct advantage over MOS-C measurements of providing a zero-gap guard ring, reducing perimeter generation.

The gate and oxide voltages of an MOS-C or COS-C are

$$V_G = V_{FB} + V_{ox} + \phi_s; V_{ox} = \frac{Q_G}{C_{ox}} = -\frac{Q_S}{C_{ox}} \qquad (7.128)$$

with Q_G the gate charge density and Q_S the semiconductor charge density. Equation (7.128) becomes

$$V_G - V_{FB} = \phi_s - \frac{Q_S}{C_{ox}} = \phi_s - \frac{Q_b - Q_n}{C_{ox}} \qquad (7.129)$$

where Q_n is the inversion charge density and Q_b the bulk charge density.

After corona charge is deposited, Q_G and V_{ox} remain constant. Differentiating Eq. (7.128) leads to

$$\frac{dV_G}{dt} = \frac{d\phi_s}{dt} \qquad (7.130)$$

assuming $dV_{FB}/dt = 0$. The bulk charge density is

$$Q_b = -qN_A W = -\sqrt{2qK_s\varepsilon_o N_A \phi_s} \qquad (7.131)$$

where W is the space-charge region width. With Q_G and Q_S constant with time

$$\frac{dQ_S}{dt} = 0 = -\frac{dQ_n}{dt} + \frac{dQ_b}{dt} = -\frac{dQ_n}{dt} - qN_A\frac{dW}{dt} \qquad (7.132)$$

or, using Eq. (7.131),

$$\frac{dQ_n}{dt} = -\sqrt{\frac{qK_s\varepsilon_o N_A}{2\phi_s}}\frac{d\phi_s}{dt} = -\frac{K_s\varepsilon_o}{W}\frac{d\phi_s}{dt} = -\frac{K_s\varepsilon_o}{W}\frac{dV_G}{dt} \qquad (7.133)$$

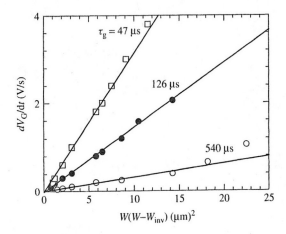

Fig. 7.35   COS generation lifetime plot.

With dQ_n/dt given by Eq. (7.112)

$$\frac{dV_G}{dt} = \frac{qn_iW}{K_s\varepsilon_o}\left(\frac{W - W_{inv}}{\tau_{g,eff}} + s_{eff}\right)\qquad(7.134)$$

The dependence of the voltage rate of change on $W(W - W_{inv})$ is shown in Fig. 7.35. The linear relationship is in good agreement with the prediction of Eq. (7.134). The slopes of these lines give $\tau_{g,eff}$. All lines intersect at the origin, implying s_{eff} to be negligibly small.

7.7   STRENGTHS AND WEAKNESSES

Recombination Lifetime: Recombination lifetime or diffusion length measurements have become ubiquitous in the semiconductor industry, because they are a good indicator of wafer contamination. Among the optical recombination lifetime measuring methods, the microwave reflection or inductive coupling photoconductance decay technique is commonly used. Its major strength is the contactless nature and rapid measurement. Its major weakness is the unknown surface recombination velocity. If the sample thickness can be changed, then both the bulk lifetime and the surface recombination velocity can be extracted. The *quasi-steady-state photoconductance method* is a more recent method and has found wide acceptance in the photovoltaic community. Its main strength is a measure of the lifetime as a function of injection level in one measurement. One of its disadvantages is the large sample area (several cm^2) precluding high density mapping. *Free carrier absorption* through the use of black body emitters is an interesting method allowing a lifetime map to be obtained in a very short time through the use of two dimensional imagers.

Another common optical technique is surface photovoltage. A chief application is the detection of iron in *p*-Si. It is a low injection level method and is not subject to trapping. The open-circuit voltage decay method is the most common electrical recombination lifetime method. It is easy to interpret, but a junction diode is required. Measured τ_r or L_n mean little for thin layers, *e.g.*, epitaxial layers on highly doped substrates, denuded zones on heavily precipitated substrates, or silicon-on-insulator films. Such layers are best characterized through generation lifetime measurements.[94]

Generation Lifetime: The generation lifetime is commonly determined with the pulsed MOS capacitor. The Zerbst plot implementation is the most common, but the current versus inverse capacitance is easier to interpret because the doping density of the sample need not be known. Since τ_g is measured in the space-charge region of a reverse-biased device (diode or MOS device), it lends itself easily for the characterization of thin layers, *e.g.*, epitaxial layers on highly doped substrates,[94] denuded zones on heavily precipitated substrates, or silicon-on-insulator films. Furthermore, since the scr width can be varied by an applied voltage, it is possible to generate a τ_g depth profile, that is difficult to do with τ_r measurements, because the measurement depth for τ_r and L_n measurements is the minority carrier diffusion length. To avoid contact formation, one can use the corona-oxide-semiconductor approach, replacing the metal or poly-Si gate with corona charge.

APPENDIX 7.1

Optical Excitation

Steady State: We consider the p-type semiconductor of Fig. A7.1. It is a wafer of thickness d, reflectivity R, minority carrier lifetime τ, minority carrier diffusion coefficient D, minority carrier diffusion length L, and surface recombination velocities s_{r1} and s_{r2} at the two surfaces. Monochromatic light of photon flux density Φ, wavelength λ, and absorption coefficient α, is incident on one side of this wafer. Carriers generated by absorbed photons diffuse in the x-direction and the wafer is infinite in the $y - z$ plane, allowing edge effects to be neglected. The steady-state, small-signal excess minority carrier density $\Delta n(x)$ is obtained from a solution of the one-dimensional continuity equation

$$D\frac{d^2 \Delta n(x)}{dx^2} - \frac{\Delta n(x)}{\tau} + G(x) = 0 \qquad (A7.1)$$

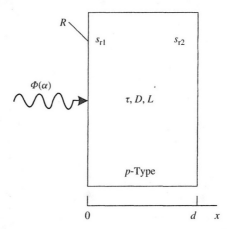

Fig. A7.1   Homogeneous p-type sample geometry with optical excitation.

subject to the boundary conditions

$$\frac{d\Delta n(x)}{dx}\bigg|_{x=0} = s_{r1}\frac{\Delta n(0)}{D} \quad \text{and} \quad \frac{d\Delta n(x)}{dx}\bigg|_{x=d} = -s_{r2}\frac{\Delta n(d)}{D} \tag{A7.2}$$

The generation rate is

$$G(x,\lambda) = \Phi(\lambda)\alpha(\lambda)[1 - R(\lambda)]\exp(-\alpha(\lambda)x) \tag{A7.3}$$

An implicit assumption in this expression is that each *absorbed* photon generates one ehp.

The solution to Eq. (A7.1) using (A7.2) and (A7.3) is[151]

$$\Delta n(x) = \frac{(1-R)\Phi\alpha\tau}{(\alpha^2 L^2 - 1)}\left(\frac{A_1 + B_1 e^{-\alpha d}}{D_1} - \exp(-\alpha x)\right) \tag{A7.4}$$

where

$$A_1 = \left(\frac{s_{r1}s_{r2}L}{D} + s_{r2}\alpha L\right)\sinh\left(\frac{d-x}{L}\right) + (s_{r1} + \alpha D)\cosh\left(\frac{d-x}{L}\right)$$

$$B_1 = \left(\frac{s_{r1}s_{r2}L}{D} - s_{r1}\alpha L\right)\sinh\left(\frac{x}{L}\right) + (s_{r2} - \alpha D)\cosh\left(\frac{x}{L}\right)$$

$$D_1 = \left(\frac{s_{r1}s_{r2}L}{D} + \frac{D}{L}\right)\sinh\left(\frac{d}{L}\right) + (s_{r1} + s_{r2})\cosh\left(\frac{d}{L}\right)$$

For some measurement methods, the excess carrier density is required, for others, the current density.

In the derivation of Eq. (A7.4) only diffusion was considered. The electric fields are assumed to be sufficiently small that drift is negligible. The diffusion current density is

$$J_n(x) = qD\frac{d\Delta n(x)}{dx} \tag{A7.5}$$

From Eq. (A7.4), $J_n(x)$ can be written as

$$J_n(x) = \frac{q(1-R)\Phi\alpha L}{(\alpha^2 L^2 - 1)}\left(\frac{A_2 - B_2 e^{-\alpha d}}{D_1} - \alpha L \exp(-\alpha x)\right) \tag{A7.6}$$

where

$$A_2 = \left(\frac{s_{r1}s_{r2}L}{D} + s_{r2}\alpha L\right)\cosh\left(\frac{d-x}{L}\right) + (s_{r1} + \alpha D)\sinh\left(\frac{d-x}{L}\right)$$

$$B_2 = \left(\frac{s_{r1}s_{r2}L}{D} - s_{r1}\alpha L\right)\cosh\left(\frac{x}{L}\right) + (s_{r2} - \alpha D)\sinh\left(\frac{x}{L}\right)$$

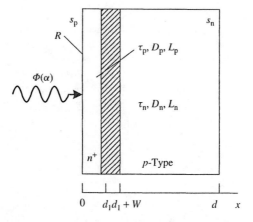

Fig. A7.2 Junction geometry for optical excitation.

For the n^+p junction of Fig. A7.2 we derive the excess carrier density and the current density expressions by some modification to Eqs. (A7.4) and (A7.6). Hovel gives an excellent discussion.[152] For the n^+ layer we are concerned with the thin top layer of thickness d_1. Hence in Eq. (A7.4): $d \to d_1$ and $s_{r1} \to s_p$. We are especially interested in the excess carrier densities under short-circuit current conditions, where the excess carrier density is zero at the edge of the space-charge region ($x = d_1$). From a surface recombination point of view, this means $s_{r2} = \infty$, resulting in

$$\Delta p(x) = \frac{(1 - R)\Phi\alpha\tau_p}{(\alpha^2 L_p^2 - 1)} \left(\frac{A_3 + B_3 e^{-\alpha d_1}}{D_3} - \exp(-\alpha x) \right) \tag{A7.7}$$

with

$$A_3 = \left(\frac{s_p L_p}{D_p} + \alpha L_p \right) \sinh\left(\frac{d_1 - x}{L_p} \right)$$

$$B_3 = \left(\frac{s_p L_p}{D_p} \right) \sinh\left(\frac{x}{L_p} \right) + \cosh\left(\frac{x}{L_p} \right)$$

$$D_3 = \left(\frac{s_p L_p}{D_p} \right) \sinh\left(\frac{d_1}{L_p} \right) + \cosh\left(\frac{d_1}{L_p} \right)$$

Similar arguments for the p-substrate, using $x' = (x - d_1 - W)$, $d' = (d - d_1 - W)$, and $s_{r1} = \infty$, give

$$\Delta n(x') = \frac{(1 - R)\Phi\alpha\tau_n}{(\alpha^2 L_n^2 - 1)} \left(\frac{A_4 + B_4 e^{-\alpha d'}}{D_4} - \exp(-\alpha x') \right) \exp(-\alpha(d_1 + W)) \tag{A7.8}$$

with

$$A_4 = \left(\frac{s_n L_n}{D_n} \right) \sinh\left(\frac{d' - x'}{L_n} \right) + \cosh\left(\frac{d' - x'}{L_n} \right)$$

$$B_4 = \left(\frac{s_n L_n}{D_n} - \alpha L_n \right) \sinh\left(\frac{x'}{L_n} \right)$$

$$D_4 = \left(\frac{s_n L_n}{D_n} \right) \sinh\left(\frac{d'}{L_n} \right) + \cosh\left(\frac{d'}{L_n} \right)$$

The additional term $\exp[-\alpha(d_1 + W)]$ in Eq. (A7.8) accounts for carrier generation beyond $x = d_1 + W$. The absorbed photon flux density is already diminished by this factor when the photons enter the p-substrate. The current density for the short-circuited structure of Fig. A7.2 is obtained by considering the diffusion current only, as in Eq. (A7.5). An implicit assumption is that there are no voltage drops across the n^+ and p regions and that drift currents are negligible in these two regions. In the scr the electric field is dominant, and recombination is negligible. With these assumptions, the short-circuit current density is

$$\boxed{J_{sc} = J_p + J_n + J_{scr}} \tag{A7.9}$$

The *hole* current density is

$$J_p = \frac{q(1 - R)\Phi\alpha L_p}{(\alpha^2 L_p^2 - 1)} \left(\frac{A_5 - B_5 e^{-\alpha d_1}}{D_5} - \alpha L_p \exp(-\alpha d_1) \right) \tag{A7.10}$$

where

$$A_5 = \frac{s_p L_p}{D_p} + \alpha L_p$$

$$B_5 = \left(\frac{s_p L_p}{D_p} \right) \cosh\left(\frac{d_1}{L_p} \right) + \sinh\left(\frac{d_1}{L_p} \right)$$

$$D_5 = \left(\frac{s_p L_p}{D_p} \right) \sinh\left(\frac{d_1}{L_p} \right) + \cosh\left(\frac{d_1}{L_p} \right)$$

The *electron* current density is

$$J_n = \frac{q(1 - R)\Phi\alpha L_n}{(\alpha^2 L_n^2 - 1)} \left(\frac{-A_6 + B_6 e^{-\alpha d'}}{D_6} + \alpha L_n \right) \exp(-\alpha(d_1 + W)) \tag{A7.11}$$

where

$$A_6 = \left(\frac{s_n L_n}{D_n} \right) \cosh\left(\frac{d'}{L_n} \right) + \sinh\left(\frac{d'}{L_n} \right)$$

$$B_6 = \frac{s_n L_n}{D_n} - \alpha L_n$$

$$D_6 = \left(\frac{s_n L_n}{D_n} \right) \sinh\left(\frac{d'}{L_n} \right) + \cosh\left(\frac{d'}{L_n} \right)$$

and the *space-charge region* current density is

$$J_{scr} = q(1 - R)\Phi \exp(-\alpha d)(1 - \exp(-\alpha W)) \tag{A7.12}$$

Transient: The transient one-dimensional continuity equation for the sample geometry of Fig. A7.1 is

$$\frac{\partial \Delta n(x, t)}{\partial t} = D \frac{\partial^2 \Delta n(x, t)}{\partial x^2} - \frac{\Delta n(x, t)}{\tau_B} + G(x, t) \tag{A7.13}$$

Generally, during transient measurements, the carrier decay is monitored after the excitation source is turned off; that is, $G(x, t) = 0$ during the measurement.

A solution to Eq. (A7.13) with $G(x, t) = 0$ and the boundary conditions

$$\frac{\partial \Delta n(x, t)}{\partial x}\Big|_{x=0} = s_{r1}\frac{\Delta n(0, t)}{D} \quad \text{and} \quad \frac{\partial \Delta n(x, t)}{\partial x}\Big|_{x=d} = -s_{r2}\frac{\Delta n(d, t)}{D} \tag{A7.14}$$

gives[21, 153]

$$\Delta n(x, t) = \sum_{m=1}^{\infty} A_m \exp(-t/\tau_m) \tag{A7.15}$$

where the coefficients A_m depend upon the initial conditions. For optical carrier generation with light of wavelength λ and optical absorption coefficient α, A_m is given by[22]

$$A_m = \frac{8G_o \exp(-\alpha d/2)}{d} \frac{\sin(\beta_m d/2)}{(\alpha^2 + \beta^2)(\beta_m d + \sin(\beta_m d))}$$

$$\times \left(\alpha \sinh\left(\frac{\alpha d}{2}\right) \cos\left(\frac{\beta_m d}{2}\right) + \beta_m \cosh\left(\frac{\alpha d}{2}\right) \sin\left(\frac{\beta_m d}{2}\right)\right)$$

$$\times \exp\left(-\left(\frac{1}{\tau_B} + \beta_m^2 D\right)t\right) \tag{A7.16}$$

where G_o is the generation rate. Equation (A7.16) holds for $s_{r1} = s_{r2}$. For the more general case of $s_{r1} \neq s_{r2}$ the expression becomes slightly more complicated.[154] The appropriate expressions for excess carriers generated by an electron beam are given in ref. 21.

The decay time constants τ_m are given by

$$\frac{1}{\tau_m} = \frac{1}{\tau_B} + D\beta_m^2 \tag{A7.17}$$

with β_m being the mth root of

$$\tan(\beta_m d) = \frac{\beta_m(s_{r1} + s_{r2})D}{\beta_m^2 D^2 - s_{r1}s_{r2}} \tag{A7.18}$$

These equations are similar to Eqs. (7.28) and (7.29). Equation (A7.18) becomes Eq. (7.29) for $s_{r1} = s_{r2} = s_r$. The excess carrier decay curve is a sum of exponentials with the higher-order solutions decaying more rapidly with time than the first and may be neglected after an initial transient period as shown in Fig. A7.3. The dominant mode decays exponentially with a time constant τ_{eff}

$$\frac{1}{\tau_{eff}} = \frac{1}{\tau_B} + D\beta_1^2 \tag{A7.19}$$

with β_1 being the first real root of Eq. (A7.18).

For low surface recombination velocity ($s_{r1} = s_{r2} = s_r \to 0$)

$$\frac{1}{\tau_{eff}} = \frac{1}{\tau_B} + \frac{2s_r}{d} \tag{A7.20a}$$

while for high s_r ($s_{r1} = s_{r2} \to \infty$)

$$\frac{1}{\tau_{eff}} = \frac{1}{\tau_B} + \frac{\pi^2 D}{d^2} \tag{A7.20b}$$

The measured lifetime is always less than the true recombination lifetime according to Eq. (A7.20). The discrepancy of the measured lifetime from the true lifetime depends on s_r, τ_B, and d. A more detailed discussion of the decay rate is given in refs. 21 and 22.

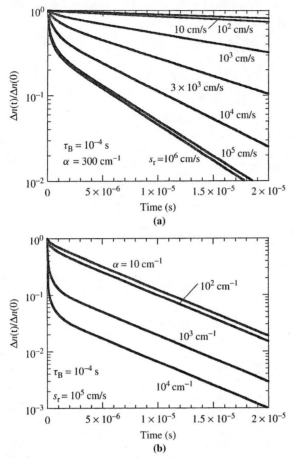

Fig. A7.3 Calculated normalized excess carrier density versus time as a function of (a) surface recombination velocity, (b) absorption coefficient. $d = 400 \ \mu$m.

The curves in Fig. A7.3 exhibit an initial rapid decay that depends on s_r and α. While it is difficult to vary s_r reproducibly, it is easy to vary α by changing the incident light wavelength, allowing s_r to be extracted.[51, 155]

All of the above theories are valid for low-level injection, where the SRH, radiative, and Auger lifetimes can be treated as constants and, aside from surface effects, the transient decay can be considered to be of an exponential form. This is no longer true for high-level injection, especially for radiative and Auger recombination, because the lifetimes themselves are functions of the excess carrier densities and the decay is no longer exponential. The equations become very complex, and a detailed discussion is given by Blakemore.[156]

In some measurement techniques, a phase shift between the optical excitation source and the detected parameter is measured. For a sinusoidally varying generation rate,

$$G(x, t) = (G_0 + G_1 e^{j\omega t}) \exp(-\alpha x) = (\Phi_0 + \Phi_1 e^{j\omega t})\alpha(1 - R)\exp(-\alpha x) \quad \text{(A7.21)}$$

the fundamental component of the variation of the excess minority carrier density $\Delta n_1(x)\exp(j\omega t)$ is determined from the equation

$$D\frac{d^2\Delta n_1(x)}{dx^2} - \frac{\Delta n_1(x)}{\tau_B} + G_1\exp(-\alpha x) = j\omega\Delta n_1(x) \qquad (A7.22)$$

The solution to this equation, subject to the same boundary conditions as Eq. (A7.1), is

$$\Delta n_1(x) = \frac{(1-R)\Phi_1\alpha\tau_B}{(\alpha^2 L^2 - 1 - j\omega\tau_B)}\left(\frac{A' + B'e^{-\alpha d}}{D'} - \exp(-\alpha x)\right) \qquad (A7.23)$$

where A', B', and D' are similar to A, B and D in Eq. (A7.4), except that in those equations L is replaced by $L/(1 + j\omega t)^{1/2}$, the frequency-dependent diffusion length.

Trapping: For low-level injection and low trap density ($N_T \ll N_A$), the above analysis holds. For high N_T, $\Delta n \neq \Delta p$ and the transient decay is not a simple exponential. There may also be trapping centers which capture the carriers and then release them back to the band from which they were captured, illustrated in Fig. A7.4. An excess ehp is introduced into the semiconductor. Instead of recombining directly, the electron is temporarily captured or trapped onto level E_{T2} (Fig. A7.4(a)). It is subsequently re-emitted into the conduction band (Fig. A7.4(b)), and finally it recombines with the hole (Fig. A7.4(c)). Clearly the electron "lives" longer in this case by the length of time that it is trapped, before it "dies" by recombination and a lifetime measurement gives an erroneously high value. Quite the opposite happens for diffusion length measurements. Due to trapping, the electron distribution is determined by where the electrons are generated rather than by the diffusion process. The minority carrier distribution "frozen" by trapping is responsible for shortening of the minority carrier diffusion length observed in SPV measurements on Si-wafers with traps.[157]

The resultant effective lifetime with trapping is

$$\tau_n' = \tau_n\frac{1 + b + b\tau_2/\tau_1}{1 + b} \qquad (A7.24)$$

where $b = \mu_n/\mu_p$, τ_1 is the average time the minority electron spends in the conduction band before it is trapped by a trapping center, and τ_2 is the mean time the electron spends in the trap before being emitted back into the conduction band. With no trapping $\tau_n' = \tau_n$; with trapping $\tau_n' > \tau_n$, and τ_n' can be very long. For example, certain wide

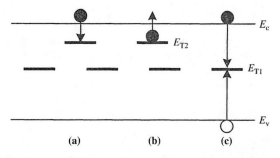

Fig. A7.4   Band diagram showing trapping and recombination.

band-gap phosphors exhibit afterglow effects lasting minutes following cessation of the excitation caused by trapping effects in these materials. However, even Si samples can exhibit significant trapping.[158]

Trapping can be much reduced by illuminating the sample with a steady-state bias light that continually creates ehps, keeping the traps filled, and any additional ehps created by a light flash will tend to recombine with reduced trapping. Another alternative is to use a very short, intense light pulse. If the pulse width is much less than τ_1, the trap density will not change appreciably during the pulse and will play a negligible part during the carrier decay.

APPENDIX 7.2

Electrical Excitation

Optical excitation as a means to create ehps in semiconductors for lifetime measuring is non-contacting. The $\alpha - \lambda$ relationship must be accurately known for some methods, *e.g.*, surface photovoltage. Electrical injection is easier to control, and it is a planar source of minority carriers injected at the edges of the scr in a pn junction. The main disadvantage is the requirement of a junction as the source of minority carriers. In most electrical lifetime methods, a junction is forward biased to inject minority carriers into both quasi-neutral regions. The injection can be thought of as proceeding from a plane located at the edge of the scr. Consider the p-substrate of an n^+p junction. The spatial distribution of electrons injected from $x = 0$ into the base is given by

$$\Delta n(x) = n_{po}\left(\exp\left(\frac{qV_f}{kT}\right) - 1\right)\frac{A}{B} \tag{A7.25}$$

where

$$A = \left(\frac{s_n L_n}{D_n}\right)\sinh\left(\frac{d-x}{L_n}\right) + \cosh\left(\frac{d-x}{L_n}\right)$$

$$B = \left(\frac{s_n L_n}{D_n}\right)\sinh\left(\frac{d}{L_n}\right) + \cosh\left(\frac{d}{L_n}\right)$$

where d is the p-substrate thickness. Equation (A7.25) resembles Eq. (A7.4) if in the latter we let $\alpha \to \infty$, which is similar to confining the optical carrier generation to the plane at $x = 0$.

One of the key differences between optical and electrical injection is that during optical injection excess carriers are generated in the sample volume, with the generation depth controlled by the absorption coefficient. Electrical injection proceeds from a plane. Excess carriers exist beyond that plane because they diffuse there, *not* because they are generated there.

REFERENCES

1. R.N. Hall, "Electron-Hole Recombination in Germanium," *Phys. Rev.* **87**, 387, July 1952.

2. W. Shockley and W.T. Read, "Statistics of the Recombinations of Holes and Electrons," *Phys. Rev.* **87**, 835–842, Sept. 1952.

3. R.N. Hall, "Recombination Processes in Semiconductors," *Proc. IEE* **106B**, 923–931, March 1960.

4. D.K. Schroder, "The Concept of Generation and Recombination Lifetimes in Semiconductors," *IEEE Trans. Electron Dev.* **ED-29**, 1336–1338, Aug. 1982.

5. Y.P. Varshni, "Band-to-Band Radiative Recombination in Groups IV, VI and III-V Semiconductors (I) and (II)," *Phys. Stat. Sol.* **19**, 459–514, Feb. 1967; *ibid.* **20**, 9–36, March 1967.

6. G. Augustine, A Rohatgi, and N.M. Jokerst, "Base Doping Optimization for Radiation-Hard Si, GaAs, and InP Solar Cells," *IEEE Trans. Electron Dev.* **39**, 2395–2400, Oct. 1992.

7. Y. Rosenwaks, Y. Shapira, and D. Huppert, "Picosecond Time-resolved Luminescence Studies of Surface and Bulk Recombination Processes in InP," *Phys. Rev.* **B 45**, 9108–9119, April 1992; I. Tsimberova, Y. Rosenwaks, and M. Molotskii, "Minority Carriers Recombination in n-InP Single Crystals," *J. Appl. Phys.* **93**, 9797–9802, June 2003.

8. U. Strauss and W.W. Rühle, "Auger Recombination in GaAs," *Appl. Phys. Lett.* **62**, 55–57, Jan. 1993; G.W. 't Hooft, "The Radiative Recombination Coefficient of GaAs from Laser Delay Measurements and Effective Nonradiative Lifetimes," *Appl. Phys. Lett.* **39**, 389–390, Sept. 1981.

9. J. Pietzsch and T. Kamiya, "Determination of Carrier Density Dependent Lifetime and Quantum Efficiency in Semiconductors with a Photoluminescence Method (Application to InGaAsP/InP Heterostructures," *Appl. Phys.* **A42**, 91–102, Jan. 1987.

10. T. Trupke, M.A. Green, P. Würfel, P.P. Altermatt, A. Wang, J. Zhao, and R. Corkish, "Temperature Dependence of the Radiative Recombination Coefficient of Intrinsic Crystalline Silicon," *J. Appl. Phys.* **94**, 4930–4937, Oct. 2003.

11. D.K. Schroder, "Carrier Lifetimes in Silicon," in *Handbook of Silicon Technology* (W.C. O'Mara and R.B. Herring, eds.) Noyes Publ., Park Ridge, NJ, 1987; J. Burtscher, F. Dannhäuser and J. Krausse, "The Recombination in Thyristors and Rectifiers in Silicon: Its Influence on the Forward-Bias Characteristic and the Turn-Off Time," (in German) *Solid-State Electron.* **18**, 35–63, Jan. 1975; J. Dziewior and W. Schmid, "Auger Coefficients for Highly Doped and Highly Excited Silicon," *Appl. Phys. Lett.* **31**, 346–348, Sept. 1977; I.V. Grekhov and L.A. Delimova, "Auger Recombination in Silicon," *Sov. Phys. Semicond.* **14**, 529–532, May 1980; L.A. Delimova, "Auger Recombination in Silicon at Low Temperatures," *Sov. Phys. Semicond.* **15**, 778–780, July 1981; L. Passari and E. Susi, "Recombination Mechanisms and Doping Density in Silicon," *J. Appl. Phys.* **54**, 3935–3937, July 1983; D. Huber, A. Bachmeier, R. Wahlich and H. Herzer, "Minority Carrier Diffusion Length and Doping Density in Nondegenerate Silicon," in *Semiconductor Silicon/1986* (H.R. Huff, T. Abe and B. Kolbesen, eds.) Electrochem. Soc., Pennington, NJ, 1986, pp. 1022–1032; E.K. Banghart and J.L. Gray, "Extension of the Open-Circuit Voltage Decay Technique to Include Plasma-Induced Bandgap Narrowing," *IEEE Trans. Electron Dev.* **39**, 1108–1114, May 1992; T.F. Ciszek, T. Wang, T. Schuyler and A. Rohatgi, "Some Effects of Crystal Growth Parameters on Minority Carrier Lifetime in Float-Zoned Silicon," *J. Electrochem. Soc.* **136**, 230–234, Jan. 1989; S.K. Pang and A. Rohatgi, "Record High Recombination Lifetime in Oxidized Magnetic Czochralski Silicon," *Appl. Phys. Lett.* **59**, 195–197, July 1991 and citations in these references.

12. D.K. Schroder, "Carrier Lifetimes in Silicon," *IEEE Trans. Electron Dev.* **44**, 160–170, Jan. 1997.

13. M.J. Kerr and A. Cuevas, "General Parameterization of Auger Recombination in Crystalline Silicon," *J. Appl. Phys.* **91**, 2473–2480, Feb. 2002.

14. D.J. Fitzgerald and A.S. Grove, "Surface Recombination in Semiconductors," *Surf. Sci.* **9**, 347–369, Feb. 1968.

15. A.G. Aberle, S. Glunz, and W. Warta, "Impact of Illumination Level and Oxide Parameters on Shockley-Read-Hall Recombination at the $Si-SiO_2$ Interface," *J. Appl. Phys.* **71**, 4422–4431, May 1992; S.J. Robinson, S.R. Wenham, P.P. Altermatt, A.G. Aberle, G. Heiser, and M.A. Green, "Recombination Rate Saturation Mechanisms at Oxidized Surfaces of High-Efficiency Silicon Solar Cells," *J. Appl. Phys.* **78**, 4740–4754, Oct. 1995.

16. D.K. Schroder, "The Concept of Generation and Recombination Lifetimes in Semiconductors," *IEEE Trans. Electron Dev.* **ED-29**, 1336–1338, Aug. 1982.

17. A.S. Grove in *Physics and Technology of Semiconductor Devices* (Wiley, New York, 1967). Grove introduced τ_0 and s_0 as bulk and surface generation parameters. He assumes that $\sigma_n = \sigma_p$ and $E_T = E_i$ and finds $\tau_0 = \tau_n = \tau_p$ and $G = n_i/2\tau_0$. This places undue restrictions on τ_0. I prefer the more general definition of Eq. (7.20) which requires no assumptions regarding τ_g. By similar arguments Grove defines the surface generation rate as $G_S = n_i(s_0/2)$. Again, I prefer the more general definition of Eq. (7.22) with no assumptions.

18. H. Nagel, C. Berge, and A.G. Aberle, "Generalized Analysis of Quasi-Steady-State and Quasi-Transient Measurements of Carrier Lifetimes in Semiconductors," *J. Appl. Phys.* **86**, 6218–6221, Dec. 1999.

19. S.K. Pang and A. Rohatgi, "A New Methodology for Separating Shockley-Read-Hall Lifetime and Auger Recombination Coefficients from the Photoconductivity Decay Technique," *J. Appl. Phys.* **74**, 5554–5560, Nov. 1993; T. Maekawa and K. Fujiwara, "Measurable Range of Bulk Carrier Lifetime for a Thick Silicon Wafer by Induced Eddy Current Method," *Japan. J. Appl. Phys.* **35**, 3955–3964, Aug. 1995.

20. ASTM Standard F28-91, "Standard Method for Measuring the Minority-Carrier Lifetime in Bulk Germanium and Silicon," *1996 Annual Book of ASTM Standards*, Am. Soc. Test. Mat., West Conshohocken, PA, 1996.

21. M. Boulou and D. Bois, "Cathodoluminescence Measurements of the Minority-Carrier Lifetime in Semiconductors," *J. Appl. Phys.* **48**, 4713–4721, Nov. 1977.

22. K.L. Luke and L.J. Cheng, "Analysis of the Interaction of a Laser Pulse with a Silicon Wafer: Determination of Bulk Lifetime and Surface Recombination Velocity," *J. Appl. Phys.* **61**, 2282–2293, March 1987.

23. D.T. Stevenson and R.J. Keyes, "Measurement of Carrier Lifetimes in Germanium and Silicon," *J. Appl. Phys.* **26**, 190–195, Feb. 1955.

24. S.M. Ryvkin, *Photoelectric Effects in Semiconductors* Consultants Bureau, New York, 1964, 19–22.

25. T.S. Horányi, T. Pavelka, and P. Tüttö, "In Situ Bulk Lifetime Measurement on Silicon with a Chemically Passivated Surface," *Appl. Surf. Sci.*, **63**, 306–311, Jan. 1993; H. M'saad, J. Michel, J.J. Lappe, and L.C. Kimerling, "Electronic Passivation of Silicon Surfaces by Halogens," *J. Electron. Mat.* **23**, 487–491, May 1994.

26. E. Yablonovitch and T.J. Gmitter, "A Contactless Minority Carrier Lifetime Probe of Heterostructures, Surfaces, Interfaces, and Bulk Wafers," *Solid-State Electron.* **35**, 261–267, March 1992.

27. A. Sanders and M. Kunst, "Characterization of Silicon Wafers by Transient Microwave Photoconductivity Measurements," *Solid-State Electron.* **34**, 1007–1015, Sept. 1991; E. Gaubas and A. Kaniava, "Determination of Recombination Parameters in Silicon Wafers by Transient Microwave Absorption," *Rev. Sci. Instrum.* **67**, 2339–2345, June 1996; ASTM Standard F1535-94, "Standard Test Method for Carrier Recombination Lifetime in Silicon Wafers by Noncontact Measurement of Photoconductivity Decay by Microwave Reflectance," *1996 Annual Book of ASTM Standards*, Am. Soc. Test. Mat., West Conshohocken, PA, 1996.

28. E. Yablonovitch, R.M. Swanson, W.D. Eades, and B.R. Weinberger, "Electron-Hole Recombination at the Si-SiO$_2$ Interface," *Appl. Phys. Lett.* **48**, 245–247, Jan. 1986.

29. E. Yablonovitch, D.L. Allara, C.C. Chang, T. Gmitter and T.B. Bright, "Unusually Low Surface-Recombination Velocity on Silicon and Germanium Surfaces," *Phys. Rev. Lett.* **57**, 249–252, July 1986.

30. J. Schmidt and A.G. Aberle, "Accurate Method for the Determination of Bulk Minority-Carrier Lifetimes in Mono- and Multicrystalline Silicon Wafers," *J. Appl. Phys.* **81**, 6186–6196, May 1997.

31. E. Yablonovitch, C.J. Sandroff, R. Bhat and T. Gmitter, "Nearly Ideal Electronic Properties of Sulfide Coated GaAs Surfaces," *Appl. Phys. Lett.* **51**, 439–441, Aug. 1987.

32. Y. Mada, "A Nondestructive Method for Measuring the Spatial Distribution of Minority Carrier Lifetime in Si Wafer," *Japan. J. Appl. Phys.* **18**, 2171–2172, Nov. 1979.

33. M. Kunst and G. Beck, "The Study of Charge Carrier Kinetics in Semiconductors by Microwave Conductivity Measurements," *J. Appl. Phys.* **60**, 3558–3566, Nov. 1986; J.M. Borrego, R.J. Gutmann, N. Jensen and O. Paz, "Non-Destructive Lifetime Measurement in Silicon Wafers by Microwave Reflection," *Solid-State Electron.* **30**, 195–203, Feb. 1987.

34. R.J. Deri and J.P. Spoonhower, "Microwave Photoconductivity Lifetime Measurements: Experimental Limitations," *Rev. Sci. Instrum.* **55**, 1343–1347, Aug. 1984.

35. R.A. Sinton and A. Cuevas, "Contactless Determination of Current-Voltage Characteristics and Minority-carrier Lifetimes in Semiconductors from Quasi-steady-state Photoconductance Data," *Appl. Phys. Lett.* **69**, 2510–2512, Oct. 1996.

36. A. Cuevas and R.A. Sinton, "Characterisation and Diagnosis of Silicon Wafers and Devices," in *Practical Handbook of Photovoltaics: Fundamentals and Applications* (T. Markvart and L. Castaner, eds.) Elsevier, Oxford, 2003.

37. M.J. Kerr, A. Cuevas, and R.A. Sinton, "Generalized Analysis of Quasi-Steady-State and Transient Decay Open Circuit Voltage," *J. Appl. Phys.* **91**, 399–404, Jan. 2002.

38. J.E. Mahan, T.W. Ekstedt, R.I. Frank and R. Kaplow, "Measurement of Minority Carrier Lifetime in Solar Cells from Photo-Induced Open Circuit Voltage Decay," *IEEE Trans. Electron Dev.* **ED-26**, 733–739, May 1979; S.R. Dhariwal and N.K. Vasu, "Mathematical Formulation for the Photo-Induced Open Circuit Voltage Decay Method for Measurement of Minority Carrier Lifetime in Solar Cells," *IEEE Electron Dev. Lett.* **EDL-2**, 53–55, Feb. 1981.

39. O. von Roos, "Analysis of the Photo Voltage Decay (PVD) Method for Measuring Minority Carrier Lifetimes in PN Junction Solar Cells," *J. Appl. Phys.* **52**, 5833–5837, Sept. 1981.

40. B.H. Rose and H.T. Weaver, "Determination of Effective Surface Recombination Velocity and Minority-Carrier Lifetime in High-Efficiency Si Solar Cells," *J. Appl. Phys.* **54**, 238–247, Jan. 1983; Corrections *J. Appl. Phys.* **55**, 607, Jan. 1984.

41. B.H. Rose, "Minority-Carrier Lifetime Measurements on Si Solar Cells Using I_{sc} and V_{oc} Transient Decay," *IEEE Trans. Electron Dev.* **ED-31**, 559–565, May 1984.

42. S.C. Jain, "Theory of Photo Induced Open Circuit Voltage Decay in a Solar Cell," *Solid-State Electron.* **24**, 179–183, Feb. 1981; S.C. Jain and U.C. Ray, "Photovoltage Decay in PN Junction Solar Cells Including the Effects of Recombination in the Emitter," *J. Appl. Phys.* **54**, 2079–2085, April 1983.

43. A.R. Moore, "Carrier Lifetime in Photovoltaic Solar Concentrator Cells by the Small Signal Open Circuit Decay Method," *RCA Rev.* **40**, 549–562, Dec. 1980.

44. R.K. Ahrenkiel, "Measurement of Minority-Carrier Lifetime by Time-Resolved Photoluminescence," *Solid-State Electron.* **35**, 239–250, March 1992.

45. J. Dziewior and W. Schmid, "Auger Coefficients for Highly Doped and Highly Excited Silicon," *Appl. Phys. Let.*. **31**, 346–348, Sept. 1977.

46. G. Bohnert, R. Häcker and A. Hangleiter, "Position Resolved Carrier Lifetime Measurement in Silicon Power Devices by Time Resolved Photoluminescence Spectroscopy," *J. Physique* **C4**, 617–620, Sept. 1988.

47. E.O. Johnson, "Measurement of Minority Carrier Lifetime with the Surface Photovoltage," *J. Appl. Phys.* **28**, 1349–1353, Nov. 1957.

48. A. Quilliet and P. Gosar, "The Surface Photovoltaic Effect in Silicon and Its Application to Measure the Minority Carrier Lifetime (in French)," *J. Phys. Rad.* **21**, 575–580, July 1960.

49. A.M. Goodman, "A Method for the Measurement of Short Minority Carrier Diffusion Lengths in Semiconductors," *J. Appl. Phys.* **32**, 2550–2552, Dec. 1961; A.M. Goodman, L.A. Goodman and H.F. Gossenberger, "Silicon-Wafer Process Evaluation Using Minority-Carrier Diffusion Length Measurements by the SPV Method," *RCA Rev.* **44**, 326–341, June 1983.

50. S.C. Choo, L.S. Tan, and K.B. Quek, "Theory of the Photovoltage at Semiconductor Surfaces and Its Application to Diffusion Length Measurements," *Solid-State Electron.* **35**, 269–283, March 1992.

51. A. Buczkowski, G. Rozgonyi, F. Shimura, and K. Mishra, "Photoconductance Minority Carrier Lifetime vs. Surface Photovoltage Diffusion Length in Silicon," *J. Electrochem. Soc.* **140**, 3240–3245, Nov. 1993.

52. A.R. Moore, "Theory and Experiment on the Surface-Photovoltage Diffusion-Length Measurement as Applied to Amorphous Silicon," *J. Appl. Phys.* **54**, 222–228, Jan. 1983; C.L. Chiang, R. Schwarz, D.E. Slobodin, J. Kolodzey and S. Wagner, "Measurement of the Minority-Carrier Diffusion Length in Thin Semiconductor Films," *IEEE Trans. Electron Dev.* **ED-33**, 1587–1592, Oct. 1986; C.L. Chiang and S. Wagner, "On the Theoretical Basis of the Surface Photovoltage Technique," *IEEE Trans. Electron Dev.* **ED-32**, 1722–1726, Sept. 1985.

53. M.A. Green and M.J. Keevers, "Optical Properties of Intrinsic Silicon at 300 K," *Progr. Photovolt.* **3**, 189–192, May/June 1995.

54. M.D. Sturge, "Optical Absorption of Gallium Arsenide Between 0.6 and 2.75 eV," *Phys. Rev.* **127**, 768–773, Aug. 1962; D.D. Sell and H.C. Casey, Jr., "Optical Absorption and Photoluminescence Studies of Thin GaAs Layers in GaAs-AlGaAs Double Heterostructures," *J. Appl. Phys.* **45**, 800–807, Feb. 1974; D.E. Aspnes and A.A. Studna, "Dielectric Functions and Optical Parameters of Si, Ge, GaP, GaAs, GaSb, InP, InAs and InSb from 1.5 to 6 eV," *Phys. Rev.* **B27**, 985–1009, Jan. 1983.

55. S.S. Li, "Determination of Minority-Carrier Diffusion Length in Indium Phosphide by Surface Photovoltage Measurement," *Appl. Phys. Lett.* **29**, 126–127, July 1976; H. Burkhard, H.W. Dinges and E. Kuphal, "Optical Properties of InGaPAs, InP, GaAs, and GaP Determined by Ellipsometry," *J. Appl. Phys.* **53**, 655–662, Jan. 1982.

56. ASTM Standard F391-90a, "Standard Test Method for Minority-Carrier Diffusion Length in Silicon by Measurement of Steady-State Surface Photovoltage," *1996 Annual Book of ASTM Standards,* Am. Soc. Test. Mat., West Conshohocken, PA, 1996.

57. L. Jastrzebski, O. Milic, M. Dexter, J. Lagowski, D. DeBusk, P. Edelman, and K. Nauka, "Monitoring Heavy Metal Contamination During Chemical Cleaning With Surface Photovoltage," *J. Electrochem. Soc.* **140**, 1152–1159, April 1993.

58. W. Kern and D.A. Puotinen, "Cleaning Solutions Based on Hydrogen Peroxide for Use in Silicon Semiconductor Technology," *RCA Rev.* **31**, 187–206, June 1970.

59. S.C. Choo, "Theory of Surface Photovoltage in a Semiconductor with a Schottky Contact," *Solid-State Electron.* **38**, 2085–2093, Dec. 1995.

60. W.H. Howland and S.J. Fonash, "Errors and Error-Avoidance in the Schottky Coupled Surface Photovoltage Technique," *J. Electrochem. Soc.* **142**, 4262–4268, Dec. 1995.

61. R.H. Micheels and R.D. Rauh, "Use of a Liquid Electrolyte Junction for the Measurement of Diffusion Length in Silicon Ribbon," *J. Electrochem. Soc.* **131**, 217–219, Jan. 1984; A.R. Moore and H.S. Lin, "Improvement in the Surface Photovoltage Method of Determining Diffusion Length in Thin Films of Hydrogenated Amorphous Silicon," *J. Appl. Phys.* **61**, 4816–4819, May 1987.

62. B.L. Sopori, R.W. Gurtler and I.A. Lesk, "Effects of Optical Beam Size on Diffusion Length Measured by the Surface Photovoltage Method," *Solid-State Electron.* **23**, 139–142, Feb. 1980.

63. W.E. Phillips, "Interpretation of Steady-State Surface Photovoltage Measurements in Epitaxial Semiconductor Layers," *Solid-State Electron.* **15**, 1097–1102, Oct. 1972.

64. O.J. Antilla and S.K. Hahn, "Study on Surface Photovoltage Measurement of Long Diffusion Length Silicon: Simulation Results," *J. Appl. Phys.* **74**, 558–569, July 1993.

65. D.K. Schroder, "Effective Lifetimes in High Quality Silicon Devices," *Solid-State Electron.* **27**, 247–251, March 1984; T.I. Chappell, P.W. Chye and M.A. Tavel, "Determination of the Oxygen Precipitate-Free Zone Width in Silicon Wafers from Surface Photovoltage Measurements," *Solid-State Electron.* **26**, 33–36, Jan. 1983.

66. H. Shimizu and C. Munakata, "Nondestructive Diagnostic Method Using ac Surface Photovoltage for Detecting Metallic Contaminants in Silicon Wafers," *J. Appl. Phys.* **73**, 8336–8339, June 1993; "AC Photovoltaic Images of Thermally Oxidized p-Type Silicon Wafers Contaminated With Metals," *Japan. J. Appl. Phys.* **31**, 2319–2321, Aug. 1992.

67. E.D. Stokes and T.L. Chu, "Diffusion Lengths in Solar Cells from Short-Circuit Current Measurements," *Appl. Phys. Lett.* **30**, 425–426, April 1977.

68. N.D. Arora, S.G. Chamberlain and D.J. Roulston, "Diffusion Length Determination in pn Junction Diodes and Solar Cells," *Appl. Phys. Lett.* **37**, 325–327, Aug. 1980.

69. E. Suzuki and Y. Hayashi, "A Measurement of a Minority-Carrier Lifetime in a p-Type Silicon Wafer by a Two-Mercury Probe Method," *J. Appl. Phys.* **66**, 5398–5403, Dec. 1989; "A Method of Determining the Lifetime and Diffusion Coefficient of Minority Carriers in a Semiconductor Wafer," *IEEE Trans Electron Dev.* **36**, 1150–1154, June 1989.

70. V. Lehmann and H. Föll, "Minority Carrier Diffusion Length Mapping in Silicon Wafers Using a Si-Electrolyte-Contact," *J. Electrochem. Soc.* **135**, 2831–2835, Nov. 1988; J. Carstensen, W. Lippik, and H. Föll, "Mapping of Defect Related Silicon Bulk and Surface Properties With the ELYMAT Technique," in *Semiconductor Silicon/94* (H. Huff, W. Bergholz, and K. Sumino, eds.), Electrochem. Soc, Pennington, NJ, 1994, 1105–1116.

71. M.L. Polignano, A. Giussani, D. Caputo, C. Clementi, G. Pavia, and F. Priolo, "Detection of Metal Segregation at the Oxide-Silicon Interface," *J. Electrochem. Soc.* **149**, G429–G439, July 2002.

72. D.K. Schroder, R.N. Thomas and J.C. Swartz, "Free Carrier Absorption in Silicon," *IEEE Trans Electron Dev.* **ED-25**, 254–261, Feb. 1978.

73. L. Jastrzebski, J. Lagowski and H.C. Gatos, "Quantitative Determination of the Carrier Concentration Distribution in Semiconductors by Scanning Infrared Absorption: Si," *J. Electrochem. Soc.* **126**, 260–263, Feb. 1979.

74. J. Isenberg and W. Warta, "Free Carrier Absorption in Heavily Doped Silicon Layers," *Appl. Phys. Lett.* **84**, 2265–2267, March 2004.

75. S.W. Glunz and W. Warta, "High-Resolution Lifetime Mapping Using Modulated Free-Carrier Absorption," *J. Appl. Phys.* **77**, 3243–3247, April 1995.

76. D.L. Polla, "Determination of Carrier Lifetime in Silicon by Optical Modulation," *IEEE Electron Dev. Lett.* **EDL-4**, 185–187, June 1983.

77. J. Waldmeyer, "A Contactless Method for Determination of Carrier Lifetime, Surface Recombination Velocity, and Diffusion Constant in Semiconductors," *J. Appl. Phys.* **63**, 1977–1983, March 1988.

78. J. Linnros, P. Norlin, and A. Hallén, "A New Technique for Depth Resolved Carrier Recombination Measurements Applied to Proton Irradiated Thyristors," *IEEE Trans Electron Dev.* **40**, 2065–2073, Nov. 1993; H.J. Schulze, A. Frohnmeyer, F.J. Niedernostheide, F. Hille, P. Tüttö, T. Pavelka, and G. Wachutka, "Carrier Lifetime Analysis by Photoconductance Decay and Free Carrier Absorption Measurements," *J. Electrochem. Soc.* **148**, G655–G661, Nov. 2001.

79. M. Bail, J. Kentsch, R. Brendel, and M. Schulz, "Lifetime Mapping of Si Wafers by an Infrared Camera," *Proc. 28th IEEE Photovolt. Conf.* 99–103, 2000; R. Brendel, M. Bail, B. Bodmann, J. Kentsch, and M. Schulz, "Analysis of Photoexcited Charge Carrier Density Profiles in Si Wafers by Using an Infrared Camera," *Appl. Phys. Lett.* **80**, 437–439, Jan. 2002; J. Isenberg, S. Riepe, S.W. Glunz, and W. Warta, "Imaging Method for Laterally Resolved Measurement of Minority Carrier Densities and Lifetimes: Measurement Principle and First Applications," *J. Appl. Phys.* **93**, 4268–4275, April 2003.

80. M.C. Schubert, J. Isenberg, and W. Warta, "Spatially Resolved Lifetime Imaging of Silicon Wafers by Measurement of Infrared Emission," *J. Appl. Phys.* **94**, 4139–4143, Sept. 2003.

81. J.F. Bresse, "Quantitative Investigations in Semiconductor Devices by Electron Beam Induced Current Mode: A Review," in *Scanning Electron Microscopy 1*, 717–725, 1978.

82. C.A. Klein, "Band Gap Dependence and Related Features of Radiation Ionization Energies in Semiconductors," *J. Appl. Phys.* **39**, 2029–2038, March 1968; F. Scholze, H. Rabus, and G. Ulm, "Measurement of the Mean Electron-Hole Pair Creation Energy in Crystalline Silicon for Photons in the 50–1500 eV Spectral Range," *Appl. Phys. Lett.* **69**, 2974–2976, Nov. 1996.

83. H.J. Leamy, "Charge Collection Scanning Electron Microscopy," *J. Appl. Phys.* **53**, R51–R80, June 1982.

84. D.E. Ioannou and C.A. Dimitriadis, "A SEM-EBIC Minority Carrier Diffusion Length Measurement Technique," *IEEE Trans. Electron Dev.* **ED-29**, 445–450, March 1982.

85. D.S.H. Chan, V.K.S. Ong, and J.C.H. Phang, "A Direct Method for the Extraction of Diffusion Length and Surface Recombination Velocity from an EBIC Line Scan: Planar Junction Configuration," *IEEE Trans. Electron Dev.* **42**, 963–968, May 1995.

86. J.D. Zook, "Theory of Beam-Induced Currents in Semiconductors," *Appl. Phys. Lett.* **42**, 602–604, April 1983.

87. C. Van Opdorp, "Methods of Evaluating Diffusion Lengths and Near-Junction Luminescence-Efficiency Profiles from SEM Scans," *Phil. Res. Rep.* **32**, 192–249, 1977; F. Berz and H.K. Kuiken, "Theory of Lifetime Measurements with the Scanning Electron Microscope: Steady State," *Solid-State Electron.* **19**, 437–445, June 1976.

88. H.K. Kuiken, "Theory of Lifetime Measurements with the Scanning Electron Microscope: Transient Analysis," *Solid-State Electron.* **19**, 447–450, June 1976; C.H. Seager, "The Determination of Grain-Boundary Recombination Rates by Scanned Spot Excitation Methods," *J. Appl. Phys.* **53**, 5968–5971, Aug. 1982.

89. C.M. Hu and C. Drowley, "Determination of Diffusion Length and Surface Recombination Velocity by Light Excitation," *Solid-State Electron.* **21**, 965–968, July 1978.

90. J.D. Zook, "Effects of Grain Boundaries in Polycrystalline Solar Cells," *Appl. Phys. Lett.* **37**, 223–226, July 1980.

91. W.H. Hackett, "Electron-Beam Excited Minority-Carrier Diffusion Profiles in Semiconductors," *J. Appl. Phys.* **43**, 1649–1654, April 1972.

92. Y. Murakami, H. Abe, and T. Shingyouji, "Calculation of Diffusion Component of Leakage Current in pn Junctions Formed in Various Types of Silicon Wafers (Intrinsic Gettering, Epitaxial, Silicon-on-Insulator) *Japan. J. Appl. Phys.,* **34,** 1477–1482, March 1995.

93. C. Claeys, E. Simoen, A. Poyai, and A. Czerwinski, "Electrical Quality Assessment of Epitaxial Wafers Based on p-n Junction Diagnostics," *J. Electrochem. Soc.* **146**, 3429–3434, Sept. 1999.

94. D.K. Schroder, B.D. Choi, S.G. Kang, W. Ohashi, K. Kitahara, G. Opposits, T. Pavelka, and J.L. Benton, "Silicon Epitaxial Layer Recombination and Generation Lifetime Characterization," *IEEE Trans. Electron Dev.* **50**, 906–912, April 2003.

95. Y. Murakami and T. Shingyouji, "Separation and Analysis of Diffusion and Generation Components of pn Junction Leakage Current in Various Silicon Wafers," *J. Appl. Phys.* **75**, 3548–3552, April 1994.

96. E. Simoen, C. Claeys, A. Czerwinski, and J. Katcki "Accurate Extraction of the Diffusion Current in Silicon p-n Junction Diodes," *Appl. Phys. Lett.* **72**, 1054–1056, March 1998; J. Vanhellemont, E. Simoen, A. Kaniava, M. Libezny, and C. Claeys, "Impact of Oxygen Related Extended Defects on Silicon Diode Characteristics," *J. Appl. Phys.* **77**, 5669–5676, June 1995.

97. A. Czerwinski, E. Simoen, C. Claeys, K. Klima, D. Tomaszewski, J. Gibki, and J. Katcki, "Optimized Diode Analysis of Electrical Silicon Substrate Properties," *J. Electrochem. Soc.,* **145**, 2107–2113, June 1998.

98. E.M. Pell, "Recombination Rate in Germanium by Observation of Pulsed Reverse Characteristic," *Phys. Rev.* **90**, 278–279, April 1953.

99. R.H. Kingston, "Switching Time in Junction Diodes and Junction Transistors," *Proc. IRE* **42**, 829–834, May 1954.

100. B. Lax and S.F. Neustadter, "Transient Response of a PN Junction," *J. Appl. Phys.* **25**, 1148–1154, Sept. 1954.

101. K. Schuster and E. Spenke, "The Voltage Step at the Switching of Alloyed PIN Rectifiers," *Solid-State Electron.* **8**, 881–882, Nov. 1965.

102. H.J. Kuno, "Analysis and Characterization of PN Junction Diode Switching," *IEEE Trans. Electron Dev.* **ED-11**, 8–14, Jan. 1964.

103. R.H. Dean and C.J. Nuese, "A Refined Step-Recovery Technique for Measuring Minority Carrier Lifetimes and Related Parameters in Asymmetric PN Junction Diodes," *IEEE Trans. Electron Dev.* **ED-18**, 151–158, March 1971.

104. S.C. Jain and R. Van Overstraeten, "The Influence of Heavy Doping Effects on the Reverse Recovery Storage Time of a Diode," *Solid-State Electron.* **26**, 473–481, May 1983.

105. B. Tien and C. Hu, "Determination of Carrier Lifetime from Rectifier Ramp Recovery Waveform," *IEEE Trans. Electron Dev. Lett.* **9**, 553–555, Oct. 1988; S.R. Dhariwal and R.C. Sharma, "Determination of Carrier Lifetime in p-i-n Diodes by Ramp Recovery," *IEEE Trans. Electron Dev. Lett.* **13**, 98–101, Feb. 1992.

106. L. De Smet and R. Van Overstraeten, "Calculation of the Switching Time in Junction Diodes," *Solid-State Electron.* **18**, 557–562, June 1975; F. Berz, "Step Recovery of pin Diodes," *Solid-State Electron.* **22**, 927–932, Nov. 1979.

107. M. Derdouri, P. Leturcq and A. Muñoz-Yague, "A Comparative Study of Methods of Measuring Carrier Lifetime in pin Devices," *IEEE Trans. Electron Dev.* **ED-27**, 2097–2101, Nov. 1980.

108. S.C. Jain, S.K. Agarwal and Harsh, "Importance of Emitter Recombinations in Interpretation of Reverse-Recovery Experiments at High Injections," *J. Appl. Phys.* **54**, 3618–3619, June 1983.

109. B.R. Gossick, "Post-Injection Barrier Electromotive Force of PN Junctions," *Phys. Rev.* **91**, 1012–1013, Aug. 1953; "On the Transient Behavior of Semiconductor Rectifiers," *J. Appl. Phys.* **26**, 1356–1365, Nov. 1955.

110. S.R. Lederhandler and L.J. Giacoletto, "Measurement of Minority Carrier Lifetime and Surface Effects in Junction Devices," *Proc. IRE* **43**, 477–483, April 1955.

111. S.C. Choo and R.G. Mazur, "Open Circuit Voltage Decay Behavior of Junction Devices," *Solid-State Electron.* **13**, 553–564, May 1970.

112. S.C. Jain and R. Muralidharan, "Effect of Emitter Recombinations on the Open Circuit Voltage Decay of a Junction Diode," *Solid-State Electron.* **24**, 1147–1154, Dec. 1981.

113. R.J. Basset, W. Fulop and C.A. Hogarth, "Determination of the Bulk Carrier Lifetime in Low-Doped Region of a Silicon Power Diode by the Method of Open Circuit Voltage Decay," *Int. J. Electron.* **35**, 177–192, Aug. 1973; P.G. Wilson, "Recombination in Silicon p-π-n Diodes," *Solid-State Electron.* **10**, 145–154, Feb. 1967.

114. J.E. Mahan and D.L. Barnes, "Depletion Layer Effects in the Open-Circuit-Voltage-Decay Lifetime Measurement," *Solid-State Electron.* **24**, 989–994, Oct. 1981.

115. M.A. Green, "Minority Carrier Lifetimes Using Compensated Differential Open Circuit Voltage Decay," *Solid-State Electron.* **26**, 1117–1122, Nov. 1983; "Solar Cell Minority Carrier Lifetime Using Open-Circuit Voltage Decay," *Solar Cells* **11**, 147–161, March 1984.

116. D.H.J. Totterdell, J.W. Leake and S.C. Jain, "High-Injection Open-Circuit Voltage Decay in pn-Junction Diodes with Lightly Doped Bases," *IEE Proc. Pt. I* **133**, 181–184, Oct. 1986.

117. K. Joardar, R.C. Dondero and D.K. Schroder, "A Critical Analysis of the Small-Signal Voltage Decay Technique for Minority-Carrier Lifetime Measurement in Solar Cells," *Solid-State Electron.* **32**, 479–483, June 1989.

118. P. Tomanek, "Measuring the Lifetime of Minority Carriers in MIS Structures," *Solid-State Electron.* **12**, 301–303, April 1969

119. J. Müller and B. Schiek, "Transient Responses of a Pulsed MIS-Capacitor," *Solid-State Electron.* **13**, 1319–1332, Oct. 1970.

120. A.C. Wang and C.T. Sah, "New Method for Complete Electrical Characterization of Recombination Properties of Traps in Semiconductors," *J. Appl. Phys.* **57**, 4645–4656, May 1985.

121. E. Soutschek, W. Müller and G. Dorda, "Determination of Recombination Lifetime in MOS-FET's," *Appl. Phys. Lett.* **36**, 437–438, March 1980.

122. D.K. Schroder, J.D. Whitfield and C.J. Varker, "Recombination Lifetime Using the Pulsed MOS Capacitor," *IEEE Trans. Electron Dev.* **ED-31**, 462–467, April 1984.

123. D.K. Schroder, "Bulk and Optical Generation Parameters Measured with the Pulsed MOS Capacitor," *IEEE Trans. Electron Dev.* **ED-19**, 1018–1023, Sept. 1972.

124. T.W. Jung, F.A. Lindholm and A. Neugroschel, "Unifying View of Transient Responses for Determining Lifetime and Surface Recombination Velocity in Silicon Diodes and Back-Surface-Field Solar Cells," *IEEE Trans. Electron Dev.* **ED-31**, 588–595, May 1984; T.W. Jung, F.A. Lindholm and A. Neugroschel, "Variations in the Electrical Short-Circuit Current Decay for Recombination Lifetime and Velocity Measurements," *Solar Cells* **22**, 81–96, Oct. 1987.

125. A. Zondervan, L.A. Verhoef and F.A. Lindholm, "Measurement Circuits for Silicon-Diode and Solar-Cell Lifetime and Surface Recombination Velocity by Electrical Short-Circuit Current Delay," *IEEE Trans. Electron Dev.* **ED-35**, 85–88, Jan. 1988.

126. P. Spirito and G. Cocorullo, "Measurement of Recombination Lifetime Profiles in Epilayers Using a Conductivity Modulation Technique," *IEEE Trans. Electron Dev.* **ED-32**, 1708–1713, Sept. 1985; P. Spirito, S. Bellone, C.M. Ransom, G. Busatto and G. Cocorullo, "Recombination Lifetime Profiling in Very Thin Si Epitaxial Layers Used for Bipolar VLSI," *IEEE Electron Dev. Lett.* **EDL-10**, 23–24, Jan. 1989.

127. P.C.T. Roberts and J.D.E. Beynon, "An Experimental Determination of the Carrier Lifetime Near the Si-SiO$_2$ Interface," *Solid-State Electron.* **16**, 221–227, Feb. 1973; "Effect of a Modified Theory of Generation Currents on an Experimental Determination of Carrier Lifetime," *Solid-State Electron.* **17**, 403–404, April 1974.

128. A.S. Grove and D.J. Fitzgerald, "Surface Effects on pn Junctions: Characteristics of Surface Space-Charge Regions Under Non-Equilibrium Conditions," *Solid-State Electron.* **9**, 783–806, Aug. 1966; D.J. Fitzgerald and A.S. Grove, "Surface Recombination in Semiconductors," *Surf. Sci.* **9**, 347–369, Feb. 1968.

129. R.F. Pierret, "The Gate-Controlled Diode s_o Measurement and Steady-State Lateral Current Flow in Deeply Depleted MOS Structures," *Solid-State Electron.* **17**, 1257–1269, Dec. 1974.

130. G.A. Hawkins, E.A. Trabka, R.L. Nielsen and B.C. Burkey, "Characterization of Generation Currents in Solid-State Imagers," *IEEE Trans. Electron Dev.* **ED-32**, 1806–1816, Sept. 1985; G.A. Hawkins, "Generation Currents from Interface States in Selectively Implanted MOS Structures," *Solid-State Electron.* **31**, 181–196, Feb. 1988.

131. M. Zerbst, "Relaxation Effects at Semiconductor-Insulator Interfaces" (in German), *Z. Angew. Phys.* **22**, 30–33, May 1966.

132. J.S. Kang and D.K. Schroder, "The Pulsed MIS Capacitor—A Critical Review," *Phys. Stat. Sol.* **89a**, 13–43, May 1985.

133. P.U. Calzolari, S. Graffi and C. Morandi, "Field-Enhanced Carrier Generation in MOS Capacitors," *Solid-State Electron.* **17**, 1001–1011, Oct. 1974; K.S. Rabbani, "Investigations on Field Enhanced Generation in Semiconductors," *Solid-State Electron.* **30**, 607–613, June 1987.

134. J. van der Spiegel and G.J. Declerck, "Theoretical and Practical Investigation of the Thermal Generation in Gate Controlled Diodes," *Solid-State Electron.* **24**, 869–877, Sept. 1981.

135. D.K. Schroder and J. Guldberg, "Interpretation of Surface and Bulk Effects Using the Pulsed MIS Capacitor," *Solid-State Electron.* **14**, 1285–1297, Dec. 1971.

136. W.R. Fahrner, D. Braeunig, C.P. Schneider and M. Briere, "Reduction of Measurement Time of Lifetime Profiles by Applying High Temperatures," *J. Electrochem. Soc.* **134**, 1291–1296, May 1987.

137. D.K. Schroder, "Bulk and Optical Generation Parameters Measured with the Pulsed MOS Capacitor," *IEEE Trans. Electron Dev.* **ED-19**, 1018–1023, Sept. 1972; R.F. Pierret and

W.M. Au, "Photo-Accelerated MOS-C C-t Transient Measurements," *Solid-State Electron.* **30**, 983–984, Sept. 1987.

138. W.W. Keller, "The Rapid Measurement of Generation Lifetime in MOS Capacitors with Long Relaxation Times," *IEEE Trans. Electron Dev.* **ED-34**, 1141–1146, May 1987.

139. C.S. Yue, H. Vyas, M. Holt and J. Borowick, "A Fast Extrapolation Technique for Measuring Minority-Carrier Generation Lifetime," *Solid-State Electron.* **28**, 403–406, April 1985.

140. M. Xu, C. Tan, Y. He, and Y. Wang, "Analysis of the Rate of Change of Inversion Charge in Thin Insulator p-Type Metal-Oxide-Insulator Structures," *Solid-State Electron.* **38**, 1045–1049, May 1995.

141. A. Vercik and A.N. Faigon, "Modeling Tunneling and Generation Mechanisms Governing the Nonequilibrium Transient in Pulsed Metal–Oxide–Semiconductor Diodes," *J. Appl. Phys.* **88**, 6768–6774, Dec. 2000.

142. M. Kohno, S. Hirae, H. Okada, H. Matsubara, I. Nakatani, Y. Imaoka, T. Kusuda, and T. Sakai, "Noncontact Measurement of Generation Lifetime," *Japan. J. Appl. Phys.* **35**, 5539–5544, Oct. 1996; T. Sakai, M. Kohno, H. Okada, H. Matsubara, and S. Hirae, "Improvement of Sensor for Noncontact Capacitance/Voltage Measurement and Lifetime Measurement of Bare Silicon (100)," *Japan. J. Appl. Phys.* **36**, 935–942, Feb. 1997.

143. D.E. Ioannou, S. Cristoloveanu, M. Mukherjee and B. Mazhari, "Characterization of Carrier Generation in Enhancement Mode SOI MOSFET's," *IEEE Electron Dev. Lett.* **11**, 409–411, Sept. 1990; A.M. Ionescu and S. Cristoloveanu, "Carrier Generation in Thin SIMOX Films by Deep-depletion Pulsing of MOS Transistors," *Nucl. Instrum. Meth. Phys. Res.* **B84**, 265–269, 1994; H. Shin, M. Racanelli, W.M. Huang, J. Foerstner, S. Choi, and D.K. Schroder, "A Simple Technique to Measure Generation Lifetime in Partially Depleted SOI MOSFETs," *IEEE Trans. Electron Dev.* **45**, 2378–2380, Nov. 1998.

144. D. Munteanu and A-M Ionescu, "Modeling of Drain Current Overshoot and Recombination Lifetime Extraction in Floating-Body Submicron SOI MOSFETs," *IEEE Trans. Electron Dev.* **49**, 1198–1205, July 2002.

145. R. Sorge, "Double-Sweep LF-CV Technique for Generation Rate Determination in MOS Capacitors," *Solid-State Electron.* **38**, 1479–1484, Aug. 1995; R. Sorge, P. Schley, J. Grabmeier, G. Obermeier, D. Huber, and H. Richter, "Rapid MOS-CV Generation Lifetime Mapping Technique for the Characterisation of High Quality Silicon," *Proc. ESSDERC* 1998, 296–299.

146. R.F. Pierret, "A Linear Sweep MOS-C Technique for Determining Minority Carrier Lifetimes," *IEEE Trans. Electron Dev.* **ED-19**, 869–873, July 1972.

147. R.F. Pierret and D.W. Small, "A Modified Linear Sweep Technique for MOS-C Generation Rate Measurements," *IEEE Trans. Electron Dev.* **ED-22**, 1051–1052, Nov. 1975.

148. W.D. Eades, J.D. Shott and R.M. Swanson, "Refinements in the Measurement of Depleted Generation Lifetime," *IEEE Trans. Electron Dev.* **ED-30**, 1274–1277, Oct. 1983.

149. S. Venkatesan, R.F. Pierret, and G.W. Neudeck, "A New Lifetime Sweep Technique to Measure Generation Lifetimes in Thin-Film SOI MOSFET's," *IEEE Trans. Electron Dev.* **41**, 567–574, April 1994.

150. D.K. Schroder, M.S. Fung, R.L. Verkuil, S. Pandey, W.C. Howland, and M. Kleefstra, "Corona-Oxide-Semiconductor Generation Lifetime Characterization," *Solid-State Electron.* **42**, 505–512, April 1998.

151. G. Duggan and G.B. Scott, "The Efficiency of Photoluminescence of Thin Epitaxial Semiconductors," *J. Appl. Phys.* **52**, 407–411, Jan. 1981.

152. H.J. Hovel, "Solar Cells," in *Semiconductors and Semimetals* (R.K. Willardson and A.C. Beer, eds.) **11**, Academic Press, New York, 1975, 17–20.

153. H.S. Carslaw and J.C. Jaeger, *Conduction of Heat in Solids*, Oxford University Press, Oxford, 1959.

154. Y.I. Ogita, "Bulk Lifetime and Surface Recombination Velocity Measurement Method in Semiconductor Wafers," *J. Appl. Phys.* **79**, 6954–6960, May 1996.

155. E. Gaubas and J. Vanhellemont, "A Simple Technique for the Separation of Bulk and Surface Recombination Parameters in Silicon," *J. Appl. Phys.* **80**, 6293–6297, Dec. 1996.

156. J.S. Blakemore, *Semiconductor Statistics*, Pergamon Press, New York, 1962.

157. Private communication by J. Lagowski, Semiconductor Diagnostics, Inc.

158. D. Macdonald, R.A. Sinton, and A. Cuevas, "On the Use of a Bias-light Correction for Trapping Effects in Photoconductance-based Lifetime Measurements of Silicon," *J. Appl. Phys.* **89**, 2772–2778, March 2001.

PROBLEMS

7.1 Calculate and plot the SRH, the radiative, and the Auger low-level recombination lifetimes in Si and GaAs using the parameters in Table 7.1. Use the D/S Auger coefficients for Si and the S/R coefficients for GaAs. Plot $\log(\tau_{SRH})$, $\log(\tau_{rad})$, $\log(\tau_{Auger})$, and the resultant overall $\log(\tau_r)$ over the hole concentration range of 10^{15} to 10^{20} cm^{-3} all on the same figure. $T = 300$ K, $\sigma_n = 10^{-16}$ cm^2, $\sigma_p = 10^{-15}$ cm^2, $N_T = 10^{13}$ cm^{-3}, $E_T = E_i + 0.15$ eV, $v_{th} = 10^7$ cm/s.

7.2 The effective recombination lifetime τ_{eff} is given by

$$\frac{1}{\tau_{eff}} = \frac{1}{\tau_B} + D\beta^2, \text{ with } \tan\left(\frac{\beta d}{2}\right) = \frac{s_r}{\beta D}$$

Determine and plot τ_{eff} versus thickness d for $\tau_B = 9 \times 10^{-4}$ s and $s_r = 10, 10^3$, 10^5, and 10^7 cm/s and $0.001 \le d \le 1$ cm. $D = 30$ cm^2/s. Plot $\log(\tau_{eff})$ versus $\log(d)$. *Hint:* It may be easier to solve the equation: $\beta d/2 = \arctan(s_r/\beta D)$

7.3 Plots of $1/\tau_{eff}$ versus $1/d$ and $1/d^2$ are shown in Fig. P7.3. (a) is for $s_r \to 0$ and (b) is for $s_r \to \infty$.
Determine τ_B for each case and also determine s_r for (a) and D for (b).

7.4 The recombination lifetime τ_r is shown in Fig. P7.4 and given by

$$\frac{1}{\tau_r} = \frac{1}{\tau_{SRH}} + \frac{1}{\tau_{rad}} + \frac{1}{\tau_{Auger}}; \tau_{SRH} = \frac{1}{\sigma_p v_{th} N_T}, \tau_{rad} = \frac{1}{B n_o}, \tau_{Auger} = \frac{1}{C n_o^2}$$

Determine $\sigma_p N_T$ and C for device (i) and $\sigma_p N_T$ and B for device (ii). $v_{th} = 10^7$ cm/s.

7.5 The effective recombination lifetime is shown in Fig. P7.5 as a function of wafer thickness; all samples have identical τ_B and s_r.

$$\frac{1}{\tau_{reff}} = \frac{1}{\tau_B} + \frac{1}{\tau_S}; \tau_S = \frac{d}{2s_r}$$

Determine τ_B and s_r.

7.6 The effective recombination lifetime, given by

$$\frac{1}{\tau_{eff}} = \frac{1}{\tau_B} + \frac{1}{\tau_S}; \tau_B = \frac{1}{\sigma_n v_{th} N_T}, \tau_S = \frac{d}{2s_r}$$

is plotted in Fig. P7.6 as a function of impurity density N_T. Determine σ_n and s_r. $v_{th} = 10^7$ cm/s.

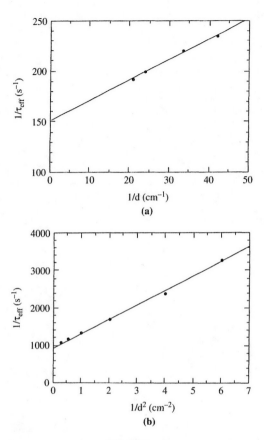

Fig. P7.3

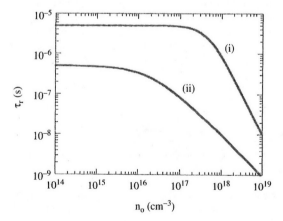

Fig. P7.4

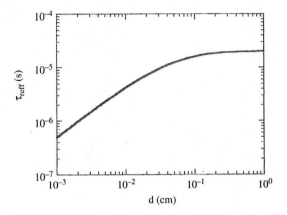

Fig. P7.5

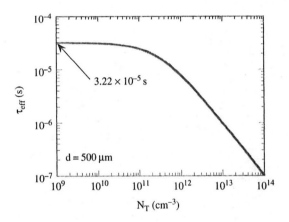

Fig. P7.6

7.7 Calculate and plot $\log(\tau_{eff})$ versus $\log(N_T)$ for $10^9 \leq N_T \leq 10^{14}$ cm^{-3}.

$$\tau_{eff} = \frac{\tau_B}{1 + \tau_B D_n \beta^2} \text{ where } \beta \text{ is determined from the relationship } \tan\left(\frac{\beta d}{2}\right) = \frac{s_r}{\beta D_n}.$$

Use: $\tau_B = \dfrac{1}{\sigma_n v_{th} N_T}$, $\sigma_n = 2 \times 10^{-14}$ cm^2, $v_{th} = 10^7$ cm/s, $d = 650$ μm, $D_n = 30$ cm^2/s. Plot three curves on the same figure for $s_r = 1$, 100, and 10,000 cm/s.

7.8 Calculate and plot the photoconductive decay curves according to Eq. (7.41) for: p-type Si, $N_A = 10^{15}$ cm^{-3}, load resistor $R = 10$ Ω, sample length = 0.3 cm, sample area = 0.01 cm^2, $\tau = 5$ μs, $T = 300$ K. Use Eq. (A8.3) for the mobilities. Plot $\log(\Delta V / V_o)$ versus t for $0 \leq t \leq 10^{-4}$ s for: (a) $\Delta n(0) = 10^{14}$ cm^{-3}, (b) $\Delta n(0) = 10^{16}$ cm^{-3}, and (c) $\Delta n(0) = 10^{18}$ cm^{-3} using $\Delta n(t) = \Delta n(0) \exp(-t/\tau)$. Discuss the relevant features of the curves.

7.9 The normalized photoconductance decay curve ($\Delta n(t)/\Delta n(0)$ versus t) is shown in Fig. P7.9 for wafer thicknesses $d = 0.025$ and $d = 0.05$ cm. From these curves, determine the effective recombination lifetimes τ_{eff}, the bulk lifetime τ_B and the surface recombination velocity s_r. It is best if you plot $1/\tau_{eff}$ versus $1/d$ and then use the equation

$$\frac{1}{\tau_{eff}} = \frac{1}{\tau_B} + \frac{2s_r}{d}$$

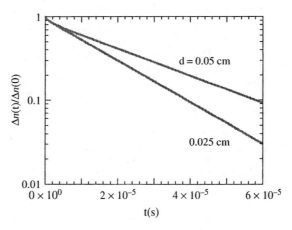

Fig. P7.9

7.10 The surface photovoltage data of an SPV measurement on a p-type Si substrate are given.

Determine the minority carrier diffusion length L_n and the surface recombination velocity s_1.

$D_n = 32$ cm^2/s, $R = 0.3$, $\Delta n(W) = 10^{10}$ cm^{-3}, use the λ to α conversion of Eq. (7.60).

λ (μm)		0.7	0.725	0.75	0.775	0.8	0.825	0.85	0.875	0.9
Φ (10^{15} Photons/s $\cdot$ cm^2)	2.08	2.09	2.11	2.13	2.15	2.19	2.23	2.29	2.38

λ (μm)		0.925	0.95	0.975	1.0	1.025	1.05
Φ (10^{15} Photons/s $\cdot$ cm^2)	2.51	2.71	3.05	3.65	4.91	8.17

7.11 Generate and plot surface photovoltage (SPV) curves of Φ versus $1/\alpha$ for a p-Si substrate using the following parameters: (i) $L_n = 100$ μm, $s_1 = 100$ cm/s; (ii) $L_n = 100$ μm, $s_1 = 10,000$ cm/s; (iii) $L_n = 10$ μm, $s_1 = 100$ cm/s. Plot all three curves on the same figure. Give appropriate units to Φ and $1/\alpha$. Use Eqs. (7.54), (7.60), and (7.63), $D_n = 30$ cm^2/s, $\Delta n(W) = 10^{10}$ cm^{-3}, $0.7 \le \lambda \le 1.05$ μm. Then determine L_n and s_1 from these SPV plots to see if you obtain the starting values.

7.12 Calculate and plot the SPV curves for Fe and Fe-B in p-Si, *i.e.*, plot Φ versus $1/\alpha$. After plotting the curves, extrapolate to $\Phi = 0$ and determine the minority carrier diffusion length L_n from the plots.

$$V_{SPV} = \frac{(kT/q)(1-R)\Phi}{n_{po}(s_1 + D_n/L_n)} \frac{L_n}{(L_n + 1/\alpha)},$$

$$\text{where} \quad \alpha = \left(\frac{83.15}{\lambda} - 74.87\right)^2 \text{ cm}^{-1} \quad (\lambda \text{ in } \mu\text{m})$$

for $0.7 \leq \lambda \leq 1 \, \mu\text{m}$. Fe: $N_T = 10^{12} \text{ cm}^{-3}$, $\sigma_n = 5.5 \times 10^{-14} \text{ cm}^2$, Fe-B: $N_T = 10^{12} \text{ cm}^{-3}$, $\sigma_n = 5 \times 10^{-15} \text{ cm}^2$. Use the following parameters: $s_1 = 1000 \text{ cm/s}$, $D_n = 30 \text{ cm}^2/\text{s}$, $R = 0.3$, $T = 300 \text{ K}$, $p_o = 10^{15} \text{ cm}^{-3}$, $n_i = 10^{10} \text{ cm}^{-3}$, $V_{SPV} = 5 \text{ mV}$.

7.13 The surface photovoltage plot of an iron-contaminated sample is shown in Fig. P7.13. Determine the iron density, N_{Fe}.

$$N_{Fe} = 1.05 \times 10^{16} \left(\frac{1}{L_{n,final}^2} - \frac{1}{L_{n,initial}^2}\right) \text{ cm}^{-3}$$

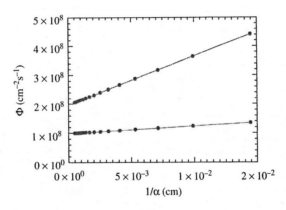

Fig. P7.13

7.14 (a) Calculate and plot $\log(\tau_{eff})$ versus $\log(N_{Fe})$ for $10^9 \leq N_{Fe} \leq 10^{14} \text{ cm}^{-3}$. Use: $s_r = 100 \text{ cm/s}$, electron diffusion coefficient $D_n = 30 \text{ cm}^2/\text{s}$, wafer thickness $d = 650 \, \mu\text{m}$. For Fe-B: $\sigma_n = 3 \times 10^{-15} \text{ cm}^2$; for interstitial iron Fe$_i$: $\sigma_n = 3 \times 10^{-14} \text{ cm}^2$; $v_{th} = 10^7 \text{ cm/s}$. The effective recombination lifetime τ_{eff} is given by

$$\frac{1}{\tau_{eff}} = \frac{1}{\tau_B} + D_n\beta^2, \quad \text{with} \quad \tan\left(\frac{\beta d}{2}\right) = \frac{s_r}{\beta D_n} \text{ and } \tau_B = \frac{1}{\sigma_n v_{th} N_T}$$

where N_T is N_{Fe-B} or N_{Fei}. *Hint:* It may be easier to solve the equation: $\beta d/2 = \arctan(s_r/D_n\beta)$; try whatever works.

(b) Next determine and plot Φ versus $1/\alpha$ for $N_T = 10^{12} \text{ cm}^{-3}$ for Fe-B and Fe$_i$ and determine the minority carrier diffusion length L_n from these surface photovoltage plots. In a real experiment, you would, of course, measure and plot

Φ versus $1/\alpha$. Here you calculate and plot it to determine L_n which should be the same as the starting values. The relevant equation is

$$\Phi = \frac{V_{SPV}\, n_{po}(s_1 + D_n/L_n)(L_n + 1/\alpha)}{(kT/q)(1 - R)L_n}$$

Use: $R = 0.3$, $s_1 = 1000$ cm/s, $T = 300$ K, $N_A = 10^{15}$ cm^{-3}, $n_i = 10^{10}$ cm^{-3}, $V_{SPV} = 10$ mV, $\alpha = 64, 157, 310, 540, 870, 1340, 2000$, and 3020 cm^{-1}. These absorption coefficients correspond to certain photon wavelengths in Si.

(c) Determine the iron density from the expression

$$N_{Fe} = 1.05 \times 10^{16} \left(\frac{1}{L_n^2(Fe_i)} - \frac{1}{L_n^2(Fe\text{-}B)} \right) \text{cm}^{-3}$$

with the diffusion lengths in units of micrometers. N_{Fe} should be very similar to the iron density starting value.

7.15 The $C - t$ data from a pulsed MOS capacitor measurement ($V_G: 0 \to V_{G1}$) give:

t(s)	C(pF)	t(s)	C(pF)	t(s)	C(pF)	t(s)	C(pF)	t(s)	C(pF)
0	5.53	60	6.94	120	9.08	180	12.36	240	16.81
5	5.62	65	7.10	125	9.29	185	12.71	245	17.09
10	5.72	70	7.24	130	9.53	190	13.05	250	17.29
15	5.83	75	7.40	135	9.76	195	13.42	255	17.43
20	5.95	80	7.55	140	10.01	200	13.79	260	17.53
25	6.06	85	7.72	145	10.26	205	14.17	265	17.58
30	6.18	90	7.89	150	10.53	210	14.55	270	17.62
35	6.30	95	8.08	155	10.81	215	14.94	275	17.64
40	6.42	100	8.26	160	11.09	220	15.34	280	17.64
45	6.54	105	8.45	165	11.39	225	15.72	285	17.64
50	6.68	110	8.65	170	11.71	230	16.10	290	17.65
55	6.81	115	8.85	175	12.03	235	16.48	295	17.65

Determine the generation lifetime $\tau_{g,eff}$ (defined in Eq. (7.116)) using the Zerbst technique.
Use $t_{ox} = 110$ nm, $N_A = 3.5 \times 10^{14}$ cm^{-3}, $n_i = 10^{10}$ cm^{-3}, $A = 3.45 \times 10^{-3}$ cm^2, $T = 300$ K, $K_{ox} = 3.9$, $K_s = 11.7$, $V_{FB} = 0$.

7.16 The pulsed MOS capacitor $C - t$ curve is shown in Fig. P7.16 when the gate voltage is pulsed from 0 to V_{G1}. Draw on the same figure the $I - t$ curve that is measured during this transient if an ammeter is placed in the circuit where the arrow is shown.

7.17 Switch S_1 in Fig. P7.17 is switched from ground to V_{G1} at $t = 0$. For the three cases of switch S_2 in position: (i) A (ground), (ii) B (open circuit), and (iii) C ($V_{D1} = V_{G1}$), draw the $C-V_G$ curves immediately after S_1 is closed *and* as $t \to \infty$. Switch S_1 does not close instantly; there is a certain rise time to reach V_{G1}. Also draw the $C - t$ curves. $V_{FB} = 0$, $V_{G1} > V_T$, where V_T is the threshold voltage when the n$^+$ p diode is grounded. The capacitance is measured at high frequencies and the measurement circuit is not shown for simplicity. The gate overlaps the diode slightly. C at $V_G = 0$ is C_{FB}.

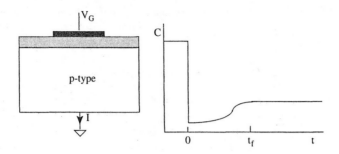

Fig. P7.16

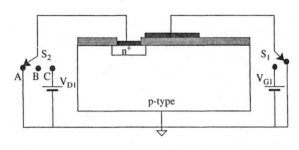

Fig. P7.17

REVIEW QUESTIONS

- Name the three recombination mechanisms.
- What is the difference between *recombination* and *generation* lifetimes?
- How does *photoconductance decay* work?
- How does *quasi-steady-state photoconductance* work?
- How does *surface photovoltage* work?
- What is special about iron in *p*-Si?
- How is *free carrier absorption* used for lifetime determination?
- How does *surface recombination* affect the effective recombination lifetime?
- How does the *diode reverse recovery* technique work?
- What techniques give the generation lifetime?
- What recombination/generation parameters can be determined from *gate-controlled diode* measurements?
- How does *corona oxide charge* method work and what recombination/generation parameters can be determined with it?

8

MOBILITY

8.1 INTRODUCTION

The *carrier mobility* influences the device performance through its frequency or time response in two ways. First, at low electric fields the carrier velocity is proportional to the mobility with higher mobility material leading to higher frequency response, because carriers take less time to travel through the device. Second, higher mobility devices have higher currents that charge capacitances more rapidly resulting in a higher frequency response.

There are several mobilities in use. The fundamental mobility is the *microscopic mobility*, calculated from basic concepts. It describes the mobility of the carriers in their respective band. The *conductivity mobility* is derived from the conductivity or the resistivity of a semiconducting material. The *Hall mobility* is determined from the Hall effect and differs from the conductivity mobility by the Hall factor. The *drift mobility* refers to the mobility when minority carriers drift in an electric field. The *effective mobility* refers to the MOSFET mobility. In addition there are considerations that cause further division between *majority carrier mobility* and *minority carrier mobility*. Momentum considerations show that electron-electron or hole-hole scattering have no first-order effect on the mobility. However, electron-hole scattering does reduce the mobility, since electrons and holes have opposite average drift velocities. Hence minority carriers experience ionized impurity and electron-hole scattering, while majority carriers experience ionized impurity scattering.

8.2 CONDUCTIVITY MOBILITY

The conductivity σ of a semiconductor is given by

$$\sigma = q(\mu_n n + \mu_p p) \tag{8.1}$$

Semiconductor Material and Device Characterization, Third Edition, by Dieter K. Schroder
Copyright © 2006 John Wiley & Sons, Inc.

For reasonably extrinsic p-type semiconductors $p \gg n$, and the hole or *conductivity mobility* is

$$\mu_p = \frac{\sigma}{qp} = \frac{1}{q\rho p} \tag{8.2}$$

Measuring the conductivity and carrier density was one of the first means of determining the semiconductor mobility, namely, the conductivity mobility.[1,2] The main reasons for its use are ease of measurement and the fact that the Hall scattering coefficient need not be known. To determine the conductivity mobility, it suffices to measure the majority carrier density and either the conductivity or the resistivity of the sample independently.

8.3 HALL EFFECT AND MOBILITY

8.3.1 Basic Equations for Uniform Layers or Wafers

The Hall effect was discovered by Hall in 1879 when he investigated the nature of the force acting on a conductor carrying a current in a magnetic field.[3] In particular, he measured the transverse voltage on gold foils. Suspecting the magnet may tend to deflect the current, he wrote "... that in this case there would exist a state of stress in the conductor, the electricity pressing, as it were, toward one side of the wire... I thought it necessary to test for a difference of potential between points on opposite sides of the conductor". Sopka gives a nice discussion of the discovery of the Hall effect including excerpts from Hall's unpublished notebook.[4]

Discussions of the Hall effect can be found in many solid state and semiconductor books. A comprehensive treatment is given by Putley.[5] The Hall effect measurement technique has found wide application in the characterization of semiconductor materials because it gives the *resistivity*, the *carrier density,* and the *mobility*. Resistivity measurements are discussed in Chapter 1 and carrier density in Chapter 2. In this chapter we give a more detailed discussion of the Hall effect and its application to mobility measurements.

Hall found that a magnetic field applied to a conductor perpendicular to the current flow direction produces an electric field perpendicular to the magnetic field and the current. Consider the p-type semiconductor sample in Fig. 8.1. A current I flows in the x-direction, indicated by the holes flowing to the right and a magnetic field B is applied in the z-direction. The current is given by

$$I = qApv_x = qwdpv_x \tag{8.3}$$

The voltage along the x-direction, indicated by V_ρ, is

$$V_\rho = \frac{\rho s I}{wd} \tag{8.4}$$

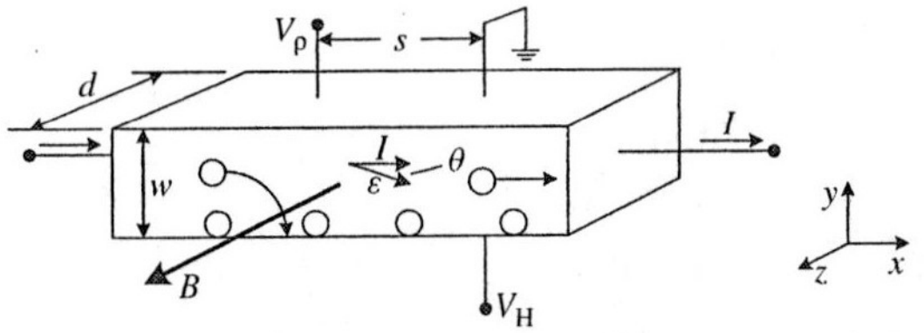

Fig. 8.1 Schematic illustrating the Hall effect in a p-type sample.

from which the resistivity is derived as

$$\rho = \frac{wd}{s}\frac{V_\rho}{I} \tag{8.5}$$

Consider now the motion of holes in a uniform magnetic field strength B. The force on the holes is given by the vector expression

$$F = q(\mathscr{E} + v \times B) \tag{8.6}$$

The magnetic field in conjunction with the current deflects some holes to the bottom of the sample, as indicated in Fig. 8.1. For n-type samples, the electrons are also deflected to the bottom of the sample for the same current direction as that in Fig. 8.1, because they flow in the opposite direction to holes and have opposite charge. In the y-direction there is no net force on the holes since no current can flow in that direction and $F_y = 0$. Combining Eqs. (8.6) and (8.3) gives

$$\mathscr{E}_y = Bv_x = \frac{BI}{qwdp} \tag{8.7}$$

The electric field in the y-direction produces the Hall voltage V_H

$$\int_0^{V_H} dV = V_H = -\int_w^0 \mathscr{E}_y\, dy = -\int_w^0 \frac{BI}{qwtp}dy = \frac{BI}{qtp} \tag{8.8}$$

The Hall coefficient R_H is defined as

$$R_H = \frac{dV_H}{BI} \tag{8.9}$$

The Hall angle θ between the current and the net electric field is

$$\tan(\theta) = \frac{\mathscr{E}_y}{\mathscr{E}_x} = B\mu_p \tag{8.10}$$

using Eq. (8.7) and $I = qp\mu_p\mathscr{E}_x wd$.

Exercise 8.1

Problem: How is R_H converted from mks to cgs units?

Solution: For the mks system, the units of R_H are m^3/C for d in m, V_H in V, B in T (1 T = 1 Tesla = 1 Weber/m^2 = 1 V·s/m^2), and I in A.
What are the cgs units? One way to determine this is to use Eq. (8.9), *i.e.*,

$$V_H = \frac{R_H BI}{d} = \frac{R_H(cm^3/C) \times 10^{-6}(m^3/cm^3)B(G) \times 10^{-4}(T/G) \times I(A)}{d(cm) \times 10^{-2}(m/cm)}$$

$$= 10^{-8}\frac{R_H BI}{d} \quad \text{or} \quad R_H = 10^8\frac{dV_H}{BI}$$

for R_H in cm^3/C, d in cm, V_H in V, B in G (Gauss; 10,000 G = 1 T), and I in A.

For $B = 5000$ G, $I = 0.1$ mA, and $p = 10^{15}$ cm^{-3}, we find $V_H = 3.1/d$. For a wafer of thickness $d = 5 \times 10^{-2}$ cm, this gives a Hall voltage $V_H \approx 6$ mV and $R_H \approx 60,000$ cm^3/C.

Combining Eqs. (8.8) and (8.9) gives

$$p = \frac{1}{qR_H}; n = -\frac{1}{qR_H} \tag{8.11}$$

When both holes and electrons are present, the Hall coefficient becomes[6]

$$R_H = \frac{(p - b^2 n) + (\mu_N B)^2 (p - n)}{q[(p + bn)^2 + (\mu_N B)^2 (p - n)^2]} \tag{8.12}$$

This expression is relatively complex and depends on the mobility ratio $b = \mu_n/\mu_p$ and on the magnetic field strength B. In the limit of low and high magnetic field strength, the Hall coefficient becomes

$$B \Rightarrow 0: R_H = \frac{(p - b^2 n)}{q(p + bn)^2}; B \Rightarrow \infty: R_H = \frac{1}{q(p - n)} \tag{8.13}$$

For Eq. (8.13) to hold in the low field limit, $B \ll 1/\mu_n$ for $p \gg n$ and $B \ll 1/\mu_p$ for $p \ll n$. For a mobility of 1000 cm^2/V·s this requires $B \ll 10$ T. For mobilities of 10^5 cm^2/V·s, this requirement becomes more severe, with $B \ll 0.1$ T. The high-field limit requires $B \gg 1/\mu_n$ for $p \gg n$ and $B \gg 1/\mu_p$ for $p \ll n$. Hence magnetic fields much larger than 10 T or 0.1 T, respectively, are necessary in this example.

For semiconductors with modest mobilities in the 100 to 1000 cm^2/V·s range and with mobility ratios of $b \approx 3$ to 10, the Hall coefficient is generally found to vary little with magnetic field and Eq. (8.13) with $B \Rightarrow \infty$ is used. However, for semiconductors with high mobilities and high b the Hall coefficient is found to vary with magnetic field and changes sign as a function of temperature. Such behavior is found in semiconductors like HgCdTe, as shown in Fig. 8.2(a) for a p-type HgCdTe with $E_G = 0.15$ eV.[7] Electron conduction dominates for temperatures of 220 to 300 K, with $n = n_i^2/p \gg p$, because n_i^2 is high for narrow band gap materials. $R_H = -1/qn$ in this temperature range, and it is independent of B. For $T \approx 100$ to 200 K holes begin to participate and mixed conduction causes R_H to decrease and be magnetic field dependent. Hole conduction dominates at lower temperatures. The Hall coefficient becomes positive and is magnetic field independent. This figure exhibits the temperature and magnetic field dependent behavior of mixed conduction very nicely. Figure 8.2(b) shows the Hall coefficient for GaAs, with neither magnetic field dependence nor mixed conduction[8] and the electron density derived from the Hall coefficient using Eq. (8.11). Sometimes it is necessary to consider the contribution of light and heavy holes.[9]

Equations (8.11) to (8.13) are derived under simplifying assumptions of energy-independent scattering mechanisms. With this assumption relaxed, the expressions for the hole and electron densities become[5-6]

$$p = \frac{r}{qR_H}; n = -\frac{r}{qR_H} \tag{8.14}$$

where r is the Hall scattering factor, defined by $r = \langle \tau^2 \rangle / \langle \tau \rangle^2$, with τ being the mean time between carrier collisions. The scattering factor depends on the type of scattering

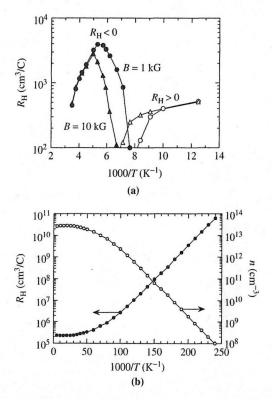

Fig. 8.2 (a) Temperature and magnetic field dependent Hall coefficient for HgCdTe showing typical mixed conduction behavior. Reprinted with permission after Zemel et al.[7] (b) Hall coefficient and electron density for GaAs adapted from Stillman and Wolfe.[8]

mechanism in the semiconductor and generally lies between 1 and 2. For lattice scattering, $r = 3\pi/8 = 1.18$, for impurity scattering $r = 315\pi/512 = 1.93$, and for neutral impurity scattering $r = 1$.[6, 10] The scattering factor is also a function of magnetic field and temperature and can be determined by measuring R_H in the high magnetic field limit, *i.e.*, $r = R_H(B)/R_H(B = \infty)$. In the high field limit $r \to 1$. The scattering factor in n-type GaAs as a function of magnetic field and was found to vary from 1.17 at $B = 0.01$ T, as expected from lattice scattering, to 1.006 at $B = 83$ kG.[11] The high fields necessary for r to approach unity are difficult to achieve, and $r > 1$ for most Hall measurements. Typical magnetic fields used for Hall measurements lie between 0.05 and 1 T.

The *Hall mobility* μ_H, defined by

$$\mu_H = \frac{|R_H|}{\rho} = |R_H|\sigma \tag{8.15}$$

differs from the conductivity mobility. Substituting Eq. (8.1) into Eq. (8.15) gives

$$\mu_H = r\mu_p; \mu_H = r\mu_n \tag{8.16}$$

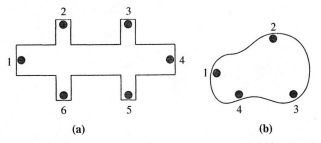

Fig. 8.3   (a) Bridge-type Hall sample, (b) lamella-type van der Pauw Hall sample.

for extrinsic p- and n-type semiconductors, respectively. Hall mobilities can differ significantly from conductivity mobilities since r is generally larger than unity. For most Hall-determined mobilities, r is taken as unity, but this assumption should be specified.

The schematic Hall sample of Fig. 8.1 has a variety of practical implementations. One of these is the bridge-type Hall bar in Fig. 8.3(a). The current flows into 1 and out of 4, the Hall voltage is measured between 2 and 6 or between 3 and 5 in the presence of a magnetic field. The resistivity is determined in the absence of the magnetic field by measuring the voltage between 2 and 3 or between 6 and 5. The equations above apply for this geometry.

A more general geometry is the irregularly shaped sample in Fig. 8.3(b). The theoretical foundation of Hall measurement evaluation for irregularly shaped samples is based on conformal mapping developed by van der Pauw.[12-13] He showed how the resistivity, carrier density, and mobility of a flat sample of arbitrary shape can be determined without knowing the current pattern if the following conditions are met: the contacts are at the circumference of the sample and are sufficiently small, the sample is uniformly thick, and does not contain isolated holes.

For the sample of Fig. 8.3(b) the resistivity is given by[12]

$$\rho = \frac{\pi t}{\ln(2)} \frac{R_{12,34} + R_{23,41}}{2} F \qquad (8.17)$$

where $R_{12,34} = V_{34}/I$. The current I enters the sample through contact 1 and leaves through contact 2 and $V_{34} = V_3 - V_4$ is the voltage between contacts 3 and 4. $R_{23,41}$ is similarly defined. Current enters the sample through two adjacent terminals and the voltage is measured across the other two adjacent terminals. F is a function of the ratio $R_r = R_{12,34}/R_{23,41}$ only, satisfying the relation

$$\frac{R_r - 1}{R_r + 1} = \frac{F}{\ln(2)} \text{ arcosh } \left(\frac{\exp(\ln(2)/F)}{2} \right) \qquad (8.18)$$

and is plotted in Fig. 8.4. For symmetric samples (circles or squares) $F = 1$.

The van der Pauw Hall mobility is determined by measuring the resistance $R_{24,13}$ with and without a magnetic field. $R_{24,13}$ is measured by forcing the current into one and out of the opposite terminal, $e.g.$, terminals 2 and 4 in Fig. 8.3, with the voltage measured

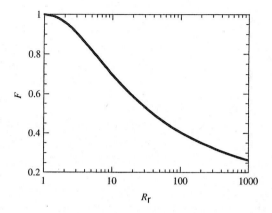

Fig. 8.4   The van der Pauw F factor plotted against R_r.

across terminals 1 and 3. The Hall mobility is then given by

$$\mu_H = \frac{d\Delta R_{24,13}}{B\rho} \tag{8.19}$$

where $\Delta R_{24,13}$ is the resistance change of $R_{24,13}$ due to the magnetic field.

Equations (8.14) and (8.17) are for carrier densities per unit volume and for resistivity ρ (ohm·cm). Occasionally it is useful to determine carrier densities per unit area and sheet resistance R_{sh} (ohms/square). For uniformly doped samples of thickness d, the *sheet Hall coefficient* R_{Hsh} is defined as

$$R_{Hsh} = \frac{R_H}{t} \tag{8.20}$$

and

$$\mu_H = \frac{|R_{Hsh}|}{R_{sh}} \tag{8.21}$$

where $R_{sh} = \rho/d$.

The thickness is well defined for bulk samples. For thin layers on substrates of opposite conductivity or on semi-insulating substrates, the active film thickness is not necessarily the total film thickness. Depletion effects caused by Fermi level pinned band bending or surface charges and by band bending at the layer-substrate interface lead to errors in the Hall coefficient.[14-15] For sufficiently lightly doped films it is possible for the surface-induced space-charge region to deplete the entire film. Hall effect measurements then indicate a semi-insulating film. For semiconducting films on insulating substrates, the mobility is frequently observed to decrease toward the substrate. Surface depletion forces the current to flow in the low-mobility portion of the film, giving apparent mobilities lower than true mobilities.[15] Even the temperature dependence of the surface and interface space-charge regions should be considered for unambiguous temperature-dependent mobility and carrier density measurements.[16]

8.3.2   Non-uniform Layers

Hall effect measurements are simple to interpret for uniformly doped samples. Non-uniformly doped layer measurements are more difficult to interpret. If the doping density varies with film thickness, then its resistivity and mobility also vary with thickness. A Hall effect measurement gives the *average* resistivity, carrier density, and mobility. For spatially varying mobility $\mu_p(x)$ and carrier density $p(x)$, the Hall sheet coefficient R_{Hsh}, the sheet resistance R_{sh}, and the average Hall mobility $\langle \mu_H \rangle$ for a p-type film of thickness d are given by[17-18]

$$R_{Hsh} = \frac{\int_0^t p(x)\mu_p^2(x)\,dx}{q\left(\int_0^t p(x)\mu_p(x)\,dx\right)^2}; R_{sh} = \frac{1}{q\int_0^t p(x)\mu_p(x)\,dx}; \langle \mu_H \rangle = \frac{\int_0^d p(x)\mu_p^2(x)\,dx}{\int_0^d p(x)\mu_p(x)\,dx}$$

(8.22)

assuming $r = 1$. Here x specifies the distance into the sample. To determine resistivity and mobility *profiles*, Hall measurements are made as a function of film thickness. The film thickness is varied by either removing thin portions of the film by etching and measuring the Hall coefficient repeatedly, or by making portions of the film electrically inactive by a reverse-biased space-charge region.

In principle, one can use chemical etching to remove thin layers of the film. In practice, it is difficult to remove thin layers reproducibly by chemical etching. The electrochemical profiler, discussed in Section 2.2.6, has been successfully used to remove thin layers by electrolytic etching of GaAs in Tiron (1,2 dihydroxybenzene-3,5 disulphonic acid, disodium salt in an aqueous solution).[19] Hall effect measurements are made after each etch. A more common method for reliable layer removal is anodic oxidation and subsequent oxide etch.[17-18, 20-24] Anodic oxidation consumes a fraction of the semiconductor during oxidation. When the oxide is subsequently etched, that portion of the semiconductor consumed during oxidation is also removed, providing for reproducible semiconductor removal without altering the doping profile. A discussion of anodic oxidation is given in Section 1.4.1.

A second method uses a junction formed on the upper surface of the film to be profiled. The film must be sufficiently thin for the reverse-biased space-charge region to be able to deplete it and it must be bounded at its lower surface by an insulator or a junction. The upper junction may be a pn junction, a Schottky barrier junction, or an MOS capacitor. An example in Fig. 8.5 consists of a p-layer on an insulator. The layer is provided with a Schottky gate. The zero-biased metal-semiconductor junction induces a space-charge region of width W under the metal. The insulator could be replaced by a semi-insulating substrate or by an n-type substrate. The square sample is laterally isolated by etching but could be isolated by surrounding it with an n-type film. Four contacts provide for current and voltage probes.

Van der Pauw measurements provide information on the undepleted film of thickness $d-W$, where d is the total film thickness. A single measurement gives the mobility, the

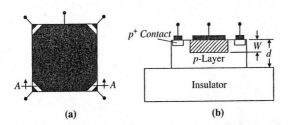

(a)   (b)

Fig. 8.5 Schottky-gated thin film van der Pauw sample, (a) top view, (b) cross-section along line A-A showing the gate, two contacts and the space-charge region of width W.

resistivity, and the carrier density averaged over d-W. When the Schottky barrier junction is reverse biased, its space-charge region extends into the film, reducing the thickness of the neutral portion of the film. By measuring the Hall effect as a function of reverse-bias voltage, one can determine mobility, resistivity, and carrier density profiles of the underlying layer. This method has been implemented with MOSFETs for thin Si films on sapphire,[25-27] for Si-on-insulator,[28] and with Schottky diodes for GaAs on semi-insulating substrates.[29-30] A comparison of the destructive "anodize-etch-measure" with the "gated" technique has shown the "gated" method to give more reliable mobilities and to have higher spatial resolution.[31]

The spatially varying Hall mobility is determined from the spatially varying sheet Hall coefficient and sheet conductance $G_{sh} = 1/R_{sh}$ by the relationship[17-18, 32]

$$\mu_H = \frac{d(R_{Hsh}G_{sh}^2)/dx}{dG_{sh}/dx} \tag{8.23}$$

and the spatially varying carrier density is

$$p(x) = \frac{r}{q}\frac{(dG_{sh}/dx)^2}{d(R_{Hsh}G_{sh}^2)/dx} \tag{8.24}$$

In the differential Hall effect (DHE) discussed in Chapter 1, the mobility and carrier density profiles are determined after each layer removal step by making Hall measurements and using the Hall measured values of adjacent layers in the calculations. The average values of mobility and carrier density may differ from the true values if there are large inhomogeneities in the sample. To reduce this effect, it is necessary to make Δx_i, where Δx_i is the thickness of the i th layer, small to approximate the non-uniform film by a uniform film. For ion-implanted and fully annealed samples with no mobility anomalies, the error between the measured and real mobility and carrier density is less than 1% if $\Delta x_i < 0.5\Delta R_p$, where ΔR_p is the standard deviation of the implanted profile.[33] A density profile of a boron layer implanted into Si is shown in Fig. 8.6, where the Hall measured

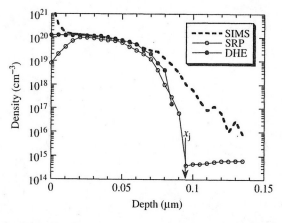

Fig. 8.6 Dopant density profiles determined by DHE, spreading resistance profiling, and secondary ion mass spectrometry. Data after ref. 34.

profile is compared with the profile determined by secondary ion mass spectrometry and spreading resistance profiling.[34]

Difficulties can arise when there are large mobility variations through the film. Consider a film consisting of two layers of equal thickness. The upper layer has a carrier density of P_1 holes/cm^2 with mobility μ_1 and the lower one has P_2 holes/cm^2 and μ_2.[35] The total hole density is $P_1 + P_2$. The Hall effect measures the weighted averages[18]

$$P = \frac{(P_1\mu_1 + P_2\mu_2)^2}{P_1\mu_1^2 + P_2\mu_2^2} \tag{8.25}$$

$$\mu_H = \frac{P_1\mu_1^2 + P_2\mu_2^2}{P_1\mu_1 + P_2\mu_2} \tag{8.26}$$

Here P will be significantly less than $(P_1 + P_2)$ and μ_H will lie between μ_1 and μ_2 for $P_1 > P_2$ and $P_1\mu_1^2 < P_2\mu_2^2$. For example, for $P_1 = 10P_2$ and $\mu_2 = 10\,\mu_1$ we find $P \approx 4P_2$ and $\mu_H = 0.55\,\mu_2$. For inhomogeneous samples it is possible for the mobility to be higher than the expected bulk mobility. One cause of abnormally high mobilities is the inclusion of metallic precipitates in the crystal. A thorough discussion of this effect is given by Wolfe and Stillman.[36]

8.3.3 Multi Layers

The measurement of non-uniform films on an "inert" substrate, *i.e.*, a substrate that does not contribute to the measurement, was addressed in the previous section. A *p*-film on an *n*-substrate or an *n*-film on a *p*-substrate might be thought to be in the same category, with the space-charge region (scr) between two semiconductors of opposite conductivity considered an insulating boundary. But this is a more precarious situation. For example, a leaky junction can no longer be considered an insulator. Even if the insulating properties of the scr are sufficiently good, there may be leakage paths along the surface, or, even worse, the heavily doped contacts may be diffused into the substrate, providing a leakage path. Film characterization is then no longer unique to the film, and the substrate properties are reflected in the measurements.

This problem was originally addressed by Neduloha and Koch[37] and by Petritz,[38] who considered a substrate whose surface is inverted by surface charges, *e.g.*, an *n*-type inversion layer on a *p*-type substrate. The two-layer interacting configuration was later extended.[39-40] For a simple two-layer structure with an upper layer having thickness d_1 and conductivity σ_1 and a substrate of thickness d_2 and conductivity σ_2, the Hall constant is given by[37]

$$R_H = \frac{d[(R_{H1}\sigma_1^2 d_1 + R_{H2}\sigma_2^2 d_2) + R_{H1}\sigma_1^2 R_{H2}\sigma_2^2 (R_{H1}d_2 + R_{H2}d_1)B^2]}{(\sigma_1 d_1 + \sigma_2 d_2)^2 + \sigma_1^2\sigma_2^2(R_{H1}d_2 + R_{H2}d_1)^2 B^2} \tag{8.27}$$

which becomes[38, 40]

$$R_H = \frac{d(R_{H1}\sigma_1^2 d_1 + R_{H2}\sigma_2^2 d_2)}{(\sigma_1 d_1 + \sigma_2 d_2)^2} = R_{H1}\frac{d_1}{d}\left(\frac{\sigma_1}{\sigma}\right)^2 + R_{H2}\frac{d_2}{d}\left(\frac{\sigma_2}{\sigma}\right)^2 \tag{8.28}$$

in the low magnetic field limit, and

$$R_H = \frac{R_{H1}R_{H2}d}{R_{H1}d_2 + R_{H2}d_1} \tag{8.29}$$

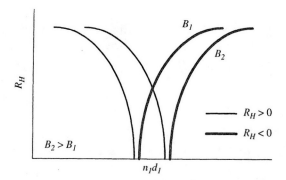

Fig. 8.7 Hall coefficient of a p-type substrate with an n-type layer as a function of $n_1 t_1$ for two magnetic fields.

in the high magnetic field limit. In these equations R_{H1} is the Hall constant of layer 1, R_{H2} is the Hall constant of substrate 2, $d = d_1 + d_2$ and σ is given by

$$\sigma = \frac{d_1}{d}\sigma_1 + \frac{d_2}{d}\sigma_2 \tag{8.30}$$

The magnetic field dependence of Eq. (8.27) gives additional information by measuring the Hall coefficient as a function of magnetic field, illustrated in Fig. 8.7 for an n-layer on a p-substrate where $R_{H1} = -1/qn_1$ and $R_{H2} = 1/qp_2$. The Hall coefficients are of opposite sign, making it possible for the measured Hall coefficient to reverse its sign with magnetic field. The Hall coefficient is plotted against the $n_1 d_1$ product. For low $n_1 d_1$ the Hall coefficient is dominated by the p-substrate and is magnetic field independent. Both p_2 and μ_2 can be determined from R_H. For intermediate values of $n_1 d_1$, the Hall coefficient becomes field dependent. Conduction is initially dominated by holes, and then by electrons as the Hall coefficient changes its sign. The carrier density and mobility of both the n-layer and the p-substrate can be deduced from an analysis of the field-dependent R_H using the two-layer model of Eq. (8.27). For high $n_1 d_1$ values the Hall coefficient is negative, conduction is dominated by the n-layer and R_H becomes again magnetic field independent. A good discussion can be found in Zemel et al.[7]

If the upper layer is more heavily doped than the substrate or is formed by inversion through surface states, for example, and the carriers in the substrate freeze out at low temperatures making σ_2 very small. Examples of an n-type skin on a p-type bulk, an n-type film on p-type bulk, and an n-type skin on n-type bulk are given for HgCdTe and InSb.[40–41]

8.3.4 Sample Shapes and Measurement Circuits

Hall samples come in two basic geometries: bridge type and lamella type. The parallelepiped sample shape of Fig. 8.1 is not recommended because contacts have to be directly soldered to the sample. To ease the contact problem, the Hall bridge has extended arms as shown in Fig. 8.8(a).[42] Both six- and eight-arm geometries can be used with dimensions given in ASTM Standard F76.[42] The lamellar specimen may be of arbitrary shape, but a symmetrical configuration is preferred. The sample must be free of geometrical holes; typical shapes are shown in Fig. 8.8(b) to (d). For the lamella-type specimen

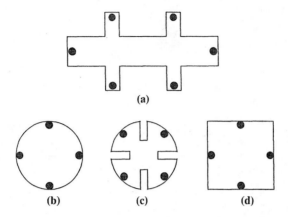

Fig. 8.8 (a) Bridge-type Hall configuration, (b)-(d) lamella-type Hall configuration.

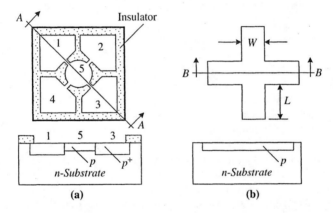

Fig. 8.9 van der Pauw Hall sample shapes.

it is important for the contacts to be small and to be placed as close to the periphery as possible.

Some common lamella or van der Pauw shapes are shown in Fig. 8.9. During the early days of ion implantation development, implant uniformity was often characterized by sample shapes of the type in Fig. 8.9. In Fig. 8.9(a), photolithography is used to provide the patterns for the p^+ contact diffusions 1 to 4. The area to be measured is region 5. A transfer length contact resistance test pattern has also been used for Hall measurements. In addition to the contact resistance, specific contact resistance, and sheet resistance, the mobility in the implanted layer and under the contacts, as well as the sheet carrier density, were extracted by applying a magnetic field.[43]

The size and placement of the contacts is important. For van der Pauw samples the contacts should be point contacts located symmetrically on the periphery. This is not achievable in practice, and some error is introduced thereby. A few cases were treated by van der Pauw.[12] He considered circular samples with contacts spaced at 90° intervals. The contacts are equipotential areas with three cases shown in Fig. 8.10. In each case

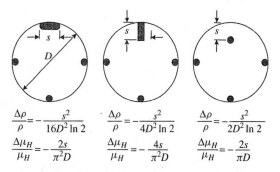

$$\frac{\Delta\rho}{\rho} = -\frac{s^2}{16D^2 \ln 2} \qquad \frac{\Delta\rho}{\rho} = -\frac{s^2}{4D^2 \ln 2} \qquad \frac{\Delta\rho}{\rho} = -\frac{s^2}{2D^2 \ln 2}$$

$$\frac{\Delta\mu_H}{\mu_H} = -\frac{2s}{\pi^2 D} \qquad \frac{\Delta\mu_H}{\mu_H} = -\frac{4s}{\pi^2 D} \qquad \frac{\Delta\mu_H}{\mu_H} = -\frac{2s}{\pi D}$$

Fig. 8.10 Effect of non ideal contact length or contact placement on the resistivity and mobility for van der Pauw samples. Reprinted with permission from van der Pauw.[12]

there are three ideal contacts, with the fourth being non-ideal. The fourth contact is either of length s and larger than a point contact or is a point contact displaced a distance s from the periphery. Also indicated for each geometry is the relative error in resistivity $\Delta\rho/\rho$ and mobility $\Delta\mu_H/\mu_H$ introduced by the non-ideal contact, valid for small s/D and low $\mu_H B$. The errors are additive to first order if more than one contact is not ideal.

One implementation is the use of some form of cloverleaf geometry shown in Fig. 8.8 (c) and Fig. 8.9. The errors due to displaced contacts on square specimen are discussed in refs. 44–45. The placement of the contacts on square samples is better at the midpoint of the sides than at the corners.[44] The *Greek Cross* in Fig. 8.9(b) makes use of this type of geometry, where for $L = 1.02W$ less than 0.1% error is introduced.[45] For square samples with sides of length L having square and triangular contacts of contact length s in the four corners, less than 10% error is introduced for Hall measurements as long as $s/L < 0.1$.[46] The contacts need not be exactly opposite one another, since the magnetic field reversal routinely made during Hall measurements tends to cancel any unbalanced voltage. But for an unbalanced voltage higher than the Hall voltage, the Hall voltage is the difference of two large numbers, and errors are likely to be introduced.

Some samples use the geometry of a semiconductor device. For example, a MOSFET fabricated in a thin film on an insulating substrate has the general shape of the Hall sample in Fig. 8.11, where the p^+ regions 1 and 2 are the source and the drain and 7 is the gate. The contact regions 3–6 are added for Hall measurements. The Hall voltage is developed between contacts 3 and 5 and 4 and 6. In some cases there are only two contacts, *e.g.*, 4 and 6, and they should be about halfway between source and drain and $W/L \leq 3$.[47] However, the sample is shorted at the ends by the source and the drain with a significant influence on the measured Hall voltage V_{Hm}. For $L = 3W$ in Fig. 8.12(a), V_{Hm} is less than the Hall voltage for samples with $L > 3W$. The Hall voltage V_H for sample dimensions of $L \gg W$ used in the earlier equations in this chapter is related to the measured Hall voltage for short samples by $V_H = V_{Hm}/G$, where G is shown in Fig. 8.12(b).[48] The curves in Fig. 8.12(b) are calculated for the Hall voltage measured across the sample at $x = L/2$. For sample lengths $L = 3W$, the shorting effect is negligible, and the measured voltage is the usual Hall voltage.

ASTM Standard F76 gives a detailed discussion of the measurement procedure and measurement precautions.[42] The current and the magnetic field are reversed and the readings averaged for more accurate measurements. Special precautions are necessary when the specimen resistance is very high to eliminate current leakage paths and sample

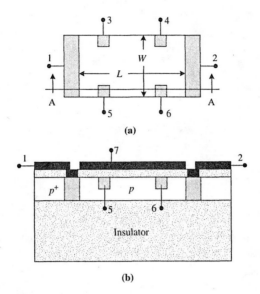

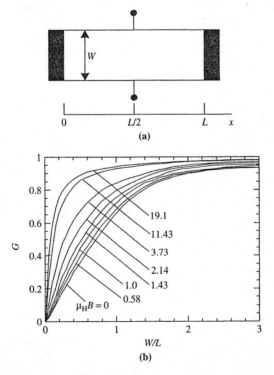

Fig. 8.11 Hall sample with electrically shorted regions at the ends; (a) top view with the gate not shown, (b) cross section along cut A-A.

Fig. 8.12 (a) Hall sample with electrically shorted end regions; (b) ratio of measured voltage V_{Hm} to Hall voltage V_H. $G = V_{Hm}/V_H$. Reprinted with permission after Lippmann and Kuhrt.[48]

loading by the voltmeter. The *guarded* approach utilizes high input impedance unity gain amplifiers between each probe on the sample and the external circuitry.[49] The unity gain outputs drive the shields on the leads between the amplifier and the sample to reduce leakage currents and system time constant by effectively eliminating the stray capacitance in the leads. Measurements of resistances up to 10^{12} ohms have been made with such a system.[50] Measurements on semi-insulating GaAs have been made by illuminating a slit across a "dark" wafer and introducing a dark spot within the illuminated slit.[51] A resistance measurement along the slit determines essentially the resistance of the small dark spot since the dark spot resistance is much higher than the resistance of the illuminated strip. A resistance map can be obtained by moving the dark spot.

Hall effect profiling measurements have other possible errors. For example, the bottom pn junction may be leaky, causing smaller Hall voltages than would be measured for perfect isolation of the film from the substrate. The upper junction in a Schottky contact configuration may also be leaky. Junction leakage currents can be reduced by sample cooling.[21] If the upper junction is forward biased to reduce the space-charge region width to profile closer to the surface, considerable error is introduced due to the forward-biased junction current.[29] Although the effect of injected gate current can be corrected,[52] the correction is large and the accuracy may be questionable. Instead of conventional dc measurement circuits, ac circuits can be employed,[29-30] where the device is driven with an ac current at one frequency and a gate voltage containing both a dc bias to vary the scr width and an ac component of a frequency different from the current. The appropriate ac voltages are measured with a lock-in amplifier without interference from the dc leakage current. In one implementation the magnetic field and the current frequencies were 60 Hz and 200 Hz, respectively.[53] The Hall voltage is detected by a lock-in amplifier at 260 Hz, eliminating most thermoelectric and thermomagnetic errors allowing Hall voltages as low as 10 μV to be measured.

8.4 MAGNETORESISTANCE MOBILITY

Typical Hall-effect structures are either long or of the van der Pauw variety. They require four or more contacts. A long Hall bar is shown schematically in Fig. 8.13(a) with $L \gg W$. Field-effect transistors (FETs) are short with $L \ll W$, shown in Fig. 8.13(b). The Hall electric field, resulting from an applied magnetic field, is nearly shorted by the long contacts and FET structures do not lend themselves well to Hall measurements. The extreme of this short geometry is when one contact is in the center of a circular sample and the other contact is at the periphery, shown in Fig. 8.13(c). The Hall electric field in this *Corbino disk*[54] is shorted, and no Hall voltage exists. The geometries of Fig. 8.13(b) and (c), however, lend themselves well to *magnetoresistance* measurements.

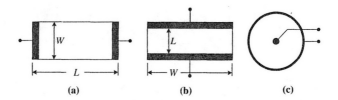

Fig. 8.13 (a) Hall sample, (b) short, wide sample, (c) Corbino disk.

The *resistivity* of a semiconductor generally increases when the sample is placed in a magnetic field. This is the *physical magnetoresistance effect* (PMR), if the conduction is anisotropic, if conduction involves more than one type of carrier, and if carrier scattering is energy dependent. The *resistance* of a semiconductor is also influenced by magnetic fields.[55] The magnetic field causes the path of the charge carriers to deviate from a straight line, raising the sample resistance. This depends on the sample geometry and is known as the *geometrical magnetoresistance* (GMR). The magnetic field induced resistance change is due to resistivity changes of the semiconductor as well as geometrical effects and is larger the higher the sample mobility is. Geometric effects usually dominate. For example, in GaAs at room temperature and in a magnetic field of $1\ T$, the *PMR* is about 2%, whereas the GMR is about 50%. The geometric magnetoresistance mobility μ_{GMR} is related to the Hall mobility μ_H by

$$\mu_{GMR} = \xi \mu_H \tag{8.31}$$

where ξ is the magnetoresistance scattering factor given by $\xi = (\langle \tau^3 \rangle \langle \tau \rangle / \langle \tau^2 \rangle^2)^2$.[10] For τ independent of energy, the mean time between collisions becomes isotropic, $\xi = 1$ and $\mu_{GMR} = \mu_H$. The physical magnetoresistivity change ratio $\Delta\rho_{PMR} = (\rho_B - \rho_0)/\rho_0$ becomes zero under those conditions, where ρ_B is the resistivity in the presence and ρ_0 in the absence a magnetic field.

The dependence of the resistance ratio R_B/R_0 is shown in Fig. 8.14 as a function of $\mu_{GMR}B$ for rectangular samples of varying L/W ratios.[56] Here R_B is the resistance with $B \neq 0$ and R_0 is the resistance with $B = 0$. For long rectangular samples with contacts at the ends of the long sample as in Fig. 8.13(a), the ratio is near unity and the magnetoresistance effect is very small. The ratio is higher for short, wide samples. The highest ratio is obtained for the Corbino disk with $L/W = 0$. Figure 8.14 shows the magnetoresistance and the Hall effect to be complementary. When one decreases, the other increases. For example, in Fig. 8.12 the Hall voltage is reduced for short, wide samples. But those same sample shapes produce maximum magnetoresistance. Magnetoresistance measurements are suitable for field-effect transistors that are short and wide. The current flow in a Corbino disk is radial from the center to the periphery for $B = 0$. With a magnetic field perpendicular to the sample, the current streamlines become logarithmic

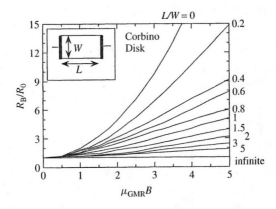

Fig. 8.14 Geometric magnetoresistance ratio of rectangular samples versus $\mu_{GMR}B$ as a function of the length/width ratio. Reprinted with permission after Lippmann and Kuhrt.[56]

spirals and the resistance ratio becomes

$$\frac{R_B}{R_0} = \frac{\rho_B}{\rho_0}[1 + (\mu_{\text{GMR}}B)^2]$$ (8.32)

Equation (8.32) represents the Corbino disk's curve in Fig. 8.14. Generally, the magnetoresistance scattering factor ξ is taken as unity just as the Hall scattering factor is generally taken to be unity for simplicity and because the scattering mechanisms are not known precisely. Measurements of μ_{GMR} on a modified Corbino disk geometry and of μ_H on Hall samples from the Corbino disk showed ξ to be unity for GaAs within experimental error.[57-58] The measurements were performed for magnetic fields up to 0.7 T and temperatures from 77 to 400 K.[58] Under those conditions $\rho_B \approx \rho_0$ and $\mu_{\text{GMR}} \approx \mu_H$. Making the additional assumption of $\mu_H \approx \mu_p$, the mobility is given by

$$\mu_p \approx \frac{1}{B}\sqrt{\frac{R_B}{R_0} - 1}$$ (8.33)

The mobility is obtained from the slope of a plot of $(R_B/R_0 - 1)^{1/2}$ versus B and can be profiled by using a Corbino disk with a Schottky gate and measuring the resistance as a function of the gate voltage.[59]

The use of Corbino disks is inconvenient because of its special geometrical configuration. However, as is evident from Fig. 8.14, rectangular sample shapes with low L/W ratios are equally suitable for magnetoresistance measurements.[60] For rectangular samples with low L/W ratios and $\mu_{\text{GMR}}B < 1$, Eq. (8.36) is replaced by[56-57]

$$\frac{R_B}{R_0} = \frac{\rho_B}{\rho_0}[1 + (\mu_{\text{GMR}}B)^2(1 - 0.54L/W)]$$ (8.34)

If the error in the determination of μ_{GMR} is to be less than 10%, then the aspect ratio L/W must be less than 0.4. For typical FET structures with $L/W \ll 1$, Eq. (8.34) is a close approximation to Eq. (8.32), and it is for that reason that Eq. (8.32) is generally used in GMR measurements. Magnetoresistance measurements were first used for GaAs Gunn effect devices.[57, 61] It is a rapid technique that can be used for functional devices, requiring no special test structures. Instead of measuring the resistance as a function of the FET gate voltage, it also possible to determine the mobility from transconductance measurements with and without a magnetic field.[62]

The magnetoresistance mobility measurement method has been applied to metal-semiconductor FETs (MESFETs) as well as to modulation-doped FETs (MODFETs). By using the magnetic field dependence of the GMR effect, it is possible to extract the mobilities of the various conducting regions and sub-bands in MODFETs.[63] The method has been used to determine the mobility dependence on gate electric field.[64] Effects of gate currents for Schottky-gate devices and series resistance effects must be corrected.[52, 62, 65] Gate current corrections are particularly important when the gate becomes forward biased. Contact resistance, which is of only secondary importance for Hall measurements, is very important for GMR measurements because it adds to the measured resistance and contact resistance is relatively independent of magnetic field. When the mobility is measured as a function of gate bias, the average mobility is measured for each value of gate voltage. Both the *average* and the *differential mobilities* can be determined from transconductance measurements.[66]

The GMR effect is not universally applicable the way the Hall effect is, shown by Eq. (8.32). Assuming that $\rho_B/\rho_0 \approx 1$, which is a reasonable assumption, to observe a resistance change, $\Delta R/R_0 = (R_B - R_0)/R_0$ of, say, 10%, the condition $\mu_{GMR} = 0.3/B$ must be met. For typical magnetic fields of 0.1 to 1 T, this requires $\mu_{GMR} = 30,000$ to 3000 cm^2/V·s, mobilities found in MESFETs and MODFETs made in III-V materials, especially at low temperatures. These are the very materials that have been successfully characterized by GMR. For higher magnetic fields, as obtained with superconducting magnets, lower mobilities can be determined. Silicon, whose mobility lies in the 500–1300 cm^2/V·s range, is unsuitable for magnetoresistance measurements because its GMR is negligibly small for typical laboratory magnet fields.

8.5 TIME-OF-FLIGHT DRIFT MOBILITY

The *time-of-flight* method to determine the *minority carrier mobility* was first demonstrated in the *Haynes-Shockley experiment*.[67–69] The first comprehensive mobility measurements for Ge and Si were made with this technique by Prince.[70] The principle of the method is demonstrated with the *p*-type semiconductor bar in Fig. 8.15(a). A drift voltage $-V_{dr}$

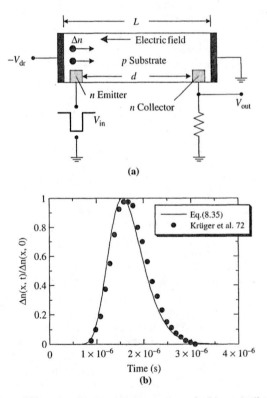

(a)

(b)

Fig. 8.15 (a) Drift mobility measurement arrangement and (b) normalized output voltage pulse ($\mu_p = 180$ cm^2/V·s, $\tau_n = 0.67$ μs, $T = 423$ K, $\mathscr{E} = 60$ V/cm), (c) output voltage pulses ($\mu_n = 1000$ cm^2/V·s, $\tau_n = 1$ μs, $T = 300$ K, $\mathscr{E} = 100$ V/cm, $N = 10^{11}$ cm^{-2}), (d) output voltage pulses ($\mu_n = 1000$ cm^2/V·s, $d = 0.075$ cm, $T = 300$ K, $\mathscr{E} = 100$ V/cm, $N = 10^{11}$ cm^{-2}).

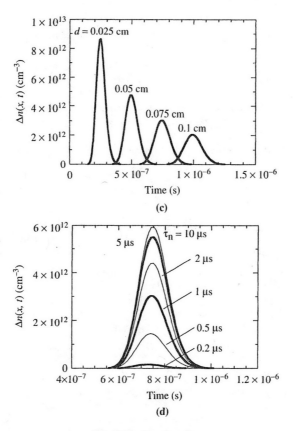

Fig. 8.15   (*continued*)

produces an electric field $\mathscr{E} = V_{dr}/L$ along the bar. Minority electrons are injected by negative polarity pulses at the n-emitter. The injected electron packet drifts from the emitter to the collector in the applied electric field to be collected by the collector.

The electrons are injected as a narrow pulse at $t = 0$ and diffuse and recombine with majority holes as they drift along the bar. Consequently, the minority carrier pulse broadens by diffusion and its area decreases by recombination. The pulse shape is given as a function of space and time by[71]

$$\Delta n(x, t) = \Delta n(x, 0) \exp\left(-\frac{(x - vt)^2}{4D_n t} - \frac{t}{\tau_n}\right) = \frac{N}{\sqrt{4\pi D_n t}} \exp\left(-\frac{(x - vt)^2}{4D_n t} - \frac{t}{\tau_n}\right) \tag{8.35}$$

where N is the electron density (electrons/cm^2) in the packet at $t = 0$ at the point of injection. The first term in the exponent describes diffusion and drift, and the second term describes recombination.

The time for the electron packet to drift from emitter to collector is $t_d = d/v$, where d is the spacing between contacts in Fig. 8.15(a) and $v = \mu_n \mathscr{E}$ is the electron packet velocity. The normalized output voltage waveform according to Eq. (8.35) is shown in Fig. 8.15(b) along with data points from Ref. 72. Calculated output voltages are shown in Figs. 8.15(c) and (d) as a function of spacing d and lifetime τ_n. Note the area decrease and pulse broadening with time in (c) and the pulse amplitude dependence on lifetime in (d).

The delay time t_d is determined by measuring the output pulses versus time for varying amplitude input pulses and extrapolating to zero injection or the injection pulse amplitude can be reduced until the peak position of the output pulse no longer shifts in time. This ensures low-level injection with the injected carrier density well below the majority carrier equilibrium density, eliminating any local disturbance of the electric field by the minority carrier pulse.

With the velocity given by $v = \mu_n \mathscr{E}$, the drift mobility is determined from the relationship

$$\mu_n = \frac{d}{t_d \mathscr{E}} \tag{8.36}$$

The time-of-flight method actually measures the *minority carrier velocity* or the *minority carrier mobility*. It is therefore useful for the determination of the carrier velocity-electric field behavior. This relationship is difficult to determine with other mobility measurement techniques.

To determine the diffusion constant D_n, the collected pulse width is measured at half its maximum amplitude. It can be shown (see Problem 8.10) that D_n is given by

$$D_n = \frac{(d\Delta t)^2}{16\ln(2)t_d^3} \tag{8.37}$$

where Δt is the pulse width.

The lifetime is determined by measuring the collected electron packet pulse at times t_{d1} and t_{d2}, corresponding to the two drift voltages, V_{dr1} and V_{dr2}. In the ideal case with no minority carrier trapping, the collected pulse has the predicted Gaussian shape and the lifetime is obtained by comparing the corresponding output pulse amplitudes V_{01} and V_{02}. The electron lifetime is then[72]

$$\tau_n = \frac{t_{d2} - t_{d1}}{\ln(V_{01}/V_{02}) - 0.5\ln(t_{d2}/t_{d1})} \tag{8.38}$$

Electrical injection can be replaced by *optical injection* with the basic method unchanged. For example, in one technique, electron-hole pairs (ehps) are created optically near the front in the p region of a pn junction.[73] Electrons diffuse and are collected by the junction acting as an integrator. The resulting voltage is measured, allowing the mobility to be extracted. A variation combines optical injection with *optical detection*. Laser pulse created ehps drift and diffuse. Electron-hole pair recombination is accompanied by photon emission, especially in III-V materials where radiative recombination dominates. It is this radiative recombination that is detected in the "photon in–photon out" time-of-flight method. In one particular scheme, quantum wells are used as time markers for both GaAs/AlGaAs[74] and InGaAs/InP.[75] The ehps can also be created by a pulsed electron beam[76] or by placing the sample into a microwave circuit, with the electron beam deflected at microwave frequencies across the sample, and the resulting microwave current is detected.[77] The drift velocity is determined from the amplitude and phase of the microwave current.[77, 78] The region between emitter and collector can be oxidized and provided with a gate, if surface recombination is of concern.[72] An appropriate gate voltage biases the surface into accumulation, effectively reducing surface recombination. Surface recombination is discussed in Chapter 7.

In addition to mobility, the carrier drift velocity in the form of the velocity—electric field curve is important. There are two techniques to determine the drift velocity as a function of electric field: the *current* and the *time-of-flight* technique. In the former the electron drift velocity is determined from the current of an *n*-type neutral region at high electric fields. The electron drift velocity is given by

$$v = \frac{I}{qwtn} \tag{8.39}$$

where I is the current, w the sample width, t the thickness, and n the electron density. Accurate determination of the drift velocity requires precise knowledge of the physical dimensions and the carrier density. To minimize heating, pulses are applied to the sample to generate appropriate electric fields with pulse widths in the 50–100 ns range.[79] The sample may need to be bathed in an inert ambient to prevent arcing. The technique has been used to determine the $v - \mathscr{E}$ curve of SiC.[79]

The principle of the time-of-flight method is illustrated in Fig. 8.16(a). Voltage $-V_1$ is applied to the cathode of two parallel plates. Electrons, liberated at the cathode by UV light, for example, drift with velocity v_n from the cathode to the anode in the electric field generated by $-V_1$. The electron charge $Q_N = qN$ C/cm^2 induces charges Q_C and Q_A in the cathode and anode, respectively, with $Q_N = Q_C + Q_A$. The arrows represent

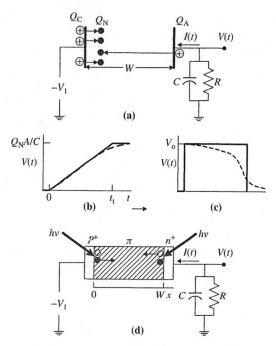

Fig. 8.16   (a) Time-of-flight measurement schematic, (b) output voltage for $t_t \ll RC$, (c) output voltage for $t_t \gg RC$, (d) implementation with a $p^+\pi n^+$ diode.

electric field lines from Q_C and Q_A terminating on Q_N. The electric field lines due to the applied voltage are not shown.

The charge on both plates redistributes itself continuously as the charge between the plates drifts from the cathode to the anode. The anode charge is $Q_A = 0$ at $t = 0$ and $Q_A = Q_N$ at $t = t_t$, where t_t is the transit time defined by

$$t_t = \frac{W}{v_n} \tag{8.40}$$

When Q_A changes from zero to Q_N, the charge flows through the external circuit as current $I(t)$ during the transit time t_t, given by[80-81]

$$I(t) = \frac{Q_N A}{t_t} = \frac{Q_N A v_n}{W}, 0 \leq t \leq t_t \tag{8.41a}$$

$$I(t) = 0, t > t_t \tag{8.41b}$$

where A is the electrode area.

The sample, connecting leads, and input to the voltage-sensing circuit all contain capacitances that are lumped into C. R is the load resistance in Fig. 8.16(a). The output voltage is

$$V(t) = \frac{Q_N A v_n R}{W}(1 - e^{-t/RC}) = V_o(1 - e^{-t/RC}) \tag{8.42}$$

Exercise 8.2

Problem: Derive Eq. (8.42).

Solution: In the frequency domain

$$V(\omega) = Z(\omega)I = \frac{R}{1 + j\omega RC}I$$

Taking the Laplace transform gives

$$V(s) = Z(s)I(s) = \frac{R}{1 + sRC}I(s) = \frac{R}{s(1 + sRC)}\frac{Q_N A v_n}{W}$$

using a step current of $I(s) = I/s = (Q_N A v_n/W)(1/s)$, where "s" is the Laplace operator. Taking the inverse Laplace transform gives

$$V(t) = \frac{Q_N A v_n R}{W}(1 - e^{-t/RC}) = V_o(1 - e^{-t/RC})$$

Equation (8.42) has two limits that are of interest for transit time measurements:

1. For $t_t \ll RC$, the voltage becomes

$$V(t) \approx \frac{V_o t}{RC} = \frac{Q_N A v_n t}{WC}, 0 \leq t \leq t_t \tag{8.43}$$

$$V(t) = \frac{Q_N A}{C}, t > t_t \tag{8.44}$$

In this approximation the RC circuit acts as an integrator, and the voltage is shown in Fig. 8.16(b) by the solid line.

2. For $t_t \gg RC$, the voltage becomes

$$V(t) \approx V_o = \frac{Q_N A v_n R}{W}, 0 \le t \le t_t \tag{8.45}$$

$$V(t) = 0, t > t_t \tag{8.46}$$

The RC time constant in this approximation is so small that the capacitor never charges and $V(t) \approx RI(t)$. The voltage is shown in Fig. 8.16(c) by the solid line. The transit time can be determined for either case and the carrier velocity is extracted from t_t.

This time-of-flight method can be implemented with the $p^+\pi n^+$ junction in Fig. 8.16(d). The π region is a lightly doped p-region in this figure. Bias voltage $-V_1$ depletes the π region entirely. Shallow penetration excitation (high energy light or an electron beam) from the *left* creates ehps near $x = 0$. The holes flow into the p^+ contact layer and the electrons drift to $x = W$, allowing the *electron velocity* to be determined. With excitation from the *right*, holes drift to the left, and the *hole velocity* is measured.

Two slightly different implementations of time-of-flight measurement geometries are shown in Fig. 8.17. Both use pn diodes combined with MOS structures. Figure 8.17(a) shows a gate-controlled diode with diode and gate biased to V_1, ensuring deep depletion under the gate so that an inversion layer cannot form.[82-83] The gate of the gate-controlled diode is a high-resistivity poly-silicon film with sheet resistance around 10 kohms/square. The voltage pulse V_2 with 200 ns pulse length and 10 kHz repetition rate creates a periodic voltage along the gate as well as along the semiconductor, leading to a lateral electric

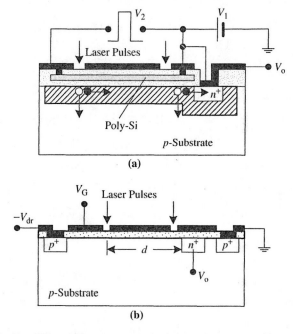

Fig. 8.17 Two drift mobility measurement implementations as discussed in the text.

field. Optical pulses, from a mode-locked Nd:YAG laser, are directed to two openings in a metal gate creating electron-hole pairs in the semiconductor. The holes drift into the substrate and the electrons drift along the surface to the collecting diode to produce a current pulse in the output circuit. By injecting minority carriers into two locations, defined by optical apertures, the difference in arrival times is used to determine the drift velocity. To obtain the field dependence of the mobility, the *lateral* or *tangential* electric field is varied by changing V_2. To determine the gate voltage dependence of the mobility, the *normal* or *vertical* electric field is varied by adjusting V_1.[82]

The electric field in the semiconductor in Fig. 8.17(b) is obtained from a voltage applied between two p^+ contacts in the semiconductor itself.[84] The Al gate sets the surface potential, but the lateral electric field is independent of the vertical electric field since the lateral field does not originate from a gate voltage. The continuous gate is also a light shield with two slits for the laser pulses to generate ehps. Optical pulses with 70 ps pulse widths from a mode-locked Nd-YAG laser have been used. The minority carrier packets are collected by the n^+ collector and displayed on a sampling oscilloscope. The circuits of Figs. 8.15 and 8.17 are similar. The chief difference lies in the method of minority carrier injection. In Fig. 8.15 minority carriers are injected electrically, in Fig. 8.17 optically. In all of these techniques, it is important for carrier trapping to be eliminated or accounted for in the data analysis.[75, 85] The dashed lines in Fig. 8.16(b) and (c) indicate the effects of trapping.[81]

The saturation velocity can also be determined from MOSFET current-voltage data. For short channel MOSFETs, the drain current under saturation conditions in the presence of source resistance R_S, can be written as[86]

$$I_{D,sat} = \frac{W_{eff} v_{sat} \mu_{eff} C_{ox} (V_{GS} - V_T - I_{D,sat} R_S)^2}{2v_{sat} L_{eff} + \mu_{eff} (V_{GS} - V_T - I_{D,sat} R_S)} \qquad (8.47)$$

Solving for $I_{D,sat}$ and dropping higher order $I_{D,sat}$ terms, allows Eq. (8.47) to be written as

$$\frac{1}{I_{D,sat}} = \frac{2R_S W_{eff} v_{sat} C_{ox} + 1}{W_{eff} v_{sat} C_{ox} (V_{GS} - V_T)} + \frac{2(L_m - \Delta L)}{W_{eff} \mu_{eff} C_{ox} (V_{GS} - V_T)^2} \qquad (8.48)$$

A plot of $1/I_{D,sat}$ versus L_m has intercepts $(1/I_{D,sat})_{int}$ and $L_{m,int}$ given by

$$\left(\frac{1}{I_{D,sat}}\right)_{int} = \frac{2R_S W_{eff} v_{sat} C_{ox} + 1}{W_{eff} v_{sat} C_{ox} (V_{GS} - V_T)} - \frac{2\Delta L}{W_{eff} \mu_{eff} C_{ox} (V_{GS} - V_T)^2} \qquad (8.49)$$

$$L_{m,int} = \Delta L - \frac{\mu_{eff} (V_{GS} - V_T)(2R_S W_{eff} v_{sat} C_{ox} + 1)}{2v_{sat}} \qquad (8.50)$$

Substituting Eq. (8.49) into (8.50) gives

$$L_{m,int} = \Delta L + \frac{2R_S W_{eff} v_{sat} C_{ox} + 1}{W_{eff} v_{sat} C_{ox} (V_{GS} - V_T)} \frac{L_{m,int}}{(1/I_{D,sat})_{int}} = \Delta L + A \frac{L_{m,int}}{(1/I_{D,sat})_{int}} \qquad (8.51)$$

Note that Eq. (8.51) no longer contains μ_{eff}. Plotting $L_{m,int}$ versus $L_{m,int}/(1/I_{D,sat})_{int}$ has the slope A. Plotting A versus $1/(V_{GS} - V_T)$ gives a line with slope S, which leads to v_{sat} through the expression

$$v_{sat} = \frac{1}{W_{eff} C_{ox} (S - 2R_S)} \qquad (8.52)$$

8.6   MOSFET MOBILITY

The conductivity, Hall, and magnetoresistance mobilities are *bulk* mobilities. Surfaces play a relatively minor role in their determination. The carriers are free to move throughout the sample and a mobility, averaged over the sample thickness, is measured. The main scattering mechanisms are *lattice* or *phonon scattering* and *ionized impurity scattering*. *Neutral impurity scattering* is important at low temperatures, where ionized impurities become neutral due to carrier freeze out. For some semiconductors there is *piezoelectric scattering*. Each scattering mechanism is associated with a mobility. According to Mathiessen's rule, the net mobility μ depends on the various mobilities as[87]

$$\frac{1}{\mu} = \frac{1}{\mu_1} + \frac{1}{\mu_2} + \dots \dots \tag{8.53}$$

and the lowest mobility dominates.

In this section we are concerned with additional scattering mechanisms that occur when the current carriers are confined within a narrow region as in a MOSFET channel. The location of the carriers near the oxide-semiconductor interface introduces additional scattering mechanisms like *Coulomb scattering* from oxide charges and interface states, as well as *surface roughness scattering*, reducing the MOSFET mobility below the bulk mobility.[88] Quantization of carriers in inversion layers further reduces the mobility.[89-91]

8.6.1   Effective Mobility

We consider an n-channel MOSFET of gate length L and width W. The considerations for p-channel devices are similar. The drain current I_D is a combination drift and diffusion currents

$$I_D = \frac{W \mu_{eff} Q_n V_{DS}}{L} - W \mu_{eff} \frac{kT}{q} \frac{dQ_n}{dx} \tag{8.54}$$

where Q_n is the mobile channel charge density (C/cm^2), and μ_{eff} the effective mobility, usually measured at drain voltages of typically 50–100 mV. Lower V_{DS} is better, because then the channel charge is more uniform from source to drain, allowing the diffusive second term in Eq. (8.54) to be dropped. Solving for the *effective mobility* μ_{eff} gives

$$\boxed{\mu_{eff} = \frac{g_d L}{W Q_n}} \tag{8.55}$$

where the drain conductance g_d is defined as

$$g_d = \frac{\partial I_D}{\partial V_{DS}} \Big|_{V_{GS}} = \text{constant} \tag{8.56}$$

How is Q_n determined? Two approaches are commonly used. In the first, the mobile channel charge density is approximated by

$$Q_n = C_{ox}(V_{GS} - V_T) \tag{8.57}$$

Although channel charge exists in the sub-threshold region below V_T, the expression $V_{GS} - V_T$ ensures the device operation in the above-threshold, drift-limited regime. Nevertheless, this approach has some deficiencies. The first is that the channel charge density is not exactly given by Eq. (8.57). Second, the threshold voltage is not necessarily well known and C_{ox} is not strictly the oxide capacitance/unit area. It is an effective oxide capacitance taking into account poly-Si gate depletion and the fact that the inversion layer resides slightly below the SiO$_2$/Si interface. Both effects introduce additional series capacitances.

When μ_{eff} is determined with Eqs. (8.55) and (8.57), one usually observes a significant mobility drop near $V_{GS} = V_T$. The reasons for this are that Eq. (8.57) is only an approximation to the true value of Q_n, the threshold voltage is not precisely known, and the channel charge density decreases as the gate voltage is decreased and ionized impurity scattering becomes more important. It is less significant at higher gate voltages, because the inversion charge screens the ionized impurities.

The approach giving better results is based on a direct measure of Q_n from capacitance measurements, with the mobile channel charge density determined from the gate-to-channel capacitance/unit area, C_{GC}, according to

$$Q_n = \int_{-\infty}^{V_{GS}} C_{GC} \, dV_{GS} \tag{8.58}$$

Then C_{GC} is measured using the connection of Fig. 8.18. The capacitance meter is connected between the gate and the source/drain connected together (not shown) with the substrate grounded. For a more detailed discussion see Appendix 6.1. For negative gate voltage (Fig. 8.18(a)), the channel region is accumulated and the overlap capacitances $2C_{ov}$ are measured. For $V_{GS} > V_T$ (Fig. 8.18(b)), the surface is inverted and all three capacitances, $2C_{ov} + C_{ch}$, are measured. A $C_{GC} - V_{GS}$ curve is shown in Fig. 8.19(a). Subtracting $2C_{ov}$ from this curve and integrating gives the $Q_n - V_{GS}$ curve of Fig. 8.19(a). Figure 8.19(b) gives the drain output characteristics. These curves give the drain conductance g_d from the slope at low V_{DS}. Extracting the mobility from Fig. 8.19 through Eq. (8.55) gives the mobility shown in Fig. 8.20.

Even if the mobility is determined with Eqs. (8.55) and (8.57), there are still some frequently ignored sources of error. We will briefly mention them. C_{GC} is most commonly measured as shown in Fig. 8.18(a). In this configuration, $V_{DS} = 0$, but the drain current to determine g_d is obviously measured with $V_{DS} > 0$. It is very common to use $V_{DS} = 100$ mV for I_D measurements. A better choice is to use as small a drain voltage as possible. However, if V_{DS} is too low, the measurement becomes noisy, but $V_{DS} \approx 20$–50 mV is reasonable. $V_{DS} > 0$ introduces an error in $\mu_{eff} - V_{GS}$ data, primarily near $V_{GS} = V_T$, because Q_n reduces as V_{DS} is increased for a given V_{GS}.[92-94] Modifying the measurement circuit slightly allows a drain bias to be applied during the C_{GC} measurement, with the capacitance measured between G and S (C_{GS}) with the drain reverse biased.[94] Then the

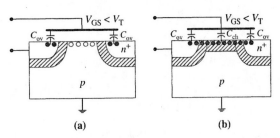

Fig. 8.18 Schematic for gate-to-channel capacitance measurements for (a) $V_{GS} < V_T$, (b) $V_{GS} > V_T$.

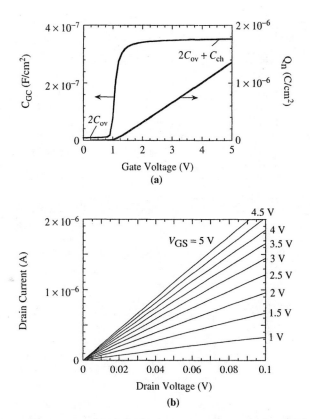

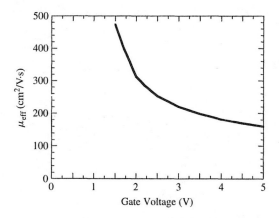

Fig. 8.19 (a) C_{GC} and Q_n versus V_{GS}; (b) I_D versus V_{DS}. W/L $= 10$ μm/10 μm, $t_{ox} = 10$ nm, $N_A = 1.6 \times 10^{17}$ cm^{-3}.

Fig. 8.20 μ_{eff} versus V_{GS} for the data of Fig. 8.19. $V_T = 0.5$ V.

G to D capacitance (C_{GD}) is measured. C_{GC} is $C_{GS} + C_{GD}$. Another error is the neglect of the overlap capacitances C_{ov} in Fig. 8.18, although it may be permissible to neglect these capacitances for large MOSFETs with gate lengths of 100 μm or so. Nevertheless some error is introduced if these capacitances are not considered in the analysis.

A further error is introduced by assuming the drain current to be drift current only. While this may be a good approximation for operation above threshold, for V_{GS} near V_T, diffusion current begins to be important. In fact, as is well known, for $V_{GS} < V_T$, *i.e.*, in the sub-threshold region the drain current is mainly due to diffusion. The capacitance should be measured at a sufficiently high frequency for interface traps to be unable to follow the ac signal, typically 100 kHz to 1 MHz. For low frequencies, the interface traps contribute a capacitive component.

The mobile channel charge density measurement technique is known as the "split C-V" technique, with the capacitance measured between the gate and source-drain and between the gate and the substrate, as illustrated in Fig. 8.18. The method was originally proposed by Koomen to measure the interface trapped charge density and the substrate doping density.[95] It was later adapted to mobility measurements.[96]

To understand the split C-V technique, consider Fig. 8.21. A time-varying gate voltage gives rise to currents I_1 and I_2. With the substrate grounded, I_1 is

$$I_1 = \frac{dQ_n}{dV_{GS}} \frac{dV_{GS}}{dt} = C_n \frac{dV_{GS}}{dt} = C_{GC} \frac{dV_{GS}}{dt} \tag{8.59}$$

Similarly,

$$I_2 = \frac{dQ_b}{dV_{GS}} \frac{dV_{GS}}{dt} = C_b \frac{dV_{GS}}{dt} = C_{GB} \frac{dV_{GS}}{dt} \tag{8.60}$$

The mobile channel charge density Q_n is derived from Eq. (8.59) and from (8.60) one derives the bulk charge density Q_b or the substrate doping density. Typical C_{GC} and C_{GB} curves are shown in Fig. 8.22.

The effective mobility depends on lattice scattering, ionized impurity scattering, and surface scattering. Ionized impurity and surface scattering depend on the substrate doping density and the gate voltage. The dependence of the effective mobility on gate voltage, illustrated in Fig. 8.23, is sometimes expressed as the dependence of μ_{eff} on the vertical

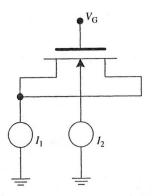

Fig. 8.21   Split C-V measurement arrangement.

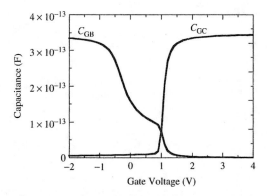

Fig. 8.22   Capacitance as a function of gate voltage.

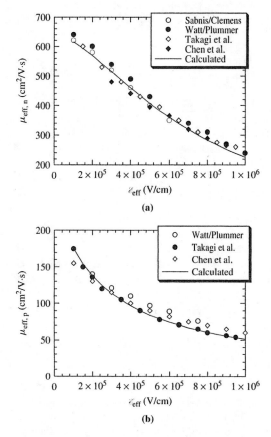

Fig. 8.23   (a) Electron and (b) hole effective mobility as a function of effective electric field. Data taken from the references in the inserts.

surface electric field $\mathscr{E}_{eff}$, according to

$$\mu_{eff} = \frac{\mu_o}{1 + (\alpha \mathscr{E}_{eff})^{\gamma}} \tag{8.61}$$

where α and γ are constants. Equation (8.61) produces "universal" mobility-electric field curves if the electric field produced by the gate voltage is expressed as the electric field due to the space-charge region and the inversion layer charges as[97-98]

$$\mathscr{E}_{eff} = \frac{Q_b + \eta Q_n}{K_s \mathscr{E}_o} \tag{8.62}$$

where Q_b and Q_n are the charge densities (C/cm^2) in the space-charge region and the inversion layer, respectively. The η in the inversion layer charge accounts for averaging of the electric field over the electron distribution in the inversion layer, usually taken as $\eta = 1/2$ for the electron mobility and 1/3 for the hole mobility.[97-98] The "universal" μ_{eff} versus $\mathscr{E}_{eff}$ plot contains the gate voltage and the bulk/inversion charge dependence.

A large body of experimental room temperature Si data agrees closely with the empirical expressions[99-102]

$$\mu_{eff,n} = \frac{638}{1 + (\mathscr{E}_{eff}/7 \times 10^5)^{1.69}}; \mu_{eff,p} = \frac{240}{1 + (\mathscr{E}_{eff}/2.7 \times 10^5)} \tag{8.63}$$

Electron and hole effective mobilities for SiO_2/Si devices calculated with Eq. (8.63) are shown in Fig. 8.23. Also shown are experimental data, showing that Eq. (8.63) is a good predictor of the experimental effective mobility data.

This form of presentation is appealing from the point of view of universality, but it is more difficult to arrive at from an operational point of view. It is, after all, the gate voltage that is measured experimentally, not the electric field. Conversion of measured voltages to electric field requires the doping density under the gate and the inversion charge density to be known. "Universal" mobility curves as a function of gate voltage are shown in Appendix 8.2.

The effective mobility reduction with effective electric field or with gate voltage has been attributed to enhanced surface roughness scattering with increased gate voltage and to quantization effects and it is common practice to express the effective mobility through an empirical relationship. The effective mobility is sometimes given by

$$\mu_{eff} = \frac{\mu_o}{1 + \theta(V_{GS} - V_T)} \tag{8.64}$$

The mobility degradation factor θ varies with gate oxide thickness and with doping density.[103] The low-field mobility μ_o is the intercept of the μ_{eff} versus $(V_{GS} - V_T)$ curve shown in Fig. 8.24(a). The constant θ is obtained from the slope of the μ_o/μ_{eff} versus $(V_{GS} - V_T)$ plot in Fig. 8.24(b).

A number of variations of the expression in Eq. (8.64) have been proposed to agree with experimental data. Some expressions include series resistance;[104] others include mobility reduction due to lateral electric fields,[105-106] important only for short-channel devices in which the drain voltage or the lateral electric field affects the mobility.

Effect of Gate Depletion and Channel Location: As discussed in Chapter 6, n^+ poly-silicon gates in n-channel MOSFETs and p^+-Si gates in p-channel MOSFETs are partially

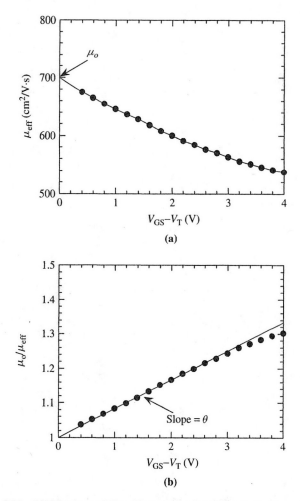

Fig. 8.24 (a) Effective mobility, (b) normalized mobility versus $V_{GS} - V_T$.

depleted when the device is biased in inversion, introducing a gate capacitance C_G in series with the oxide capacitance, thereby lowering C_{ox}. Location of the inversion channel slightly below the Si surface introduces an additional channel capacitance, C_{ch}, lowering the oxide capacitance further. The gate-to-channel capacitance due to these two parasitic capacitances is

$$C_{GC} = \frac{C_{ox}}{1 + C_{ox}/C_G + C_{ox}/C_{ch}} \tag{8.65}$$

C_{GC} for metal gate and n^+ poly-Si gate in Fig. 8.25 clearly shows the C_{GC} droop for positive gate voltages. Such reduced capacitance, if not properly accounted for, yields artificially high mobilities according to Eqs. (8.57) and (8.58). For $N_D = 10^{19}$ cm^{-3}, C_{GC} increases when the gate inverts at $V_G = 1.6$ V.

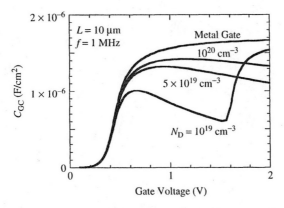

Fig. 8.25 Simulated gate-to-channel capacitance versus gate voltage as a function of poly-Si gate doping density. Oxide leakage current not considered. $t_{ox} = 2$ nm, $N_A = 10^{17}$ cm^{-3}, $\mu_{eff} = 300$ cm^2/V-s.

Effect of Gate Current: The gate current influences the drain current for sufficiently thin gate insulators, as illustrated in Fig. 8.26. For low drain voltage, *e.g.*, $V_{DS} = 10\text{-}20$ mV, the channel region can be considered an equipotential and approximately half the *gate* current flows to the source and half to the drain. The gate current flowing to the drain opposes the current from source to drain. This reduced drain current leads to a drain conductance reduction and, according to Eq. (8.55), to lower effective mobility. Experimental drain and gate currents are shown in Fig. 8.27 for two drain voltages. It is obvious that as the gate current increases the drain current decreases.

Several approaches have been proposed to correct this problem. In one of these, it is assumed that half the gate current flows to the drain giving the effective drain current[107]

$$I_{D,eff} = I_D + I_G/2 \qquad (8.66)$$

where I_D and I_G are the measured drain and gate currents. $I_{D,eff}$ should be used to determine the drain conductance. In another approach, the drain current is measured at two drain voltages and the effective drain current is[108]

$$I_{D,eff} = \Delta I_D = I_D(V_{DS2}) - I_D(V_{DS1}) \qquad (8.67)$$

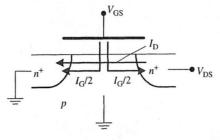

Fig. 8.26 MOSFET cross section showing drain and gate currents. Gate current adds to source current and subtracts from drain current.

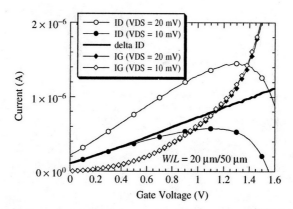

Fig. 8.27 Drain and gate currents versus gate voltage for an n-channel MOSFET. Gate insulator: HfO$_2$ $\sim$ 2 nm thick. With permission of W. Zhu and T.P. Ma, Yale University.

For example, in Fig. 8.27 the drain current is measured at 10 mV and 20 mV. For gate voltages slightly higher than 1 V, the drain current drops. Subtracting one from the other yields the lower current, assuming that the gate current is not affected by such low drain voltages. As seen in Fig. 8.27 that is a good assumption with the gate current only weakly dependent on drain voltage.

Effect of Inversion Charge Frequency Response: Errors can occur in the determination of Q_n due to the channel frequency response. Consider the cross-sectional MOSFET diagram in Fig. 8.28, consisting of C_{ox}, C_{ov}, C_{ch} the channel capacitance, C_b the space-charge region or bulk capacitance, R_S the source resistance, R_D the drain resistance, and R_{ch} the channel resistance. The capacitances are in F/unit area. When the gate-to-channel capacitance C_{GC} is measured, electrons are supplied by the source and drain. They encounter resistance R_{ch} and various capacitances limiting the frequency response of the inversion charge. The gate-to-channel capacitance is given by[92]

$$C_{GC} = \frac{C_{ox} C_{ch}}{C_{ox} + C_{ch} + C_b} \text{Re} \left(\frac{\tanh(\lambda)}{\lambda} \right) \tag{8.68}$$

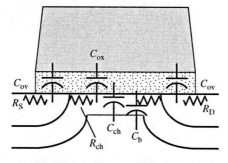

Fig. 8.28 MOSFET cross-section showing source and drain resistances (R_S and R_D), channel resistance R_{ch}, overlap, oxide, channel and bulk capacitances (C_{ov}, C_{ox}, C_{ch}, and C_b).

where

$$\lambda = \sqrt{j0.25\omega C' R_{ch} L^2} \tag{8.69}$$

The derivation is given in Appendix 8.3.

Figure 8.29 gives C_{GC} as a function of frequency and channel length, showing that the frequency must be sufficiently low or the gate sufficiently short for the channel frequency effects to be negligible. If those conditions are not satisfied, the integrated C_{GC} leads to an incorrect mobility.

Effect of Interface Trapped Charge: Interface traps affect mobility measurements in several ways. They provide scattering sites, reducing the mobility. Electrons trapped in interface traps contribute to Q_n but not to drain current or drain conductance.[109] Interface traps can contribute a capacitance, leading to a Q_n increase. Hence, Q_n increases and g_d decreases, both leading to reduced mobility, according to Eq. (8.55). Let us consider two of these effects. Interface traps contribute a capacitance if they can respond to the applied

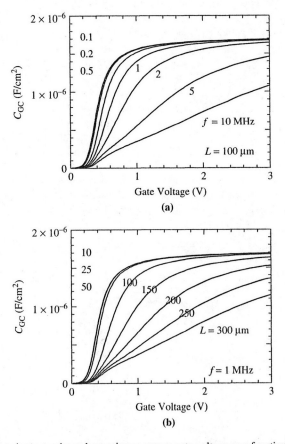

Fig. 8.29 Simulated gate-to-channel capacitance versus gate voltage as a function of (a) frequency and (b) channel length. Gate depletion and oxide leakage current not considered. $t_{ox} = 2$ nm, $N_A = 10^{17}$ cm^{-3}, $\mu_n = 300$ cm^2/V-s.

frequency. Their response is determined by the capture and emission time constants. From Eq. (6.60) the interface trap electron emission time constant is

$$\tau_{it} = \frac{\exp(\Delta E/kT)}{\sigma_n v_{th} N_c} = 4 \times 10^{-11} \exp(\Delta E/kT) \ [s] \tag{8.70}$$

where ΔE is the interface trap energy interval measured from the bottom of the conduction band to the interface trap energy of interest. The numerical value is calculated for $\sigma_n = 10^{-16}$ cm^2, $v_{th} = 10^7$ cm/s, and $N_c = 2.5 \times 10^{19}$ cm^{-3}. The frequency dependent capacitance is then

$$C_{it} = \frac{q^2 D_{it}}{1 + \omega^2 \tau_{it}^2} \tag{8.71}$$

where D_{it} is the interface trap density.

The interface trap capacitance is in parallel with C_{ch} and the effective gate capacitance is

$$C_{GC} = \frac{C_{ox}(C_{ch} + C_{it})}{C_{ox} + C_{ch} + C_b + C_{it}} \tag{8.72}$$

C_{it} only contributes a noticeable component for low C_{ch}, which typically occurs near the threshold voltage where the channel is weak. Furthermore, C_{it} is only significant for high interface trap densities, typically observed for insulators other than SiO$_2$ on Si. The measurement frequency is, of course, also important, according to Eq. (8.71). This frequency depends on ΔE, which, in turn depends on the doping density, as discussed in Appendix 8.4.

A second effect of interface traps is its influence on the gate voltage. The effect of D_{it} on V_T is given by

$$V_G = V_{FB} + \phi_s + \frac{Q_s}{C_{ox}} \pm \frac{Q_{it}}{C_{ox}} \tag{8.73}$$

The "$\pm$" sign accounts for the interface trap charge. It is generally accepted that for SiO$_2$/Si interfaces, the interface traps in the upper half of the band gap are acceptors and in the lower half are donors. For inverted n-channel MOSFETs, donors occupied by electrons are neutral and acceptors occupied by electrons are negatively charged, leading to positive threshold voltage shifts, illustrated in Fig. 8.30. The increased capacitance in Fig. 8.30(a) is seen only at low C_{GC} and the shifted gate voltage is quite obvious. Interface traps lead to an incorrect integrated C_{GC} and incorrect mobility. Fig. 8.30(b) includes the effects of gate depletion and interface traps.

Effect of Series Resistance: We discussed in Chapter 4 how series resistance degrades the MOSFET current-voltage behavior. Source, drain and contact resistances also affect the mobility, since the effective mobility depends on drain conductance g_d. As I_D depends on series resistance R_{SD}, μ_{eff} also depends on R_{SD}. It should be understood that μ_{eff} itself does, of course, not depend on R_{SD}. The drain conductance becomes

$$g_d(R_{SD}) = \frac{g_{d0}}{1 + g_{d0}R_{SD}} \tag{8.74}$$

where g_{d0} is the drain conductance for $R_{SD} = 0$. With g_d reduced, the effective mobility is also reduced as evident in Eq. (8.55).

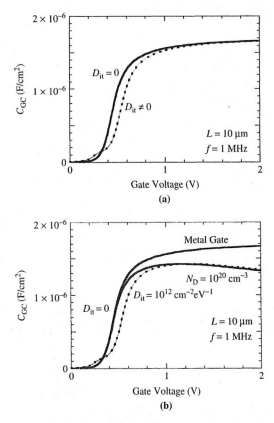

Fig. 8.30 Simulated gate-to-channel capacitance versus gate voltage as a function of interface trap density. (a) Gate depletion and oxide leakage current not considered, (b) oxide leakage current not considered $t_{ox} = 2$ nm, $N_A = 10^{17}$ cm^{-3}, $\mu_n = 300$ cm^2/V-s, $D_{it} = 10^{12}$ cm^{-2} eV^{-1}, $\tau_{it} = 5 \times 10^{-8}$ s.

The effective mobility of *depletion-mode* devices can be measured by the same drain conductance method. In depletion-mode devices mobility profiles are obtained by varying the gate voltage. In order to extract depth-dependent mobilities, it is necessary to determine the carrier density independently, by capacitance-voltage measurements for example.[110]

8.6.2 Field-Effect Mobility

While the effective mobility is derived from the drain conductance, the *field-effect mobility* is determined from the transconductance, defined by

$$g_m = \frac{\partial I_D}{\partial V_{GS}}|_{V_{DS}} = \text{constant} \tag{8.75}$$

The drift component of the drain current with $Q_n = C_{ox}(V_{GS} - V_T)$ is

$$I_D = \frac{W}{L}\mu_{eff}C_{ox}(V_{GS} - V_T)V_{DS} \tag{8.76}$$

When the field-effect mobility is determined, the transconductance is usually taken to be

$$g_m = \frac{W}{L} \mu_{eff} C_{ox} V_{DS} \tag{8.77}$$

When this expression is solved for the mobility, it is known as the *field-effect mobility*

$$\boxed{\mu_{FE} = \frac{L g_m}{W C_{ox} V_{DS}}} \tag{8.78}$$

The field-effect mobility, defined by Eq. (8.78), is generally lower than the effective mobility, as shown in Fig. 8.31. This is rather disturbing, since it is the same device measured under identical bias conditions. This discrepancy between μ_{eff} and μ_{FE} is due to the neglect of the electric field dependence of the mobility in the derivation of Eq. (8.78).[111–112] Considering the μ_{eff} dependence on gate voltage, gives the transconductance

$$g_m = \frac{W}{L} \mu_{eff} C_{ox} V_{DS} \left(1 + \frac{(V_{GS} - V_T)}{\mu_{eff}} \frac{d\mu_{eff}}{dV_{GS}} \right) \tag{8.79}$$

and the field-effect mobility becomes

$$\boxed{\mu_{FE} = \frac{L g_m}{W C_{ox} V_{DS} \left(1 + \dfrac{(V_{GS} - V_T)}{\mu_{eff}} \dfrac{d\mu_{eff}}{dV_{GS}} \right)}} \tag{8.80}$$

Since $d\mu_{eff}/dV_{GS} < 0$, it is obvious that Eq. (8.80) gives a higher μ_{FE} than Eq. (8.78).

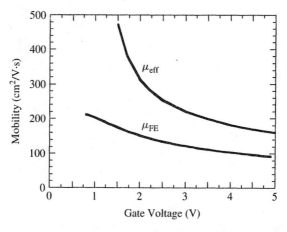

Fig. 8.31 Effective and field-effect mobilities.

8.6.3   Saturation Mobility

Sometimes the MOSFET mobility is derived from the drain current–drain voltage curves with the device in saturation. The saturation drain current can be expressed as

$$I_{D,sat} = \frac{BW\bar{\mu}_n C_{ox}}{2L}(V_{GS} - V_T)^2 \tag{8.81}$$

where B represents the body effect which is weakly dependent on the gate voltage. When Eq. (8.81) is solved for the mobility, this mobility, sometimes called the *saturation mobility*, is

$$\mu_{sat} = \frac{2Lm^2}{BWC_{ox}} \tag{8.82}$$

where m is the slope of the $(I_{D,sat})^{1/2}$ versus $(V_{GS} - V_T)$ plot. The saturation mobility in Eq. (8.82) is usually lower than μ_{eff} because the gate voltage dependence of the mobility is also neglected in Eq. (8.82). Additional error is introduced because the factor B is not well known and usually assumed as unity. The saturation mobility is only valid when the drain current is governed by mobility, not by velocity saturation.

8.7   CONTACTLESS MOBILITY

Infrared (IR) reflectance, discussed in Section 2.6.1 for carrier density characterization, can also be used for mobility measurements. In this technique, the infrared reflectance is measured over a wide wavelength range and the data are fitted to obtain the mobility. Long wavelength IR reflectance data have a characteristic plasma frequency

$$\omega_p = \sqrt{\frac{q^2 p}{K_s \varepsilon_o m^*}} \tag{8.83}$$

The mobility is given by

$$\mu = \frac{q}{\gamma_p m^*} \tag{8.84}$$

where γ_p is the free carrier damping constant. The SiC mobility over the 10^{17}–10^{19} cm^{-3} doping density range was obtained by a fitting procedure involving ω_p and γ_p.[113]

8.8   STRENGTHS AND WEAKNESSES

Conductivity Mobility: The weakness of the conductivity mobility method is the requirement for both sample resistivity and carrier density, requiring independent measurements. Its strength lies in that it is directly defined from the sample resistivity or conductivity and no correction factors are required in its analysis.

Hall Effect Mobility: The weakness of the Hall method lies in the special sample requirements and the inability to predict a precise value for the Hall scattering factor. The usual assumption of $r = 1$ introduces an error into the measured mobility. Although appropriate sample geometries exist for profiling, the method is awkward for mobility profiling. The strength of the Hall technique lies in its common use and the availability of mobilities determined by this method for the common semiconductors.

Magnetoresistance Mobility: The weakness of the magnetoresistance technique lies in its limited use and the inability to characterize low-mobility semiconductors. For example, it does not work well for Si. In common with the Hall effect, it is difficult to determine the magnetoresistance scattering factor, and the assumption $\xi = 1$ introduces an error. Its strength is the ability to measure devices requiring no special test structures. MESFETs and MESFET-like devices can be easily characterized.

Time-of-Flight or Drift Mobility: The weakness of this method is the requirement for special test structures and high speed electronics and/or optics. This puts the method into the hands of specialists in a few laboratories. Its strength lies in the ability to measure the mobility and the carrier velocity at high electric fields. Many of the experimental data of velocity-electric field curves were generated by this method.

MOSFET Mobility: This method is only suitable for MOSFETs, MESFETs, and MODFETs and an operational mobility is extracted. Depending on how the mobility is measured, different experimental values are obtained. The effective mobility is the most common and the least ambiguous. Both the field-effect and saturation mobility, as usually defined, yield lower mobilities than μ_{eff} and should not be used to characterize a device, unless appropriately modified equations are used in their derivation.

APPENDIX 8.1

Semiconductor Bulk Mobilities

Silicon: The dependence of the mobility on carrier density giving good agreement with experiment at room temperature is given by the empirical expressions[114]

$$\mu_n = \mu_o + \frac{\mu_{max} - \mu_o}{1 + (n/C_r)^a} - \frac{\mu_1}{1 + (C_s/n)^b} \tag{A8.1}$$

$$\mu_n = \mu_o e^{-p_c/p} + \frac{\mu_{max}}{1 + (p/C_r)^a} - \frac{\mu_1}{1 + (C_s/n)^b} \tag{A8.2}$$

The parameters that give the best fit to experimental data are given in Table A8.1. These two expressions are plotted in Figs. A8.1 and A8.2. For clarity, the experimental points are not shown, but are found in Masetti et al.[114]

The doping density dependence of the mobility is often expressed as[115]

$$\mu = \mu_{min} + \frac{\mu_o}{1 + (N/N_{ref})^\alpha} \tag{A8.3}$$

where μ is either the electron or hole mobility and N is the donor or acceptor doping density. The temperature dependence of the various parameters in Eq. (A8.3) has the form

$$A = A_o(T/300)^n \tag{A8.4}$$

The parameters that give the best fit to experimental data are given in Table A8.2.

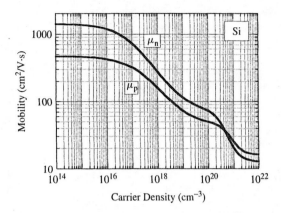

Fig. A8.1 Room temperature electron and hole mobilities in Si.

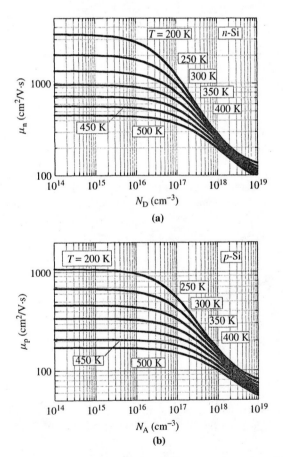

Fig. A8.2 (a) Electron and (b) hole mobilities in silicon as a function of temperature calculated from Eqs. (A8.3) and (A8.4).

TABLE A8.1   Mobility Fit Parameters for Silicon.

Parameter	Arsenic	Phosphorus	Boron
μ_o (cm²/V·s)	52.2	68.5	44.9
μ_{max} (cm²/V·s)	1417	1414	470.5
μ_1 (cm²/V·s)	43.4	56.1	29.0
C_r (cm⁻³)	9.68×10^{16}	9.20×10^{16}	2.23×10^{17}
C_s (cm⁻³)	3.43×10^{20}	3.41×10^{20}	6.10×10^{20}
a	0.680	0.711	0.719
b	2.00	1.98	2.00
p_c (cm⁻³)	—	—	9.23×10^{16}

Source: Masetti et al. Ref. 114.

TABLE A8.2   Mobility Fit Parameters for Silicon.

	Temperature-Independent Prefactors		Temperature
Parameter	Electrons	Holes	Exponent
μ_o (cm²/V·s)	1268	406.9	−2.33 electrons
			−2.23 holes
μ_{min} (cm²/V·s)	92	54.3	−0.57
N_{ref} (cm⁻³)	1.3×10^{17}	2.35×10^{17}	2.4
α	0.91	0.88	−0.146

Source: Baccarani and Ostoja, Arora et al., and Li and Thurber, Refs. 116–118.

Equation (A8.3) is plotted in Fig. A8.2 for n-Si and p-Si as a function of temperature. Experimental data from Li and Thurber[118] and Li[119] agree reasonably well with the mobilities in Fig. A8.2. Other mobility expressions have also been proposed.[120–121]

Gallium Arsenide: The mobilities of $n-$ and p-type GaAs are shown in Fig. A8.3 at $T = 300$ K.

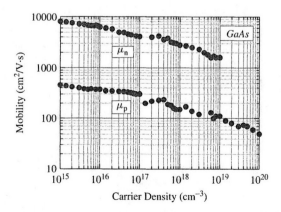

Fig. A8.3   Room temperature electron and hole mobilities in GaAs. Data adapted from ref. 122.

APPENDIX 8.2

Semiconductor Surface Mobilities

SiO₂/Si: The dependence of the effective mobility on effective electric field is shown in Section 8.6.1. However, what is measured is the gate voltage. Hence it is useful to determine the effective mobility as a function of gate voltage. Such curves are shown in Fig. A8.4 for electron and hole mobilities, described by the equations[123]

$$\mu_{n,eff} = \frac{540}{1 + (\mathscr{E}_{eff}/9 \times 10^5)^{1.85}} \tag{A8.5a}$$

$$\mu_{p,eff} = \frac{180}{1 + (\mathscr{E}_{eff}/4.5 \times 10^5)} \tag{A8.5b}$$

with $\mathscr{E}_{eff}$ in V/cm. These curves resemble those in Section 8.6.1. The effective electric field is converted to gate voltage through[123]

$$\mathscr{E}_{eff}(\mu_{n,eff}) = \frac{V_G + V_T}{6t_{ox}} \tag{A8.6a}$$

$$\mathscr{E}_{eff}(\mu_{p,eff}) = \frac{V_G + 1.5V_T - \alpha}{7.5t_{ox}} \tag{A8.6b}$$

with V_G and V_T in V and t_{ox} in cm. In Eq. (A8.6b) $\alpha = 0$ for p^+ poly-Si gate surface channel p-MOSFETs, $\alpha = 2.3$ for n^+ poly-Si gate buried channel p-MOSFETs, and $\alpha = 2.7$ for p^+ poly-Si gate p-MOSFETs.

APPENDIX 8.3

Effect of Channel Frequency Response

The effective MOSFET mobility is

$$\mu_{eff} = \frac{g_d L}{W Q_n} \tag{A8.7}$$

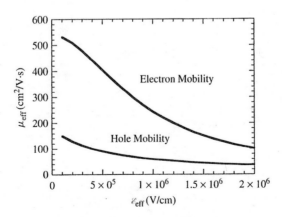

Fig. A8.4 Room temperature effective mobilities for surface channel MOSFETs.

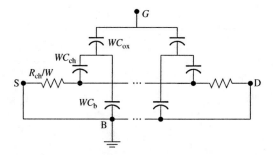

Fig. A8.5   MOSFET transmission line equivalent circuit.

Errors can occur in the determination of g_d and Q_n. Consider the cross-sectional MOSFET diagram in Fig. 8.28, consisting of overlap, oxide, channel, and bulk capacitances, as well as source, drain, and channel resistances. The MOSFET is represented by the equivalent circuit in Fig. A8.5.[92] The gate-to-channel capacitance is

$$C_{GC} = \frac{C_{ox} C_{ch}}{C_{ox} + C_{ch} + C_b} \operatorname{Re} \frac{tanh(\lambda)}{\lambda}; \lambda = \sqrt{j\omega\tau_{GC}} \tag{A8.8}$$

where the channel time constant is

$$\tau_{GC} = \frac{C_{GC0} L^2}{4 R_{sh,ch}}; C_{GC0} = \frac{C_{ch}(C_{ox} + C_b)}{C_{ox} + C_{ch} + C_b}; R_{sh,ch} = \frac{1}{Q_n \mu_n} \tag{A8.9}$$

with $C_{ch} = dQ_n/d\phi_s$ the inversion capacitance, $C_b = dQ_b/d\phi_s$ the space-charge region or bulk capacitance, $R_{sh,ch}$ the channel sheet resistance, and μ_n the channel mobility. During capacitance measurements, the channel charge is supplied by the source and drain junctions. To avoid distortion of the capacitance-voltage curve due to channel charging effects, the measurement frequency f must satisfy the criterion

$$f \ll \frac{1}{2\pi \tau_{GC}} = \frac{4 R_{sh,ch}}{2\pi C_{GC0} L^2} \tag{A8.10}$$

placing limits on C_{GC}, $R_{sh,ch}$, f, and L as shown in Fig. 8.29. For oxides thinner than 2 nm, L should be 10 μm or less for $f = 1$ MHz.[124] For thinner oxides or longer channels, the source and drain are unable to supply the required channel. This will clearly be important for effective mobility extraction since the channel charge density Q_n in Eq. (A8.7) depends on C_{GC}.

APPENDIX 8.4

Effect of Interface Trapped Charge

From Eq. (6.57 the interface trap electron emission time constant is

$$\tau_{it} = \frac{\exp(\Delta E/kT)}{\sigma_n v_{th} N_c} = 4 \times 10^{-11} \exp(\Delta E/kT) \ [s] \tag{A8.11}$$

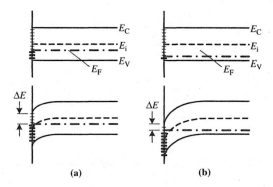

Fig. A8.6   Semiconductor band diagram of a MOSFET or (a) low N_A and (b) high N_A showing ΔE variation.

where ΔE is the interface trap energy interval from the bottom of the conduction band to the interface trap energy of interest. The numerical value is calculated for $\sigma_n = 10^{-16}$ cm^2, $v_{th} = 10^7$ cm/s, and $N_c = 2.5 \times 10^{19}$ cm^{-3}. With $f_{it} = 1/2\pi\tau_{it}$

$$f_{it} = 4 \times 10^9 \exp(-\Delta E/kT) \qquad (A8.12)$$

Figure A8.6 shows the band diagrams for two n-channel MOSFETs at flatband and at the onset of inversion where the surface potential is $(\phi_s = 2\phi_F)$. Figure A8.6(a) has a lightly-doped and (b) a heavily-doped substrate. Only the semiconductor band diagram is shown for simplicity. The vertical line represents the insulator/semiconductor interface. The small horizontal lines at the interface denote interface traps. The heavy lines represent interface traps below the Fermi energy E_F, occupied by electrons, and the light lines are unoccupied traps. Those interface traps responding to the external ac signal are around the Fermi energy. ΔE is lower for highly-doped substrates, since it takes a higher surface potential to invert the surface. However, according to Eq. (A8.12), a lower ΔE corresponds to a higher frequency response. For lightly-doped substrates ($N_A = 10^{16}$ cm^{-3}), $\Delta E = 0.41$ eV and $f = 500$ Hz, while for $N_A = 10^{18}$ cm^{-3}, $\Delta E = 0.17$ eV and $f = 5.5$ MHz. Interface traps responding to the ac signal contribute the interface trap capacitance C_{it}. The resulting gate-to-channel capacitance becomes

$$C_{GC} = \frac{C_{ox}(C_{ch} + C_{it})}{C_{ox} + C_{ch} + C_b + C_{it}} \qquad (A8.13)$$

REFERENCES

1. F.J. Morin, "Lattice-Scattering Mobility in Germanium," *Phys. Rev.* **93**, 62–63, Jan. 1954.

2. F.J. Morin and J.P. Maita, "Electrical Properties of Silicon Containing Arsenic and Boron," *Phys. Rev.* **96**, 28–35, Oct. 1954.

3. E.H. Hall, "On a New Action of the Magnet on Electric Currents," *Amer. J. Math.* **2**, 287–292, 1879.

4. K.R. Sopka, "The Discovery of the Hall Effect: Edwin Hall's Hitherto Unpublished Account," in *The Hall Effect and Its Applications* (C.L. Chien and C.R. Westgate, eds.), Plenum Press, New York, 1980, 523–545.

5. E.H. Putley, *The Hall Effect and Related Phenomena*, Butterworths, London, 1960; "The Hall Effect and Its Application," *Contemp. Phys.* **16**, 101–126, March 1975.

6. R.A. Smith, *Semiconductors*, Cambridge University Press, Cambridge, 1959, Ch. 5.

7. A. Zemel, A. Sher, and D. Eger, "Anomalous Hall Effect in p-Type $Hg_{1-x}Cd_x$ Te Liquid-Phase-Epitaxial Layers," *J. Appl. Phys.* **62**, 1861–1868, Sept. 1987.

8. G.E. Stillman and C.M. Wolfe, "Electrical Characterization of Epitaxial Layers," *Thin Solid Films* **31**, 69–88, Jan. 1976.

9. M.C. Gold and D.A. Nelson, "Variable Magnetic Field Hall Effect Measurements and Analyses of High Purity, Hg Vacancy (p-Type) HgCdTe," *J. Vac. Sci. Technol.* **A4**, 2040–2046, July/Aug. 1986.

10. A.C. Beer, *Galvanomagnetic Effects in Semiconductors*, Academic Press, New York, 1963, p. 308.

11. D.L. Rode, C.M. Wolfe, and G.E. Stillman, "Magnetic-Field Dependence of the Hall Factor of Gallium Arsenide," in *GaAs and Related Compounds* (G.E. Stillman, ed.) Conf. Ser. No. 65, Inst. Phys., Bristol, 1983, pp. 569–572.

12. L.J. van der Pauw, "A Method of Measuring Specific Resistivity and Hall Effect of Discs of Arbitrary Shape," *Phil. Res. Rep.* **13**, 1–9, Feb. 1958.

13. L.J. van der Pauw, "A Method of Measuring the Resistivity and Hall Coefficient on Lamellae of Arbitrary Shape," *Phil. Tech. Rev.* **20**, 220–224, Aug. 1958.

14. A. Chandra, C.E.C. Wood, D.W. Woodard, and L.F. Eastman, "Surface and Interface Depletion Corrections to Free Carrier-Density Determinations by Hall Measurements," *Solid-State Electron.* **22**, 645–650, July 1979.

15. W.E. Ham, "Surface Charge Effects on the Resistivity and Hall Coefficient of Thin Silicon-On-Sapphire Films," *Appl. Phys. Lett.* **21**, 440–443, Nov. 1972.

16. T.R. Lepkowski, R.Y. DeJule, N.C. Tien, M.H. Kim, and G.E. Stillman, "Depletion Corrections in Variable Temperature Hall Measurements," *J. Appl. Phys.* **61**, 4808–4811, May 1987.

17. R. Baron, G.A. Shifrin, O.J. Marsh, and J.W. Mayer, "Electrical Behavior of Group III and V Implanted Dopants in Silicon," *J. Appl. Phys.* **40**, 3702–3719, Aug. 1969.

18. H. Maes, W. Vandervorst, and R. Van Overstraeten, "Impurity Profile of Implanted Ions in Silicon," in *Impurity Doping Processes in Silicon* (F.F.Y. Wang, ed.) North-Holland, Amsterdam, 1981, 443–638.

19. T. Ambridge and C.J. Allen, "Automatic Electrochemical Profiling of Hall Mobility in Semiconductors," *Electron. Lett.* **15**, 648–650, Sept. 1979.

20. J.W. Mayer, O.J. Marsh, G.A. Shifrin, and R. Baron, "Ion Implantation of Silicon; II Electrical Evaluation Using Hall-Effect Measurements," *Can. J. Phys.* **45**, 4073–4089, Dec. 1967.

21. N.G.E. Johannson, J.W. Mayer, and O.J. Marsh, "Technique Used in Hall Effect Analysis of Ion Implanted Si and Ge," *Solid-State Electron.* **13**, 317–335, March 1970.

22. N.D. Young and M.J. Hight, "Automated Hall Effect Profiler for Electrical Characterisation of Semiconductors," *Electron. Lett.* **21**, 1044–1046, Oct. 1985.

23. H. Müller, F.H. Eisen, and J.W. Mayer, "Anodic Oxidation of GaAs as a Technique to Evaluate Electrical Carrier Concentration Profiles," *J. Electrochem. Soc.* **122**, 651–655, May 1975.

24. L. Bouro and D. Tsoukalas, "Determination of Doping and Mobility Profiles by Automated Electrical Measurements and Anodic Stripping," *Phys. E: Sci. Instrum.* **20**, 541–544, May 1987.

25. A.C. Ipri, "Variation in Electrical Properties of Silicon Films on Sapphire Using the MOS Hall Technique," *Appl. Phys. Lett.* **20**, 1–2, Jan. 1972.

26. A.B.M. Elliot and J.C. Anderson, "An Investigation of Carrier Transport in Thin Silicon-On-Sapphire Films Using MIS Deep Depletion Hall Effect Structures," *Solid-State Electron.* **15**, 531–545, May 1972.

27. P.A. Crossley and W.E. Ham, "Use of Test Structures and Results of Electrical Tests for Silicon-On-Sapphire Integrated Circuit Processes," *J. Electron. Mat.* **2**, 465–483, Aug. 1973.

28. S. Cristoloveanu, J.H. Lee, J. Pumfrey, J.R. Davies, R.P. Arrowsmith, and P.L.F. Hemment, "Profiling of Inhomogeneous Carrier Transport Properties with the Influence of Temperature in Silicon-On-Insulator Films Formed by Oxygen Implantation," *J. Appl. Phys.* **60**, 3199–3203, Nov. 1986.

29. T.L. Tansley, "AC Profiling by Schottky-Gated Cloverleaf," *J. Phys. E: Sci. Instrum.* **8**, 52–54, Jan. 1975.

30. C.W. Farley and B.G. Streetman, "The Schottky-Gated Hall-Effect Transistor and Its Application to Carrier Concentration and Mobility Profiling in GaAs MESFET's," *IEEE Trans. Electron. Dev.* **ED-34**, 1781–1787, Aug. 1987.

31. P.R. Jay, I. Crossley, and M.J. Caldwell, "Mobility Profiling of FET Structures," *Electron. Lett.* **14**, 190–191, March 1978.

32. H.H. Wieder, *Laboratory Notes on Electrical and Galvanomagnetic Measurements*, Elsevier, Amsterdam, 1979, Ch. 5–6.

33. H. Ryssel, K. Schmid, and H. Müller, "A Sample Holder for Measurement and Anodic Oxidation of Ion Implanted Silicon," *J. Phys. E: Sci. Instrum.* **6**, 492–494, May 1973.

34. S.B. Felch, R. Brennan, S.F. Corcoran, and G. Webster, "A Comparison of Three Techniques for Profiling Ultrashallow p^+n Junctions," *Solid State Technol.* **36**, 45–51, Jan. 1993.

35. J.W. Mayer, L. Eriksson, and J.A. Davies, *Ion Implantation in Semiconductors; Silicon and Germanium*, Academic Press, New York, 1970.

36. C.M. Wolfe and G.E. Stillman, "Apparent Mobility Enhancement in Inhomogeneous Crystals," in *Semiconductors and Semimetals* (R.K. Willardson and A.C. Beer, eds.) Academic Press, New York, **10**, 175–220, 1975.

37. A. Neduloha and K.M. Koch, "On the Mechanism of the Resistance Change in a Magnetic Field (in German)," *Z. Phys.* **132**, 608–620, 1952.

38. R.L. Petritz, "Theory of an Experiment for Measuring the Mobility and Density of Carriers in the Space-Charge Region of a Semiconductor Surface," *Phys. Rev.* **110**, 1254–1262, June 1958.

39. R.D. Larrabee and W.R. Thurber, "Theory and Application of a Two-Layer Hall Technique," *IEEE Trans. Electron Dev.* **ED-27**, 32–36, Jan. 1980.

40. L.F. Lou and W.H. Frye, "Hall Effect and Resistivity in Liquid-Phase-Epitaxial Layers of HgCdTe," *J. Appl. Phys.* **56**, 2253–2267, Oct. 1984.

41. A. Zemel and J.R. Sites, "Electronic Transport Near the Surface of Indium Antimonide Films," *Thin Solid Films* **41**, 297–305, March 1977.

42. ASTM Standard F76-86, "Standard Method for Measuring Hall Mobility and Hall Coefficient in Extrinsic Semiconductor Single Crystals," *1996 Annual Book of ASTM Standards*, Am. Soc. Test. Mat., West Conshohocken, PA, 1996.

43. D.C. Look, "Bulk and Contact Electrical Properties by the Magneto-Transmission-Line Method: Application to GaAs," *Solid-State Electron.* **30**, 615–618, June 1987.

44. D.S. Perloff, "Four-Point Sheet Resistance Correction Factors for Thin Rectangular Samples," *Solid-State Electron.* **20**, 681–687, Aug. 1977.

45. J.M. David and M.G. Buehler, "A Numerical Analysis of Various Cross Sheet Resistor Test Structures," *Solid-State Electron.* **20**, 539–543, June 1977.

46. R. Chwang, B.J. Smith, and C.R. Crowell, "Contact Size Effects on the van der Pauw Method for Resistivity and Hall Coefficient Measurement," *Solid-State Electron.* **17**, 1217–1227, Dec. 1974.

47. H.P. Baltes and R.S. Popović, "Integrated Semiconductor Magnetic Field Sensors," *Proc. IEEE*, **74**, 1107–1132, Aug. 1986.

48. H.J. Lippmann and F. Kuhrt, "The Geometrical Influence of Rectangular Semiconductor Plates on the Hall Effect, (in German)" *Z. Naturforsch.* **13a**, 474–483, 1958; I. Isenberg, B.R. Russell and R.F. Greene, "Improved Method for Measuring Hall Coefficients," *Rev. Sci. Instrum.* **19**, 685–688, Oct. 1948.

49. P.M. Hemenger, "Measurement of High Resistivity Semiconductors Using the van der Pauw Method," *Rev. Sci. Instrum.* **44**, 698–700, June 1973.

50. L. Forbes, J. Tillinghast, B. Hughes, and C. Li, "Automated System for the Characterization of High Resistivity Semiconductors by the van der Pauw Method," *Rev. Sci. Instrum.* **52**, 1047–1050, July 1981.

51. R.T. Blunt, S. Clark, and D.J. Stirland, "Dislocation Density and Sheet Resistance Variations Across Semi-Insulating GaAs Wafers," *IEEE Trans. Electron Dev.* **ED-29**, 1038–1045, July 1982; K. Kitahara and M. Ozeki, "Nondestructive Resistivity Measurement of Semi-Insulating GaAs Using Illuminated n^+-GaAs Contacts," *Japan. J. Appl. Phys.* **23**, 1655–1656, Dec. 1984.

52. D.C. Look, "Schottky-Barrier Profiling Techniques in Semiconductors: Gate Current and Parasitic Resistance Effects," *J. Appl. Phys.* **57**, 377–383, Jan. 1985.

53. P. Chu, S. Niki, J.W. Roach, and H.H. Wieder, "Simple, Inexpensive Double ac Hall Measurement System for Routine Semiconductor Characterization," *Rev. Sci. Instrum.* **58**, 1764–1766, Sept. 1987.

54. O.M. Corbino, "Electromagnetic Effects Resulting from the Distortion of the Path of Ions in Metals Produced by a Field, (in German)" *Physik. Zeitschr.* **12**, 561–568, July 1911.

55. H. Weiss, "Magnetoresistance," in *Semiconductors and Semimetals* (R.K. Willardson and A.C. Beer, eds.) Academic Press, New York, **1**, 315–376, 1966.

56. H.J. Lippmann and F. Kuhrt, "The Geometrical Influence on the Transverse Magnetoresistance Effect for Rectangular Semiconductor Plates, (in German)" *Z. Naturforsch.* **13a**, 462–474, 1958.

57. T.R. Jervis and E.F. Johnson, "Geometrical Magnetoresistance and Hall Mobility in Gunn Effect Devices," *Solid-State Electron.* **13**, 181–189, Feb. 1970.

58. P. Blood and R.J. Tree, "The Scattering Factor for Geometrical Magnetoresistance in GaAs," *J. Phys. D: Appl. Phys.* **4**, L29–L31, Sept. 1971.

59. H. Poth, "Measurement of Mobility Profiles in GaAs at Room Temperature by the Corbino Effect," *Solid-State Electron.* **21**, 801–805, June 1978.

60. J.R. Sites and H.H. Wieder, "Magnetoresistance Mobility Profiling of MESFET Channels," *IEEE Trans. Electron Dev.* **ED-27**, 2277–2281, Dec. 1980.

61. R.D. Larrabee, W.A. Hicinbothem, Jr., and M.C. Steele, "A Rapid Evaluation Technique for Functional Gunn Diodes," *IEEE Trans. Electron Dev.* **ED-17**, 271–274, April 1970.

62. F. Kharabi and D.R. Decker, "Magnetotransconductance Profiling of Mobility and Doping in GaAs MESFET's," *IEEE Electron. Dev. Lett.* **11**, 137–139, April 1990.

63. D.C. Look and G.B. Norris, "Classical Magnetoresistance Measurements in $Al_xGa_{1-x}As$/GaAs MODFET Structures: Determination of Mobilities," *Solid-State Electron.* **29**, 159–165, Feb. 1986.

64. W.T. Masselink, T.S. Henderson, J. Klem, W.F. Kopp, and H. Morkoç, "The Dependence of 77 K Electron Velocity-Field Characteristics on Low-Field Mobility in AlGaAs-GaAs Modulation-Doped Structures," *IEEE Trans. Electron Dev.* **ED-33**, 639–645, May 1986.

65. D.C. Look and T.A. Cooper, "Schottky-Barrier Mobility Profiling Measurements with Gate-Current Corrections," *Solid-State Electron.* **28**, 521–527, May 1985.

66. S.M.J. Liu and M.B. Das, "Determination of Mobility in Modulation-Doped FET's Using Magnetoresistance Effect," *IEEE Electron. Dev. Lett.* **EDL-8**, 355–357, Aug. 1987.

67. J.R. Haynes and W. Shockley, "Investigation of Hole Injection in Transistor Action," *Phys. Rev.* **75**, 691, Feb. 1949.

68. J.R. Haynes and W. Shockley, "The Mobility and Life of Injected Holes and Electrons in Germanium," *Phys. Rev.* **81**, 835–843, March 1951.

69. J.R. Haynes and W.C. Westphal, "The Drift Mobility of Electrons in Silicon," *Phys. Rev.* **85**, 680–681, Feb. 1952.

70. M.B. Prince, "Drift Mobilities in Semiconductors. I Germanium," *Phys. Rev.* **92**, 681–687, Nov. 1953; "Drift Mobilities in Semiconductors. II Silicon," *Phys. Rev.* **93**, 1204–1206, March 1954.

71. J.P. McKelvey, *Solid State and Semiconductor Physics*, Harper & Row, New York, 1966, 342.

72. B. Krüger, Th. Armbrecht, Th. Friese, B. Tierock, and H.G. Wagemann, "The Shockley-Haynes Experiment Applied to MOS Structures," *Solid-State Electron.* **39**, 891–896, June 1996.

73. R.K. Ahrenkiel, D.J. Dunlavy, D. Greenberg, J. Schlupmann, H.C. Hamaker, and H.F. MacMillan, "Electron Mobility in p-GaAs by Time of Flight," *Appl. Phys. Lett.* **51**, 776–779, Sept. 1987; M.L. Lovejoy, M.R. Melloch, R.K. Ahrenkiel, and M.S. Lundstrom, "Measurement Considerations for Zero-Field Time-of-Flight Studies of Minority Carrier Diffusion in III-V Semiconductors," *Solid-State Electron.* **35**, 251–259, March 1992.

74. H. Hillmer, G. Mayer, A. Forchel, K.S. Löchner, and E. Bauser, "Optical Time-of-Flight Investigation of Ambipolar Carrier Transport in GaAlAs Using GaAs/GaAlAs Double Quantum Well Structures," *Appl. Phys. Lett.* **49**, 948–950, Oct. 1986.

75. D.J. Westland, D. Mihailovic, J.F. Ryan, and M.D. Scott, "Optical Time-of-Flight Measurement of Carrier Diffusion and Trapping in an InGaAs/InP Heterostructure," *Appl. Phys. Lett.* **51**, 590–592, Aug. 1987.

76. C.B. Norris, Jr. and J.F. Gibbons, "Measurement of High-Field Carrier Drift Velocities in Silicon by Time-of-Flight Technique," *IEEE Trans. Electron Dev.* **ED-14**, 38–43, Jan. 1967.

77. A.G.R. Evans and P.N. Robson, "Drift Mobility Measurements in Thin Epitaxial Semiconductor Layers Using Time-of-Flight Techniques," *Solid-State Electron.* **17**, 805–812, Aug. 1974; P.M. Smith, M. Inoue, and J. Frey, "Electron Velocity in Si and GaAs at Very High Electric Fields," *Appl. Phys. Lett.* **37**, 797–798, Nov. 1980.

78. T.H. Windhorn, L.W. Cook, and G.E. Stillman, "High-Field Electron Transport in InGaAsP ($\lambda_g = 1.2$ μm)," *Appl. Phys. Lett.* **41**, 1065–1067, Dec. 1982.

79. W. von Münch and E. Pettenpaul, "Saturated Electron Drift Velocity in 6H Silicon Carbide," *J. Appl. Phys.* **48**, 4823–4825, Nov. 1977; I.A. Khan and J.A. Cooper, Jr., "Measurement of High-Field Electron Transport in Silicon Carbide," *IEEE Trans. Electron Dev.* **47**, 269–273, Feb. 2000.

80. W. Shockley, "Currents to Conductors Induced by a Moving Point Charge," *J. Appl. Phys.* **9**, 635–636, Oct. 1938.

81. W.E. Spear, "Drift Mobility Techniques for the Study of Electrical Transport Properties in Insulating Solids," *J. Non-Cryst. Sol.* **1**, 197–214, April 1969.

82. J.A. Cooper, Jr. and D.F. Nelson, "High-Field Drift Velocity of Electrons at the Si-SiO$_2$ Interface as Determined by a Time-of-Flight Technique," *J. Appl. Phys.* **54**, 1445–1456, March 1983.

83. J.A. Cooper, Jr., D.F. Nelson, S.A. Schwarz, and K.K. Thornber, "Carrier Transport at the Si-SiO$_2$ Interface," in *VLSI Electronics Microstructure Science* (N.G. Einspruch and R.S. Bauer, eds.), Academic Press, Orlando, FL, **10**, 1985, 323–361.

84. D.D. Tang, F.F. Fang, M. Scheuermann, and T.C. Chen, "Time-of-Flight Measurements of Minority-Carrier Transport in p-Silicon," *Appl. Phys. Lett.* **49**, 1540–1541, Dec. 1986.

85. C. Canali, M. Martini, G. Ottaviani, and K.R. Zanio, "Transport Properties of CdTe," *Phys. Rev.* **B4**, 422–431, July 1971.

86. R.J. Schreutelkamp and L. Deferm, "A New Method for Measuring the Saturation Velocity of Submicron CMOS Transistors," *Solid-State Electron.* **38**, 791–793, April 1995.

87. C. Kittel, *Introduction to Solid State Physics*, 4th ed., Wiley, New York, 1975, 261.

88. J.R. Schrieffer, "Effective Carrier Mobility in Surface-Space Charge Layers," *Phys. Rev.* **97**, 641–646, Feb. 1955.

89. M.S. Lin, "The Classical Versus the Quantum Mechanical Model of Mobility Degradation Due to the Gate Field in MOSFET Inversion Layers," *IEEE Trans. Electron Dev.* **ED-32**, 700–710, March 1985.

90. A. Rothwarf, "A New Quantum Mechanical Channel Mobility Model for Si MOSFET's," *IEEE Electron Dev. Lett.* **EDL-8**, 499–502, Oct. 1987.

91. M.S. Liang, J.Y. Choi, P.K. Ko, and C. Hu, "Inversion Layer Capacitance and Mobility of Very Thin Gate-Oxide MOSFET's," *IEEE Trans. Electron Dev.* **ED-33**, 409–413, March 1986.

92. P.M.D. Chow and K.L. Wang, "A New ac Technique for Accurate Determination of Channel Charge and Mobility in Very Thin Gate MOSFET's," *IEEE Trans. Electron Dev.* **ED-33**, 1299–1304, Sept. 1986; U. Lieneweg, "Frequency Response of Charge Transfer in MOS Inversion Layers," *Solid-State Electron.* **23**, 577–583, June 1980.

93. C.L. Huang and G.Sh. Gildenblat, "Correction Factor in the Split C-V Method for Mobility Measurements," *Solid-State Electron.* **36**, 611–615, April 1993.

94. C.L. Huang, J.V. Faricelli, and N.D. Arora, "A New Technique for Measuring MOSFET Inversion Layer Mobility," *IEEE Trans. Electron Dev.* **40**, 1134–1139, June 1993.

95. J. Koomen, "Investigation of the MOST Channel Conductance in Weak Inversion," *Solid-State Electron.* **16**, 801–810, July 1973.

96. C.G. Sodini, T.W. Ekstedt, and J.L. Moll, "Charge Accumulation and Mobility in Thin Dielectric MOS Transistors," *Solid-State Electron.* **25**, 833–841, Sept. 1982.

97. A.G. Sabnis and J.T. Clemens, "Characterization of the Electron Mobility in the Inverted (100) Si Surface," *IEEE Int. Electron Dev. Meet.,* Washington, DC, 1979, 18–21.

98. S.C. Sun and J.D. Plummer, "Electron Mobility in Inversion and Accumulation Layers on Thermally Oxidized Silicon Surfaces," *IEEE Trans. Electron Dev.* **ED-27**, 1497–1508, Aug. 1980.

99. S. Selberherr, W. Hänsch, M. Seavey, and J. Slotboom, "The Evolution of the MINIMOS Mobility Model," *Solid-State Electron.* **33**, 1425–1436, Nov. 1990.

100. S.I. Takagi, A. Toriumi, M. Iwase, and H. Tango, "On the Universality in Si MOSFET's: Part I—Effects of Substrate Impurity Concentration," *IEEE Trans. Electron Dev.* **41**, 2357–2362, Dec. 1994.

101. K. Chen, H.C. Wann, P.K. Ko, and C. Hu, "The Impact of Device Scaling and Power Supply Change on CMOS Gate Performance," *IEEE Electron Dev. Lett.* **17**, 202–204, May 1996.

102. J.T. Watt and J.D. Plummer, "Universal Mobility-Field Curves for Electrons and Holes in MOS Inversion Layers," *Proc. VLSI Symp.* 81, 1987.

103. K.Y. Fu, "Mobility Degradation Due to the Gate Field in the Inversion Layer of MOSFET's," *IEEE Electron Dev. Lett.* **EDL-3**, 292–293, Oct. 1982.

104. L. Risch, "Electron Mobility in Short-Channel MOSFET's with Series Resistances," *IEEE Trans. Electron Dev.* **ED-30**, 959–961, Aug. 1983.

105. N. Herr and J.J. Barnes, "Statistical Circuit Simulation Modeling of CMOS VLSI," *IEEE Trans. Comp.-Aided Des.* **CAD-5**, 15–22, Jan. 1986.

106. M.H. White, F. van de Wiele, and J.P. Lambot, "High-Accuracy MOS Models for Computer-Aided Design," *IEEE Trans. Electron Dev.* **ED-27**, 899–906, May 1980.

107. P.M. Zeitzoff, C.D. Young, G.A. Brown, and Y. Kim, "Correcting Effective Mobility Measurements for the Presence of Significant Gate Leakage Current," *IEEE Electron Dev. Lett.* **24**, 275–277, April 2003.

108. W. Zhu, J.P. Han, and T.P. Ma, "Mobility Measurement and Degradation Mechanisms of MOSFETs Made With Ultrathin High-K Dielectrics," *IEEE Trans. Electron Dev.* **51**, 98–105, Jan. 2004.

109. L. Perron, A.L. Lacaita, A. Pacelli, and R. Bez, "Electron Mobility in ULSI MOSFET's: Effect of Interface Traps and Oxide Nitridation," *IEEE Electron Dev. Lett.* **18**, 235–237, May 1997.

110. S.T. Hsu and J.H. Scott, Jr., "Mobility of Current Carriers in Silicon-on-Sapphire (SOS) Films," *RCA Rev.* **36**, 240–253, June 1975; R.A. Pucel and C.A. Krumm, "Simple Method of Measuring Drift-Mobility Profiles in Thin Semiconductor Films," *Electron. Lett.* **12**, 240–242, May 1976.

111. F.F. Fang and A.B. Fowler, "Transport Properties of Electrons in Inverted Silicon Surfaces," *Phys. Rev.* **169**, 619–631, May 1968.

112. J.S. Kang, D.K. Schroder, and A.R. Alvarez, "Effective and Field-Effect Mobilities in Si MOSFET's," *Solid-State Electron.* **32**, 679–681, Aug. 1989.

113. K. Narita, Y. Hijakata, H. Yaguchi, S. Yoshida, and S. Nakashima, "Characterization of Carrier Concentration and Mobility in n-type SiC Wafers Using Infrared Reflectance Spectroscopy," *Japan. J. Appl. Phys.* **43**, 5151–5156, Aug. 2004.

114. G. Masetti, M. Severi, and S. Solmi, "Modeling of Carrier Mobility Against Carrier Concentration in Arsenic-, Phosphorus-, and Boron-Doped Silicon," *IEEE Trans. Electron Dev.* **ED-30**, 764–769, July 1983 and references therein.

115. D.M. Caughey and R.E. Thomas, "Carrier Mobilities in Silicon Empirically Related to Doping and Field," *Proc. IEEE* **55**, 2192–2193, Dec. 1967.

116. G. Baccarani and P. Ostoja, "Electron Mobility Empirically Related to the Phosphorus Concentration in Silicon," *Solid-State Electron.* **18**, 579–580, June 1975.

117. N.D. Arora, J.R. Hauser, and D.J. Roulston, "Electron and Hole Mobilities in Silicon as a Function of Concentration and Temperature," *IEEE Trans. Electron Dev.* **ED-29**, 292–295, Feb. 1982.

118. S.S. Li and W.R. Thurber, "The Dopant Density and Temperature Dependence of Electron Mobility and Resistivity in n-Type Silicon," *Solid-State Electron.* **20**, 609–616, July 1977.

119. S.S. Li, "The Dopant Density and Temperature Dependence of Hole Mobility and Resistivity in Boron-Doped Silicon," *Solid-State Electron.* **21**, 1109–1117, Sept. 1978.

120. J.M. Dorkel and Ph. Leturcq, "Carrier Mobilities in Silicon Semi-Empirically Related to Temperature, Doping and Injection Level," *Solid-State Electron.* **24**, 821–825, Sept. 1981.

121. Y. Sasaki, K. Itoh, E. Inoue, S. Kishi, and T. Mitsuishi, "A New Experimental Determination of the Relationship Between the Hall Mobility and the Hole Concentration in Heavily Doped p-Type Silicon," *Solid-State Electron.* **31**, 5–12, Jan. 1988.

122. J.R. Lowney and H.S. Bennett, "Majority and Minority Electron and Hole Mobilities in Heavily Doped GaAs," *J. Appl. Phys.* **69**, 7102–7110, May 1991.

123. K. Chen, C. Hu, J. Dunster, P. Fang, M.R. Lin, and D.L. Wolleson, "Predicting CMOS Speed with Gate Oxide and Voltage Scaling and Interconnect Loading Effects," *IEEE Trans. Electron Dev.* **44**, 1951–1957, Nov. 1997.

124. K. Ahmed, E. Ibok, G.C.F. Yeap, Q. Xiang, B. Ogle, J.J. Wortman, and J.R. Hauser, "Impact of Tunnel Currents and Channel Resistance on the Characterization of Channel Inversion Layer Charge and Polysilicon-Gate Depletion of Sub-20-Å Gate Oxide MOSFET's," *IEEE Trans. Electron Dev.* **46**, 1650–1655, Aug. 1999.

PROBLEMS

8.1 The excess carrier density in a Haynes-Shockley experiment follows the equation

$$\Delta n(x, t) = \frac{N}{\sqrt{4\pi D_n t}} \exp\left(-\frac{(x - vt)^2}{4 D_n t} - \frac{t}{\tau_n}\right) ; D_n = \frac{(d\Delta t)^2}{16 \ln(2) t_d^3} ;$$

$$\tau_n = \frac{t_{d2} - t_{d1}}{\ln(\Delta n_1 / \Delta n_2) - 0.5 \ln(t_{d2}/t_{d1})}$$

where t_d is the delay time (the time at the peak in the curves) and Δt is the pulse width at half its maximum amplitude. From the $\mathscr{E} = 75$ V/cm curve in Fig. P8.1 determine the velocity v, the mobility μ_n ($v = \mu_n \mathscr{E}$, where $\mathscr{E}$ is the electric field), and the diffusion constant D_n. From the $\mathscr{E} = 75$ V/cm and $\mathscr{E} = 150$ V/cm curves, determine the lifetime τ_n. $d = 2.5 \times 10^{-2}$ cm, $T = 300$ K.

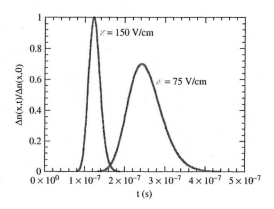

Fig. P8.1

8.2 The $I_D - V_D$ and $I_D - V_G$ curves of a MOSFET are shown in Fig. P8.2(a) and (b).

(a) Determine and plot μ_{eff} versus V_G for this device to $V_G = 5$ V.

(b) Determine θ and μ_o.

(c) Determine and plot μ_{FE} versus V_G, using Eq. (8.78).

(d) Derive a modified version of Eq. (8.78) taking into account the dependence of μ_{eff} on V_G. Then use this modified equation to determine and plot new and more accurate values of μ_{FE}.

Plot all mobilities on the same figure. W/L = 20, $C_{ox} = 1.7 \times 10^{-7}$ F/cm^2, effective mobility can be represented by $\mu_{eff} = \mu_o/[1 + \theta(V_G - V_T)]$.

8.3 The $I_D - V_D$ curves of a MOSFET are shown in Fig. P 8.3.

(a) Determine and plot μ_{eff} versus V_G for this device to $V_G = 6$ V.

(b) Determine V_T, μ_o, and θ. $W = 20$ μm, $L = 2$ μm, $t_{ox} = 12.5$ nm, the effective mobility can be represented by

$$\mu_{eff} = \mu_o/[1 + \theta(V_G - V_T)].$$

8.4 The $I_D - V_{GS}$ curve of a MOSFET with $\mu_{eff} = $ constant and $R_{SD} = 0$, measured at $V_{DS} = 100$ mV, is shown in Fig. P 8.4.

(a) Determine the threshold voltage.

(b) Draw the transconductance curve g_m on the same diagram and label the g_m axis.

(c) Draw the $I_D - V_{GS}$ curve if μ_{eff} is not constant, but depends on gate voltage according to $\mu_{eff} = \mu_o[1 + \theta(V_{GS} - V_T)]$; show how the g_m curve changes.

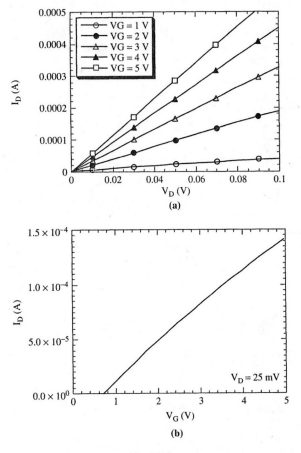

(a)

(b)

Fig. P8.2

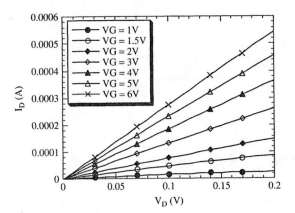

Fig. P8.3

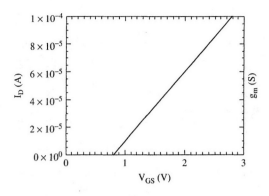

Fig. P8.4

8.5 A short duration light flash creates excess electron-hole pairs, $\Delta n = \Delta p$, at $x = 0$ at $t = 0$ as shown in Fig. P8.5. Minority carriers drift in the electric field. Draw Δn at $t = t_1$ corresponding to $x = x_1$. What material parameters can be determined with this experiment? Explain.

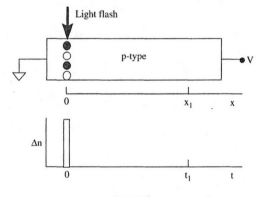

Fig. P8.5

8.6 A sheet of electrons and holes is generated at $x = 0.5$ mm at $t = 0$ as shown in Fig. P8.6. These carriers drift in the electric field. The electron velocity is 10^7 cm/s and the hole velocity is 5×10^6 cm/s. Draw the electron current, the hole current, and the total current (in arbitrary units, but all currents on the same scale) on the diagram provided.

8.7 In the time-of-flight measurement, electron-hole pairs (ehp) are generated in the depleted space-charge region of width W of a reverse-biased junction as shown in Fig. P8.7. At $t = 0$ there is a brief light flash creating ehp. The flash has a negligible pulse width and it creates 6.25×10^6 ehp.

(a) Determine the transit time for electrons or holes for the flash at positions (a) and (c).

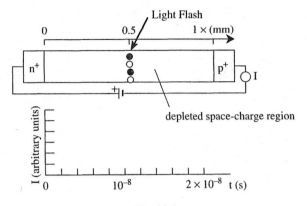

Fig. P8.6

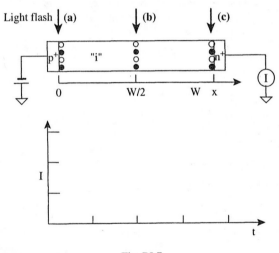

Fig. P8.7

(b) Determine the current for positions (a) and (c).

(c) Draw the I-*t* curves for the light flash at positions (a) $x = 0$, (b) $x = W/2$, and (c) $x = W$.

$v_n = 2v_p = 10^7$ cm/s, $W = 100$ μm. There are no capacitances in the output circuit, *i.e.*, the RC time constant is zero. *Label the axes with numerical values.*

8.8 In a time-of-flight measurement, electron-hole pairs (ehp) are generated in the depleted space-charge region of width W of a reverse-biased p^+in^+ junction as shown in Fig. P8.8. At $t = 0$ there is a brief light flash creating ehps *uniformly* throughout the entire space-charge region. Draw the resulting $V(t)$ versus t curve. Use $v_n = 2v_p$, where v_n is the electron velocity and v_p is the hole velocity. Neglect any capacitance in the output circuit.

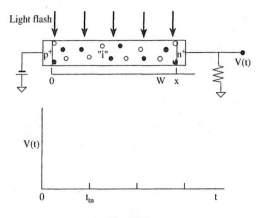

Fig. P8.8

8.9 In a time-of-flight measurement, carriers are generated in the depleted space-charge region of width W of a reverse-biased p^+in^+ junction, shown in Fig. P8.9. "i" stands for intrinsic region. At $t = 0$ there is a brief light flash creating *electrons only* uniformly throughout the entire space-charge region. This is strange light that only creates electrons. Draw the I-t curve. t_t is the transit time of electrons to drift from $x = 0$ to $x = W$. There are no capacitances in the output circuit, *i.e.*, the RC time constant is zero.

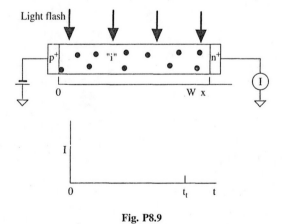

Fig. P8.9

8.10 Derive Eq. (8.37). Consider the pulse width, which has spread due to diffusion, at its 50% maximum value in the solution.

8.11 The current expression for a MOSFET with series resistance (Fig. P8.11) is given by

$$I_D = (W/L)\mu_{eff}C_{ox}(V'_G - V_T - V'_D/2)V'_D \approx \beta(V'_G - V_T)V'_D \qquad (P8.11)$$

where

$$V'_G = V_G - I_D R_S; \quad V'_D = V_D - I_D R_{SD}; \quad R_{SD} = R_S + R_D$$

Show that the measured transconductance $g_m = \partial I_D / \partial V_{GS|VDS=constant}$ is given by

$$g_m = \frac{g_{mo}}{1 + g_{do} R_{SD} + g_{mo} R_s}$$

where $g_{mo} = \partial I_D / \partial V'_{GS|VDS=constant}$ and $g_{do} = \partial I_D / \partial V'_{DS|VGS=constant}$.

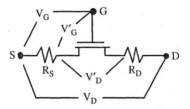

Fig. P8.11

8.12 From the $I_D - V_D$ and the $C_{GC} - V_G$ curves in Fig. P8.12, determine and plot μ_{eff} versus $V_G - V_T$. Plot both on the same figure. To integrate the $C_{GC} - V_G$ curve to determine Q_n use a simple graphical integration

$$\mu_{eff} = \frac{g_d L}{W C_{ox}(V_G - V_T)}; \quad \mu_{eff} = \frac{g_d L}{W Q_n}$$

$W = 1$ μm, $L = 0.18$ μm, $t_{ox} = 2.5$ nm, $V_T = 0.5$ V, $R_{SD} = 0$, $K_{ox} = 3.9$.

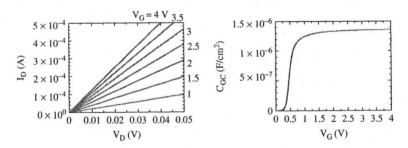

Fig. P8.12

8.13 The current-voltage relationship of a MOSFET in the presence of series resistance is (source and substrate are grounded):

$$I_D \approx \frac{W_{eff} C_{ox}}{L_{eff}} \frac{\mu_o}{[1 + \theta(V_G - V_T)]} (V_G - V_T - 0.5 V_D) V'_D$$

where $V'_D = V_D - I_D(R_S + R_D)$, $W_{eff} = W - \Delta W$, and $L_{eff} = L - \Delta L$. Using the $I_D - V_G$ curves determine V_T, μ_o, θ, ΔL, and $R_{SD} = R_S + R_D$; assume $\Delta W = 0$. $t_{ox} = 10$ nm, $W = 50$ μm, $V_D = 50$ mV.

The drain currents for various channel lengths and various gate voltages are:

I_D (A)

V_G (V)	$L = 20$ μm	12 μm	7 μm	1 μm
1.325	1.145e-05	1.876e-05	3.119e-05	0.0001527
1.625	1.636e-05	2.645e-05	4.304e-05	0.0001740
1.925	2.094e-05	3.345e-05	5.339e-05	0.0001873
2.225	2.523e-05	3.985e-05	6.250e-05	0.0001964
2.525	2.924e-05	4.572e-05	7.058e-05	0.0002031
2.825	3.301e-05	5.113e-05	7.781e-05	0.0002081
3.125	3.656e-05	5.612e-05	8.430e-05	0.0002121
3.425	3.991e-05	6.075e-05	9.017e-05	0.0002153
3.725	4.307e-05	6.504e-05	9.550e-05	0.0002179
4.025	4.606e-05	6.905e-05	0.0001004	0.0002202
4.325	4.889e-05	7.278e-05	0.0001048	0.0002220
4.625	5.157e-05	7.628e-05	0.0001089	0.0002237
4.925	5.412e-05	7.957e-05	0.0001127	0.0002251
5.225	5.655e-05	8.265e-05	0.0001162	0.0002263

8.14 In a Haynes Shockley experiment, the output voltage versus time is measured and shown in Fig. P8.14. The semiconductor is then flown in outer space, where it is bombarded with high-energy particles, creating damage in the semiconductor. After returning to earth, the V_{out} versus t curve is remeasured. Draw and justify the new curve on the same figure. Mobility and diffusion coefficient are unchanged.

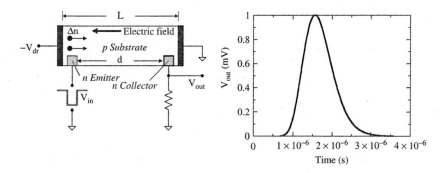

Fig. P8.14

8.15 In a Haynes Shockley experiment, the output voltage versus time is measured and shown in Fig. P8.14. Draw and the curve when the electric field is increased. Mobility and diffusion coefficient are unchanged.

REVIEW QUESTIONS

- What are the different mobilities?

- Why is the MOS effective mobility less than the bulk mobility?
- Why does the Hall mobility differ from the conductivity mobility?
- How does the Haynes-Shockley experiment work?
- What is determined with the Haynes-Shockley experiment?
- For what is the time-of-flight technique used?
- What precaution must be observed when the time-of-flight method is used to determine the mobility at high electric fields?
- How is μ_{eff} most commonly determined?
- Why is μ_{FE} usually lower then μ_{eff}?
- What are the effects of gate depletion and gate current on effective mobility measurements?
- Why is the channel frequency response important in effective mobility measurements?

9

CHARGE-BASED AND PROBE CHARACTERIZATION

9.1 INTRODUCTION

Many semiconductor characterization techniques are based on current, voltage, and capacitance measurements. They generally require some device fabrication or at least temporary contacts, *e.g.*, mercury probe $C-V$ measurements. For example, to determine the oxide charge and interface trap density of an MOS device, it is necessary to make an MOS capacitor, traditionally done by evaporating a metal gate, depositing a poly-Si gate, or using a mercury probe for the gate on an oxidized wafer. It is sometimes useful to make measurements without device fabrication. One way is to deposit charge on an oxidized wafer and measure the voltage contactless with a Kelvin or Monroe probe. The charge in this configuration becomes the "gate". After all, applying a gate voltage to an MOS capacitor is equivalent to placing a charge on the gate. Depositing the charge directly on the oxide circumvents the gate formation with the additional advantage of being contactless. The charge can be removed with a water rinse. Some of the material/device parameters that can be determined with charge-based measurements are illustrated in Fig. 9.1.

Charge-based measurements lend themselves to measurements during the development of integrated circuits (ICs) and for manufacturing control. To be effective, such test structures should provide rapid feedback to the pilot or manufacturing line. Surface voltage (SV) and surface photovoltage (SPV) semiconductor characterization techniques are suitable for such rapid feedback and have become powerful and convenient methods for a variety of material/device parameter measurements.[1] The introduction of commercial equipment led to widespread adoption by the semiconductor industry for initially measuring the minority carrier diffusion length,[2] later expanded to encompass routine characterization of surface voltage, surface barrier height, flatband voltage, oxide thickness, oxide leakage current, interface trap density, mobile charge density, oxide integrity, generation

Semiconductor Material and Device Characterization, Third Edition, by Dieter K. Schroder
Copyright © 2006 John Wiley & Sons, Inc.

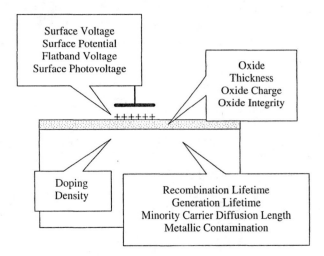

Fig. 9.1 Schematic illustration of the various material/device parameters measurable with charge/ probe/light techniques.

lifetime, recombination lifetime, and doping density. Charge, in these measurements, is used in two basic ways: as the "gate" in MOS-type measurements, where the charge replaces the metal or poly-silicon gate, and as a surface modifying method, where the charge controls the surface potential.

IBM developed corona charge for semiconductor characterization during the period 1983–1992.[3] However, due to lack of commercial instruments, the technique was initially only sparingly used. Later, it was developed into commercial products. We give an introduction to this technique here, review the relevant theory and compare the technique to the well-established MOS technique and illustrate it with several examples.

9.2. BACKGROUND

Bardeen and Brattain first described the SPV technique in 1953.[4] They characterized the light-induced surface potential variation in Ge samples with a mechanically vibrating reed. In 1955 Garret and Brattain presented the basic theory of the photo-induced change of the surface potential in a semiconductor when illuminated with light.[5] In the same year, Moss considered diffusion of photo-generated carriers during surface photovoltage measurements.[6] He called it "photovoltage" and the "photovoltaic effect". The name "surface photovoltage" appears to have been used first by Brattain and Garret in 1956 using continuous illumination.[7] Morrison used a chopped light signal for capacitive voltage detection.[8] The use of SPV for minority carrier diffusion length determination was proposed by Moss in 1955, by Johnson in 1957,[9] by Quilliet and Gosar in 1960,[10] and by Goodman in 1961.[11] It was Goodman's SPV approach that led to the first full-scale implementation of the technique in the semiconductor industry at RCA,[12] where it was employed during semiconductor production by placing high-diffusion length wafers into critical furnaces and measuring the diffusion length after heating the wafers. Through this relatively simple, contactless method, they were able to detect cracked furnace tubes, contaminated solid source diffusion sources, metallic contact contamination, and other contamination sources. Instead of dc surface voltage or photovoltage measurements, lifetimes or diffusion lengths can also be extracted from frequency-dependent, charge-based measurements. Nakhmanson introduced frequency-dependent optically induced lifetime measurements.[13] The equivalent circuit concept has proven to be very powerful for the analysis of such measurements.[14]

During charge-based measurements, charge is deposited on the wafer and the semiconductor response is measured with a Kelvin probe. To understand charge-based measurements, it is necessary to understand Kelvin probes, first proposed by Kelvin in 1881.[15] Kronik and Shapira give an excellent explanation of such probes and applications.[16]

9.3 SURFACE CHARGING

Charge is deposited *chemically* or as a *corona charge*. During chemical treatment, for *n*-type silicon, the oxide on the sample surface should be removed and the sample should be boiled in H_2O_2 or in water for about 15 min and then rinsed in deionized water (DI).[17] Alternately, one can soak the sample in $KMnO_4$ for 1–2 min and then rinse in DI water. These treatments produce a stable depletion surface potential barrier. For *p*-type silicon very little treatment is required. In case of very low V_{SPV}, etching in buffered HF followed by a DI water rinse is recommended.

Corona charging is used in copying processes using xerographic techniques where the charge is deposited on a photoconductive drum.[18] One of the first uses of deposited charge for semiconductors was in the characterization of ZnO in 1968.[19] Williams and Woods[20] and later Weinberg[21] expanded the approach to the characterization of oxide leakage current and mobile charge drift.[22] Ions are deposited on a surface at atmospheric pressure through an electric field applied to a source of ions. The corona source consists of a wire, a series of wires, a single point, or multiple points located a few mm or cm above the sample surface.[23] The substrate may be moved during charging or between charging cycles and the sample may be charged uniformly or in well-defined areas through a mask. It is even possible to deposit positive (negative) charge in a given area and surround the area with negative (positive) charge, to act as a zero-gap guard ring.[24]

A potential of 5,000–10,000 V of either polarity is applied to the corona source, as illustrated in Fig. 9.2. Ions are generated close to the electrode, where a faint glow may be observed in a darkened room. For a negative source potential, positive ions bombard the source while free electrons are rapidly captured by ambient molecules to form negative ions. For a positive source potential, electrons are attracted to the source and positive ions follow the electric field lines to the substrate. The negative and positive corona ionic species are predominantly CO^{-3} and H_3O^+ (hydrated protons), respectively. The corona source forces a uniform flow of ionized air molecules toward the surface. The very short (approximately 0.1 μm) atmospheric mean free path of the ionized gas ensures collision-dominated ion transport with the molecules retaining very little kinetic energy. Typically a few seconds are required to charge an insulating surface to a saturation potential.

One of the advantages for oxide thickness and oxide integrity measurements using corona charge "gates" rather than conductive gates is the low surface mobility of the "corona" ions on the sample surface. A charge deposited on the surface of an oxidized wafer, creates an oxide electric field. The oxide breaks down at its weakest spot, with the current confined to the breakdown spot, because the surface corona charge does not readily drift or diffuse along the surface. By contrast, for a conductive gate with applied gate voltage, the breakdown area may be the same as for the corona charge method, but

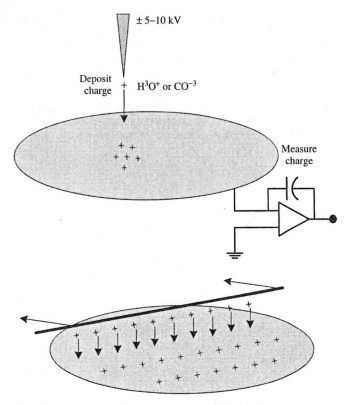

Fig. 9.2 Schematic illustration of point and wire electrode corona charging methods. The deposited charge is precisely measured with the op-amp charge meter.

the current from the entire gate area will be channeled into the weak spot, possibly leading to catastrophic breakdown.

9.4 THE KELVIN PROBE

How does a surface voltage or photovoltage come about and how is it measured? A surface voltage is generated by a surface or insulator charge or work function difference and is most commonly detected with a non-contacting probe. The probe is a small plate, 2–4 mm in diameter, held typically 0.1–1 mm above the sample. Two types of probes are used, as illustrated in Fig. 9.3. In the Kelvin probe, the electrode is vibrated vertically changing the capacitance between probe and sample. In the Monroe probe, the electrode is fixed and a grounded shutter, mounted in front of the electrode, is vibrated horizontally thereby modulating the probe to wafer capacitance. The vibrational frequencies are typically 500–600 Hz. Two modes are used to determine the voltage: measuring the current and measuring the voltage.

To understand the operation of a Kelvin probe, let us start with the band diagrams in Fig. 9.4, consisting of two metals with differing work functions spaced a distance d_1 and forming a capacitor. In Fig. 9.4(a) the two metals are not connected and there is a voltage

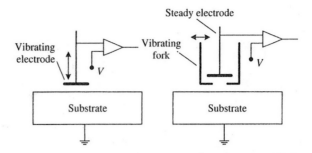

Fig. 9.3   Kelvin probe (*left*) and Monroe probe (*right*) for contact potential difference measurements.

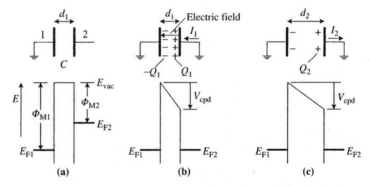

Fig. 9.4   Two metal plates and their band diagrams (a) plate 2 is floating electrically, (b) plate 2 is grounded with spacing d_1, and (c) plate 2 is grounded with spacing $d_2 > d_1$.

difference between them due to the work function difference $(\Phi_{M2} - \Phi_{M1})/q$, illustrated by its band diagram. This voltage, indicated by the different metal Fermi energies E_{F1} and E_{F2}, can be measured with a voltmeter. There is, however, no electric field in the gap between the metal plates, since neither metal plate is charged. In Fig. 9.4(b), the metals are both grounded and the Fermi energies equalize. Electrons flow from plate 2 to plate 1 establishing a net charge Q_1 and $-Q_1$ on the plates and an electric field in the gap. The external voltage is now zero, but there is an internal voltage, the contact potential difference indicated by V_{cpd}. Clearly, as plate 2 is grounded, current I_1 flows momentarily. The charge is related to the voltage and capacitance by

$$Q = VC = V\varepsilon_o/d \tag{9.1}$$

where C is the capacitance and V is the internal voltage between the plates, d the spacing between plates and ε_o the permittivity of free space. If now the plates are pulled apart to distance d_2 on Fig. 9.4(c) while remaining grounded, the charge on each plate must decrease, since the voltage is constant at V_{cpd}, but the electric field is reduced. Electrons flowing from plate 1 to plate 2 give rise to current I_2. In the *vibrating Kelvin probe* the current is[25]

$$I = \frac{dQ}{dt} = V\frac{dC}{dt} = -V\frac{\varepsilon_o}{d^2}\frac{dd}{dt} \tag{9.2}$$

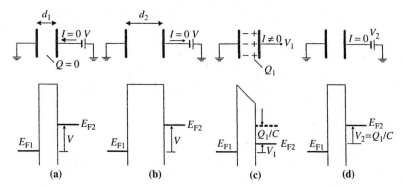

Fig. 9.5 Two metal plates and their band diagrams, (a) voltage $-V$ is applied to plate 2 with spacing d_1, (b) voltage $-V$ is applied to plate 2 with spacing $d_2 > d_1$, (c) charge Q_1 on plate 2, and (d) voltage $-V_2$ reduces the current to zero.

For the example in Fig. 9.4, the current is

$$I = V_{cpd} \frac{dC}{dt} \sim V_{cpd} \qquad (9.3)$$

The contact potential difference is determined by calibrating the current. V_{cpd} depends on the work function, adsorption layers, oxides, doping density in semiconductors, and sample temperature variations.

The *zero current* mode is illustrated in Fig. 9.5. In Fig. 9.5(a) a negative voltage $V = \Phi_{M2}/q - \Phi_{M1}/q$ is applied to plate 2 reducing the charge on the plates and the electric field between the plates to zero and there is no current flow. When the plates are pulled further apart (9.5(b)) there is still no current flow since there is no charge on the plates. The applied voltage V is adjusted until the current reaches zero when one of the plates is vibrated. This voltage is equal to V_{cpd} and is also shown on Fig. 9.3 as V. The method in Fig. 9.4 is faster than that in Fig. 9.5 and is typically used for mapping purposes.

Let us now consider the case when charge Q_1 is deposited on plate 2 as shown on Fig. 9.5(c) with charge $-Q_1$ induced on plate 1. Initially the voltage on floating gate 2 is V_{cpd} (Fig. 9.4(a)). The voltage Q_1/C alters floating plate 2 voltage to V_1 and a current pulse flows when the charge is deposited. Applying an external voltage $V_2 = Q_1/C$ in Fig. 9.5(d) reduces the charge and the current to zero. Hence, knowing the capacitance one can determine the charge from the voltage.

Having established the basic operation of the Kelvin probe with two metal plates, we now turn to semiconductors. The potential band diagram of the probe-air-semiconductor system is shown in Fig. 9.6, where Φ_M/q and Φ_S/q are the metal and semiconductor work function *potentials, i.e.,* the potentials between the vacuum potential E_{vac}/q and the Fermi potential ϕ_F. E_c and E_v are the conduction band and valence band energies and E_c/q and E_v/q their potentials. The potential of the intrinsic energy level in the neutral bulk semiconductor, ϕ, is taken as the reference potential. The *semiconductor* surface potential ϕ_s (ϕ_s is ϕ at $x = 0$) is zero for flatband, positive for depletion and inversion, and negative for accumulation for *p*-substrates.

The potential on the sample surface is the surface voltage V_S. For a bare sample $V_S = \phi_s$, but $V_S \neq \phi_s$ for oxidized wafers with charge in or on the oxide. The potential

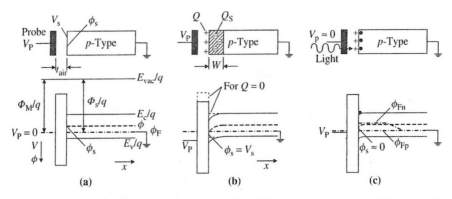

Fig. 9.6 Cross-section and band diagram of a metal-air-semiconductor system with zero work function difference; (a) no surface charge, (b) positive surface charge, (c) strong light excitation.

measured at the probe is the contact potential difference V_{cpd} also called the contact potential, denoted as the probe potential V_P from now on. All potentials are measured with respect to the grounded substrate. The probe voltage is the difference in Fermi potential between probe and substrate.

First we consider the bare, grounded p-type semiconductor in Fig. 9.6 with the metal probe placed a distance t_{air} above the sample. There is no surface charge and Φ_M and Φ_S are equal, leading to the work function difference $\Phi_{MS} = \Phi_M - \Phi_S = 0$ and $V_P = 0$ in Fig. 9.6(a). The band diagram is very similar to that of an MOS capacitor, with the oxide replaced by air. Next, positive charge density Q (C/cm^2) is deposited on the semiconductor surface in Fig. 9.6(b), inducing charge density Q_S in the semiconductor. The dashed lines on the energy band diagram obtain for zero charge and the solid lines for charge density Q, inducing charge only in the semiconductor, not in the probe, because the probe is floating electrically. Hence, no electric field exists between the sample and the probe making $V_P = V_S = \phi_s$.

The induced semiconductor charge density Q_S, in the absence of an inversion layer, consists of ionized acceptors in the space-charge region (scr) and is

$$Q = -Q_S = q N_A W \tag{9.4}$$

where W is the scr width and N_A the acceptor doping density. The scr width W is

$$W = \sqrt{\frac{2 K_s \varepsilon_o \phi_s}{q N_A}} = \frac{Q}{q N_A} \tag{9.5}$$

Solving for the surface potential ϕ_s gives

$$\phi_s = \frac{Q^2}{2 K_s \varepsilon_o q N_A} = \frac{(q N)^2}{2 K_s \varepsilon_o q N_A} = 9.07 \times 10^{-7} \frac{N^2}{K_s N_A} \tag{9.6}$$

where N is the surface charge atom density (cm^{-2}). For example, for Si with $N_A = 10^{16}$ cm^{-3}, $K_s = 11.7$, and a surface charge atom density $N = 10^{11}$ cm^{-2}, the surface potential is $\phi_s = 0.077$ V.

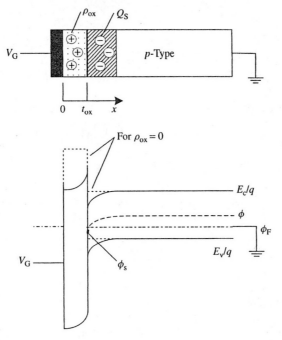

Fig. 9.7 MOS capacitor cross-section showing oxide charge ρ_{ox} and the potential band diagram.

Many semiconductor samples characterized by SV or SPV techniques are oxidized and contain charges and work function differences. To understand the effect of work function differences *and* charge density on surface voltage, we will first consider the simpler and well-known MOS capacitor (MOS-C) in Fig. 9.7, containing work function difference Φ_{MS} and uniform positive oxide charge density ρ_{ox} (C/cm³). From Chapter 6, the gate voltage is

$$V_G = V_{FB} + V_{ox} + \phi_s \tag{9.7}$$

where V_{ox} is the potential across the oxide. The flatband voltage is

$$V_{FB} = \Phi_{MS}/q - \frac{1}{C_{ox}} \int_0^{t_{ox}} \frac{x}{t_{ox}} \rho_{ox}\, dx \tag{9.8}$$

With the gate floating electrically, there is no charge on the gate and zero electric field in the oxide at the gate, shown by the zero slope of the oxide band diagram at $x = 0$.

Let us now extend this example to an electrically floating Kelvin probe held above a semiconductor covered with an insulator and a probe-semiconductor work function difference Φ_{MS} leading to the negative probe potential $V_P = \Phi_{MS}/q$ in Fig. 9.8(a). Next uniform oxide charge density ρ_{ox} (C/cm³) and surface charge density Q (C/cm²) are added in Fig. 9.8(b). These charges induce charge density qN_AW in the semiconductor (indicated by the negative charges). The probe voltage, calculated with the same approach as for MOS capacitors, is

$$V_P = V_{FB} + V_{air} + V_{ox} + \phi_s \tag{9.9}$$

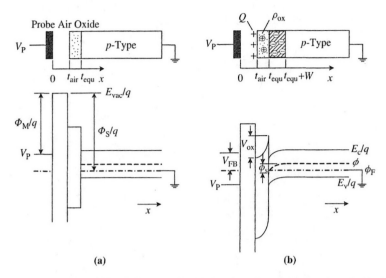

Fig. 9.8 Cross-sections and band diagrams for (a) $\Phi_{MS}/q < 0$, and (b) $\Phi_{MS}/q < 0$, $Q > 0$, and $\rho_{ox} > 0$.

For the floating gate configuration in Fig. 9.8(b), $V_{air} = 0$, since there is no charge on the probe and no electric field in the air gap. The flatband voltage is

$$V_{FB} = \Phi_{MS}/q - \frac{t_{air}}{t_{equ}} \frac{Q}{C_{equ}} - \frac{1}{C_{equ}} \int_{t_{air}}^{t_{equ}} \frac{x}{t_{equ}} \rho_{ox}\, dx \tag{9.10}$$

where C_{equ} is the equivalent capacitance and t_{equ} the equivalent thickness given by

$$C_{equ} = \frac{C_{air} C_{ox}}{C_{air} + C_{ox}} = \frac{\varepsilon_o}{t_{equ}}; t_{equ} = t_{air} + t_{ox}/K_{ox} \tag{9.11}$$

Equations (9.9)–(9.11) show the probe voltage to be due to Φ_{MS}, Q, and ρ_{ox}. A single measurement is unable to distinguish between these three parameters.

Next we consider the effect of light on the sample. For simplicity, we will use the bare sample in Fig. 9.6. Fig. 9.6(a) shows the band diagram with surface charge density Q in the dark and in Fig. 9.6(c) the sample is strongly illuminated driving the semiconductor to the flatband condition and the probe potential approaches zero. The Fermi level splits into two quasi-Fermi levels and measuring the surface voltage without and with light yields the surface potential and thus the charge density from Eq. (9.5). To understand how this comes about, we must look at the flatband condition in more detail.

The semiconductor charge density Q_S for a p-type semiconductor in depletion or inversion is

$$Q_S = -\sqrt{2kT K_s \varepsilon_o n_i}\, F(U_S, K) \tag{9.12}$$

where F is the normalized surface electric field (for more details see Chapter 6), defined as[26]

$$F(U_S, K) = \sqrt{K(e^{-U_S} + U_S - 1) + K^{-1}(e^{U_S} - U_S - 1) + K(e^{U_S} + e^{-U_S} - 2)\Delta} \tag{9.13}$$

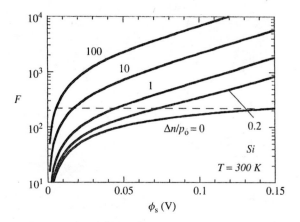

Fig. 9.9 Normalized surface electric field, F, function versus surface potential as a function of normalized excess carrier density or light intensity.

where $K = p_o/n_i$ (p_o is the equilibrium majority carrier density and n_i the intrinsic carrier density), $U_s = q\phi_s/kT$ is the normalized surface potential, ϕ_s the surface potential, and Δ the normalized excess carrier density ($\Delta = \Delta n/p_o$, where $\Delta p = \Delta n$ is the excess carrier density). In the absence of excess carriers, *i.e.*, in equilibrium, the last term in Eq. (9.13) vanishes.

F is plotted versus ϕ_s in Fig. 9.9 as a function of the normalized illumination-induced excess carrier density. The electric field is related to the charge density through Eq. (9.12). Constant charge implies constant electric field or constant F. Hence, as Δn increases, the surface potential decreases, because the locus of the F-ϕ_s plot is along a horizontal line such as the dashed line. In the limit of intense illumination, $\phi_s \to 0$ and the semiconductor approaches flatband.

The probe potential is

$$V_P = V_{FB} + V_{air} + V_{ox} + \phi_s; V_{ox} = Q/C_{ox} = -Q_S/C_{ox} \tag{9.14}$$

The voltages in the dark and under intense illumination ($\phi_s \to 0$) are

$$V_{P,dark} = V_{FB} + V_{air} + Q/C_{ox} + \phi_s; V_{P,light} \approx V_{FB} + V_{air} + Q/C_{ox} \tag{9.15}$$

The charge density Q remains constant with illumination and the change in the surface voltage becomes

$$\Delta V_P = V_{P,dark} - V_{P,light} \approx \phi_s \tag{9.16}$$

showing that the surface potential is determined in this method. The flatband voltage corresponds to $\phi_s = 0$, *i.e.*, $V_{SPV} \approx 0$, as illustrated in the surface photovoltage versus probe voltage plot measured under intense illumination in Fig. 9.10. Flatband voltage is indicated as the point where $V_{SPV} = 0$. Note that determination of V_{FB} in this way requires

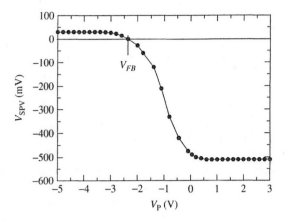

Fig. 9.10 Surface photovoltage versus gate voltage. $N_A = 2.6 \times 10^{14}$ cm^{-3}.

neither oxide thickness nor substrate doping density to be known, in contrast to MOS-C flatband voltage determination, where both must be known.

9.5 APPLICATIONS

9.5.1 Surface Photovoltage (SPV)

Surface photovoltage was one of the first characterization techniques using surface charge as discussed in Chapter 7 and is commonly used to determine the minority carrier diffusion length.[27] The concept of surface photovoltage can be understood with the band diagram in Fig. 9.11. Surface charge density Q induces charge density Q_S in the semiconductor with $Q + Q_S = 0$ shown in Fig. 9.11(a). The surface charge must be of a polarity to drive the semiconductor into depletion. The band diagram in the dark is shown in Fig. 9.11(b). Incident light creates electron-hole pairs (ehps). Some ehps recombine in the neutral p-substrate, some diffuse toward the surface. If they reach the edge of the space-charge region (scr), the holes neutralize acceptor atoms, thereby reducing the scr width and the electrons drift in the scr electric field to the surface exchanging negatively electrons for

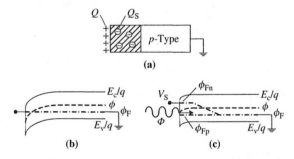

Fig. 9.11 (a) Cross-section with surface charge Q and semiconductor charge density Q_s, (b) band diagram in the dark, (c) illuminated band diagram.

negatively charged acceptors. This generates a forward bias, reducing the band bending and splitting the Fermi level into the quasi-Fermi levels ϕ_{Fn} and ϕ_{Fp} giving the surface photovoltage $V_S = \phi_{Fn} - \phi_{Fp}$ in Fig. 9.11(c). The SPV voltage, being a surface voltage, is named V_S here to be consistent with the nomenclature in this chapter, even though it is V_{SPV} in Chapter 7. For constant photon flux density Φ, the diffusion length is extracted form a plot of $1/V_S$ versus $1/\alpha$.

9.5.2 Carrier Lifetimes

Carrier lifetimes are discussed in Chapter 7. Here we outline the use of corona charge in lifetime measurements. Lifetimes are divided into *recombination lifetime* and *generation lifetime*.[28] For *recombination* lifetime determination, charge is deposited onto an oxidized wafer to invert the semiconductor surface forming a surface charge-induced *np* junction in a *p*-type substrate, shown in Fig. 9.12. A brief light pulse injects excess carriers into the sample thereby *forward biasing* this *np* junction. The junction bias changes as ehps recombine, leading to a time-dependent surface voltage. This method is very similar to the open-circuit voltage decay technique with the recombination lifetime determined by[29]

$$\tau_r = \frac{kT/q}{dV_P/dt} \tag{9.17}$$

The *generation* lifetime is determined by depositing a charge pulse onto an oxidized wafer, driving the corona-oxide-semiconductor (COS) device into deep depletion, illustrated in Fig. 9.13(a). The space-charge region width is controlled by the amount of corona charge. The wafer is then quickly transported under a Kelvin probe and the time-varying probe voltage due to ehp generation is measured as a function of time (Fig. 9.13(b)).

The generation lifetime is extracted from the probe voltage transient through the expression[24]

$$\frac{dV_P}{dt} = \frac{qn_i}{C_{ox}} \left(\frac{(W - W_{min})}{\tau_{g,eff}} - s_{g,eff} \right) \tag{9.18}$$

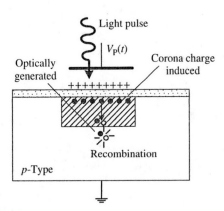

Fig. 9.12 Corona charge forms a charge-induced np junction. Pulsed light modifies the junction voltage measured with the contactless probe.

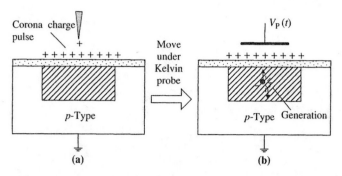

Fig. 9.13 Corona charge pulse forms a deep-depleted space-charge region. Thermal electron-hole pair generation leads to time varying probe voltage.

The gate voltage of a metal-oxide-semiconductor (MOS-C) or corona-oxide-semiconductor (COS-C) is

$$V_G = V_S = V_{FB} + V_{ox} + \phi_s; \; V_{ox} = Q_G/C_{ox} = -Q_S/C_{ox} = (Q_n + Q_b)/C_{ox} \quad (9.19)$$

V_G is the gate voltage for the MOS-C or the surface voltage V_S for the COS-C. After corona charge is deposited on the gate, Q_G remains constant throughout the measurement, causing V_{ox} to remain constant too. Differentiating Eq. (9.19) leads to

$$\frac{dV_S}{dt} = \frac{d\phi_s}{dt} \quad (9.20)$$

assuming the flatband voltage to be invariant with time, *i.e.*, $dV_{FB}/dt = 0$—a good approximation for room temperature measurements.

With the bulk charge density

$$Q_b = qN_A W = \sqrt{2qK_s\varepsilon_o N_A\phi_s} \quad (9.21)$$

and Q_G and Q_S constant with time, we get

$$\frac{dQ_S}{dt} = 0 = -\frac{dQ_n}{dt} - \frac{dQ_b}{dt} = -\frac{dQ_n}{dt} - qN_A\frac{dW}{dt} \quad (9.22)$$

or, using Eq. (9.21),

$$-\frac{dQ_n}{dt} = \sqrt{\frac{qK_s\varepsilon_o N_A}{2\phi_s}}\frac{d\phi_s}{dt} = \frac{K_s\varepsilon_o}{W}\frac{d\phi_s}{dt} = \frac{K_s\varepsilon_o}{W}\frac{dV_S}{dt} \quad (9.23)$$

For the pulsed MOS-C with *constant gate voltage* after the device is driven into deep depletion, the capacitance is measured as a function of time. In that case, dQ_n/dt is given by (see Chapter 7)

$$-\frac{dQ_n}{dt} = \frac{qK_s\varepsilon_o N_A C_{ox}}{C^3}\frac{dC}{dt} \quad (9.24)$$

For COS measurements, the surface voltage is monitored as a function of time.

dQ_n/dt, the rate at which inversion carriers are generated in the non-equilibrium, deep-depleted semiconductor, is

$$-\frac{dQ_n}{dt} = \frac{qn_i(W - W_{inv})}{\tau_{g,eff}} + qn_i s_{g,eff} \tag{9.25}$$

Equations (9.23) to (9.24) now become

$$\frac{dV_S}{dt} = \frac{qn_i W}{K_s \varepsilon_o} \left(\frac{(W - W_{min})}{\tau_{g,eff}} - s_{g,eff} \right) \tag{9.26}$$

$$\frac{1}{C^3} \frac{dC}{dt} = \frac{n_i}{K_s \varepsilon_o N_A C_{ox}} \left(\frac{(W - W_{min})}{\tau_{g,eff}} - s_{g,eff} \right) \tag{9.27}$$

One of the advantages of the COS approach is the constancy of the surface charge. With Q_G constant, V_{ox} also remains constant in contrast to conventional MOS-C measurements, where V_{ox} increases with time, limiting the gate voltage because of oxide breakdown or oxide current. It is possible for V_{ox} to become sufficiently high for appreciable oxide current to flow or for the oxide to break down. Gate current for a p-type substrate consists of electrons from the thermally generated inversion layer. As some of these electrons are injected into the oxide, it will take longer to build up the inversion layer. In other words, it will appear as if the generation lifetime is longer than it actually is.[30] This problem is reduced in the COS method because the oxide voltage remains constant.

COS generation and recombination lifetime measurements were used to characterize epitaxial films and their substrates.[31] The epitaxial layer is characterized through generation lifetime measurements, with the thermal carrier generation confined to the charge-induced space-charge region, which is typically on the order of 1 μm below the semiconductor surface. The recombination lifetime, on the other hand, characterizes a depth determined by the minority carrier diffusion length. Figure 9.14 illustrates corona-induced generation and recombination lifetime measurements of n-epitaxial layers on n-substrates. The figure shows the results for both "good" and "bad" epi-layers and "good"

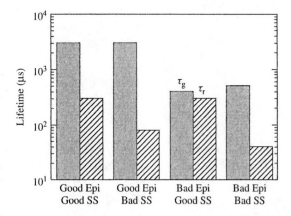

Fig. 9.14 Corona-induced generation and recombination lifetimes of n-epitaxial layers on n-substrates. Data adapted from ref. 31.

and "bad" substrates (SS). This is a good example of how these two complementary charge-based techniques yield information that neither one alone can provide.

9.5.3   Surface Modification

Surface charge can be used to control the surface potential and surface recombination by driving the sample into accumulation, depletion, or inversion, illustrated in Fig. 9.15. Positive surface charge in Fig. 9.15(a) leads to a depleted surface. Excess minority carriers are attracted to the surface to recombine there with high surface recombination velocity. In contrast, the accumulated surface in Fig. 9.15(b) repels excess minority carriers with a concomitant low surface recombination velocity. The effective lifetime and surface recombination velocity are plotted in Fig. 9.16 as a function of surface charge density.[32] The effective lifetime is measured with the photoconductance decay/microwave reflectivity technique. The wafer surface is slightly inverted for zero surface charge. As negative corona charge is deposited, the surface initially depletes. The effective lifetime decreases because surface recombination increases. With more negative corona charge, the surface becomes accumulated, surface recombination is reduced and the lifetime increases. In this case, corona charge is used to modify the surface recombination velocity by controlling the surface condition.

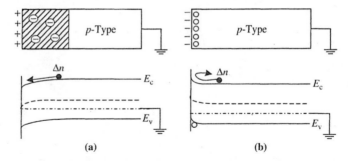

(a)   (b)

Fig. 9.15 Band diagrams for (a) attractive potential, (b) repulsive potential.

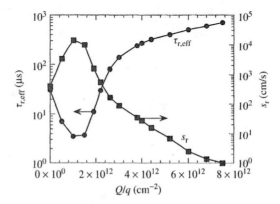

Fig. 9.16 Effective recombination lifetime and surface recombination velocity as a function of negative surface corona density. $N_A = 4.2 \times 10^{16}$ cm^{-3}, wafer thickness = 280 μm. After ref. 32.

9.5.4 Near-Surface Doping Density

Near-surface doping density is a measure of the average doping density in the top few microns of the semiconductor. This near-surface doping density is determined in the COS technique by forming a field-induced junction and pulsing it into deep depletion, similar to pulsed MOS measurements. The data analysis is similar to that of MOS measurements. The junction is formed by first creating an accumulation region at the test site. Then, an inverted region is created in the center of this test site. The accumulation region acts as a guard ring, suppressing lateral conduction to give the junction a well-defined area.

The junction is then pulsed into deep depletion with an additional charge ΔQ and the resulting voltage transient is recorded. The deposited charge is imaged in the substrate by repelling majority carriers to a space-charge region width W. As minority carriers are generated, the scr collapses in time and returns to its equilibrium width W_{inv} and equilibrium voltage V_{Si}. During the measurement, the charge increment ΔQ and the transient voltage increase ΔV_{Si} are measured and the doping density is a function of these two variables.

W and ΔQ are given by

$$W = W_{inv} + \Delta W; \Delta Q = q N_A W \tag{9.28}$$

The voltage during the depleting pulse is

$$\Delta V_{Si} + V_{Si} = \frac{q N_A W^2}{2 K_s \varepsilon_o} \tag{9.29}$$

where V_{Si} and space-charge region width W_{inv} are related by

$$V_{Si} = \frac{q N_A W_{inv}^2}{2 K_s \varepsilon_o} \tag{9.30}$$

V_{Si} is also given by[33]

$$V_{Si} = \frac{kT}{q} \left[2.1 \ln \left(\frac{N_A}{n_i} \right) + 2.08 \right] \tag{9.31}$$

Equations (9.28) to (9.31) are solved iteratively for N_A. A comparison between N_A measured by COS and MOS-C for n-epitaxial layers is shown in Fig. 9.17. For the MOS-C, the max-min MOS-C method was used to determine N_A (see Chapter 2).

9.5.5 Oxide Charge

The surface voltage dependence on surface charge lends itself to measurements of charge *in* the insulator on a semiconductor wafer or charge *on* the wafer. This charge can be oxide charge, interface trapped charge, plasma damage charge, or other charge. Let us illustrate this by considering the mobile charge density Q_m of an oxidized wafer.[34] One way to measure such a mobile charge is to combine SV measurements with corona charge techniques by depositing corona charge on an oxidized semiconductor surface. First deposit positive corona charge, heat the wafer to a moderate temperature of around 200°C for a few minutes, driving the mobile charge to the oxide-semiconductor interface. Cool the sample and determine the flatband voltage V_{FB1}. Next repeat the procedure with a negative corona charge and drive the mobile charge to the oxide-air interface determining

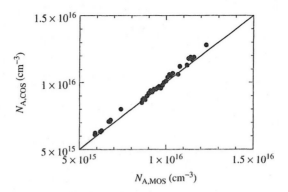

Fig. 9.17 Doping density determined by COS and MOS techniques. The line represents perfect correlation between the two.

V_{FB2}. Q_m is then determined by the flatband voltage difference $\Delta V_{FB} = V_{FB2} - V_{FB1}$ through the relation

$$Q_m = C_{ox} \Delta V_{FB} \tag{9.32}$$

The sensitivity of the measurement can be enhanced by decreasing the oxide capacitance through thicker oxides, but that is inconsistent with today's thin gate oxides. The flatband voltage due to oxide charge density ρ_{ox} alone is $V_{FB} = -\rho_{ox} t_{ox}^2 / 2K_{ox}\varepsilon_o$. For a charge density of $\rho_{ox} t_{ox}/q = 10^{10}$ cm^{-2}, we find $V_{FB} = -2.3 \times 10^3 t_{ox}$. For example, for $t_{ox} = 5$ nm, $V_{FB} = -1.1$ mV, illustrating that voltage measurements become impractical for thin oxides. A solution to this problem is to measure the surface potential of an oxidized wafer by measuring the surface voltage without and with intense light. Then deposit corona charge until the surface potential becomes zero. The deposited corona charge is equal in magnitude but opposite in sign to the original oxide charge.[35] The accuracy and precision of this charge-based measurement is identical for thin and thick oxides.

Other charges that have been determined with SV measurements are plasma-induced charge and damage as well as hydrogen-stabilized silicon surfaces.[36] When Si surfaces were exposed to two hydrogen treatments: annealing in hydrogen or immersing in HF, hydrogen-annealed surfaces were more stable, determined by measuring the surface barrier as a function of time. The measurement of oxide charge in buried oxides of silicon-on-insulator materials is also feasible.[37] An example of plasma charge induced surface voltage is shown in Fig. 9.18.

Charge-based oxide charge measurements have an advantage over voltage-based measurements. For example, to determine the oxide charge of an MOS device one can measure the *charge* or the *voltage*. The relationship between the oxide voltage uncertainty ΔV_{ox} and oxide charge uncertainty ΔQ_{ox} is

$$\Delta Q_{ox} = C_{ox} \Delta V_{ox} = K_{ox} \varepsilon_o \Delta V_{ox}/t_{ox} \tag{9.33}$$

Equation (9.33) is plotted in Fig. 9.19. Suppose the oxide charge is determined from a voltage measurement with an uncertainty of $\Delta V_{ox} = 1$ mV. ΔQ_{ox} varies from 2.2×10^{10} to 2.2×10^{11} cm^{-2} for oxide thicknesses from 10 nm to 1 nm. In voltage-based measurements, there is a large uncertainty in oxide charge. For charge-based measurements,

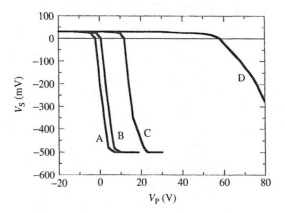

Fig. 9.18 Surface voltage versus probe voltage as a function of plasma charging. 15 nm SiO_2, 1000 nm PSG glass, power 700 W, A: 16 torr, B: 12 torr, C: 8.5 torr with anneal, D: 8.5 torr. After M.S. Fung, "Monitoring PSG Plasma Damage with COS," *Semicond. Int.* **20**, July 1997.

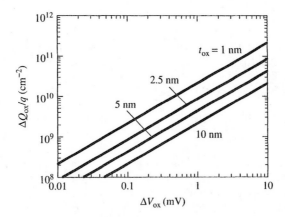

Fig. 9.19 Oxide charge density uncertainty versus oxide voltage uncertainty as a function of oxide thickness.

there is a charge uncertainty, but that is independent of oxide thickness and is on the order of $\Delta Q_{ox}/q = 10^9$ cm^{-2} or less. An example of charge-based and voltage-based measurements is shown in Fig. 9.20.

9.5.6 Oxide Thickness and Interface Trap Density

To determine the oxide thickness, corona charge density Q is deposited on the oxidized wafer and the surface voltages are measured in the dark and under intense light,[38] giving the surface voltage V_S, that is plotted versus deposited charge density as in Fig. 9.21.[39] In accumulation or inversion the curves are linear and the oxide thickness is

$$C_{ox} = \frac{dQ}{dV_S}; t_{ox} = \frac{K_{ox}\varepsilon_o}{C_{ox}} = K_{ox}\varepsilon_o \frac{dV_S}{dQ} \tag{9.34}$$

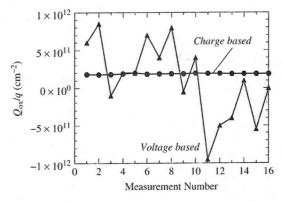

Fig. 9.20 Charge-based and voltage-based oxide charge repeatability for 3 nm oxides. After Weinzierl and Miller.[35]

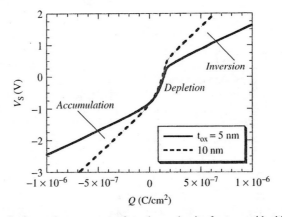

Fig. 9.21 Surface voltage versus surface charge density for two oxide thicknesses.

This method is not subject to the poly-Si gate depletion effects of MOS-C measurements.[40] It is also not affected by probe punchthrough and is relatively insensitive to oxide pinhole leakage currents. Interface traps distort the low-frequency $C_{lf} - V_S$ curve, as discussed in Chapter 6 (V_S is V_G in Ch. 6). Similarly, interface traps distort the $V_S - Q$ curve and the interface trap density is determined from that distortion.

9.5.7   Oxide Leakage Current

To determine oxide leakage current, known as gate current in MOS devices, corona charge is deposited on the surface of an oxidized wafer and the Kelvin probe voltage is measured as a function of time. If the charge leaks through the oxide, the voltage decreases with time. The device is biased into accumulation or inversion and the oxide leakage current is related to the voltage through the relationship[41]

$$I_{leak} = C_{ox}\frac{dV_P(t)}{dt} \Rightarrow V_P(t) = \frac{I_{leak}}{C_{ox}}t \tag{9.35}$$

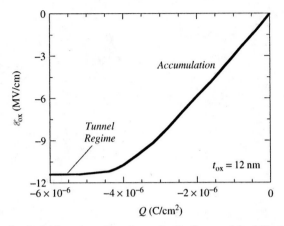

Fig. 9.22 Oxide electric field versus surface charge density for an oxidized Si wafer. $t_{ox} = 12$ nm. After Roy et al.[38]

The device should be biased into accumulation. When biased into inversion, some inversion electrons tunnel to the gate and have to be supplied through thermal ehp generation. If the generation rate is lower than the oxide leakage rate, the leakage current is limited by thermal generation giving erroneous leakage currents.

When the device is biased into accumulation, charge builds up on the oxide. However, when the charge density is too high, it leaks through the oxide by Fowler-Nordheim or direct tunneling and the surface voltage becomes clamped. The deposited charge density is related to the oxide electric field $\mathscr{E}_{ox}$ through the relationship

$$Q = K_{ox}\varepsilon_o\mathscr{E}_{ox} = 3.45 \times 10^{-13}\mathscr{E}_{ox} \tag{9.36}$$

for SiO_2. Silicon dioxide breaks down at electric fields of $10-14$ MV/cm. For $\mathscr{E}_{ox} = 12$ MV/cm, the charge is 4.1×10^{-6} C/cm². The $\mathscr{E}_{ox}$ versus Q plot in Fig. 9.22 clearly shows the electric field saturation at a charge density of around 4.4×10^{-6} C/cm², corresponding to a breakdown electric field of 12.8 MV/cm.

9.6 SCANNING PROBE MICROSCOPY (SPM)

Scanning probe microscopy refers to techniques in which a sharp tip is scanned across a sample surface at very small distances to obtain two- or three-dimensional images of the surface at nanometer or better lateral and/or vertical resolution.[42] In the extreme, one can obtain lateral resolution on the order of 0.1 nm and vertical resolution of 0.01 nm. The original application of SPM was the scanning tunneling microscope (STM), invented in 1982[43] based on the earlier topografiner.[44] It is the only technique for imaging at atomic resolution other than transmission electron microscopy. A myriad of SPM instruments has been developed over the past decade, and one can sense current, voltage, resistance, force, temperature, magnetic field, work function, and so on with these instruments at high resolution as outlined in Table 9.1.[45] The operation of the instruments is generally based on detecting the near-field image, as described for the near-field optical microscope in Chapter 10. We briefly describe several scanning microscopy techniques. For a more

TABLE 9.1   Scanning Probe Techniques and Their Abbreviations and Acronyms.

AFM	Atomic Force Microscopy
BEEM	Ballistic Electron Emission Microscopy
CAFM	Conducting AFM
CFM	Chemical Force Microscopy
IFM	Interfacial Force Microscopy
MFM	Magnetic Force Microscopy
MRFM	Magnetic Resonance Force Microscopy
MSMS	Micromagnetic Scanning Microprobe System
Nano-Field	Nanometer Electric Field Gradient
Nano-NMR	Nanometer Nuclear Magnetic Resonance
NSOM	Near Field Optical Microscopy
SCM	Scanning Capacitance Microscopy
SCPM	Scanning Chemical Potential Microscopy
SEcM	Scanning Electrochemical Microscopy
SICM	Scanning Ion-Conductance Microscopy
SKPM	Scanning Kelvin Probe Microscopy
SSRM	Scanning Spreading Resistance Microscopy
SThM	Scanning Thermal Microscopy
STOS	Scanning Tunneling Optical Spectroscopy
STM	Scanning Tunneling Microscopy
TUNA	Tunneling AFM

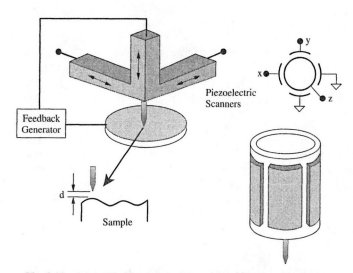

Fig. 9.23   Schematic illustration of a scanning tunneling microscope.

detailed description of these and other probe techniques, the reader is referred to the extensive published literature.

9.6.1   Scanning Tunneling Microscopy (STM)

The schematic in Fig. 9.23 shows the major features of a *scanning tunneling microscope*,[46] consisting of a very sharp metallic probe that is scanned across the sample at distances of about 1 nm, with a bias voltage between the tip and the sample that is less than the work function of the tip or the sample. The probe is usually made from tungsten or Pt-Ir. It is not trivial to make such probes with radii on the order of 100–1000 nm. Experimental evidence suggests "mini tips" of <10 nm radii form at the tip of the probes.[47] Piezoelectric elements provide the scanning mechanism. A piezoelectric material is one that changes dimension upon application of a voltage. By applying voltages to x, y, and z-oriented piezoelectric elements, the tip or the sample can be scanned in all three directions. Early implementations used the three-arm tripod arrangement in Fig. 9.23 that is subject to low resonance frequencies and was later changed to the tubular implementation. The outside of the tube contains four symmetric electrodes. Applying equal but opposite voltages to opposing electrodes causes the tube to bend due to contraction and expansion. The inner wall is contacted by a single electrode for actuation voltages for vertical movement.[48]

Since the probe tip is very close to the sample surface, a tunnel current of typically 1 nA flows across the gap. Clearly, both probe and sample must be conducting for this technique. For high-resolution images it is very important that the tip be extremely sharp and it is believed that a single atom at the probe tip primarily determines the device operation. The current is given by[49]

$$ I = \frac{C_1 V}{d} \exp\left(-2d\sqrt{\frac{8\pi^2 m \Phi_B}{h^2}} \right) = \frac{C_1 V}{d} \exp(-1.025 d \sqrt{\Phi_B}) \qquad (9.37) $$

for d in Å and Φ_B in eV, where C_1 is a constant, V is the voltage, d the gap spacing between tip and sample, and Φ_B an effective work function defined by $\Phi_B = (\Phi_{B1} + \Phi_{B2})/2$ with Φ_{B1} and Φ_{B2} the work functions of the tip and sample, respectively. For $\Phi_B \approx 4$ eV, a typical work function, a gap spacing change from 10 Å to 11 Å, changes the current density by about a factor of eight. Hence, small variations in gap spacing produce large current changes, suggesting application for surface flatness characterization.

There are two modes of operation. In the first the gap spacing is held constant, as the probe is scanned in the x and y dimensions, through a feedback circuit holding the current constant. The voltage on the piezoelectric transducer is then proportional to the vertical displacement giving a contour plot. In the second mode, the probe is scanned across the sample with varying gap and current. The current is now used to determine the wafer flatness. Equation (9.37) is somewhat simplified, because the tunnel current is actually a measure of the overlap of the electronic wave functions of probe and sample in the gap and the probe actually images surface wave functions rather than just atomic positions. However, the current is largely determined by the gap spacing or sample topography. Holding the probe above a given location of the sample and varying the probe voltage gives the tunneling spectroscopy current, allowing the band gap and the density of states to be probed. By using the STM in its spectroscopic mode, the instrument probes the electronic states of a surface located within a few electron volts on either side of the Fermi energy. The sensitivity of STM to electronic structure can lead to undesirable artifacts. For example, a region of lower conductivity appears as a dip in the image.

9.6.2 Atomic Force Microscopy (AFM)

The *atomic force microscope* was introduced in 1986 to examine the surface of insulating samples. There was a clear implication in the first paper that it was capable of resolving single atoms.[50] However, unambiguous evidence for atomic resolution with the AFM did

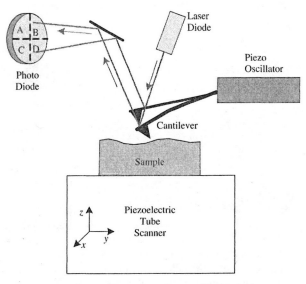

Fig. 9.24   Schematic illustration of an atomic force microscope.

not appear until 1993. In the intervening years the AFM evolved into a mature instrument providing new insights in the fields of surface science, electrochemistry biology and technology.[51] Atomic force microscopy operates by measuring the force between a probe and the sample. This force depends on the nature of the sample, the distance between the probe and the sample, the probe geometry, and sample surface contamination. In contrast to scanning tunneling microscopy, which requires electrically conducting samples, AFM is suitable for conducting as well as insulating samples.

The AFM principle is illustrated in Fig. 9.24. The instrument consists of a cantilever with a sharp tip mounted on its end. The cantilever is usually formed from silicon, silicon oxide or silicon nitride and is typically 100 μm long, 20 μm wide, and 0.1 μm thick, but other dimensions are used. The vertical sensitivity depends on the cantilever length. For topographic imaging, the tip is brought into continuous or intermittent contact with the sample and scanned across the sample surface. Depending on the design, piezoelectric scanners translate either the sample under the cantilever or the cantilever over the sample. Moving the sample is simpler because the optical detection system need not move. The motion of the cantilever can be sensed by one of several methods.[52] It can be one mirror of an optical laser interferometer or the cantilever deflection can be sensed by a capacitance change between the cantilever and a reference electrode. A common technique is to sense the light reflected from the cantilever into a two-segment or four-segment, position sensitive photodiode in Fig. 9.24.[53] The cantilever motion causes the reflected light to impinge on different segments of the photodiode. Vertical motion is detected by $z = (A + C) - (B + D)$ and horizontal motion by $x = (A + B) - (C + D)$. Holding the signal constant, equivalent to constant cantilever deflection, by varying the sample height through a feedback arrangement, gives the sample height variation. Cantilevers come in various shapes. Two common shapes are shown in Fig. 9.25. For the beam cantilever, the resonance frequency is given by

$$f_o = \frac{1}{2\pi}\sqrt{\frac{k}{m}} \tag{9.38}$$

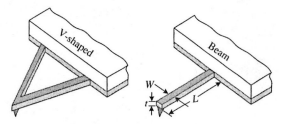

Fig. 9.25 AFM cantilevers.

where k is the spring constant and m the mass of the cantilever. Typical resonance frequencies lie in the 50–500 kHz range.

AFMs can operate in several modes. In the *contact mode*, the probe tip is dragged across the surface and the resulting image is a topographical map of the sample surface. While this technique has been very successful for many samples, it has some drawbacks. The dragging motion of the probe tip, combined with adhesive forces between the tip and the surface, can damage both sample and probe and create artifacts in the data. Under ambient air conditions, most surfaces are covered by a layer of condensed water vapor and other contaminants. When the scanning tip touches this layer, capillary action causes a meniscus to form and surface tension pulls the cantilever into the layer. Trapped electrostatic charge on the tip and sample contribute additional adhesive forces. These downward forces increase the overall force on the sample and, when combined with lateral shear forces caused by the scanning motion, can distort measurement data and damage the sample.

In the *non-contact mode*, the instrument senses van der Waal attractive forces between the surface and the probe tip held above the sample surface. Unfortunately, these forces are substantially weaker than the contact mode forces—so weak in fact that the tip must be given a small oscillation and ac detection methods are used to detect the small forces between tip and sample. The attractive forces also extend only a short distance from the surface, where the adsorbed gas layer may occupy a large fraction of their useful range. Hence, even when the sample-tip separation is successfully maintained, non-contact mode provides lower resolution than either contact or tapping mode.

Tapping mode imaging overcomes the limitations of the conventional scanning modes by alternately placing the tip in contact with the surface to provide high resolution and then lifting the tip off the surface to avoid dragging the tip across the surface.[54] It is implemented in ambient air by oscillating the cantilever assembly at or near the cantilever's resonant frequency with a piezoelectric crystal. The piezo motion causes the cantilever to oscillate when the tip does not contact the surface. The oscillating tip is then moved toward the surface until it begins to lightly touch, or "tap" the surface. During scanning, the vertically oscillating tip alternately contacts the surface and lifts off, generally at a frequency of 50 to 500 kHz. As the oscillating cantilever contacts the surface intermittently, energy loss caused by the tip contacting the surface reduces the oscillation amplitude that is then used to identify and measure surface features. When the tip passes over a bump in the surface, the cantilever has less room to oscillate and the amplitude of oscillation decreases. Conversely, when the tip passes over a depression, the cantilever has more room to oscillate and the amplitude increases approaching the maximum free air amplitude. The oscillation amplitude of the tip is measured and the feedback loop adjusts the tip-sample separation maintaining a constant amplitude and force on the sample.

Fig. 9.26 Non-contact AFM image of metal lines showing the grains and grain boundaries. 10 μm × 10 μm scan area. Courtesy of Veeco Corp.

Tapping mode imaging works well for soft, adhesive, or fragile samples, allowing high resolution topographic imaging of sample surfaces that are easily damaged or otherwise difficult to image by other AFM techniques. It overcomes problems associated with friction, adhesion, electrostatic forces, and other difficulties that can plague conventional AFM scanning methods. An AFM image is shown in Fig. 9.26.

9.6.3 Scanning Capacitance Microscopy (SCM)

The two main techniques that have emerged for *lateral doping density* profiling are *scanning capacitance microscopy* and *scanning spreading resistance microscopy*.[55] Scanning capacitance microscopy has received much attention as a lateral profiling tool.[56] A small-area capacitive probe measures the capacitance of a metal/semiconductor or an MOS contact, similar to techniques described in Chapter 2. Scanning capacitance microscopy combines atomic force microscopy with highly sensitive capacitance measurements. SCM is able to measure the local capacitance-voltage characteristics between the SCM tip and a semiconductor with nanometer resolution. SCM images have been used to extract two dimensional carrier profiles and to locate electrical $p - n$ junctions. The original SCM used an insulating stylus.[57] Later a metallized tip was used in combination with AFM.[58] The metallized AFM tip is used for imaging the wafer topography in conventional contact mode and also serves as an electrode for simultaneously measuring the MOS capacitance. SCM images of actively biased cross-sectional MOSFETs and of operating *pn* junctions allow visualization of the operation of semiconductor devices.

The semiconductor device is usually cleaved or polished so that the device cross section is exposed, as shown in Fig. 9.27, although the sample top, without cleaving, can also be measured. An oxide is deposited on the cross-sectional area and the probe is scanned across the area in the contact mode, measuring the capacitance variations in the nanometric probe/oxide/silicon MOS capacitor by applying a high-frequency ac voltage between the probe and the semiconductor. For constant electrical bias, the space-charge region in the MOS capacitor is wider for lower doping densities. Dedicated simulation

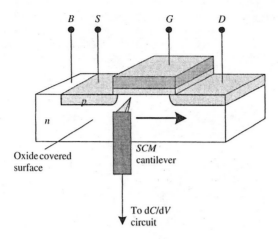

Fig. 9.27   Scanning capacitance schematic.

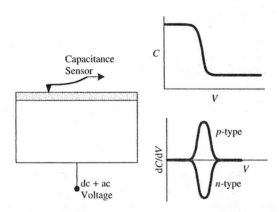

Fig. 9.28   Schematic of the AFM/SCM design. $C-V$ curve of n-type substrate with bias applied to the substrate and dC/dV curve. The sign identifies the dopant type.

models are necessary to obtain a realistic conversion curve which relates the local SCM signal with the local carrier density. A schematic of the measurement in Fig. 9.28 shows the conducting AFM tip on the oxidized sample, the $C-V$ and dC/dV curves. The voltage is applied to the substrate in this case. In some cases, it is applied to the tip. The shape of the dC/dV curve identifies the doping type. SCM is sensitive to carrier density densities from 10^{15} to 10^{20} cm^{-3}, with a lateral resolution of 20–150 nm, depending on tip geometry and dopant density. Extraction of absolute dopant densities requires reverse simulation incorporating tip geometry and sample oxide thickness. Example SCM maps are given in Fig. 9.29, showing the formation of a channel in a MOSFET with increasing gate voltage.[59]

The capacitance between the tip and sample is measured with a capacitance sensor from an RCA Video Disk player,[60] which is electrically connected to the tip. The capacitance measurement is made independently of and simultaneously with the AFM measurement

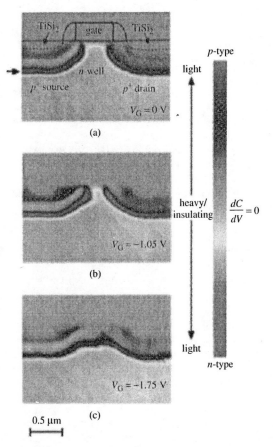

Fig. 9.29 Sequence of SCM images of a Si, *p*-channel MOSFET with $V_D = -0.1$ V, $V_S = V_B = 0$ V, and $V_G = $ (a) 0, (b) -1.05, and (c) 1.75 V. The progression of the SCM images shows the formation of a conducting channel between the source and drain. The schematic drawing in (a) shows the approximate locations of the polysilicon gate, titanium nitride spacers, the titanium silicide contacts. Images were acquired with $V_{ac} = 2.0$ V peak to peak and $V_{dc} = 0$ applied to the SCM tip. After Nakakura et al.[59]

of topography. The sensor measures the capacitance at a frequency of 915 MHz, allowing small variations in capacitance to be resolved. This ultra sensitive capacitance sensor can detect *relative* variations in capacitance in the range of 10^{-18} F around an input capacitance of about 0.1 pF. The conducting tips are made by coating commercially available cantilevered AFM tips with metal. Silicon nitride cantilevers coated with approximately 20 nm Cr or Ti have useful lifetimes during probe scanning. Commercially available Co/Cr-coated, highly doped silicon cantilevers, such as those used for magnetic force microscopy have also been successful. A degree of electrical isolation of the SCM from the environment is achieved by enclosing the entire microscope in a grounded acoustic isolation hood.[61]

Two standard SCM methods have been developed for two-dimensional dopant profiling: in the ΔC *mode*, a constant amplitude ac bias voltage is applied between tip and sample, and in the ΔV *mode*, a feedback loop adjusts the applied ac bias voltage to keep the change in capacitance, ΔC, constant as the tip is moved from one region to another.[62] In the former, the ac bias voltage produces a corresponding change in capacitance measured by a lock-in amplifier. As the tip moves from a region of high dopant density to a more lightly doped region, the lock-in amplifier output increases owing to the larger $C-V$ curve slope in the lightly doped region. In the latter, a feedback loop adjusts the applied ac bias voltage to keep ΔC constant as the tip is moved from one region to another. In this case, the magnitude of the required ac bias voltage is measured to determine the dopant density.

The advantage of the ΔC mode is simplicity. The disadvantage of this system is that a large ac bias voltage (several volts ac) is needed to measure finite SCM signals at high doping densities. When this same voltage is applied to lightly-doped silicon, it creates a larger depletion volume, reducing the spatial resolution and making accurate modeling more difficult. The advantage of the ΔV method is that the physical geometry of the depletion problem remains relatively constant as the tip is scanned from a lightly to heavily-doped region. The disadvantage is that an additional feedback loop is required.

For reproducible measurements, samples must be prepared carefully.[61] Factors that influence the repeatability and the reproducibility of SCM measurements, include: sample-related problems (mobile and fixed oxide charges, interface states, non-uniform oxide thickness, surface humidity and contamination, sample aging, water-related oxide traps), tip-related problems (increase of the tip radius, fracture of the tip-apex, mechanical wear of the metal coating, contaminants on the tip picked up from the sample), and problems related to the electrical operating conditions (amplitude of the ac probing signal in the capacitance sensor, scanning rate, compensation of the stray capacitance, electric field induced oxide growth, dc tip-bias voltage).

9.6.4 Scanning Kelvin Probe Microscopy (SKPM)

Scanning Kelvin probe microscopy falls in the category of *electrostatic force microscopy* (EFM) techniques. EFM can be divided into three regimes based on tip-sample separation: long range, intermediate, and short range.[63] Additional regimes can be described depending on whether the tip is driven mechanically or electrostatically. The SKPM probe, typically held 30–50 nm above the sample, is scanned across the surface and the potential is measured. Frequently this measurement is combined with AFM measurements. During the first AFM scan the sample topography is measured and during the second scan, in the SKPM mode, the surface potential is determined.[64]

The conducting probe and conducting substrate can be treated as a capacitor with the gap spacing being the spacing between probe and sample surface. A dc and ac voltage is applied to the tip (sometimes the voltage is applied to the sample with the tip held at ground potential). This leads to an oscillating electrostatic force between tip and sample from which the surface potential can de determined. The method is similar to the Kelvin probe discussed earlier in this chapter, except that a force is measured instead of a current. An advantage of force over current measurements is that the latter is proportional to the probe size while the former is independent of it. The frequency is chosen equal or close to the cantilever resonance frequency, which it typically around several 100 kHz.

Let us consider a capacitance C, a voltage V, and a charge Q. The capacitance and energy stored in the capacitor are

$$C = \frac{Q}{V}; E = \frac{1}{2}CV^2 = \frac{1}{2}\frac{Q^2}{C} \tag{9.39}$$

A voltage across the capacitor leads to an attractive force between the tip and the sample. The relationship between energy and force is

$$F = \frac{dE}{dz} = -\frac{1}{2}\frac{Q^2}{C^2}\frac{dC}{dz} = -\frac{1}{2}V^2\frac{dC}{dz} \tag{9.40}$$

for constant charge and constant voltage where z is the tip-to-sample spacing.[65] The tip potential is

$$V_{tip} = V_{dc} + V_{ac}\sin(\omega t) \tag{9.41}$$

Substituting into Eq. (9.40) gives

$$F = \frac{1}{2}\frac{dC}{dz}\left[(V_{dc} - V_{surf})^2 + \frac{1}{2}V_{ac}^2(1 - \cos(2\omega t)) + 2(V_{dc} - V_{surf})V_{ac}\sin(\omega t)\right] \tag{9.42}$$

where V_{surf} is the surface potential. The force between the tip and surface consists of static, first harmonic, and second harmonic components

$$F_{dc} = \frac{1}{2}\frac{dC}{dz}\left[(V_{dc} - V_{surf})^2 + \frac{1}{2}V_{ac}^2\right] \tag{9.43}$$

$$F_{\omega} = \frac{dC}{dz}(V_{dc} - V_{surf})V_{ac}\sin(\omega t) \tag{9.44}$$

$$F_{2\omega} = -\frac{1}{4}\frac{dC}{dz}V_{ac}^2\cos(2\omega t) \tag{9.45}$$

Using an ac signal *without* the dc component yields dc and 2ω force components, but none at ω. Equation (9.44) shows F_{ω} going to zero when $V_{dc} = V_{surf}$.

The method consists of applying an ac voltage of constant amplitude together with a dc voltage. A lock-in technique allows extraction of the first harmonic signal in the form of the first harmonic tip deflection proportional to F_{ω}. Using a feedback loop the oscillation amplitude is minimized by adjusting V_{dc}. The detection technique is the AFM method with a measure of the feedback voltage V_{dc} being a measure of the surface potential. The null technique renders the measurement independent of dC/dz or to variations in the sensitivity of the system to applied forces. SKPM has also been combined with optical excitation, similar to the surface photovoltage measurements in Fig. 9.6.[66]

The spatial resolution depends on the tip shape, illustrated for the sample in Fig. 9.30 consisting of two regions with surface potentials V_{surf1} and V_{surf2}. The force now is

$$F_{\omega} = \left(\frac{dC_1}{dz}(V_{dc} - V_{surf1}) + \frac{dC_2}{dz}(V_{dc} - V_{surf2})\right)V_{ac}\sin(\omega t) \tag{9.46}$$

The dc tip potential to null the F_{ω} force becomes

$$V_{dc} = \frac{V_{surf1}dC_1/dz + V_{surf2}dC_2/dz}{dC_1/dz + dC_2/dz} \tag{9.47}$$

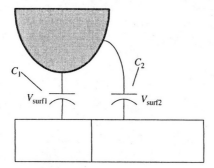

Fig. 9.30   Schematic illustration of a tip near a sample with two surface potentials.

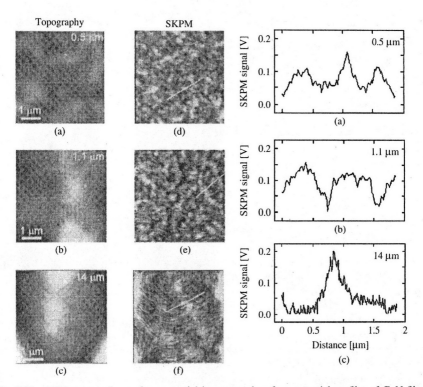

Fig. 9.31   AFM topographs, surface potential images, and surface potential profiles of GaN films 0.5, 1.1, and 14 μm thick. The grey scales correspond to 15 nm for the AFM and 0.1–0.2 V for the surface potential images. After Simpkins et al.[68]

The measured potential depends on the capacitances and surface potentials of the two regions. The measured potential of an area approaches the value of the surrounding surface potential as the area decreases in size.[67] Example AFM and SKPM plots are shown in Figs. 9.31 and 9.32. Fig. 9.31 gives AFM, surface potential maps and surface potential line scans of GaN showing the effect of dislocations.[68] Figure 9.32 is an effective illustration of surface potentials.[69] The AFM topograph (9.32(a)) exhibits no differences associated

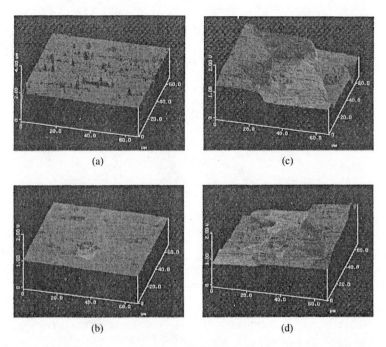

(a) (c)

(b) (d)

Fig. 9.32 (a) ZnO AFM surface topography. (b) SKPM image on grounded surface shows local work-function variations, under lateral (c) positive and (d) negative bias exhibit potential drops at grain boundaries. The direction of potential drops inverts with bias. After ref. 69.

with multiple phases or grain boundaries in this ZnO sample. In the surface potential map with no external perturbation (9.32(b)), a depression of approximately 60 mV is observed due to the difference in work functions of the ZnO surface and pyrochlore phase. The surface potential map with the sample under applied lateral bias shows a potential drop at the grain boundaries in (c) and (d).

9.6.5 Scanning Spreading Resistance Microscopy (SSRM)

Scanning Spreading Resistance Microscopy, based on the atomic force microscope, uses a small conductive tip to measure the local spreading resistance.[70] The resistance is measured between a sharp conductive tip and a large back surface contact. A precisely controlled force is used while the tip is stepped across the sample. SSRM sensitivity and dynamic range are similar to conventional spreading resistance (SRP is discussed in Chapter 1). The small contact size and small stepping distance allow measurements on the device cross section with no probe conditioning. The high spatial resolution allows direct two-dimensional nano-SRP measurements, without the need for special test structures. Spatial resolution of 3 nm has been demonstrated.[71]

For one- or two-dimensional carrier density profile measurements the sample is cleaved to obtain a cross section. The cleavage plane is polished using decreasing grit-size abrasive paper and finally colloidal silica to obtain a flat silicon surface. After polishing the sample is cleaned to eliminate contaminants and finally rinsed in deionized water. The sole limitation is the requirement that the structure be sufficiently wide for the profile in the direction perpendicular to the cross-section of the sample to be uniform.

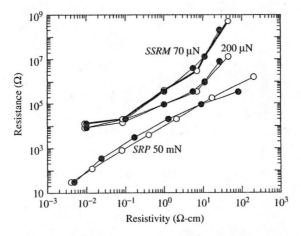

Fig. 9.33 Calibration curves for a W-coated diamond tip on *n*-type (open circles) and *p*-type (filled circles) (100) silicon at a load of 70 μN and 200 μN. Calibration curves for a conventional W/Os SRP probe at a load of 50 mN are given for comparison. After DeWolf et al.[72]

The AFM equipment is standard commercially available equipment. A conductive cantilever with a highly-doped ion-implanted diamond tip can be used as a resistance probe. Diamond protects the tip from deformation due to the rather high loads (~50–100 μN) required to penetrate the native oxide layer and make good electrical contact. Coating the tip with a thin tungsten layer improves the conductivity. Like conventional SRP, nano-SRP needs a calibration curve to convert the measured resistances into carrier densities. The resistance is measured at a bias of ~5 mV, as in conventional SRP. Scanning the tip over the cross section of the sample provides a two-dimensional map of the local spreading resistance with a spatial resolution set by the tip radius of typically 10–15 nm. A straight conversion of spreading resistance to local resistivity is made. Example calibration curves in Fig. 9.33 show the dependence of the measured resistance on probe pressure and their deviation from conventional spreading resistance calibration.[72] To compensate for non-linearities the experimental calibration curve and quantification is based on a look-up procedure using calibration curves. More refined data treatment is required to correct for two-dimensional current spreading effects induced by nearby layers, which are, however, second-order corrections.[73]

As in conventional spreading resistance measurements, a proper model must be used to interpret the experimental data. It is frequently assumed that the contact between the probe and the sample is ohmic. However, it has been shown that the contact is not ohmic.[74] The *I-V* curves vary from an ohmic-like shape in heavily doped areas to a rectifying in lightly doped areas and that surface states induced by the sample preparation influence the *I-V* curves. The presence of surface states due to sample polishing reduces the current, particularly pronounced in lightly-doped areas.

9.6.6 Ballistic Electron Emission Microscopy (BEEM)

Ballistic Electron Emission Microscopy, based on scanning tunneling microscopy is a powerful low-energy tool for non-destructive local characterization of semiconductor heterostructures, such as Schottky diodes.[75] We follow the discussions in ref. 76. A schematic

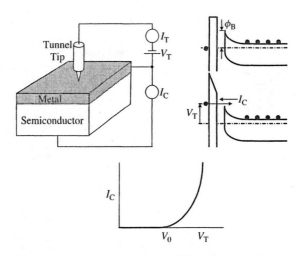

Fig. 9.34 A schematic BEEM experimental setup, band diagrams showing electron emission, and a typical BEEM spectrum, with a threshold voltage V_o corresponding to the Schottky barrier height ϕ_B.

of the BEEM experimental set-up is shown in Fig. 9.34. The BEEM structure is analogous to a bipolar junction transistor. The metal tip, the emitter, injects electrons across the tunneling gap into the metal, the base, deposited on a semiconductor. The substrate, the collector, collects those electrons that have traversed the interface. The emitter or tunnel current I_T is on the order of 1 nA and the collector current I_C is on the order of 10 pA.

A fine metal tip is brought close to the Schottky diode metal and a negative voltage V_T between the tip and the metal gate allows tunnel current I_T to flow by electron tunneling from the negatively biased tip to the metal. This current is the conventional STM current. I_T is held constant while I_C is measured as a function of V_T. Since the scattering mean-free path of the electrons in the metal film is on the order of several nm, some electrons reach the metal/semiconductor interface ballistically for metal films on the order of 10 nm thick. A sufficiently high V_T lifts the tip Fermi level above the barrier height ϕ_B allowing electrons to tunnel through the air gap into the semiconductor leading to BEEM current I_C that depends strongly on the local properties of the interface as well as the scattering properties of the metal film. Varying the tip voltage allows a spectroscopic determination of the Schottky-barrier height with high accuracy from the threshold voltage V_0, in a plot of I_C versus V_T. The lateral resolution is determined by the conditions of tunneling, scattering processes in the metal, and interface transmission. Values below 1 nm can be achieved, providing information on the homogeneity of the interface electronic structure. BEEM can also yield energy-resolved information on hot-electron transport in the metal film, at the interface, and in the semiconductor.

Although BEEM was originally used as a unique microscopic and spectroscopic method to probe Schottky barriers on a local scale, it has been successfully used for heterojunction offsets, resonant transport through single barrier, double-barrier and super-lattice resonant tunneling heterostructures, for investigation of hot carrier transport in low-dimensional nanostructures such as quantum wires and quantum dots, as well as for imaging of defects buried below the surface.

Special care has to be taken when designing the instrument: The main objective for a usual STM is vibration isolation to obtain mechanical-noise levels of less than 0.01 nm for atomic resolution. Such low-level mechanical noise is also required for BEEM, although it does not seem to be required for the typical lateral resolution of BEEM. However, a difference in tip-to-sample spacing by only 0.1 nm results in a variation of the tunnel current by about a factor 10, so that mechanical vibrations yield strong tunnel-current variations.

9.7   STRENGTHS AND WEAKNESSES

Corona Charge: The strength of corona charge based systems is the contactless nature of the measurements allowing some semiconductor processes to be monitored without having to fabricate test structures as well as the variety of semiconductor parameters that can be determined. A weakness is the specialized nature of the equipment not as routinely found as current-voltage or capacitance-voltage systems.

Probe Microscopy: The strength of probe microscopy lies in the variety of possible measurements (topography, electric field, temperature, magnetic field, *etc.*) and their high resolution to atomic scale. Weaknesses include the measurement time and the fragility of the probes, although recent equipment has become automated and is more rugged than early versions.

REFERENCES

1. D.K. Schroder, "Surface Voltage and Surface Photovoltage: History, Theory and Applications," *Meas. Sci. Technol.* **12**, R16–R31, 2001; D.K. Schroder, "Contactless Surface Charge Semiconductor Characterization," *Mat. Sci. Eng.*, **B91-92**, 196–210, 2002.

2. J. Lagowski, P. Edelman, M. Dexter, and W. Henley, "Non-contact Mapping of Heavy Metal Contamination for Silicon IC Fabrication, *Semicond. Sci. Technol.* **7**, A185–A192, 1982.

3. M.S. Fung and R.L. Verkuil, "Contactless Measurement of Silicon Generation Leakage and Crystal Defects by a Corona-Pulsed Deep-Depletion Potential Transient Technique," *Extended Abstracts*, Electrochem. Soc. Meet. Chicago, IL, 1988; R.L. Verkuil and M.S. Fung, "Contactless Silicon Doping Measurements by Means of a Corona-Oxide-Semiconductor (COS) Technique," *Extended Abstracts*, Electrochem. Soc. Meet. Chicago, IL, 1988; M.S. Fung and R.L. Verkuil, "Process Learning by Nondestructive Lifetime Testing," in *Semiconductor Silicon 1990* (H.R. Huff, K.G. Barraclough, and J.I. Chikawa, eds.), Electrochem. Soc., Pennington, NJ, 1990, 924–950; R.L. Verkuil and M.S. Fung, "A Contactless Alternative to MOS Charge Measurements by Means of a Corona-Oxide-Semiconductor (COS) Technique," *Extended Abstracts*, Electrochem. Soc. Meet. Chicago, IL, 1988.

4. W.H. Brattain and J. Bardeen, "Surface Properties of Germanium," *Bell Syst. Tech. J.* **32**, 1–41, Jan. 1953.

5. C.G.B. Garrett and W.H. Brattain, "Physical Theory of Semiconductor Surfaces," *Phys. Rev.* **99**, 376–387, July 1955.

6. T.S. Moss, "Photovoltaic and Photoconductive Theory Applied to InSb," *J. Electron. Ctl.* **1**, 126–138, 1955.

7. W.H. Brattain and C.G.B. Garrett, "Combined Measurements of Field-Effect, Surface Photo-Voltage and Photoconductivity," *Bell Syst. Tech. J.* **35**, 1019–1040, Sept. 1956.

8. S.R. Morrison, "Changes of Surface Conductivity of Germanium with Ambient *J. Phys. Chem.* **57**, 860–863, Nov. 1953.

9. E.O. Johnson, "Measurement of Minority Carrier Lifetime with the Surface Photovoltage," *J. Appl. Phys.* **28**, 1349–1353, Nov. 1957.

10. A. Quilliet and P. Gosar, "The Surface Photovoltaic Effect in Silicon and Its Application to Measure the Minority Carrier Lifetime (in French)," *J. Phys. Rad.* **21**, 575–580, July 1960.

11. A.M. Goodman, "A Method for the Measurement of Short Minority Carrier Diffusion Lengths in Semiconductors," *J. Appl. Phys.* **32**, 2550–2552, Dec. 1961.

12. A.M. Goodman, L.A. Goodman and H.F. Gossenberger, "Silicon-Wafer Process Evaluation Using Minority-Carrier Diffusion Length Measurements by the SPV Method," *RCA Rev.* **44**, 326–341, June 1983.

13. R.S. Nakhmanson, "Frequency Dependence of the Photo-EMF of Strongly Inverted Ge and Si MIS Structures—I. Theory," *Solid-State Electron.* **18**, 617–626, 1975; "Frequency Dependence of the Photo-EMF of Strongly Inverted Ge and Si MIS Structures—II. Experiment," *Solid-State Electron.* **18**, 627–634, July/Aug. 1975.

14. K. Lehovec and A. Slobodskoy, "Impedance of Semiconductor-Insulator-Metal Capacitors," *Solid-State Electron.* **7**, 59–79, Jan. 1964; S.R. Hofstein and G. Warfield, "Physical Limitations on the Frequency Response of a Semiconductor Surface Inversion Layer," *Solid-State Electron.* **8**, 321–341, March 1965; D.K. Schroder, J.E. Park, S.E. Tan, B.D. Choi, S. Kishino, and H. Yoshida, "Frequency-Domain Lifetime Characterization," *IEEE Trans. Electron Dev.* **47**, 1653–1661, Aug. 2000.

15. Lord Kelvin, "On a Method of Measuring Contact Electricity," *Nature*, April 1881; "Contact Electricity of Metals," *Phil. Mag.* **46**, 82–121, 1898.

16. L. Kronik and Y. Shapira, "Surface Photovoltage Phenomena: Theory, Experiment, and Applications", *Surf. Sci. Rep.* **37**, 1–206, Dec. 1999.

17. Semiconductor Diagnostics, Inc. Manual "Contamination Monitoring System Based on SPV Diffusion Length Measurements," SDI, 1993.

18. R.M. Shaffert, *Electrophotography*, Wiley, New York, 1975.

19. R. Williams and A. Willis, "Electron Multiplication and Surface Charge on Zinc Oxide Single Crystals," *J. Appl. Phys.* **39**, 3731–3736, July 1968.

20. R. Williams and M.H. Woods, "High Electric Fields in Silicon Dioxide Produced by Corona Charging," *J. Appl. Phys.* **44**, 1026–1028, March 1973.

21. Z.A. Weinberg, "Tunneling of Electrons from Si into Thermally Grown SiO_2," *Solid-State Electron.* **20**, 11–18, Jan. 1977.

22. M.H. Woods and R. Williams, "Injection and Removal of Ionic Charge at Room Temperature Through the Interface of Air with SiO_2," *J. Appl. Phys.* **44**, 5506–5510, Dec. 1973.

23. R.B. Comizzoli, "Uses of Corona Discharges in the Semiconductor Industry," *J. Electrochem. Soc.* **134**, 424–429, Feb. 1987.

24. D.K. Schroder, M.S. Fung, R.L. Verkuil, S. Pandey, W.H. Howland, and M. Kleefstra, "Corona-Oxide-Semiconductor Device Characterization," *Solid-State Electron.* **42**, 505–512, April 1998.

25. J. Lagowski and P. Edelman, "Contact Potential Difference Methods for Full Wafer Characterization of Oxidized Silicon," presented 7th *Int. Conf. on Defect Recognition and Image Proc.*, 1997.

26. E.O. Johnson, "Large-Signal Surface Photovoltage Studies with Germanium," *Phys. Rev.* **111**, 153–166, July 1958.

27. J. Lagowski, P. Edelman, M. Dexter, and W. Henley, "Non-contact Mapping of Heavy Metal Contamination for Silicon IC Fabrication, *Semicond. Sci. Technol.* **7**, A185–A192, 1982.

28. D.K. Schroder, "The Concept of Generation and Recombination Lifetimes in Semiconductors" *IEEE Trans. Electron Dev.* **ED-29**, 1336–1338, Aug. 1982.

29. S.C. Choo and R.G. Mazur, "Open Circuit Voltage Decay Behavior of Junction Devices," *Solid-State Electron.* **13**, 553–564, May 1970.

30. M.Z. Xu, C.H. Tan, Y.D. He, and Y.Y. Wang, "Analysis of the Rate of Change of Inversion Charge in Thin Insulator p-Type Metal-Oxide-Semiconductor Structures," *Solid-State Electron.* **38**, 1045–1049, May 1995.

31. P. Renaud and A. Walker, "Measurement of Carrier Lifetime: Monitoring Epitaxy Quality," *Solid State Technol.* **43**, 143–146, June 2000.

32. M. Schöfthaler, R. Brendel, G. Langguth, and J.H. Werner, First WCPEC, 1994, 1509.

33. E.H. Nicollian and J.R. Brews, *MOS Physics and Technology*, Wiley, New York, 1982, 63.

34. D.K. DeBusk and A.M. Hoff, "Fast Noncontact Diffusion-Process Monitoring," *Solid State Technol.* **42**, 67–74, April 1999.

35. S.R. Weinzierl and T.G. Miller, "Non-Contact Corona-Based Process Control Measurements: Where We've Been and Where We're Headed," in *Analytical and Diagnostic Techniques for Semiconductor Materials, Devices, and Processes* (B.O. Kolbesen, C. Claeys, P. Stallhofer, F. Tardif, J. Benton, T. Shaffner, D. Schroder, S. Kishino, and P. Rai-Choudhury, eds.), Electrochem. Soc. **ECS 99-16**, 342–350, 1999.

36. K. Nauka and J. Lagowski, "Advances in Surface Photovoltage Techniques for Monitoring of the IC Processes," in *Characterization and Metrology for ULSI Technology: 1998 Int. Conf.* (D.G. Seiler, A.C. Diebold, W.M. Bullis, T.J. Shaffner, R. McDonald, and E.J. Walters, eds.), Am. Inst. Phys. 245–249, 1998; M.S. Fung, "Monitoring PSG Plasma Damage with COS," *Semicond. Int.* **20**, 211–218, July 1997.

37. K. Nauka, "Contactless Measurement of the Si-Buried Oxide Interfacial Charges in SOI Wafers with Surface Photovoltage Technique," *Microelectron. Eng.* **36**, 351–357, June 1997.

38. P.K. Roy, C. Chacon, Y. Ma, I.C. Kizilyalli, G.S. Horner, R.L. Verkuil, and T.G. Miller, "Non-Contact Characterization of Ultrathin Dielectrics for the Gigabit Era," in *Diagnostic Techniques for Semiconductor Materials and Devices* (P. Rai-Choudhury, J.L. Benton, D.K. Schroder, and T.J. Shaffner, eds.), Electrochem. Soc. **PV97-12**, 280–294, 1997.

39. T.G. Miller, "A New Approach for Measuring Oxide Thickness," *Semicond. Int.* **18**, 147–148, 1995.

40. S.H. Lo, D.A. Buchanan, and Y. Taur, "Modeling and Characterization of Quantization, Polysilicon Depletion, and Direct Tunneling Effects in MOSFETs with Ultrathin Oxides," *IBM J. Res. Dev.* **43**, 327–337, May 1999.

41. Z.A. Weinberg, W.C. Johnson, and M.A. Lampert, "High-Field Transport in SiO_2 on Silicon Induced by Corona Charging of the Unmetallized Surface," *J. Appl. Phys.* **47**, 248–255, Jan. 1976.

42. D.A. Bonnell, *Scanning Probe Microscopy and Spectroscopy*, 2nd Ed., Wiley-VCH, New York, 2001.

43. G. Binnig, H. Rohrer, C. Gerber, and E. Weibel, "Surface Studies by Scanning Tunneling Microscopy," *Phys. Rev. Lett.* **49**, 57–60, July 1982; G. Binnig and H. Rohrer, "Scanning Tunneling Microscopy," *Surf. Sci.* **126**, 236–244, March 1983.

44. R. Young, J. Ward, and F. Scire, "The Topografiner: An Instrument for Measuring Surface Microtopography," *Rev. Sci. Instrum.* **43**, 999–1011, July 1972.

45. T.J. Shaffner, "Characterization Challenges for the ULSI Era," in *Diagnostic Techniques for Semiconductor Materials and Devices* (P. Rai-Choudhury, J.L. Benton, D.K. Schroder, and T.J. Shaffner, eds.), Electrochem. Soc., Pennington, NJ, 1997, 1–15.

46. R.J. Hamers and D.F. Padowitz, "Methods of Tunneling Spectroscopy with the STM," in *Scanning Probe Microscopy and Spectroscopy*, 2nd Ed., (D. Bonnell, ed.), Wiley-VCH, New York, 2001, Ch. 4.

47. R.L. Smith and G.S. Rohrer, "The Preparation of Tip and Sample Surfaces for Scanning Probe Experiments," in *Scanning Probe Microscopy and Spectroscopy*, 2nd Ed., (D. Bonnell, ed.), Wiley-VCH, New York, 2001, Ch. 6.

48. E. Meyer, H.J. Hug, and R. Bennewitz, *Scanning Probe Microscopy*, Springer, Berlin, 2004.

49. J. Simmons, "Generalized Formula for the Electric Tunnel Effect Between Similar Electrodes Separated by a Thin Insulating Film," *J. Appl. Phys.* **34**, 1793–1803, June 1963.

50. G. Binnig, C.F. Quate, and Ch. Gerber, "Atomic Force Microscope," *Phys. Rev. Lett.* **56**, 930–933, March 1986.

51. C.F. Quate, "The AFM as a Tool for Surface Imaging," *Surf. Sci.* **299–300**, 980–95, Jan. 1994.

52. D. Sarid, *Scanning Force Microscopy with Applications to Electric, Magnetic, and Atomic Forces*, Revised Edition, Oxford University Press, New York, 1994.

53. G. Meyer and N.M. Amer, "Novel Optical Approach to Atomic Force Microscopy," *Appl. Phys. Lett.* **53**, 1045–1047, Sept. 1988.

54. Q. Zhong, D. Inniss, K. Kjoller, and V.B. Elings, "Fractured Polymer/Silica Fiber Surface Studied by Tapping Mode Atomic Force Microscopy," *Surf. Sci. Lett.* **290**, L668–L692, 1993.

55. Y. Huang and C.C. Williams, "Capacitance-Voltage Measurement and Modeling on a Nanometer Scale by Scanning C-V Microscopy," *J. Vac. Sci. Technol.* **B12**, 369–372, Jan./Feb. 1994.

56. G. Neubauer, A. Erickson, C.C. Williams, J.J. Kopanski, M. Rodgers, and D. Adderton, "Two-Dimensional Scanning Capacitance Microscopy Measurements of Cross-Sectioned Very Large Scale Integration Test Structures," *J. Vac. Sci. Technol.* **B14**, 426–432, Jan./Feb. 1996; J.S. McMurray, J. Kim, and C.C. Williams, "Quantitative Measurement of Two-dimensional Dopant Profile by Cross-sectional Scanning Capacitance Microscopy," *J. Vac. Sci. Technol.* **B15**, 1011–1014, July/Aug. 1997.

57. J.R. Matey and J. Blanc, "Scanning Capacitance Microscopy," *J. Appl. Phys.* **57**, 1437–1444, March 1985.

58. C.C. Williams, W.P. Hough, and S.A. Rishton, "Scanning Capacitance Microscopy on a 25 nm Scale," *Appl. Phys. Lett.* **55**, 203–205, July 1989.

59. C.Y. Nakakura, P. Tangyunyong, D.L. Hetherington, and M.R. Shaneyfelt, "Method for the Study of Semiconductor Device Operation Using Scanning Capacitance Microscopy," *Rev. Sci. Instrum.* **74**, 127–133, Jan. 2003.

60. J.K. Clemens, "Capacitive Pickup and Buried Subcarrier Encoding System for RCA Videodisc," *RCA Rev.* **39**, 33–59, Jan. 1978; R.C. Palmer, E.J. Denlinger, and H. Kawamoto, "Capacitive Pickup Circuitry for Videodiscs," *RCA Rev.* **43**, 194–211, Jan. 1982.

61. J.J. Kopanski, J.F. Marchiando, and J.R. Lowney, "Scanning Capacitance Microscopy Measurements and Modeling: Progress Towards Dopant Profiling of Silicon," *J. Vac. Sci. Technol.* **B14**, 242–247, Jan./Feb. 1996.

62. C.C. Williams, Two-Dimensional Dopant Profiling by Scanning Capacitance Microscopy," *Annu. Rev. Mater. Sci.* **29**, 471–504, 1999.

63. S.V. Kalinin and D.A. Bonnell, "Electrostatic and Magnetic Force Microscopy," in *Scanning Probe Microscopy and Spectroscopy*, 2nd Ed., (D. Bonnell, ed.), Wiley-VCH, New York, 2001, Ch. 7.

64. M. Nonnenmacher, M.P. Boyle, and H.K. Wickramasinghe, "Kelvin Probe Microscopy," *Appl. Phys. Lett.* **58**, 2921–2923, June 1991.

65. R.P. Feynman, R.B. Leighton, and M. Sands, *The Feynman Lectures on Physics*, Vol. 2, Addison-Wesley, Reading, MA, 1964, 8-2–8-4.

66. J.M.R. Weaver and H.K. Wickramasinghe, "Semiconductor Characterization by Scanning Force Microscope Surface Photovoltage Microscopy," *J. Vac. Sci. Technol.* **B9**, 1562–1565, May/June 1991.

67. H.O. Jacobs, H.F. Knapp, S. Müller, and A. Stemmer, "Surface Potential Mapping: A Qualitative Material Contrast in SPM," *Ultramicroscopy*, **69**, 39–49, 1997.

68. B.S. Simpkins, D.M. Schaadt, E.T. Yu, and R.J. Molner, "Scanning Kelvin Probe Microscopy of Surface Electronic Structure in GaN Grown by Hydride Vapor Phase Epitaxy," *J. Appl. Phys.* **91**, 9924–9929, June 2002.

69. D.A. Bonnell and S. Kalinin, "Local Potential at Atomically Abrupt Oxide Grain Boundaries by Scanning Probe Microscopy," *Proc. Int. Meet. on Polycryst. Semicond.* (O. Bonnaud, T. Mohammed-Brahim, H.P. Strunk, and J.H. Werner, eds.) in *Solid State Phenomena*, Scitech Publ. Uettikon am See, Switzerland, 33–47, 2001.

70. W. Vandervorst, P. Eyben, S. Callewaert, T. Hantschel, N. Duhayon, M. Xu, T. Trenkler and T. Clarysse, "Towards Routine, Quantitative Two-dimensional Carrier Profiling with Scanning

Spreading Resistance Microscopy," in *Characterization and Metrology for ULSI Technology,* (D.G. Seiler, A.C. Diebold, T.J. Shaffner, R. McDonald, W.M. Bullis, P.J. Smith, and E.M. Secula, eds.), Am. Inst. Phys. **550**, 613–619, 2000.

71. P. Eyben, N. Duhayon, D. Alvarez, and W. Vandervorst, "Assessing the Resolution Limits of Scanning Spreading Resistance Microscopy and Scanning Capacitance Microscopy," in *Characterization and Metrology for VLSI Technology: 2003 Int. Conf.,* (D.G. Seiler, A.C. Diebold, T.J. Shaffner, R. McDonald, S. Zollner, R.P. Khosla, and E.M. Secula, eds.) Am. Inst. of Phys. **683**, 678–684, 2003.

72. P. De Wolf, T. Clarysse, W. Vandervorst, J. Snauwaert and L. Hellemans, "One- and Two-dimensional Carrier Profiling in Semiconductors by Nanospreading Resistance Profiling," *J. Vac. Sci. Technol.* **B14**, 380–385, Jan-Feb. 1996.

73. P. De Wolf, T. Clarysse and W. Vandervorst, "Quantification of Nanospreading Resistance Profiling Data," *J. Vac. Sci. Technol.* **B16**, 320–326, Jan./Feb. 1998.

74. P. Eyben, S. Denis, T. Clarysse, and W. Vandervorst, "Progress Towards a Physical Contact Model for Scanning Spreading Resistance Microscopy," *Mat. Sci. Eng.* **B102**, 132–137, 2003.

75. W.J. Kaiser and L.D. Bell, "Direct Investigation of Subsurface Interface Electronic Structure by Ballistic-Electron-Emission Microscopy," *Phys. Rev. Lett.* **60**, 1406–1410, April 1988.

76. M. Prietsch, "Ballistic-Electron Emission Microscopy (BEEM): Studies of Metal/Semiconductor Interfaces With Nanometer Resolution," *Phys. Rep.* **253**, 163–233, 1995; L.D. Bell and W.J. Kaiser, "Ballistic Electron Emission Microscopy: A Nanometer-Scale Probe of Interfaces and Carrier Transport," *Ann. Rev. Mater. Sci.* **26**, 189–222, 1996; V. Narayanamurti and M. Kozhevnikov, BEEM Imaging and Spectroscopy of Buried Structures in Semiconductors," *Phys. Rep.* **349**, 447–514, 2001.

PROBLEMS

9.1 A positive charge density of 3×10^{-7} C/cm^2 is deposited on a p-type semiconductor surface doped to $N_A = 10^{15}$ cm^{-3}. Compute the surface potential and the charge-induced space-charge region width.

9.2 Does the potential measured with a Kelvin probe change as the Kelvin probe spacing above the sample surface is changed? Discuss.

9.3 A capacitor consisting of two parallel metal plates with an air dielectric has a voltage V_1 applied across the plates. Charge Q_1 and $-Q_1$ exists on the plates. Then a dielectric with dielectric constant >1 is inserted into the gap between the plates. With the voltage remaining at V_1, does the charge and the capacitance change? Does it increase, decrease, or remain the same?

9.4 Draw a band diagram similar to Fig. 9.7 for positive and negative surface charge on an n-type substrate.

9.5 Charge is pulse deposited onto a p-type wafer as in Fig. 9.15(b). The band diagram immediately after the charge pulse is shown in Fig. P9.5. Draw the band diagram when, due to thermal electron-hole pair generation, the space-charge region width has decreased, but has not yet reached equilibrium.

9.6 Compute the maximum charge density, in C/cm^2 and in cm^{-2}, that can be deposited onto an oxidized Si wafer. The limit is the oxide breakdown electric field of 1.5×10^7 V/cm. Does it matter how thick the oxide is?

9.7 10^{12} cm^{-2} positive charges are deposited on gates 1, 2, and 3 in Fig. P9.7. The oxide has a dielectric constant of 4. Compute the oxide electric field and the oxide voltage

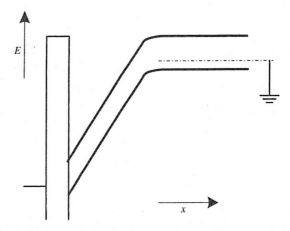

Fig. P9.5   Schematic illustration of a tip near a sample with two surface potentials.

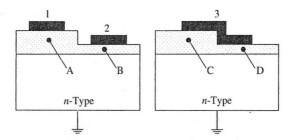

Fig. P9.7   Schematic illustration of a tip near a sample with two surface potentials.

in regions A, B, C, and D. $t_{ox}(A) = t_{ox}(C) = 100$ nm, $t_{ox}(B) = t_{ox}(D) = 50$ nm. Neglect any voltage drop across the semiconductor. The gate areas on thick and thin oxide are identical.

9.8 Compute and plot the scanning tunneling microscope tunnel current for an air gap width of 0.2 to 2 nm. Plot is as $I/C_1 V$ versus d, according to Eq. (9.37). The work function of the probe and the sample is 4 eV.

REVIEW QUESTIONS

- How is surface charging done?
- How does a Kelvin probe work?
- Does the Kelvin probe voltage depend on the distance between the probe and the surface?
- Name an advantage of generation lifetime measurement using corona charge compared to a conventional gate.
- How can charge be used to vary the effective recombination lifetime?
- How is the oxide thickness determined with the charge-based method?

- How does a scanning tunneling microscope work?
- How does an atomic force microscope work?
- What is "tapping" mode in an AFM?
- How is the force determined in scanning Kelvin probe microscopy?
- What is ballistic electron emission microscopy (BEEM)?
- What is measured with BEEM?

10

OPTICAL CHARACTERIZATION

10.1 INTRODUCTION

Optical measurements are attractive because they are almost always non-contacting with minimal sample preparation—a major advantage when contacts are detrimental. The instrumentation for many optical techniques is commercially available and is often automated. The measurements can have very high sensitivity. The main concepts are discussed in this chapter with some of the details left to the published literature. An overview of optical measurements is given by Herman.[1]

Optical measurements fall into three broad categories (1) *photometric* measurements (amplitude of reflected or transmitted light is measured), (2) *interference* measurements (phase of reflected or transmitted light is measured), and (3) *polarization* measurements (ellipticity of reflected light is measured). The main optical techniques are summarized in Fig. 10.1. Light is either reflected, absorbed, emitted, or transmitted. Most of the techniques in that figure are discussed here; some have been discussed in earlier chapters (*e.g.*, photoconductance) and some are not discussed at all (*e.g.*, ultraviolet photoelectron spectroscopy). For completeness we also discuss several non-optical film thickness and line width methods in this chapter.

Optical measurements use the ultraviolet to the far infrared region of the electromagnetic spectrum. Parameters are wavelength (λ), energy (E or $h\nu$), and wavenumber (WN). The most common units are: *wavelength* in nanometer (1 nm = 10^{-9} m = 10^{-7} cm = 10^{-3} μm), Ångström (1 Å = 10^{-10} m = 10^{-8} cm = 10^{-4} μm) or micrometer (1 μm = 10^{-6} m = 10^{-4} cm); *energy* in electron volt (1 eV = 1.6×10^{-19} J), and *wavenumber* in inverse wavelength (1 WN = $1/\lambda$). The relationship between energy and wavelength is

$$E = h\nu = \frac{hc}{\lambda} = \frac{1.2397 \times 10^3}{\lambda(nm)} = \frac{1.2397 \times 10^4}{\lambda(\text{Å})} = \frac{1.2397}{\lambda(\mu\text{m})} \ [eV] \qquad (10.1)$$

Semiconductor Material and Device Characterization, Third Edition, by Dieter K. Schroder
Copyright © 2006 John Wiley & Sons, Inc.

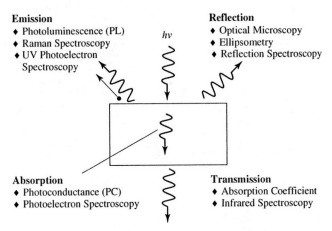

Fig. 10.1   Optical characterization techniques.

10.2   OPTICAL MICROSCOPY

The compound optical microscope is one of the most versatile and useful instruments in a semiconductor laboratory. Many of the features of integrated circuits and other semi-conductor devices are sufficiently gross to be seen through such a microscope. However, optical microscopy becomes useless as feature sizes shrink to the sub-micron regime. Typically optical microscopy remains useful for feature sizes above about 0.5 μm. For smaller sizes, electron beam microscopes become useful. The basic optical microscope can be enhanced by adding phase and differential interference contrast as well as polariz-ing filters. Optical microscopy is not only used to view the features of integrated circuits; it is also useful for analyzing particles found on such circuits. To identify and analyze particles requires a skilled and practiced microscopist. The technique is most useful for particles larger than one micron and the analysis depends on matching the unknown with data on known particles. Particle atlases are available to aid in identification.[2]

The essential elements of a compound optical microscope are illustrated in Fig. 10.2. Its optical elements, the *objective* and the *ocular* or *eyepiece,* are shown as simple lenses; in modern microscopes they consist of six or more highly corrected compound lenses. Object *O* is placed just beyond the first focal point f_{obj} of the objective lens that forms a real and enlarged image *I*. This image lies just within the first focal point f_{oc} of the ocular, forming a virtual image of *I* at *I'*. A virtual image is an image that does not actually exist and cannot be observed on a screen, for example. The position of *I'* may lie anywhere between near and far points of the eye. The objective merely forms an enlarged real image which is examined by the eye looking through the ocular. The overall magnification *M* is a product of the lateral magnification of the objective and the angular magnification of the ocular. The simplest microscope is the monocular microscope, with only one eyepiece. The binocular instrument has two eyepieces to make viewing of the sample more convenient. When one objective is used with a binocular microscope, the observed image is generally not stereoscopic. For stereoscopic viewing, one uses a stereo microscope consisting of two compound microscopes arranged so that each eye has its own individual view of the same field. Through the use of prisms in each microscope, erect images are presented to the eyes.

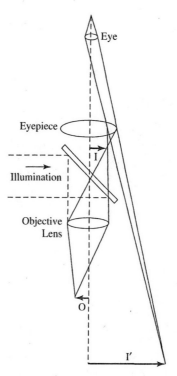

Fig. 10.2 Simplified representation of a compound microscope's optical paths.

10.2.1   Resolution, Magnification, Contrast

Light can be thought of as waves as well as particles. To explain some experimental results it is easier to use the wave concept, while for others the particle concept is more useful. Waves interfere with one another placing certain limits on the performance of microscopes. Airy[3] first computed the diffracted image and showed in 1834 that for diffraction at a circular aperture of diameter d, the angular position of the first minimum (measured from the center) is given by (see Fig. 10.3(a))[4]

$$\sin(\alpha) = \frac{1.22\lambda}{d} \tag{10.2}$$

where λ is the wavelength of light in free space. The central spot containing most of the light is called the Airy or diffraction disc. You can do your own experiment by looking at a bright point source at a distance of several meters, for example, a microscope lamp, through a small pinhole in a cardboard sheet. The same kind of pattern is formed when a point object is imaged by a microscope. There is no lower limit to the size of an object that can be detected *in isolation*, given adequate illumination.

Generally one is not interested in detecting a point object, but a two- or three-dimensional object. Two point objects, a distance s apart, produce overlapping images, as shown in Fig. 10.3(b). If they are too close, it is impossible to resolve them. Raleigh suggested that two objects can be distinguished when the central maximum of one coincides

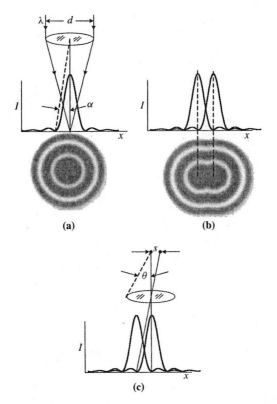

Fig. 10.3 (a) Diffraction at the aperture of a lens showing the Airy disc and (b) the Raleigh criterion for resolution, (c) the resolution limit of an optical microscope. I represents the intensity. Reprinted with permission after Spencer.[3]

with the first minimum of the other. The intensity between the two peaks then decreases to 80% of the peak height, as shown in Fig. 10.3(c). The equation

$$s = \frac{0.61\lambda}{n \sin(\theta)} = \frac{0.61\lambda}{NA} \tag{10.3}$$

gives the *resolution* (the minimum distance between points or parts of an object) that satisfies Raleigh's criterion, where n is the refractive index of the medium separating the object from the objective and θ is the half angle subtended by the lens at the object. Sometimes the intensity at 50% of the peak height is used as the resolution limit and the "0.61" in Eq. (10.3) becomes "0.5".

The *numerical aperture* (*NA*), usually engraved on the objective mount, is a number that expresses the resolving power of the lens and the brightness of the image it forms. Equation (10.3) is sometimes expressed in terms of the $f/\#$ of the lens as

$$s = \frac{1.22\lambda f/\#}{n} \tag{10.4}$$

The higher the NA, the higher is the lens quality. For high resolution, *i.e.*, small s, NA should be made as large as possible. However, high NA corresponds not only to high resolution, but also to shallow depth of field and shallow working distance—the distance from the focus point of the object plane to the front surface of the objective. The *depth of focus* D_{focus}—the thickness of the image space that is simultaneously in focus—is given by

$$D_{focus} = \frac{\lambda}{4NA^2} \tag{10.5}$$

D_{focus} is insufficient for both the top and bottom surfaces of an integrated circuit to be in focus simultaneously at 200× magnification. The *depth of field* D_{field}—the thickness of the object space that is simultaneously in focus—is given by

$$D_{field} = \frac{\sqrt{n^2 - NA^2}}{NA^2}\lambda = \frac{\sqrt{(n/NA)^2 - 1}}{NA}\lambda \tag{10.6}$$

Both D_{focus} and D_{field} decrease with increasing NA, but the resolution increases.

According to Eq. (10.3) three variables may be adjusted to reduce s or increase the resolution. The wavelength may be reduced Blue light has higher resolution than red light. One frequently uses a green filter with its transmission peak at the wavelength for which the objective is chromatically corrected and the eye is most sensitive and leads to the least eye fatigue. The resolution may be improved by increasing the angle θ toward the theoretical maximum of 90°. $NA \approx 0.95$ is the upper practical limit. Beyond this, further gain in resolution is achieved by use of immersion objectives in which a fluid with higher index of refraction than air is placed between the sample and the front lens of the objective. With air as the immersion medium, the numerical aperture is sometimes referred to as "dry" NA. Immersion fluids can be water ($n = 1.33$), glycerin ($n = 1.44$), oil ($n = 1.5$–1.6), cargille ($n = 1.52$), or monobromonaphthalene ($n = 1.66$). Water is frequently used in immersion optical lithography. Practical limits of $NA \approx 1.3$–1.4 for oil-immersion optics limit the resolution to $s \approx 0.25$ μm for green light with $\lambda \approx 0.5$ μm.

Magnification M is related to the resolving power of the microscope objective and the eye. However, the image must be magnified sufficiently for detail to be visible to the eye. The *resolving power* is the ability to reveal detail in an object by means of the eye, microscope, camera, or photograph. An approximate relationship for the magnification is[5]

$$M = \frac{maximum\ NA\ of\ microscope}{minimum\ NA\ of\ eye} \approx \frac{1.4}{0.002} = 700 \tag{10.7}$$

Magnification is sometimes expressed as the ratio of the resolution limits

$$M = \frac{limit\ of\ resolution\ (eye)}{limit\ of\ resolution\ (microscope)} \approx \frac{200\mu m}{0.61\lambda/NA} \approx \frac{200\mu m}{0.25\mu m}NA = 800\ NA \tag{10.8}$$

where the eye resolution is related to the distance between the rod and cone receptors on the retina of the eye. The maximum magnification of a microscope when the image is viewed by the eye is around 750×. Magnification above this is *empty magnification*, giving no additional information. It is useful when the light detector is not the eye, but photographic film or photodetectors; then higher magnification than that implied by the equations above is possible.

The eye fatigues easily if used at its limits of resolving power so it is desirable to supply more magnification than the minimum required for convenience. A reasonable rule is to make the magnification about 750 *NA*, but one should always use the lowest magnification that permits comfortable viewing. Excessive magnification produces images of lower brilliance and poorer definition and the object detail that can be seen is frequently reduced.

Contrast—the ability to distinguish between parts of an object—depends on many factors. Dirty eyepieces or objectives degrade image quality. Glare will reduce contrast, especially if the sample is highly reflecting. It is most serious when viewing samples with little contrast and can be controlled to some extent by controlling the field diaphragm, the opening that controls the area of the lighted region. The diaphragm should never be open more than just enough to illuminate the complete field of the microscope. For critical cases it may be reduced to illuminate only a small portion of the normal field.

10.2.2 Dark-Field, Phase, and Interference Contrast Microscopy

In *bright field microscopy*, the light impinges vertically on the sample. Horizontal surfaces reflect most of the light while slanted or vertical surfaces reflect less, illustrated in Fig. 10.4(a), resulting in the intensity *I* scan. In *dark-field microscopy*, the light impinges at a shallow angle on the sample as illustrated in Fig. 10.4(b). Light from horizontal sample surfaces does not reach the lens, but light from slanted and vertical surfaces does. The image contrast is the reverse of that of bright light imaging. Dark field microscopy is especially useful to bring out small surface irregularities that are difficult or impossible to see with bright light microscopy. Dark field microscopy is akin to seeing dust in the air in a darkened room when bright sunlight falls into the room, scattering light from dust particles.

In *phase contrast microscopy* one makes use of the phase shift which can occur when light is transmitted through or reflected from a sample. Phase shifts arise when the sample consists of regions of differing refractive indices, when the path length through the sample varies (in transmission), or when there are changes in sample surface height (in reflection). We illustrate the principle of phase contrast microscopy in Fig. 10.5.[6] Let us first consider amplitude contrast microscopy in Fig. 10.5(a). Light of amplitude A_1 is incident on a sample. Some of the light is absorbed, which can be considered as being scattered or

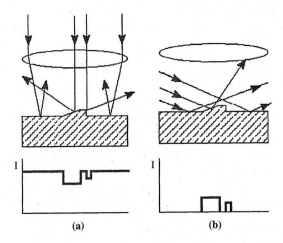

Fig. 10.4 Schematic illustration of (a) bright and (b) dark field imaging. The light intensity is shown in the lower part.

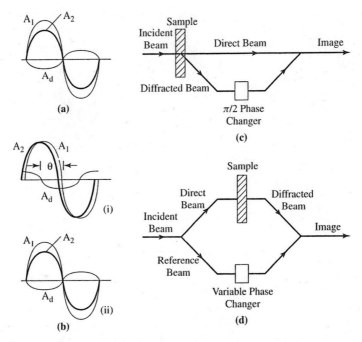

Fig. 10.5 (a) Amplitude contrast, (b) phase contrast, (c) phase contrast, and (d) interference contrast.

diffracted. Diffracted light is π or $\lambda/2$ out of phase with the incident light. The diffracted light is shown as A_d. These two waves interfere, giving the resultant wave of amplitude $A_2 = A_1 - A_d$. Now consider the case of phase contrast microscopy in Fig. 10.5(b). The incident light has amplitude A_1 and the diffracted wave is A_d. The reflected wave has amplitude A_2, identical to A_1 (we assume zero absorption in this case) but retarded by phase angle $\theta = \pi/2$, as shown in Fig. 10.5(b)(i). Since there is only a phase change, but not an amplitude change the eye cannot detect the difference between A_1 and A_2. If now A_d is retarded by a further $\pi/2$, we get the case of Fig. 10.5(b)(ii) which is identical to Fig. 10.5(a). Now there is a difference in amplitude between A_1 and A_2. In other words, the phase difference has become an amplitude difference that can be observed by the eye or other detector. Phase contrast microscopy is schematically illustrated in Fig. 10.5(c).

The fundamental differences between phase contrast and interference contrast microscopy are shown in Fig. 10.5(c) and (d). In phase contrast, the incident light is split by the sample into direct and diffracted beams. The phase of the diffracted beam is changed by $\pi/2$, the beams recombine in the image plane and their interference gives amplitude contrast relative to the background. In interference contrast, the incident beam is split into a direct and a reference beam. The direct beam is altered by the sample and the phase of the reference beam is adjusted. When the reference and diffracted beams recombine, interference between them gives an amplitude contrast image. The system allows for appropriate phase adjustment for optimum contrast. Both edges of a raised portion of a sample can appear bright against a dark background or dark against a light background for monochromatic light. For white light, edges of one color appear against a

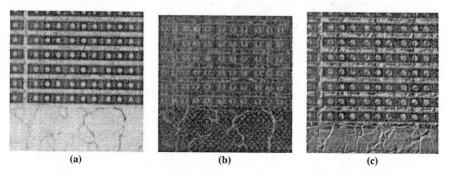

(a)   (b)   (c)

Fig. 10.6 Micrographs of an integrated circuit for reflected light (a) bright field, (b) dark field, and (c) differential interference contrast. $100 \times$ objective, $1.5 \times$ zoom, $10 \times$ magnification to camera. Courtesy of T. Wetteroth, Motorola Semiconductor.

different colored background. Other effects can be produced with one edge brighter and the other darker than the background, giving an interesting but strictly spurious impression of "shadowing." Step heights as small as 3 nm can be observed, making this technique suitable for measurements of planarity of wafer surface and etch pit studies.

Interference contrast images are generally sharper than phase contrast images. Interference contrast is also more sensitive to gradual topographical sample changes. The technique is also known as differential interference contrast. There are various implementations of this technique, generally based on the *Nomarski* system.[7] Figure 10.6 shows a comparison of bright-field, dark-field, and differential interference contrast micrographs. Richardson gives a good discussion of microscopy with many examples.[8]

10.2.3 Confocal Optical Microscopy

Confocal optical microscopy, invented in 1955,[9] is a method of generating three-dimensional images of an object and increasing the contrast of microscopic images.[10-11] By restricting the observed volume, compared to conventional microscopy, it keeps overlying or nearby scatterers from contributing to the detected signal. However, it only images one point at a time and a complete image is built up by scanning a light beam across the sample. The resolution of a confocal microscope is[12]

$$s = \frac{0.44\lambda}{n \sin(\theta)} = \frac{0.44\lambda}{NA} \tag{10.9}$$

which is slightly better than Eq. (10.3) because the confocal diffraction pattern has less energy outside the central peak than does the single lens pattern, and the resolution is less degraded by sample contrast variations.

To see how a confocal microscope works, consider image formation of a point in Fig. 10.7(a). Point A is focused in focal plane A, while point B is focused in plane B. The microscope objective lens forms an image at the pinhole plane, *i.e.*, the sample plane and the pinhole plane are conjugate planes. The pinhole is conjugate to the **focal** point of the lens, *i.e.*, it is a **confocal** pinhole. When a pinhole is placed at plane A (Fig. 10.7(b)), the light from point A passes through the pinhole, while most of the light from point B does not. In order for light from point B to pass through the pinhole, one raises the

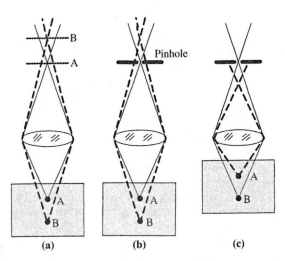

Fig. 10.7 (a) Point A is focused at plane A and point B at plane B, (b) pinhole is placed at plane A, (c) raising the sample places point B at the pinhole.

sample for point B to be in focus, in Fig. 10.7(c). Now, of course, most of the light from point A does not pass through the pinhole. In other words, only one plane is in focus at a time. By scanning the light across the sample surface, a two-dimensional image of one plane is generated. Then by raising the sample, the next plane is imaged, and so on, until a three-dimensional image of the entire sample is created. Instead of moving the sample, one can also move the objective lens with a piezoelectric transducer. At the same time, conventional optical microscopy allows the eye to view the sample.

Two systems are used to generate two-dimensional pictures. In the first, the laser light source deflects off the dichroic mirror and is then deflected by two scanning mirrors across the sample in Fig. 10.8(a) The light reflected from the sample is deflected by the same scanning mirrors, passes through the dichroic mirror and then through the pinhole to be detected by a photomultiplier or charge-coupled device. The image is built up one pixel

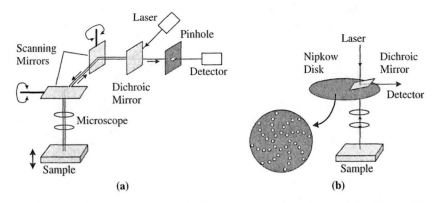

Fig. 10.8 Schematic scanning confocal microscope (a) scanning mirrors and (b) Nipkow disk.

at a time. Instead of using scanning mirrors, one can also use acousto-optic deflection for the fast scan direction. Scanning the sample has the advantage that no light reaches the detector from areas outside the beam and the contrast is not degraded by unwanted background light. It has the disadvantage that only a small portion of the sample is illuminated.

The second method uses the Nipkow disk, invented in 1884[13] and later adapted to the confocal microscope.[14] The disk is a mechanically spinning disk with a series of equally distanced holes in Fig. 10.8(b). When the disk rotates, the holes trace circular ring surfaces. Each hole in the spiral takes a horizontal "slice" through the image which is picked up as a pattern of light and dark by a sensor. To increase the light through the pinholes, one can use two disks with the second disk containing microlenses to focus the excitation light. The disk contains thousands of pinholes that are on the order of 20 µm in diameter. Confocal microscopy is used for scanning through the sample depth in biological specimen and for investigating various heights in integrated circuits, for example.[15]

10.2.4 Interferometric Microscopy

Interferometric microscopy is a contactless method for determining horizontal and vertical features of a sample. Quantitative vertical features are determined through phase-shift interferometry (PSI). In a typical application, the sample is imaged through a microscope giving a maximum lateral, *i.e.*, x and y, resolution of conventional microscopy of $\sim 0.5\lambda/NA$. The z resolution, however, is determined by the ability to interpret fringes using phase modulation techniques. The vertical resolution is around 1 nm. A good source for interference microscopy and other optical measurements is the book *Optical Shop Testing*.[16]

Exercise 10.1

Question: What is interference?

Answer: Consider the two optically flat pieces of glass forming an angle α in Fig. E10.1 Monochromatic light of wavelength λ is incident. The air gap spacing is αx, where x is the distance from the line of intersection and the optical path distance is $2\alpha x$, since the light travels twice through the air gap. When light travels from a low-n to a high-n

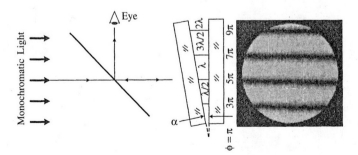

Fig. E10.1

material, there is a phase shift π, *e.g.*, at the bottom plane, leading to an optical path difference of $2\alpha x + \lambda/2$. Dark fringes occur at

$$2\alpha x = m\lambda$$

where m is an integer, and bright fringes occur at

$$2\alpha x + \lambda/2 = m\lambda$$

In each case the fringe separation is

$$d = \frac{\lambda}{2\alpha}$$

In interferometric microscopy, the reflected light passes through an interference objective, giving an image containing height contours. For monochromatic light of wavelength λ, the resultant intensity I of two interfering waves of intensity I_o in the image plane is[17]

$$I = KI_o\left[1 + \cos\left(\frac{4\pi}{\lambda}h(x, y) + \delta\phi\right)\right] \tag{10.10}$$

where K is a constant, $h(x,y)$ is a comparison of the sample height with a reference mirror and $\delta\phi$ is a phase change introduced into one of the optical paths to help in fringe analysis. The phase is changed by varying the vertical displacement of the sample by a piezoelectric crystal or a stepping motor. Using three phase steps of $-120°$, $0°$, and $120°$, the resulting three images become[18]

$$I_1 = C[1 + \cos(\phi - 120°)], I_2 = C[1 + \cos(\phi)], I_3 = C[1 + \cos(\phi + 120°)] \tag{10.11}$$

where C is a constant. From these three equations the height $h(x,y)$ is

$$h(x, y) = \frac{1}{4\pi}\arctan\left(\frac{-\sqrt{3}(I_1 - I_3)}{2I_2 - I_1 - I_3}\right) \tag{10.12}$$

The resulting interferogram consists of light and dark fringes, which when aligned parallel to the sample surface, represent height contours separated by $\lambda/2$. Phase modulation techniques allow phase calculations to better than 0.01 of two neighboring fringes or a vertical resolution of 0.1–1 nm. From Eq. (10.12), the height of the surface at each x,y location can be calculated and represented as a gray scale. Because the solution to the arctan() expression has values between $-\pi/2$ and $\pi/2$, a discontinuity occurs every change in phase of π in the interferometer or every change in height of $\lambda/4$ (165 nm for $\lambda = 660$ nm, red light). Hence, PSI is unable to determine heights beyond $\lambda/4$ unambiguously. An independent measurement must be used to determine which order applies, *i.e.*, how many $\lambda/4$ increments, to determine sample heights greater than $\lambda/4$.

One method of removing this height ambiguity is to make these measurements at two wavelengths λ_1 and λ_2 and subtract the two. Now the limitation in height difference is $\lambda_e/4$, where λ_e is an effective wavelength given by $\lambda_e = \lambda_1\lambda_2/|\lambda_1 - \lambda_2|$. A disadvantage of this approach is a degradation in the precision of the measurement by the ratio λ_e/λ. Multiple wavelength interferometry works well on some samples, but for rough samples noisy data points lead to errors. For microscopes with high magnification and high numerical aperture, the upper and lower surface of a sample discontinuity may not both be in focus at the same time. This is a further source of errors.

Interferometric microscopy is implemented in several ways; two of these are: the *Mirau interference microscope* and the *Linnik interference microscope*. The Mirau interference microscope is schematically illustrated in Fig. 10.9(a). Light is incident on the microscope objective. Some is transmitted to the sample, the remainder is reflected by the beam splitter to the reference surface. The light reflected by the sample and the reference surface are combined at the beam splitter and interfere. The resulting interference fringes give the difference between the sample surface and the reference plane. The reference surface, objective lens, and beam splitter are attached to a piezoelectric transducer translating the reference surface to vary the phase of the reference beam.[19]

The Linnik interference microscope in Fig. 10.9(b) is a *Michelson interferometer*. A plane wave front from a coherent, monochromatic light of wavelength λ is incident on a beam splitter. Part of the beam is transmitted to a stationary reference mirror and part is transmitted to the sample. Both beams are reflected to the beam splitter where they combine and are transmitted to the detector. The phase change is introduced by

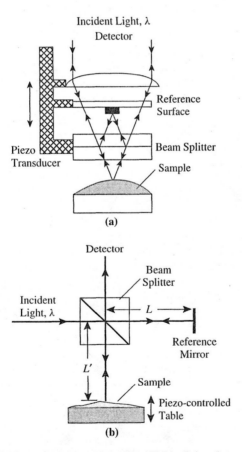

Fig. 10.9   (a) Mirau interference microscope, (b) Linnik interference microscope.

varying the sample-beam splitter distance with a piezoelectrically controlled table. The Mirau objective is typically used at magnifications between $10\times$ and $50\times$, and numerical apertures of $0.25-0.55$. The Linnik objective is suitable for any magnification, but is used primarily at high magnifications (*e.g.*, $100\times$) and high numerical apertures up to 0.95.

An extension of optical sample height measurements uses white light in a Linnik interferometer.[20] With a white light source, the interference fringes with the best contrast obtain only when the two paths in the interferometer are equal. Hence, if the path length to the reference mirror in the interferometer is varied to give maximum fringe contrast, that path length corresponds to the distance to the sample surface. There are no height ambiguities, especially since the sample is in focus when the maximum fringe contrast is observed. Height variations of many wavelengths, up to $100~\mu m$, can be measured in this manner. In an interference microscope, where height measurements are the main concern, lateral resolution limits take on a somewhat different meaning. The blurring of small objects leads to edge smoothing at height steps, reducing the accuracy of the height measurement.

Optical properties of the sample must be considered in interpreting interferometric microscopy measurements, because the height depends on the phase change upon reflection. For example, consider a step of height h_1 in a metal line on a semiconductor wafer. This wafer is subsequently covered with a glass to produce better planarity. The resulting step height of the glass is h_2, which may be much less than h_1. Optical interference measurements are likely to ignore the glass layer and measure the step height of the underlying metal film, giving a step height of h_1. Materials with different optical constants also affect height measurements. Coating the material with a reflective material eliminates this problem.

10.2.5   Defect Etches

Optical microscopy is frequently used to determine defect size, type, and density in semiconductors delineated by particular defect etches. The sample is subjected to an etch that renders particular defects visible through etch pits of particular shapes. Table 10.1 lists some etches. Instructions on their use are given in ref. 21. To count the defects, it is recommended to use an optical microscope with $100\times$ magnification and count the number of defects within a known area. Do this on nine locations on the wafer and average the readings. A very detailed series of photographs showing many examples of defects is given in Ref. 22. A cross-section and top view of some defects in silicon are shown in Fig. 10.10. The effect of the type of etch is illustrated in Fig. 10.11, showing a Si wafer with vacancy-type defects etched in Secco, Wright, and HF-HNO_3 etches. It is quite obvious that typical defect etches in (a) and (b) bring out the defects much more than the HF-HNO_3 polishing etch.

10.2.6   Near-Field Optical Microscopy (NFOM)

In *near-field optical microscopy* the resolution is not related to the wavelength of the exciting radiation forming the image, but rather determined by the geometry of the imaging system. The physician's stethoscope is a very practical demonstration of near-field imaging. The stethoscope has an aperture of several cm, while the acoustic wavelength is around 100 m, giving the stethoscope a resolving power of roughly $\lambda/1000$.

Conventional wisdom holds that the lower limit to imaging is related to the radiation wavelength. Abbé pointed out in 1873 that light focused by a converging, aberration-free

TABLE 10.1   Etches for Semiconductor Defect Delineation.

Semiconductor	Etch	Chemical Composition	Application
Si	Sirtl[23]	Dissolve 50 g CrO_3 in 100 ml H_2O immediately before using add 1 part HF to 1 part of the solution by volume	Best applicable to {111}-oriented surfaces
Si	Dash[24]	HF:HNO_3:CH_3COOH 1:3:10	Generally applicable for both n-Si and p-Si of {111} and {100} orientation; but works best for p-Si
		Dissolve 55 g $CuSO_4$ 5 H_2O in 950 ml H_2O, add 50 ml HF	Cu displacement etch; delineates defects by Cu decoration
Si	Secco[25]	HF:$K_2Cr_2O_7$ (0.15 M) (11 g $K_2Cr_2O_7$ in 250 ml H_2O) i.e. 2:1 or HF:CrO_3 (0.15 M) 2:1	Generally applicable, but is particularly suitable for {100} orientation
Si	Schimmel[26]	Add 75 g CrO_3 to H_2O to make 1000 ml sol'n (0.75 M sol'n)	For n-Si, p-Si, {100}, {111} For $\rho > 0.2$ Ω-cm add 2 pts. HF to 1 pt. sol'n; For $\rho < 0.2$ Ω-cm add 2 pts. HF to 1 pt. sol'n and 1.5 pts. H_2O
Si	Wright[27]	HF:HNO_3:5MCrO_3:$Cu(NO_3)_2$·3H_2O:{111}; CH_3COOH:H_2O 2:1:1:2g:2:2 Best results obtained by first dissolving the $Cu(NO_3)_2$ in the H_2O; otherwise order of mixing not critical	For n-Si and p-Si, {100} and {111}; defect-free regions are not roughened following etching
Si	Yang[28]	Add 150 g CrO_3 to 1000 ml H_2O (1.5 M); add 1 part sol'n to 1 part HF	Delineates various defects on {100}, {111}, and {110} surfaces without agitation
Si	Seiter[29]	9 parts by volume of a solution of 120 g CrO_3 in 100 ml H_2O and 1 part HF (49%)	Etches {100} planes 0.5–1 μm/min; 20–60 s etch time; delineates dislocations, stacking faults, swirl defects
Si	MEMC[30]	Add 1 g of $Cu(NO_3)_2$:3H_2O to 100 ml of HF:HNO_3: CH_3COOH: H_2O	Similar to Sirtl or Wright etch, without chromium; etches dislocations and slip
GaAs	KOH[31]	Molten KOH	Sample immersed in molten KOH for 3 h at 350°C in covered Ni crucible.
InP	Huo et al.[32]	HBr:H_2O_2:H_2O:HCl 20:2:20:20	Reveals dislocations on {100} and {111} surfaces

CH_3 COOH: glacial acetic acid.

lens onto an object surface can be focused to a spot diameter no smaller than about $\lambda/2$,[33] limited by diffraction effects and usually expressed by the Raleigh limit of Eq. (10.3). Microscopy that follows the Abbé or Raleigh limit, known as far-field microscopy because it is the far field of the radiation that is imaged, applies to conventional optical, electron, and acoustic imaging.

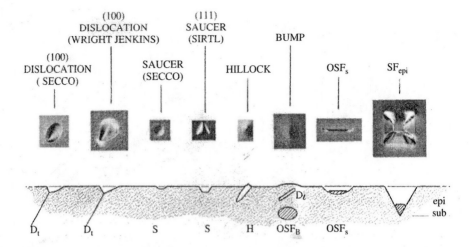

Fig. 10.10 Some common etch patterns in silicon when etched with some of the etches in Table 10.1. Reprinted with permission after Miller and Rozgonyi, ref. 7.

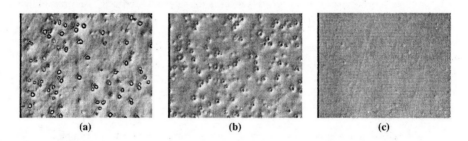

Fig. 10.11 D-type vacancy defects in silicon etched with (a) Secco, (b) Wright, and (c) HF-HNO$_3$. Micrographs courtesy of M.S. Kulkarni, MEMC.

NFOM relies on physical dimensions to image at resolutions much better than the Abbé or Raleigh limits. The idea of near-field microscopy was suggested in 1928[34] and demonstrated with microwaves in 1972, achieving a resolution of $\lambda/200$.[35] More recently it has also been demonstrated optically.[36] Later it was extended to infrared and μm and mm waves.[37] The principle of a NFOM is based on the concept that by illuminating an object through an aperture smaller than the wavelength of the exciting radiation, and detecting the reflected or transmitted radiation very close to the object, closer than the wavelength, it is possible to record a scanned image with a resolution determined by the aperture size and not by the wavelength. It was the development of nanometer positioning technology that has led to the successful implementation of NFOM.

The principle of near-field optical microscopy is illustrated in Fig. 10.12. In conventional microscopy shown in Fig. 10.12(a), the focused light spot diameter is approximately

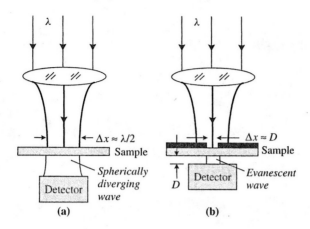

Fig. 10.12   (a) Conventional far-field optical imaging and (b) near-field optical imaging.

$\lambda/2$. Photons transmitted through the aperture in Fig. 10.12(b) have their transverse position defined to an accuracy $\Delta x \approx D$, where D is the aperture diameter. The transmitted photons have two distinct wave components. One is the diverging wave with the spatial distribution of the radiation related to the Fourier transform—the far-field region. The other is the *evanescent* or vanishing wave in proximity to the aperture exit—the near-field region. This wave is collimated to the aperture size and has a rapidly decreasing intensity and must be detected very close to the aperture—typically a few nm. If the aperture is scanned across the sample with the sample always in the near-field region, it is possible to generate an image with spatial resolution determined by the aperture size.

In the reflection mode the detection system must contain some means of implementing a very narrow aperture or "optical receiver." One method is to thin a glass fiber by etching or drawing it to a very fine tip whose outer surface is coated with metal to prevent extraneous light from entering through the walls of the pipette. Aperture sizes of 100 nm or less are practical and routine for light of wavelength between 400 nm and 1.5 μm. Another example of near-field imaging is the scanning tunneling microscope, discussed in Chapter 9, where the probe tip defines the aperture. Its diameter of around 0.2 nm allows imaging of atoms at that resolution, even though the wavelength of electrons accelerated to a potential of 1 V is around 1.2 nm.

For high spatial resolution, the sample must be placed in the near-field zone of the tip. For typical apertures of 100 nm, the tip-sample separation should be ~20 nm. As a rule of thumb, the tip-sample separation should be less than one third of the aperture size.[38] A feedback mechanism keeps the tip-sample separation constant. A shear force mechanism has been widely adapted to regulate tip-sample separation. The NFOM tip is attached to a piezoelectric element (the dither piezo) and held vertically above the sample surface. By applying a time-dependent ac voltage to the dither piezo at the resonant frequency, the fiber tip vibrates parallel to the sample surface. As the tip approaches the sample, the amplitude decreases due to interaction with the sample and the tip-sample separation is regulated through a feedback loop by monitoring the changes in the tip dithering motion.

10.3   ELLIPSOMETRY

10.3.1   Theory

Ellipsometry is a contactless, non-invasive technique measuring changes in the polarization state of light reflected from a surface.[39] It deals with intensity-dependent complex quantities compared to intensities for reflectance or transmittance measurements. Ellipsometry can be thought of as an *impedance measurement*, while reflectance or transmittance can be viewed as *power measurements*. Impedance measurements give the amplitude and phase, whereas power measurements only give amplitudes. One determines the complex reflection coefficient ratio of the sample that depends on the ratio of the complex reflection coefficient for light polarized parallel and perpendicular to the plane of incidence.

Ellipsometry is used predominantly to determine the thickness of thin dielectric films on absorbing substrates, line width, and optical constants of films or substrates.[40] It does not measure the film directly, rather it measures certain optical properties from which thickness and other sample parameters are derived. Recent additions to basic ellipsometry include variable angle and variable wavelength ellipsometry, allowing thickness measurements at least an order of magnitude smaller than interferometric methods. Before going into the details of ellipsometry, it is important to understand the properties of *polarized light*.

When light is reflected from a single surface it will generally be reduced in amplitude and shifted in phase. For multiple reflecting surfaces, the various reflecting beams interact and give maxima and minima as a function of wavelength or incident angle. Since ellipsometry depends on angle measurements, optical variables can be measured with great precision, being independent of light intensity, reflectance, and detector-amplitude sensitivity.

Consider plane-polarized light incident on a plane surface, illustrated in Fig. 10.13. The light spot is typically on the order of millimeters in diameter, but can be focused to about 100 μm. The incident polarized light can be resolved into a component *p*, parallel to the plane of incidence and a component *s* perpendicular to the plane of incidence ("*s*" is the first letter of the German word *senkrecht*, meaning vertical). For zero absorption material, only the amplitude of the reflected wave is affected. Linearly polarized light is reflected as linearly polarized light. However, the two components experience different amplitudes *and* phase shifts upon reflection for absorbing materials and for multiple reflections in a thin layer between air and the substrate. The parallel component reflectance is always less than the vertical component for angles of incidence other than 0° and 90°. The two are equal at those two angles. The phase-shift difference introduces an additional component

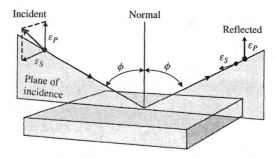

Fig. 10.13   Schematic of polarized light reflection from a plane surface. ϕ is the angle of incidence.

polarized 90° to the incident beam rendering the reflected light *elliptically* polarized. The key property of polarized light for ellipsometry is the change of plane polarized light into elliptically polarized light or elliptically polarized light into plane polarized light upon reflection.

Exercise 10.2

Question: What is Polarization?

Answer: Light is an electromagnetic wave with electric and magnetic field components propagating in the z direction perpendicular to each other as shown in Fig. E10.2(a). The polarization is defined by the orientation and the phase of the electric field vector. To describe the polarization of the wave, the wave is projected onto the x and y axes with the two components, $\mathscr{E}_x$ and $\mathscr{E}_y$. When these components propagate in the same direction, are orthogonal and in phase with each other, a linearly polarized wave results as in (b). When the two components are equal in amplitude and 90° out of phase as in (c), the result is circularly polarized light. Elliptically polarized light in (d) is the result of the two components having arbitrary phase and amplitude. For a nice discussion of polarization and ellipsometry see the web site for the J.A. Woollam Co. at http://jawoollam.com/Tutorial_1.html. Figure E10.2 is adapted from that site.

Light propagates as a fluctuation in electric and magnetic fields at right angles to the direction of propagation. The total electric field consists of the parallel component $\mathscr{E}_p$ and the vertical component $\mathscr{E}_s$. The reflection coefficients

$$R_p = \frac{\mathscr{E}_p(\text{reflected})}{\mathscr{E}_p(\text{incident})}; R_s = \frac{\mathscr{E}_s(\text{reflected})}{\mathscr{E}_s(\text{incident})} \tag{10.13}$$

are not separately measurable. However, the complex reflection ratio, ρ, defined in terms of the reflection coefficients R_p and R_s or the *ellipsometric angles* Ψ and Δ is measurable

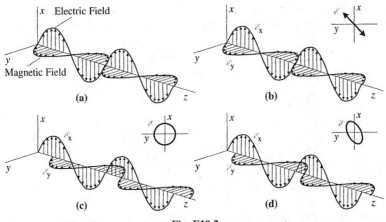

Fig. E10.2

and given by

$$\rho = \frac{R_p}{R_s} = \tan(\Psi)e^{j\Delta} \qquad (10.14)$$

where $j = (-1)^{1/2}$. Since ρ is the *ratio* of reflection coefficients, *i.e.*, the ratio of the intensities and the relative phase difference, it is not necessary to make absolute intensity and phase measurements.

The ellipsometric angles $\Psi(0° \leq \Psi \leq 90°)$ and $\Delta(0° \leq \Delta \leq 360°)$ are the most commonly used variables in ellipsometry and are defined as

$$\Psi = \tan^{-1}|\rho|; \Delta = \textit{differential phase change} = \Delta_p - \Delta_s \qquad (10.15)$$

The angles Ψ and Δ determine the differential changes in amplitude and phase, respectively, experienced upon reflection by the vibrations of the parallel and perpendicular electric field vector components.

How are Ψ and Δ used to determine the sample's optical parameters? Consider the example of light reflected at an air-solid absorbing substrate interface in Fig. 10.14. The air is characterized by its index of refraction n_0 and the sample by $n_1 - jk_1$, where n_1 is the index of refraction and k_1 the extinction coefficient. From Fresnel's equations[41]

$$n_1^2 - k_1^2 = n_0^2 \sin^2(\phi)\left[1 + \frac{\tan^2(\phi)[\cos^2(2\Psi) - \sin^2(2\Psi)\sin^2(\Delta)]}{[1 + \sin(2\Psi)\cos(\Delta)]^2}\right] \qquad (10.16)$$

$$2n_1k_1 = \frac{n_0^2 \sin^2(\phi)\tan^2(\phi)\sin(4\Psi)\sin(\Delta)}{[1 + \sin(2\Psi)\cos(\Delta)]^2} \qquad (10.17)$$

Of considerable importance in ellipsometric measurements is a substrate covered by a thin film, *e.g.*, an insulator. For the air (n_0)—thin film (n_1)—substrate $(n_2 - jk_2)$ system, the equations become more complex because they depend on the refractive indices, the film thickness, the angle of incidence, and the wavelength. If n_2 and k_2 are known from an independent measurement and if the film is transparent, then n_1 and film thickness may be calculated from the results of a single Ψ and Δ measurement, but the computation becomes very tedious. For example, an entire book is devoted to ellipsometric tables and curves showing the dependence of Ψ and Δ on the oxide thickness and oxide refractive index for the air-SiO_2-Si system at selected mercury and He-Ne laser spectral lines.[42]

10.3.2 Null Ellipsometry

In the PCSA (Polarizer—Compensator—Sample—Analyzer) null ellipsometer configuration in Fig. 10.14, a collimated beam of unpolarized monochromatic light, typically from a laser, is linearly polarized by the *polarizer*.[43] The Glan-Thompson prism, consisting of two sections of calcite cemented together, is a common polarizer. When unpolarized light is incident on such a polarizer, total internal reflection allows only linearly polarized light to exit. The *compensator* or *retarder* changes the linearly polarized light to elliptically polarized light. The compensator contains a fast and a slow optical axis perpendicular to the direction of transmission. The component of incident polarized light with electric field parallel to the slow axis is retarded in phase relative to the component parallel to the fast axis as the light passes through the compensator. When the relative retardation is $\pi/2$, the compensator is called a *quarter-wave retarder* or *quarter-wave plate*.

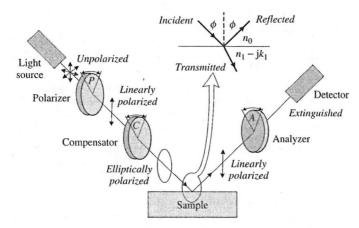

Fig. 10.14   Ellipsometer schematic.

The angles P and C of the polarizer and the compensator can be adjusted to any state of polarization ranging from linear to circular. The aim of ellipsometry measurements is a null at the detector, attained by choosing P and C to give light of elliptical polarization which, when reflected from the sample, becomes linearly polarized to be extinguished by the *analyzer*. The linearly polarized light is passed through the analyzer, which is similar to the polarizer, and the angle A is adjusted for minimum detector output. Stepping motors adjust the polarizer and analyzer angles sequentially for minimum detector signal. The angular convention is that all angles are measured as positive counterclockwise from the plane of incidence when looking into the beam and the polarizer angle is adjusted to zero when the plane of transmission is in the plane of incidence.

There are 32 combinations of P, C, and A that can result in a given pair of Ψ and Δ. Because any two angles of the polarizer, compensator, and analyzer that are 180° apart are optically identical, the number of combinations of P, C, and A settings giving any pair of Ψ and Δ can be reduced to 16 if all angles are restricted to less than 180°. The 16 equation pairs can be further reduced to two pairs by restricting the compensator to one angle, for example 45°, and the ranges of P and A to two zones.

10.3.3   Rotating Analyzer Ellipsometry

The *rotating analyzer ellipsometer* falls within a class of ellipsometers known as *photometric ellipsometers* that speed up the measurement, because null ellipsometers are too slow for real-time and spectroscopic ellipsometry measurements. In the rotating analyzer ellipsometer linearly polarized light is incident on the sample and becomes elliptically polarized upon reflection.[44-45] The reflected beam passes through the analyzer, rotating around the beam axis at a constant angular velocity (typically 50–100 Hz), to be detected. If the light incident on the analyzer were linearly polarized, the detected light would be a sine-squared function with a maximum and *zero* minimum per half rotation of the analyzer. Unmodulated, uniform output for circularly polarized light and sinusoidal output variations, similar to linearly polarized light, is observed for elliptically polarized light, but the maxima are smaller and the minima larger, reducing the amplitude variation. The amplitude variation of the sinusoidal detector output is a function of the ellipticity of the reflected light. The output is generally Fourier analyzed to yield Ψ and Δ. A single frequency measurement can be made in a few ms for an angular velocity of 100 Hz. In some systems, a rotating polarizer is used.

The light intensity at the detector is[41]

$$I(\theta) = I_o[1 + a_2 \cos(2\theta) + b_2 \sin(2\theta)] \tag{10.18}$$

where θ is the angle between the polarizing plane of the analyzer and the plane of incidence of the reflected light, I_o the average intensity of one complete revolution of the analyzer. Ψ and Δ are determined from a_2 and b_2, the parameters describing the state of polarization of the reflected light as

$$\Psi = \frac{1}{2} \operatorname{arcosh} (-a_2); \Delta = \operatorname{arcosh} \left(\frac{b_2}{\sqrt{1 - a_2^2}} \right) \tag{10.19}$$

The major advantages of rotating analyzer ellipsometers lie in their higher speed and their increased accuracies. Effects of noise and random errors are reduced, since hundreds or thousands of light intensity samples constitute a single measurement. The lack of a compensator improves the measurement, since errors associated with commercial compensators do not affect the measurement. The demands on the optical system, however, are more stringent. Stray light must be carefully controlled and the light source intensity should not change with time. The detector response must be linear to avoid generation of harmonics. Rotating analyzer ellipsometers are particularly suited to spectroscopic ellipsometric measurements, because any wavelength-dispersive properties of the compensator play no role since there is no compensator, and because the data acquisition time is short.

10.3.4 Spectroscopic Ellipsometry (SE)

A common application of single wavelength ellipsometry is in film thickness measurements. But it can also be used for other applications, because the ellipsometric angles Ψ and Δ are sensitive not only to layer thickness, but also to composition, microstructure, and optical constant of the sample surface. *Spectroscopic* ellipsometric measurements have extended the range of ellipsometry by using more than one wavelength.[46] Furthermore, it is possible to vary not only the wavelength but also the angle of incidence, providing yet another degree of freedom. This allows non-invasive, real-time process measurements such as layer growth monitoring during MBE,[47] and *in situ* diagnostic and process control.[48] Variable wavelength and angle allows optimization for a material parameter of interest, something that is not usually possible with fixed-angle, constant-wavelength ellipsometry.

An ellipsometer is sensitive to surface changes on the order of a monolayer. Film thickness and alloy composition can be determined during growth or during etch. During etch measurements it is possible to stop the etch before reaching an interface, in contrast to most other *in situ* sensors, which only give a signal after an interface has been reached. Optical measurements are ideal for real-time measurements because they are non-invasive and can be used in any transparent ambient including ambients associated with plasma processing and chemical vapor deposition. Spectroscopic ellipsometry has also been used to measure the temperature during semiconductor processing.[41]

10.3.5   Applications

Film Thickness: Measurements of thickness and index of refraction of thin, non-absorbing films on semiconductor substrates is a major application of ellipsometry. There is, in principle, no limit to the thickness of the layer that can be determined. Films as thin as 1 nm have been measured. Although ellipsometry gives numeric values, the results for very thin films are questionable, because the model assumes uniform optical properties and a sharp planar film-substrate boundary and the ellipsometric equations are based on the macroscopic Maxwell's equations which may not apply to layers only a few atomic layers thick. Nevertheless, the measurements appear to give reasonable average thicknesses.

Thick layers have a different problem. The interpretation becomes more difficult due to optical path lengths. In the thin transparent layer on a substrate of Fig. 10.15, the two reflected rays interfere with one another, going from being completely in phase to completely out of phase. This interference causes the cyclical nature of thickness measurements, where Ψ and Δ are cyclic functions of film thickness. They repeat for the full-cycle film thickness

$$d = \frac{\lambda}{2\sqrt{n_1^2 - \sin^2(\phi)}} \tag{10.20}$$

For example, at $\phi = 70°$ the full-cycle thickness of SiO_2 films with $n_1 = 1.465$ at $\lambda = 632.8$ nm is 281.5 nm. If a 10 nm thick SiO_2 film gives certain ellipsometric angles, the same angles will be measured for films of $(10 + 281.5)$ nm, $(10 + 563)$ nm, and so on. Hence for films thicker than the full-cycle thickness, one must have independent knowledge of the film thickness to within one full-cycle thickness.

Substrates, Layer Growth: Although the major use of ellipsometry is for the analysis of non-absorbing, insulating films, the method is used to characterize semiconductors, *e.g.*, to study semiconducting materials that have been modified in some way. For example, ion implantation damage has been correlated with ellipsometric measurements for Si, GaAs, and InP.[49-50] It is believed that the implant-induced damage, not the doping density, changes the refractive indices. Although quantitative results are difficult to obtain, the measurements do allow a rapid, non-destructive measurement of the crystal damage and the behavior of this damage with annealing.

Ellipsometry has also found applications during crystal growth, where its non-contacting, real-time nature is particularly useful when used *in situ*. For example, it has been used to monitor the growth of superlattice structures grown by molecular beam epitaxy (MBE) and metalorganic chemical vapor deposition (MOCVD).[51-52] Ellipsometry is a benign

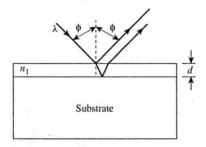

Fig. 10.15   Schematic showing multiple reflections.

technique that is little affected by deposition methods, if used to monitor the growth of a layer. It does not influence the deposition process if used as an *in situ* process tool to monitor the growth of MBE films or the growth of insulating layers.

Line Width or *Critical Dimension:* The use of fixed angle SE or spectral reflectance measurements from periodic structures shows strong promise for high speed topography measurements. In many cases, SE-based measurements have proven to be more detailed and more accurate than top-down scanning electron microscopy critical dimension (CD) measurements and this method is used as an in-line process control tool. The emergence of this approach is directly analogous to that of conventional thin film ellipsometry when low cost computers enabled the accurate solution of thin film reflection models. For structures for which the diffraction problem can be numerically solved nearly exactly, the advantages of spectroscopic ellipsometry for patterned structures are now being realized.[53] Line width measurements are discussed in Section 10.8.

10.4 TRANSMISSION

10.4.1 Theory

Optical *transmission* or *absorption* measurements are used to determine optical absorption coefficients and certain impurities. Shallow-level impurities respond to optical measurements as discussed in Sections 2.6.3 and 2.6.4. Certain impurities possess characteristic absorption lines due to vibrational modes, for example oxygen and carbon in silicon. Photons absorbed in a semiconductor can change the immediate environment around certain impurities producing local vibrational modes. We discuss in this chapter the appropriate theory of optical transmission measurements and give some examples.

During transmission measurements light is incident on the sample and the transmitted light is measured as a function of wavelength as illustrated in Fig. 10.16(a). The sample is characterized by reflection coefficient R, absorption coefficient α, complex refractive index $(n_1 - jk_1)$, and thickness d. Light of intensity I_i is incident from the left. The absorption coefficient is related to the extinction coefficient k_1 by $\alpha = 4\pi k_1/\lambda$. Absorption coefficients and refractive indices are given in Appendix 10.2 for selected semiconductors. The transmitted light I_t can be measured absolutely or the ratio of transmitted to incident light can be formed. As shown in Appendix 10.1, the transmittance T of a sample with identical front and back reflection coefficient and light incident normal to the sample surface is

$$T = \frac{(1-R)^2 e^{-\alpha d}}{1 + R^2 e^{-2\alpha d} - 2Re^{-\alpha d}\cos(\phi)} \tag{10.21}$$

where $\phi = 4\pi n_1 d/\lambda$ and the reflectance R is given by

$$R = \frac{(n_0 - n_1)^2 + k_1^2}{(n_0 + n_1)^2 + k_1^2} \tag{10.22}$$

A normalized curve of I_t for polished Si is shown in Fig. 10.16(b).

The semiconductor band gap can be determined by measuring the absorption coefficient as a function of the photon energy. Light, with energy higher than the band gap, is absorbed. However, absorption is low to moderate for $h\nu$ near E_G. For indirect band-gap

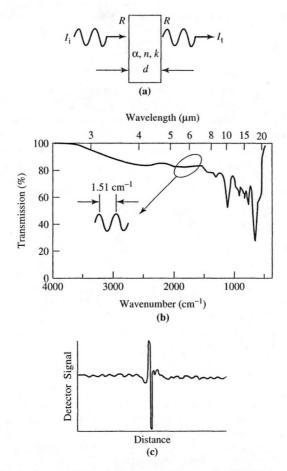

Fig. 10.16 (a) Schematic transmittance measurement, (b) normalized FTIR transmittance curve for a double-side polished Si wafer ($\Delta f = 4$ cm^{-1} for transmittance curve, $\Delta f = 1$ cm^{-1} for inset), (c) interferogram for the same wafer, $\Delta f = 4$ cm^{-1}. From the period of 1.51 cm^{-1} in (b) the wafer thickness is 970 μm. Courtesy of N.S. Kang, Arizona State University.

semiconductors, $\alpha^{1/2}$ is plotted against $h\nu$, the extrapolated intercept on the $h\nu$ axis yields the semiconductor band gap. Such a plot is sometimes referred to as a *Tauc plot*. For direct band-gap semiconductors like GaAs, for example, α^2 is plotted against $h\nu$ and the band gap is again determined from the extrapolated intercept.

Semiconductors are generally transparent ($\alpha \approx 0$) for photon energies less than the band gap energy and the transmittance becomes

$$T = \frac{(1-R)^2}{1+R^2-2R\cos(\phi)} \tag{10.23}$$

The "cos" term can be written as $\cos(f/f_1)$, where $f = 2\pi/\lambda$ and $f_1 = 1/2n_1d$ is a characteristic spatial frequency. If the resolution of the detector is sufficiently high, $\Delta f \leq$

$1/2n_1d$, then an oscillatory transmittance curve is observed. For example, $\Delta f \leq 4.9$ cm^{-1} for a Si wafer with thickness $d = 300$ μm and refractive index $n_1 = 3.42$. If the resolution of the instrument is insufficient to resolve these fine-structure oscillations, then Eq. (10.23) becomes

$$T = \frac{(1 - R)^2}{1 - R^2} = \frac{1 - R}{1 + R} \tag{10.24}$$

For the Si example this becomes $T \approx 0.54$ for $R = 0.3$. As shown in Appendix 10.1, the wafer thickness can be determined from the period of the oscillatory transmittance versus wavenumber curve with

$$d = \frac{1}{2n_1 \Delta(1/\lambda)} \tag{10.25}$$

where $\Delta(1/\lambda)$ is the wavenumber interval between two maxima or two minima. The transmittance curves can be plotted as a function of wavelength or wavenumber (wavenumber $= 1/$wavelength).

Certain impurities in a semiconductor sample exhibit absorption. Examples are interstitial oxygen and substitutional carbon in silicon. Their density is proportional to the absorption coefficient at those wavelengths. The transmittance with absorption, but no "cos" oscillations, is

$$T = \frac{(1 - R)^2 e^{-\alpha d}}{1 - R^2 e^{-2\alpha d}} \tag{10.26}$$

The absorption coefficient from Eq. (10.26) is[54]

$$\alpha = -\frac{1}{d} \ln \left(\frac{\sqrt{(1 - R)^4 + 4T^2 R^2} - (1 - R)^2}{2T R^2} \right) \tag{10.27}$$

R can be determined from that part of the transmittance curve where $\alpha \approx 0$. In some spectral regions there may be absorption due to lattice vibrations and free-carrier absorption for heavily-doped substrates. The lattice absorption coefficient is about $0.85-1$ cm^{-1} for oxygen in Si and about 6 cm^{-1} for carbon in Si. This must be considered in the analysis.[55]

Complications in interpretation of transmission data occur when both surfaces are not polished. Due to surface roughness, the transmittance becomes wavelength dependent and T can vary significantly from wafer to wafer. If transmission is severely impaired, the signal-to-noise ratio can be so poor that the measurement becomes meaningless.[56]

10.4.2 Instrumentation

Monochromator: There are two instruments for transmission measurements: *monochromator* and *interferometer*. The monochromator, illustrated in Fig. 10.17(a), selects a narrow band of wavelengths $\Delta\lambda$ from a source of radiation. The spectral band is centered on a wavelength λ that can be varied. The monochromator can be thought of as a tunable filter with a band pass $\Delta\lambda$ and resolution $\Delta\lambda/\lambda$. Light enters the monochromator through a narrow entrance slit. Light falling on the prism or grating is dispersed, breaking the light into its spectral components, by virtue of having a wavelength-dependent refractive index. Short wavelength light is refracted more than long wavelength light. A grating consists of many equidistant parallel lines inscribed on a polished substrate (glass or metal film

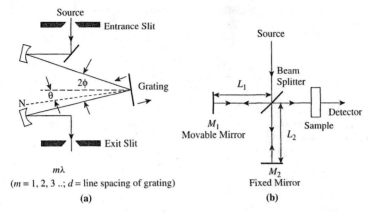

Fig. 10.17 (a) Monochromator and (b) FTIR schematics.

on glass) with typically between 4,000 and 20,000 lines or grooves per cm. The dispersed light depends on the groove spacing and on the incident angle.

The dispersed light passes through a narrow exit slit that largely controls the spectral resolution; the narrower the slit, the narrower the wavelength range that reaches the detector. The slits can be thought of as spectral bandpass filters. As the slit becomes narrower, however, the amount of light reaching the sample is likewise reduced. The wavelengths are varied by changing the angular position of the prism or the grating. In a monochromator only a narrow band of wavelengths is selected for the transmission measurement avoiding simultaneous excitation of competing processes that can result from other wavelengths. For example, above band gap light creates electron-hole pairs, which may interfere with measurements using below band gap light. Monochromator transmission measurements avoid this problem by eliminating above band gap light. A disadvantage of the monochromator is that only a small portion of the total spectrum is available at one time leading to low signals. This can be overcome through lock-in or signal averaging techniques. For greater sensitivity and minimization of atmospheric attenuation, double beam instruments are frequently used, splitting the beam into two similar paths with the sample placed in one of these paths and a reference sample in the reference beam and the sample transmission is compared to that of the reference.

A monochromator is inserted between the light source and the sample, ensuring that only selected wavelengths are incident on the sample at one time. It is also possible for all wavelengths to be incident on the sample at one time and spectrally resolve the light *after* it has been transmitted through the sample. Then the instrument is known as a *spectrometer*. A spectrometer is more commonly used when the light is emitted from the sample, while a monochromator is used to decompose white light into its spectral components for a subsequent spectral response measurement.

Fourier Transform Infrared Spectroscopy: The foundations of modern *Fourier Transform Infrared Spectroscopy* (FTIR) were laid in the latter part of the nineteenth century by Michelson[57] and Lord Raleigh who recognized the relationship of an interferogram to its spectrum by a Fourier transformation.[58] It was not until the advent of computers and the fast Fourier algorithm[59] that interferometry began to be applied to spectroscopic measurements in the 1970s.

The basic optical component of Fourier transform spectrometers is the *Michelson interferometer* shown in simplified form in Fig. 10.17(b).[60] Light from an infrared source—a heated element or a glowbar—is collimated and directed onto a beam splitter, creating two separate optical paths by reflecting 50% of the incident light and transmitting the remaining 50%. In one path the beam is reflected back to the beam splitter by a fixed-position mirror, where it is partially transmitted to the source and partially reflected to the detector. In the other leg of the interferometer, the beam is reflected by the movable mirror that is translated back and forth while maintained parallel to itself. The movable mirror rides on an air bearing for good stability. The beam from the movable mirror is also returned to the beam splitter where it, too, is partially reflected back to the source and partially transmitted to the detector. Although the light from the source is incoherent, when it is split into two components by the beam splitter, the components are *coherent* and can produce interference phenomena when the beams are combined.

The light intensity reaching the detector is the sum of the two beams. The two beams are in phase when $L_1 = L_2$. When M_1 is moved, the optical path lengths are unequal and an optical path difference δ is introduced. If M_1 is moved a distance x, the retardation is $\delta = 2x$ since the light has to travel an additional distance x to reach the mirror and the same distance to reach the beam splitter.

Consider the output signal from the detector for a *single wavelength* source. For $L_1 = L_2$ the two beams reinforce each other because they are in phase, $\delta = 0$, and the detector output is a maximum. If M_1 is moved by $x = \lambda/4$, the retardation becomes $\delta = 2x = \lambda/2$. The two wave fronts reach the detector 180° out of phase, resulting in destructive interference or zero output. For an additional $\lambda/4$ movement by M_1, $\delta = \lambda$ and constructive interference results again. The detector output—the interferogram—consists of a series of maxima and minima that can be described by the equation

$$I(x) = B(f)[1 + \cos(2\pi x f)] \tag{10.28}$$

where $B(f)$ is the source intensity modified by the sample. $B(f)$ and $I(x)$ are shown in Fig. 10.18(a) for this simple case. When the source emits more than one frequency, Eq. (10.28) is replaced by the integration

$$I(x) = \int_0^f B(f)[1 + \cos(2\pi x f)]\, df \tag{10.29}$$

For example, consider the source spectral distribution, $B(f) = A$ for $0 \le f \le f_1$ in Fig. 10.18(b). The interferogram is obtained by eliminating the unmodulated term from Eq. (10.29):

$$I(x) = \int_0^{f_1} A \cos(2\pi x f)\, df = A f_1 \frac{\sin(2\pi x f_1)}{2\pi x f_1} \tag{10.30}$$

shown in Fig. 10.18(b). The interferogram becomes narrower as f_1 is increased.

The interferogram always retains its maximum at $x = 0$ where $L_1 = L_2$, because *all* wavelengths interfere constructively for that mirror position. For $x \ne 0$, waves interfere destructively and the interferogram amplitude decreases from its maximum as shown in the interferogram for a Si wafer in Fig. 10.16(c). The strong maximum at $x = 0$ is the *centerburst*. The higher resolution spectral information is contained in the wings of the interferogram, corresponding to larger mirror travel. There is a practical limit to the mirror displacement, represented by $x = L$. The best spectral resolution is $\Delta f = 1/L$. In practice

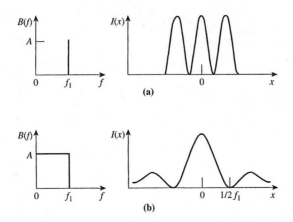

Fig. 10.18   Spectrum and interferogram for (a) cosine wave and (b) band-limited signal.

other practical considerations reduce Δf below this value. In most FTIR instruments, numerous movable mirror sweeps are averaged for enhanced signal-to-noise ratio.

What is measured in FTIR is the interferogram, containing not only the spectral information of the source, which we have considered so far, but also the transmittance characteristics of the sample. The interferogram, however, is of little direct interest. It is the spectral response, calculated from the interferogram using the Fourier transformation, that is of interest

$$B(f) = \int_{-\omega}^{\omega} I(x)\cos(2\pi x f)\,dx \qquad (10.31)$$

$B(f)$ contains the spectral content of the source, the sample, and the ambient in the path of the measurement. It is common practice to reduce atmospheric H_2O and CO_2 absorption lines by purging the apparatus with dry nitrogen. The effect of the source is eliminated by making one measurement without the sample—the background measurement—and one with the sample. The ratio of the two eliminates the background. Since the mirror travel is finite, irregularities are introduced into the interferogram. Some of these irregularities can be subsequently reduced by using *weighting* or *apodization* schemes.[61]

FTIR has two major advantages over monochromators. One is the multiplex gain or the *Fellget* advantage. In monochromator transmission measurements only a small fraction of the entire spectrum is observed at a given time while in FTIR the entire spectrum is observed over the measurement period of a second or less. With N spectral elements, each $\Delta\lambda$ wide, the FTIR has a signal-to-noise advantage of $N^{1/2}$ over the monochromator when the detector is limited by noise other than photon noise.[62] A second major advantage is the optical throughput gain or *Jacquinot* advantage, referring to the amount of light one is able to pass through the instrument. Monochromators are limited by the entrance and exit slits while FTIRs have relatively large entrance apertures. The optical throughput gain is typically about 100.

10.4.3   Applications

Transmittance spectroscopy is primarily used to detect certain impurities, *e.g.*, oxygen and carbon in Si. Interstitial oxygen in silicon causes absorption at $\lambda = 9.05$ μm (1105 cm^{-1})

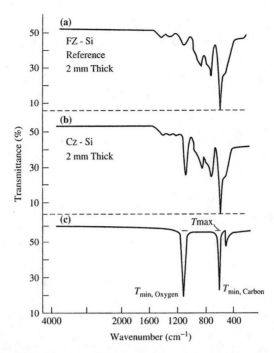

Fig. 10.19 Transmission spectra for (a) Si wafer low in oxygen and carbon, (b) Si wafer with more oxygen and carbon, (c) difference between spectra in (a) and (b). Data after ref. 55. Reprinted from the Aug. 1983 edition of *Solid State Technology*. Copyright 1983 by Penn Well Publishing Company.

at 300 K and at 8.87 μm (1227.6 cm^{-1}) at 77 K due to the antisymmetric vibration of the SiO$_2$ complex.[63] Substitutional carbon has absorption peaks at $\lambda = 16.47$ μm (607.2 cm^{-1}) at 300 K and at $\lambda = 16.46$ μm (607.5 cm^{-1}) at 77 K due to a local vibrational mode.[64] These absorption peaks are superimposed on phonon excitations of the silicon substrate and should be subtracted from the spectrum of a carbon- and oxygen-free reference sample. An example is shown in Fig. 10.19, where the transmission spectrum of a low oxygen and carbon Si wafer is subtracted from the spectrum of a sample containing oxygen and carbon, giving the spectrum of oxygen and carbon alone. Nitrogen has shown an absorption peak at 963 cm^{-1}.[65]

The optical absorption coefficients are converted to densities by

$$N = C_1\alpha \text{ [cm}^{-3}]; N = C_2\alpha \text{ [ppma]} \tag{10.32}$$

ppma is parts per million atomic. C_1 and C_2 are given in Table 10.2. Also shown is the full width at half maximum line width (FWHM) dictating the bandwidth of the measuring system. The oxygen conversion factors were obtained by calibrating the IR transmittance against oxygen densities determined by charged particle activation analysis, gas fusion analysis,[66] and photon activation analysis. For oxygen in silicon measurements it is necessary to specify the conversion factor, due to the diversity of these values. A good discussion of the state of oxygen-in-silicon measurements is given in ref. 67.

The lower detection limit for oxygen in silicon by the IR technique is around 5×10^{15}cm^{-3}; for carbon in silicon it is around 10^{16}cm^{-3} at room temperature and

TABLE 10.2   Conversion Factors of α to Densities.

Impurity	C_1 (cm^{-2})	C_2 (cm^{-2})	FWHM (cm^{-1})	Ref.
Oxygen in Si (300 K)	4.81×10^{17}	9.62	34	"Old ASTM" 63
Oxygen in Si (300 K)	2.45×10^{17}	4.9	34	"New ASTM" 63
Oxygen in Si (77 K)	0.95×10^{17}	1.9	19	"New ASTM" 63
Oxygen in Si (300 K)$^{@}$	3.03×10^{17}	6.06	34	"JEIDA" 71
Oxygen in Si (300 K)	2.45×10^{17}	4.9	34	"DIN" 72
Oxygen in Si (300 K)$^{\#}$	3.14×10^{17}	6.28	34	IOC-88 73
Carbon in Si (300 K)	8.2×10^{16}	1.64	6	64, 74
Carbon in Si (77 K)	3.7×10^{16}	0.74	3	64
Nitrogen in Si (300 K)	4.07×10^{17}	8.14		65
EL2 in GaAs (300 K)*	1.25×10^{16}	0.25		75

$^{@}$ JEIDA: Japan Electronic Industry Development Association.
$^{\#}$ International Oxygen Coefficient 1988.
* at $\lambda = 1.1$ μm; EL2, being a deep-level impurity, has a broad absorption band.

5×10^{15} cm^{-3} at 77 K. Low carbon densities are particularly difficult to measure because there is a strong two-phonon lattice absorption band near the carbon band at $\lambda = 16$ μm. Separation of these absorption bands requires either sample cooling to "freeze out" the lattice band or a comparison with a "carbon-lean" reference sample. A measurement method based on low-temperature photoluminescence of samples subjected to a CF$_4$ reactive ion etch, suggests detection limits for C in Si as low as 10^{13} cm^{-3}.[68] Transmittance measurements are, of course, also used to determine the optical absorption coefficients of semiconductors[69] and have been used to determine the boron and phosphorus content of deposited glasses.[70] Microspot FTIR measurements use beams as small as 1 μm.

10.5   REFLECTION

10.5.1   Theory

Reflection or *reflectivity* measurements are commonly made to determine layer thicknesses, both for insulating layers on semiconducting substrates and for epitaxial semiconductor films. The reflectance for the structure in Fig. 10.20(a), consisting of an absorbing layer of thickness d_1 on a non-absorbing substrate, is given by[76]

$$R = \frac{r_1^2 e^{\alpha d_1} + r_2^2 e^{-\alpha d_1} + 2r_1 r_2 \cos(\varphi_1)}{e^{\alpha d_1} + r_1^2 r_2^2 e^{-\alpha d_1} + 2r_1 r_2 \cos(\varphi_1)} \tag{10.33}$$

where

$$r_1 = \frac{n_0 - n_1}{n_0 + n_1}; r_2 = \frac{n_1 - n_2}{n_1 + n_2}; \varphi_1 = \frac{4\pi n_1 d_1 \cos(\phi')}{\lambda}; \phi' = \arcsin\left[\frac{n_0 \sin(\phi)}{n_1}\right] \tag{10.34}$$

For a non-absorbing layer, $\alpha = 0$ in Eq. (10.33).

The reflectance exhibits maxima at the wavelengths

$$\lambda(\max) = \frac{2n_1 d_1 \cos(\phi')}{m} \tag{10.35}$$

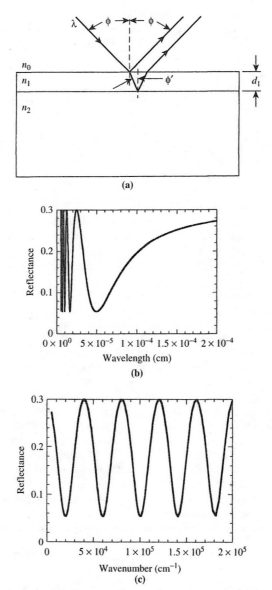

Fig. 10.20 (a) Reflection spectroscopy schematic, theoretical reflectance for SiO_2 on Si versus (b) wavelength and (c) wavenumber. $t_{ox} = 10^{-5}$ cm, $n_0 = 1$, $n_1 = 1.46$, and $n_2 = 3.42$, $\phi = 50°$.

where $m = 1, 2, 3, \ldots$ Taking the wavelengths at two maxima and subtracting one from the other using Eq. (10.35) gives the layer thickness[77]

$$d_1 = \frac{i\lambda_0\lambda_i}{2n_1(\lambda_i - \lambda_0)\cos(\phi')} = \frac{i}{2n_1(1/\lambda_o - 1/\lambda_i)\cos(\phi')} \qquad (10.36)$$

where i = number of complete cycles from λ_0 to λ_i, the two wavelength peaks that bracket the i cycles. For two adjacent maxima $i = 1$, for a maximum and an adjacent minimum $i = 1/2$, for two adjacent minima $i = 1$ and so on. As evident from Fig. 10.20(b), the wavelengths are difficult to determine from R versus λ plots. However, R versus $1/\lambda$ (wavenumber) plots yield values that are easier to extract. For example, in Fig. 10.20(c) for the first two peaks, $i = 1$, $1/\lambda_0 = 1.62 \times 10^5$ cm^{-1} and $1/\lambda_1 = 1.22 \times 10^5$ cm^{-1} giving $d_1 = 10^{-5}$ cm. The same thickness is obtained by choosing any other two adjacent peaks or by using $i = 3$, $1/\lambda_0 = 1.62 \times 10^5$ cm^{-1} and $1/\lambda_3 = 4.2 \times 10^4$ cm^{-1}. $2n_1 cos(\phi')$ is determined by the experimental arrangement and the film's index of refraction, sometimes written in terms of the incidence angle ϕ as

$$2n_1 \cos(\phi') = 2\sqrt{n_1^2 - n_0^2 \sin^2(\phi)} \tag{10.37}$$

Instead of illuminating the sample with monochromatic light and varying the wavelength, it is possible to shine white light, containing many wavelengths, onto the sample and analyze the reflected light by passing it through a spectrometer. Small areas can be characterized by shining the light through a microscope. Once the various wavelengths have been dispersed by the spectrometer, they can be detected by a photodiode array with the various wavelengths falling on different diodes in the array, for automatic data acquisition.[78] Reflectance measurements are also used to determine the thickness of epitaxial semiconductor layers, but it only works if there is a substantial doping density change at the epitaxial-substrate interface, because there must be a measurable index of refraction change at that interface.

Dielectric film thicknesses can also be measured using white light without a spectrometer. The white light is reflected onto a detector from a reference variable-thickness film and the unknown sample onto a detector. The detector output is a maximum for $n_r d_r = n_x d_x$ where n_r, d_r are the refractive index and thickness of the reference and n_x, d_x are those of the unknown.[79] The variable-thickness reference can be a semicircular wedge of oxidized Si. For $n_r = n_x$ the maximum detector output corresponds to the unknown film thickness when it is equal to the reference film thickness.

An alternate method uses FTIR. As described in Section 10.4.2, a maximum in the interferogram is observed when both optical paths from the beam splitter to the mirrors are identical. For thickness measurements of a layer on a substrate, a secondary maximum is observed when the movable mirror has moved by a distance equal to the optical path through the layer. The thickness is determined from the location x of this secondary maximum relative to the centerburst on the interferogram by the relation[80]

$$d_1 = \frac{x}{2n_1 \cos(\phi)} \tag{10.38}$$

This relationship is not strictly correct, as phase shifts in the reflected beam alter the shape and position of the side burst peaks. These phase shifts are not easy to include in the analysis, because the detector sees a broad range of wavelengths and thus a broad range of phase shifts. In practice, an empirical relationship is established between side burst position and layer thickness.

10.5.2 Applications

Dielectrics: The reflectance method lends itself to dielectric film thickness measurements; for SiO$_2$ films on Si thicker than 50 nm or so. For thinner films ($d < 50$ nm), it is easier to use ellipsometry. Instead of varying the wavelength, it is also possible to keep the wavelength constant and vary the incident angle. The technique is then known as *variable-angle monochromator fringe observation* (VAMFO).[81] When dielectric films on semiconductor substrates are viewed by eye or through a microscope, interference effects give the layer a characteristic color determined by the film thickness, its index of refraction, and the spectral distribution of the light source. Using calibrated color charts, thicknesses can be judged accurate to 10 to 20 nm. Such color charts are useful for oxide films thicker than about 80 nm.

A potential difficulty arises for films thicker than about 300 nm for SiO_2 or about 200 nm for Si_3N_4 since different orders have substantially the same color. A trained eye will be able to detect slight color changes for different orders. However, a more definite approach is to view the sample at an angle and compare it with the calibrated samples held at the same angle. The colors will not match unless they are both of the same order. A guide to the colors is given in Tables 10.3 and 10.4. The charts may also be used for films other than SiO_2 or Si_3N_4. In that case $d_x = d_0 n_0/n_f$, where d_x = unknown film thickness, d_0 = film thickness from color chart, n_0 = index of refraction of the original film (*e.g.*, SiO_2), and n_x = index of refraction of the film to be measured.

Semiconductors: Two types of semiconductor layers are of interest for thickness measurements: epitaxial layers and diffused or ion-implanted layers. Spectrophotometer reflectance measurements as given in Eqs. (10.35) and (10.36) pose several difficulties. There is only a small difference in the refractive index between the epitaxial layer and the substrate, leading to low-amplitude reflection from that interface. The index difference increases at longer wavelengths, giving enhanced interference patterns at longer wavelengths. Typical wavelengths for epitaxial layer thickness measurements lie in the 2 to 50 μm range. The index of refraction difference also increases with substrate doping density increase. The ASTM recommendation calls for Si epitaxial layer resistivity $\rho_{epi} > 0.1$ Ω·cm and substrate resistivity $\rho_{subst} < 0.02$ Ω·cm.[82] Additional complications arise because the phase shift at the air-semiconductor interface is different from that at the epitaxial-substrate interface, leading to the modified thickness equation[82–83]

$$d_{epi} = \frac{(m - 1/2 + \theta_i/2\pi)\lambda_i}{2\sqrt{n_1^2 - \sin^2(\phi)}} \tag{10.39}$$

where m = order of the maxima or minima in the spectrum, θ_i = phase shift at the epitaxial-substrate interface, and λ_i = wavelength of ith extrema in the spectrum. The 1/2 comes from the phase shift term. The phase shift at the epitaxial-substrate interface must be accurately known. Tabulated values for both n-Si and p-Si are given in Ref. 80. For very thin layers or layers on very thin buried structures, these phase shift values are crucial.[84]

Exercise 10.3

Question: What is a *magic mirror?*

Answer: A magic mirror is a contactless optical characterization method based upon the *Makyoh* concept. It is used to detect small changes in radius of curvature of a nominal flat

TABLE 10.3   Color Chart for Thermally Grown SiO$_2$ Films Observed Perpendicularly under Daylight Fluorescent Lighting.[81].

Film Thickness (μm)	Color	Film Thickness (μm)	Color
0.05	Tan	0.63	Violet-red
0.07	Brown dark violet to red violet	0.68	"Bluish" (Not blue but borderline 0.10
0.12	Royal blue		between violet and blue-green. It
0.15	Light blue to metallic blue		appears more like a mixture
0.17	Metallic to very light yellow green		between violet-red and blue-green
0.20	Light gold or		and looks grayish)
	yellow slightly metallic	0.72	Blue-green to green (quite broad)
0.22	Gold with slight yellow-orange	0.77	"Yellowish"
0.25	Orange to melon	0.80	Orange (rather broad for orange)
0.27	Red-violet	0.82	Salmon
0.30	Blue to violet-blue	0.85	Dull, light red-violet
0.31	Blue	0.86	Violet
0.32	Blue to blue-green	0.87	Blue-violet
0.34	Light green	0.89	Blue
0.35	Green to yellow-green	0.92	Blue-green
0.36	Yellow-green	0.95	Dull yellow-green
0.37	Green-yellow	0.97	Yellow to "yellowish"
0.39	Yellow	0.99	Orange
0.41	Light orange	1.00	Carnation pink
0.42	Carnation-pink	1.02	Violet-red
0.44	Violet-red	1.05	Red-violet
0.46	Red-violet	1.06	Violet
0.47	Violet	1.07	Blue-violet
0.48	Blue-violet	1.10	Green
0.49	Blue	1.11	Yellow-green
0.50	Blue-green	1.12	Green
0.52	Green (broad)	1.18	Violet
0.54	Yellow-green	1.19	Red-violet
0.56	Green-yellow	1.21	Violet-red
0.57	Yellow to "yellowish" (not yellow	1.24	Carnation pink to salmon
	but is in the position where	1.25	Orange
	yellow is to be expected. At	1.28	"Yellowish"
	times it appears to be light	1.32	Sky blue to green-blue
	creamy gray or metallic)	1.40	Orange
0.58	Light orange or yellow to pink	1.45	Violet
	borderline	1.46	Blue-violet
0.60	Carnation pink	1.50	Blue
		1.54	Dull yellow-green

surface and is based on an ancient Chinese mysterious mirror. It was a simple, feature-less, flat mirror made of bronze. However, the image of a feature (sometimes a Buddha) engraved on the back of this mirror appeared on a wall when sunlight was reflected from the front of the mirror onto a wall. The ancient Chinese gave it the name *light penetrating mirror*, the Japanese call it *Makyoh* or *magic mirror*.

TABLE 10.4   Color Chart for Deposited Si$_3$N$_4$ Films Observed Perpendicularly under Daylight Fluorescent Lighting.[85].

Film Thickness (μm)	Color	Film Thickness (μm)	Color
0.01	Very light brown	0.095	light blue
0.017	Medium brown	0.105	Very light blue
0.025	Brown	0.115	Light blue - brownish
0.034	Brown-pink	0.125	Light brown-yellow
0.035	Pink-purple	0.135	Very light yellow
0.043	Intense purple	0.145	Light yellow
0.0525	Intense dark blue	0.155	Light to medium yellow
0.06	Dark blue	0.165	Medium yellow
0.069	Medium blue	0.175	Intense yellow

The technique is illustrated in Fig. E10.3. A light beam is shone onto the sample and the reflected beam is projected onto a screen or video detector to form a slightly defocused image of the sample surface. The detail illustration shows that if the sample surface contains a flaw, such as a depression, the reflected image at the image plane, not the focal plane, shows this defect. It has been used in the semiconductor field for transforming latent damage, scratches, waviness, and other flaws on mirror-like semiconductor wafer surfaces into visual images. It can detect undulations of a few nm over distances of 0.5 mm.

For further discussions, see K. Kugimiya, S. Hahn, M. Yamashita, P. R. Blaustein, and K. Tanahashi, "Characterization of Mirror Polished Silicon Wafers Using the "Makyoh", the Magic Mirror Method," in *Semiconductor Silicon/1990* (H. R. Huff, K. G. Barraclough, and J. I. Chikawa, Eds.), Electrochem. Soc., Pennington, NJ, 1990, 1052–1067; K. Kugimiya, "Makyoh Topography: Comparison with X-Ray Topography," *Semicond. Sci. Technol.* **7**, A91-A94, Jan. 1992; I.E Lukács and F. Riesz, "Imaging-limiting Effects

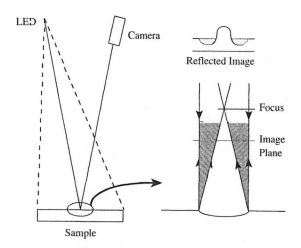

Fig. E10.3

of Apertures in Makyoh-topography Instruments," *Meas. Sci. Technol.* **12**, N29-N33, Aug. 2001.

10.5.3 Internal Reflection Infrared Spectroscopy

When light is incident on the interface between two media with differing refractive indices, *e.g.*, n_0 and n_1, some light is reflected. If both media are transparent, the light not reflected is transmitted and refracted according to Snell's law

$$n_1 \sin(\theta) = n_0 \sin(\varphi) \tag{10.40}$$

where θ, φ, n_0, and n_1 are indicated on Fig. 10.21(a) with $n_0 < n_1$. For $\varphi = 90°$, the light is totally reflected. For this condition, θ becomes the *critical angle* θ_c, given by

$$\sin(\theta_c) = \frac{n_0}{n_1} \tag{10.41}$$

Total internal reflection occurs for $\theta \leq \theta_c$. It is evident from Eq. (10.41) that total internal reflection calls for medium 0 to be less optically dense than medium 1 as found for air and a solid. For the air-Si interface with $n_0 = 1$ and $n_1 = 3.42$, $\theta_c = 17°$.

Internal reflection infrared spectroscopy probes the chemical nature of surfaces, films, and interfaces by relying on total internal reflection,[86] with the special geometry sample shown in Fig. 10.21(b). Infrared light is incident on one surface. The energy of the light must be less than the band-gap energy for there to be minimum absorption in the semiconductor. To be totally internally reflected, the incidence angle θ must be smaller than the critical angle. Once the light enters the solid sample, it encounters multiple reflections as it travels through the sample by total internal reflection before being detected. The large number of reflections give this technique its high sensitivity, *i.e.*, the light samples the surface many times as it traverses the sample.

The number of internal reflections N is given by

$$N = \frac{L}{d} \frac{1}{\tan(\theta)} \tag{10.42}$$

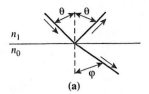

(a)

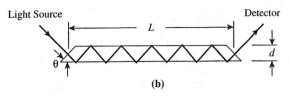

(b)

Fig. 10.21   (a) The behavior of light at the interface between two media, (b) fixed-angle, multiple pass internal reflection plate.

If only one surface is of interest then $N \to N/2$. This large number of reflections allows detection of surface species that are not amenable to IR analysis with just a single pass. Applications to semiconductors have been to study Si oxidation, Si/SiO$_2$ fluorination, hydrogen passivation of Si and a number of other applications.[87] The technique can also be used for real-time *in situ* semiconductor surface cleaning monitoring, for example.

10.6 LIGHT SCATTERING

One form of light scattering is *scatterometry*, which is the elastic scattering of light from particles and from surfaces with random or periodic variations. The particle may be much smaller than the wavelength of light. Elastic light scattering can detect particles in the gas phase, on surfaces, and in liquids. It is most commonly used in semiconductor characterization for detecting particles on surfaces and for critical dimension measurements. A useful rule of thumb is that particles with diameter of one-third the minimum feature size (usually the MOSFET gate length) can be killer defects, *i.e.*, have a substantial effect on circuit yield.

The schematic measurement arrangement is shown in Fig. 10.22. A focused laser beam is scanned across the sample surface and the scattered light is detected. The specular (directly reflected) light leaves the system to prevent "blinding" the detector. Although a semiconductor surface is very flat, it can have some microroughness or haze, scattering some light around the direction of the specularly reflected light. Particles scatter light in all directions. Optical detectors are placed at various locations around the system to capture as much light as possible and the light may be conditioned by polarizing it. For a particle in isolation, the scattered light is proportional to the optical scattering cross-section[88]

$$\sigma = \frac{\pi^4}{18} \frac{D^6}{\lambda^4} \left(\frac{K - 1}{K + 2} \right)^2 \tag{10.43}$$

where D is the particle diameter, λ the laser wavelength, and K the relative dielectric constant of the particle. Equation (10.43) is valid for $D \ll \lambda$. Scattering is not very sensitive to particle shape. It has been suggested that scattering from a particle on a surface is proportional to D^8.[89]

The particle density is detected by scattered light pulses as the laser is scanned across the sample. The particle size is determined through the size dependence of scattered light

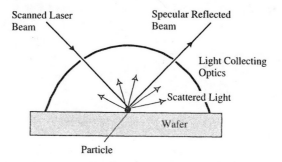

Fig. 10.22 Light scattering experimental schematic.

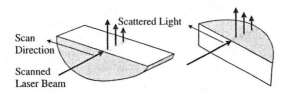

Fig. 10.23   Light scattering tomography schematic.

as given by Eq. (10.43). Smaller particles scatter less light than larger particles. Particles can be smaller than the wavelength of the light and still be detected. Their size, however, cannot be determined without recourse to calibrated standards, which are usually calibrated latex or Si spheres. Detection of small particles is akin to seeing smoke with a bare eye. We know where the smoke is, without being able to tell the size of the smoke particles. There is, however, a lower limit to the particle size that can be detected, because of wafer surface roughness. The wafer itself has a certain amount of scattered light and if the particle scatter falls within this surface scatter, it cannot be detected. Some of the surface interference can be overcome by using angle-resolved scatter measurements.[90]

Light scattering is also used in *light scattering tomography*, illustrated in Fig. 10.23. Light is incident on one surface of a sample. Because the wavelength is sufficiently high for the material to be reasonably transparent and owing to the high refractive index of semiconductors, the laser beam enters the material as a quasi-parallel beam. For incident light of 1060 nm wavelength, the light penetrates approximately 1000 μm into the Si wafer. The scattered light is detected as a linear image at right angles. The laser or sample is then moved and the next image is acquired. This is repeated until a two-dimensional image of the scattering centers is built up, giving a virtual "tomographic plane" parallel to the surface.[91] By appropriate sample preparation, one can obtain wafer cross-section or surface images.

10.7   MODULATION SPECTROSCOPY

Modulation spectroscopy is a sensitive technique to determine fine details of interband transitions in semiconductors through the derivative of the response function, *e.g.*, derivative of the optical reflectance or transmittance instead of the response function itself.[92] The derivative amplifies weak features in the response function and suppresses high background signals, giving the method high sensitivity to small spectral features not detectable by conventional means. The measurement is implemented by varying a property of the sample, *e.g.*, the electric field, or of the measuring system, *e.g.*, the light wavelength or polarization, and measuring the resultant signal.

The reflectance R of a sample depends on the dielectric function, which depends on a number of physical properties, *e.g.*, electric field. For an electric field with a *dc* ($\mathscr{E}_0$) and a small ac component ($\mathscr{E}_1 \cos(\omega t)$), the reflectance is[93]

$$R(\mathscr{E}) = R[\mathscr{E}_0 + \mathscr{E}_1 \cos(\omega t)] \approx R(\mathscr{E}_0) + \frac{dR}{d\mathscr{E}}(\mathscr{E}_1 \cos(\omega t)) \tag{10.44}$$

provided $\mathscr{E}_1 \ll \mathscr{E}_0$. The second term is a periodic function of time at the modulation frequency ω and small features in the optical spectrum are enhanced. In *electroreflectance*, the periodic perturbation is an applied electric field and in *photoreflectance* carriers are optically injected by a modulated laser. These injected carriers modulate the internal electric field and the reflectance. Electric fields can be applied by contacting a junction device or using electrolyte-semiconductor junctions. Other excitation sources are electron beams, heat pulses, and stress.

10.8   LINE WIDTH

Line widths are frequently called *critical dimensions* (*CD*) and their measurements are referred to as *CD* measurements. They are measured electrically, optically, by scanning probe techniques, and by scanning electron microscopy. A line width measurement system should be able to measure the width of the line and be repeatable to less than the tolerance—typically 10%. The measurement error should be three to ten times smaller than the process error. There are several terms related to line width measurements. *Accuracy* is the deviation of a measured line width from the true line width; *short-term precision* is the distribution of errors due to the instrument in repeated measurements; *long-term stability* is the variation of the average measured line width over time.

10.8.1   Optical-Physical Methods

Scatterometry: Optical techniques are capable of *CD* and overlay measurements of both conducting and insulating lines. Their strength is versatility, speed, and simplicity. Early optical techniques were: video scan, slit scan, laser scan, and image shearing.[94] Angle-resolved laser scattering from grating structures has been used for dimensional measurements.[95] The scattered/diffracted light depends on the structure and composition of the features. In a strict physical sense, this light 'scattered' from a periodic sample is due to *diffraction*, but in a general sense it is termed *scatter*. The scattered light from periodic features is sensitive to the geometry of the scattering features. The distribution of the energy pattern can be thought of as a scattering "signature." The technique is rapid, non-destructive and has demonstrated excellent precision, making it an attractive alternative to other metrologies in semiconductor manufacturing.

Scatterometry can be divided into the "forward problem" and the "inverse problem".[96] In the "forward problem" the scatter signature is measured, with the grating illuminated and the light detected to determine the "signature". In the "inverse problem" the line width of the scattering structure is quantified by model-based analyses, where the optical scatter data are compared to simulations from a theoretical model derived from Maxwell's equations. Traditionally, the model has been used *a priori* to generate a series of signatures that correspond to discrete iterations of various grating parameters, such as its thickness and the width of the grating lines. The resulting signature is known as a signature "library" or database. When the scatter signature is measured in the forward problem, it is compared against the library to find the closest match. The parameters of the modeled signature that agree most closely with the measured signature are taken to be the parameters of this measured signature.

Spectroscopic Ellipsometry (SE):   Spectroscopic ellipsometry measurements from period structures shows strong promise for high-speed topography measurements. SE-based scatterometry has proven to be more detailed than top-down scanning electron microscopy *CD* measurements. The emergence of this approach is directly analogous that of conventional thin-film ellipsometry when low-cost computers enabled the accurate solution of thin film reflection models. Data from complex thin film stacks could be analyzed very rapidly to yield-film thicknesses that compare favorably to cross-sectional transmission electron microscopy. By applying structures for which the diffraction problem can be numerically solved nearly exactly, the advantages of spectroscopic ellipsometry for patterned structures are being realized.

In SE for topography extraction the spectral reflectance is collected in the specular mode on reflection from a sample with a one-dimensional grating. The reflection problem from the grating is modeled using high-accuracy numerical simulation of Maxwell's equations. The optical dielectric functions of all materials in the lines and any underlying smooth thin films are assumed to be known, and are usually obtained by spectroscopic ellipsometry measurements of similarly prepared unpatterned thin films. Either a pattern matching procedure using a large, pre-simulated library of line shapes or a parameterized non-linear regression procedure is used to find the best fit between theory and experimental reflection data.[97]

Scanning Electron Microscopy: In scanning electron microscopy *CD* measurements, a focused electron beam is scanned across the sample and an image is built up by detecting secondary electrons.[98] The yield of secondary electrons depends on the sample geometry with higher secondary electron yield on sloped than on flat surfaces shown in Fig. 10.24(a) and (b). The line width is determined by measuring the distance between two edges on a line scan through the image, but it depends on the choice of the line width definition shown as W_1, W_2, and W_3. A typical width is taken at the 50% point. SEM *CD* metrology is routinely used and its main strength is the excellent SEM resolution, but the sample must be placed in a vacuum and is subject to charging. It is also subject to *line slimming* where the *e*-beam radiation can cause the line width to shrink due to photoresist cross linking as in Fig. 10.24(c).

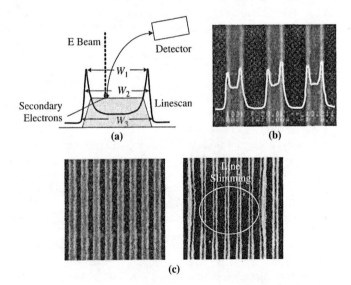

Fig. 10.24   Line width by scanning electron microscopy, (a) schematic showing the sample and line scan, (b) experimental curve courtesy of M. Postek, NIST, $W = 0.21$ μm (c) effect of line slimming.

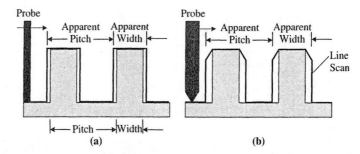

Fig. 10.25 Line width measurement by a mechanical probe. (a) Blunt probe (b) pointed probe. Shown are the resulting line widths modified by the probe shape.

Atomic Force Microscopy: Among the physical line width characterization techniques, one of the most sensitive is atomic force microscopy (AFM), discussed in Chapter 9. A mechanical stylus is scanned across the sample and the resulting profile is measured. AFM is very sensitive in the vertical dimension and somewhat less sensitive in the horizontal dimension. Nevertheless, horizontal resolution of tens of Ångströms is possible. AFM can also trace out the contours of lines and trenches. In spite of the high sensitivity of AFMs, one must use care in interpreting the experimental data, as illustrated in Fig. 10.25.[99] In Fig. 10.25(a) we show a line cross section on a semiconductor substrate, consisting of width W and pitch P. The probe scan depends on the probe shape, as illustrated in Figs. 10.25(a) and (b). The pitch is measured correctly in either case, but the line width is in error, even for the "ideal" rectangular probe shape. Not only is the line width in error, but so is the line shape. A knowledge of the probe geometry allows this to be corrected.

10.8.2 Electrical Methods

Electrical line width measurements, suitable only for conducting lines, are based on the test structure in Fig. 10.26.[100] Such measurements have shown high levels of repeatability. For a line width of 1 μm, the repeatability has been demonstrated to be on the order of 1 nm.[101] Precisions of 0.005 μm and lines as narrow as 0.1 μm have been measured. The left portion of the test structure is a cross resistor for van der Pauw sheet resistance measurements and the right portion is a bridge resistor. The cross resistor, discussed in Section 1.2.2, gives the sheet resistance as

$$R_{sh} = \frac{\pi}{\ln(2)} \frac{V_{34}}{I_{12}} \tag{10.45}$$

where $V_{34} = V_3 - V_4$ and I_{12} is the current flowing into contact I_1 and out of contact I_2. The voltage is measured between two adjacent contacts with the current flowing between the two opposite adjacent contacts. Averaging is usually done by changing the voltage and current contacts. The sheet resistance is determined in the shaded area.

The line width W is determined from the bridge resistor by

$$W = \frac{R_{sh} L}{V_{45}/I_{26}} \tag{10.46}$$

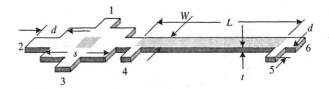

Fig. 10.26   Cross-bridge line width test structure.

where $V_{45} = V_4 - V_5$ and I_{26} is the current flowing into contact I_2 and out of contact I_6 and L is the length between voltage taps 4 and 5, known from the test structure layout.

An assumption in Eq. (10.46) is that the sheet resistance in the bridge portion of the test structure is the same as that in the cross portion, *i.e.*, in both shaded areas. If that is not true, W will be in error.[102] What exactly is L? Is it the center-to-center spacing as illustrated in Fig. 10.26? That depends on the exact layout of the structure. With arms 4 and 5 extending only below the measured line as in Fig. 10.26, L is approximately as shown. For symmetrical structures, *i.e.*, arms 4 and 5 extending above as well as below the line, an effective length is $L_{eff} \approx L - W_1$, where W_1 is the arm width. For long structures, *i.e.*, $L \approx 20\ W$, this correction is negligible, but for short lines, it must be considered, because the contact arms distort the current path. Other considerations are: $t \leq W$, $W \leq 0.005L$, $d \geq 2t$, $t \leq 0.03s$, $s \leq d$.[103]

10.9 PHOTOLUMINESCENCE (PL)

Photoluminescence, also known as *fluorometry*, provides a non-destructive technique for the determination of certain impurities in semiconductors.[104] It is particularly suited for the detection of shallow-level impurities, but can also be applied to certain deep-level impurities, provided their recombination is radiative.[105] Photoluminescence is also used in other applications. For example, ultraviolet light in fluorescent tubes, generated by an electric discharge, is absorbed by a phosphor inside the tube and visible light is emitted by photoluminescence. We discuss PL only briefly by giving the main concepts and a few examples. *Identification* of impurities is easy with PL, but measurement of the impurity *density* is more difficult. PL can provide simultaneous information on many types of impurities in a sample, but only those impurities that produce radiative recombination processes can be detected.

Photoluminescence has been largely the domain of III–V semiconductor characterization in the past with high internal efficiency. Internal efficiency is a measure of optically generated electron-hole pairs recombining radiatively thereby emitting light. Silicon, being an indirect band gap semiconductor, has low internal efficiency, because most recombination takes places through Shockley-Read-Hall or Auger recombination, neither of which emits light. In spite of the low internal efficiency, PL is now used to characterize Si, because the emitted light intensity depends on the defect and the doping density and is used to map either one.

A typical PL set-up is illustrated in Fig. 10.27. The sample is placed in a cryostat and cooled to temperatures near liquid helium. It is important that the sample be mounted in a strain-free manner, as strain affects the emitted light. Low temperature measurements are desirable to obtain the fullest spectroscopic information by minimizing thermally activated non-radiative recombination processes and thermal line broadening. The thermal

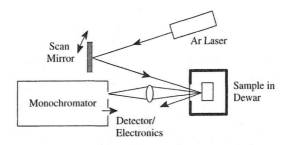

Fig. 10.27 Schematic photoluminescence arrangement.

distribution of carriers excited into a band contributes a width of approximately $kT/2$ to an emission line originating from that band. This makes it necessary to cool the sample to reduce the width. The thermal energy $kT/2$ is only 1.8 meV at $T = 4.2$ K. For many measurements this is sufficiently low, but occasionally it is necessary to reduce this broadening further by reducing the sample temperature below 4.2 K. Room temperature PL measurements, especially for Si, have become routine recently. They are used not to identify impurities but to provide PL maps of doping and trap densities.

The sample is excited with an optical source, typically a laser with energy $h\nu > E_G$, generating electron-hole pairs (ehps) which recombine by one of several mechanisms, discussed in Chapter 7. Photons are emitted for *radiative* recombination, but not for *non-radiative* recombination bulk or surface recombination. Some of the photons may be reabsorbed in the sample, provided they are directed at the surface within the critical angle. The emitted light is focused onto either a dispersive or a Fourier transform spectrometer and then a detector.

The internal PL efficiency is[106]

$$\eta_{int} = \int_0^d \frac{\Delta n}{\tau_{rad}} \exp(-\beta x) dx \approx \int_0^d \frac{\Delta n}{\tau_{rad}} dx \tag{10.49}$$

where d is the sample thickness, Δn the excess minority carrier density, and β the absorption coefficient of the *generated* light within the sample. The emitted light in Si has a wavelength near the band gap making the absorption coefficient β is very low ($\alpha \approx 2$ cm^{-1} for $h\nu = 1.12$ eV) and $\exp(-\beta x)$ is often neglected. This is not the case in general for other semiconductors. Δn depends on reflectance, photon flux density, and the various recombination mechanisms discussed in Appendix 7.1.

The photon energy depends on the recombination process, illustrated in Fig. 10.28, where five commonly observed PL transitions are shown.[107] Band-to-band recombination (Fig. 10.28(a)) dominates at room temperature but is rarely observed at low temperatures in materials with small effective masses due to the large electron orbital radii. Excitonic recombination is commonly observed, but what are excitons? When a photon generates an ehp, Coulombic attraction can lead to the formation of an excited state in which an electron and a hole remain bound to each other in a hydrogen-like state.[108] This excited state is referred to as a free *exciton* (FE). Its energy, shown in Fig. 10.28(b), is slightly less than the band gap energy required to create a *separated* ehp. An exciton can move through the crystal, but because it is a *bound* ehp, both electron and hole move together and neither photoconductivity nor current results. A free hole can combine with a neutral donor (Fig. 10.28(c)) to form a positively charged excitonic ion or *bound exciton* (BE).[109]

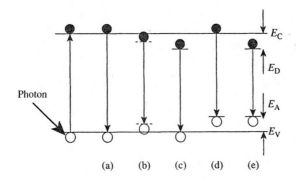

Fig. 10.28 Radiative transitions observed with photoluminescence.

The electron bound to the donor travels in a wide orbit about the donor. Similarly electrons combining with neutral acceptors also form bound excitons.

If the material is sufficiently pure, free excitons form and recombine by emitting photons. The photon energy in direct band-gap semiconductors of band-gap energy E_G is[109]

$$hv = E_G - E_x \tag{10.50}$$

where E_x is the excitonic binding energy. In indirect band gap semiconductors, momentum conservation requires the emission of a phonon, giving[109]

$$hv = E_G - E_x - E_p \tag{10.51}$$

where E_p is the phonon energy. Bound exciton recombination dominates over free exciton recombination for less pure material. A free electron can also recombine with a hole on a neutral acceptor (Fig. 10.28(d)), and similarly a free hole can recombine with an electron on a neutral donor.

Lastly, an electron on a neutral donor can recombine with a hole on a neutral acceptor, the well-known donor-acceptor (*D*-A) recombination, illustrated in Fig. 10.28(e). The emission line has an energy modified by the Coulombic interaction between donors and acceptors[105]

$$hv = E_G - (E_A + E_D) + \frac{q^2}{K_s \varepsilon_o r} \tag{10.52}$$

where r is the distance between donor and acceptor. The photon energy in Eq. (10.52) can be higher than the band gap for low $(E_A + E_D)$. Such photons are generally reabsorbed in the sample. The full widths at half maximum for bound exciton transitions are typically $\leq kT/2$ and resemble slightly broadened delta functions. This distinguishes them from donor-valence band transitions which are usually a few kT wide. Energies for these two transitions are frequently similar and the line widths are used to determine the transition type.

The optics in a PL apparatus are designed to ensure maximum light collection. The PL-emitted light from the sample can be analyzed by a grating monochromator and detected by a photodetector. A Michelson interferometer leads to enhanced sensitivity and reduced measurement time. One can also vary the wavelength of the incident light using a tunable dye laser. For wide band gap semiconductors it may be necessary to use electron beam excitation, since the excitation energy has to exceed the semiconductor band gap. PL radiation from shallow-level impurities in Si and GaAs can be detected with a photomultiplier tube with an S-1 photocathode able to detect wavelengths from about 0.4 to 1.1 μm. Lower-energy light from deeper levels requires a PbS (1–3 μm) or doped germanium detector.

The volume analyzed in PL measurements is determined by the absorption depth of the exciting laser light and the diffusion length of the minority carriers. Usually the absorption depth is on the order of microns or so. It is possible, however, to confine the absorbed light to a very thin layer near the surface by using ultraviolet light. This is useful in such materials as silicon-on-insulator, in which the active Si layer is only about 0.1 μm thick.[110] It is generally difficult to correlate the intensity of a given PL spectral line with the density of the impurity due to non-radiative bulk and surface recombination that vary from sample to sample and from location to location on a given sample. A novel approach to this problem is due to Tajima.[111] For Si samples of different resistivity, he found spectra with both intrinsic and extrinsic peaks as shown in Fig. 10.29. Higher resistivity samples showed higher intrinsic peaks. The ratio $X_{TO}(BE)/I_{TO}$ (FE) is proportional to the doping density, where X_{TO} (BE) is the transverse optical phonon PL intensity peak of the bound exciton for element X (boron or phosphorus) and I_{TO} (FE) is the transverse optical phonon intrinsic PL intensity peak of the free exciton.

Figure 10.30 shows two PL maps of an n/n^+ epitaxial Si wafer. The excitation wavelength for Fig. 10.30(a) was 0.532 nm with an absorption depth of about 1 μm and the figure shows quite uniform PL response indicative of a uniform epi layer. For Fig. 10.30(b), $\lambda = 827$ nm with an absorption depth around 9 μm, which probes the substrate, showing doping density variations of the heavily-doped substrate.

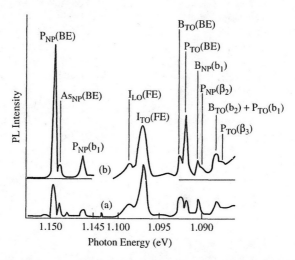

Fig. 10.29 Photoluminescence spectra for Si at $T = 4.2$ K. (a) Starting material, (b) after neutron transmutation doping. Base lines for measuring the peak heights are shown by the horizontal lines. Symbols: I = intrinsic, TO = transverse optical phonon, LO = longitudinal optical phonon, BE = bound exciton, FE = free exciton. The sample contains residual arsenic. Components labeled b_n and β_n are due to recombination of multiple bound excitons. Reprinted with permission after Tajima et al.[111] This paper was originally presented at the Spring 1981 Meeting of the Electrochemical Society, Inc. held in Minneapolis, MN.

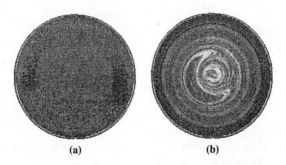

Fig. 10.30   Room temperature PL maps of an n/n^+ Si epitaxial wafer, $t_{epi} = 5$ μm; (a) $\lambda = 532$ nm, $1/\alpha \approx 1$ μm, (b) $\lambda = 827$ nm, $1/\alpha \approx 9$ μm. Courtesy of A. Buczkowski, SUMCO USA.

Calibration curves of photoluminescence intensity ratio versus impurity density for Si are shown in Fig. 2.29. Good agreement is found between the resistivity measured electrically and the resistivity calculated from the carrier density measured by photoluminescence. Very pure float-zone Si was used and varying amounts of phosphorus were introduced using neutron transmutation doping to generate calibration curves for the PL data.[112] It is estimated that for samples with areas of 0.3 cm^2 and 300 μm thickness, the detection limits for P, B, Al and As in Si are around 5×10^{10}, 10^{11}, 2×10^{11}, and 5×10^{11} cm^{-3}, respectively. Various impurities in Si have been catalogued.[113] The interpretation has also been applied to InP, where the donor density as well as the compensation ratio was determined.[114]

The ionization energies of donors in GaAs are typically around 6 meV and the energy difference between the various donor impurities is too small to be observable by conventional PL. However, acceptors with their wider spread of ionization energies can be detected by using the transitions: free electron to neutral acceptor (Fig. 10.28(d)) and electron on a neutral donor to hole on a neutral acceptor (Fig. 10.28(e)). Acceptors in GaAs determined with PL have also been catalogued.[115] Complications arise when the energy difference between the ground states of two or more acceptors is identical to the difference between their band-acceptor and donor-acceptor pair transitions. Such transitions can often be differentiated through variable temperature or variable excitation power measurements that cause a shift of the donor-acceptor pair transition to higher energies.[116] Donors in GaAs can be detected by *magneto-photoluminescence* measurements. The magnetic field splits some of the spectral lines into several components by splitting of the bound exciton initial states.[117] *Photothermal ionization spectroscopy*, discussed in Section 2.6.3, also allows donors to be identified.

10.10 RAMAN SPECTROSCOPY

Raman spectroscopy is a vibrational spectroscopic technique that can detect both organic and inorganic species and measure the crystallinity of solids.[118] It is free from charging effects. We mention it here because it is finding increased use in semiconductor characterization. For example, it is sensitive to strain, allowing it to be used to detect stress in a semiconductor material or device. Since the light beam can be focused to a small diameter, one can measure stress in small areas.

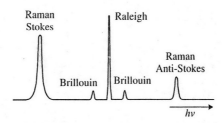

Fig. 10.31   Energy distribution of scattered light.

When light is scattered from the surface of a sample, the scattered light is found to contain mainly wavelengths that were incident on the sample (*Raleigh* scattering) but also at different wavelengths at very low intensities (few parts per million or less) that represent an interaction of the incident light with the material. The interaction of the incident light with optical phonons is called *Raman* scattering while the interaction with acoustic phonons results in *Brillouin* scattering. Optical phonons have higher energies than acoustic phonons giving larger photon energy shifts, illustrated in Fig. 10.31, but even for Raman scattering the energy shift is small. For example, the optical phonon energy in Si is about 0.067 eV, while the exciting photon energy is several eV (Ar laser light with $\lambda = 488$ nm has an energy of $h\nu = 2.54$ eV). Since the intensity of Raman scattered light is very weak (about 1 in 10^8 parts), Raman spectroscopy is only practical when an intense monochromatic light source like a laser is used.

Raman spectroscopy is based on the Raman effect first reported by Raman in 1928.[119] If the incident photon imparts energy to the lattice in the form of a phonon (phonon emission) it emerges as a lower-energy photon. This down-converted frequency shift is known as *Stokes-shifted* scattering. In *Anti-Stokes-shifted* scattering the photon absorbs a phonon and emerges with higher energy. The anti-Stokes mode is much weaker than the Stokes mode and it is Stokes-mode that is usually monitored.

During Raman spectroscopy measurements a laser pump beam is incident on the sample. The weak scattered light or signal is passed through a double monochromator to reject the Raleigh scattered light and the Raman-shifted wavelengths are detected by a photodetector. In the Raman microprobe, a laser illuminates the sample through a commercial microscope. Laser power is usually held below 5 mW to reduce sample heating and specimen decomposition. In order to separate the signal from the pump it is necessary that the pump be a bright, monochromatic source. Detection is made difficult by the weak signal against an intense background of scattered pump radiation. The signal-to-noise ratio is enhanced if the Raman radiation is observed at right angles to the pump beam. A major limitation in Raman spectroscopy is the interference caused by fluorescence, either of impurities or the sample itself. The fluorescent background problem is eliminated by combining Raman spectroscopy with FTIR.[120] Advances in FTIR and dispersive Raman measurements as well as lasers and detectors have been summarized.[121]

By using lasers with varying wavelengths and hence different absorption depths, it is possible to profile the sample to some depth. The technique is non-destructive and requires no contacts to the sample. Most semiconductors can be characterized by Raman spectroscopy. The wavelengths of the scattered light are analyzed and matched to known wavelengths for identification.

Various properties of the sample can be characterized. Its composition can be determined. Raman spectroscopy is also sensitive to crystal structure. For example, different

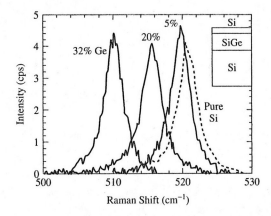

Fig. 10.32   Raman spectra of Si and Si grown on SiGe grown on Si. The percentages represent the germanium content in the SiGe layer. Courtesy of M. Canonico, Freescale Semiconductor.

crystal orientations give slightly different Raman shifts. However, damage and structural imperfections induce scattering by the forbidden TO phonons, allowing implant damage to be monitored, for example. The Stokes line shifts, broadens and becomes asymmetric for microcrystalline Si with grain sizes below 100 Å.[122] The lines become very broad for amorphous semiconductors, allowing a distinction to be made between single crystal, polycrystalline, and amorphous materials. The frequency is also shifted by stress and strain in thin film.[123] The strain in Si MOS technology introduced by SiGe and other approaches is eminently suitable for Raman characterization.[124] Both compressive and tensile stress can be determined with compressive stress giving an upward and tensile stress a downward shift from the unstressed 520 cm^{-1} Si shift. $1/\lambda \approx 520$ cm^{-1} corresponds to the optical phonon energy of 0.067 eV. The plots in Fig. 10.32 show the Raman spectra of Si, and Si on SiGe on Si. What is shown in these plots is the *wavenumber shift* from the incident light by the sample. SiGe has larger lattice constant than Si. Si grown on SiGe is under tensile stress leading to downward shift. The higher the Ge content, the higher the stress and the larger the shift. Regions as small as 200 nm have been characterized.[125]

The Raman microprobe is able to identify organic contaminants that appear as particles as small as 2 μm or as films as thin as 1 μm. The technique is most successful for organic materials because organic spectral data bases exist. For example, silicone films, teflon, cellulose, and other contaminants have been detected.[126] Raman spectroscopy is very effective, when coupled with other characterization techniques, for problem solving in semiconductor processing.[127]

10.11   STRENGTHS AND WEAKNESSES

Optical Microscopy: The strength of optical microscopy lies in its simplicity and well-established nature. It has been used for many years, is well developed and can be used for many applications from defect determination to IC inspection. Augmenting the basic technique with differential interference contrast, confocal microscopy, and near-field microscopy has expanded the technique further. The contactless nature of the measurement is a definite advantage. One of its main weaknesses is its resolution limit of about 0.25 μm. Near-field microscopy overcomes this limit, but is not easy to use.

Interference optical microscopes can measure areas from 50 μm^2 to 5 mm^2 by changing the microscope objective lens. The main disadvantage is that the height depends on the phase shift upon reflection. If a single material is measured, there is no problem. However, samples with differing optical properties yield erroneous results. Coating the material with a reflective material eliminates this problem.

Ellipsometry: The strength of ellipsometry is its widespread application for film thickness measurements. The addition of variable angle and multiple wavelength features has expanded the application of ellipsometry further, including such uses as *in-situ* process control due to its contactless nature. It measures the optical thickness of films, not the physical thickness. Knowing the refractive index allows the physical thickness to be determined. But the refractive index is not always known, especially for thin films, because the composition of the film may differ from a thicker film.

Transmission is primarily used for absorption coefficient and impurity density (*e.g.*, oxygen and carbon in silicon) determination. For α measurements there is no alternative. Impurities can, of course, be determined by such methods as secondary ion mass spectrometry, but optical transmission measurements are contactless and non-destructive. A weakness for impurity density determination is the sensitivity at low densities. For example, the density of carbon in silicon is around 10^{16} cm^{-3} or less, but the measurement sensitivity is around 10^{16} cm^{-3}, making it difficult to determine this impurity.

Reflection has traditionally been used for insulator thickness measurements where its contactless nature is a definite advantage. The use of internal reflection infrared spectroscopy has extended the method significantly by allowing the state of the surface to be monitored. As with ellipsometry, a disadvantage for thickness determination is that an optical thickness is measured.

Photoluminescence: This technique has the advantage of very high sensitivity. It is one of the most sensitive techniques to determine doping densities. It can also give defect information although defect densities are more difficult, since bulk and surface recombination are usually not known and they are difficult to separate. The disadvantages include the need for low temperature measurements for best sensitivity and the unknown nature of bulk and surface recombination.

Raman spectroscopy has become important for stress measurements, *e.g.*, for strained Si devices with stress introduced by one of several methods. It can also be used to measure stress in trenches, for example.

APPENDIX 10.1

Transmission Equations

Consider the sample of Fig. A10.1, characterized by reflection coefficients R_1, R_2, absorption coefficient α, complex refractive index $(n_1 - jk_1)$, and thickness d. Light of intensity I_i is incident from the left. $I_{r1} = R_1 I_i$ is reflected at point A and $(1 - R_1)I_i$ is transmitted into the sample, where it is attenuated as it traverses the sample. At point B, just inside the sample at $x = d$, the intensity is $(1 - R_1)\exp(-\alpha d)I_i$. The fraction $R_2(1 - R_1)\exp(-\alpha d)I_i$ is reflected back into the sample at point B and the fraction $I_{t1} = (1 - R_2)(1 - R_1)\exp(-\alpha d)I_i$ is transmitted through the sample. Some of the light reflected at B is reflected back into the sample at C and the component I_{r2} is reflected

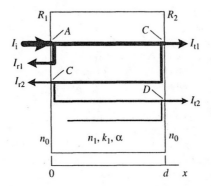

Fig. A10.1 Schematic showing the various reflected and transmitted light components.

back. Light is reflected back and forth and each time some of it is reflected, some is absorbed and some is transmitted.

When all the components are summed, the transmittance T is[76]

$$T = \frac{I_t}{I_i} = \frac{(1 - R_1)(1 - R_2)e^{-\alpha d}}{1 + R_1 R_2 e^{-2\alpha d} - 2\sqrt{R_1 R_2} e^{-\alpha d} \cos(\varphi)} \tag{A10.1}$$

where $\varphi = 4\pi n_1 d/\lambda$. For symmetrical samples, $R_1 = R_2 = R$, allowing Eq. (A10.1) to become

$$T = \frac{(1 - R)^2 e^{-\alpha d}}{1 + R^2 e^{-2\alpha d} - 2R e^{-\alpha d} \cos(\varphi)} \tag{A10.2}$$

The "cos" term can be written as $\cos(f/f_1)$, where $f = 2\pi/\lambda$ and $f_1 = 1/2n_1 d$ is a spatial frequency. If the detector does not have sufficient spectral resolution, then the oscillations due to the "$\cos(\varphi)$" term average to zero, calculated by averaging the transmitted intensity over a period of the cosine term as[128]

$$T = \frac{1}{2\pi} \int_{-\pi}^{\pi} \frac{(1 - R)^2 e^{-\alpha d}}{1 + R^2 e^{-2\alpha d} - 2R e^{-\alpha d} \cos(\varphi)} \, d\varphi \tag{A10.3}$$

Assuming α and n_1 to be constant over the wavelength interval, the transmittance becomes

$$T = \frac{(1 - R)^2 e^{-\alpha d}}{1 - R^2 e^{-2\alpha d}} \tag{A10.4}$$

R is the reflectance given by

$$R = \frac{(n_0 - n_1)^2 + k_1^2}{(n_0 + n_1)^2 + k_1^2} \tag{A10.5}$$

and the absorption coefficient α is related to the extinction coefficient k_1 by

$$\alpha = \frac{4\pi k_1}{\lambda} \tag{A10.6}$$

$\cos(\varphi)$ has maxima when $m\lambda_0 = 2n_1 d$, where $m = 1, 2, 3 \ldots$ and can be used to determine the sample thickness through the relationship

$$d = \frac{m\lambda_0}{2n_1} = \frac{(m+1)\lambda_1}{2n_1} = \frac{(m+i)\lambda_i}{2n_1} \tag{A10.7}$$

or $m = i\lambda_i/[\lambda_0 - \lambda_i]$, where i = number of complete cycles from λ_0 to λ_i. For one cycle $i = 1$ and

$$d = \frac{1}{2n_1(1/\lambda_0 - 1/\lambda_1)} = \frac{1}{2n_1\Delta(1/\lambda)} \tag{A10.8}$$

where $1/\lambda$ = wavenumber and $\Delta(1/\lambda)$ = wavenumber interval between two maxima or minima.

APPENDIX 10.2

Absorption Coefficients and Refractive Indices for Selected Semiconductors

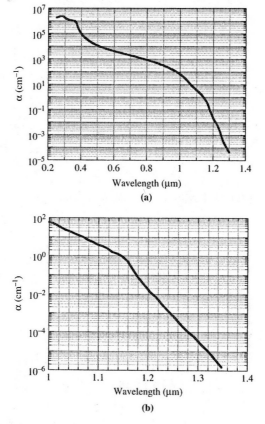

Fig. A10.2 Absorption coefficient as a function of wavelength for Si. Adapted from data in Green[129] and Daub/Würfel.[130]

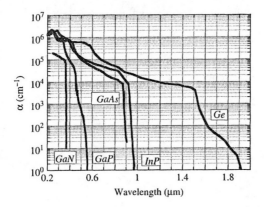

Fig. A10.3   Absorption coefficient as a function of wavelength for selected semiconductors. Adapted from data in Palik[131] and Muth et al.[132]

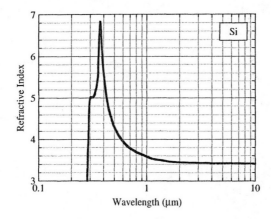

Fig. A10.4   Refractive index as a function of wavelength for silicon. Adapted from data in Palik.[131]

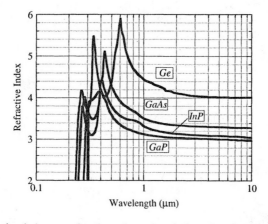

Fig. A10.5   Refractive index as a function of wavelength for selected semiconductors. Adapted from data in Palik.[131]

REFERENCES

1. I.P. Herman, *Optical Diagnostics for Thin Film Processing*, Academic Press, San Diego, 1996.

2. W.C. McCrone and J.G. Delly, *The Particie Atlas*, Ann Arbor Science Publ., Ann Arbor, MI, 1973; J.K. Beddow (ed.) *Particle Characterization in Technology*, Vols. I and II, CRC Press, Boca Raton, FL, 1984.

3. G. Airy, *Mathematical Transactions*, 2nd ed., Cambridge, 1836.

4. M. Spencer, *Fundamentals of Light Microscopy*, Cambridge University Press, Cambridge, 1982.

5. T.G. Rochow and E.G. Rochow, *An Introduction to Microscopy by Means of Light, Electrons, X-Rays, or Ultrasound*, Plenum Press, New York, 1978.

6. H.N. Southworth, *Introduction to Modern Microscopy*, Wykeham Publ., London, 1975.

7. G. Nomarski, "Microinterferometre Differentiel a Ondes Polarisés," *J. Phys. Radium* **16**, 9S-13S, 1955. French Patents Nos. 1059124 and 1056361; D.C. Miller and G.A. Rozgonyi, "Defect Characterization by Etching, Optical Microscopy and X-Ray Topography," in *Handbook on Semiconductors*, **3** (S.P. Keller, ed.) North-Holland, Amsterdam, 1980, 217–246.

8. J.H. Richardson, *Handbook for the Light Microscope, A User's Guide*, Noyes Publ., Park Ridge, NJ, 1991.

9. M. Minsky, "Memoir on Inventing the Confocal Scanning Microscope," *Scanning*, **10**, 128–138, 1988.

10. T. Wilson and C.J.R. Sheppard, *Theory and Practice of Scanning Optical Microscopy*, Academic Press, London, 1984; T. Wilson (ed.), *Confocal Microscopy*, Academic Press, London, 1990.

11. T.R. Corle and G.S. Kino, *Confocal Scanning Optical Microscopy and Related Imaging Systems*, Academic Press, San Diego, 1996.

12. R.H. Webb, "Confocal Optical Microscopy," *Rep. Progr. Phys.* **59**, 427–471, March 1996.

13. P. Nipkow, Electrical Telescope (in German), German Patent # 30105, 1884.

14. M. Petran, M. Hadravsky, M.D. Egger and R. Galambos, "Tandem-scanning Reflected Light Microscope," *J. Opt. Soc.* **58**, 661–664, May 1968; M. Petran, M. Hadravsky, and A. Boyde, "The Tandem-scanning Reflected Light Microscope," *Scanning,* **7**, 97–108, March/April 1985.

15. G. Udupa, M. Singaperumal, R.S. Sirohi, and M.P. Kothiyal, "Characterization of Surface Topography by Confocal Microscopy: I. Principles and the Measurement System," *Meas. Sci. Technol.* **11**, 305–314, March 2000.

16. D. Malacara (ed.), *Optical Shop Testing*, 2nd ed., Wiley, New York, 1992.

17. P.C. Montgomery, J.P. Fillard, M. Castagné, and D. Montaner, "Phase-Stepping Microscopy (PSM): A Qualification Tool for Electronic and Optoelectronic Devices," *Semicond. Sci. Technol.* **7**, A237–A242, Jan. 1992.

18. K. Creath, "Step Height Measurement Using Two-Wavelength Phase-Shifting Interferometry," *Appl. Opt.* **26**, 2810–2816, July 1987.

19. B. Bhushan, J.C. Wyant, and C.L. Koliopoulos, "Measurement of Surface Topography of Magnetic Tapes by Mirau Interferometry," *Appl. Opt.* **24**, 1489–1497, May 1985.

20. P.J. Caber, S.J. Martinek, and R.J. Niemann, "A New Interferometric Profiler for Smooth and Rough Surfaces," *Proc. SPIE* **2088**, 195–203, 1993.

21. ASTM Standard F 47–94, "Standard Test Method for Crystallographic Perfection of Silicon by Preferential Etch Techniques," *1996 Annual Book of ASTM Standards*, Am. Soc. Test. Mat., West Conshohocken, PA, 1996.

22. ASTM Standard F 154–94, "Standard Practices and Nomenclature for Identification of Structures and Contaminants Seen on Specular Silicon Surfaces," *1996 Annual Book of ASTM Standards*, Am. Soc. Test. Mat., West Conshohocken, PA, 1996.

23. E. Sirtl and A. Adler, "Chromic Acid-Hydrofluoric Acid as Specific Reagents for the Development of Etching Pits in Silicon," *Z. Metallkd.* **52**, 529–534, Aug. 1961.

24. W.C. Dash, "Copper Precipitation on Dislocations in Silicon," *J. Appl. Phys.* **27**, 1193–1195, Oct. 1956; "Evidence of Dislocation Jogs in Deformed Silicon," *J. Appl. Phys.* **29**, 705–709, April 1958.

25. F. Secco d'Aragona, "Dislocation Etch for (100) Planes in Silicon," *J. Electrochem. Soc.* **119**, 948–951, July 1972.

26. D.G. Schimmel, "Defect Etch for < 100 > Silicon Ingot Evaluation," *J. Electrochem. Soc.* **126**, 479–483, March 1979; D.G. Schimmel and M.J. Elkind, "An Examination of the Chemical Staining of Silicon," *J. Electrochem. Soc.* **125**, 152–155, Jan. 1978.

27. M. Wright-Jenkins, "A New Preferential Etch for Defects in Silicon Crystals," *J. Electrochem. Soc.* **124**, 757–762, May 1977.

28. K.H. Yang, "An Etch for Delineation of Defects in Silicon," *J. Electrochem. Soc.* **131**, 1140–1145, May 1984.

29. H. Seiter, "Integrational Etching Methods," in *Semiconductor Silicon/1977* (H.R. Huff and E. Sirtl, eds,), Electrochem. Soc., Princeton, NJ, 1977, 187–195.

30. T.C. Chandler, "MEMC Etch - A Chromium Trioxide-Free Etchant for Delineating Dislocations and Slip in Silicon," *J. Electrochem. Soc.* **137**, 944–948, March 1990.

31. M. Ishii, R. Hirano, H. Kan, and A. Ito, "Etch Pit Observation of Very Thin {001}-GaAs Layer by Molten KOH," *Japan. J. Appl. Phys.* **15**, 645–650, April 1976; for a more detailed discussion of GaAs Etching see D.J. Stirland and B.W. Straughan, "A Review of Etching and Defect Characterisation of Gallium Arsenide Substrate Material," *Thin Solid Films* **31**, 139–170, Jan. 1976.

32. D.T.C. Huo, J.D. Wynn, M.Y. Fan and D.P. Witt, "InP Etch Pit Morphologies Revealed by Novel HCl-Based Etchants," *J. Electrochem. Soc.* **136**, 1804–1806, June 1989.

33. E. Abbé, *Archiv. Mikroskopische Anat. Entwicklungsmech.* **9**, 413, 1873; E. Abbé, *J. R. Microsc. Soc.* **4**, 348, 1884.

34. E.H. Synge, "A Suggested Method for Extending Microscopic Resolution into the Ultra-Microscopic Region," *Phil. Mag.* **6**, 356–362, 1928.

35. E.A. Ash and G. Nicholls, "Super-Resolution Aperture Scanning Microscope," *Nature* **237**, 510–512, June 1972.

36. D.W. Pohl, W. Denk, and M. Lanz, "Optical Stethoscopy: Image Recording with Resolution $\lambda/20$," *Appl. Phys. Lett.* **44**, 651–653, Apr. 1984; A. Lewis, M. Isaacson, A. Harootunian, and A. Muray, "Development of a 500 Å Spatial Resolution Light Microscope: I. Light Is Efficiently Transmitted Through $\lambda/6$ Diameter Apertures," *Ultramicroscopy*, **13**, 227–231, 1984; E. Betzig and J.K. Trautman, "Near-Field Optics: Microscopy, Spectroscopy, and Surface Modification Beyond the Diffraction Limit," *Science* **257**, 189–195, July 1992.

37. B.T. Rosner and D.W. van der Weide, "High-frequency Near-field Microscopy," *Rev. Sci. Instrum.* **73**, 2502–2525, July 2002.

38. J.W.P. Hsu, "Near-field Scanning Optical Microscopy Studies of Electronic and Photonic Materials and Devices," *Mat. Sci. Eng. Rep.* **33**, 1–50, May 2001.

39. H.G. Tompkins, *A User's Guide to Ellipsometry,"* Academic Press, Boston, 1993; ASTM Standard F576-90, "Standard Test Method for Measurement of Insulator Thickness and Refractive Index on Silicon Substrates by Ellipsometry," *1996 Annual Book of ASTM Standards*, Am. Soc. Test. Mat., West Conshohocken, PA, 1996.

40. R.M.A. Azzam and N.M. Bashara, *Ellipsometry and Polarized Light*, North-Holland, Amsterdam, 1989.

41. R.K. Sampson and H.Z. Massoud, "Resolution of Silicon Wafer Temperature Measurement by *In situ* Ellipsometry in a Rapid Thermal Processor," *J. Electrochem. Soc.* **140**, 2673–2678, Sept. 1993.

42. G. Gergely, ed., Ellipsometric Tables of the Si–SiO$_2$ System for Mercury and He-Ne Laser Spectral Lines, Akadémiai Kiadó, Budapest, 1971.

43. R.H. Muller, "Principles of Ellipsometry," in *Adv. in Electrochem. and Electrochem. Eng.* **9**, (R.H. Muller, ed.), Wiley, New York, 1973, 167–226.

44. D.E. Aspnes and A.A. Studna, "High Precision Scanning Ellipsometer," *Appl. Opt.* **14**, 220–228, Jan. 1975.

45. K. Riedling, *Ellipsometry for Industrial Applications*, Springer, Vienna, 1988.

46. D.E. Aspnes, "New Developments in Spectroellipsometry: The Challenge of Surfaces," *Thin Solid Films* **233**, 1–8, Oct. 1993.

47. G.N. Maracas and C.H. Kuo, "Real Time Analysis and Control of Epitaxial Growth," in *Semiconductor Characterization: Present Status and Future Needs* (W.M. Bullis, D.G. Seiler, and A.C. Diebold, Eds.), Am. Inst. Phys., Woodbury, NY, 476–484, 1996.

48. W.M. Duncan and S.A. Henck, "*In situ* Spectral Ellipsometry for Real-Time Measurement and Control," *Appl. Surf. Sci.* **63**, 9–16, Jan. 1993.

49. A. Moritani and C. Hamaguchi, "High-Speed Ellipsometry of Arsenic-Implanted Si During CW Laser Annealing," *Appl. Phys. Lett.* **46**, 746–748, April 1985.

50. M. Erman and J.B. Theeten, "Multilayer Analysis of Ion Implanted GaAs Using Spectroscopic Ellipsometry," *Surf. and Interf. Analys.* **4**, 98–108, June 1982.

51. F. Hottier, J. Hallais and F. Simondet, "*In situ* Monitoring by Ellipsometry of Metalorganic Epitaxy of GaAlAs-GaAs Superlattice," *J. Appl. Phys.* **51**, 1599–1602, March 1980.

52. D.E. Aspnes, "The Characterization of Materials by Spectroscopic Ellipsometry," *Proc. SPIE* **452**, 60–70, 1983.

53. H-T Huang and F.L. Terry, Jr., "Spectroscopic Ellipsometry and Reflectometry from Gratings (Scatterometry) for Critical Dimension Measurement and *In situ*, Real-time Process Monitoring," *Thin Solid Films*, **455/456**, 828–836, May 2004.

54. ASTM Standard F 120, "Standard Practices for Determination of the Concentration of Impurities in Single Crystal Semiconductor Materials by Infrared Absorption Spectroscopy," *1988 Annual Book of ASTM Standards*, Am. Soc. Test. Mat., Philadelphia, 1988.

55. P. Stallhofer and D. Huber, "Oxygen and Carbon Measurements on Silicon Slices by the IR Method," *Solid State Technol.* **26**, 233–237, Aug. 1983; H.J. Rath, P. Stallhofer, D. Huber and B.F. Schmitt, "Determination of Oxygen in Silicon by Photon Activation Analysis for Calibration of the Infrared Absorption," *J. Electrochem. Soc.* **131**, 1920–1923, Aug. 1984.

56. K.L. Chiang, C.J. Dell'Oca and F.N. Schwettmann "Optical Evaluation of Polycrystalline Silicon Surface Roughness," *J. Electrochem. Soc.* **126**, 2267–2269, Dec. 1979.

57. A.A. Michelson, "Visibility of Interference Fringes in the Focus of a Telescope," *Phil Mag.* **31**, 256–259, 1891; "On the Application of Interference Methods to Spectroscopic Measurements, *Phil. Mag.* **31**, 338–346, 1891; **34**, 280–299, 1892.

58. Lord Raleigh, "On the Interference Bands of Approximately Homogeneous Light; in a Letter to Prof. A. Michelson," *Phil Mag.* **34**, 407–411, 1892.

59. J.W. Cooley and J.W. Tukey, "An Algorithm for the Machine Calculation of Complex Fourier Series," *Math. Comput.* **19**, 297–301, April 1965.

60. P.R. Griffith and J.A. de Haseth, *Fourier Transform Infrared Spectrometry*, Wiley, New York, 1986.

61. W.D. Perkins, "Fourier Transform-Infrared Spectroscopy," *J. Chem Educ.* **63**, A5–A10, Jan. 1986.

62. G. Horlick, "Introduction to Fourier Transform Spectroscopy," *Appl. Spectrosc.* **22**, 617–626, Nov./Dec. 1968.

63. ASTM Standard F 121, "Standard Test Method for Interstitial Atomic Oxygen Content of Silicon by Infrared Absorption," *1988 Annual Book of ASTM Standards*, Am. Soc. Test. Mat., Philadelphia, 1988.

64. ASTM Standard F 1391-93, "Standard Test Method for Substitutional Atomic Carbon Content of Silicon by Infrared Absorption," *1996 Annual Book of ASTM Standards*, Am. Soc. Test. Mat., West Conshohocken, PA, 1996.

65. K. Tanahashi and H. Yamada-Kaneta, "Technique for Determination of Nitrogen Concentration in Czochralski Silicon by Infrared Absorption Measurement," *Japan. J. Appl. Phys.* **42**, L 223–L 225, March 2003.

66. R.W. Shaw, R. Bredeweg, and P. Rossetto, "Gas Fusion Analysis of Oxygen in Silicon: Separation of Components," *J. Electrochem. Soc.* **138**, 582–585, Feb. 1991.

67. W.M. Bullis, M. Watanabe, A. Baghdadi, Y.Z. Li, R.I. Scace, R.W. Series and P. Stallhofer, "Calibration of Infrared Absorption Measurements of Interstitial Oxygen Concentration in Silicon," in *Semiconductor Silicon/1986* (H.R. Huff, T. Abe and B.O. Kolbesen, eds.), Electrochem. Soc., Pennington, NJ, 1986, 166–180; W.M. Bullis, "Oxygen Concentration Measurements" in *Oxygen in Silicon* (F. Shimura, ed.), Academic Press, Boston, 1994, Ch. 4.

68. J. Weber and M. Singh, "New Method to Determine the Carbon Concentration in Silicon," *Appl. Phys. Lett.* **49**, 1617–1619, Dec. 1986.

69. G.G. MacFarlane, T.P. McClean, J.E. Quarrington and V. Roberts, "Fine Structure in the Absorption-Edge Spectrum of Si," *Phys. Rev.* **111**, 1245–1254, Sept. 1958.

70. W. Kern and G.L. Schnable, "Chemically Vapor-Deposited Borophosphosilicate Glasses for Silicon Device Applications," *RCA Rev.* **43**, 423–457, Sept. 1982.

71. T. Iizuka, S. Takasu, M. Tajima, T. Arai, N. Inoue, and M. Watanabe, "Determination of Conversion Factor for Infrared Measurement of Oxygen in Silicon," *J. Electrochem. Soc.* **132**, 1707–1713, July 1985.

72. K. Graff, E. Grallath, S. Ades, G. Goldbach, and G. Tolg, "Determination of Parts Per Billion of Oxygen in Silicon by Measurement of the IR-Absorption of 77 K," *Solid-State Electron.* **16**, 887–893, Aug. 1973; Deutsche Normen DIN 50 438/1 "Determination of the Contamination Level in Silicon Through IR Absorption: O_2 in Si," (in German) Beuth Verlag, Berlin, 1978.

73. A. Baghdadi, W.M. Bullis, M.C. Croarkin, Y-Z. Li, R.I. Scace, R.W. Series, P. Stallhofer, and M. Watanabe, "Interlaboratory Determination of the Calibration Factor for the Measurement of the Interstitial Oxygen Content of Silicon by Infrared Absorption," *J. Electrochem. Soc.* **136**, 2015–2024, July 1989; ASTM Standard F 1188-93a, "Standard Test Method for Interstitial Atomic Oxygen Content of Silicon by Infrared Absorption," *1996 Annual Book of ASTM Standards*, Am. Soc. Test. Mat., West Conshohocken, PA, 1996.

74. J.L. Regolini, J.P. Stoquert, C. Ganter, and P. Siffert, "Determination of the Conversion Factor for Infrared Measurements of Carbon in Silicon," *J. Electrochem. Soc.* **133**, 2165–2168, Oct. 1986.

75. G.M. Martin, "Optical Assessment of the Main Electron Trap in Bulk Semi-Insulating GaAs," *Appl. Phys. Lett.* **39**, 747–748, Nov. 1981.

76. H. Anders, *Thin Films in Optics*, The Focal Press, London, 1967, Ch.1.

77. W.R. Runyan and T.J. Shaffner, *Semiconductor Measurements and Instrumentation*, McGraw-Hill, New York, 1997.

78. P. Burggraaf, "How Thick Are Your Thin Films?" *Semicond. Int.* **11**, 96–103, Sept. 1988.

79. J.R. Sandercock, "Film Thickness Monitor Based on White Light Interference," *J. Phys. E: Sci. Instrum.* **16**, 866–870, Sept. 1983.

80. W.E. Beadle, J.C.C. Tsai and R.D. Plummer, *Quick Reference Manual for Silicon Integrated Circuit Technology*, Wiley-Interscience, New York, 1985, 4–23.

81. W.A. Pliskin and E.E. Conrad, "Nondestructive Determination of Thickness and Refractive Index of Transparent Films," *IBM J. Res. Develop.* **8**, 43–51, Jan. 1964; W.A. Pliskin and R.P. Resch, "Refractive Index of SiO_2 Films Grown on Silicon," *J. Appl. Phys.* **36**, 2011–2013, June 1965.

82. ASTM Standard F 95-89, "Standard Test Method for Thickness of Lightly Doped Silicon Epitaxial Layers on Heavily Doped Silicon Substrates by an Infrared Dispersive Spectrophotometer," *1997 Annual Book of ASTM Standards*, Am. Soc. Test. Mat., West Conshohocken 1997.

83. P.A. Schumann, Jr., "The Infrared Interference Method of Measuring Epitaxial Layer Thickness," *J. Electrochem. Soc.* **116**, 409–413, March 1969.

84. B. Senitsky and S.P. Weeks, "Infrared Reflectance Spectra of Thin Epitaxial Silicon Layers," *J. Appl. Phys.* **52**, 5308–5313, Aug. 1981.

85. F. Reizman and W.E. van Gelder, "Optical Thickness Measurement of SiO_2-Si_3N_4 Films on Si," *Solid-State Electron.* **10**, 625–632, July 1967.

86. Y.J. Chabal, "Surface Infrared Spectroscopy," *Surf. Sci. Rep.* **8**, 211–357, May 1988.

87. V.A. Burrows, "Internal Reflection Infrared Spectroscopy for Chemical Analysis of Surfaces and Thin Films," *Solid-State Electron.* **35**, 231–238, March 1992.

88. J. Stover, *Optical Scattering: Measurement and Analysis*, McGraw-Hill, New York, 1990.

89. H.R. Huff, R.K. Goodall, E. Williams, K.S. Woo, B.Y.H. Liu, T. Warner, D. Hirleman, K. Gildersleeve, W.M. Bullis, B.W. Scheer, and J. Stover, "Measurement of Silicon Particles by Laser Surface Scanning and Angle-Resolved Light Scattering," *J. Electrochem. Soc.* **144**, 243–250, Jan. 1997.

90. T.L. Warner and E.J. Bawolek, "Reviewing Angle-Resolved Methods for Improved Surface Particle Detection," *Microcont.* **11**, 35–39, Sept./Oct. 1993.

91. K. Moriya and T. Ogawa, "Observation of Lattice Defects in GaAs and Heat-Treated Si Crystals by Infrared Light Scattering Tomography," *Japan. J. Appl. Phys.* **22**, L207–L209, April 1983; J.P. Fillard, P. Gall, J. Bonnafé, M. Castagné, and T. Ogawa, "Laser-Scanning Tomography: A Survey of Recent Investigations in Semiconductor Materials," *Semicond. Sci. Technol.* **7**, A283–A287, Jan. 1992; G. Kissinger, D. Gräf, U. Lambert, and H. Richter, "A Method for Studying the Grown-in Defect Density Spectra in Czochralski Silicon Wafers," *J. Electrochem. Soc.* **144**, 1447–1456, April 1997.

92. F.H. Pollack and H. Shen, "Modulation Spectroscopy of Semiconductors: Bulk/Thin Film, Microstructures, Surfaces/Interfaces, and Devices," *Mat. Sci. Eng.* **R10**, 275–374, Oct. 1993.

93. S. Perkowitz, D.G. Seiler, and W.M. Duncan, "Optical Characterization in Microelectronics Manufacturing," *J. Res. Natl. Inst. Stand. Technol.* **99**, 605–639, Sept./Oct. 1994.

94. P.H. Singer, "Linewidth Measurement Aids Process Control," *Semicond. Int.* **8**, 66–73, Feb. 1985.

95. S.A. Coulombe, B.K. Minhas, C.J. Raymond, S.S.H. Naqvi, and J.R. McNeil, "Scatterometry Measurement of Sub-0.1 μm Linewidth Gratings," *J. Vac. Sci. Technol.* **B16**, 80–87, Jan. 1998.

96. C.J. Raymond, "Scatterometry for Semiconductor Metrology," in *Handbook of Silicon Semiconductor Technology* (A.C. Diebold, ed.), Dekker, New York, 2001, Ch. 18.

97. H-T Huang and F.L. Terry, Jr., "Spectroscopic Ellipsometry and Reflectometry from Gratings (Scatterometry) for Critical Dimension Measurement and *In situ*, Real-time Process Monitoring," *Thin Solid Films*, **455/456**, 828–836, May 2004.

98. M.T. Postek, "Scanning Electron Microscope Metrology," in *Handbook of Critical Dimension Metrology and Process Control* (K.M. Monahan, ed.), SPIE Optical Engineering Press, Bellingham, WA, 1994, 46–90.

99. J.E. Griffith and D.A. Grigg, "Dimensional Metrology With Scanning Probe Microscopes," *J. Appl. Phys.* **74**, R83–R109, Nov. 1993.

100. M.G. Buehler and C.W. Hershey, "The Split-Cross-Bridge Resistor for Measuring the Sheet Resistance, Linewidth, and Line Spacing of Conducting Layers," *IEEE Trans. Electron Dev.* **ED-33**, 1572–1579, Oct. 1986; ASTM Standard F1261M-95, "Standard Test Method for Determining the Average Electrical Width of a Straight, Thin-Film Metal line," *1996 Annual Book of ASTM Standards*, Am. Soc. Test. Mat., West Conshohocken, PA, 1996.

101. M.W. Cresswell, J.J. Sniegowski, R.N. Goshtagore, R.A. Allen, W.F. Guthrie, and L.W. Linholm, "Electrical Linewidth Test Structures Fabricated in Mono-Crystalline Films for Reference-Material Applications," in *Proc. Int. Conf. Microelectron. Test. Struct.*, Monterey, CA, 1997, 16–24.

102. R.A. Allen, M.W. Cresswell, and L.M. Buck, "A New Test Structure for the Electrical Measurement of the Width of Short Features With Arbitrarily Wide Voltage Taps," *IEEE Electron Dev. Lett.* **13**, 322–324, June 1992.

103. G. Storms, S. Cheng, and I. Pollentier, "Electrical Linewidth Metrology for Sub-65 nm Applications," *Proc. SPIE*, **5375**, 614–628, 2004.

104. H.B. Bebb and E.W. Williams, "Photoluminescence I: Theory," in *Semiconductors and Semimetals* (R.K. Willardson and A.C. Beer, eds.) Academic Press, New York, **8**, 181–320, 1972; E.W. Williams and H.B. Bebb, "Photoluminescence II: Gallium Arsenide," *ibid.* 321–392.

105. P.J. Dean, "Photoluminescence as a Diagnostic of Semiconductors," *Prog. Crystal Growth Charact.* **5**, 89–174, 1982.

106. J. Vilms and W.E. Spicer, "Quantum Efficiency and Radiative Lifetime in p-Type Gallium Arsenide," *J. Appl. Phys.* **36**, 2815–2821, Sept. 1965; H.J. Hovel, "Scanned Photoluminescence of Semiconductors," *Semicond. Sci. Technol.* **7**, A1–A9, Jan. 1992.

107. K.K. Smith, "Photoluminescence of Semiconductor Materials," *Thin Solid Films* **84**, 171–182, Oct. 1981.

108. J.P. Wolfe and A. Mysyrowicz, "Excitonic Matter," *Sci. Am.* **250**, 98–107, March 1984.

109. J.I. Pankove, *Optical Processes in Semiconductors*, Dover Publications, New York, 1975.

110. M. Tajima, S. Ibuka, H. Aga, and T. Abe, "Characterization of Bond and Etch-Back Silicon-on-Insulator Wafers by Photoluminescence Under Ultraviolet Excitation," *Appl. Phys. Lett.* **70**, 231–233, Jan. 1997.

111. M. Tajima, "Determination of Boron and Phosphorus Concentration in Silicon by Photoluminescence Analysis," *Appl. Phys. Lett.* **32**, 719–721, June 1978; M. Tajima, T. Masui, T. Abe and T. Iizuka, "Photoluminescence Analysis of Silicon Crystals," in *Semiconductor Silicon/1981* (H.R. Huff, R.J. Kriegler, and Y. Takeishi, eds.), Electrochem. Soc., Pennington, NJ, 1981, pp. 72–89.

112. M. Tajima, "Recent Advances in Photoluminescence Analysis of Si: Application to an Epitaxial Layer and Nitrogen in Si," *Japan. J. Appl. Phys.* **21**, Supplement 21–1, 113–119, 1982.

113. P.J. Dean, R.J. Haynes, and W.F. Flood, "New Radiative Recombination Processes Involving Neutral Donors and Acceptors in Silicon and Germanium," *Phys. Rev.* **161**, 711–729, Sept. 1967.

114. G. Pickering, P.R. Tapster, P.J. Dean, and D.J. Ashen, "Determination of Impurity Concentration in n-Type InP by a Photoluminescence Technique," in *GaAs and Related Compounds* (G.E. Stillman, ed.) Conf. Ser. No. 65, Inst. Phys., Bristol, 1983, 469–476.

115. D.J. Ashen, P.J. Dean, D.T.J. Hurle, J.B. Mullin and A.M. White, "The Incorporation and Characterization of Acceptors in Epitaxial GaAs," *J. Phys. Chem. Solids* **36**, 1041–1053, Oct. 1975.

116. G.E. Stillman, B. Lee, M.H. Kim, and S.S. Bose, "Quantitative Analysis of Residual Impurities in High Purity Compound Semiconductors," in *Diagnostic Techniques for Semiconductor Materials and Devices* (T.J. Shaffner and D.K. Schroder, eds.), Electrochem. Soc., Pennington, NJ, 1988, 56–70.

117. S.S. Bose, B. Lee, M.H. Kim, and G.E. Stillman, "Identification of Residual Donors in High-Purity GaAs by Photoluminescence," *Appl. Phys. Lett.* **51**, 937–939, Sept. 1987.

118. D.A. Long, *Raman Spectroscopy*, McGraw-Hill, New York, 1977.

119. C.V. Raman and K.S. Krishna, "A New Type of Secondary Radiation," *Nature* **121**, 501–502, March 1928.

120. B.D. Chase, "Fourier Transform Raman Spectroscopy," *J. Am. Chem. Soc.* **108**, 7485–7488, Nov. 1986.

121. B.D. Chase, "A New Generation of Raman Instrumentation," *Appl. Spectrosc.* **48**, 14A-19A, July 1994.

122. H. Richter, Z.P. Wang and L. Ley, "The One Phonon Raman Spectrum in Microcrystalline Silicon," *Solid State Commun.* **39**, 625–629, Aug. 1981.

123. G.H. Loechelt, N.G. Cave, and J. Menéndez, "Measuring the Tensor Nature of Stress in Silicon Using Polarized Off-Axis Raman Spectroscopy," *Appl. Phys. Lett.* **66**, 3639–3641, June 1995.

124. R. Liu and M. Canonico, "Applications of UV–Raman Spectroscopy and High-resolution X-ray Diffraction to Microelectronic Materials and Devices," *Microelectron. Eng.* **75**, 243–251, Sept. 2004.

125. B. Dietrich, V. Bukalo, A. Fischer, K.F. Dombrowski, E. Bugiel, B. Kuck, and H.H. Richter, "Raman-spectroscopic Determination of Inhomogeneous Stress in Submicron Silicon Devices," *Appl. Phys. Lett.* **82**, 1176–1178, Feb. 2003.

126. F. Adar, "Application of the Raman Microprobe to Analytical Problems in Microelectronics," in *Microelectronic Processing: Inorganic Materials Characterization* (L.A. Casper, ed.), American Chemical Soc., ACS Symp. Series 295, 1986, 230–239.

127. I. De Wolf, "Micro-Raman Spectroscopy to Study Local Mechanical Stress in Silicon Integrated Circuits," *Semicond. Sci. Technol.* **11**, 139–154, Feb. 1996.

128. A. Baghdadi, "Multiple-Reflection Corrections in Fourier Transform Spectroscopy," in *Defects in Silicon* (W.M. Bullis and L.C. Kimerling, eds.) Electrochem. Soc., Pennington, NJ, 1983, 293–302.

129. M.A. Green, *High Efficiency Silicon Solar Cells*, Trans. Tech. Publ., Switzerland, 1987.

130. E. Daub and P. Würfel, "Ultralow Values for the Absorption Coefficient of Si Obtained from Luminescence," *Phys. Rev. Lett.* **74**, 1020–1023, Feb. 1995.

131. E.D. Palik (ed.), *Handbook of Optical Constants of Solids*, Academic Press, Orlando, FL, 1985.

132. J.F. Muth, J.H. Lee, I.K. Shmagin, R.M. Kolbas, H.C. Casey, Jr., B.P. Keller, U.K. Mishra, and S.P. DenBaars, "Absorption Coefficient, Energy Gap, Exciton Binding Energy, and Recombination Lifetime of GaN Obtained from Transmission Measurements," *Appl. Phys. Lett.* **71**, 2572–2574, Nov. 1997.

PROBLEMS

10.1   In an optical transmission measurement, the transmittance of a thin semiconductor sample of thickness d is shown by the curve of Fig. P10.1(a).

$$T = \frac{(1 - R)^2 e^{-\alpha d}}{1 + R^2 e^{-2\alpha d} - 2R e^{-\alpha d} \cos \phi} \qquad (1)$$

where α can be assumed negligibly small, *i.e.*, $\alpha \approx 0$. For the curve in Fig. P10.1(b) the appropriate equation is

$$T = \frac{(1 - R)^2 e^{-\alpha d}}{1 - R^2 e^{-2\alpha d}} \qquad (2)$$

The transmittance dip at $\lambda = 0.0004$ cm is due to an impurity of density $N_1 = 4 \times 10^{15} \alpha_1 \mathrm{cm}^{-3}$, where α_1 is the absorption coefficient of that impurity. Determine R, n_1, d, α_1, N_1, and E_G. k_1 is negligibly small for $\lambda > 0.0002$ cm, $n_o = 1$. E_G can be determined by plotting $\alpha^{1/2}$ versus E, where $E = h\nu = hc/\lambda$.

λ (μm)	0.883	0.879	0.876	0.873	0.870	0.867
T	0.5359	0.4251	0.2610	0.1455	0.0726	0.0338

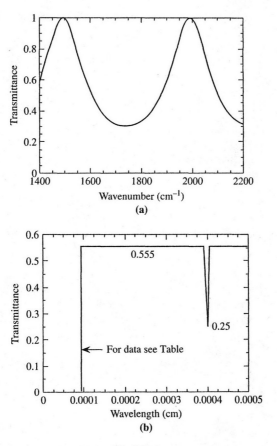

Fig. P10.1

10.2 In an optical transmission measurement, the transmittance T of a semiconductor sample of thickness d and index of refraction n_1 is determined. The appropriate equation is Eq. (10.21), where α can be assumed negligibly small, *i.e.*, $\alpha \approx 0$. In this measurement $\Delta(1/\lambda) = 14.3$ cm^{-1}. Then another measurement is made which can be described by Eq. (10.26). For most of that curve $\alpha = 0$ and $T = 0.504$. At one particular wavelength there is a dip in that T versus λ curve to $T = 0.482$ due to an impurity of density $N_i = 3 \times 10^{17}\alpha_i$cm^{-3}, where α_i is the absorption coefficient of the impurity. Determine R, n_1, d, α_i, and N_i. k_1 is negligibly small, $n_0 = 1$.

10.3 In an optical transmission measurement, the transmittance of a semiconductor sample of thickness d is shown by the curve in Fig. P10.3. The appropriate equation is

$$T = \frac{(1-R)^2 e^{-\alpha d}}{1 + R^2 e^{-2\alpha d} - 2Re^{-\alpha d}\cos\phi} \tag{1}$$

where α can be assumed negligibly small, *i.e.*, $\alpha \approx 0$. Then another measurement is made which can be described by

$$T = \frac{(1-R)^2 e^{-\alpha d}}{1 - R^2 e^{-2\alpha d}} \qquad (2)$$

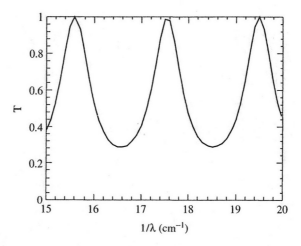

Fig. P10.3

For most of that curve $\alpha = 0$ and $T = 0.516$. At one particular wavelength there is a dip in that T versus λ curve to $T = 0.407$ due to an impurity of density $N_i = 3 \times 10^{17} \alpha_i \mathrm{cm}^{-3}$, where α_i is the absorption coefficient of the impurity. Determine R, n_1, d, α_i, and N_i. k_1 is negligibly small, $n_0 = 1$.

10.4   The reflectance curve of an insulator on a Si substrate was measured according to the diagram in Fig. 10.20 for $\phi = 50°$. This is a common method to determine insulator thickness.

(a) Determine d_1 using Eqs. (10.33) to (10.36). Use $n_0 = 1$, $n_1 = 1.46$, $n_2 = 4$. The wavelengths for R_{max} and R_{min} are:

R_{max}	λ (μm)	R_{min}	λ (μm)
0.36	0.066	0.093	0.08
0.36	0.1	0.093	0.132
0.36	0.199	0.093	0.398

(b) Plot R versus λ and R versus $1/\lambda$ for $n_0 = 1$, $n_1 = 1.46$, $n_2 = 4$, $d_1 = 1000$ Å, and $\phi = 70°$ over the wavelength range $0.1 \leq \lambda \leq 0.8$ μm.

10.5   The reflectance curve of an insulator on a Si substrate is shown in Fig. P10.5. It was measured according to the diagram in Fig. 10.20 for the angle indicated on the figure. This is a common method to determine insulator thickness.

(a) Determine n_1 using Eqs. (10.33) to (10.36). Use $n_0 = 1$, $n_2 = 4$, $d_1 = 1000$ Å.

(b) Plot R versus angle ϕ for $\lambda = 6000$ Å. The wavelengths for R_{max} and R_{min} for the curve below are:

R_{max}	λ (μm)	R_{min}	λ (μm)
0.36	0.094	0.026	0.113
0.36	0.141	0.026	0.189
0.36	0.283	0.026	0.567

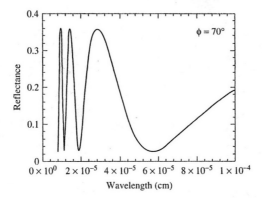

Fig. P10.5

10.6 Show that if the resolution of the detector in the FTIR instrument is insufficient, then

$$T = \frac{(1 - R)^2 e^{-\alpha d}}{1 + R^2 e^{-2\alpha d} - 2Re^{-\alpha d} \cos(\phi)} \quad \text{becomes} \quad T = \frac{(1 - R)^2 e^{-\alpha d}}{1 - R^2 e^{-2\alpha d}}$$

Hint: See Appendix 10.1.

10.7 You are flying in a plane at 30,000 feet altitude and are looking out of the window. How large must an object be on the ground for you to be able to see it? Give answer in meters. Use $s(\text{eye}) = 2.5 \times 10^{-3}$ cm and the focal length of the eye is 2 cm.

10.8 The function $B(f)$ in an FTIR system is a bandpass function with constant amplitude A containing only frequencies between f_1 and f_2. Calculate and plot $I(x)$ versus x for this function for $0 \le x \le 1/f_1$ for: $f_2 = 2f_1$ and (b) $f_2 = 10f_1$.

10.9 The reflectance curves of an insulator on a Si substrate are shown in Fig. P10.9 for $\phi = 60°$. $\alpha = 0$. This is a method to determine insulator thicknesses, provided they are not too thin. Determine the thickness d_1. Use $n_0 = 1$, $n_1 = 1.46$, $n_2 = 3.4$.

10.10 The optical transmittance T of a semiconductor wafer of thickness d is shown in Fig. P10.10 for $\lambda > \lambda_G = hc/E_G$. Draw, on the *same figure*, the $T - 1/\lambda$ curve when the semiconductor wafer is *thinned*, i.e., d is reduced. Justify your answer. The reflectance R remains the same:

$$T = \frac{(1 - R)^2 e^{-\alpha d}}{1 - R^2 e^{-2\alpha d}}$$

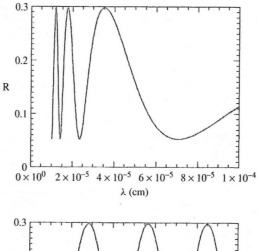

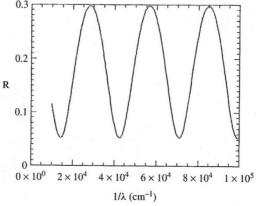

Fig. P10.9

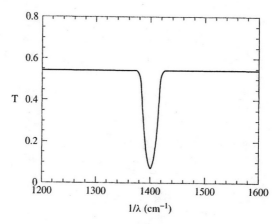

Fig. P10.10

10.11 Light is incident perpendicular on a Si wafer covered with a dielectric of thickness d_1 and refractive index n_1. The reflectance versus wavenumber is shown in Fig. P10.11. Determine the dielectric thickness. $n_1 = 2$.

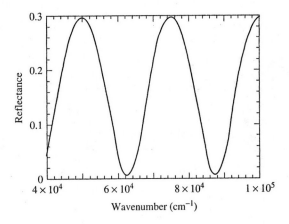

Fig. P10.11

10.12 Give two methods for better spatial resolution in an optical microscope *without* changing the lens and *without* using a near-field optical microscope.

10.13 Give three material parameters that can be determined from a transmission versus $1/\lambda$ spectrum.

10.14 Briefly discuss the main application of Raman spectroscopy.

10.15 Why does confocal microscopy give depth resolution?

REVIEW QUESTIONS

- What determines the resolution limit in conventional optical microscopy?
- What is near field optical microscopy?
- Explain interferometric microscopy.
- What is confocal optical microscopy?
- Why does near field optical microscopy give higher resolution then conventional optical microscopy?
- What are the basic elements of ellipsometry?
- How does FTIR work?
- Where are transmission measurements used?
- Where are reflection measurements used?
- Why does an oil slick on wet pavement show different colors?
- What is luminescence?
- How can photoluminescence be used in Si characterization?
- What are line width measurement methods?
- Name two applications for Raman spectroscopy.

11

CHEMICAL AND PHYSICAL CHARACTERIZATION

11.1 INTRODUCTION

The chemical and physical characterization techniques in this chapter include electron beam, ion beam, X-ray, as well as a few others. These techniques are generally more specialized and require more sophisticated and more expensive equipment than those of the previous chapters. Some methods are used a great deal by a few specialists or are offered as services. For example, secondary ion mass spectrometry is a common characterization method. Because of the specialized nature of the methods, only a brief description is given of the principles, the instrumentation, and the most important areas of application. The specialist using any of the methods is already familiar with the details and the non-specialist is usually not interested in the details, but may be interested in an overview, in the detection limits, the required sample size, and so on.

Many papers, review papers, chapters in books, and books have been written describing these characterization techniques in great detail. I will give reference to these publications as they occur throughout this chapter. Books that give good overviews and offer more practical aspects are: *Metals Handbook, 9th Ed., Vol. 10 Materials Characterization* (R. E. Whan, coord.), Am. Soc. Metals, Metals Park, OH, 1986; *Encyclopedia of Materials Characterization* (C. R. Brundle, C. A. Evans, Jr., and S. Wilson, eds.), Butterworth-Heinemann, Boston, 1992; *Surface Analysis: The Principal Techniques* (J. C. Vickerman, ed.), Wiley, Chichester, 1997; *Handbook of Surface and Interface Analysis* (J. C. Riviere and S. Myhra, eds.), Marcel Dekker, New York, 1998; W. R. Runyan and T. J. Shaffner, *Semiconductor Measurements and Instrumentation*, McGraw-Hill, New York, 1998; D. Brandon and W.D. Kaplan, *Microstructural Characterization of Materials*, Wiley, Chichester, 1999; *Surface Analysis Methods in Materials Science* (D.J. O'Connor, B.A. Sexton and R. St. C. Smart, eds.), Springer, Berlin, 2003.

Semiconductor Material and Device Characterization, Third Edition, by Dieter K. Schroder
Copyright © 2006 John Wiley & Sons, Inc.

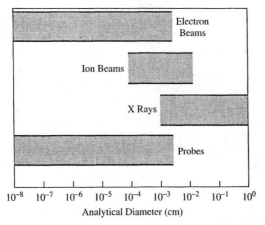

Fig. 11.1   Diameter capabilities of electron-beam, ion-beam, X-ray, and probe characterization techniques.

The capabilities of surface-analysis instruments, especially spatial resolution, sample handling and treatment, speed of data acquisition, and data processing and analysis have been greatly enhanced during the past three decades. The applications of these instruments have now advanced from research and development to problem solving, failure analysis, and quality control and the instruments have become more reliable. The most salient features of the characterization techniques are summarized in Appendix 11.1.

The characterization of semiconductor materials and devices frequently requires a measurement of an impurity spatially in the x and y as well as in the z-dimension. Typical x-y resolution capabilities are shown in Fig. 11.1. Electron beams can be focused to diameters as small as 0.1 nm. Ion beams cover the 1 to 100 μm range and X-rays typically have diameters of 100 μm and above. There is a dichotomy in the characterization of materials at small dimensions: high sensitivity and small volume sampling are mutually exclusive. Generally decreasing beam diameter results in poorer sensitivity. High sensitivity requires large excitation beam diameters.

All analytical techniques are based on similar principles. A primary electron, ion, or photon beam causes backscattering or transmission of the incident particles-waves or the emission of secondary particles/waves. The mass, energy, or wavelength of the emitted entities is characteristic of the target element or compound from which it originated. The distribution of the unknown can be mapped in the x-y plane and frequently also in depth. Each of the techniques has particular strengths and weaknesses, and frequently more than one method must be used for unambiguous identification. Differences between the various techniques include sensitivity, elemental or molecular information, spatial resolution, destructiveness, matrix effects, speed, imaging capability, and cost.

Spectroscopy is used for techniques that are primarily qualitative in their ability to determine densities even though they may be quantitative for identifying impurities; *spectrometry* is used for quantitative methods.

11.2   ELECTRON BEAM TECHNIQUES

Electron beam techniques are summarized in Fig. 11.2. Incident electrons are absorbed, emitted, reflected, or transmitted and can, in turn, cause light or X-ray emission. An

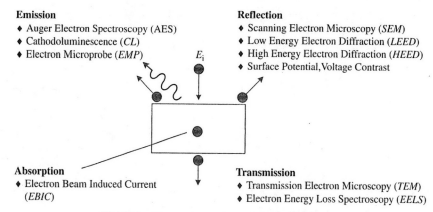

Fig. 11.2   Electron beam characterization techniques.

electron beam of energy E_i causes emission of electrons from the surface over a wide range of energies, as illustrated in Fig. 11.3, where the electron yield $N(E)$ is plotted against the electron energy. Three groups of electrons can be distinguished: *secondary*, *Auger* and *backscattered electrons*. $N(E)$ shows a maximum for *secondary electrons*. The interaction of an electron beam with a solid can lead to the ejection of loosely bound electrons from the conduction band. These are the secondary electrons with energies below about 50 eV with a maximum $N(E)$ at 2 to 3 eV. The secondary electron yield depends on the material and its topography.[1] Auger electrons are emitted in an intermediate energy range. Backscattered electrons, having undergone large-angle elastic collisions, leave the sample with essentially the same energy as the incident electrons.

Electrons can be focused, deflected, and accelerated by appropriate potentials; they can be efficiently detected and counted, their energy and angular distribution can be measured, and they do not contaminate the sample or the vacuum system. However, because they may cause sample charging that may distort the measurement.

11.2.1 Scanning Electron Microscopy (SEM)

Principle: An *electron microscope* uses an electron beam (e-beam) to produce a magnified image of the sample. The three principal electron microscopes are: *scanning, transmission*, and *emission*. In the scanning and transmission electron microscope, an electron beam incident on the sample produces an image while in the field-emission microscope the specimen itself is the source of electrons. A good discussion of the history of electron microscopy is given by Cosslett.[2] A scanning electron microscope consists of an electron gun, a lens system, scanning coils, an electron collector, and a cathode ray display tube (CRT). The electron energy is typically 10–30 keV for most samples, but for insulating samples the energy can be as low as several hundred eV. The use of electrons has two main advantages over optical microscopes: much larger magnifications are possible since electron wavelengths are much smaller than photon wavelengths and the depth of field is much higher.

De Broglie proposed in 1923 that particles can also behave as waves.[3] The electron wavelength λ_e depends on the electron velocity v or the accelerating voltage V as

$$\lambda_e = \frac{h}{mv} = \frac{h}{\sqrt{2qmV}} = \frac{1.22}{\sqrt{V}} \text{ [nm]} \tag{11.1}$$

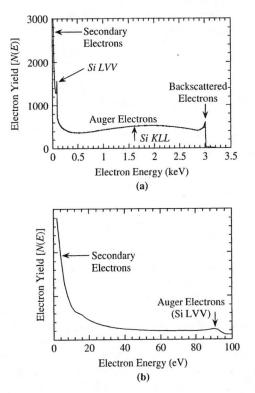

Fig. 11.3 Electron yield $N(E)$ as a function of electron energy for silicon. (a) Entire electron energy range, (b) restricted energy range. Incident energy is 3 keV. For Auger electrons the Si LVV and KLL transitions are shown. Data courtesy of M.J. Rack, Arizona State University.

$\lambda_e = 0.012$ nm for $V = 10,000$ V—a wavelength significantly below the 400 to 700 nm wavelengths of visible light—making the resolution of an SEM much better than that of an optical microscope.

The image in an SEM is produced by scanning the sample with a focused electron beam and detecting the secondary and/or backscattered electrons. We will not concern ourselves with the details of focusing electrons because this is discussed in appropriate books and papers.[4] Electrons and photons are emitted at each beam location and subsequently detected. Secondary electrons form the conventional SEM image, backscattered electrons can also form an image, X-rays are used in the electron microprobe, emitted light is known as cathodoluminescence, and absorbed electrons are measured as electron-beam induced current. All of these signals can be detected and amplified to control the brightness of a CRT scanned in synchronism with the sample beam scan in the SEM. A one-to-one correspondence is thus established between each point on the display and each point on the sample. Magnification M results from the mapping process according to the ratio of the dimension scanned on the CRT to the dimension of the scanned sample

$$M = \frac{Length\ of\ CRT\ display}{Length\ of\ sample\ scan} \qquad (11.2)$$

For a 10-cm-wide CRT displaying a sample scanned over a 100-μm length, the magnification is 1000×. Magnifications of 100,000× or higher are possible in SEMs, but low magnifications are more difficult. An SEM typically has one large viewing CRT and a high-resolution CRT with typically 2500 lines resolution for photography.

The contrast in an SEM depends on a number of factors. For a flat, uniform sample the image shows no contrast. If, however, the sample consists of materials with different atomic numbers, a contrast is observed if the signal is obtained from backscattered electrons, because the backscattering coefficient increases with the atomic number Z. The secondary electron emission coefficient, however, is not a strong function of Z and atomic number variations give no appreciable contrast. Contrast is also influenced by surface conditions and by local electric fields. But the main SEM contrast-enhancing feature is the sample topography. Secondary electrons are emitted from the top 10 nm or so of the sample surface. When the sample surface is tilted from normal beam incidence, the electron beam path lying within this 10 nm is increased by the factor $1/\cos\theta$ where θ is the angle from normal incidence ($\theta = 0°$ for normal incidence). The interaction of the incident beam with the sample increases with path length and the secondary electron emission coefficient increases. The contrast C depends on the angle as[4]

$$C = \tan(\theta)\,d\theta \qquad (11.3)$$

For $\theta = 45°$ a change in angle of $d\theta = 1°$ produces a contrast of 1.75% while at 60° the contrast increases to 3% for $d\theta = 1°$.

The sample stage is an important component in SEMs. It must allow precise movement in tilt and rotation for the sample to be viewed at the appropriate angle. The angle effect is responsible for the three-dimensional nature of SEM images, but the striking pictures come about also due to the signal collection. Secondary electrons are attracted and collected by the detector even if they leave the sample in a direction away from the detector. This does not happen in optical microscopes, where light reflected away from the detector (the eye) is not observed. An SEM forms its picture in an entirely different manner than an optical microscope, where light reflected from a sample passes through a lens and is formed into an image. In an SEM no true image exists. The secondary electrons that make up the conventional SEM image are collected, and their density is amplified and displayed on a CRT. Image formation is produced by mapping, which transforms information from specimen space to CRT space.

Instrumentation: A schematic representation of an SEM is shown in Fig. 11.4. Electrons emitted from an electron gun pass through a series of lenses to be focused and scanned across the sample. The electron beam should be bright with small energy spread. The tungsten "hairpin" filament gun emits electrons thermionically with an energy spread of around 2 eV. Tungsten sources have been largely replaced by lanthanum hexaboride (LaB_6) sources with higher brightness, lower energy spread ($\sim$1 eV) and longer life and by field-emission guns with an energy spread of about 0.2 to 0.3 eV. Field-emission guns are about 100× brighter than LaB_6 sources and 1000× brighter than tungsten sources with longer lifetimes.

The incident or primary electron beam causes secondary electrons to be emitted from the sample and these are ultimately accelerated to 10 to 12 kV. They are most commonly detected with an Everhart-Thornley (ET) detector.[5] The basic component of this detector is a scintillation material that emits light when struck by energetic electrons accelerated from the sample to the detector. The light from the scintillator is channeled through a light pipe to a photomultiplier, where the light incident on a photocathode produces electrons that

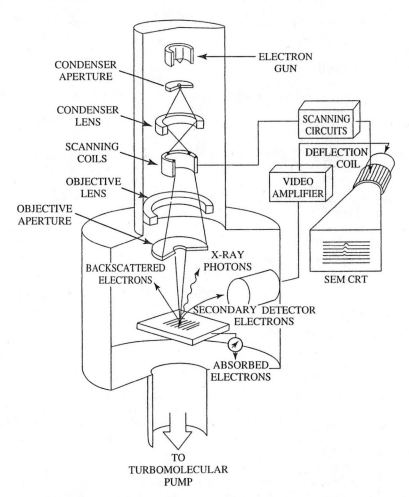

Fig. 11.4 Schematic of a scanning electron microscope. Reprinted with permission from Micro-electronics Processing: Inorganic Materials Characterization Fig. 1, p.51, Copyright 1986 American Chemical Society, after Young and Kalin, ref. 6.

are multiplied, creating the very high gains necessary to drive the CRT. High potentials of 10 to 12 kV are necessary for efficient light emission by the scintillator.

The beam diameter in lens-based SEMs is around 0.4 nm and in field-emission SEMs around 0.1 nm. Yet the resolution of e-beam measurements is not always that good. Why is that? It has to do with the shape of the electron-hole cloud generated in the semiconductor. When electrons impinge on a solid, they lose energy by elastic scattering (change of direction with negligible energy loss) and inelastic scattering (energy loss with negligible change in direction). Elastic scattering is caused mainly by interactions of electrons with nuclei and is more probable in high atomic number materials and at low beam energies. Inelastic scattering is caused mainly by scattering from valence and core electrons. The result of these scattering events is a broadening of the original nearly collimated, well focused electron beam within the sample.

The generation volume is a function of the e-beam energy and the atomic number Z of the sample. Secondary electrons, backscattered electrons, characteristic and continuum X-rays, Auger electrons, photons, and electron-hole pairs are produced. For low-Z samples most electrons penetrate deeply into the sample and are absorbed. For high-Z samples there is considerable scattering near the surface and a large fraction of the incident electrons is backscattered. The shape of the electron distribution within the sample depends on the atomic number. For low-Z material ($Z \leq 15$) the distribution has the "teardrop" shape in Fig. 11.5. For $15 < Z < 40$ the shape becomes more spherical and for $Z \geq 40$ it becomes hemispherical. "Teardrop" shapes have been observed by exposing polymethylmethacrylate to an electron beam and etching the exposed portion of the material.[4] Electron trajectories, calculated with Monte Carlo techniques, also agree with these shapes.

The electron penetration depth is the *electron range* R_e, defined as the average distance from the sample surface that an electron travels in the sample along a trajectory. A number of empirical expressions have been derived for R_e. One such expression is[7]

$$R_e = \frac{4.28 \times 10^{-6} E^{1.75}}{\rho} \text{ (cm)} \qquad (11.4)$$

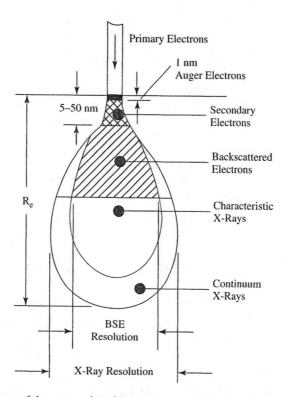

Fig. 11.5 Summary of the range and spatial resolution of backscattered electrons, secondary electrons, X-rays, and Auger electrons for electrons incident on a solid. Reprinted with permission after ref. 4.

where ρ is the sample density (g/cm^3) and E the electron energy (keV). The electron ranges for Si ($\rho = 2.33$ g/cm^3), Ge (5.32 g/cm^3), GaAs (5.35 g/cm^3), and InP (4.7 g/cm^3) are

$$R_e(Si) = 1.84 \times 10^{-6} E^{1.75}; R_e(Ge) = 8.05 \times 10^{-7} E^{1.75};$$

$$R_e(GaAs) = 8.0 \times 10^{-7} E^{1.75}; R_e(InP) = 9.1 \times 10^{-7} E^{1.75}.$$

Equation (11.4) is sufficiently accurate for $20 < E < 200$ keV, but underestimates the range slightly compared to a more accurate expression.[8]

Applications: The most common use of SEMs for semiconductor applications, when used as a microscope, is to view the surface of the device, frequently during failure analysis and for cross-sectional analysis to determine device dimensions, for example MOSFET channel length, junction depth, and so on. SEMs are also used in wafer processing production lines for on-line inspection and line width measurement (see Chapter 10). When inspecting integrated circuits, it is important to reduce or eliminate surface charging by coating the surface with a thin conductive layer (Au, AuPd, Pt, PtPd, and Ag provide an oxide-free surface) or by reducing the beam energy until the number of primary electrons is roughly equal to the number of secondary and backscattered electrons. The energy for this balance is around 1 keV and is sufficiently low to minimize electron beam damage to devices. The reduced signal-to-noise ratio of low-energy beams is optimized by using high beam brightness and digital frame storage for signal enhancement.

11.2.2 Auger Electron Spectroscopy (AES)

Principle: *Auger electron spectroscopy* is based on the Auger (pronounced something like "O-J") effect, discovered by Auger in 1925.[9] It has become a powerful surface characterization method for the study of chemical and compositional properties of materials. All elements except hydrogen and helium can be detected. Data interpretation is simplified by the large data base of available literature for identification of elemental species.[10] Spectra from individual elements do not interfere with one another and chemical binding state information is obtained from Auger transition energy shifts as well. There can be interferences from different elements, but these can usually be eliminated by measuring Auger peaks at various energies corresponding to different electronic transitions. Although the basic Auger technique samples a depth of typically 0.5 to 5 nm, sputter etching the sample provides depth information.

Auger electron emission is illustrated in Fig. 11.6 for a semiconductor. The energy band diagram of the semiconductor shows the vacuum level, the conduction and valence bands, and the lower lying core levels not usually shown on semiconductor energy band diagrams. In particular, we assume a material with a K level at energy E_K and two L levels (E_{L1} and E_{L2}).[11] A primary electron with typically 3–5 keV energy from an electron gun ejects an electron from the K shell. The K-shell vacancy is filled by an outer shell electron (L_1 in this case) or by a valence band electron. The energy $E = E_{L1} - E_K$ is transferred to a third electron—the Auger electron—originating in this case at the $L_{2,3}$ level. The energy can also result in the emission of an X-ray photon, discussed in the next section. Auger emission dominates over X-ray emission for the lower Z elements.

The atom remains in a doubly ionized state and the entire process is labeled "$KL_1L_{2,3}$" or simply as "KLL." In the KLL transition, the L shell ends up with two vacancies. These can be filled by valence band electrons, resulting in LVV Auger electrons. Since the Auger process is a three-electron process, it is obvious why hydrogen and helium cannot be detected; both have less than three electrons. The dominant Auger energy

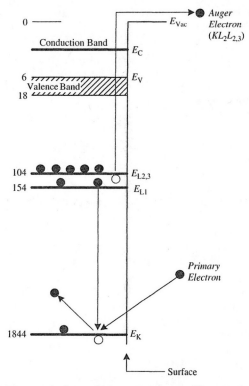

Fig. 11.6 Electronic processes in Auger electron spectroscopy. The numeric values on the energy axis are for silicon.

transitions are for $3 < Z < 14$: KLL transitions, for $14 < Z < 40$: LMM transitions, and for $40 < Z < 82$: MNN transitions.[12] Transitions between the valence band and the K shell are KVV and between the valence band and the L level are LVV. For example, LVV is the dominant transition in Si.

Exercise 11.1

Problem: In AES an electron falls from the L to the K shell, giving its energy to another electron on the L shell. For Li there is no second electron on the L shell (Li has three electrons). Yet Li can be detected by AES. Why?

Solution: In atomic lithium there can be no Auger transition, because there is no second electron on the L shell. For solid lithium, however, Auger emission is possible because of the ready availability of conduction band electrons, which can supply the "extra" electron required for the Auger transition. The L shell and the conduction band are effectively the same. A. J. Jackson, C. Tate, T. E. Gallon, P. J. Bassett, and J. A. D. Matthew, "The KVV Auger Spectrum of Lithium Metal'" *J. Phys. F: Metal Phys.* **5**, 363–374, Feb. 1975.

The Auger electron energy, characteristic of the emitting atom of atomic number Z, for the $KL_1L_{2,3}$ transition is[13]

$$E_{KL_1L_{2,3}} = E_K(Z) - E_{L_1}(Z) - E_{L_{2,3}}(Z+1) - q\phi \qquad (11.5a)$$

where $q\phi$ is the sample work function. In general, when an electron is excited from level A, the vacancy is filled by an electron from level B and an electron from level C is ejected, the kinetic energy of the Auger electron is[14]

$$E_{ABC} = E_A(Z) - E_B(Z) - E_C(Z+\Delta) - q\phi \qquad (11.5b)$$

where Δ is included to account for the energy of the final doubly ionized state being higher than the sum of the energies for individual ionization of the same levels. Δ lies between 0.5 and 0.75.[10] Auger electron energies are characteristic of the sample and are independent of the incident electron energy.

Instrumentation: Auger electron spectroscopy instrumentation consists of an electron gun, electron beam control, an electron energy analyzer, and data analysis electronics. The incident electron beam energy is typically 1 to 5 keV. Higher beam energies produce Auger electrons deeper within the sample with little chance of escaping. The focused electron beam diameter depends on the electron source, the beam energy, the electron optics, and the beam current. For non-scanning AES, the beam diameter is on the order of 100 μm, for scanning systems it is smaller. Field emission electron sources can achieve beam diameters of 10 nm at 1 nA beam current. The emitted Auger electrons are detected with a retarding potential analyzer, a cylindrical mirror analyzer, or a hemispherical analyzer. A common analyzer is the cylindrical mirror analyzer (CMA) in Fig. 11.7.[15] A coaxial configuration with the analyzer wrapped around the electron gun reduces shadowing and allows room for positioning the ion sputter gun. Auger electrons, entering the inlet aperture between the two concentric cylinders, are focused by a negative potential creating a cylindrical electric field between the coaxial electrodes. The CMA allows electrons with $E \sim V_a$ and energy spread ΔE to pass through the exit slit. Ramping the analyzing potential V_a provides the electron energy spectrum. The energy resolution is defined by

$$R = \frac{\Delta E}{E} \qquad (11.6)$$

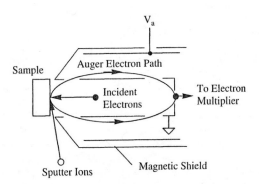

Fig. 11.7 Layout of an AES system with a cylindrical mirror analyzer detector. Reprinted with permission after ref. 15.

where ΔE is the pass energy of the analyzer and E the electron energy. $R \approx 0.005$ for the CMA. When the voltage is ramped, $\Delta E/E$ is constant due to the CMA design. According to Eq. (11.6) this requires ΔE to increase with E. Hence, the number of electrons passing through the analyzer must be proportional to E. This is why Auger spectra are usually displayed as $EN(E)$ instead of $N(E)$, the number of electrons.

Auger electrons have energies of typically 30 to 3000 eV. The Si LVV transition occurs at 92 eV. The analysis area is chiefly determined by the diameter of the primary electron beam. Although the interaction volume of an electron beam in a solid is large compared to the electron beam diameter, as illustrated in Fig. 11.5, Auger electrons escape from a very shallow surface layer. Metals tend to have the shortest escape depth, followed by semiconductors and insulators. AES can be operated in a point analysis mode, detecting many elements in a small sample area, or a map of a selected element can be generated by scanning the beam across the sample with the detector tuned to one element. A high, oil-free vacuum (10^{-9} torr or lower) is required to protect the sample from contamination as the presence of surface contamination interferes with the Auger signal.

$EN(E)$ is plotted in Fig. 11.8 as a function of energy for silicon. The Auger electron peaks appear as small perturbations on a high background (Fig. 11.3) consisting of beam electrons that have lost varying amounts of energy before being backscattered, Auger electrons that have lost energy propagating through the sample, as well as secondary electrons produced by a cascade process at the low energy end. Those Auger electrons formed deeper in the sample will lose energy and be unrecognizable in the background. In order to enhance the Auger peaks, it is common practice to differentiate the Auger signal and present it as $d[EN(E)]/dE$ versus E, also shown in Fig. 11.8. The introduction of signal differentiation led to the rapid growth of AES.[16] The Auger energy position is indicated by the peak in the $EN(E)$ spectrum or by the maximum negative excursion of the $d[EN(E)]/dE$ peak in the differentiated spectrum. Differentiation enhances the signal-to-background ratio, but degrades the signal-to-noise ratio. One can also use the undifferentiated $EN(E)$ versus E curve with background suppression to lift the signal above the background.

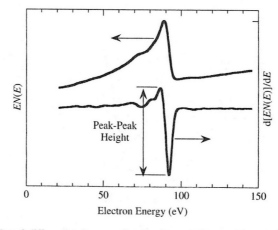

Fig. 11.8 $EN(E)$ and differentiated curves for the silicon LVV transition. Data courtesy of M.J. Rack, Arizona State University.

The AES detection limit is about 0.1% but it varies greatly from element to element. Davies et al. give the relative Auger sensitivities for the elements for 3 keV, 5 keV, and 10 keV primary electron energies.[10] Detection limits are also influenced by the beam current and by the analysis time, making quantitative AES analysis difficult. Reported accuracies are about 10% with calibrated standards with 5% precision for simple semi-conductor samples.[17] The most common correction scheme relies on published Auger intensities or sensitivity factors for the elements. The analyst corrects measured peak intensities by weighting the spectrum with each element's sensitivity factor. Peak-to-peak values of differentiated spectra are commonly used for intensities, a practice that has come under criticism.[18] It is more accurate to measure the area under the peaks in the integrated, not the differentiated, spectrum. Early applications of surface analysis by AES and XPS made extensive use of reference spectra in handbooks and published papers. Powell recently summarized a wide array of analytical resources for applications in AES, XPS, and SIMS.[18]

It is important to ascertain that the incident electron beam does not alter the sample. Insulators can exhibit sample charging artifacts.[19] Depth profiles, generated by alternately sputtering with an inert ion beam and acquiring the Auger signal, are displayed as Auger electron intensity versus sputtering time in Fig. 11.9. Depth can be correlated with sputtering time by measuring the crater depth after the analysis. When the surface is sputtered during AES depth profiling, sputter-induced artifacts may appear.[20] These include crater wall effects, redeposition of sputtered material, surface roughness, preferential sputtering, varying sputter rates, atomic mixing, charging effects, and specimen damage, *e.g.*, decomposition and desorption and oxygen loss in SiO_2.[21] Depth resolution is on the order of 10 nm.[20]

Applications: AES has found applications in measuring semiconductor composition, oxide film composition, phosphorus-doped glasses, silicides, metallization, bonding pad contamination, lead frame failures, particle analysis, and the effects of surface cleaning.[22] AES measurements are made in a high vacuum environment ($10^{-12}-10^{-10}$ torr) to retard the formation of contamination films on the sample surface. Elemental scans give a rapid means of identifying surface elements. *Scanning Auger microscopy* (SAM) allows the

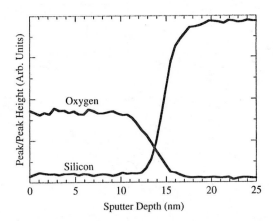

Fig. 11.9   Auger depth profile of a 14.8 nm SiO_2 film on Si. Argon ion beam sputter at 2 keV. Data courtesy of M.J. Rack, Arizona State University.

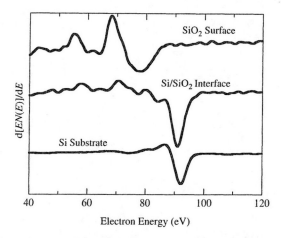

Fig. 11.10 Auger spectra for various forms of Si. Data courtesy of M.J. Rack, Arizona State University.

sample to be mapped for one selected element at a time. AES is not suitable for trace element analysis, because its sensitivity lies in the 0.1 to 1% range. Originally only elements were detected with AES, but today's AES systems allow chemical information to be obtained. When elements combine to form compounds, there is an energy shift and shape change in the Auger spectra as shown in Fig. 11.10. In this example, there is a clear difference in the AES signal between Si in elemental form, Si at the Si/SiO_2 interface and Si in bulk SiO_2. X-ray photoelectron spectroscopy (XPS) is generally considered to be more appropriate for chemical analysis, because XPS lines are narrower than AES lines, but AES energy shifts are usually larger than XPS energy shifts.

11.2.3   Electron Microprobe (EMP)

Principle: The *electron microprobe*, also known as *electron probe microanalysis* (EPM or EPMA) was first described by Castaing in his doctoral thesis in 1948.[23] The method consists of electron bombardment of the sample and X-ray emission from the sample. An EMP is usually a part of a scanning electron microscope equipped with appropriate X-ray detectors.[24] Of all the signals generated by the interaction of the primary electron beam with the sample in the SEM, X-rays are most commonly used for material characterization. The X-rays have energies characteristic of the element from which they originate, leading to elemental identification. The X-ray intensity can be compared with intensities from known samples and the ratio of the sample intensity to the intensity of the standard can be considered a measure of the amount of the element in the sample. The correlation, however, is not entirely straightforward. Other factors complicate the interpretation, *e.g.*, the influence of other elements in the sample that absorb some of the X-rays generated by the primary electron beam and release other X-rays of their own characteristic energy, known as *secondary fluorescence*. If the energy of the characteristic radiation from element A exceeds the absorption energy for element *B* in a sample containing A and *B*, a characteristic fluorescence of *B* by A will occur. Additionally, not all X-rays leaving the sample are captured by the detector. Best accuracy in quantitative density determination is obtained if the standards are identical in composition to the unknown. Pure elemental

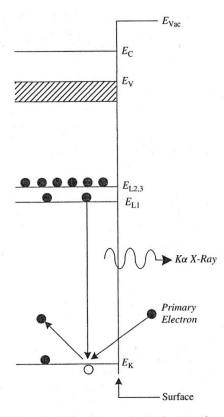

Fig. 11.11   Electronic processes in the electron microprobe.

standards can also be used, but may lead to inaccuracies. Fortunately, quantitative analysis is not always necessary.

EMP is not a true surface technique, because X-rays are emitted from within the sample volume as shown in Fig. 11.5. The method is illustrated with the aid of the band diagram in Fig. 11.11. A primary electron beam of typically 5 to 20 keV strikes the sample. The electron beam energy should be approximately three times the X-ray energy, known as *overvoltage*. X-rays are generated by electron bombardment of a target by two distinctly different processes: (1) Deceleration of electrons in the Coulombic field of the atom core, leading to formation of a continuous spectrum of X-ray energies from zero to the incident electron energy. This is the X-ray continuum or Bremsstrahlung (German for "braking radiation") extending from zero to the incident electron energy sometimes called white radiation by analogy with white light of the visible spectrum. (2) The interaction of the primary electrons with inner-shell electrons. Incident electrons eject electrons from one of the inner atomic shells with electrons from higher-lying shells dropping into the vacancies created by the ejected electrons. These are the characteristic X-rays with wavelengths practically independent of the physical or chemical state of the emitting atom.

If the X-ray emission is the result of an $L \rightarrow K$ transition, the X-rays are known as $K\alpha$ X-rays. $K\beta$ X-rays are the result of $M \rightarrow K$ transitions, $L\alpha$ X-rays are due to $M \rightarrow L$ transitions, and so on. There is but one K level, but the other levels are subdivided. The

L shell is split into a triple fine structure and the M shell has five levels. This leads to further sub-divisions. For example, the $L_2 \rightarrow K$ transition is known as $K\alpha_2$ and the $L_3 \rightarrow K$ transition results in $K\alpha_1$ X-rays.[25] Not all possible transitions occur with equal probability and some are so improbable to have earned the name "forbidden" transitions, for example, the $L_1 \rightarrow K$ transition. The ionization efficiency is low, with typically only one electron in a thousand producing a K shell vacancy.

X-ray detectors frequently lack the resolving power to separate X-ray lines close to one another (doublets). The unresolved doublets are measured in such cases as if they were a single line. This is indicated by dropping the subscript; the notation $K\alpha$ refers to the unresolved doublet $K\alpha_1 + K\alpha_2$. Sometimes the term $K\alpha_{1,2}$ is used. The X-ray photon energy for an $L \rightarrow K$ transition in the EMP is

$$E_{EMP} = E_K(Z) - E_{L2,3}(Z) \tag{11.7}$$

The energy between the K and L levels is much higher than that between the L and M levels, which in turn is higher than that between the M and N levels. For example, for silicon $E(K\alpha_1) = 1.74$ keV; for copper $E(K\alpha_1) = 8.04$ keV and $E(L\alpha_1) = 0.93$ keV, while for gold $E(K\alpha_1) = 68.79$ keV, $E(L\alpha_1) = 9.71$ keV, and $E(M\alpha_1) = 2.12$ keV. The most common EMP X-ray lines are $K\alpha_{1,2}$, $K\beta_1$, $L\alpha_{1,2}$ and $M\alpha_{1,2}$. A graphical representation of all X-ray lines observed in high quality X-ray spectra in the 0.7 to 10 keV energy range is given by Fiori and Newbury.[26] A detailed discussion of both qualitative and quantitative EMP spectra interpretation is given in Ref. 4. The relationship between X-ray energy E and wavelength λ is

$$\lambda = \frac{hc}{E} = \frac{1.2398}{E[keV]} \text{ [nm]} \tag{11.8}$$

The only possible outcomes of an ionization event involving the K shell are the emission of a K-line X-ray photon or of an Auger electron. The fraction of the total number of ionizations leading to the emission of X-rays is the *fluorescence yield*. The sum of the probability of X-ray emission and that of Auger electron emission is unity. For low-Z material Auger emission is dominant while X-ray emission dominates for high-Z material. The two probabilities are about equal for $Z \approx 30$ for K shell ionization, as shown in Fig. 11.12.

Although the electron beam diameter in the EMP is on the order of 1 μm or less, the emitted X-rays originate from a larger area and EMP does not have the high resolution associated with AES, nor is it a true surface-sensitive technique. EMP is a good example where the size of the exciting beam bears little relation to the resolution of the measurement. Reduction of the beam diameter leads to slightly higher resolution at the expense of signal reduction. As in AES, the electron beam can be stationary and an elemental scan gives the sample impurities. The use of beam rastering to scan a limited sample area leads to imaging and elemental mapping. The EMP sensitivity is better than that of AES because a larger volume is probed. The sensitivity is 10^3 to 10^4 ppm (parts per million) for $Z = 4$ to 10, 10^3 ppm for $Z = 11$ to 22, and 100 ppm for $Z = 23$ to 100 but varies with instrumental and sample parameters.[12]

Instrumentation: EMP uses the electron beam, focusing lenses, and deflection coils of an SEM. Only the X-ray detector is added and many SEMs have EMP capability. Several types of detectors are used. The most common are: *energy-dispersive* spectrometers (EDS) and *wavelength-dispersive* spectrometers (WDS), illustrated in Fig. 11.13.

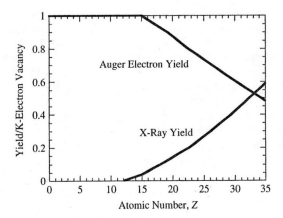

Fig. 11.12 Auger electron and X-ray yields per K vacancy as a function of atomic number. Reprinted with permission from ref. 25.

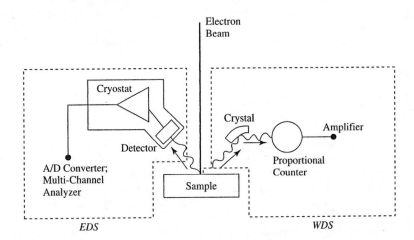

Fig. 11.13 EDS and WDS X-ray detector systems. Reprinted with permission after ref. 6.

The two spectrometers complement each other. EDS is commonly used for rapid sample analysis and WDS for high resolution measurements. A more recent detector is the *microcalorimeter*.

The X-ray detector in the EDS is a reverse-biased semiconductor (usually Si or Ge) pin or Schottky diode. X-rays are absorbed in a solid according to the equation[24]

$$I(x) = I_0 \ \exp[-(\mu/\rho)\rho x] \tag{11.9}$$

with (μ/ρ) the mass absorption coefficient, ρ the detector material density, $I(x)$ the X-ray intensity in the detector, and I_0 the incident X-ray intensity. The mass absorption coefficient is characteristic of a given element at specified X-ray energies. Its value varies with the photon wavelength and with the atomic number of the target element generally decreasing smoothly with energy. It exhibits discontinuities in the energy region immediately above the "absorption edge" energy, corresponding to the energy to eject an electron from a shell. For Si $\mu/\rho = 6.533$ cm^2/g for Mo $K\alpha$ ($E = 17.44$ keV) and 65.32 cm^2/g for Cu $K\alpha$ ($E = 8.05$ keV) X-rays.[11] The absorption equation for Cu $K\alpha$ X-rays incident on a Si detector with $\rho(Si) = 2.33$ g/cm^3 is

$$I(x) = I_0 \exp(-152.2x) \tag{11.10}$$

with x in cm. The thickness for 50% absorption is 46 μm and for 90% absorption it is 151 μm. With X-rays penetrating deeply in Si, the space-charge region (scr) of the reverse-biased diode must be sufficiently wide to absorb the X-rays. With the scr width $W \sim 1/N_D^{1/2}$ this requires either very pure Si or *lithium drifting* to produce an effectively intrinsic region.[27]

The X-rays from the sample pass through a thin beryllium window onto a lithium-drifted Si detector, which should be liquid nitrogen cooled at all times to prevent lithium diffusion and to reduce the diode leakage current. Detector bias voltage should never be applied to a non-cooled Li-drifted detector because the electric field causes the Li ions to drift even at room temperature. Each absorbed X-ray creates many electron-hole pairs, which are swept out of the diode by the high electric field in the space-charge region. The charge pulse is converted to a voltage pulse by a charge-sensitive preamplifier; the signal is further amplified and shaped and then passed to a multichannel analyzer, that measures and sorts the pulses from the preamplifier and assigns them to the appropriate channel (memory location) in the display, with the channel location or number calibrated to correspond to X-ray energy. The pulse from each absorbed X-ray should not interfere with the pulse from the next absorbed X-ray. If, say, two 5 keV pulses coincide, the detector output will be that of one 10 keV pulse. The likelihood of such an occurrence is rare, however. *Pulse pile-up* can occur if the spacing between pulses is so small that they overlap and cause erroneous amplitude measurements.

An energetic particle or photon of energy E absorbed in a semiconductor generates N_{ehp} electron-hole pairs (ehp), given by[28]

$$N_{ehp} = \frac{E}{E_{ehp}} \left(1 - \frac{\alpha E_{bs}}{E} \right) \tag{11.11}$$

where E_{ehp} is the average energy necessary to create one ehp, E_{bs} the mean energy of the backscattered electrons, and α the backscattering coefficient ($\alpha \approx 0.1$ for Si in the 2 to 60 keV energy range). $E_{ehp} \approx 3.2 E_g$ and for Si it is 3.64 eV.[29] A 5 keV X-ray photon generates about 1350 ehps or a charge of 2.2×10^{-16} C in Si. The energy of incident X-rays is determined in such semiconductor detectors by the number of ehps those X-rays produce. Elements from Na to U can be detected with EDS. However, lower-Z elements are difficult to detect due to the Be window that isolates the cooled detector from the vacuum system. Windowless systems allow lower-Z elements to be detected. It is possible for X-rays from the sample absorbed in the Si detector to generate Si $K\alpha$ X-rays that are subsequently absorbed in the detector. These X-rays, which do not originate from the sample, appear in the spectrum as a silicon internal *fluorescence peak*. A good discussion of the factors affecting EDS is found in ref. 4.

X-rays from the sample are directed onto an analyzing crystal in WDS. Only those X-rays that strike the crystal at the proper angle are diffracted through a polypropylene window into the detector, usually a gas proportional counter. The proportional counter consists of a gas-filled tube with a thin tungsten wire in the center of the tube held at a 1 to 3 kV potential. The gas (usually 90% argon, 10% methane) flows through the tube because it is difficult to seal the thin entrance window. An absorbed X-ray creates a shower of electrons and positive ions. The electrons are attracted to the wire and produce a charge pulse, much as ehps are generated and collected in a semiconductor detector.

X-ray diffraction is determined by Bragg's law

$$n\lambda = 2d \sin(\theta_B) \qquad (11.12)$$

where $n = 1, 2, 3, \ldots,$ λ is the X-ray wavelength, d the interplanar spacing of the ana-lyzing crystal, and θ_B the Bragg angle. The detector signal is amplified, converted to a standard pulse size by a single-channel analyzer, and then counted or displayed. The analyzing crystals are curved to focus the X-rays onto the detector. More than one crystal is necessary to span an appreciable wavelength range. Common crystal materials with varying lattice spacing are α-quartz, LiF, pentaerythritol (PET), potassium acid phthalate (KAP), ammonium dihydrogen phosphate (ADP) and others.

WDS detectors have larger collection areas than other types of detectors and are located at longer distances from the sample giving them lower collection efficiencies. WDS has higher energy resolution since only a small range of wavelengths is detected at one time allowing greater peak-to-background ratios, and higher count rates for individual elements. This gives approximately 1 to 2 orders of magnitude better sensitivity, but makes the method slow and requires typically 10–100 times the electron beam current of EDS. Table 11.1 summarizes the major features of the two techniques. EDS and WDS spectra are shown in Fig. 11.14 clearly illustrating the higher resolution of WDS. Typical EDS peaks have about 100 times the natural peak width, limited by the ehp statistics and electronic noise.

In the superconducting *microcalorimeter* the small temperature change (typically less then 1 K) in a metal finger due to X-ray absorption is measured,[30] allowing identifica-tion of elements at low energies (< 3 keV), not possible with EDS. Such low energies result if the e-beam energy is reduced for small X-ray emitting volume. In this energy range, the K-lines of light elements overlap the L and M-lines of heavy elements and the individual peaks cannot be resolved by EDS. Due to the small temperature change in the microcalorimeter, the finger is held at very low temperatures, around 100–200 mK. In one implementation the temperature is detected by a superconducting phase transition of an Ir/Au film with the temperature change proportional to X-ray energy.[31] In the transition

TABLE 11.1 Comparison between X-ray Spectrometers.

Operating Characteristics	WDS Crystal Diffraction	EDS Si Energy Dispersive
Quantum efficiency	Variable, <30%	~100% for 2–16 keV
Elements detected	$Z \geq 5$ (B)	$Z \geq 11$ (Na) for Be window
		$Z \geq 6$ (C) windowless
Resolution	Crystal dependent ~5 eV	Energy dependent 150 eV at 5.9 keV
Data collection time	Minutes to hours	Minutes
Sensitivity	0.01–0.1%	0.1–1%

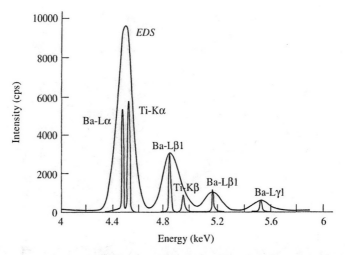

Fig. 11.14 EDS and WDS spectra of $BaTiO_3$. The EDS spectrum was obtained with a 135 eV resolution detector. The WDS spectrum resolves the overlapped peaks of the EDS spectrum. Reprinted with permission after R.H. Geiss, "Energy-Dispersive X-Ray Spectroscopy," in *Encyclopedia of Materials Characterization* (C.R. Brundle, C.A. Evans, Jr., and S. Wilson, eds.), Butterworth-Heinemann, Boston, 1992.

range between the normal and superconducting state, the film resistance depends strongly on temperature. A superconducting quantum interference device converts the resistance change to voltage. The energy resolution is about 10 eV, similar to WDS resolution and about 10 times better than EDS. The microcalorimeter combines the analysis speed of EDS with the WDS energy resolution at the expense of more stringent cooling requirements.

Applications: Electron microprobe analysis with EDS spectrometers is used for quick surveys and for spatial maps of individual elements. It is frequently one of the first techniques to solve a problem or diagnose a failure. Impurities are identified from either EDS or WDS by matching the experimental spectra to known X-ray energies. The comparison can be done automatically by appropriate software, or it is possible to display the experimental spectrum and also known spectra for best match with the experimental data. EMP is not a trace analysis method, due to its poor sensitivity; it is particularly insensitive to light elements in a heavy matrix (see Fig. 11.12 and Table 11.1). It has reasonably good spatial resolution of 1 to 10 μm determined by the electron interaction volume in the sample and is well suited for quantitative measures of metals on semiconductors, alloy compositions, and so on. Detection of carbon, oxygen, and nitrogen is difficult because of the low X-ray yield and the fact that these are common contaminants in vacuum systems. An elemental map is shown in Fig. 11.15.

11.2.4 Transmission Electron Microscopy (TEM)

Transmission electron microscopy was originally used for highly magnified sample images. Later, analytical capabilities such as electron energy loss detectors and light and X-ray detectors were added to the instrument and the technique is now also known as *analytical transmission electron microscopy* (AEM).[32-33] The "M" in TEM, SEM and

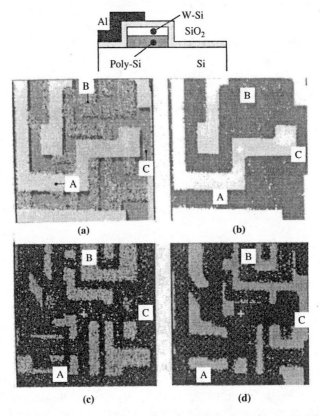

Fig. 11.15 EDS map of a Si circuit. (a) Composite EDS map of Al (A), W (B), and Si (C), (b) aluminum map from the Al line, (c) tungsten map from the W-Si line, (d) silicon map from the substrate. A schematic cross-section is shown on top. Courtesy of J.B Mohr, Arizona State University.

AEM stands for either "microscopy" or "microscope." Transmission electron microscopes are, in principle, similar to optical microscopes; both contain a series of lenses to magnify the sample. The main strength of TEM lies in its extremely high resolution, approaching 0.08 nm. The reason for this high resolution can be found in the resolution equation, $s = 0.61\lambda/NA$. In optical microscopy the numerical aperture $NA \approx 1$ and $\lambda \approx 500$ nm, giving $s \approx 300$ nm. In electron microscopy NA is approximately 0.01 due to larger electron lens imperfections, but the wavelength is much shorter. According to Eq. (11.1) $\lambda_e \approx 0.004$ nm for $V = 100$ kV giving a resolution of $s \approx 0.25$ nm and magnifications of several hundred thousand—much better than optical microscopy. The actual resolution expression is more complicated and this simple calculation should only be taken as a coarse estimate. A shortcoming of TEM is its limited depth resolution.

A schematic of a transmission electron microscope is shown in Fig. 11.16. Electrons from an electron gun are accelerated to high voltages—typically 100 to 400 kV—and focused on the sample by the condenser lenses. The sample is placed on a small copper grid a few mm in diameter. The static beam has a diameter of a few microns. The sample must be sufficiently thin (a few tens to a few hundred nm) to be transparent to electrons.

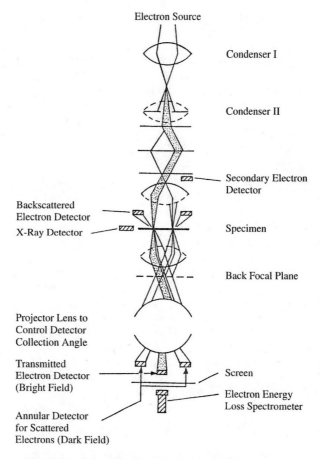

Fig. 11.16 Schematic of a transmission electron microscope.

This circumvents the resolution problem of Fig. 11.5, because the beam does not have a chance to spread when the sample is so thin. The transmitted and forward scattered electrons form a diffraction pattern in the back focal plane and a magnified image in the image plane. With additional lenses, either the image or the diffraction pattern is projected onto a fluorescent screen for viewing or electronic or photographic recording. The ability to form a diffraction pattern yields structural information.

The three primary imaging modes are *bright-field, dark-field*, and *high-resolution* microscopy. Image contrast does not depend very much on absorption, as it does in optical transmission microscopy, but rather on scattering and diffraction of electrons in the sample. Images formed with only the transmitted electrons are bright-field images and images formed with a specific diffracted beam are dark-field images. Few electrons are absorbed in the sample. Absorbed electrons lead to sample heating.

Consider an amorphous sample consisting of atoms A with inclusions of atoms B, where $Z_B > Z_A$ (Z = atomic number). Electrons experience very little scattering from atoms A, but are more strongly scattered by atoms B. The more strongly scattered electrons are not transmitted by the image forming lenses and do not reach the fluorescent screen but the weakly scattered electrons do. Hence, the heavier elements do not appear on the screen and the image brightness is determined by the intensity of those electrons transmitted through the sample that pass through the image forming lenses. For crystalline specimen, the wave nature of electrons must be considered and Bragg diffraction of electrons by the sample crystal planes occurs. Electrons "make it" to the screen if they are not deflected by Bragg diffraction. Contrast comes about by mass, thickness, diffraction, and phase contrast.

A stationary, parallel, coherent electron beam passes through the sample forming a magnified image in the image plane that is projected onto a fluorescent screen. In *scanning transmission electron microscopy* (STEM) a fine beam (diameter ≈ 0.1 nm) is rastered across the sample. The objective lens recombines the transmitted electrons from all points scanned by the probe beam to a fixed region in the back focal plane to be detected. The detector output modulates the CRT brightness, much as secondary electrons do in an SEM. The primary electrons in an STEM also produce secondary electrons, backscattered electrons, X-rays, and light (cathodoluminescence) above the sample much as in SEMs. Below the sample, inelastically scattered transmitted electrons can be analyzed for electron energy loss, making the instrument truly an analytical electron microscope. X-ray analysis has become an important aspect of transmission electron microscopy at magnifications much higher than possible for EMP in an SEM. However, the volume for X-ray generation is much smaller, giving weaker X-ray signals. The integration time for each picture element in STEM is limited since the data are collected serially.

Electron energy loss spectroscopy (EELS), an absorption spectroscopy, is the analysis of the distribution of electron energies for electrons transmitted through the sample.[34] EELS complements EDS as it is more sensitive to low-Z elements ($Z \leq 10$) while EDS detects elements with $Z > 10$. Theoretically hydrogen should be detectable, but boron is a more practical limit. EELS is concerned with the measurement of electron energy loss due to inelastic collisions and is more sensitive than EDS for several reasons: It is a primary event that does not rely on a secondary event of X-ray emission when an excited atom returns to its ground state making it a more efficient process, especially for low-Z elements. Also only a fraction of the emitted X-rays are detected, while most of the transmitted electrons are detected. EELS is primarily used to provide microanalytical and structural information approaching the very high resolution of TEM. The EELS spectrum generally consists of three distinct groupings of spectral peaks: the zero loss peak containing no useful analytical information, the low energy loss peaks due primarily to plasmons, and the high energy loss peaks due to inner shell ionization. EELS maps can be generated by displaying the intensity of a particular energy of the spectrum.

In addition to structural information, diffraction information is also available in AEM. This is very important for crystalline samples, where *selected area diffraction* may be used to identify crystalline phases, amorphous regions, crystal orientations, and defects such as stacking faults or dislocations.

High resolution TEM (HREM) gives structural information on the atomic size level, is known as *lattice imaging,* and has become very important for interface analysis and TEM micrographs have become important in semiconductor integrated circuit development.[35] For example, oxide-semiconductor, metal-semiconductor, and semiconductor-semiconductor interface studies have benefited from HREM images. In lattice imaging a number of different diffracted beams are combined to give an interference image. Many HREM examples are found in ref. 36. In Fig. 11.17 we show a cross-section of a poly-Si/SiO$_2$/Si structure with a 1.5 nm thick oxide. The white dots in the lower

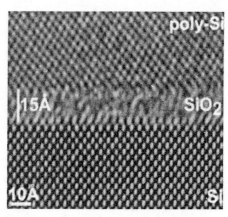

Fig. 11.17   TEM micrograph of a poly-Si/SiO₂/Si structure. Courtesy of M.A. Gribelyuk, IBM.

Si region represent Si atoms (actually columns of Si atoms), clearly showing atomic resolution.

Sample preparation has been the weak point of TEM, because the sample must be very thin. Traditionally mechanical lapping-polishing and ion milling has been used. Sheng gives a good discussion of the difficulties for early successful sample preparation.[37] A recent addition to sample preparation is the focused ion beam (FIB) instrument.[33] FIB is similar to an SEM in design and operation, except the beam consists of Ga⁺ ions instead of electrons. The Ga beam diameter is around 10 nm. These ions are rastered over a given portion of a sample and the ions mill a hole into the sample. This hole can be precisely located, allowing one particular part of an integrated circuit to be examined. For example, once a failure in an IC is located, FIB can be used to cut a hole at precisely that failure location. Once the hole is cut, one can then view the wall of the hole with an SEM. Alternately, it is possible to mill on two sides of a failed IC, leaving a free-standing film (Fig. 12.29). This film can then be taken to a TEM to be studied. The availability of FIB has made TEM much more routine and in some cases it has become a production environment tool—a far cry from the very specialized instrument only a few years ago. Free-standing films can be cut in about 20 minutes compared with many hours for traditional sample thinning. There is, however, a question of whether the ion bombardment of the FIB perturbs the sample. The surface of the sample may become amorphous and contain a high density of Ga. Nevertheless, FIB has become a routine sample preparation technique.

11.2.5   Electron Beam Induced Current (EBIC)

Electron beam induced current for minority carrier diffusion length measurements is discussed in Chapter 7. Here we extend the EBIC discussion to other applications.[38] The term EBIC was coined by Everhart.[39] The technique is also known as *charge collection scanning electron microscopy*.[7] In contrast to most of the techniques in this chapter, EBIC does not identify impurities but measures electrically active impurities. The method relies on collection of minority carriers generated by a scanned electron beam in a junction device. The electron beam generates N_{ehp} electron-hole pairs, where the dependence of N_{ehp} on the beam energy is given in Eq. (11.11). The generation rate of ehps is, according to Eq. (7.69),

$$G = \frac{I_b N_{ehp}}{q Vol} \qquad (11.13)$$

where Vol is the volume in which ehps are generated. For short minority carrier diffusion length, the ehps are generated within a volume of $(4/3)\pi R_e^3$. For semiconductors like Si, with minority carrier diffusion length $L \gg R_e$, the volume is $(4/3)\pi L^3$. The minority electron density in a p-substrate is approximately

$$n = G\tau_n \qquad (11.14)$$

Equation (11.14) expresses the essence of EBIC measurements. Minority carriers are generated by a scanned electron beam. They are collected by a junction (pn junction, Schottky barrier, MOSFET, MOS capacitor, electrolyte-semiconductor junction) and measured as a current—the *electron beam induced current*. The carrier density is dependent on the minority carrier lifetime, which in turn depends on the defect distribution of the sample. The interaction of an electron beam with the semiconductor sample can take place in a variety of geometries as shown in Chapter 7. Changes in the photocurrent collected by the junction, can be effected by moving the beam in the x- and/or y-direction. Changes in the z-direction are produced by changing the beam energy. The e-beam creates ehps at a distance d from the edge of the scr. Some of the minority carriers diffuse to the junction to be collected. We showed in Chapter 7 how the diffusion length is determined by measuring the current as the electron beam is moved away from the collecting junction.

To determine the defect or recombination center distribution, one generally forms a large area collecting junction and scans the e-beam along the junction as shown in Fig. 11.18. The current is constant for uniform material. In this figure we assume a

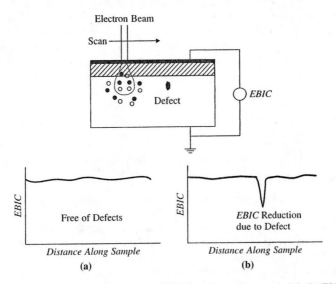

Fig. 11.18   EBIC measurement schematic; (a) EBIC scan for uniform material, (b) EBIC scan for non-uniform material.

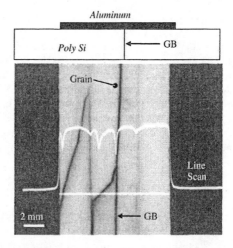

Fig. 11.19 EBIC map of polycrystalline Si showing high recombination grain boundaries. A line scan is taken along the horizontal marker line and the EBIC along that line is displayed. Courtesy of J.B Mohr, Arizona State University.

recombination defect at some depth. For low beam energies, where the ehps are generated near the top surface, most of the minority carriers are collected by the space-charge region and the current does not vary with distance (Fig. 11.18(a)). The beam penetration at higher energies is sufficient for some ehps to recombine at the defect causing the collected current to decrease in the vicinity of the defect (Fig. 11.18(b)). This example illustrates lateral as well as depth uniformity measurements by sweeping the beam and by varying the beam energy.

The current can be displayed as a line scan or as a brightness map on the SEM CRT. It can also be displayed as a pseudo three-dimensional plot. An EBIC brightness map and a line scan of a poly-Si wafer with an Al Schottky contact are shown in Fig. 11.19. The upper sketch is the cross-section and the EBIC map is the top view. The grains and the high recombination activity at the grain boundaries (GB) are clearly visible. The line scan shows the EBIC along the horizontal white line.

Typical applications for EBIC include the measurement of minority carrier diffusion length and lifetime, recombination sites (dislocations, precipitates, grain boundaries), doping density inhomogeneities, and junction location. Due to the contactless nature of the electron beam, it is possible to scan small regions of the sample. For example, in a study of recombination behavior of twin planes in dendritic web silicon, the Schottky contact was formed on the 100-μm thick wafer cross-section, and the beam was scanned across it to reveal recombination activity at the twin planes.[40] EBIC can also be used to detect oxide defects, where an oxidized wafer is provided with a conducting gate.[41] The EBIC current amplifier is connected between the gate and substrate and the electron beam is swept across the sample. Due to the high resistivity oxide between substrate and gate, the EBIC should be near zero. However, current can flow if the oxide contains defects.

11.2.6 Cathodoluminescence (CL)

Cathodoluminescence is a contactless technique based on the emission of light from a sample excited by an electron beam.[42] The most common application of CL is in television receivers, oscilloscopes, and computer monitors where an electron incident on the phosphor inside the picture tube generates light producing the image. CL is related to both EMP (e-beam excitation, X-ray emission) and photoluminescence (light excitation, light emission). Its strength lies in its imaging capability. The electron beam is scanned across the sample and the emitted light is detected and displayed on a CRT. The chief difference between EMP and CL, both employing e-beams for excitation, is that EMP X-rays originate from electronic transitions between inner core energy levels while CL photons originate from transitions between conduction and valence bands.

CL brightness maps can be related to sample recombination behavior through the external photon quantum efficiency η (emitted photons per incident electron)[43]

$$\eta = \frac{(1 - R)(1 - \cos\theta_c)}{(1 + \tau_{rad}/\tau_{non\text{-}rad})} e^{-\alpha d} \qquad (11.15)$$

where $(1 - R)$ accounts for reflection losses at the semiconductor-vacuum interface, $(1 - \cos\theta_c)$ accounts for internal reflection losses, $\exp(-\alpha d)$ accounts for internal absorption losses (d = photon path length), and τ_{rad}, $\tau_{non\text{-}rad}$ are the radiative and non-radiative minority carrier lifetimes.

All of the factors in Eq. (11.15) can be spatially dependent and can contribute to CL image contrast making quantitative interpretation difficult.[44] For example, there may be local reflectance variations, and surface morphology can produce shadowing or enhanced light emission through changes in the $(1 - \cos\theta_c)$ term. Mechanisms causing enhanced or reduced light emission include doping densities, temperature, and recombination centers (metallic impurities, dislocations, stacking faults, precipitates), as well as the presence of electric fields.

In the simplest implementation, the sample is kept at room temperature and the CL light is collected. This panchromatic technique is useful for quick data collection. Distinct advantages are gained by cooling the sample and by resolving the light spectrally. Cooling to liquid helium temperature reduces thermal line broadening and raises the signal/noise ratio. Resolving the light into its spectral components can lead to impurity identification. CL resolution is determined by a combination electron beam diameter, electron range R_e, and minority carrier diffusion length L. For $L \ll R_e$, the resolution is essentially R_e and for $L \gg R_e$ it is essentially L.

CL is mainly used for $III-V$ materials with its high radiative recombination. It is more difficult to use for Si due to its low luminescence efficiency. Of course, CL emission can be enhanced with higher beam current, but that leads to sample heating. Time-resolved CL is useful for lifetime measurements, with both bulk and surface recombination contributing to the effective lifetime.[45] The technique can be combined with other methods in the SEM (EBIC, scanning electron microscopy, EMP) for a more complete analysis. CL implementation is also possible in transmission electron microscopes, but space limitations in the instrument and small sample size make light collection more difficult.

11.2.7   Low-Energy, High-Energy Electron Diffraction (LEED)

Low energy electron diffraction, first demonstrated in 1927 by Davisson and Germer,[46] is one of the oldest surface characterization techniques for investigating the crystallography of sample surfaces.[47] It provides structural, not elemental information, and is illustrated in Fig. 11.20(a). A low-energy (10 to 1000 eV), narrow-energy spread electron beam incident on the sample penetrates only the first few atomic layers. Electrons are diffracted by the periodic atomic arrangement of the atoms. The elastically scattered, diffracted electrons emerge from the surface in directions satisfying interference conditions from the crystal periodicity and strike a fluorescent screen, forming a distinct array of diffraction spots due to the orientation of the crystal lattice of the sample. The diffraction pattern is viewed through a window behind the screen. A series of grids filter the scattered electrons.

LEED provides information on the atomic arrangement and is sensitive to crystallographic defects. It is typically used to determine surface atomic structure, surface structural disorder, surface morphology, and surface changes with time. The diffraction conditions can be most easily studied using a reciprocal lattice and an *Ewald sphere*.[48] To study the properties of the surface, it is important for the surface to be clean, as contaminated surfaces generally do not give diffraction patterns. Consequently, LEED measurements are generally made in an ultra-high vacuum (UHV) of less than 10^{-10} torr. A monolayer of contamination takes about one second to form at a pressure of 10^{-6} torr, but takes about one day at 10^{-10} torr. Even a fraction of a monolayer is sufficient to prevent accurate surface crystallography measurements. Samples should be cleaved in vacuum to expose the appropriate surfaces that have not been subjected to ambient contamination.

Electron diffraction by high-energy electrons is known as *reflection high-energy electron diffraction* (RHEED).[48] As shown in Fig. 11.20(b), 1 to 100 keV electrons are incident on the sample, but because such energetic electrons penetrate deeply, they are made to strike the sample at a shallow, glancing angle of typically less than 5°. Forward scattered electrons are utilized as there is little backscattering. RHEED gives information on surface crystal structure, surface orientation, and surface roughness. Molecular beam epitaxial growth (MBE) has done much to foster the use of RHEED by allowing continuous monitoring of the growth of epitaxial films.[49] The experimental arrangement of Fig. 11.20(b) leaves the front of the sample clear for growth beams. Additionally, since the electron beam strikes the sample at a glancing angle, it is a more critical characterization method, because it picks out surface irregularities more effectively than LEED.

11.3 ION BEAM TECHNIQUES

Ion beam characterization techniques are illustrated in Fig. 11.21. Incident ions are absorbed, emitted, scattered, or reflected leading to light, electron or X-ray emission. Aside from characterization, ion beams are also used for ion implantation. We discuss two main ion beam material characterization methods: *secondary ion mass spectrometry* and *Rutherford backscattering spectrometry*.

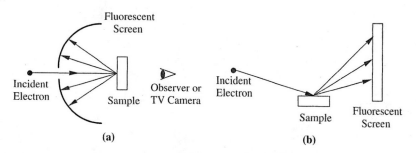

Fig. 11.20   (a) LEED diffractometer, (b) RHEED diffractometer.

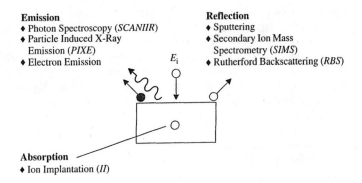

Emission
- Photon Spectroscopy (*SCANIIR*)
- Particle Induced X-Ray Emission (*PIXE*)
- Electron Emission

Reflection
- Sputtering
- Secondary Ion Mass Spectrometry (*SIMS*)
- Rutherford Backscattering (*RBS*)

E_i

Absorption
- Ion Implantation (*II*)

Fig. 11.21   Ion beam characterization techniques.

11.3.1   Secondary Ion Mass Spectrometry (SIMS)

Principle: Secondary ion mass spectrometry, also known as *ion microprobe* and *ion microscope,* is one of the most powerful and versatile analytical techniques for semiconductor characterization.[50–51] It was developed independently by Castaing and Slodzian at the University of Paris[52] and by Herzog and collaborators at the GCA Corp.[53] in the United States in the early 1960s, but did not become practical until Benninghoven showed that it was possible to maintain the surface integrity for periods well in excess of the analysis time.[54] Benninghoven did much to further the evolution and advances of SIMS. The technique is element specific and is capable of detecting all elements as well as isotopes and molecular species. Of all the beam techniques it is the most sensitive with detection limits for some elements in the 10^{14} to 10^{15} cm^{-3} range if there is very little background interference signal. Lateral resolution is typically 100 μm but can be as small as 0.5 μm with depth resolution of 5 to 10 nm.

The basis of SIMS, shown in Fig. 11.22, is the destructive removal of material from the sample by sputtering and the analysis of the ejected material by a mass analyzer. A primary

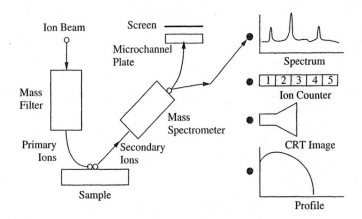

Fig. 11.22   SIMS schematic.

ion beam impinges on the sample and atoms from the sample are sputtered or ejected from the sample. Most of the ejected atoms are neutral and cannot be detected by conventional SIMS, but some are positively or negatively charged. This fraction was estimated as about 1% of the total in 1910,[55] an estimate that is still considered reasonable.[56] The mass/charge ratio of the ions is analyzed, detected as a mass spectrum, as a count, or displayed on a fluorescent screen. The detection of the mass/charge ratio can be problematic, since various complex molecules form during the sputtering process between the sputtered ions and light elements like H, C, O, and N typically found in SIMS vacuum systems. The mass spectrometer only recognizes the total mass/charge ratio and can mistake one ion for another.

Sputtering is a process in which incident ions lose their energy mainly by momentum transfer as they come to rest within the solid. In the process they displace atoms within the sample. Sputtering takes place when atoms near the surface receive sufficient energy from the incident ion to be ejected from the sample. The escape depth of the sputtered atoms is generally a few monolayers for primary energies of 10 to 20 keV typically used in SIMS. The primary ion loses its energy in the process and comes to rest tens of nm below the sample surface. Ion bombardment leads not only to sputtering, but also to ion implantation and lattice damage. The sputtering yield is the average number of atoms sputtered per incident primary ion; it depends on the sample or target material, its crystallographic orientation, and the nature, energy, and incidence angle of the primary ions. Selective or preferential sputtering can occur in multi-component or polycrystalline targets when the components have different sputtering yields. The component with lowest yield becomes enriched at the surface while that with the highest yield becomes depleted. However, once equilibrium is reached, the sputtered material leaving the surface has the same composition as the bulk material and preferential sputtering is not a problem in SIMS analysis.[57]

The yield for SIMS measurements with Cs^+, O_2^+, O^-, and Ar^+ ions of 1 to 20 keV energy ranges from 1 to 20. What is important, however, is the not the total yield, but the yield of ionized ejected atoms or the *secondary ion yield*, because only ions can be detected. The secondary ion yield is significantly lower than the total yield, but can be influenced by the type of primary ion. Electronegative oxygen (O_2^+) enhances species for electropositive elements (*e.g.*, B and Al in Si) which produce predominantly positive secondary ions. Electronegative elements (*e.g.*, P, As and Sb in Si) have higher yields when sputtered with electropositive ions like cesium (Cs^+). The secondary ion yield for the elements varies over five to six orders of magnitude.[58]

SIMS has not only a wide variation in secondary ion yield among different elements, it also shows strong variations in the secondary ion yield from the same element in different samples or matrices - the *matrix effect*. For example, the secondary ion yield for oxidized surfaces is higher than for bare surfaces by as much as 1000.[58] A striking example is a SIMS profile of B or P implanted into oxidized Si obtained by sputtering through an oxidized Si wafer. The yield of Si in SiO_2 is about 100 times higher than the yield of Si from the Si substrate. A plot of yield versus sputtering time shows a sharp drop when the sample is sputtered through the SiO_2-Si interface.

SIMS can give three types of results. For low incident ion beam current or low sputtering rate ($\sim$0.1 nm per hour), a complete mass spectrum can be recorded for surface analysis of the outer 0.5 nm or so. This mode of operation is known as *static* SIMS. In *dynamic* SIMS, the intensity of one peak for one particular mass is recorded as a function of time as the sample is sputtered at a higher sputter rate ($\sim$10 μm per hour), yielding a depth profile. It is also possible to display the intensity of one peak as a two-dimensional image. The various output signals are illustrated in Fig. 11.22.

Quantitative depth profiling is unquestionably the major strength of SIMS, with one selected mass plotted as secondary ion yield versus sputtering time. Such plots must be converted to density versus depth. The conversion of signal intensity to density can, in principle, be calculated knowing the primary ion beam current, the sputter yield, the ionization efficiency, the atomic fraction of the ion to be analyzed, and an instrumental factor. Some of these factors are generally poorly known and a successful technique for routine quantitative SIMS analysis has not yet emerged. The usual approach is one of using standards with composition and matrices identical or similar to the unknown. Ion implanted standards are very convenient and also very accurate. The implant dose of an ion-implanted standard can be controlled to an accuracy of 5% or better. When such a standard is measured, one calibrates the SIMS system by integrating the secondary ion yield signal over the entire profile. Calibrated standards are, therefore, very important for accurate SIMS measurements. The time-to-depth conversion is usually made by measuring the sputter crater depth after the analysis is completed. An example of the conversion of yield or intensity versus time to density versus depth profile is given in Fig. 11.23, showing both the raw SIMS plot and the dopant density profile.

Instrumentation: There are two instrumentation approaches to SIMS; (i) the *ion microprobe* and (ii) the *ion microscope*. A good discussion of SIMS instrumentation is given by Bernius and Morrison.[59] The ion *microprobe* is an ion analog of the electron microprobe. The primary ion beam is focused to a fine spot and rastered over the sample

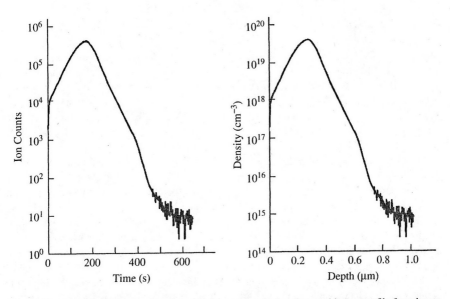

Fig. 11.23   (a) Raw $^{11}B^+$ secondary ion signal versus sputtering time and boron profile for a boron implant into a silicon substrate. Reprinted with permission after P.K. Chu, "Dynamic SIMS," in *Encyclopedia of Materials Characterization* (C.R. Brundle, C.A. Evans, Jr., and S. Wilson, eds.), Butterworth-Heinemann, Boston, 1992, 532–548.

surface. The secondary ions are mass analyzed and the mass spectrometer output signal is displayed on a CRT in synchronism with the primary beam to produce a map of secondary ion intensity across the surface. The spatial resolution is determined by the spot size of the primary ion beam and resolutions lower than 1 μm are possible. The mass spectrometer consists of electrostatic and magnetic sector analyzers in tandem.[33] In the electrostatic analyzer, the ions travel between two parallel plates separated a distance d with a radius of curvature r_V. A potential V between the two plates permits only those ions with the proper energy E to be transmitted without striking either plate, where E is

$$E = \frac{qVr_V}{2d} \tag{11.16}$$

In the magnetic sector spectrometer, a magnetic field B curves the ion of mass m, charge q, and energy E into a path of radius r_B according to

$$\frac{m}{q} = \frac{qB^2r_B^2}{2E} \tag{11.17}$$

Substituting Eq. (11.16) into (11.17) gives

$$\frac{m}{q} = \frac{B^2r_B^2d}{Vr_V} \tag{11.18}$$

The mass resolution can be as high as 40,000, equivalent to resolving two masses differing by only 0.003%. Such high mass resolution is required for detecting ions for which there are interferences. For example, ^{31}P (31.9738 amu) has a very similar mass/charge ratio as $^{30}Si^1H$ (31.9816 amu); $^{29}Si^1H_2$ (31.9921 amu) and ^{54}Fe are similar to the $^{28}Si_2$ dimer.

The ion *microscope* is a direct imaging system, analogous to an optical microscope or a TEM. The primary ion flood beam illuminates the sample and secondary ions are simultaneously collected over the entire imaged area with a resolution on the order of 1 μm. The spatial distribution of the secondary image is preserved through the system using an electrostatic and magnetic sector analyzer in tandem, amplified by a microchannel plate, and displayed on a fluorescent screen. A small aperture may be inserted to select an area for analysis. Ion imaging is also done by raster scanning the ion beam across the sample and measuring and displaying the secondary ion intensity as a function of the lateral position of the small spot scanning ion beam. The lateral resolution of this imaging method is dependent on the beam size, which can be as small as 50 nm.

Proper mass resolution is essential for unambiguous SIMS analysis. For example, a SIMS mass/charge (m/e) spectrum for high-purity Si obtained with an O_2^+ primary ion beam contains $^{28}Si^+$, $^{29}Si^+$, and $^{30}Si^+$ isotopes, polyatomic Si_2^+ and Si_3^+ as well as many molecular species involving oxygen. The latter are not from the sample itself, but are due to the oxygen primary beam causing oxygen implantation and subsequent sputtering. This plethora of signals requires a high resolution spectrometer. Another instrumentation effect that complicates SIMS analysis is the *edge* or *wall effect*. To obtain good depth resolution it is important that only the signal from the flat, bottom portion of the sputtered crater be analyzed. Atoms are also ejected from the crater bottom as well as from the sidewalls during sputtering. But the sidewalls of an ion-implanted sample, especially near the top surface, contain a much higher doping density than the crater bottom. Using electronic gating of the secondary ion yield signal or a lens system, it is possible to detect only those ions from the central part of the crater.[21]

A quadrupole mass analyzer consists of four parallel rods with an oscillating electric field through which the ions pass. It is the basis of *quadrupole SIMS*, which is robust and less expensive than the electrostatic-magnetic sector analyzers, but has lower resolution. Due to lower extraction potentials, it is suitable for analyzing insulating samples, but it cannot distinguish between ions with close mass/charge ratios. Quadrupole SIMS can also switch rapidly between different mass peaks, enabling depth profiles with more data points, thereby increasing the depth resolution.

Electrostatic or magnetic spectrometers depend on serial scanning of an electrostatic or magnetic field, requiring narrow slits for only those ions with the correct mass/charge ratio to be transmitted. This reduces the transmittance of the spectrometer substantially to values as low as 0.001%. A SIMS approach without this limitation is *time-of-flight SIMS* (TOF-SIMS). Instead of continuous sputtering by an ion beam, in TOF-SIMS the incident beam consists of pulsed ions from a liquid Ga^+ gun with beam diameters as small as 0.3 μm. For pulse widths on the order of nanoseconds, ions are sputtered in brief bursts and the time for these ions to travel to the detector is measured. Equating the kinetic and potential energy gives

$$\frac{mv^2}{2} = qV \tag{11.19}$$

where v is the ion velocity. The transit time t_t is simply v/L, where L is the path length from sample to detector, leading to the expression

$$\frac{m}{q} = \frac{2Vt_t^2}{L^2} \tag{11.20}$$

A major advantage of TOF-SIMS is the absence of narrow slits in the spectrometer increasing the ion collection by 10–50%. This allows the incident beam current to be reduced significantly compared to conventional SIMS, which, in turn, reduces the sputtering rate greatly. In fact, the sputtering rate is so low that it may take an hour to remove a fraction of a monolayer. Such low sputtering rates allow characterization of organic surface layers. Furthermore, since m/q is determined by time of flight, very large and small ion fragments can be detected, much larger than in other SIMS approaches. Consequently, a TOF-SIMS spectrum of an organic layer contains hundreds of peaks. TOF-SIMS has also proven very sensitive to surface metals. Surface densities as low as 10^8 cm^{-2} have been detected for Fe, Cr, and Ni on Si.[60]

A major source of the limited sensitivity of SIMS is the fact that most of the sputtered material is neutral and cannot be detected. In *secondary neutral mass spectrometry* (SNMS) or *resonance ionization spectroscopy* (RIS), the neutral atoms are ionized by a laser or by an electron gas and then detected.[61] Significant sensitivity enhancements over conventional SIMS are achieved. The primary ion beam in SIMS is replaced by a pulsed laser in *laser microprobe mass spectrometry* (LAMMA) or *laser ionization mass spectrometry* (LIMS).[62] The pulsed laser volatizes and ionizes a small volume of the sample and the ions are analyzed in a time-of-flight mass spectrometer. LAMMA has high sensitivity, high speed of operation, is applicable to inorganic as well as organic samples and has microbeam capability with a spatial resolution of ~1 μm. It is primarily used in failure analysis where chemical differences between contaminated and control samples must be rapidly assessed.

Applications: SIMS has found its greatest utility in semiconductor characterization, especially for dopant profiling. For a more detailed discussion and comparison with

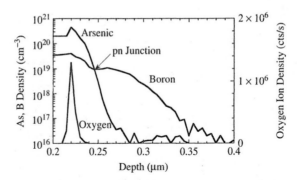

Fig. 11.24 SIMS depth profile of a shallow Si pn junction. Both As and B were measured with 3 keV Cs ions at 60° incidence. Reprinted with permission after ref. 63.

spreading resistance measurements, see Chapter 2. SIMS measurements are well suited for semiconductor applications, because matrix effects are minor and ion yields can be assumed to be linearly proportional to densities up to 1%. Furthermore, the substrate sputters very uniformly, at least for Si. An example profile in Fig. 11.24 shows that arsenic, boron, and oxygen can be determined in a single measurement. This sample was formed by diffusing As and B from a poly-Si layer deposited on the Si substrate. The plot shows the location of the junction ($N_{As} = N_B$) and the location of the poly-Si/substrate interface (oxygen peak).

Factors that need to be considered in data analysis are crater wall effects, ion knock-on, atomic mixing, diffusion, preferential sputtering, and surface roughening. Some of these are instrumental and can be alleviated to some extent, but others are intrinsic to the sputtering process. For SIMS, the most important type of atomic mixing is "cascade mixing," resulting from primary ions striking sample atoms and displacing them from their lattice positions, leading to homogenization of all atoms within the depth affected by the collision cascade. Dopant atoms originally present at a given depth in the sample will distribute throughout this "mixing depth" as sputtering proceeds and the dopant profile will give a deeper distribution than the true distribution. It is important that the primary ion penetration depth be kept to a minimum for shallow dopant profiling. Deeper junctions are often observed when SIMS doping profiles are compared to spreading resistance profiles.[64] A high vacuum is very important for SIMS. The arrival rate of gaseous species from the vacuum chamber should be less than that of the primary ion beam; otherwise it is vacuum contamination that is measured, not the sample. This is particularly important for low mass species like hydrogen. A very thorough discussion of these effects can be found in the paper by Zinner listing 35 factors affecting SIMS depth profiling.[65]

11.3.2 Rutherford Backscattering Spectrometry (RBS)

Principle: *Rutherford backscattering spectrometry*, also known as *high-energy ion (back)-scattering spectrometry* (HEIS), is based on backscattering of ions incident on a sample.[66] It is quantitative without recourse to calibrated standards. Experiments by Rutherford and his students in the early 1900s proved the existence of nuclei and scattering from these nuclei.[67] The field of ion interactions in solids was very intensively researched and developed

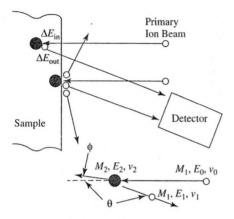

Fig. 11.25 Rutherford backscattering schematic.

following the discovery of fission and nuclear weapons development. But it was not until the late 1950s that nuclear backscattering was put to practical use.[68] Further developments in the 1960s led to identification of minerals[69] and determination of properties of thin films as well as thick samples.

RBS is based on bombarding a sample with energetic ions—typically He ions of 1 to 3 MeV energy—and measuring the energy of the backscattered He ions. It allows determination of the *masses* of the elements in a sample, their *depth distribution* over distances from 10 nm to a few microns from the surface, their areal density, and the *crystalline structure* in a non-destructive manner. The depth resolution is on the order of 10 nm. The use of ion backscattering as a quantitative materials analysis tool depends on an accurate knowledge of well known nuclear and the atomic scattering processes.

The method is illustrated in Fig. 11.25. Ions of mass M_1, atomic number Z_1, energy E_0, and velocity v_0 are incident on a solid sample or target composed of atoms of mass M_2 and atomic number Z_2. Most of the incident ions come to rest within the solid, losing their energy through interactions with valence electrons. A small fraction—around 10^{-6} of the number of incident ions—undergoes elastic collisions and is backscattered from the sample at various angles. The incident ions lose energy traversing the sample until they experience a scattering event and then lose energy again as they travel back to the surface, leaving the sample with reduced energy.

After scattering, atom M_2 has energy E_2 and velocity v_2 and ion M_1 has energy E_1 and velocity v_1. Conservation of energy gives

$$E_0 = M_1 v_0^2/2 = E_1 + E_2 = M_1 v_1^2/2 + M_2 v_2^2/2 \qquad (11.21)$$

Conservation of momentum in the directions parallel and perpendicular to the incidence direction gives

$$M_1 v_0 = M_1 v_1 \cos(\theta) + M_2 v_2 \cos(\phi); 0 = M_1 v_1 \sin(\theta) - M_2 v_2 \sin(\phi) \qquad (11.22)$$

Eliminating ϕ and v_2 and taking the ratio $E_1/E_0 = (M_1 v_1^2/2)/(M_1 v_0^2/2)$, gives the *kinematic factor* K[70]

$$K = \frac{E_1}{E_0} = \frac{[\sqrt{1 - (R \sin\theta)^2} + R \cos\theta]^2}{(1 + R)^2} \approx 1 - \frac{2R(1 - \cos\theta)}{(1 + R)^2} \qquad (11.23)$$

where $R = M_1/M_2$ and θ is the scattering angle. The approximation in Eq. (11.23) holds for $R \ll 1$ and θ close to $180°$. Equation (11.23) is the key RBS equation. The kinematic factor is a measure of the primary ion energy loss. The scattering angle should be as large as possible and angles around $170°$ are commonly used. The unknown mass M_2 is calculated from the measured energy E_1 through the kinematic factor.

We illustrate RBS with the two examples in Fig. 11.26. Fig. 11.26(a) consists of a silicon substrate with a very thin film of nitrogen, silver, and gold. The atomic weight and calculated R, K, and E_1 in Table 11.2 are for $\theta = 170°$ and incident helium ions ($M_1 = 4$) with $E_0 = 2.5$ MeV. Helium ions have energies of 0.78, 1.41, 2.16 and 2.31 MeV after scattering from the N, Si, Ag, and Au atoms at the sample surface. Since N, Ag, and Au are only at the surface in this example, RBS signals from these elements have narrow spectral distributions confirmed by experimental data. The yield is not to scale on this figure. Figure 11.26(a) brings out two important properties of RBS plots: the RBS yield increases with element atomic number and the RBS signal of elements lighter than the substrate rides on the matrix background while heavier elements are displayed by themselves. This makes the nitrogen signal more difficult to detect because it rides on the Si signal. The Si background count represents the "noise" and the signal-to-noise ratio is degraded compared to heavy elements on a light matrix.

RBS plots are more complicated for layers of finite thicknesses. In Fig. 11.26(b) we consider a gold film of thickness d on a silicon substrate. The He ions are backscattered from surface gold atoms with $E_{1,\text{Au}} = 2.31$ MeV as in Fig. 11.26(a). However, those ions backscattered from deeper within the Au film emerge with lower energies, due to additional losses within the film. These losses come from Coulombic interactions between helium ions and electrons. Consider a scattering event from those Au atoms at the Si-Au interface at $x = t$. The He ion loses energy ΔE_{in} traveling through the Au film before the scattering event at the back gold surface. Upon scattering, it loses additional energy

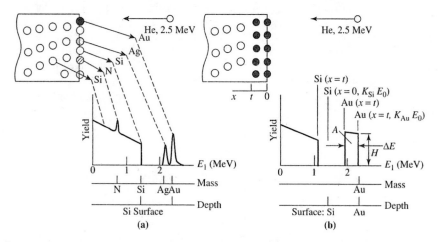

Fig. 11.26 (a) RBS calculated spectrum for N, Ag, and Au on Si, (b) schematic spectrum for a Au film on Si. "A" is the area under the curve.

TABLE 11.2 Calculated R, K and E_1 (For 2.5 MeV He ions, $\theta = 170°$).

Target Atom (M_2)	Atomic Weight	R	K	E_1 (MeV)
N	14	0.256	0.311	0.78
O	16	0.25	0.363	0.91
Si	28.1	0.142	0.566	1.41
Cu	63.6	0.063	0.779	1.95
Ag	107.9	0.037	0.863	2.16
Au	197	0.020	0.923	2.31

$(E_0 - \Delta E_{in})(1 - K_{Au})$. To reach the detector it must traverse the film a second time, losing energy ΔE_{out}. The total energy loss is the sum of these three losses. The energy of He ions scattered from the sample at depth d is

$$E_1(d) = (E_0 - \Delta E_{in})K_{Au} - \Delta E_{out} \tag{11.24}$$

The energy losses are slightly energy dependent and are listed in tables of stopping powers.[71] The energy difference of the ions backscattered from the surface and from the interface ΔE can be related to the film thickness d

$$\Delta E = \Delta E_{in} K_{Au} + \Delta E_{out} = [S_0] d \tag{11.25}$$

where $[S_0]$ is the *backscattering energy loss factor*; it has units of eV/Å and is tabulated for pure-element samples, e.g., $[S_0] = 133.6$ eV/Å for gold films with a 2 MeV beam energy.

The *backscattering yield* A, also designated as the total number of detected ions or counts

$$A = \sigma \Omega Q N_s, \tag{11.26}$$

where σ = average scattering cross-section in cm²/sr, Ω = detector solid angle in steradians [detector area-(detector-sample distance)²], Q = total number of ions incident on the sample, and N_s = sample atoms/cm². The total count A is also the area under the experimental yield-energy curve or the total number of detected He ions backscattered from the element of interest or the sum of the counts in each channel, shown on Fig. 11.26(b) as "A". $N_s = Nd$ for a thin film where N is atoms/cm³. Q is determined by the time integration of the current of charged particles incident on the target, but it is difficult to determine accurately due to secondary sample electron emission. The average scattering cross-section is

$$\sigma = \frac{1}{\Omega} \int \left(\frac{d\sigma}{d\Omega}\right) d\Omega \tag{11.27}$$

with the differential scattering cross-section given by[72]

$$\frac{d\sigma}{d\Omega} = \left(\frac{q^2 Z_1 Z_2}{2 E_0 \sin^2 \theta}\right)^2 \frac{\left[\sqrt{1 - (R \sin \theta)^2} + \cos \theta\right]^2}{\sqrt{1 - (R \sin \theta)^2}} \tag{11.28}$$

E_0 is the energy of the projectile immediately before scattering. Values for $d\sigma/d\Omega$ are tabulated for all elements for He probe ions. Typical values of the differential scattering cross-section are 1 to 10×10^{-24} cm^2/sr. The yield increases with increasing atomic number leading to higher RBS sensitivity for high-Z elements. However, due to the kinematics of scattering, high-mass elements are more difficult to distinguish from one another than low-mass elements.

The areal density N_s is determined from the yield according to Eq. (11.26), which can be cast in a different form because it may be difficult to determine Q accurately. Furthermore the detector solid angle Ω may change if the detector develops "dead" spots after prolonged exposure to energetic projectiles.

An unknown impurity on a known substrate, for example impurity "X" on a Si substrate, is determined by[73]

$$(N_s)_X = \frac{A_X}{H_{Si}} \frac{\sigma_{Si}}{\sigma_X} \frac{\delta E_1}{[\varepsilon]_{Si}} \tag{11.29}$$

where A is the total count, H the height (count/channel) of the spectrum, and $[\varepsilon] = (1/N)\, dE/dx$ the backscattering stopping cross-section.[74] The energy width of a single channel in the multichannel analyzer δE_1 corresponds to a depth uncertainty δx as

$$\delta E_1 = [S_0]\delta x \tag{11.30}$$

δE_1 is determined by the detector and the electronic system and is typically 2 to 5 keV. To find the unknown density it is only necessary to determine the RBS spectrum area, the height of the Si spectrum, and to look up the two cross-sections and the Si stopping cross-section. Typical values for the stopping cross-section lie in the 10 to 100 eV/(10^{15} atoms/cm^2) range with $[\varepsilon]_{Si} = 49.3$ eV/(10^{15} atoms/cm^2) and $[\varepsilon]_{Au} = 115.5$ eV/(10^{15} atoms/cm^2) for 2 MeV He ions.[11]

The RBS spectrum of the thick Si substrate in Fig. 11.26 has a characteristic slope with the yield increasing at lower energies due to scattering within the target. The yield is inversely proportional to the ion energy at depth d_1. The yield at energy E_1, the energy of those ions backscattered from atoms at depth d_1, is proportional to $(E_0 + E_1)^{-2}$, i.e., the yield increases as E_1 decreases deeper into the target.

The RBS sensitivity can be enhanced by changing the differential scattering cross-section in Eq. (11.28) by increasing the atomic number Z_1 of the incident ion from He to C, for example, and/or decreasing the energy E of the incident ion from several MeV to hundreds of keV, known as *heavy ion backscattering spectroscopy* (HIBS).[75] For example, replacing 3 MeV ^{4}He with 400 keV ^{12}C increases the backscattering yield by a factor of 1000. In contrast to conventional RBS with a sensitivity of around 10^{13} cm^{-2}, HIBS can reduce that to the 10^9–10^{10} cm^{-2} range. The lower energy, heavier ions, however, have the potential of inducing surface sputter damage.

Instrumentation: An RBS system consists of an evacuated chamber containing the He ion generator, the accelerator, the sample, and the detector. Negative He ions are generated in the ion accelerator at close to ground potential. In a tandem accelerator, these ions are accelerated to 1 MeV, traversing a gas-filled tube or "stripper canal," where either two or three electrons are stripped from the He$^-$ to form He$^+$ or He^{2+}, respectively.[76] These ions with energies of around 1 MeV are accelerated a second time to ground potential at which point the He$^+$ ions have 2 MeV and the He^{2+} ions have 3 MeV energy. A magnet separates the two high-energy species.

In the sample chamber, the He ions are incident on the sample and the backscattered ions are detected by a Si surface barrier detector that operates much like the X-ray EDS detector described in Section 11.2.3. The energetic ions generate many electron-hole pairs in the detector, resulting in output voltage pulses from the detector. The pulse height, proportional to the incident energy, is detected by a pulse height or multichannel analyzer that stores pulses of a given magnitude in a given voltage bin or channel. The spectrum is displayed as yield or counts versus channel number with channel number proportional to energy. The energy resolution of Si detectors, set by statistical fluctuations, is around 10 to 20 keV for typical RBS energies. The sample is mounted on a goniometer for precise sample-beam alignment or channeling measurements RBS runs take 15–30 minutes.

Applications: Typical semiconductor applications include measurements of thickness, thickness uniformity, stoichiometry, nature, amount and distribution of impurities in thin films, such as silicides and Si- and Cu-doped Al. The technique is also very useful to investigate the crystallinity of a sample. Backscattering is strongly affected by the alignment of atoms in a single crystal sample with the incident He ion beam. If the atoms are well aligned with the beam, those He ions falling between atoms in the channels penetrate deeply into the sample and have a low probability of being backscattered. Those He ions that encounter sample atoms "head-on" are, of course, scattered. The yield from a well-aligned single crystal sample can be two orders of magnitude less than that from a randomly aligned sample. This effect is referred to as *channeling* and has been extensively used to study ion implantation damage in semiconductors with the yield decreasing as the single crystal nature of an implanted sample is restored by annealing.[77]

RBS is particularly suited for heavy elements on light substrates, *e.g.*, contacts to semiconductors. Consequently RBS has been used extensively in the study of such contacts. For example, Fig. 11.27 shows RBS spectra for platinum and platinum silicide on silicon. Initially a Pt film is deposited on a Si substrate. The "no anneal" RBS spectrum clearly shows the Pt film. The Si signal is consistent with E_1 taking into account the loss into and out of the Pt film. As the film is heated, PtSi forms. Note the formation from the Pt-Si interface indicated by the Pt yield decrease for that part of the film near the Si substrate. At the same time, the Si signal moves to higher energies, indicative of Si moving into the

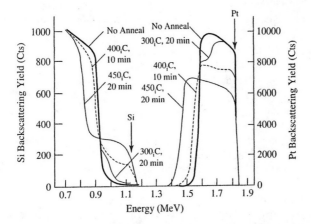

Fig. 11.27   RBS spectra for a 2000 Å Pt film on Si before and after heat treatment. Platinum silicide is formed first at the interface and then throughout the film. $E_0 = 2$ MeV. Reprinted with permission after ref. 78.

Pt film. When stoichiometry is attained, the Pt signal is uniform, but reduced and the Si signal has risen. It is difficult to obtain these data with other techniques non-destructively.

RBS can provide both atomic composition and depth scales to accuracies of 5% or better. The detection limit is 10^{17} to 10^{20} cm^{-3}, but depends on the element and on energy. The sensitivity to light elements, *e.g.*, oxygen, carbon, and nitrogen, in the presence of heavier elements is poor, because the differential scattering cross-section is low for such elements, according to Eq. (11.28). However, the cross-section can be enhanced by using ion beams for which the elastic scattering is resonant.[79] For example, the resonance at 3.08 MeV for oxygen enhances the cross-section 25 times compared to its corresponding Rutherford cross-section. Typical RBS depth resolutions are 10 to 20 nm for film thicknesses $\leq$ 200 nm. The penetration depth of 2 MeV He ions is about 10 μm in silicon and 3 μm in gold. Beam diameters are commonly around 1 to 2 mm but microbeam backscattering with beam diameters as small as 1 μm is possible.[80] Lateral non-uniformities over the area of the analyzing beam cannot be resolved.

A particular difficulty is the ambiguity of RBS spectra, because the horizontal axis is simultaneously a depth and a mass scale. A light mass at the surface of a sample generates a signal that may be indistinguishable from that of a heavier mass located within the sample. Through the use of tabulated constants, experimental techniques such as beam tilting, detector angle changes, and incident energy variations as well as good analytical reasoning, sample analysis is usually successful, but additional information may have to be provided to resolve ambiguities. Computer programs are extensively used in spectrum analysis.[81] As with other physical and chemical characterization techniques, the more is known about the sample before the analysis, the less ambiguous are the results. A comparison of RBS with SIMS is given by Magee.[82]

11.4   X-RAY AND GAMMA-RAY TECHNIQUES

X-ray interactions with a solid are illustrated in Fig. 11.28. Incident X-rays are absorbed, emitted, reflected, or transmitted and can, in turn, cause electron emission. We will discuss *X-ray fluorescence* and *X-ray photoelectron spectroscopy*, useful for chemical characterization, and briefly mention *X-ray topography*, used for structural characterization. Gamma rays, detected in *neutron activation analysis,* are included in this chapter for completeness.

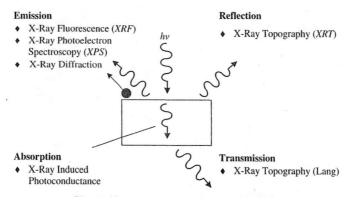

Fig. 11.28   X-ray characterization techniques.

11.4.1 X-Ray Fluorescence (XRF)

Principle: In *X-ray fluorescence*, also known as *X-ray fluorescence spectroscopy* (XRFS), *X-ray fluorescence analysis* (XRFA), and *X-ray secondary emission spectroscopy*, primary X-rays incident on the sample are absorbed by ejecting electrons from the atomic *K*-shell as illustrated in Fig. 11.29.[83] Electrons from higher-lying levels, for example the *L* shell, drop into the *K*-shell vacancies and the energy liberated in the process is given off as characteristic secondary X-rays with energy

$$E_{XRF} = E_K(Z) - E_{L2,3}(Z) \tag{11.31}$$

The X-ray energy *identifies* the impurity and the intensity gives its *density*. XRF allows non-destructive elemental analysis of solids and liquids with quantitative thin film analysis readily obtained. It is not a high-resolution method, as X-rays are difficult to focus. Typical analysis areas are 1 cm^2, although in recent instruments it is possible to analyze areas as small as 10^{-6} to 10^{-4} cm^2.[84] Microspot XRF with beam diameter around 25 μm has been used to characterize metal lines and voids in such lines.[85] The method is suitable for conductors as well as for insulators, since X-rays are uncharged.

Conventional XRF is not a surface-sensitive technique. As discussed in Section 11.2.3, X-ray penetration into a sample is governed by the X-ray absorption coefficient. In Si the penetration depth is typically microns or tens of microns. For example, to detect X-rays emerging from the sample, it is reasonable to find the depth for 50% absorption, since X-rays have to penetrate the sample and generate characteristic X-rays which in turn have

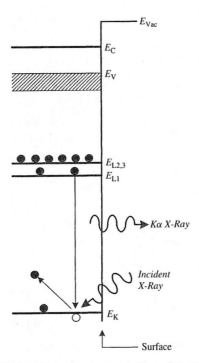

Fig. 11.29 Electronic processes in X-ray fluorescence.

to be emitted to be detected. The 50% penetration depth of Cu $K\alpha$ primary X-rays is 46 μm in Si according to Eq. (11.10).

Total reflection XRF (TXRF) is a surface sensitive technique in which X-rays strike the sample at a very shallow angle and penetrate only a small distance into the sample.[86] Theoretical penetration is on the order of several nm, actual penetration may be deeper due to surface roughness, wafer warpage, and beam divergence. In contrast to XRF, which uses angles of around 45°, TXRF uses a primary beam with a grazing incidence angle of less than 0.1°, which is below the critical angle θ_c. For Mo $K\alpha$ X-rays incident on Si, $\theta_c = 1.8$ mrad. The instrument is schematically shown in Fig. 11.30. X-rays from an X-ray tube in the shape of a strip of about 1 mm by 1 cm are monochromatized and then directed at the sample at a shallow angle. Standing waves are formed above the sample, leading to detection in a lithium-drifted Si detector. The detector is located within about 1 mm of the sample surface. Due to the total reflection property of the system, the substrate contributes little to the spectrum, *i.e.*, matrix absorption and enhancement effects are avoided, in contrast to conventional XRF where matrix effects can be significant. This leads to the high sensitivity of the technique. The instrument must be calibrated against known standards.

TXRF allows metallic surface density determination of 10^9–10^{10} cm^{-2}. With *synchrotron* TXRF (*S*-TXRF) the sensitivity is around 10^7–10^8 cm^{-2}.[87] The sensitivity can be enhanced by HF (hydrofluoric acid) condensation or *vapor phase decomposition* TXRF (VPD-TXRF),[88] where the wafer with a native or thermal oxide is exposed to HF vapor. A byproduct of HF etching is water. The HF etches the oxide with the impurities contained in the resulting water droplet. The impurities are collected *in situ* by scanning the surface with an additional water droplet. The VPD residue is allowed to dry and measured by TXRF. The assumption is that the water droplet carries all surface contaminants with it and the gain is area(wafer)/area(droplet). For a 200 mm diameter wafer and a 10 mm diameter droplet, the gain is 400, enhancing the sensitivity to about 10^8 cm^{-2} for Fe, for example. Impurities like Fe, Ni, Zn, Ca are condensed to about 80%, whereas only about 15% to 20% Cu is collected.[88] The technique is used by Si wafer producers and by IC manufacturers, with the latter often using it to determine the efficacy of cleaning methods and contamination by various IC processes. A recent study of Fe measurements by TXRF, *S*-TXRF, TOF-SIMS, surface photovoltage, ELYMAT, and DLTS showed reasonable agreement among these techniques.[89]

We will briefly mention two other methods that are used for the detection of low-density metallic contamination on wafer surfaces. *Inductively coupled plasma mass spectroscopy* (ICP-MS) is a surface sensitive technique.[90] Trace elements on a semiconductor surface are removed by etching the oxide layer that is invariably present on the wafer. The assumption is that the trace elements are removed with the oxide. The liquid is nebulized and ionized in an inductively coupled plasma. Once in ion form, the ions are analyzed in a mass spectrometer, most commonly a quadrupole mass spectrometer. It is sensitive to about 10^9–10^{10} atoms/cm^2. In *atomic absorption spectroscopy* (AAS), light from a radiation source is absorbed by the sample and detected.[91] The light source generates characteristic narrow emission lines of a selected metal. The sample is atomized in a flame

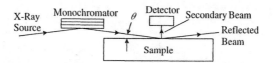

Fig. 11.30 Schematic of a TXRF instrument.

cell or graphite furnace causing a broadening of the absorption lines. Due to the narrow emission lines from the light source along with the broad absorption lines, the wavelength selector (monochromator) only needs to isolate the line of interest from other source lines.

Instrumentation: In XRF, a beam of primary X-rays illuminates the sample and secondary X-rays are detected by an energy dispersive (EDS) or a wavelength dispersive spectrometer (WDS). Energy dispersive XRF uses low-power excitation sources and provides a cost effective way for qualitative as well as quantitative detection of elements starting at $Z \approx 11$. Wavelength dispersive XRF requires higher power excitation sources, typically 3–4 kW, and offers high-precision determination of elements down to $Z \approx 4$. Detection of these light elements requires vacuum as the analysis environment. Conventional XRF sensitivity is around 0.01% or 5×10^{18} cm^{-3}, the analysis area is on the order of 1 cm^2 and typical measurement times are around 50–100 s. Total reflection XRF is sensitive to surface contamination of about 10^{10} cm^{-2} and as low as 10^8 cm^{-2} when coupled with vapor phase decomposition.

Applications: XRF is ideally suited for rapid initial sample survey to define subsequent, more detailed analyses. It is non-destructive and can be used in air for conductors, semiconductors as well as insulators. It gives the average sample composition over the X-ray absorption depth rapidly, but has no profiling capability. The technique has also found use in film thickness measurements.[92] By establishing standards of a given film in which the thickness is measured independently, thicknesses of unknown films are easily determined by measuring the intensity of the secondary X-rays. Films as thin as 10 nm can be measured. Standards are important for quantitative measurements, since XRF is subject to a *matrix effect*, which is the absorption of secondary X-rays by the sample itself. The standards should be well matched to the sample matrix. XRF requires no standards in the thin film approximation.[93]

XRF has also found application in determining the constituents of mixed conductors. For example, it is common practice in Si technology to add a small fraction of copper to aluminum to increase its electromigration resistance. The Cu fraction can be easily detected by XRF. Similarly, glasses to passivate Si chips are frequently doped with boron and phosphorus to increase the ability to "flow" at moderate process temperatures. The phosphorus content of such glasses can be determined by XRF. In contamination problems, XRF has been used to determine chlorine and fluorine contaminants in aluminum metallization after plasma etching.[94]

11.4.2 X-Ray Photoelectron Spectroscopy (XPS)

Principle: X-ray photoelectron spectroscopy, also known as *electron spectroscopy for chemical analysis* (ESCA), is the high-energy version of the photoelectric effect discovered by Hertz in 1887. It is primarily used for identifying chemical species at the sample surface, allowing all elements except hydrogen and helium to be detected. Hydrogen and helium can, in principle, also be detected, but that requires a very good spectrometer. When photons of low energy ($\leq$ 50 eV) are incident on a solid, they can eject electrons from the valence band; the effect is known as *ultraviolet photoelectron spectroscopy* (UPS). In XPS the photons that interact with core level electrons are X-rays.[95] Electrons can be emitted from any orbital with photoemission occurring for X-ray energies exceeding the binding energy. Although the principle of XPS had been known for a long time, implementation had to await the introduction of a high-resolution spectrometer for the detection of the low-energy XPS electrons in the 1960s by Siegbahn and coworkers in

Sweden.[96] He coined the term "electron spectroscopy for chemical analysis", but since other methods also give chemical information, it is more commonly known as XPS today. The early history and development of XPS has been well chronicled by Jenkin et al.[97]

The method is illustrated with the energy band diagram in Fig. 11.31 and the schematic in Fig. 11.32. Primary X-rays of 1 to 2 keV energy eject photoelectrons from the sample. The measured energy of the ejected electron at the spectrometer E_{sp} is related to the binding energy E_b, referenced to the Fermi energy E_F, by

$$E_b = h\nu - E_{sp} - q\phi_{sp} \tag{11.32}$$

where $h\nu$ is the energy of the primary X-rays and ϕ_{sp} the work function of the spectrometer (3 to 4 eV). With E_b depending on the X-ray energy, it is important that the incident X-ray energy be monochromatic. The spectrometer and the sample are connected forcing their Fermi levels to line up. The Fermi energy of metals is well defined. Care must be

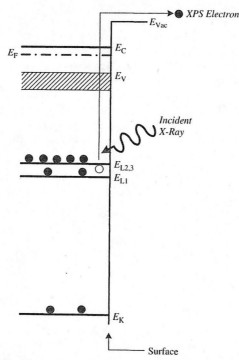

Fig. 11.31 Electronic processes in X-ray photoelectron spectroscopy.

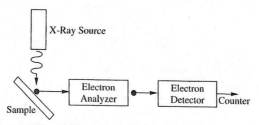

Fig. 11.32 XPS measurement schematic.

taken in analyzing XPS data from semiconductors and insulators because E_F can vary from sample to sample.

The electron binding energy is influenced by its chemical surroundings making E_b suitable for determining chemical states. This is a major strength of XPS; it allows *chemical* and *elemental* identification. Handbooks and graphs of binding energies for elements and compounds are available.[98] X-rays tend to be less destructive, making the method more suitable for organics and oxides than AES. It is sometimes claimed that XPS causes no charging. While it is true that X-rays possess no charge, electron emission from the sample may cause positive sample charging, especially for insulators. This can be compensated with an electron flood gun. X-ray induced Auger electron emission also occurs during XPS. While such Auger lines can interfere with XPS lines, they can also be used to advantage. For example, varying the incident X-ray energy changes the energy of XPS electrons but not the energy of Auger electrons.

XPS is surface sensitive because the emitted photoelectrons originate from the upper 0.5–5 nm of the sample, just as Auger electrons do, despite the deeper penetration of the primary X-rays compared to a primary electron beam.[99] The depth is governed by the electron escape depth or the related electron mean free path. Those electrons excited deeper within the sample are unable to exit the surface. Depth profiling is possible by ion beam sputtering or by sample tilting.[100] However, sputtering can alter oxidation states of the compound. Sample tilting is the basis of *angle-resolved* XPS in which the sampling depth is $\lambda \sin \theta$, where θ is the angle between the sample surface and the trajectory of the emitted photoelectrons.[101]

The major use of XPS is for identification of compounds using energy shifts due to changes in the chemical structure of the sample atoms. For example, an oxide exhibits a different spectrum than a pure element. Care must be exercised in correctly interpreting the data. Unexpected peaks may appear for a variety of reasons. XPS has a more developed chemical state analysis than AES.[18]

Instrumentation: The three basic components of XPS, shown in Fig. 11.32, are (1) the X-ray source, (2) the spectrometer, and (3) a high vacuum, even though such beam-induced chemistry as carbonization is minimized. X-ray line widths are proportional to the atomic number of the target in the X-ray tube. The X-ray line width in XPS should be as narrow as possible; hence, light elements like Al ($E_{K\alpha} = 1.4866$ keV) or Mg ($E_{K\alpha} = 1.2566$ keV) are common X-ray sources. Some XPS systems come equipped with multiple anode X-ray sources. X-ray generation from low-Z materials also has reduced background radiation. The primary X-rays may be filtered by crystal dispersion to remove X-ray satellites and continuum radiation, but filtering reduces the X-ray intensity substantially. The XPS electrons are detected by one of several types of detectors. The hemispherical sector analyzer consists of two concentric hemispheres with a voltage applied between them. A spectrum is generated by varying the voltage so that the electron trajectories with different energies are brought to a focus at the analyzer exit slit. An electron multiplier amplifies the signal.

Chemical compounds or elements are identified by the location of energy peaks on the undifferentiated XPS spectrum. Density determination is more difficult. Peak heights and peak areas can be used with appropriate correction factors to obtain densities, but the method is primarily used for identification. X-ray techniques are generally large-area methods with typically 1 cm^2 area. The analyzed sample area in XPS has been reduced over the years. Today 10 μm spot size is about the smallest that can be analyzed. This has come about by either focusing the X-rays with a monochromator crystal, or by using a large-area X-ray beam but only allowing electrons from a small sample area to enter the

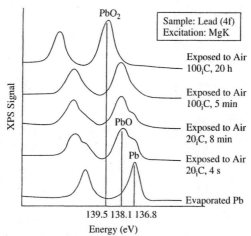

Fig. 11.33 XPS binding energy shifts of lead as an oxide forms. Reprinted with permission after ref. 102.

electron analyzer. XPS sensitivity is around 0.1% or 5×10^{19} cm^{-3} and depth resolution is around 10 nm.[20]

Applications: XPS is used primarily for chemical surface information. It is particularly useful for analyzing organics, polymers, and oxides. For example, it has been used to follow the oxidation of elements. In Fig. 11.33 we show XPS spectra of lead in its pure form and the spectral changes when the Pb oxidizes to PbO and PbO_2. XPS has been extensively used in the semiconductor industry for a variety of problem solving. It has played a major role in understanding the chemistry and reaction mechanisms in the development of plasma etching. XPS has been applied to die attachment problems, adhesion of resins to metal surfaces, and interdiffusion of nickel through gold.[103] Recently, XPS has found use in oxide thickness measurements. The intensity of the 2p peak associated with silicon dioxide is proportional to oxide thickness when it is normalized to the Si 2p peak from the un-oxidized silicon substrate. X-ray photoelectron spectroscopy shows the presence of at least a monolayer of incompletely oxidized silicon - the sub-oxide layer.[104]

11.4.3 X-Ray Topography (XRT)

X-Ray topography or *X-ray diffraction* is a non-destructive technique for determining structural crystal defects.[105] It requires little sample preparation and gives structural information over entire semiconductor wafers but it does not identify impurities. The XRT image is not magnified because no lenses are used. It is, therefore, not a high-resolution technique, but does give microscopic information through photographic enlargement of the topograph.

Consider a perfect crystal arranged to diffract monochromatic X-rays of wavelength λ from lattice planes spaced d. The X-rays are incident on the sample at an angle α, as shown in Fig. 11.34(a). The primary beam is absorbed by or transmitted through the sample; only the diffracted beam is recorded on the film. The diffracted beam emerges at twice the Bragg angle θ_B

$$\theta_B = \arcsin (\lambda/2d) \tag{11.33}$$

The diffracted X-rays are detected on a high-resolution, fine-grained photographic plate or film or solid state detector held as close as possible to the sample without intercepting

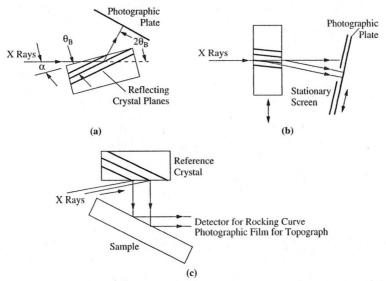

Fig. 11.34   (a) Berg-Barrett reflection topography, (b) Lang transmission topography, (c) double-crystal topography with a rocking curve.

the incident beam. The plate should be held perpendicular to the secondary X-rays for highest resolution. If the lattice spacing or lattice plane orientation vary locally due to structural defects, Eq. (11.33) no longer applies simultaneously to the perfect and the distorted regions. Consequently there is a difference in X-ray intensity from the two regions. For example, the diffracted beam from dislocations is more intense than from an area without defects caused by the mitigation of extinction and by Bragg defocusing. Dislocations produce a more heavily exposed image on the film. The image is formed as a result of diffraction from an anomaly such as strain in the crystal but does not image the defect directly. Strain S is the amount of elastic deformation defined by

$$S = \frac{d_{unstrained} - d_{strained}}{d_{unstrained}} \tag{11.34}$$

By determining d in unstrained and strained regions, using Eq. (11.33), one can determine S.

The reflection method illustrated in Fig. 11.34(a), known as the *Berg-Barrett* method, is based on the original work of Berg, modified by Barrett and further refined by Newkirk.[106] It is the simplest X-ray topography method. There are neither lenses nor moving parts except for the sample alignment goniometer. Reflection XRT probes a thin sample region near the surface, since the shallow incident angle α confines X-ray penetration to the near-surface region. This method is used to determine dislocations, for example, and is useful for dislocation densities up to about 10^6 cm^{-2}. The resolution is about 10^{-4} cm, and entire wafers can be examined.

Transmission XRT, illustrated in Fig. 11.34(b), introduced by Lang, is the most popular XRT technique.[107] Monochromatic X-rays pass through a narrow slit and strike the sample aligned to an appropriate Bragg angle. The tall and narrow primary beam is transmitted through the sample and strikes a lead screen. The diffracted beam falls on the photographic plate through a slit in the screen. X-rays are absorbed in a solid according to Eq. (11.9). However, absorption is considerably reduced when the X-rays are aligned for diffraction along certain crystal planes.[108] A topograph is generated by scanning the sample and the film in synchronism holding the screen stationary. Scanning combined with oscillation is effective when extreme sample warpage prevents large area imaging. While the crystal is scanned, both crystal and film are oscillating simultaneously around the normal to the plane containing the incident and reflected beam.[109] Entire large-area wafers can be imaged. Large-diameter wafers become warped during processing, making it necessary to adjust the specimen continuously during topography measurements to ensure that it stays on the chosen Bragg angle.

To "photograph" defects, one usually chooses a weakly diffracting plane. A uniform sample gives a featureless image. Structural defects cause stronger X-ray diffraction, thus providing film contrast or topographic features. The Lang technique has also been adapted to *reflection topography*. Scanning provides for considerably more flexibility than is possible with the Berg-Barrett technique. For semiconductors, the Lang method is used primarily to study defects introduced during crystal growth or during wafer processing.[110] Transmission topographs provide information on defects through the entire sample; reflection topographs provide information of 10 to 30 μm depth from the surface. X-ray topographs of a (100) oriented silicon epitaxial wafer are shown in Fig. 11.35. Fig. 11.35(a) shows a Lang and (b) a double crystal topograph. Clearly the double crystal image is more detailed.

In *section topography* the sample and film are stationary and a narrow "section" of the sample—the cross-section—is imaged.[112] The stationary sample is illuminated by a narrow X-ray beam and the sample cross-section is imaged on the film. The method is like that in Fig. 11.34(b), except both sample and photographic plate are stationary. Section topography has proven to be very valuable for defect depth information. For example, it is common in integrated fabrication to precipitate oxygen in silicon wafers. Section topography is a convenient method to obtain a non-destructive cross-sectional picture through the wafer clearly showing the precipitated regions.[113]

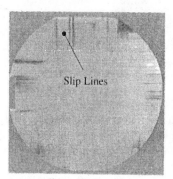
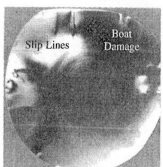

LANG TOPOGRAPHY   DOUBLE CRYSTAL
                              TOPOGRAPHY

(a)   (b)

Fig. 11.35 X-ray topographs of a 7 μm thick epitaxial layer on a (100)-oriented silicon wafer using the Lang and double crystal topography methods. The Lang topograph shows slip lines at the epi-substrate interface and the double crystal topograph shows warpage, thermal memory effects, and swirl in the substrate from grown-in defects. Reprinted with permission after ref. 111. Courtesy of T.J. Shaffner, Texas Instruments.

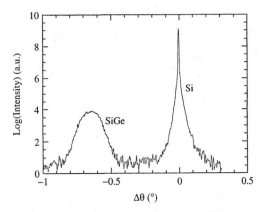

Fig. 11.36   Rocking curve of a heteroepitaxial $Si_{0.80} Ge_{0.20}$ film (150 nm) on (100) Si. The film is diffracting at a smaller angle than the substrate. From Bragg's law, this implies that the film has a large d-spacing and thus has a larger lattice parameter than the substrate. Data courtesy of T.L. Alford, Arizona State University.

Double-crystal diffraction provides higher accuracy because the beam is more highly collimated than is possible with single crystal topography.[113] The technique, shown in Fig. 11.34(c), consists of two successive Bragg reflections from reference and sample crystals. Reflection from the first, carefully selected "perfect" crystal produces a monochromatic and highly parallel beam to probe the sample. The double crystal technique is used not only for topography, but also for *rocking curve* determination. To record a rocking curve, the sample is slowly rotated or "rocked" about an axis normal to the diffraction plane and the scattered intensity is recorded as a function of the angle as shown in Fig. 11.34(c). Such a rocking curve is shown in Fig. 11.36. Its width is a measure of crystal perfection. The narrower the curve, the more perfect is the material. For epitaxial layers it provides data on lattice mismatch, layer thickness, layer and substrate perfection, and wafer curvature. Double crystal diffraction has been extended to four-crystal diffraction where four crystals are used to collimate the X-ray beam further.[114]

11.4.4   Neutron Activation Analysis (NAA)

Neutron activation analysis is a trace analysis method in which nuclear reactions lead to the production of radioactive isotopes from stable isotopes of the elements in the sample, followed by measurement of the radiation emitted by the desired radio isotopes.[115] When an element captures a neutron it emits a prompt γ ray within $\sim 10^{-14}$ s and becomes radioactive. Subsequently the nucleus emits β rays, α particles or γ rays with a half life characteristic of the element. Prompt γ rays are detected in *prompt gamma neutron activation analysis* and β rays, γ rays, etc., from decaying radionuclides are measured in NAA.[116] We mention it because it has high sensitivity to certain elements important to semiconductors. The technique has not found wide use in the semiconductor community. It is generally offered as a service, since few semiconductor laboratories have the required nuclear reactor.

The sample is sealed in a high-purity quartz vial and placed into a nuclear reactor. Those elements that absorb neutrons find themselves in a highly excited state that relaxes by beta and gamma-ray emission. The sample may also become radioactive.

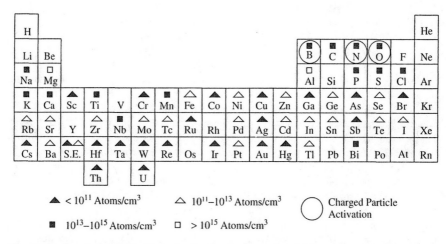

$\blacktriangle$ $< 10^{11}$ Atoms/cm^3 $\triangle$ $10^{11}-10^{13}$ Atoms/cm^3 $\bigcirc$ Charged Particle Activation

$\blacksquare$ $10^{13}-10^{15}$ Atoms/cm^3 $\square$ $> 10^{15}$ Atoms/cm^3

Fig. 11.37 Practical detection limits for elements in silicon detected by neutron activation analysis. Radionuclides have half lives > 2 h; sample volume 1 cm^3; neutron flux 10^{14} thermal and 3×10^{13} fast neutrons/cm$^2 \cdot s$; irradiation time: 1 to 5 days. Reprinted with permission after ref. 119.

Gamma-ray emission is analogous to X-ray emission from orbital electron transitions. Beta rays have a continuous spectrum and are not an attractive tool for elemental determination. Gamma rays have well defined, tabulated energies that are usually measured with a germanium detector.[117] The γ-ray energy identifies the element and their intensity determines the density. The detection system is usually calibrated against standards for quantitative measurements. Typical detection limits for elements in silicon are shown in Fig. 11.37.

NAA is not a surface-sensitive technique, since uncharged incident neutrons penetrate deeply into the sample. Similarly, emitted γ-rays are also very penetrating. A disadvantage of NAA is the attendant radioactivity of the sample. The key to successful NAA for Si is the short 2.6 h half life of Si and the longer half life of many contaminating elements. It is common to irradiate a Si sample and then measure it 24 h later, when the Si activity has decayed to insignificant levels. NAA sensitivity can be extremely high. For example, gold densities as low as 10^8 to 10^9 cm^{-3} in silicon can be determined,[118] but other elements are much less sensitive, as shown in Fig. 11.37. For example, NAA cannot be used for Al in Si, since the radioactivity from the Si interferes with that from Al. The method is most sensitive if the sample is destroyed after irradiation by etching. The more convenient *instrumental NAA*, in which the sample is measured as irradiated, is a powerful survey method but is not as sensitive.

NAA measures the purity of silicon during and after crystal growth and determines impurities introduced during processing.[119-120] Usually the total impurity content of the sample is measured. Profiling is possible by etching or lapping thin layers of the sample and measuring the activity in the removed material. NAA is not sensitive to boron, carbon, or nitrogen. Phosphorus does not emit gamma rays; instead beta ray decay must be measured. The method is not suitable for heavily doped wafers. For example, Sb and As form radioactive species. For quantitative measurements, careful calibration must be performed.[121] A good summary of NAA application to semiconductor problems can be found in the work of Haas and Hofmann.[119] They have used the method for impurity monitoring during crystal growth, device processing, detection of impurities in supplies like aluminum and even in plastic pipes used for water and examined materials for uranium and thorium, that cause alpha particle upsets in memory chips. They also use *autoradiography*, in which an impurity is imaged to show its spatial distribution.

A method related to NAA is *neutron depth profiling* (NDP).[122] It is an isotope-specific, non-destructive technique for the measurement of density profiles in the near surface region of solids. A well-collimated thermal neutron beam with energy less than 0.01 eV is directed at the sample. After capturing a neutron, the material emits a charged particle such as an α-particle. The emitted alpha particles have a characteristic energy defined by the kinematics of the reaction, which serves to determine the element. Their energy depends on the depth of generation, since they lose energy while traveling through the sample, much as He ions do in RBS. By analyzing the energy of the detected α-particles it is possible to construct a depth profile of the element. NDP lends itself to only a few elements; Li, Be, B, N, and Na are the dominant ones. NDP induces negligible damage to the sample and the surface is not sputtered during the measurement. It has been used to determine the implanted boron profile in silicon for comparison with SIMS and spreading resistance measurements with the detection limit for boron in silicon being 10^{12} cm^{-2}.[123] It has also been used to determine the boron content in borophosphosilicate glass films.[124] It is not routinely used since there are few NDP facilities. In the United States facilities exist at the University of Michigan, Texas A&M University, the University of Texas/Austin, the National Institute of Standards and Technology, and North Carolina State University.

11.5 STRENGTHS AND WEAKNESSES

Scanning Electron Microscopy: Its major strength lies in the high-resolution and high depth-of-field capability of the instrument. It has evolved from a highly specialized instrument to one routinely used today for line-width measurements, for example. Its major disadvantage is the need for a vacuum, which is true for all electron beam instruments.

Auger Electron Spectroscopy: Its major strength is the ability to characterize thin layers for both elemental and molecular information. In scanning AES, the instrument provides high-resolution images. Its major disadvantages are the need for high vacuum and its relatively low sensitivity.

Electron Microprobe: Its major strength is the availability of EMP on many scanning electron microscopes and the relatively simple way of obtaining quantitative elemental information. The energy resolution of energy dispersive spectroscopy is modest, but usually sufficient. For higher energy resolution one needs to use wavelength dispersive spectroscopy, which is more difficult to use, or microcalorimeters. A weakness is the modest spatial resolution of the technique and the damage the electron beam can inflict on semiconductor samples.

Transmission Electron Microscopy: Its major strength lies in its unprecedented atomic resolution imaging ability. To achieve this, one must prepare extremely thin samples, therefore making sample preparation its major weakness. This been somewhat alleviated by focused ion beam sample preparation.

Secondary Ion Mass Spectrometry: Its major strength lies in its sensitivity (better than most beam techniques) and the ability to detect all impurities. Furthermore, it is

APPENDIX 11.1 Selected Features of Some Analytical Techniques.

Technique	Detectable Elements	Lateral Resolution	Depth Resolution	Detection Limit (atoms/cm^3)[a]	Type of Information[b]	Destructive?	Depth Profiling?	Analysis Time	Matrix Effect
AES	≥Li	100 μm	2 nm	10^{19}–10^{20}	elem/chem	Yes[c]	Yes	30 m	sputter mixing
SAM	≥Li	10 nm	2 nm	10^{21}	elem	Yes[c]	Yes	30 m	sputter mixing
EMP-EDS	≥Na	1 μm	1 μm	10^{19}–10^{20}	elem	No	No	30 m	correctable
EMP-WDS	≥Na	1 μm	1 μm	10^{18}–10^{19}	elem	No	No	2 h	correctable
SIMS	All	1 μm	1–30 nm	10^{14}–10^{18}	elem	Yes	Yes	1 h	severe
RBS	≥Li	0.1 cm	20 nm	10^{19}–10^{20}	elem	No	Yes	30 m	free
XRF	≥C	0.1–1 cm	1–10 μm	10^{17}–10^{18}	elem	No	No	30 m	correctable
TXRF	≥C	0.5 cm	5 nm	10^{10} cm^{-2}	elem	No	No	30 m	correctable
XPS	≥Li	10 μm–1 cm	2 nm	10^{19}–10^{20}	elem/chem	Yes[c]	Yes	30 m	chemical shift
XRT	—	1–10 μm	100–500 μm	—	cryst. struct.	No	Yes	45 m	strain interact.
PL	shallow level	10 μm	1–10 μm	10^{11}–10^{15}	elem/band gap	No	No	1 h	bound exciton
FTIR	funct. grps.	1–1000 μm	1–10 μm	10^{12}–10^{16}	molecule	No	No	15 m	molec. interact.
Raman	funct. grps.	1 μm	1–10 μm	10^{19}	molecule	No	No	1 h	molecular stress
EELS	≥ B	1 nm	20 nm	10^{19}–10^{20}	elem/chem	No	No	30 m	free
LEED	—	0.1–100 μm	2 nm	—	cryst. struct.	No	No	30 m	—
RHEED	—	0.1–1000 μm	2 nm	—	cryst. struct.	No	No	30 m	—
NAA	selective	1 cm	1 μm	10^8–10^{18}	elem	No	No	2 d	free
AFM	—	1 nm	1 nm	—	surface flatness	No	No	30 m	free

[a]Depends on element to be detected; [b]elem: elemental composition, chem: chemical composition; [c]If profiled, otherwise "No"

AES: Auger electron spectroscopy
EDS: Energy dispersive spectrometer
RBS: Rutherford backscattering spectrometry
XPS: X-ray photoelectron spectroscopy
FTIR: Fourier transform infrared spectroscopy
LEED: Low energy electron diffraction
AFM: Atomic force microscopy

SAM: Scanning Auger Microprobe
WDS: Wavelength dispersive spectrometer
XRF: X-ray fluorescence
XRT: X-ray topography
Raman: Raman microprobe
RHEED: Refl. high energy electron diffract.

EMP: Electron microprobe
SIMS: Secondary ion mass spectrometry
TXRF: Total reflection XRF
PL: Photoluminescence
EELS: Electron energy loss spectroscopy
NAA: Neutron activation analysis

one of the most commonly used beam techniques for dopant profiling and, with time-of-flight SIMS, for organic and surface metal contaminants. Its weaknesses include matrix effects, molecular interferences, the destructive nature of the measurement, and the need for calibrated samples.

Rutherford Backscattering: Its major strength lies in its contactless and absolute measurements without recourse to calibrated standards. Its major weakness is the specialized nature of the instrumentation and the difficulty of measuring light elements on a heavy element substrate.

X-Ray Fluorescence: It major strength is the ability for rapid, contactless survey of elements. Its weakness is the modest resolution due to the difficulty of focusing X-rays and the presence of matrix effects. The sensitivity to surface contamination is greatly extended by total reflection X-ray fluorescence.

X-Ray Photoelectron Spectroscopy: It major strength is the ability to characterize the elemental and molecular nature of thin layers. Its weakness is the modest resolution due to the difficulty of focusing X-rays, the high vacuum requirement, and its low sensitivity.

Neutron Activation Analysis: Its major strength is the ability to detect very low densities of certain impurities in common semiconductors like Si. Its weakness is the specialized equipment. Few laboratories have nuclear reactors.

APPENDIX 11.1

Selected Features of Some Analytical Techniques

REFERENCES

1. H. Seiler, "Secondary Electron Emission in the Scanning Electron Microscope," *J. Appl. Phys.* **54**, R1–R18, Nov. 1983.
2. V.E. Cosslett, "Fifty Years of Instrumental Development of the Electron Microscope," in *Advances in Optical and Electron Microscopy* (R. Barer and V.E. Cosslett, eds.), Academic Press, London, **10**, 215–267, 1988.
3. L. de Broglie, "Waves and Quanta," (in French) *Compt. Rend.* **177**, 507–510, Sept. 1923.
4. J.I. Goldstein, D.E. Newbury, P. Echlin, D.C. Joy, C. Fiori and E. Lifshin, *Scanning Electron Microscopy and X-Ray Microanalysis*, Plenum Press, New York, 1984.
5. T.E. Everhart and R.F.M. Thornley, "Wide-Band Detector for Micro-Microampere Low-Energy Electron Currents," *J. Sci. Instrum.* **37**, 246–248, July 1960.
6. R.A. Young and R.V. Kalin, "Scanning Electron Microscopic Techniques for Characterization of Semiconductor Materials," in *Microelectronic Processing: Inorganic Material Characterization* (L.A. Casper, ed.), American Chemical Soc., Symp. Series 295, Washington, DC, 1986, 49–74.
7. H.J. Leamy, "Charge Collection Scanning Electron Microscopy," *J. Appl. Phys.* **53**, R51–R80, June 1982.
8. T.E. Everhart and P.H. Hoff, "Determination of Kilovolt Electron Energy Dissipation vs. Penetration Distance in Solid Materials," *J. Appl. Phys.* **42**, 5837–5846, Dec. 1971.

9. P. Auger, "On the Compound Photoelectric Effect (in French)," *J. Phys. Radium* **6**, 205–208, June 1925.

10. L.E. Davis, N.C. MacDonald, P.W. Palmberg, G.E. Riach and R.E. Weber, *Handbook of Auger Electron Spectroscopy*, Physical Electronics Industries Inc., Eden Prairie, MN, 1976; G.E. McGuire, *Auger Electron Spectroscopy Reference Manual*, Plenum Press, New York, 1979; *Handbook of Auger Electron Spectroscopy*, JEOL Ltd., Tokyo, 1980.

11. L.C. Feldman and J.W. Mayer, *Fundamentals of Surface and Thin Film Analysis*, North Holland, New York, 1986.

12. L.L. Kazmerski, "Advanced Materials and Device Analytical Techniques," in *Advances in Solar Energy* (K.W. Böer, ed.), **3**, American Solar Energy Soc., Boulder, CO, 1986, 1–123.

13. R.E. Honig, "Surface and Thin Film Analysis of Semiconductor Materials," *Thin Solid Films* **31**, 89–122, Jan. 1976.

14. H.W. Werner and R.P.H. Garten, "A Comparative Study of Methods for Thin-Film and Surface Analysis," *Rep. Progr. Phys.* **47**, 221–344, March 1984.

15. H. Hapner, J.A. Simpson and C.E. Kuyatt, "Comparison of the Spherical Deflector and the Cylindrical Mirror Analyzers," *Rev. Sci. Instrum.* **39**, 33–35, Jan. 1968; P.W. Palmberg, G.K. Bohn and J.C. Tracy, "High Sensitivity Auger Electron Spectrometer," *Appl. Phys. Lett.* **15**, 254–255, Oct. 1969.

16. L.A. Harris, "Analysis of Materials by Electron-Excited Auger Electrons," *J. Appl. Phys.* **39**, 1419–1427, Feb. 1968.

17. E. Minni, "Assessment of Different Models for Quantitative Auger Analysis in Applied Surface Studies," *Appl. Surf. Sci.* **15**, 270–280, April 1983.

18. C.J. Powell, "Growth and Trends in Auger-electron Spectroscopy and X-ray Photoelectron Spectroscopy for Surface Analysis," *J. Vac. Sci. Technol.* **A21**, S42–S53, Sept./Oct. 2003.

19. T.J. Shaffner, "Surface Characterization for VLSI," in *VLSI Electronics: Microstructure Science* (N.G. Einspruch and G.B. Larrabee, eds.) **6**, Academic Press, New York, 1983, 497–527.

20. S. Oswald and S. Baunack, "Comparison of Depth Profiling Techniques Using Ion Sputtering from the Practical Point of View," *Thin Solid Films* **425**, 9–19, Feb. 2003.

21. E. Zinner, "Sputter Depth Profiling of Microelectronic Structures," *J. Electrochem. Soc.* **130**, 199C–222C, May 1983; P.L. King, "Artifacts in AES Microanalysis for Semiconductor Applications," *Surf. Interface Anal.* **30**, 377–382, Aug. 2000.

22. P.H. Holloway and G.E. McGuire, "Characterization of Electronic Devices and Materials by Surface-Sensitive Analytical Techniques," *Appl. Surf. Sci.* **4**, 410–444, April 1980; G.E. McGuire and P.H. Holloway, "Applications of Auger Spectroscopy in Materials Analysis," in *Electron Spectroscopy: Theory, Techniques and Applications* (C.R. Brundle and A.D. Baker, eds.), **4**, Academic Press, New York, 1981, 2–74; J. Keenan, "TiSi$_2$ Chemical Characterization by Auger Electron and Rutherford Backscattering Spectroscopy," *TI Tech. J.* **5**, 43–49, Sept./Oct. 1988; L.A. Files and J. Newsom, "Scanning Auger Microscopy: Applications to Semiconductor Analysis," *TI Tech. J.* **5**, 89–95, Sept./Oct. 1988.

23. R. Castaing, Thesis, Univ. of Paris, France, 1948; "Electron Probe Microanalysis," in *Adv. in Electronics and Electron Physics* (L. Marton, ed.), Academic Press, New York, **13**, 317–386, 1960.

24. S.J.B. Reed, *Electron Microprobe Analysis*, Cambridge University Press, 1993; K.F.J. Heinrich and D.E. Newbury, "Electron Probe Microanalysis," in *Metals Handbook*, 9th Ed. (R.E. Whan, coord.), Am. Soc. Metals, Metals Park, OH, **10**, 516–535, 1986.

25. K.F.J. Heinrich, *Electron Beam X-Ray Microanalysis*, Van Nostrand Reinhold, New York, 1981.

26. C.E. Fiori and D.E. Newbury, "Artifacts Observed in Energy-Dispersive X-Ray Spectrometry in the Scanning Electron Microscope," *Scanning Electron Microscopy*, **1**, 401–422, 1978.

27. F.S. Goulding and Y. Stone, "Semiconductor Radiation Detectors," *Science* **170**, 280–289, Oct. 1970; A.H.F. Muggleton, "Semiconductor Devices for Gamma Ray, X Ray and Nuclear Radiation Detectors," *J. Phys. E: Scient. Instrum.* **5**, 390–404, May 1972.

28. J.F. Bresse, "Quantitative Investigations in Semiconductor Devices by Electron Beam Induced Current Mode: A Review," in *Scanning Electron Microscopy* **1**, 717–725, 1978.

29. F. Scholze, H. Rabus, and G. Ulm, "Measurement of the Mean Electron-Hole Pair Creation Energy in Crystalline Silicon for Photons in the 50–1500 eV Spectral Range," *Appl. Phys. Lett.* **69**, 2974–2976, Nov. 1996.

30. M. LeGros, E. Silver, D. Schneider, J. McDonald, S. Bardin, R. Schuch, N. Madden, and J. Beeman, "The First High Resolution, Broad Band X-Ray Spectroscopy of Ion-surface Interactions Using a Microcalorimeter," *Nucl. Instrum. Meth.* **A357**, 110–114, April 1995; D.A. Wollman, K.D. Irwin, G.C. Hilton, L.L. Dulcie, D.A. Newbury, and J.M. Martinis, "High-resolution, Energy-dispersive Microcalorimeter Spectrometer for X-Ray Microanalysis, *J. Microsc.* **188**, 196–223, Dec. 1997.

31. B. Simmnacher, R. Weiland, J. Höhne, F.V. Feilitzsch, and C. Hollerith, "Semiconductor Material Analysis Based on Microcalorimeter EDS," *Microelectron. Rel.* **43**, 1675–1680, Sept./Nov. 2003.

32. M. von Heimendahl, *Electron Microscopy of Materials*, Academic Press, New York, 1980; D.B. Williams and C.B. Carter, *Transmission Electron Microscopy*, Plenum Press, New York, 1996; D.C. Joy, A.D. Romig, Jr., and J.I. Goldstein (eds.), *Principles of Analytical Electron Microscopy*, Plenum Press, New York, 1986.

33. W.R. Runyan and T.J. Shaffner, *Semiconductor Measurements and Instrumentation*, McGraw-Hill, New York, 1998.

34. R.F. Egerton, *Electron Energy-Loss Spectroscopy in the Electron Microscopy*, 2nd ed., Plenum Press, New York, 1996; C. Colliex, "Electron Energy Loss Spectroscopy in the Electron Microscope," in *Advances in Optical and Electron Microscopy* (R. Barer and V.E. Cosslett, eds.), Academic Press, London, **9**, 65–177, 1986.

35. J.C.H. Spence, *Experimental High-Resolution Electron Microscopy*, 2nd Ed., Oxford University Press, New York, 1988; D. Cherns, "High-Resolution Transmission Electron Microscopy of Surface and Interfaces," in *Analytical Techniques for Thin Film Analysis* (K.N. Tu and R. Rosenberg, eds.), Academic Press, Boston, 1988, 297–335.

36. P.E. Batson, "Scanning Transmission Electron Microscopy," in *Analytical Techniques for Thin Film Analysis* (K.N. Tu and R. Rosenberg, eds.), Academic Press, Boston, 1988, pp. 337–387; R.J. Graham, "Characterization of Semiconductor Materials and Structures by Transmission Electron Microscopy," in *Diagnostic Techniques for Semiconductor Materials and Devices* (T.J. Shaffner and D.K. Schroder, eds.), Electrochem. Soc., Pennington, NJ, 1988, 150–167.

37. T.T. Sheng, "Cross-Sectional Transmission Electron Microscopy of Electronic and Photonic Devices," in *Analytical Techniques for Thin Film Analysis* (K.N. Tu and R. Rosenberg, eds.), Academic Press, Boston, 1988, 251–296.

38. J.I. Hanoka and R.O. Bell, "Electron-Beam-Induced Currents in Semiconductors," in *Annual Review of Materials Science* (R.A. Huggins, R.H. Bube and D.A. Vermilya, eds.), Annual Reviews, Palo Alto, CA, **11**, 353–380, 1981.

39. T.E. Everhart, O.C. Wells and R.K. Matta, "A Novel Method of Semiconductor Device Measurements," *Proc. IEEE* **52**, 1642–1647, Dec. 1964.

40. K. Joardar, C.O. Jung, S. Wang, D.K. Schroder, S.J. Krause, G.H. Schwuttke and D.L. Meier, "Electrical and Structural Properties of Twin Planes in Dendritic Web Silicon," *IEEE Trans. Electron Dev.* **ED-35**, 911–918, July 1988.

41. M. Tamatsuka, S. Oka, H.R. Kirk, and G.A. Rozgonyi, "Novel GOI Failure Analysis Using SEM/MOS/EBIC With Sub-nano Ampere Current Breakdown," in *Diagnostic Techniques for Semiconductor Materials and Devices* (P. Rai-Choudhury, J.L. Benton, D.K. Schroder, and T.J. Shaffner, eds), Electrochem. Soc., Pennington, NJ, 1997, 80–91. H.R. Kirk, Z. Radzimski, A. Romanowski, and G.A. Rozgonyi, "Bias Dependent Contrast Mechanisms in EBIC Images of MOS Capacitors," *J. Electrochem. Soc.* **146**, 1529–1535, April 1999.

42. S.M. Davidson, "Semiconductor Material Assessment by Scanning Electron Microscopy," *J. Microsc.* **110**, 177–204, Aug. 1977; B.G. Yacobi and D.B. Holt, "Cathodoluminescence Scanning Electron Microscopy of Semiconductors," *J. Appl. Phys.* **59**, R1–R24, Feb. 1986.

43. G. Pfefferkorn, W. Bröcker and M. Hastenrath, "The Cathodoluminescence Method in the Scanning Electron Microscope," *Scanning Electron Microscopy*, SEM, AMF O'Hare, IL, 251–258, 1980.

44. R.J. Roedel, S. Myhajlenko, J.L. Edwards and K. Rowley, "Cathodoluminescence Characterization of Semiconductor Materials," in *Diagnostic Techniques for Semiconductor Materials and Devices* (T.J. Shaffner and D.K. Schroder, eds.), Electrochem. Soc., Pennington, NJ, 1988, 185–196.

45. B.G. Yacobi and D.B. Holt, "Cathodoluminescence Scanning Electron Microscopy of Semiconductors," *J. Appl. Phys.* **59**, R1–R24, Feb. 1986.

46. C. Davisson and L.H. Germer, "Diffraction of Electrons by a Crystal of Nickel," *Phys. Rev.* **30**, 705–740, Dec. 1927.

47. J.B. Pendry, *Low Energy Electron Diffraction*, Academic Press, New York, 1974; K. Heinz, "Structural Analysis of Surfaces by LEED," *Progr. Surf. Sci.* **27**, 239–326, 1988.

48. M.G. Lagally, "Low-Energy Electron Diffraction," in *Metals Handbook*, 9th Ed. (R.E. Whan, coord.), Am. Soc. Metals, Metals Park, OH, **10**, 536–545, 1986.

49. B.F. Lewis, F.J. Grunthaner, A. Madhukar, T.C. Lee, and R. Fernandez, "Reflection High Energy Electron Diffraction Intensity Behavior During Homoepitaxial Molecular Beam Epitaxy Growth of GaAs and Implications for Growth Kinetics," *J. Vac. Sci. Technol.* **B3**, 1317–1322, Sept./Oct. 1985.

50. L.C. Feldman and J.W. Mayer, *Fundamentals of Surface and Thin Film Analysis*, North Holland, New York 1986; J.M. Walls (ed.), *Methods of Surface Analysis*, Cambridge University Press, Cambridge, 1989.

51. C.G. Pantano, "Secondary Ion Mass Spectroscopy," in *Metals Handbook*, 9th Ed. (R.E. Whan, coord.), Am. Soc. Metals, Metals Park, OH, **10**, 610–627, 1986; A. Benninghoven, F.G. Rüdenauer and H.W. Werner, *Secondary Ion Mass Spectrometry: Basic Concepts, Instrumental Aspects, Applications and Trends*, Wiley, New York, 1987.

52. R. Castaing, B. Jouffrey, and G. Slodzian, "On the Possibility of Local Analysis of a Specimen Using its Secondary Ion Emission," (in French) *Compt. Rend.* **251**, 1010–1012, Aug. 1960; R. Castaing and G. Slodzian, "First Attempts at Microanalysis by Secondary Ion Emission," (in French) *Compt. Rend.* **255**, 1893–1895, Oct. 1962.

53. R.K. Herzog and H. Liebl, "Sputtering Ion Source for Solids," *J. Appl. Phys.* **34**, 2893–2896, Sept. 1963.

54. A. Benninghoven, "The Analysis of Monomolecular Solid State Surface Layers with the Aid of Secondary Ion Emission," (in German) *Z. Phys.* **230**, 403–417, 1970.

55. J.J. Thomson, "Rays of Positive Electricity," *Phil. Mag.* **20**, 752–767, Oct. 1910.

56. P. Williams, "Secondary Ion Mass Spectrometry," in *Applied Atomic Collision Spectroscopy*, Academic Press, New York, 1983, 327–377.

57. D.E. Sykes, "Dynamic Secondary Ion Mass Spectrometry," in *Methods of Surface Analysis* (J.M. Walls, ed.), Cambridge University Press, Cambridge, 1989, 216–262.

58. A. Benninghoven, "Surface Analysis by Means of Ion Beams," *Crit. Rev. Solid State Sci.* **6**, 291–316, 1976.

59. M.T. Bernius and G.H. Morrison, "Mass Analyzed Secondary Ion Microscopy," *Rev. Sci. Instrum.* **58**, 1789–1804, Oct. 1987.

60. M.A. Douglas and P.J. Chen, "Quantitative Trace Metal Analysis of Si Surfaces by TOF-SIMS," *Surf. Interface Anal.* **26**, 984–994, Dec. 1998.

61. S.G. Mackay and C.H. Becker, "SALI—Surface Analysis by Laser Ionization," in *Encyclopedia of Materials Characterization* (C.R. Brundle, C.A. Evans, Jr., and S. Wilson, eds.), Butterworth-Heinemann, Boston, 1992, 559–570; J.C. Huneke, "SNMS—Sputtered

Neutral Mass Spectrometry," in *Encyclopedia of Materials Characterization* (C.R. Brundle, C.A. Evans, Jr., and S. Wilson, eds.), Butterworth-Heinemann, Boston, 1992, 571–585; Y. Mitsui, F. Yano, H. Kakibayashi, H. Shichi, and T. Aoyama, "Developments of New Concept Analytical Instruments for Failure Analyses of Sub-100 nm Devices," *Microelectron. Rel.* **41**, 1171–1183, Aug. 2001.

62. F.R. di Brozolo, "LIMS - Laser Ionization Mass Spectrometry," in *Encyclopedia of Materials Characterization* (C.R. Brundle, C.A. Evans, Jr., and S. Wilson, eds.), Butterworth-Heinemann, Boston, 1992, 586–597; M.C. Arst, "Identifying Impurities in Silicon by LIMA Analysis," in *Emerging Semiconductor Technology* (D.C. Gupta and R.P. Langer, eds.), STP **960**, Am. Soc. Test. Mat., Philadelphia, 1987, 324–335.

63. C.W. Magee and M.R. Frost, "Recent Successes in the Use of SIMS in Microelectronics Materials and Processes," *Int. J. Mass Spectrom. Ion Proc.* **143**, 29–41, May 1995.

64. E. Ishida and S.B. Felch, "Study of Electrical Measurement Techniques for Ultra-Shallow Dopant Profiling," *J. Vac. Sci. Technol.* **B14**, 397–403, Jan/Feb. 1996; S.B. Felch, D.L. Chapek, S.M. Malik, P. Maillot, E. Ishida, and C.W. Magee "Comparison of Different Analytical Techniques in Measuring the Surface Region of Ultrashallow Doping Profiles," *J. Vac. Sci. Technol.* **B14**, 336–340, Jan/Feb. 1996.

65. E. Zinner, "Depth Profiling by Secondary Ion Mass Spectrometry," *Scanning* **3**, 57–78, 1980.

66. W.K. Chu, J.W. Mayer and M-A. Nicolet, *Backscattering Spectroscopy*, Academic Press, New York, 1978; W.K. Chu, "Rutherford Backscattering Spectrometry," *in Metals Handbook*, 9th Ed. (R.E. Whan, coord.), Am. Soc. Metals, Metals Park, OH, **10**, 628–636, 1986; T.G. Finstad and W.K. Chu, "Rutherford Backscattering Spectrometry on Thin Solid Films," in *Analytical Techniques for Thin Film Analysis* (K.N. Tu and R. Rosenberg, eds.), Academic Press, Boston, 1988, 391–447.

67. E. Rutherford and H. Geiger, "Transformation and Nomenclature of the Radio-Active Emanations," *Phil. Mag.* **22**, 621–629, Oct. 1911.

68. S. Rubin, T.O. Passell and L.E. Bailey, "Chemical Analysis of Surfaces by Nuclear Methods," *Anal. Chem.* **29**, 736–743, May 1957.

69. J.H. Patterson, A.L. Turkevich and E.J. Franzgrote, "Chemical Analysis of Surfaces Using Alpha Particles," *J. Geophys. Res.* **70**, 1311–1327, March 1965.

70. L.C. Feldman and J.M. Poate, "Rutherford Backscattering and Channeling Analysis of Interfaces and Epitaxial Structures," in *Annual Review of Materials Science* (R.A. Huggins, R.H. Bube and D.A. Vermilya, eds.), Annual Reviews, Palo Alto, CA, **12**, 149–176, 1982; C.W. Magee and L.R. Hewitt, "Rutherford Backscattering Spectrometry: A Quantitative Technique for Chemical and Structural Analysis of Surfaces and Thin Films," *RCA Rev.* **47**, 162–185, June 1986.

71. J.F. Ziegler, *Helium Stopping Powers and Ranges in All Elemental Matter*, Pergamon Press, New York, 1977.

72. J.F. Ziegler and R.F. Lever, "Calculations of Elastic Scattering of ^{4}He Projectiles," *Thin Solid Films* **19**, 291–296, Dec. 1973; J.W. Mayer, M-A. Nicolet, and W.K. Chu, "Backscattering Analysis with ^{4}He Ions," in *Nondestructive Evaluation of Semiconductor Materials and Devices* (J.N. Zemel, ed.), Plenum Press, New York, 1979, 333–366.

73. W.K. Chu, J.W. Mayer, M-A Nicolet, T.M. Buck, G. Amsel, and P. Eisen, "Principles and Applications of Ion Beam Techniques for the Analysis of Solids and Thin Films," *Thin Solid Films* **17**, 1–41, July 1973.

74. J.F. Ziegler and W.K. Chu, "Stopping Cross Sections and Backscattering Factors for ^{4} He Ions in Matter: $Z = 1$–92, E(^{4}He)=400-4000 keV," in *Atomic Data and Nuclear Data Tables* **13**, 463–489, May 1974; J.F. Ziegler, R.F. Lever, and J.K. Hirvonen, in *Ion Beam Surface Analysis* (O. Mayer, G. Linker, and F. Käppeler, eds.), **I**, Plenum, New York, 1976, 163.

75. A.C. Diebold, P. Maillot, M. Gordon, J. Baylis, J. Chacon, R. Witowski, H.F. Arlinghaus, J.A. Knapp, and B.L. Doyle, "Evaluation of Surface Analysis Methods for Characterization

of Trace Metal Surface Contaminants Found in Silicon Integrated Circuit Manufacturing," *J. Vac. Sci. Technol.* **A10**, 2945–2952, July/Aug. 1992; J.A. Knapp and J.C. Banks, "Heavy Ion Backscattering Spectrometry for High Sensitivity," *Nucl. Instrum. Meth.* **B79**, 457–459, June 1993.

76. C.W. Magee and L.R. Hewitt, "Rutherford Backscattering Spectrometry: A Quantitative Technique for Chemical and Structural Analysis of Surfaces and Thin Films," *RCA Rev.* **47**, 162–185, June 1986.

77. L.C. Feldman, J.W. Mayer, and S.T. Picraux, *Materials Analysis by Ion Channeling*, Academic Press, New York, 1982.

78. M.-A. Nicolet, J.W. Mayer, and I.V. Mitchell, "Microanalysis of Materials by Backscattering Spectrometry," *Science* **177**, 841–849, Sept. 1972.

79. J. Li, F. Moghadam, L.J. Matienzo, T.L. Alford, and J.W. Mayer, "Oxygen Carbon, and Nitrogen Quantification by High-Energy Resonance Backscattering," *Solid State Technol.* **38**, 61–64, May 1995.

80. W.G. Morris, H. Bakhru and A.W. Haberl, "Materials Characterization With a He^+ Microbeam," *Nucl. Instrum. and Meth.* **B10/11**, 697–699, May 1985.

81. J.A. Keenan, "Backscattering Spectroscopy for Semiconductor Materials," in *Diagnostic Techniques for Semiconductor Materials and Devices* (T.J. Shaffner and D.K. Schroder, eds.), Electrochem. Soc., Pennington, NJ, 1988, 15–26.

82. C.W. Magee, "Secondary Ion Mass Spectrometry and Its Relation to High-Energy Ion Beam Analysis Techniques," *Nucl. Instrum. and Meth.* **191**, 297–307, Dec. 1981.

83. E.P. Berlin, "X-Ray Secondary Emission (Fluorescence) Spectrometry," in *Principles and Practice of X-Ray Spectrometric Analysis*, Plenum Press, New York, 1970, Ch. 3; R.O. Muller, *Spectrochemical Analysis by X-Ray Fluorescence*, Plenum Press, New York, 1972; J.V. Gilfrich, "X-Ray Fluorescence Analysis," *in Characterization of Solid Surfaces* (P.F. Kane and G.B. Larrabee, eds.), Plenum Press, New York, 1974, Ch. 12; D.S. Urch, "X-Ray Emission Spectroscopy," *in Electron Spectroscopy: Theory, Techniques and Applications*, **3** (C.R. Brundle and A.D. Baker, eds.), Academic Press, New York, 1978, 1–39; W.E. Drummond and W.D. Stewart, "Automated Energy-Dispersive X-Ray Fluorescence Analysis," *Am. Lab.* **12**, 71–80, Nov. 1980; J.A. Keenan and G.B. Larrabee, "Characterization of Silicon Materials for VLSI," in *VLSI Electronics: Microstructure Science* (N.G. Einspruch and G.B. Larrabee, eds.) **6**, Academic Press, New York, 1983, 1–72.

84. M.C. Nichols, D.R. Boehme, R.W. Ryon, D. Wherry, B. Cross and D. Aden, "Parameters Affecting X-Ray Microfluorescence (XRMF) Analysis," in *Adv. in X-Ray Analysis* (C.S. Barrett et al.) **30**, Plenum Press, New York, 1987, 45–51.

85. L.M. van der Haar, C. Sommer, and M.G.M. Stoop, "New Developments in X-ray Fluorescence Metrology," *Thin Solid Films* **450**, 90–96, Feb. 2004.

86. R. Klockenkämper, J. Knoth, A. Prange, and H. Schwenke, "Total-Reflection X-Ray Fluorescence Spectroscopy," *Anal. Chem.* **64**, 1115A–1123A, Dec. 1992; R. Klockenkämper, *Total-Reflection X-Ray Fluorescence Analysis*, Wiley, New York, 1997.

87. P. Pianetta, K. Baur, A. Singh, S. Brennan, Jonathan Kerner, D. Werho, and J. Wang, "Application of Synchrotron Radiation to TXRF Analysis of Metal Contamination on Silicon Wafer Surfaces," *Thin Solid Films* **373**, 222–226, Sept. 2000.

88. Y. Mizokami, T. Ajioka, and N. Terada, "Chemical Analysis of Metallic Contamination on a Wafer After Wet Cleaning," *IEEE Trans. Semic. Manufact.* **7**, 447–453, Nov. 1994.

89. D. Caputo, P. Bacciaglia, C. Carpanese, M.L. Polignano, P. Lazzeri, M. Bersani, L. Vanzetti, P. Pianetta, and L. Morod, "Quantitative Evaluation of Iron at the Silicon Surface after Wet Cleaning Treatments," *J. Electrochem. Soc.* **151**, G289–G296, May 2004.

90. B.J. Streusand, "Inductively Coupled Plasma Mass Spectrometry," in *Encyclopedia of Materials Characterization* (C.R. Brundle, C.A. Evans, Jr., and S. Wilson, eds.), Butterworth-Heinemann, Boston, 1992, 624–632.

91. J.R. Dean, *Atomic Absorption and Plasma Spectroscopy*, 2nd ed., Wiley, Chichester, 1997.

92. R. Jenkins, R.W. Gould, and D. Gedcke, *Quantitative X-ray Spectrometry*, 2nd ed, Marcel Dekker, New York, 1995.

93. R.D. Giauque, F.S. Goulding, J.M. Jaklevic, and R.H. Pehl, "Trace Element Detection With Semiconductor Detector X-Ray Spectrometers," *Anal. Chem.* **45**, 671–681, April 1973.

94. N. Parekh, C. Nieuwenhuizen, J. Borstrok, and O. Elgersma, "Analysis of Thin Films in Silicon Integrated Circuit Technology by X-Ray Fluorescence Spectrometry," *J. Electrochem. Soc.* **138**, 1460–1465, May 1991.

95. P.K. Gosh, *Introduction to Photoelectron Spectroscopy*, Wiley-Interscience, New York, 1983; D. Briggs and M.P. Seah (eds.), *Practical Surface Analysis*, Vol. 1: Auger and X-Ray Photoelectron Spectroscopy, Wiley, Chichester, 1990; J.B. Lumsden, "X-Ray Photoelectron Spectroscopy," in *Metals Handbook*, 9th Ed. (R.E. Whan, coord.), Am. Soc. Metals, Metals Park, OH, **10**, 568–580, 1986; N. Mårtensson, "ESCA," in *Analytical Techniques for Thin Film Analysis* (K.N. Tu and R. Rosenberg, eds.), Academic Press, Boston, 1988, 65–109.

96. C. Nordling, S. Hagström and K. Siegbahn, "Application of Electron Spectroscopy to Chemical Analysis," *Z. Phys.* **178**, 433–438, 1964; S. Hagström, C. Nordling and K. Siegbahn, "Electron Spectroscopic Determination of the Chemical Valence State," *Z. Phys.* **178**, 439–444, 1964.

97. J.G. Jenkin, R.C.G. Leckey and J. Liesegang, "The Development of X-Ray Photoelectron Spectroscopy: 1900–1960," *J. Electron Spectr. Rel. Phen.* **12**, 1–35, Sept. 1977; J.G. Jenkin, J.D. Riley, J. Liesegang and R.C.G. Leckey, "The Development of X-Ray Photoelectron Spectroscopy (1900–1960): A Postscript," *J. Electron Spectr. Rel. Phen.* **14**, 477–485, Dec. 1978; J.G. Jenkin, "The Development of Angle-Resolved Photoelectron Spectroscopy: 1900–1960," *J. Electron Spectr. Rel. Phen.* **23**, 187–273, June 1981.

98. T.A. Carlson, *Photoelectron and Auger Spectroscopy*, Plenum Press, New York, 1975; C.D. Wagner, W.M. Riggs, L.E. Davies, J.F. Moulder, and G.E. Muilenberg, *Handbook of X-Ray Photoelectron Spectroscopy*, Perkin Elmer, Eden Prairie, MN, 1979.

99. S. Tanuma, C.J. Powell, and D.R. Penn "Proposed Formula for Electron Inelastic Mean Free Paths Based on Calculations for 31 Materials," *Surf. Sci.* **192**, L849–L857, Dec. 1987.

100. K.L. Smith and J.S. Hammond, "Destructive and Nondestructive Depth Profiling Using ESCA," *Appl. Surf. Sci.* **22/23**, 288–299, 1985.

101. C.S. Fadley, "Angle-Resolved X-Ray Photoelectron Spectroscopy," *Progr. Surf. Sci.* **16**, 275–388, 1984.

102. D.H. Buckley, *Surface Effects in Adhesion, Friction, Wear and Lubrication*, Elsevier, Amsterdam, 1981, 73–78.

103. A. Torrisi, S. Pignataro, and G. Nocerino, "Applications of ESCA to Fabrication Problems in the Semiconductor Industry," *Appl. Surf. Sci.* **13**, 389–401, Sept./Oct. 1982.

104. A.C. Diebold, D. Venables, Y. Chabal, D. Muller, M. Weldon, E. Garfunkel, "Characterization and Production Metrology of Thin Transistor Gate Oxide Films," *Mat. Sci. Semicond. Proc.* **2**, 103–147, July 1999.

105. A.R. Lang, "Recent Applications of X-Ray Topography," *in Modern Diffraction and Imaging Techniques in Materials Science* (S. Amelinckx, G. Gevers, and J. Van Landuyt, eds.), North Holland, Amsterdam, 1978, 407–479; B.K. Tanner, *X-Ray Diffraction Topography*, Pergamon Press, Oxford, 1976; R.N. Pangborn, "X-Ray Topography," in *Metals Handbook*, 9th Ed. (R.E. Whan, coord.), Am. Soc. Metals, Metals Park, OH, **10**, 365–379, 1986; B.K. Tanner, "X-Ray Topography and Precision Diffractometry of Semiconductor Materials," in *Diagnostic Techniques for Semiconductor Materials and Devices* (T.J. Shaffner and D.K. Schroder, eds.), Electrochem. Soc., Pennington, NJ, 1988, 133–149; D.K. Bowen and B.K. Tanner, *High Resolution X-Ray Diffractometry and Topography*, Taylor and Francis, 1998.

106. W.F. Berg, "An X-Ray Method for the Study of Lattice Disturbances of Crystals," (in German) *Naturwissenschaften* **19**, 391–396, 1931; C.S. Barrett, "A New Microscopy and Its

Potentialities," *Trans. AIME* **161**, 15–64, 1945; J.B. Newkirk, "Subgrain Structure in an Iron Silicon Crystal as Seen by X-Ray Extinction Contrast," *J. Appl. Phys.* **29**, 995–998, June 1958.

107. A.R. Lang, "Direct Observation of Individual Dislocations by X-Ray Diffraction," *J. Appl. Phys.* **29**, 597–598, March 1958; A.R. Lang, "Studies of Individual Dislocations in Crystals by X-Ray Diffraction Microradiography," *J. Appl. Phys.* **30**, 1748–1755, Nov. 1959.

108. D.C. Miller and G.A. Rozgonyi, "Defect Characterization by Etching, Optical Microscopy and X-Ray Topography," in *Handbook on Semiconductors 3* (S.P. Keller, ed.) North-Holland, Amsterdam, 1980, 217–246.

109. G.H. Schwuttke, "New X-Ray Diffraction Microscopy Technique for the Study of Imperfections in Semiconductor Crystals," *J. Appl. Phys.* **36**, 2712–2721, Sept. 1961.

110. B.K. Tanner and D.K. Bowen, *Characterization of Crystal Growth Defects by X-Ray Methods*, Plenum Press, New York, 1980.

111. T.J. Shaffner, "A Review of Modern Characterization Methods for Semiconductor Materials," *Scann. Electron Microsc.* 11–23, 1986.

112. B.K. Tanner, *X-Ray Diffraction Topography*, Pergamon Press, Oxford, 1976; Y. Epelboin, "Simulation of X-Ray Topographs," *Mat. Sci. Eng.* **73**, 1–43, Aug. 1985.

113. B.K. Tanner, "X-Ray Topography and Precision Diffractometry of Semiconductor Materials," in *Diagnostic Techniques for Semiconductor Materials and Devices* (T.J. Shaffner and D.K. Schroder, eds.), Electrochem. Soc., Pennington, NJ, 1988, 133–149.

114. M. Dax, "X-Ray Film Thickness Measurements," *Semicond. Int.* **19**, 91–100, Aug. 1996.

115. T.Z. Hossain, "Neutron Activation Analysis," in *Encyclopedia of Materials Characterization* (C.R. Brundle, C.A. Evans, Jr., and S. Wilson, eds.), Butterworth-Heinemann, Boston, 1992, 671–679; P. Kruger, *Principles of Activation Analysis*, Wiley-Interscience, New York, 1971; R.M. Lindstrom, "Neutron Activation Analysis in Electronic Technology," *in Diagnostic Techniques for Semiconductor Materials and Devices* (T.J. Shaffner and D.K. Schroder, eds.), Electrochem. Soc., Pennington, NJ, 1988, 3–14.

116. C. Yonezawa, "Prompt Gamma Neutron Activation Analysis With Reactor Neutrons," in *Non-Destructive Elemental Analysis* (Z.B. Alfassi, ed.), Blackwell Science, Oxford, 2001.

117. G. Erdtmann, *Neutron Activation Tables*, Verlag Chemie, Weinheim, 1976.

118. A.R. Smith, R.J. McDonald, H. Manini, D.L. Hurley, E.B. Norman, M.C. Vella, and R.W. Odom, "Low-Background Instrumental Neutron Activation Analysis of Silicon Semiconductor Materials," *J. Electrochem. Soc.* **143**, 339–346, Jan. 1996.

119. E.W. Haas and R. Hofmann, "The Application of Radioanalytical Methods in Semiconductor Technology," *Solid-State Electron.* **30**, 329–337, March 1987.

120. P.F. Schmidt and C.W. Pearce, "A Neutron Activation Analysis Study of the Sources of Transition Group Metal Contamination in the Silicon Device Manufacturing Process," *J. Electrochem. Soc.* **128**, 630–637, March 1981.

121. M. Grasserbauer, "Critical Evaluation of Calibration Procedures for Distribution Analysis of Dopant Elements in Silicon and Gallium Arsenide," *Pure Appl. Chem.* **60**, 437–444, March 1988.

122. R.G. Downing, J.T. Maki and R.F. Fleming, "Application of Neutron Depth Profiling to Microelectronic Materials Processing," *in Microelectronic Processing: Inorganic Material Characterization* (L.A. Casper, ed.), American Chemical Soc., Symp. Series 295, Washington, DC, 1986, 163–180.

123. J.R. Ehrstein, R.G. Downing, B.R. Stallard, D.S. Simons and R.F. Fleming, "Comparison of Depth Profiling ^{10}B in Silicon Using Spreading Resistance Profiling, Secondary Ion Mass Spectrometry, and Neutron Depth Profiling," in *Semiconductor Processing, ASTM STP 850* (D.C. Gupta, ed.) Am. Soc. Test. Mat., Philadelphia, 1984, 409–425.

124. R.G. Downing and G.P. Lamaze, "Nondestructive Characterization of Semiconductor Materials Using Neutron Depth Profiling," in *Semiconductor Characterization, Present Status and Future Needs* (W.M. Bullis, D.G. Seiler, and A.C. Diebold, eds.), American Institute of Physics, Woodbury, NY, 1996, 346–350.

PROBLEMS

11.1 Consider an aluminum layer on a silicon substrate. Determine the Al film thickness so that when a 10 keV electron beam is incident on this sample, no X-rays are generated in the Si substrate. Use Eq. (11.4) for the electron range and assume electrons are only generated in a volume defined by the range R_e. Remember no X-rays can be generated when no electrons reach the Si.

11.2 Using Eqs. (11.13) and (11.14) determine the generation rate G and electron density n for a 10 keV electron beam having a beam current of 10^{-9} A incident on a Si wafer with electron lifetime of 10^{-5} s.

11.3 Using Eqs. (11.21) and (11.22) derive an expression for v_1/v_0 and derive Eq. (11.23).

11.4 The Rutherford backscattering plot of layer X_1 on substrate Y is in Fig. P11.4. A similar plot exists for layer X_2 on substrate Y. Find the identity and thickness of X_1 and X_2 and find the identity of substrate Y. The He ion energy is 2 MeV and the angle θ is 164°. Assume $\Delta E_{in} = \Delta E_{out}$. For X_1 on Y: $[S_o] = 45.5$ eV/Å, $E_1 = 0.332$ MeV, $E_2 = 0.91$ MeV, $E_3 = 1.138$ MeV. For X_2 on Y: $[S_o] = 133.6$ eV/Å, $E_1 = 0.338$ MeV, $E_2 = 1.578$ MeV, $E_3 = 1.847$ MeV.

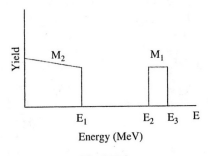

Fig. P11.4

11.5 The Rutherford backscattering plot of layer X on substrate Y is shown in Fig. P 11.4. Find the identity and thickness of X and find the identity of substrate Y. The He ion energy is 2 MeV and the angle θ is 170°. Assume $\Delta E_{in} = \Delta E_{out}$. For X on Y: $[S_o] = 193$ eV/Å, $E_1 = 0.657$ MeV, $E_2 = 1.256$ MeV, $E_3 = 1.835$ MeV.

11.6 The Rutherford backscattering plot of layer X on substrate Y is shown in Fig. P11.4. Find the identity and thickness of X and find the identity of substrate Y. The He ion energy is 2 MeV and the angle θ is 170°. Assume $\Delta E_{in} = \Delta E_{out}$. For X on Y: $[S_o] = 108$ eV/Å, $E_1 = 0.845$ MeV, $E_2 = 1.23$ MeV, $E_3 = 1.56$ MeV.

11.7 The Rutherford backscattering plot of layer X on substrate Y is shown in Fig. P11.4. Find the identity and thickness (in μm) of X and find the identity of substrate Y.

The He ion energy is 2 MeV and the angle θ is 164°. Assume $\Delta E_{in} = \Delta E_{out}$. For X on Y: $[S_o] = 109.2$ eV/Å, $E_1 = 0.954$ MeV, $E_2 = 1.51$ MeV, $E_3 = 1.73$ MeV.

11.8 Give the name for each of the characterization techniques in Fig. P 11.8.

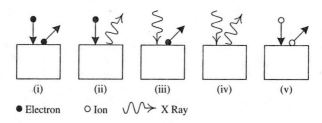

(i)   (ii)   (iii)   (iv)   (v)

● Electron   ○ Ion   〰⤳ X Ray

Fig. P11.8

11.9 Draw the Rutherford backscattering plots for He ions incident on the two samples in Fig. P11.9. Incident energy $E_0 = 2$ MeV; for substrate M_2, $K_2 = 0.6$, for layer M_3, $K_3 = 0.9$. $\Delta E_{in} = \Delta E_{out} = 0.2$ MeV.

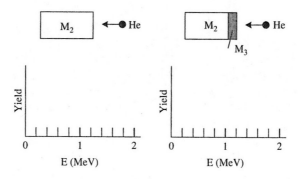

Fig. P11.9

REVIEW QUESTIONS

- What determines the magnification in an SEM?
- What is detected in *Auger electron spectroscopy*?
- What is detected in *electron microprobe*?
- What is the detection mechanism in energy dispersive spectroscopy (*EDS*)?
- What is the detection mechanism in wavelength dispersive spectroscopy (*WDS*)?
- How are X-rays generated?
- *AES* or *EMP*: Which has higher resolution? Why?
- Why can He not be detected with AES?
- What is the difference between EDS and WDS?
- What is an application for EBIC?
- What is the main application for SIMS?
- How are the SIMS vertical and horizontal data converted?

- What is TOF-SIMS?
- What is the principle for RBS?
- What is channeling?
- What is XRF?
- What is TXRF?
- Give an application for XRF.
- What is the principle of photoelectron spectroscopy?
- What makes XPS unique?
- What does X-ray topography give?
- How does neutron activation analysis work and where is it used?

12

RELIABILITY AND FAILURE ANALYSIS

12.1 INTRODUCTION

In this chapter we outline some general reliability concepts and then discuss some reliability measurements and concerns for semiconductor materials and devices. Reliability can be defined as the *probability* of operating a product for a given *time* under specified *conditions* without *failure*.[1] What does *failure* mean? Must the device cease to function? That depends. A device parameter *degradation* may be a failure. For example, if the MOSFET threshold voltage drifts from its specified value, the drain current changes and the circuit may not operate within its specification. If the line resistance of an interconnect line increases due to electromigration, the line delay time may exceed its specification. Both cases can be defined as failures, although the device or circuit still functions. Failure due to corrosion, fatigue, creep, and packages is beyond the scope of this chapter, but is discussed by Di Giacomo.[2]

Failure analysis (FA) is carried out in a number of steps.[3] Package failure can be detected with ultrasonic imaging and X-ray inspection. For semiconductor chips, the first step is usually visual inspection (optical microscope) and electrical measurements (I_{DDQ}, current-voltage, etc.). To locate the fault one uses mechanical probing, electron beams (scanning electron microscopy, voltage contrast), emission microscopy, liquid crystal, infrared microscopy, fluorescent microscopy, optical/electron beam induced current, optical beam induced resistance change, etc. For final detailed analysis, electron microprobe, Auger electron microscopy, X-ray photoelectron spectroscopy, secondary ion mass spectrometry, focused ion beam, and others are used. Increasingly FA must be carried out from the back of the chip as the front is obscured by multiple metal layers and the chip may be flip-chip mounted. Many of these characterization techniques were discussed earlier or are discussed in this chapter.

Semiconductor Material and Device Characterization, Third Edition, by Dieter K. Schroder
Copyright © 2006 John Wiley & Sons, Inc.

12.2   FAILURE TIMES AND ACCELERATION FACTORS

12.2.1   Failure Times

Various failure times are in use. Consider n products that fail after operating at times t_1, t_2, t_3, ... t_n. The *mean time to failure*, MTTF, is

$$MTTF = \frac{t_1 + t_2 + t_3 + ...t_n}{n} \tag{12.1}$$

The *median time to failure*, t_{50}, is the time when 50% of the products have failed, *i.e.*, half the products fail before t_{50} and half after. The *mean time between failures*, MTBF, is

$$MTBF = \frac{(t_2 - t_1) + (t_3 - t_2) + ...(t_n - t_{n-1})}{n} \tag{12.2}$$

Failure rate is sometimes represented by the "bathtub" curve in Fig. 12.1. During the early life of a product the failure rate is frequently high due to macro manufacturing defects, known as *infant mortality*. Such defects are usually eliminated by rigorous testing such as *burn-in*. The next region in the bathtub curve is characterized by approximately constant failure rate. This region corresponds to the working life of the component. Finally, during the last stage the failure rate increases due to *wearout*.

12.2.2   Acceleration Factors

Semiconductor circuits are designed to operate for extended times, typically 5–10 years. It is obviously impractical to test circuits over such long times and many reliability measurements are made under accelerated conditions where the test temperature and/or voltage and/or current are higher than normal operating conditions. The failure times are then extrapolated to "operating" conditions. Product life tests and long term reliability stresses are usually carried out under modest accelerated stress conditions with stress times of 10^4 to 10^6 s. Long term reliability stress is used in process qualifications for extrapolation, model parameter extraction and when new materials and/or process steps are introduced. For faster feedback to process development higher accelerated tests on

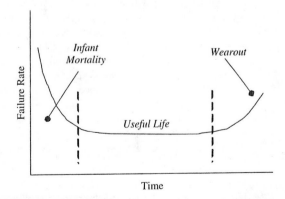

Fig. 12.1   Reliability bathtub curve showing early, intermediate and final failure.

wafer level are carried out when the degradation mechanism is known to be the same as at the package level, *e.g.*, for metal line electromigration, gate oxide degradation and device reliability.[4] Wafer level reliability (WLR) stress measurements are commonly used during process qualification in addition to package level tests. Lifetime projections based solely on data of highly accelerated WLR stresses include some degree of uncertainty. However, if they are backed up by package level stresses they are very useful and save measurement time for lifetime extrapolations.

Many failure modes are activation energy limited, *e.g.*, during electromigration atoms move in response to applied currents and the atom movement is thermally activated. Such thermally activated processes are characterized by the Arrhenius equation

$$t(T) = A \exp\left(\frac{E_A}{kT}\right) \tag{12.3}$$

where A is a constant and E_A the activation energy. The acceleration factor, AF_T, is defined as the ratio of the time at the base temperature T_0 to that at elevated temperature T_1

$$AF_T = \frac{t(T_0)}{t(T_1)} = \frac{\exp(E_A/kT_0)}{\exp(E_A/kT_1)} = \exp\left(\frac{E_A}{k}\left(\frac{1}{T_0} - \frac{1}{T_1}\right)\right) \tag{12.4}$$

assuming A and E_A are independent of temperature. When AF_T is known, $t(T_0)$ can be determined at any temperature. For some accelerated measurements, the voltage is increased above its operating value and the time to failure is

$$t(V) = B \exp(-\gamma V) \tag{12.5}$$

where B is a constant and γ the voltage factor. The acceleration factor becomes

$$AF_V = \frac{t(V_0)}{t(V_1)} = \exp(\gamma(V_1 - V_0)) \tag{12.6}$$

One of the uncertainties with AF is the assumption that elevated temperature or voltage data can be extrapolated to operating conditions. The failure mechanisms under elevated conditions may differ from those at operating condition. This suggests making the reliability measurements close to the operating condition, which, however, may incur extraordinarily long measurement times.

Elevated temperature measurements are made by placing the device to be tested into an oven, on a temperature-controlled probe station, or provide the wafer itself with a heater. Oven-heated devices are usually in a package and some ovens can hold many packaged test structures for simultaneous testing. The built-in heater can be a poly-Si resistor and temperatures up to 300°C can be achieved.[5] Diodes incorporated into or near the built-in heater allow the temperature to be measured making use of the diode temperature dependent current-voltage relationship

$$I = K n_i^2 \exp\left(\frac{qV}{kT}\right) = K_1 T^3 \exp\left(\frac{qV}{kT} - \frac{E_G}{kT}\right) \tag{12.7}$$

where K and K_1 are assumed to be constants. One can either use the current at a given voltage or the voltage at a given current. For a constant current, the temperature-dependent diode voltage is

$$\frac{dV}{dT} = \frac{1}{q}\frac{dE_G}{dT} + \frac{V - E_G/q}{T} - \frac{3k}{q} \approx -2.5 \ mV/K \tag{12.8}$$

where E_G is the band gap and V the applied forward bias voltage. The -2.5 mV/K is for Si around $T = 300$ K.

12.3 DISTRIBUTION FUNCTIONS

When a series of devices are tested, they will fail in time giving a *frequency distribution* and a *failure rate*. The failure or *hazard rate* is λ

$$\lambda = \frac{N}{t} \ \text{or} \ \lambda = \frac{f(t)}{1 - F(t)} \tag{12.9}$$

where N is the number of failures and t the total time. λ is defined as the probability of failure/unit time at time t given that the member of the original distribution has survived until time t.[6] $f(t)$ and $F(t)$ are defined below. Since the failure rate is quite low, the unit of FIT (failure unit) is used (1 FIT = 1 failure/10^9 hours).

Various functions are used to describe failures. The *cumulative distribution function* $F(t)$ also known as the *failure probability* is the probability that the device will fail at or before time t. $F(t) \rightarrow 1$ as $t \rightarrow \infty$. The *reliability function* $R(t)$ is the probability that the device will survive without failure to time t. It is

$$R(t) = 1 - F(t) \tag{12.10}$$

The *probability density function* $f(t)$ also known as the number of failures is

$$f(t) = \frac{d}{dt}F(t) \tag{12.11}$$

or

$$F(t) = \int_0^t f(t)\,dt \tag{12.12}$$

We will illustrate these concepts with an example. The number of oxide breakdown failures versus oxide electric field is shown in Fig. 12.2(a).[7] The cumulative distribution function is shown in 12.2(b). It is quite obvious that $F(t)$ provides more information by its two slopes representing defect-related and intrinsic oxide breakdown, although the information is the same in the two figures. The mean time to failure is given by

$$MTTF = \int_0^\infty t f(t)\,dt \tag{12.13}$$

Exponential Distribution: The *exponential function* is the simplest distribution function. It is characterized by a constant failure rate over the lifetime of the devices and is useful when early failures and wearout mechanisms have been eliminated. It is characterized by the functions

$$\lambda(t) = \lambda_o = \text{constant}; \ R(t) = \exp(-\lambda_o t); \ F(t) = 1 - \exp(-\lambda_o t); \ f(t) = \lambda \exp(-\lambda_o t) \tag{12.14a}$$

$$MTTF = \int_0^\infty t\lambda_o \exp(-\lambda_o t)\, dt = \lambda_0^{-1} \tag{12.14b}$$

The exponential function is frequently used in semiconductor failure analysis because it has a constant failure rate.

Weibull Distribution: In the *Weibull distribution* function[8] the failure rate varies as a power of the device age.

$$\lambda(t) = \frac{\beta}{\tau}\left(\frac{t}{\tau}\right)^\beta ; R(t) = \exp\left(-\left(\frac{t}{\tau}\right)^\beta\right) ; F(t) = 1 - \exp\left(-\left(\frac{t}{\tau}\right)^\beta\right) ;$$

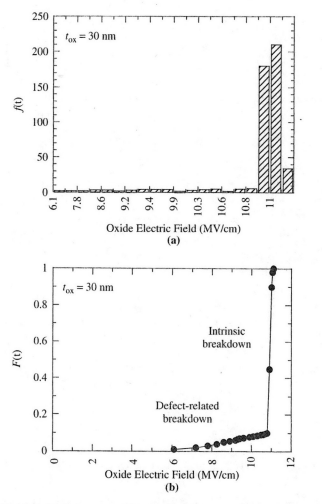

Fig. 12.2   (a) Number of failures versus oxide electric field, (b) cumulative failure versus electric field. Data adapted from ref. [49]; originally published in *Phil. J. Res.* **40**, 1985 (Philips Research).

$$f(t) = \frac{\beta}{\tau} \left(\frac{t}{\tau}\right)^{\beta} \exp\left(-\left(\frac{t}{\tau}\right)^{\beta}\right) \tag{12.15a}$$

$$MTTF = \tau \Gamma(1 + 1/\beta) \tag{12.15b}$$

where τ and β (shape parameter) are constants and Γ is the gamma function. For $\beta < 1$ the failure rate decreases and for $\beta > 1$ it increases with time. The former represents the early and the latter the wearout period. For $\beta = 1$, the Weibull becomes the exponential distribution. For experimental data to lie on straight line, they can be plotted on "Weibull" plots. Rearranging $F(t)$ in Eq. (12.15a) gives

$$\ln[-\ln(1 - F(t))] = \beta \ln(t) - \beta \ln(\tau) \tag{12.16}$$

which is linear of the form $y = mx + b$.

Normal Distribution: For the *normal distribution* function

$$F(t) = \frac{1}{\sigma\sqrt{2\pi}} \int_0^t \exp\left(-\frac{1}{2}\left(\frac{t-\tau}{\sigma}\right)^2\right) dt; f(t) = \frac{1}{\sigma\sqrt{2\pi}} \exp\left(-\frac{1}{2}\left(\frac{t-\tau}{\sigma}\right)^2\right) \tag{12.17a}$$

where the median time to failure t_{50}, the scale parameter σ, and the failure rate are

$$\sigma = \ln\left(\frac{t_{50}}{t_{15.87}}\right); \lambda(t) = \frac{f(t)}{1 - F(t)} \tag{12.17b}$$

where $t_{15..87}$ is the time when 15.87% of the devices have failed.[9] An interesting aspect of the normal distribution is the "six sigma" reliability practiced by some companies. From $f(t)$ in Eq. (12.17a) 99.999908% falls within $\pm 6\sigma$, *i.e.*, no more than 3.4 defective parts per million are tolerated.

Log-Normal Distribution: The *log-normal distribution* function is frequently used to describe the failure statistics of semiconductor devices over long times. Here

$$F(t) = \frac{1}{\sigma\sqrt{2\pi}} \int_0^t \frac{1}{t} \exp\left(-\frac{1}{2}\left(\frac{\ln(t) - \ln(t_{50})}{\sigma}\right)^2\right) dt;$$

$$f(t) = \frac{1}{\sigma t\sqrt{2\pi}} \exp\left(-\frac{1}{2}\left(\frac{\ln(t) - \ln(t_{50})}{\sigma}\right)^2\right) \tag{12.18a}$$

where the median time to failure t_{50}, the scale parameter σ, and the failure rate are

$$\sigma = \ln\left(\frac{t_{50}}{t_{15.87}}\right); \lambda(t) = \frac{f(t)}{1 - F(t)} \tag{12.18b}$$

Experimental data are displayed on log-normal plots.

Which function should be used to make lifetime predictions? One common procedure is to select the probability plotting paper (exponential, Weibull, log-normal, *etc.*) that allows the data to be graphed as a straight line, but it is not always possible to find an unambiguous model. Electromigration failures usually follow the log-normal distribution while gate oxide breakdown statistics are usually plotted with the Weibull distribution also known as an extreme value distribution in which there may be many identical and independent competing failure processes, but the first to reach some critical point determines the time to failure. For example, there may be several weak spots in an oxide, but the first to fail determines the failure time, *i.e.*, the weakest link in a chain causes it to fail.

12.4   RELIABILITY CONCERNS

12.4.1   Electromigration (EM)

In 1861 Geradin first observed that liquid solder subjected to direct electric current showed segregation of its components.[10] Skaupy in 1914 suggested the importance of the interaction between the metal atoms and the moving electrons and in 1953 Seith and Wever first measured the mass transport of alloys showing that the driving force for *electromigration* was not only influenced by the electrostatic force from the applied current, but also depended on the direction of motion of the electrons.[11] This work laid the foundation for electromigration by introducing the "electron wind" force that drives the mass transport. Huntington and Grone developed theoretical and mathematical formulations describing the driving forces during electromigration.[12] However, it was not until the late 1960s that the study of thin film electromigration gained significant attention because of its role in the failure of semiconductor integrated circuits.

Failures in conductors and contacts in integrated circuits are attributed to electromigration and *stress migration* (SM). We will briefly describe the mechanisms responsible for these failure modes and then give some of the characterization techniques to detect such failures. Failures are typically characterized by a certain percentage increase in the line resistance, by a line becoming an open circuit, or by adjacent lines becoming short circuited. Sometimes one observes voids at one end of the line and hillocks at the other end, illustrated in Fig. 12.3. Line degradation is a slow process and under normal operating conditions can take many years. Hence, measurements are made under accelerated conditions. For example, ICs normally operate at maximum temperatures of 100–175°C and line current densities of $\leq 5 \times 10^5$ A/cm^2. Accelerated tests typically use temperatures above 200°C and current densities above 10^6 A/cm^2.

Why do metal lines degrade? Metals deposited on insulators are polycrystalline, consisting of small single crystal grains having varying crystal orientations as illustrated in Fig. 12.4. Adjacent grains meet at grain boundaries—regions of imperfection. Three or more grain boundaries meet at *triple points*. Grain sizes depend on processing, but are on the order of 100 nm or so. The line is also under considerable mechanical stress due to thermal mismatch. When a potential is applied along such a line, two forces act on

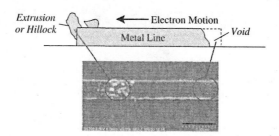

Fig. 12.3   Void and hillock formation in a Ag line stressed with $J = 23$ MA/cm^2 at $T = 160°$C. Image courtesy of T.L. Alford, Arizona State University.

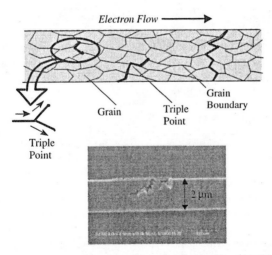

Fig. 12.4 Schematic of a polycrystalline line containing grains, grain boundaries, and triple points. The SEM micrograph shows a propagating crack. Micrograph courtesy of P. Nguyen and T.L. Alford, Arizona State University.

the metal ions: one is due to the electric field along the line and the other due to the electron "wind" effect. For an electric field pointing to the left in Fig. 12.4, the positive metal ions tend to drift to the left due to the electric field. However, the electrons flow to the right and momentum transfer from the electrons to the ions pushes the ions to the right. It is the momentum transfer that dominates in Al lines.[13] The motion of atoms is a complicated process depending on grain size, grain boundary orientation, triple point density, thermally induced stress, surface conditions, and so on.

Passivating a metal line increases its EM resistance, presumably by introducing additional stress. Voids typically form at triple points due to the accumulation of vacancies that move by diffusion. The heavy lines on Fig. 12.4 indicate the formation of cracks, leading eventually to an open circuit. Metal diffusion occurs primarily through vacancies. Electromigration alone cannot induce failures in metal lines unless there is a non-vanishing divergence of the atomic flux. Such a divergence exists at triple points and at the boundary from small to large grains. Mass build up occurs at a small-to-large grain boundary and mass depletion is observed at a large-to-small grain boundary. An experiment comparing a single crystal and a polycrystalline Al line showed that testing the lines under identical conditions (175°C, 2×10^6 A/cm^2) led to polycrystalline line failure after 30 h, whereas the single crystal line showed no degradation after 26,000 h.[14] Most EM reliability measurements are made under dc conditions. The ac lifetime for Al/Si and Cu is orders of magnitude longer than dc lifetime and proportional to frequency over the mHz to 200 MHz frequency range.[15]

While it is not possible to form single crystal lines on polycrystalline insulators, it is possible to modify the grain boundary structure by eliminating triple points. As lines become narrower, there is a high probability that no triple points exist, as in the *bamboo* structure. The higher EM resistance of such lines is due the reduced number of triple point and grain boundaries and the fact that the activation energy of intragrain diffusion is higher than for grain boundary diffusion. Adding impurities to retard diffusion along grain boundaries "strengthens" Al lines. Adding small amounts of Cu to Al lines extends their lifetime significantly. For example, adding 4 wt% Cu increased the lifetime 70 times.[16] Another way to extend the EM lifetime is through layered structures. For example, depositing Al lines on top of a TiN film allows the current to be shunted through the TiN layer if weak spots develop in the higher conductivity Al, and line lifetime is thus greatly extended. Electromigration in refractory metals like TiN is virtually non-existent. An even better solution is to replace of Al with a higher EM resistance material, *e.g.*, Cu, which has both higher conductivity and higher EM resistance than Al.[17] The line length is also important during EM measurements. A critical length for metal lines, known as the *Blech length*, exists below which electromigration is inhibited.[18] Blech found that the accumulation of Al atoms at the anode end of the line results in a stress gradient which can balance the electromigration driving force. When metal ions diffuse toward the anode end of the line a stress buildup opposes the electron wind, thus restricting the electromigration void growth.

In 1969, Black published a simple theory relating the median time to failure of a conductor to the transport and geometrical parameters of the line.[19] He assumed the rate of mass transport by momentum transfer between thermally activated ions and electrons to be directly proportional to the momentum of the electrons, to the number of activated ions, to the number of electrons/s·cm^3, and to the effective target cross-section. This resulted in the well-known "Black" equation relating the median time to failure, the current density J, and the activation energy E_A as

$$t_{50} = \frac{A e^{E_a/kT}}{J^n} \tag{12.19}$$

where A is a constant related to the line cross-sectional area. Once E_A and n are determined under accelerated conditions, one extrapolates to operating conditions through the equations

$$\frac{t_{50}(T_1)}{t_{50}(T_2)} = \exp\left[\frac{E_A}{k}\left(\frac{1}{T_1} - \frac{1}{T_2}\right)\right]; \; \frac{t_{50}(J_1)}{t_{50}(J_2)} = \left(\frac{J_2}{J_1}\right)^n \tag{12.20}$$

The underlying assumption is that those mechanisms causing degradation under accelerated stress are also active under normal operating conditions.

A standard test line, developed by the National Institute of Standards and Technology, is shown in Fig. 12.5.[20] The test structure in Fig. 12.5(a) is designed for a $2 \times N$ probe card. Electromigration test structures 1, 2, 7, 8; 10, 3, 6, 14; or 9, 10, 15, 16 consist of a straight line of length of about 800 μm. Temperature gradients due to heat sinking by the end sections are confined to the ends of such lines. Significant temperature gradients can exist for lines less than 400 μm long. The line resistance should be around 20–30 ohms. The van der Pauw test structure 2, 3, 10, 11 measures the line sheet resistance and determines the temperature coefficient of resistance, which must be accurately known.[21] The terminating end segments of the EM lines should be twice the width of the test lines and voltage taps are provided for Kelvin measurements.[22] The line in Fig. 12.5(b) has extrusion detectors. By monitoring the resistance between the line and the "extrusion" lines, one can detect shorts that might form due to metal migration. Since contacts are especially important in EM measurements, test structures usually consist of lines and contacts, illustrated in Fig. 12.5(c).

To obtain satisfactory statistical data, one usually stresses a number of test lines. With the test structure of Fig. 12.5, each line requires a power supply. A simpler approach is to use a test structure with a number of lines in parallel. One end of these lines terminates on one contact pad and the other end on another contact pad. All lines are

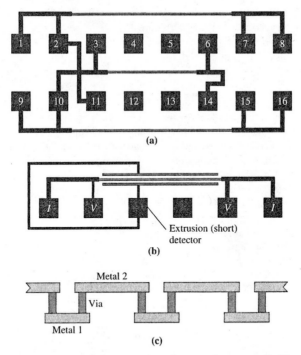

Fig. 12.5 Electromigration test structures. (a) three lines can be measured, (b) one EM line and extrusion detectors, (c) line with contacts.

tested simultaneously by applying a constant voltage with the current dividing among all of the lines. All lines are subjected to a constant current density, regardless of whether any line has failed, since the total current adjusts itself to the number of active lines. By monitoring the current through the entire test structure, the failure of individual lines can be detected.[23] In the serially connected method, the lines are connected to a single current source with each line having a current bypass circuit, consisting of a shunt relay and a Zener diode.[24] A large number of samples can be tested with the same current, making it suitable for reliability assessment. Electromigration is sometimes characterized by low frequency noise measurements, showing typically a $1/f^n$ behavior and the noise amplitude and n factor are related to EM.[25]

Electromigration failure data are usually analyzed by means of the log-normal distribution. A number of metal lines are tested at various temperatures for a given current density. The resulting data are plotted as cumulative failures as a function of test time as in Fig. 12.6(a). The median times to failure are then plotted as $\log(t_{50})$ versus $1/T$ and the activation energy is extracted (Fig. 12.6(b)). From Eq. (12.19)

$$E_A = \ln(10)k \frac{\Delta \log(t_{50})}{\Delta(1/T)} \qquad (12.21)$$

Then measurements are made for various current densities at a given temperature. The exponent n in Eq. (12.19) is determined from

$$n = -\frac{\Delta \log(t_{50})}{\Delta \log(J)}, \qquad (12.22)$$

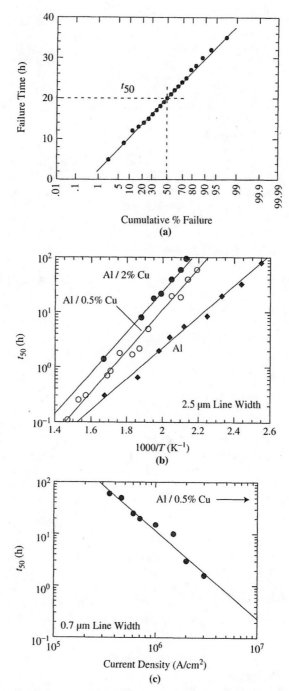

Fig. 12.6 Electromigration data representation; (a) median time to failure, (b) activation energy determination, (c) n factor determination.

as in Fig. 12.6(c). Knowing E_A and n allows t_{50} to be predicted for other temperatures or current densities according to Eq. (12.20).

Sometimes it is desirable to use EM test methods that are very fast, for a production environment, for example. One such method uses the *standard wafer-level electromigration acceleration test* (SWEAT) structure. Measurement times are reduced to 30–60 s without external heating.[26] There is no need for a hot plate or oven as the heat is supplied through Joule heating of the line by the current flowing through it. The test structure consists of alternating narrow and wide segments or simple straight lines. The transition from one to the next can be gradual or abrupt. The wide regions act as heat sinks and the transition from the narrow to the wide region creates current and stress gradients. Due to the high current densities, the measurement time is short. Good correlation between wafer level SWEAT and conventional standard package level tests on via terminated structures can be obtained.[27] Similar t_{50} values were obtained by extrapolating the data from both tests to normal conditions with the same failure mechanism for both. Copper lines require longer test times due to its higher electromigration resistance and higher acceleration factors are needed to reach reasonable test times. Self-heated test methods, such as SWEAT use Joule heating to reach stress temperatures up to 600°C and much lower test times. SWEAT measurements compare well with conventional EM test conditions.[28]

Electromigration also occurs at contacts. In fact contact EM has become the dominant metal failure mechanism. However, such EM is dependent on the type of contact. Consider the schematic in Fig. 12.7(a), consisting of two Al lines connected by a tungsten plug. Electrons flow from the upper level M_2 to the lower metal M_1. As they enter M_1, Al atoms migrate and since W migration is negligible, a void develops *under* the W plug. For electron flow in the opposite direction, the void forms in M_2. The interface between

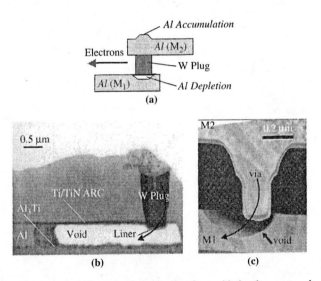

Fig. 12.7 Contact electromigration (a) schematic showing void development under the W plug with electrons flowing from M_2 to M_1; TEM cross-sections showing the void in (b) Al and (c) Cu lines. The arrows indicate the electron flow. TEM micrographs courtesy of (b) T.S. Sriram and E. Piccioli, Compaq Computer Corp.; (c) Reprinted after M. Ueki, M. Hiroi, N. Ikarashi, T. Onodera, N. Furutake, N. Inoue, and Y. Hayashi, *IEEE Trans. Electron Dev.* **51**, 1883–1891, Nov. 2004 by permission of IEEE (© 2004, IEEE).

dissimilar materials, *e.g.*, Al/Si, Al/TiN, Al/W, is highly vulnerable to electromigration, because the rate of electromigration of refractory metals is insignificant compared to Al and Al is transported from the interface.[29] Figures 12.7(b) and (c) illustrate contact electromigration. The EM voids below the tungsten plug are obvious. Another weak spot is EM in solder joints.[30]

12.4.2 Hot Carriers

Hot carriers (electrons or holes) are of concern in integrated circuits, because electrons and/or holes that gain energy in an electric field can be injected into the oxide to become oxide trapped charge, they can drift through the oxide, causing gate current, they can create interface traps, and they can generate photons, all illustrated in Fig. 12.8.[31] The term *hot carriers* is somewhat misleading. The carriers are energetic. The carrier temperature T and energy E are related through the expression $E = kT$. At room temperature, $E \approx 25$ meV for $T = 300$ K. When carriers gain energy by being accelerated in an electric field, their energy E increases. For example, $T = 1.2 \times 10^4$ K for $E = 1$ eV. Hence the name *hot carriers* means *energetic carriers*, not that the entire device is hot.

Let us briefly discuss the effects of hot carriers. As shown in Fig. 12.8, some electrons in the channel entering the drain space-charge region experience impact ionization. The resulting hot carriers can be injected into the oxide (N_{ot}), can flow through the oxide (I_G), can generate interface traps (D_{it}), flow to the substrate contact as substrate current (I_{sub}), and create photons. The photons, in turn, can propagate into the device, be absorbed, and create electron-hole pairs. N_{ot} and D_{it} lead to threshold voltage changes and mobility degradation. The substrate current causes a voltage drop in the substrate, forward biasing the source-substrate junction, leading to further impact ionization and possibly snapback breakdown. The device can be viewed as a parasitic bipolar junction transistor (BJT) in parallel with the MOSFET. The BJT has an almost open base and open base BJTs often exhibit snapback breakdown with negative differential resistance. Almost open base means the base potential is not well controlled and although the base contact is grounded, the interior base has an ill-defined potential.

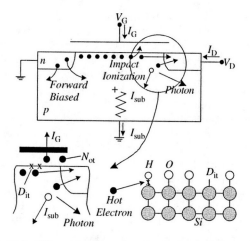

Fig. 12.8   Effect of hot electrons near the drain of MOSFETs.

One method to determine hot carrier degradation in n-channel devices is to bias the device at maximum substrate current. The substrate current dependence on gate voltage is shown in Fig. 12.9(a). The substrate current depends on the channel lateral electric field. At low V_G, with the device in saturation, the lateral electric field increases with increasing gate voltage until $V_G \approx V_D/3 - V_D/2$. I_{sub} increases to a maximum at that gate voltage for n-channel devices. For higher gate voltages, the device enters its linear region, the lateral electric field decreases as does the substrate current.

The device is biased at $I_{sub,max}$ for a certain time and a device parameter, e.g., saturation drain current, threshold voltage, mobility, transconductance, or interface trap density, is measured.[32] This process is repeated until the measured parameter has changed by some amount (typically 10–20%) as shown in Fig. 12.9(b) for I_{Dsat}. The lifetime corresponds to that time. Next the substrate current is changed by choosing a different gate voltage and the process is repeated and plotted as lifetime versus I_{sub} in Fig. 12.9(c). The data points, measured over a restricted range, are extrapolated to the IC life, typically ten years, giving the maximum I_{sub} that should not be exceeded during the device operation.

The chief degradation mechanism for n-channel MOSFETs is believed to be interface trap generation, and the *substrate* current is a good monitor of such damage. There are, of course, other measurements that can be used, such as interface trap density measurements by charge pumping, for example. Because it is simple to measure, I_{sub} is commonly used. The main degradation mechanism for p-channel devices is believed to be trapped electrons near the gate-drain interface and it manifests itself at a maximum in *gate* current. Hence, in p-channel devices I_G is usually measured.[33] Hot carrier damage can be reduced by reducing the electric field at the drain by, for example, forming lightly doped drains and by using deuterium instead of hydrogen during post-metallization anneal at temperatures around 400–450°C, since the Si-D bond is stronger than the Si-H bond.[34]

An issue related to hot carriers is *plasma induced damage* during semiconductor processing, where charge in the plasma environment lands on the device. If it lands on MOS gates, the charge produces electric fields which, in turn, can generate insulator leakage currents and their attendant damage. A common test structure is the *antenna structure*, with a large conducting area, consisting of polysilicon or metal layers, attached to a MOSFET or MOS capacitor gate.[35] Frequently the antenna resides on a thicker oxide than the MOSFET gate oxide. The ratio between antenna area and gate oxide area has typical values of 500–5000. The antenna test structure is placed into a plasma environment; charge builds up on the antenna and channels gate current through the MOSFET gate oxide where it generates damage that is subsequently detected by measuring the transconductance, drain current, threshold voltage, etc. The highest V_T sensitivity exists for gate oxides 4–5 nm or thicker. Below 4 nm the gate leakage current is a more suitable measure. Another test structure, the *charge monitor*, is based on an electrically erasable programmable read-only memory (EEPROM) structure, consisting of a MOSFET with a floating gate inserted between the substrate and the control gate. The control gate is a large-area collecting electrode.[36] The device is exposed to the plasma, charge builds up and develops a control gate voltage. Part of that control gate voltage is capacitively coupled to the floating gate. For sufficiently high floating gate voltage, charge is injected from the substrate and is trapped on the floating gate changing the device threshold voltage. The threshold voltage is subsequently measured and converted to charge generating a contour map of the plasma charge distribution. The potential sensors are implemented in pairs, where one sensor measures negative and the other positive potentials.

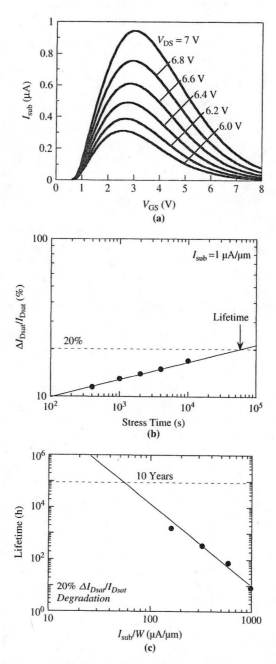

Fig. 12.9 (a) Substrate current, (b) drain current degradation, (c) lifetime plots for hot carrier degradation. Substrate current plots courtesy of L. Liu, Motorola.

12.4.3   Gate Oxide Integrity (GOI)

The gate oxide of an MOS device is one of the most important MOS device parameters. It is very sensitive to damage and can easily degrade. Although the oxide resistivity is on the order of 10^{15} $\Omega \cdot$cm, it is not infinite. Hence currents flow through a gate oxide for any gate voltage. However, for moderate gate voltages, corresponding typically to oxide electric fields $\leq 3 \times 10^6$ V/cm, the gate currents are negligible. However, for higher gate oxide electric fields, gate currents increase rapidly with voltage. To characterize the lifetime and integrity of gate oxides, voltages higher than operating voltages or temperatures higher than operating temperatures are used with appreciable current flow through an oxide. There are two main gate current flow mechanisms. The oxide voltage required for one or the other of these two mechanisms to occur is shown in the band diagrams of an MOS device with an n^+ poly-Si gate and a p-substrate in Fig. 12.10. For $V_{ox} < q\phi_B$ (the barrier height ϕ_B is in eV), as in 12.10(a), the electrons "see" the full oxide thickness and the gate current is due to *direct tunneling*. For $V_{ox} > q\phi_B$, as in 12.10(b), the electrons "see" a triangular barrier and the gate current is due to *Fowler-Nordheim* (FN) *tunneling*. The dividing oxide voltage between the two is $V_{ox} = q\phi_B$, which is approximately 3.2 V for the SiO$_2$-Si interface. Of course, the oxide electric field, $\mathscr{E}_{ox} = V_{ox}/t_{ox}$, must be sufficiently high for tunneling to take place. For oxide thicknesses of 4–5 nm and above, Fowler-Nordheim tunneling dominates and for $t_{ox} \leq 3.5$ nm or so, direct tunneling is dominant. A recent study concludes that silicon dioxide-based dielectrics provide reliable gate dielectrics, even as thin as 1 nm.[37]

The gate current is discussed in Appendix 12.1. The *Fowler-Nordheim* current density is[38]

$$J_{FN} = A\mathscr{E}_{ox}^2 \exp\left(-\frac{B}{\varepsilon_{ox}}\right) \tag{12.23}$$

where $\mathscr{E}_{ox}$ is the oxide electric field and A and B are given in Appendix 12.1. The *direct tunnel* current density expression is more difficult to derive and several versions have been published.[39] We give the empirical expression[40]

$$J_{dir} = \frac{AV_G}{t_{ox}^2}\frac{kT}{q}C\exp\left(-\frac{B(1 - (1 - qV_{ox}/\Phi_B)^{1.5})}{\mathscr{E}_{ox}}\right) \tag{12.24}$$

with C is given in Appendix 12.1.

The total gate current is the sum of FN and direct currents. The breakpoint between J_{FN} and J_{dir} occurs at gate voltages of approximately ± 4 V. Experimental gate current

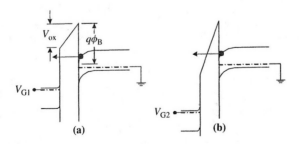

Fig. 12.10   MOS band diagrams for (a) $V_{ox} < q\phi_B$ (direct tunneling) and (b) $V_{ox} > q\phi_B$ (Fowler-Nordheim tunneling).

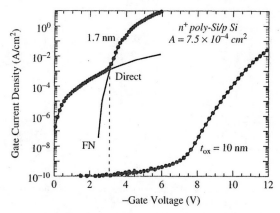

Fig. 12.11 Gate current density versus gate voltage illustrating direct and Fowler-Nordheim tunneling. Points are experimental data.

densities are plotted in Fig. 12.11. FN current dominates for the 10 nm oxide while J_{FN} dominates for $V_G > 3$ V and J_{dir} for $V_G < 3$ V for the 1.7 nm oxide and $J_{dir} \gg J_{FN}$ in the regime where J_{dir} dominates.

Oxide tunnel currents sometimes saturate at high currents, especially when measured on MOS capacitors with inverted substrates, because the tunnel electrons originate from thermal electron-hole pair generation, which can be very low for high-lifetime substrates (discussed in Chapter 7). The "tunnel" current under those conditions is actually the thermally generated leakage current. This problem does not exist in MOSFETs because the necessary electrons are supplied by the grounded source and drain. Oxide currents are very low at low gate voltages and frequently obscured by system (probe station, cables, etc.) leakage currents.

A method to measure the very low gate oxide currents is based on the floating gate configuration in Fig. 12.12(a).[41] The gate electrode of a MOSFET is connected to the capacitor (MOS-C) under test. The common gate is biased to V_G and then open circuited. As the MOS-C discharges due to oxide current the MOSFET gate potential and drain current decrease. The variation of I_D is measured and related to the gate current through the relationship

$$I_G = C\frac{dV_G}{dt} = C\frac{dV_G}{dI_D}\frac{dI_D}{dt} = \frac{C}{g_m}\frac{dI_D}{dt} \tag{12.25}$$

where C is the sum of the MOSFET (C_{MOSFET}) and MOS-C (C_{MOS-C}) capacitances. With $C_{MOS-C} \gg C_{MOSFET}$, the gate discharge is due to current flowing through the MOS-C. The I_G-V_G plot in Fig. 12.12(b) clearly shows the very low gate current measurement capability of this method.

Exercise 12.1

Problem: Oxide breakdown is often characterized as A-, B-, and C-mode. What do these designations mean?

Solution: When MOS devices are measured on a given wafer or within a given lot, the oxide breakdown voltages can exhibit a wide range of breakdown electric fields. It

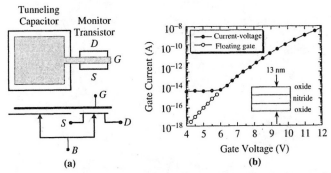

Fig. 12.12   (a) Schematic showing device for very low gate current measurements, (b) I_G-V_G for conventional and modified measurements.

is customary to divide oxide breakdowns into three distinct regions. *A*-mode failures are those oxides breaking down at very low oxide electric fields, *e.g.*, 1–2 MV/cm; oxides breaking down at intermediate electric fields, *e.g.*, 2–8 MV/cm are termed *B*-mode failures, and *C*-mode failures are those of the intrinsic oxide at typical fields of 9–12 MV/cm or higher, as illustrated in Fig. E12.1(a). *A*-mode failures are attributed to pinholes, scratches and other gross defects, as illustrated in Fig. E12.1(b). *B*-mode failures have been attributed to oxide thinning, *e.g.*, at LOCOS edges and defects. *C*-mode failures are due to the intrinsic nature of the oxide.

Oxide integrity is determined by *time-zero* and *time-dependent* measurements.[42] The time-zero method is simply an I_G-V_G MOS device measurement with increasing gate voltage until the oxide breaks down, illustrated in Fig. 12.13. The breakdown voltage is

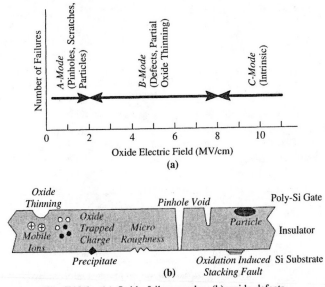

Fig. E12.1   (a) Oxide failure modes, (b) oxide defects.

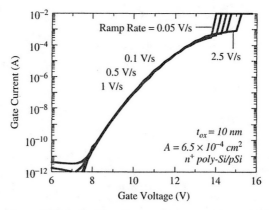

Fig. 12.13 $I_G - V_G$ plots exhibiting the effect of gate voltage ramp rate on oxide breakdown voltage. Data courtesy of P. Ku, Arizona State University.

gate voltage ramp-rate dependent. This dependence is related to the damage created in the oxide during the measurement. For low ramp rates, more time is available to create damage resulting in lower breakdown voltage than for higher ramp rates.

The time-dependent measurements are the *constant gate voltage* and *constant gate current* methods of Fig. 12.14. In the constant voltage method, a gate voltage near the breakdown voltage is applied and the gate current is measured as a function of time. The current typically decreases before rising precipitously at breakdown [Fig. 12.14(a)]. For constant current measurements, a constant current is forced through the oxide and the gate voltage is measured as a function of time. Typically, the gate voltage rises gently and drops as the device breaks down [Fig. 12.14(b)].

When the oxide is driven into breakdown, one defines a *charge-to-breakdown* Q_{BD} as

$$Q_{BD} = \int_0^{t_{BD}} J_G \, dt \tag{12.26}$$

where t_{BD} is the *time to breakdown*. Q_{BD} is the charge density flowing through the oxide necessary to break it down and it depends on the gate oxide thickness. In Fig. 12.14(a), Q_{BD} is the area under the curve, while for 12.14(b) it is simply $Q_{BD} = J_G t_{BD}$. Q_{BD} depends not only on the oxide itself, *i.e.*, how the oxide is grown, but also on how Q_{BD} is measured. For example, it depends on the gate current density and the gate voltage during the measurement. The current may be constant or it may be stepped. For constant current, the stress current density is often around 0.1 A/cm². This comes about as follows. $Q_{BD} \approx 10$ C/cm² for oxides with thicknesses around 10 nm. For reasonable measurement times of $t_{BD} \approx 100$ s and $J_G = Q_{BD}/t_{BD} \approx 0.1$ A/cm². In the stepped current technique, the current is applied for a certain time, *e.g.*, 10 s; it is then increased by a factor of ten for the same time, and so on, until the oxide breaks down.[43] Since in this method the current starts at a low value, *e.g.*, 10^{-5} A/cm², it is a more sensitive technique to bring out *B*-mode failures. One can also apply the voltage in steps. It has also been proposed to apply a certain gate voltage to stress the device, then reduce the voltage and measure the device; increase the stress voltage and return to the same original measuring voltage, and so on.[44] Sometimes one applies a constant gate voltage not near breakdown, but closer to the operating voltage of the device. Since breakdown takes inordinately

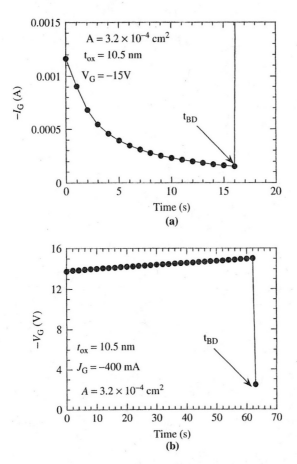

Fig. 12.14 (a) Gate current-time and (b) gate voltage-time characteristics of n^+-gate/SiO$_2$/p-substrate devices. Gate injection was used for these measurements. Data courtesy of Z. Zhou, Motorola.

long times under those conditions, the measurement temperature is increased. The *oxide breakdown mechanism* is not completely understood. In the *percolation* model, breakdown is envisioned as the formation of a connecting path of defects, as a result of random defect generation throughout the insulating film. A sufficient defect density forms a percolation path leading to oxide breakdown.[45]

Q_{BD} also depends on whether the substrate is the anode or cathode. Electron injection from the substrate typically exhibits higher Q_{BD} than injection from the poly-Si gate. This has been attributed to a rougher interface at the gate/oxide interface than at the substrate/oxide interface. The frequency of the applied voltage also plays a role with ac stress generally resulting in higher t_{BD} than dc stress.[46] One explanation is that holes need to drift through the oxide to generate oxide traps. Under ac excitation, the holes have insufficient time to drift before the oxide electric field reverses direction. The test structures have various shapes and geometries: large capacitors with rectangular gates, transistors, arrays of transistors or arrays of small unit cell capacitors, finger- and serpentine-structured capacitors. Output data can be strongly influenced by the edge-to-area ratio. A structure can also have more than one edge component. The main aim of the test structure is to reflect all critical structure issues that occur in the products.

The oxide integrity is most commonly determined from current-voltage or time-dependent measurements. Occasionally, however, it is desired to know the pinhole density of an oxide or other insulator. One can use chemical methods for this. For example, fluorescent tracers deposited on an oxidized sample can be viewed under ultraviolet light.[47] They emit light wherever there is a pinhole. Alternatively, one can use copper decoration.[48] The sample as a cathode is placed in a methanol bath with a copper mesh anode. The copper dissolves from the Cu anode to become colloidal particles. When a voltage is applied between anode and cathode, the colloidal copper precipitates at local oxide defect sites. The oxide defect structure on the sample is not disturbed due to the low bath current.

Oxide Breakdown Statistics: Oxide breakdown data are presented in a variety of ways. The simplest is to plot the number of failures versus the oxide electric field, shown in Fig. 12.2(a). Next is the cumulative failure distribution shown in Fig. 12.2(b). It is a plot of cumulative failure as a function of oxide electric field. Sometimes the cumulative failure is plotted as a function of time-to-breakdown. The statistics of oxide breakdown are usually described by extreme value distributions or Weibull statistics based on the observation that oxide breakdown usually occurs in a small area of the device.[49] If the device contains a multitude of weak spots, the first breakdown occurs at the spot with the lowest dielectric strength. The assumptions underlying the use of extreme value distribution functions are (1) a breakdown may take place at any spot out of a large number of spots, (2) the spot with the lowest dielectric strength gives rise to the breakdown event, and (3) the probability of breakdown at a given spot is independent of the occurrence of breakdown at other spots.

Consider a set of n MOS capacitors. Each of these capacitors ($i = 1, 2 \ldots n$) fails at an electric field $\mathscr{E}_i$. For a device with area A and defect density D, the cumulative failure F is[50]

$$F = 1 - \exp(-AD) \tag{12.27}$$

In Fig. 12.2(b), F is plotted versus $\mathscr{E}_{ox}$. This plot has two distinct regions. Those devices breaking down at low electric fields are due to oxide defects. The values at the higher electric fields are due to intrinsic oxide breakdown. Equation (12.27) can be written as

$$-\ln(1 - F) = AD \tag{12.28}$$

plotted in Fig. 12.15(a). A particular oxide electric field (in Fig. 12.15 it is 10 MV/cm) gives a value of $-\ln(1-F)$ equal to AD. Hence, this point gives the defect density, provided the area is known. For the example, on Fig. 12.15(a), $-\ln(1 - F) = 0.08$, giving $D = 8$ cm^{-2} for $A = 0.01$ cm^2. Choosing a different value of $\mathscr{E}_{ox}$ gives a different D, since the defect density causing oxide breakdown depends on the oxide electric field. Frequently it is Q_{BD} that is of most interest and Weibull plots are then given as in Fig. 12.15(b). This is a good example of a high quality and a low quality oxide. Oxide 1 is defect dominated while the breakdown in oxide 2 is predominantly intrinsic.

The cumulative failure is sometimes written as[51]

$$F = 1 - \exp(-x/\alpha)^\beta \tag{12.29}$$

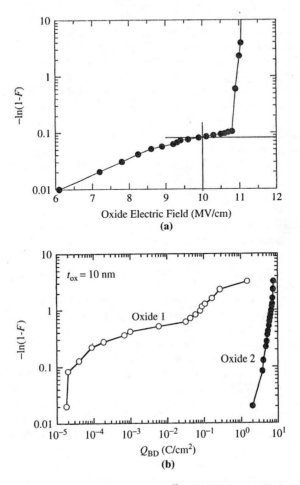

Fig. 12.15   (a) Weibull plot. Data adapted from ref. [49], (b) Weibull plot of charge-to-breakdown. Data courtesy of S. Hong, Motorola.

where x can be either charge or time. The characteristic life α is percentile 63.2, and β is the Weibull slope. Plotting $\ln(-\ln(1-F))$ versus $\ln(x)$ yields a straight line with slope β. If the sample area is increased by a factor N, the curve shifts vertically by $\ln(N)$.[52] If the desired low failure rate is F_{chip} over the product lifetime t_{life} for the total gate area A_{ox} on the chip, this is equivalent to a higher failure rate F_{test} in time t_{test} on the test structures with area A_{test}. This gives

$$\frac{t_{life}}{t_{test}} \approx \left(\frac{A_{test}}{A_{ox}} \frac{F_{chip}}{F_{test}} \right)^{1/\beta} \tag{12.30}$$

This equation is used to scale measured breakdown times to the expected product lifetime, or to estimate the chip failure rate from test-structure measurements. Since $F_{chip} < F_{test}$ and typically $A_{test} < A_{ox}$, then $t_{test} > t_{life}$, making it necessary to measure the test structure under accelerated voltage and temperature stress conditions. The Weibull parameter β is an important parameter for reliability projections.[37] It is a function of oxide thickness, decreasing for thinner oxides.

12.4.4   Negative Bias Temperature Instability (NBTI)

Negative bias temperature instability has been known since the very early days of MOS device development.[53] NBTI, occurring in *p*-channel MOS devices stressed with negative gate voltages at elevated temperatures,[54] manifests itself as absolute drain current and transconductance decrease, and absolute threshold voltage increase. Typical stress temperatures lie in the 100–250°C range with oxide electric fields typically below 6 MV/cm, *i.e.*, fields below those that lead to hot carrier degradation. Such fields and temperatures are typically encountered during *burn in*, but are also approached in high-performance ICs. Either negative gate voltages or elevated temperatures can produce NBTI, but a stronger and faster effect is produced by their combined action. It occurs primarily in *p*-channel MOSFETs with negative gate voltage bias and appears to be negligible for positive gate voltage and for either positive or negative gate voltages in *n*-channel MOSFETs.[55] In MOS circuits, it occurs most commonly during the "high" state of *p*-channel MOSFETs inverter operation. It also leads to timing shifts and potential circuit failure due to increased spreads in signal arrival in logic circuits.

NBTI degradation is believed to be caused by the creation of interface traps and fixed oxide charge in *p*-channel MOSFETs. A fraction of NBTI degradation can be recovered by annealing if the NBTI stress voltage is removed. The electric field applied during anneal can play a role in the recovery of NBTI degradation. Scaling of technology results in a significant increase in the susceptibility to NBTI degradation. Hence it may ultimately limit device lifetime, since NBTI is more severe than hot carrier stress for thin oxides at low electric fields. NBTI has also been reported for HfO_2 high-*k* insulators.[56]

Threshold voltage and transconductance degradation as a function of stress time is shown in Fig. 12.16.[57] Transconductance is related to mobility degradation during the stress. Although such plots vary from researcher to researcher, the general NBTI trends are embodied in this figure.

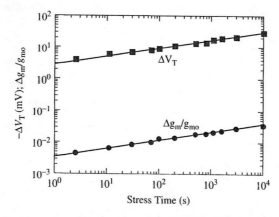

Fig. 12.16 Negative bias temperature instability effect on threshold voltage and transconductance changes. $V_G = -2.3$ V, $T = 100$C, $L = 0.1$ μm, $t_{ox} = 2.2$ nm. Reprinted after ref. 57 by permission of IEEE (© 2000, IEEE).

12.4.5 Stress Induced Leakage Current (SILC)

An effect frequently observed in thin electric field-stressed oxides is a gate oxide current increase, referred to as *stress-induced leakage current*. It is defined as the increase of oxide leakage current after high-field stress ($\mathscr{E}_{ox} \approx 10 - 12$ MV/cm) compared to before stress and first reported in 1982.[58] It is typically observed at low to moderate oxide electric fields ($\mathscr{E}_{ox} \approx 4 - 8$ MV/cm) and increases markedly as oxide thicknesses decrease. However, SILC decreases for oxides thinner than about 5 nm, believed to be due to reduced trap generation rates in thin oxides. Different models have been proposed to explain SILC: interface-state generation, bulk-oxide electron-trap generation, non-uniformities or weak spot formation in the oxide films, trapped holes injected from the anode. SILC can be best explained by the generation of neutral electron traps in the oxide, allowing more current to flow through the oxide layer by these traps acting as "stepping stones" for tunneling carriers, known as trap-assisted tunneling.[59] The generation of these neutral sites is caused mainly by the "trap creation" phenomenon related to hydrogen release by hot electrons. SILC degrades data retention of non-volatile memories that store charge on floating gates and it is usually not detected by time-zero or time-dependent breakdown measurements. Reliance on these latter characterization techniques may lead to overestimation of oxide integrity and reliability.

12.4.6 Electrostatic Discharge (ESD)

Electrostatic discharge is the transient discharge of static charge due to human handling or contact with equipment. A very good discussion is given by Amerasekera and Duvvury.[60] In a typical work environment a charge of about 0.6 µC on a body discharged through a 150 pF capacitor generates electrostatic potentials of around 4 kV. A contact by a charged human body with an IC pin can result in a discharge for about 100 ns with peak currents in the ampere range, leading to failure in electronic devices. Typically, the damage is thermally initiated in the form of device or interconnect burn-out, but the voltages can be sufficiently high to cause oxide breakdown in MOS devices. Even if a device is not destroyed, it can incur damaged that is difficult to detect resulting in latency effects known as *walking wounded*. The static voltages that can be generated are given in Table 12.1.

The three principal sources of electrostatic charging and discharging are (1) human handling, (2) automated test and (3) handling systems, and the IC is charged during transport or contact with a highly charged surface or material. The IC remains charged

TABLE 12.1 Static Voltages as a Function of Relative Humidity.

	20%	80%
Walking across vinyl floor	12 kV	0.2 kV
Walking across synthetic carpet	35	1.5
Arising from foam cushion	18	1.5
Picking up polyethylene bag	20	0.6
Sliding styrene box on carpet	18	1.5
Removing mylar tape from PC board	12	1.5
Shrinkable film on PC board	16	3
Triggering vacuum solder remover	8	1
Aerosol circuit freeze spray	15	5

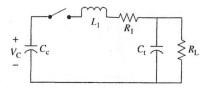

Fig. 12.17 Equivalent LCR circuit for modeling human body and machine model discharge waveforms.

until it contacts a grounded surface and is then discharged through its pins. The three ESD mechanism models are the *Human Body Model* (HBM), the *Machine Model* (MM), and the *Charged Device Model* (CDM). The HBM is the ESD testing standard and can be modeled using the LCR circuit in Figure 12.17. The discharge waveform of an HBM tester through a zero ohm load has rise and decay times of approximately 10 ns and 150 ns. The waveform is obtained by the discharge of a 100 pF capacitor with an initial voltage of 2 kV through a 1.5 k-ohm resistor. C_c is the discharge capacitor and the charging voltage is V_c. L_1 is the parasitic inductance which determines the rise time of the discharge pulse together with the resistor R_1. C_S is the parasitic stray capacitance of R_1 and the interconnect. C_t is the parasitic capacitance of the test board and R_L is the resistance of the load or device under test.

The MM discharge circuit can be defined by the LCR network in Figure 12.17. C_c is 200 pF, while $R_1 = 0$. In practice $R_1 > 0$ and during a discharge the dynamic impedance of the circuit can be much higher than zero. Hence, existing MM standards specify the output current waveform in terms of peak current and oscillating frequency for a given discharge voltage, automatically defining L_1 and R_1. The MM and the HBM tests are different forms of the same discharge mechanism, that of an externally charged object discharging through the IC. The failure modes are similar, although the severity of the damage varies between the two tests. In contrast, CDM type ESD events result in gate oxide breakdown inside the IC which are related to the internal discharge paths and voltage build-up in the chip. Hence, the CDM is a different type of ESD test and device sensitivity to the CDM cannot be inferred from results of HBM or MM tests. The increased usage of automated manufacturing and testing equipment has led to environments more suitable to CDM type ESD, rather than HBM ESD.

Gross ESD events usually leave visible craters observed in optical microscopes. Less visible defects are best detected using thermal detection techniques such as liquid crystal or fluorescence imaging.[61] For more detailed investigation, one needs focused ion beam cuts coupled with SEM or TEM. An ESD failure example is shown in Fig. 12.18.[62] The sample consists of tungsten TiN, $TiSi_2$ contacts to Si. After a 300 V pulse, the silicon in contact with the $TiSi_2$ melted with Ti dissolving in the molten silicon and diffusing through the filament and tungsten dissolving in the liquid silicon. To guard against ESD damage, sensitive devices on a chip are protected by providing relevant bonding pads with protection diodes, silicon controlled rectifiers or MOSFETs with gate connected to source directly or through a resistor.

12.5   FAILURE ANALYSIS CHARACTERIZATION TECHNIQUES

12.5.1   Quiescent Drain Current (I_{DDQ})

Quiescent drain current testing, more commonly known as I_{DDQ} testing, refers to integrated circuit testing based on measurement of the steady state current of a packaged chip

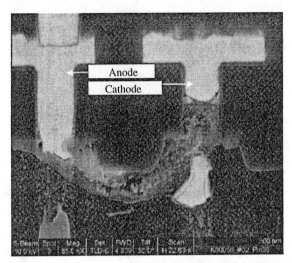

Fig. 12.18 SEM micrograph of the cross section after ESD failure. The surface has been decorated with xenon difluoride. Reprinted after ref. 62 by permission of IEEE (© 2003, IEEE).

in the quiescent mode. In steady state a CMOS circuit dissipates very low static current, typically below 1 μA. However, if the chip contains a defect such as gate-oxide short or short between metal lines, a conduction path from power supply to ground is formed and the current increases. This faulty I_{DDQ} is several orders of magnitude higher than the fault-free leakage current and monitoring this current distinguishes between faulty and fault-free circuits.[63] I_{DDQ} targets physical defects and for detailed circuit testing must be supplemented by functional testing.

The concept is illustrated in Fig. 12.19. A voltage ramp is supplied to the CMOS inverters with a gate oxide short. The current conduction path formed by this defect is highlighted. Defects typically detected by I_{DDQ} are: gate oxide shorts, metal line bridging, shorts from gate to drain or source or drain to source, *etc.* Opens are more difficult or impossible to detect. Example defects are shown in Fig. 12.20. The I_{DDQ} in each case indicated a problem and further measurements led to identification of the culprit. The speed of I_{DDQ} testing is constrained by the speed at which the measurement system can respond and by the speed at which the circuit settles after the input signal is applied. I_{DDQ} testing is typically 1–2 orders of magnitude slower than normal circuit operating speed, but requires only few measurements and has been combined with emission microscopy[64] and rear optical beam induced current[65] for failure location. I_{DDQ} measurements are also useful to determine drift mechanisms in MOSFETs. For example, sodium, potassium or hydrogen drift in the gate or field oxide can lead to drain current changes detected by I_{DDQ}.[66]

As ICs are scaled with reduced threshold voltages and oxide thicknesses, both MOSFET *off* current and oxide leakage current and increase, making I_{DDQ} interpretation more difficult. It is possible to reduce the *off* current by supplying substrate bias, lowering the supply voltage V_{DD} and/or the temperature. I_{DDQ} is very cost effective and uses root cause of problems (physical defects) to identify defective circuits. For IC manufacturers, this is an attractive, low cost supplemental test to functional testing. I_{DDQ} instrumentation is discussed by Wallquist. [67]

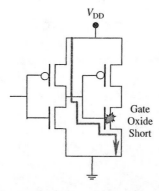

Fig. 12.19   Effect of gate oxide short on I_{DDQ}.

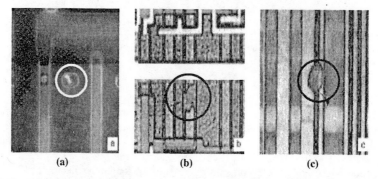

(a)   (b)   (c)

Fig. 12.20   Examples of circuit failures detected with I_{DDQ}. (a) Gate oxide short, $I_{DDQ} = 360$ μA, (b) poly Si-poly Si short, $I_{DDQ} = 5$ mA,(c) metal bridging defect, $I_{DDQ} = 5$ μA. Micrographs courtesy of IBM.

12.5.2 Mechanical Probes

Mechanical probes are used to contact parts of an IC during FA. This, of course, becomes more difficult as lines become narrower. Nevertheless, with care it is possible to contact lines on the order of a micron wide. With the availability of scanning probes, *e.g.*, conducting AFM probes that can be manipulated over sub-micron dimensions, sub-micron probing has become easier. Scanning capacitance and spreading resistance microscopy has recently been implemented in FA for ion implant monitoring and dielectric characterization.[68]

12.5.3 Emission Microscopy (EMMI)

Emission Microscopy is the emission of light in response to an electrical stimulus.[69] A familiar example is the emission of light from a forward-biased junction due to radiative recombination, as in light emitting diodes. Radiative recombination is also active in CMOS circuits under latch-up conditions when there is a high density of excess carriers in the device. An entirely different mechanism is active when carriers are accelerated to high energies in an electric field and subsequently lose their energy. Some of that energy is converted into light. This happens in reverse-biased diodes, *e.g.*, the drain of a MOSFET.

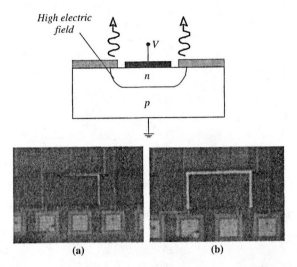

Fig. 12.21 Schematic illustrating hot carrier light emission at the high field region of an np junction, (a) weak breakdown, (b) strong breakdown condition shown by the lines. Courtesy of J.E. Park, Arizona State University.

It can also happen when carriers flow through an oxide and lose energy. An example of light emission is the reverse-biased np junction at moderate reverse bias in Fig. 12.21(a). Some light emission is observed around the junction periphery in the high electric field region. With higher reverse bias, light emission increases in (b).

Emission microscopy has become an important tool for failure location.[70] The failed chip is placed in the emission microscope and illuminated and one records an image of the chip to locate the various devices. Then the illumination is turned off, voltage is applied to the chip, and the emitted light is detected with a sensitive light amplifier, such as a charge-coupled device or photocathode image intensifier. An example of such an image is shown in Fig. 12.22. In this example, the circuit latched and EMMI located the latch-up spot.[71] Fig. 12.22(a) shows a front image with light emission at the circled location. However, it was felt that this was not the latch-up spot, rather the light emerged here because this was an open area not obscured by metal. The wafer was then observed from the back surface in (b) at a current of 30 ma with no latch-up. In (c) at 70 mA latch-up

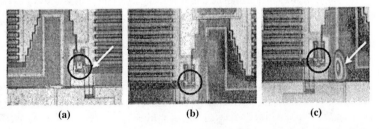

Fig. 12.22 Light emission to locate a latch-up spot. (a) Front surface view light emission in the circle, (b) back surface view with light emission (I = 30 mA), but no latch-up, (c) Latch-up light emission (arrow) at I = 70 mA. Reprinted after ref. 71 by permission of Semiconductor International.

occurred and the location, marked by the arrow, is clearly not the location of the original light spot, marked by the circle.

A common EMMI application is the detection of weak spots in gate oxides. The light can be imaged from the top or the bottom of the chip. Top imaging is obviously simpler, but can be complicated by metal layers obscuring the defective area or even light bouncing between substrate and metal layers, leaving the chip at a location different from the failure site. Both of these complications are ameliorated by back surface imaging. The back surface must, of course, be free of metallization layers and the light must traverse the sample thickness. Light near the band edge can propagate through wafers not too heavily doped. For epitaxial layers on heavily-doped substrates, the substrate must be thinned to about 50 μm by diamond milling, plasma etching or mechanical etching to be reasonably transparent.[72] Otherwise the optical signal is attenuated due to free carrier absorption. *Laser ablation* thinning uses high-intensity femtosecond pulses to ionize surface atoms and the resulting high density plasma ablates Si without thermal damage.[73]

The spectral content of the emitted light can be used to gain some insight into the failure mode.[74] A common method to characterize the hot electron behavior of MOSFETs is to measure the substrate current. Substrate current is due to impact ionization in the high electric field drain region, which is also the very region of light generation. It has been shown that light emission correlates well with substrate current and device degradation. [75]

Instead of measuring steady-state light emission, time varying light emission is detected in *picosecond imaging circuit analysis* (PICA).[76] During hot carrier light emission, there are generally few hot carriers, and the efficiency of their coupling to light is weak, making the light intensity emission very low. However, in most cases, the intensity of any background emission from the silicon devices is negligible, so that the experimental challenge for PICA is the detection of a small, background-free light pulse with the duration of the switching time of the device. For example, in CMOS circuits under static conditions only the sub-threshold leakage current flows with no detectable emission. Maximum optical emission occurs during switching. Fig. 12.23 shows the emission in a ring oscillator at various times,[77] clearly showing which device switches at what time. The time interval for image acquisition is 34 ps!

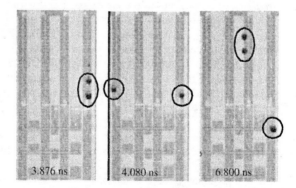

Fig. 12.23 Spatial and temporal response of light emission from a ring oscillator. The pulse nature of the emission is clearly seen, as is the ability to spatially resolve light pulses from next-nearest-neighbor gates. The emission is superimposed on an image of the circuit, and appears as the dark spots. Each *snapshot* in this shift register lasts 34 ps! Reprinted after ref. 77 by permission of IEEE (© 2000, IEEE).

12.5.4   Fluorescent Microthermography (FMT)

In *fluorescent microthermography*[78], a thin film of europium thenoyl-trifluoroacetonate dissolved in acetone is deposited onto the surface and illuminated with 340–380 nm ultraviolet light, stimulating fluorescence mainly at the bright 612 nm line.[79] Since no significant absorption—and thus no fluorescence—occurs above 500 nm, the excitation source and the fluorescence emission can be separated. The fluorescence quantum efficiency decreases exponentially with temperature, so the measurement of the fluorescence intensity gives the temperature of the device during operation with hot areas appearing darker than colder regions. The technique has a spatial resolution of 0.5 μm and a thermal resolution of about 5 mK. Specimen preparation is comparable to liquid crystal preparation. For quantitative temperature measurements, the film must be calibrated. Time-dependent measurements are, in principle, possible, since the fluorescence lifetime is about 200 μs.

12.5.5   Infrared Thermography (IRT)

Infrared thermography uses thermal radiation of a solid in thermal equilibrium with its surroundings. In contrast to a black body, real surfaces reflect a part of the incident radiation that depends on the wavelength. The total radiant emittance is not only a function of the temperature but also of the material-dependent emissivity. Hence, the temperature of real materials, known as grey bodies, cannot be determined by measuring the total radiant emittance alone.[80] To circumvent this problem, the emitting surface of the device is frequently blackened or a suitable calibration is made. The system collects emitted light by scanning the device surface at two different calibration temperatures to determine the emissivity. Most IR microscopes use this procedure. For qualitative information, a radiance image frequently suffices.

Typical temperature resolutions are around 1 K with InSb and HgCdTe infrared detectors. With silicon largely transparent in the near infrared region, IR thermography can also be used to measure from the rear of the IC when the front surface is masked by multiple metallization layers. However, highly-doped substrates with $N_A > 10^{18}$ cm^{-3} must be thinned to reduce IR absorption.

In *photothermal radiometry*, a version of IR thermography, thermal waves generated by a modulated laser lead to fluctuations in the IR emission. Thermal properties of the sample are determined from the phase measured over a larger region of the modulation frequency. The technique is material-specific and includes determination of the heat capacity/thermal conductivity of thin films and of heat transmission resistances, film thickness, detection of material inhomogeneities, and delaminations.[81] The temperature resolution is superior to conventional IR thermography and values of several 10 μK appear attainable.

12.5.6   Voltage Contrast

Voltage contrast is an electron beam technique using local electric field-induced secondary yield.[82] We illustrate it in Fig. 12.24 with three conductors. In Fig. 12.24(a) all three lines are at ground potential and a certain number of secondary electrons are collected by the detector upon e-beam excitation. Fewer electrons are collected at the detector from a line at a 5 V potential in Fig. 12.24(b) than from a line at ground potential. Similarly, a −5 V line gives a still higher signal. The reason, of course, is that electrons emitted from a line at a positive potential experience not only the attractive potential of the detector, but also the attractive potential of the emitting line, allowing line voltages to be determined. Using

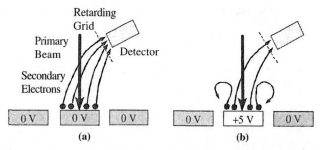

Fig. 12.24 Voltage contrast showing the effects of (a) ground potential and (b) positive potential on electron detection.

stroboscopic techniques, one can measure the transient behavior of an IC, *i.e.*, observe the circuit switch from one state to another.

Electron beams have a number of advantages over mechanical probes for IC failure analysis.[83] The beam is small, allowing narrow lines to be contacted, no capacitive loading of the circuit (important during transient analysis), high spatial resolution, and voltages can be measured to the millivolt range and switching voltage waveforms into the sub-nanosecond range. Voltage contrast measurements are illustrated in Fig. 12.25. The e-beam voltage x-y image in Fig. 12.25(a) shows the state of the various IC lines. Light corresponds to high voltage and dark to low voltage. If, for example, one of the lines had an open circuit, as might happen from electromigration, this would clearly show in such an image, but would be very difficult to detect by other means. The time-dependent behavior in Fig. 12.25(b) is obtained be setting the beam at a particular y location and scanning the beam in x and time. The transition of a line from *high* to *low* or *low* to *high* is clearly shown. In this mode, one observes whether various portions of the IC switch correctly. If, for example, the line resistance increases due to electromigration, the RC switching time may be affected and a voltage contrast measurement will display it. In a contact chain it has been used to detect high resistance contacts.

12.5.7 Laser Voltage Probe (LVP)

A method somewhat akin to voltage contrast is the *laser voltage probe* introduced in the early 1990s.[84] An infrared laser probes the electric field and the free carrier induced

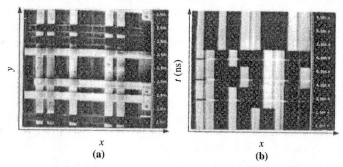

Fig. 12.25 Voltage contrast images in the (a) x-y and (b) x-time configuration. Photos courtesy of T.D. McConnell, Intel Corp.

absorption modulation in a reverse biased pn junction without requiring a vacuum. This absorption modulation is related directly to the voltage across the junction. The mode-locked laser is focused through the heavily-doped silicon onto the diffusion regions of the CMOS chip. The small modulations in laser power riding on the reflected optical beam are detected to measure the voltage across the junction. Stroboscopic measurements allow frequencies into the GHz by phase locking the mode-locked laser to the tester driving the chip.[85]

Laser voltage probe is based upon two principles. Heavily doped silicon is partially transparent to IR light with sub-band gap photon energies, and there are optoelectronic interactions in a semiconductor *pn* junction, when an optical beam is focused onto the junction: electroabsorption or Franz–Keldysh effect, electrorefraction, and free carrier absorption and refractive index changes. In the *Franz-Keldysh* effect, high electric fields (> 10^4 V/cm) reduce the band gap, allowing photons with energies near the band gap to be absorbed more in the presence of an electric field. For such high electric fields, there is also a refractive index change (electroabsorption) and there are free carrier effects as charge carriers are swept into and out of the space-charge region. The modulation in the free carrier charge density causes a modulation of both the optical absorption coefficient and of the local refractive index of the region. LVP has been used to acquire timing wave forms directly from CMOS circuits through the silicon rear side, allowing internal timing measurements from high frequency circuits packaged in flip chip packages with inaccessible front side interconnects.

12.5.8 Liquid Crystals (LC)

Liquid crystals, introduced in 1971 for FA,[86] detect small temperature changes. They were first used to map logic states of operating ICs.[87] Improved methods for cholesteric liquid crystals were used for hot spot detection in 1981.[88] Later nematic liquid crystals were used.[89] Liquid crystals exist in *isotropic, nematic, cholesteric, smectic* and *crystal* phases with transitions between the different phases induced by thermal variation. Liquid crystals are not quite liquid and not quite solid. They flow like liquids, but they have some properties of crystalline solids and can be considered to be crystals that have lost some or all of their positional order, while maintaining full orientational order.

For FA the chip is coated with a thin film of a liquid crystal and illuminated with polarized light from a white light source. The liquid is deposited on the sample with a syringe or eye dropper. It changes the polarization plane of the transmitted light above a certain temperature. The liquid crystal rotates the plane of polarization and a thin LC layer appears transparent when viewed through a polarizing microscope with perpendicular polarizers in the light source and optical path as shown in Fig. 12.26(a). If a portion of the IC is heated for the LC to change from the nematic to the isotropic stage, the polarized light is no longer rotated and appears as a dark spot (Fig. 12.26(b)). To visualize a defect location, the coated chip is heated close to the *transient* or *clearing* temperature of the liquid crystal. The additional local heating of the defect-induced current, changes the optical properties of the liquid crystal rotating the polarization plane resulting in a dark spot. Sometimes it is difficult to see the hot spot. Switching the chip current on and off creates a pulsating spot at the defect location. This technique allows defect localization down to the micrometer range. The development of highly sensitive cameras allows temperature changes of about 0.1°C or slightly lower with spatial resolution on the order of a μm to be detected. The temporal resolution is a few ms precluding measurements of high-speed circuits. An example LC image is shown in Fig. 12.27.

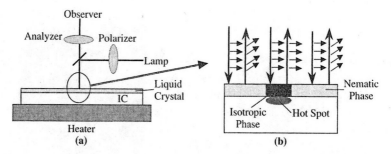

Fig. 12.26 (a) Schematic of polarizing microscope for liquid crystal measurements, (b) cool liquid crystal rotates polarized light, whereas hot liquid crystal does not and appears opaque when viewed through crossed polarizers.

Fig. 12.27 Dark spot liquid crystal image. Courtesy of D. Alavrez, Microchip.

12.5.9 Optical Beam Induced Resistance Change (OBIRCH)

The *optical beam induced resistance change* method has become an important FA technique. In OBIRCH a constant voltage or current is applied to the IC. A scanning laser irradiates the chip and some of its energy is converted into heat. The temperature coefficient of resistance (TCR) of metals is usually positive, so that a temperature increase leads to a line resistance increase and a current decrease or voltage increase, illustrated in Fig. 12.28(a). For *constant voltage*, the current change (ΔI) due to the laser heating is approximately proportional to the resistance change (ΔR), which is proportional to the temperature increase (ΔT). For *constant current*, the voltage change (ΔV) is proportional to the resistance change (ΔR).[90] When the laser beam scans, the heat is transmitted freely across defect-free areas, but heat transmission is impeded when the beam encounters defects, such as voids and Si nodules, creating differences in temperature increases

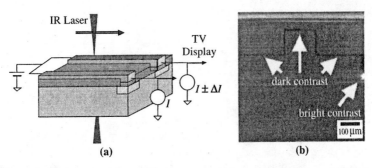

Fig. 12.28 OBIRCH (a) schematic showing the incident laser from either top or bottom and the current change, (b) uniform metal line in dark with a defect ion the bright contrast area. After Nikawa ref. [92].

between irradiated points that are near defects and those that are not. The resulting differences in ΔRs are converted to ΔIs or ΔVs and displayed on a cathode ray tube in the form of brightness changes. The chip under investigation is powered by simply applying the appropriate chip voltage to the bonding pad and then it is scanned while all lines are electrically active.

A 1.3-μm laser, with energy less than the Si band gap, does not generate electron-hole pairs, *i.e.*, it does not give an OBIC signal. OBIRCH images can also be observed by shining the 1.3-μm laser beam from the rear of a chip, because the 1.3-μm laser penetrates into the Si substrate with about 40% power loss for moderately doped Si about 500 μm thick. Materials with negative TCR include W (containing Ga), Ti (containing O) unintentionally left to be etched, and Ti-Al amorphous layers (containing O). The maximum temperature increase on the Al line is on the order of 10 K making the method non-destructive. To increase the resolution, OBIRCH has been combined with a near-field optical probe.[91] The OBIRCH image in Fig. 12.28(b) shows a leakage current path as the dark contrast and the defective part as the bright contrast.[92] FIB cross-sectioning the bright area for TEM observation revealed a short between Al lines. Energy dispersive X-ray analysis showed the existence of Ti and O in that region with a negative temperature coefficient leading to the bright contrast in the OBIRCH image.

A technique related to OBIRCH is *Thermally-Induced Voltage Alteration* (TIVA), in which a laser is scanned across the chip from the front or rear of the chip producing localized thermal gradients in the IC interconnects.[93] The effects of the thermal gradients on IC power consumption are detected by monitoring the voltage fluctuations of the IC power supply voltage when the IC is biased with a constant current power supply. TIVA images can localize shorts in a single, entire die field of view image. Shorted conductors cause increased IC power consumption that depends on the resistance of the short and its location in the circuit. As a laser is scanned over an IC with a short circuit, laser heating changes the resistance of the short when it is illuminated, changing the supply voltage. Open conductors are detected using the thermoelectric power or Seebeck effect to change the power demands of the IC. If a conductor is electrically isolated from a driving transistor or power bus, the Seebeck effect changes the conductor potential, altering the bias of transistors, whose gates are connected to the electrically open conductor, and the power dissipation. An image of the changing IC power demands displays the location of electrically floating conductors.

12.5.10 Focused Ion Beam (FIB)

Focused ion beam is not a characterization technique, but is used to prepare specimen for further analysis. It uses a finely focused probe of Ga^+ ions, extracted from a liquid droplet in the ion gun by an intense electric field, to etch selected regions of an IC.[94] The tip diameter of the liquid is about 100 nm, making it possible to form a final focused probe that is less than 10 nm in diameter at the sample's surface. The basic components in the FIB column are analogous to those used in SEMs: lenses, defining apertures, and scanning coils to raster the probe across the sample. In preparing a cross section by FIB, one can image the region of interest by collecting low-energy electrons liberated from the surface by the scanning Ga beam. The most common use of FIB in analytical work is to prepare cross sections for optical or electron microscopy. By moving the beam repeatedly along a single line or within a narrow raster pattern, the FIB cuts through metal and polysilicon interconnects as well as oxide and nitride layers, with little or no damage to adjacent structures. Typically, a high current broad beam gives an initial rough cut followed with a tighter focus low current probe for final polishing. Free standing films as shown in Fig. 12.29, illustrating the power of FIB, take about 20 minutes to prepare.

12.5.11 Noise

Noise is one characterization technique that is neither extensively discussed nor used as much as many of the other techniques in this book. We will briefly mention the main noise sources and how they can be used to characterize semiconductors. Noise increases during some device degradation. The recent review papers by Wong[95] and Claeys/Mercha/Simoen[96] give a good overview of the present state of noise theory and measurements. Earlier noise issues are covered in books by van der Ziel,[97] one of the early noise experts, Robinson,[98] and Motchenbacher and Fitchen.[99] At high frequencies, *thermal* noise and *shot* noise dominate up to frequencies beyond the gigahertz range. Both of these noises are fundamental in nature, forming an intrinsic lower noise limit. At low

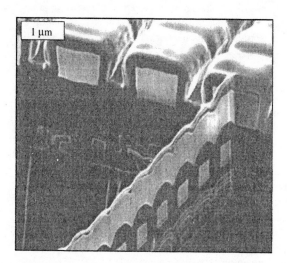

Fig. 12.29 TEM cross section prepared by FIB, showing a "rib" less than 100 nm thick left standing after FIB cut. Courtesy of H-L Tsai, Texas Instruments Inc.

frequencies, *flicker* or $1/f$ noise dominates with a $1/f^n$ frequency behavior with n close to unity. *Generation-recombination* (G-R) noise can also occur in this frequency range. It is characterized by a Lorentzian spectrum with a constant plateau at $f < f_c$ and a $1/f^2$ roll off beyond the characteristic frequency f_c. In contrast to the fundamental thermal and shot noise, 1/f and G-R noise depend on material and semiconductor processing and can be used for FA.

Thermal Noise: One of the earliest noise sources was predicted by Einstein in 1906, when he proposed that Brownian motion of charge carriers would lead to fluctuations in the potential across a resistor in thermal equilibrium.[100] This noise is known as *thermal*, *Johnson*, or *white noise*, first measured by Johnson[101] and its noise power calculated by Nyquist.[102] The noise voltage mean square value is

$$\overline{v_n^2} = \frac{4kTR\Delta f}{1 + (\omega\tau)^2} \approx 4kTR\Delta f \tag{12.31}$$

where Δf is the bandwidth of the measurement system and τ the carrier scattering time ($\sim$picoseconds). For most practical frequencies, the second term in the denominator can be neglected and the thermal noise power is then frequency independent. Thermal noise exists in almost all electronic systems and because of its fundamental nature it is frequently used to compare other noise types. Sometimes the noise is expressed as the noise power spectral density

$$S_v = 4kTR \ [\text{V}^2/\text{Hz}] \tag{12.32}$$

Thermal noise can be used for thermometry purposes, provided the resistance R is accurately known.[103] It merely requires a low-noise amplifier, spectrum analyzer, and a dedicated test structure with at least four body contacts.[96]

Shot Noise: *Shot noise* is the second fundamental noise source due to the discrete nature of charge transport. It is usually observed in devices containing barriers, *e.g.*, *pn* junctions, Schottky diodes, *etc.* Schottky gave the first explanation of this type of noise in relation to vacuum tubes.[104] Its noise current mean square value is given by

$$\overline{i_n^2} = 2qI_{dc}\Delta f \tag{12.33}$$

where I_{dc} is the dc current flowing through the device.

In his classic paper Schottky formulated this equation based on the fact that the vacuum diode plate current is not a continuum but rather a sequence of discrete increments of charge carried by each electron arriving at the plate at random times. The average rate of charge arrival constitutes the dc component of the plate current on which is superimposed a fluctuation component as each discrete charge arrives. He referred to this phenomenon as "Schrot Effekt" or "shot effect".

Generation-Recombination Noise: *Generation-recombination noise* is due to generation and recombination of electrons and holes. The dependence of this type of noise on frequency is determined by the lifetime τ of these charge carriers. The current noise spectrum density is given by

$$S_i = \frac{KI^2\tau}{1 + (\omega\tau)^2} \tag{12.34}$$

where K is a constant determined by the trap concentration, I the device current, and τ the trap time constant determined by emission and capture of carriers by the trap. Generation-recombination noise is quite well defined by well-established models and theories.

Low Frequency or Flicker Noise: *Low frequency* or *flicker noise*, first observed in vacuum tubes over eighty years ago,[105] dominates the noise spectrum at low frequencies. It gets its name from the anomalous "flicker" that was seen in the plate current. Flicker noise is also commonly called 1/f noise, because the noise spectrum varies as $1/f^n$, where the exponent n is very close to unity. Fluctuations with a $1/f$ power law have been observed in practically all electronic materials and devices, including homogenous semiconductors, junction devices, metal films, liquid metals, electrolytic solutions, super-conducting Josephson junctions, and even in mechanical, biological, geological, and even musical systems. Two competing models have been proposed to explain flicker noise: the McWhorter number fluctuation theory[106] and the Hooge mobility fluctuation theory[107] with experimental evidence to support both theories. Christensson et al. were the first to apply the McWhorter theory to MOSFETs, using the assumption that the necessary time constants are caused by the tunneling of carriers from the channel into traps located within the oxide.[108] *Popcorn noise*, sometimes called burst noise or random-telegraph-signal (RTS) noise, is a discrete modulation of the channel current caused by the capture and emission of a channel carrier. [109]

The MOSFET current is proportional to the product of mobility μ times the charge carrier density or number N. Low frequency fluctuations in charge transport are caused by stochastic changes in either of these parameters, which can be independent (uncorrelated) or dependent (correlated). In most cases, fluctuations in the current, or more specifically in the product of $\mu \times N$ are monitored, which does not allow the separation of mobility from number effects and therefore obscures the identification of the dominant $1/f$ noise source.

The voltage noise spectrum density is[110]

$$S_V(f) = \frac{q^2 kT\lambda}{\alpha WLC_{ox}^2 f}(1 + \sigma\mu_{eff}N_s)^2 N_{ot} \qquad (12.35)$$

where λ is the tunneling parameter, μ_{eff} the effective carrier mobility, σ the Coulombic scattering parameter, N_s the density of channel carriers, C_{ox} the gate oxide capacitance/unit area, WL the gate area, and N_{ot} the oxide trap density (cm^{-3} eV^{-1}) near the interface. In weak inversion, the channel carrier density N_s is very low (10^7-10^{11} cm^{-2}), so that the mobility fluctuation contribution becomes negligible and the second term in the bracket can be neglected.

It is assumed that the free carriers tunnel to traps in the oxide with a tunneling time constant, which varies with distance x from the interface. The tunneling parameter and time constant are[111]

$$\lambda = \frac{\hbar}{\sqrt{8m_t\Phi_B}}, \tau_T = \frac{\exp(x/\lambda)}{\sigma_p v_{th} p_{os}} \qquad (12.36)$$

where m_t is the oxide tunnel effective electron mass, ϕ_B the oxide-semiconductor barrier height, σ_p the hole capture cross-section, and p_{os} the surface hole density near the source. τ_T represents the time for carriers to tunnel into traps in the oxide. λ is typically around 5×10^{-9} cm and it is obvious that the tunnel times become very long for traps any appreciable distance into the oxide from the semiconductor-oxide interface. For example, for $\sigma_p = 10^{-15}$ cm^2, $v_{th} = 10^7$ cm/s, $p_{os} = 10^{17}$ cm^{-3} and for a trap at 1 nm from the

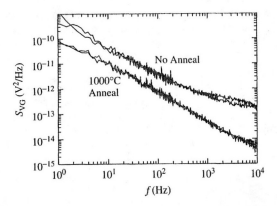

Fig. 12.30 Low-frequency noise spectra before and after annealing. $W/L = 10\ \mu\text{m}/0.8\ \mu\text{m}$, $t_{ox} = 3.3$ nm, $V_G - V_T = -1.05$ V, $V_D = -0.005$ V. Reprinted after ref. 112 by permission of IEEE (© 2004, IEEE).

semiconductor interface, $\tau_T \approx 0.5$ s or $f = 1/2\pi\tau_T \approx 0.3$ Hz. Hence, a distribution of traps in the oxide gives rise to a wide range of frequencies and can explain the $1/f$ dependence. $1/f$ noise shows sensitivity to the wafer orientation, which correlates with interface trap density. Figure 12.30 shows an example of low frequency noise before and after annealing where the anneal reduced the interface trap density and the lf noise. [112]

Noise spectroscopy has been applied to the study of deep levels in MOSFETs.[113] The main advantage of using a noise-based technique is that it can be applied even to very small area devices, which is not possible for standard capacitance-based DLTS. Low-frequency noise has become a FA characterization technique.[114]

12.6 STRENGTHS AND WEAKNESSES

Electromigration: None of the electromigration measurements are truly representative of the operating conditions of integrated circuits. The strength of conventional measurements using somewhat elevated stress conditions is that this method has been used for many years and is well accepted. Using established theory, one extrapolates the failure data to normal operating data. The weakness of this technique is the time consuming nature of the measurements and the uncertainty whether the mechanisms causing failure under elevated current and temperature are also active under normal current/voltage and temperature. Short-time measurements using test structures such as the SWEAT structure have the chief advantage of brief test times, thus lending themselves to production evaluation. Their disadvantage is the possibly different failure mode compared to conventional IC operation.

Hot Carriers: Hot carriers lead to avalanche multiplication in high electric field regions and to interface trap generation. The avalanche generated current is measured as a substrate current for n-MOSFETs and usually as a gate current for p-MOSFETs. The interface trap density can be measured directly by charge pumping, for example, or indirectly by threshold voltage, transconductance, or drain current changes. The aim is frequently to use the simplest technique that yields reliable results and that means substrate or gate currents. The weakness of this is that it is an indirect measure.

Oxide Integrity: Oxides are commonly characterized by their charge-to-breakdown behavior or time-to-breakdown and oxide integrity is commonly measured by the constant or ramped gate voltage or gate current techniques. Q_{BD}, usually determined with constant current stress, is more representative of the device physics; t_{BD} is usually determined with constant voltage stress and used for most gate oxides. The constant gate current has the advantage of simply yielding the charge-to-breakdown as a simple product $Q_{BD} = J_G t_{BD}$. However, it has the disadvantage that the current may not be uniform and most devices are not operated under constant current, but under constant gate voltage conditions. Thin oxides frequently do not exhibit well defined breakdowns partly because the gate leakage current prior to breakdown is quite high.

NBTI   is most commonly characterized by threshold voltage, transconductance, interface trap density, and drain current measurements.

ESD   is not an FA technique. Several of the techniques in this chapter are used to characterize ESD. To reduce ESD, the devices in a circuit are protected by some kind of current shunting device.

I_{DDQ}:   **Pro:** Simple to implement as only the input current to an IC needs to be monitored; very good at detecting shorts. **Con:** Does not localized the fault; difficult to determine opens.

Emission Microscopy: **Pro:** Entire die may be viewed at one time; no deprocessing except for lid removal; functional failure does not need to propagate to output. In the form of PICA it can be used to follow the switching function of an IC and allows circuit FA. **Con:** IC must be biased and toggled; ohmic defects do not emit light; no light detection through opaque layers; emission site may not be defect site. For imaging from the back of the chip: sample preparation; substrate thinning may impact device characteristics; Si is an infrared filter and limits detection bandwidth of emission sites; doping atoms serve to scatter IR photons leading to reduced sensitivity; CCD based systems have a low quantum efficiency in the required IR spectrum.

Voltage Contrast: **Pro:** Contactless method to determine the spatial and temporal voltage within an IC. The electron beam is small and can contact most lines in an IC. **Con:** Difficult when the line of interest is buried below other metallization levels.

Liquid Crystal: **Pro:** Low cost, easy to use, very good thermal and spatial resolution, useful for thermal and voltage contrast analysis, real time imaging. **Con:** Tends to "wick up" around probes and bonding wires making identification of hot spots difficult; poor thermal resolution for measurements from the back of the wafer; the number of layers between the source of the failure and the surface where the liquid crystal resides limits spatial resolution and sensitivity; liquid crystal has a set transition temperature. Multiple hot spots can be difficult to resolve if the warmer spot creates a significant temperature gradient.

Fluorescent Microthermography: **Pro:** Offers high thermal and spatial resolution. **Con:** The film must be calibrated for quantitative temperature measurements.

Infrared Thermography: **Pro:** Is a passive technique not requiring thermal excitation with good temperature resolution allowing imaging from front and rear surfaces. **Con:**

Calibration is necessary for quantitative information but is not easy since the emissivity is generally not known.

OBIRCH: **Pro:** A sensitive technique for a variety of FA investigation with high resolution. When OBIRCH does not work well, frequently EMMI does. The two are complementary. **Con:** Cannot be used for multiple metal layer chips; when used from the rear surface, wafer must be thinned to 150–200 μm.

Noise: **Pro:** Some noise measurements, *e.g.*, low-frequency and generation-recombination noise, are sensitive to surface states, interface traps, and bulk traps and noise is very sensitive. Noise measurements are not only a diagnostic tool, but give information about the performance of the device in a circuit. The measurement is made on an actual device not a test structure. **Con:** Requires specialized equipment that is not as routinely available as current-voltage equipment and the measurement is more difficult to make.

APPENDIX 12.1

Gate Currents

Consider the band diagrams in Fig. 12.10. Electrons tunnel through a triangular potential barrier in FN tunneling with $V_{ox} > q\phi_B$. In direct tunneling, the electrons tunnel through the entire oxide thickness. The transition voltage between FN and direct tunneling is $V_{ox} = q\phi_B$, which is about 3.2 eV for the SiO_2/Si system.

For n^+ poly-Si/p-substrate, the gate voltage is

$$V_G = V_{FB} + \phi_{s,G} + \phi_{s,sub} + V_{ox}; \quad V_{FB} = \phi_{MS} - \frac{Q_{ox}}{C_{ox}}; \quad \phi_{MS} = -\frac{E_G}{2q} - \phi_{F,sub} \quad (A12.1)$$

where ϕ_{MS} is the metal-semiconductor work function difference. For oxide charge densities on the order of $Q_{ox}/q \leq 10^{11}$ cm^{-2} and oxide thicknesses $t_{ox} \leq 10$ nm, the Q_{ox}/C_{ox} term is negligible. The surface potentials depend on the substrate and gate types and doping densities (*p*- or *n*-type) as well as the gate voltage polarity. To determine oxide tunnel currents, we need the oxide electric field. The electric field in the poly-Si gate is

$$\mathscr{E}_{s,G} = \frac{Q_G}{K_s \varepsilon_o} = \frac{q N_G W_G}{K_s \varepsilon_o} = \sqrt{\frac{2q N_G \phi_{s,G}}{K_s \varepsilon_o}} \quad (A12.2)$$

where N_G is the gate doping density and the W_G the gate scr width. We use a simple approach to get the main ideas across, neglecting quantization effects, for example. With

$$\mathscr{E}_{ox} = \frac{K_s}{K_{ox}} \mathscr{E}_{s,G} \text{ and } V_{ox} = \mathscr{E}_{ox} t_{ox} \quad (A12.3)$$

$\mathscr{E}_{ox}$ can be written

$$\mathscr{E}_{ox} = \frac{q K_s \varepsilon_o N_G}{(K_{ox}\varepsilon_o)^2} \left(\sqrt{t_{ox}^2 + \frac{2(K_{ox}\varepsilon_o)^2}{q K_s \varepsilon_o N_G}(V_G - V_{FB} - \phi_{s,sub})} - t_{ox} \right) \quad (A12.4)$$

where $\phi_{s,sub} \approx 2\phi_F$.

For the structure of Fig. 12.10 with $+V_G$ both p-substrate and n^+ gate are depleted/inverted. For electron tunneling from the substrate to occur, with oxide electric fields in the $5 \times 10^6 - 2 \times 10^7$ V/cm range, the substrate is strongly and the gate weakly inverted, giving

$$\phi_s \approx 2\phi_{F,sub} + 2\phi_{F,gate} \approx 2\phi_{F,sub} + \frac{E_G}{2q} \rightarrow V_G(inv) = V_{ox} - \frac{E_G}{2q} + \phi_{F,sub} + \frac{E_G}{2q}$$

$$\approx V_{ox} + \phi_{F,sub} \tag{A12.5}$$

For $-V_G$ both gate and substrate are accumulated and

$$\phi_s \approx -\phi_{s,sub} - \phi_{s,gate} \rightarrow V_G(acc) = -V_{ox} - \frac{E_G}{2q} - \phi_{F,sub} - \phi_{s,sub} - \phi_{s,gate}$$

$$\approx -V_{ox} - \frac{E_G}{q} \tag{A12.6}$$

The FN current density is[38]

$$J_{FN} = A\mathscr{E}_{ox}^2 \exp\left(-\frac{B}{\mathscr{E}_{ox}}\right) \tag{A12.7}$$

where A and B are given by

$$A = \frac{q^3}{8\pi h \Phi_B}\left(\frac{m}{m_{ox}}\right) = 1.54 \times 10^{-6}\left(\frac{m}{m_{ox}}\right)\frac{1}{\Phi_B} \ [A/V^2]$$

$$B = \frac{8\pi\sqrt{2m_{ox}\Phi_B^3}}{3qh} = 6.83 \times 10^7 \sqrt{\frac{m_{ox}\Phi_B^3}{m}} \ [V/cm] \tag{A12.8}$$

with m_{ox} is the effective electron mass in the oxide, m the free electron mass, and Φ_B (eV) the barrier height at the Si-oxide interface. Φ_B is an effective barrier height that takes into account barrier height lowering and quantization of electrons at the semiconductor surface and is not strictly constant.

The FN equation is derived under the assumptions: the electrons in the emitting electrode can be described by a free Fermi gas, the electrons in the oxide have a single effective mass m_{ox}, and the tunneling probability is derived by taking into account the component of the electron momentum normal to the interface only.

Rearranging Eq. (A12.8) gives

$$\ln\left(\frac{I_{FN}}{A_G \mathscr{E}_{ox}^2}\right) = \ln\left(\frac{J_{FN}}{\mathscr{E}_{ox}^2}\right) = \ln(A) - \frac{B}{\mathscr{E}_{ox}} \tag{A12.9}$$

A plot of $\ln(J_{FN}/\mathscr{E}_{ox}^2)$ versus $1/\mathscr{E}_{ox}$, known as a *Fowler-Nordheim* plot is linear if the oxide conduction is pure Fowler-Nordheim conduction. The intercept of this linear FN plot gives A and the slope yields B.

Tunneling currents through thin oxides contain a small oscillatory component due to quantum interference of electrons. They show a strong dependence on oxide thickness, suggesting that these oscillations can be used for a precise measurement of the oxide thickness. [115]

Direct tunneling is the flow of electrons through the entire oxide thickness illustrated in Fig. 12.10. Its current expression is more difficult to derive and several versions have been published.[39] We give the empirical expression[116]

$$J_{dir} = \frac{A V_G}{t_{ox}^2} \frac{kT}{q} C \exp\left(-\frac{B(1 - (1 - q V_{ox}/\Phi_B)^{1.5})}{\mathscr{E}_{ox}}\right) \qquad \text{(A12.10)}$$

because it is relatively simple and used in the BSIM model. In Eq. (A12.10)

$$C = N \exp\left(\frac{20}{\Phi_B}\left(1 - \frac{V_{ox}}{\Phi_B}\right)^\alpha \left(1 - \frac{V_{ox}}{\Phi_B}\right)\right) \qquad \text{(A12.11)}$$

where $\alpha = 0.6$ for the SiO_2/Si system.

$$N = \left(n_{inv} \ln\left(1 + \exp\left(\frac{V_{G,eff} - V_T}{n_{inv} kT}\right)\right)\right) + \ln\left(1 + \exp\left(\frac{V_G - V_{FB}}{kT}\right)\right) \qquad \text{(A12.12)}$$

represents the inversion or accumulation layer carrier density. $n_{inv} = qS/kT$ (typically 1.2–1.5), with S the sub-threshold swing ($n_{inv} > 0$ for n-MOSFETs and $n_{inv} < 0$ for p-MOSFETs).

$$V_{G,eff} = V_{FB} + 2\phi_F + \frac{\gamma_{Gate}^2}{2}\left(\sqrt{1 + \frac{4(V_G - V_{FB} - 2\phi_F)}{\gamma_{Gate}^2}} - 1\right);$$

$$\gamma_{Gate} = \frac{\sqrt{2q K_s \varepsilon_o N_{Gate}}}{C_{ox}} \qquad \text{(A12.13)}$$

is the effective gate potential, accounting for poly-Si gate depletion.

$$V_{ox} = V_{G,eff} - \left(\frac{\gamma}{2}\left(\sqrt{1 + \frac{4(V_G - V_{G,eff} - V_{FB})}{\gamma^2}} - 1\right)\right)^2 - V_{FB}; \quad \gamma = \frac{\sqrt{2q K_s \varepsilon_o N_A}}{C_{ox}}$$
$$\text{(A12.14)}$$

Φ_B in these equations is the barrier height for either electrons or holes, depending whether the electron or hole tunnel current is calculated.

REFERENCES

1. M. Ohring, *Reliability and Failure of Electronic Materials and Devices*, Academic Press, San Diego, 1998.

2. G. Di Giacomo, *Reliability of Electronic Packages and Semiconductor Devices*, McGraw-Hill, New York, 1997.

3. L.C. Wagner, *Failure Analysis of Integrated Circuits: Tools and Techniques*, Kluwer, Boston, 1999.

4. A. Martin and R-P Vollertsen, "An Introduction to Fast Wafer Level Reliability Monitoring for Integrated Circuit Mass Production," *Microelectron. Reliab.* **44**, 1209–1231, Aug. 2004.

5. W. Muth, A. Martin, J. von Hagen, D. Smeets, and J. Fazekas, "Polysilicon Resistive Heated Scribe Lane Test Structure for Productive Wafer Level Reliability Monitoring of NBTI," *IEEE Int. Conf. Microelectron. Test Struct.*, 155–160, 2003.

6. F.R. Nash, *Estimating Device Reliability: Assessment of Credibility*, Kluwer, Boston, 1993.

7. D.R. Wolters and J.J. van der Schoot, "Dielectric Breakdown in MOS Devices," *Phil. Res. Rep.* **40**, 115–192, 1985.

8. W. Weibull, "A Statistical Distribution Function of Wide Applicability," *J. Appl. Mech.* **18**, 293–297, Sept. 1951.

9. W.J. Bertram, "Yield and Reliability," in *VLSI Technology* 2nd ed. (S.M. Sze, ed.), McGraw-Hill, New York, 1988.

10. S. Kilgore, Freescale Semiconductor, private communication.

11. T. Kwok and P.S. Ho, "Electromigration in Metallic Thin Films," in *Diffusion Phenomena in Thin Films and Microelectronic Materials* (D. Gupta and P.S. Ho, eds.). Noyes Publ., Park Ridge, NJ, 1988.

12. H.B. Huntington and A.R. Grone. "Current-induced Marker Motion in Gold Wires," *J. Phys. Chem. Sol.* **20**, 76–87, Jan. 1961.

13. A. Scorzoni, B. Neri, C. Caprile, and F. Fantini, "Electromigration in Thin-Film Interconnection Lines: Models, Methods and Results," *Mat. Sci. Rep.* **7**, 143–220, Dec. 1991.

14. F.M. d'Heurle and I. Ames, "Electromigration in Single-Crystal Aluminum Films," *Appl. Phys. Lett.* **16**, 80–81, Jan. 1970.

15. J. Tao, N.W. Cheung, and C. Hu, "Metal Electromigration Damage Healing Under Bidirectional Current Stress," *IEEE Electron Dev. Lett.* **14**, 554–556, Dec. 1993.

16. I. Ames, F.M. d'Heurle, and R.E. Horstmann, "Reduction of Electromigration in Al Films by Cu Doping," *IBM J. Res. Dev.* **14**, 461–463, July 1970.

17. S.P. Murarka and S.W. Hymes, "Copper Metallization for ULSI and Beyond," *Crit. Rev. Solid State Mat. Sci.* **20**, 87–124, Jan. 1995.

18. I. Blech, "Electromigration in Thin Aluminum Films on Titanium Nitride", *J. Appl. Phys.* **47**, 1203–1208, April 1976.

19. J.R. Black, "Electromigration Failure Modes in Aluminum Metallization for Semiconductor Devices," *Proc. IEEE* **57**, 1587–1594, Sept. 1969.

20. ASTM Standard F1259-89, "Standard Guide for Design of Flat, Straight-Line Test Structure for Detecting Metallization Open-Circuit or Resistance-Increase Failure due to Electromigration," *1996 Annual Book of ASTM Standards*, Am. Soc. Test. Mat., West Conshohocken, PA, 1996.

21. H.A. Schafft and J.S. Suehle, "The Measurement, Use and Interpretation of the Temperature Coefficient of Resistance of Metallizations," *Solid-State Electron.* **35**, 403–410, March 1992; H.A. Schafft, T.C. Staton, J. Mandel, and J.D. Shott, "Reproducibility of Electromigration Measurements," *IEEE Trans Electron Dev.* **ED-34**, 673–681, March 1987.

22. ASTM Standard F1260-89, "Standard Test Method for Estimating Electromigration Median Time-to-Failure and Sigma of Integrated Circuit Metallizations," *1996 Annual Book of ASTM Standards*, Am. Soc. Test. Mat., West Conshohocken, PA, 1996.

23. C.V. Thompson and J. Cho, "A New Electromigration Testing Technique for Rapid Statistical Evaluation of Interconnect Technology," *IEEE Electron Dev. Lett.* **EDL-7**, 667–668, Dec. 1986.

24. C-U Kim, N.L. Michael, Q.-T. Jiang, and R. Augur, "Efficient Electromigration Testing with a Single Current Source," *Rev. Sci. Instrum.* **72**, 3962–3967, Oct. 2001.

25. B. Neri, A. Diligenti, and P.E. Bagnoli, "Electromigration and Low-Frequency Resistance Fluctuations in Aluminum Thin-Film Interconnections," *IEEE Trans Electron Dev.* **ED-34**, 2317–2322, Nov. 1987.

26. B.J. Root and T. Turner, "Wafer Level Electromigration Tests for Production Monitoring," *IEEE Int. Reliab. Phys. Symp.*, IEEE, New York, 1985, 100–107.

27. A. Zitzelsberger, A. Pietsch, and J. von Hagen, "Electromigration Testing on Via Line Structures with a SWEAT Method in Comparison to Standard Package Level Tests," *IEEE Int. Integr. Reliab. Workshop Final Rep.* 57–60, 2000.

28. J. von Hagen, R. Bauer, S. Penka, A. Pietsch, W. Walter, and A. Zitzelsberger, "Extrapolation of Highly Accelerated Electromigration Tests on Copper to Operation Conditions," *IEEE Int. Integr. Reliab. Workshop Final Rep.*, 41–44, 2002.

29. A.S. Oates, "Electromigration Failure of Al Alloy Integrated Circuit Metallizations," in *Diagnostic Techniques for Semiconductor Materials and Devices* (D.K. Schroder, J.L. Benton, and P. Rai-Choudhury, eds.), Electrochem. Soc., Pennington, NJ, 1994, 178–192.

30. K.N. Tu, "Recent Advances on Electromigration in Very-large-scale-integration of Interconnects," *J. Appl. Phys.* **94**, 5451–5473, Nov. 2003.

31. E. Takeda, C.Y. Yang, and A. Miura-Hamada, *Hot Carrier Effects in MOS Devices*, Academic Press, San Diego, 1995; A. Acovic, G. La Rosa, and Y.C. Sun, "A Review of Hot- Carrier Degradation Mechanisms in MOSFETs," *Microelectron. Reliab.* **36**, 845–869, July/Aug. 1996.

32. W.H. Chang, B. Davari, M.R. Wordeman, Y. Taur, C.C.H. Hsu, and M.D. Rodriguez, "A High-Performance 0.25 μm CMOS Technology: I - Design and Characterization," *IEEE Trans. Electron Dev.* **39**, 959–966, Apr. 1992.

33. J.T. Yue, "Reliability," in *ULSI Technology* (C.Y. Chang and S.M. Sze, eds.), McGraw-Hill, New York, 1996, Ch. 12.

34. E. Li, E. Rosenbaum, J. Tao, and P. Fang, "Projecting Lifetime of Deep Submicron MOS-FETs," *IEEE Trans. Electron Dev.* **48**, 671–678, April 2001; K. Cheng and J.W. Lyding, "An Analytical Model to Project MOS Transistor Lifetime Improvement by Deuterium Passivation of Interface Traps," *IEEE Electron Dev. Lett.* **24**, 655–657, Oct. 2003.

35. H.C. Shin and C.M. Hu, "Dependence of Plasma-Induced Oxide Charging Current on Al Antenna Geometry," *IEEE Electron Dev. Lett.* **13**, 600–602, Dec. 1992; K. Eriguchi, Y. Uraoka, H. Nakagawa, T. Tamaki, M. Kubota, and N. Nomura "Quantitative Evaluation of Gate Oxide Damage During Plasma Processing Using Antenna Structure Capacitors," *Japan. J. Appl. Phys.* **33**, 83–87, Jan. 1994.

36. J. Shideler, S. Reno, R. Bammi, C. Messick, A. Cowley, and W. Lukas., "A New Technique for Solving Wafer Charging Problems," *Semicond. Internat.* **18**, 153–158, July 1995; W. Lukaszek, "Understanding and Controlling Wafer Charging Damage," *Solid State Technol.*, **41**, 101–112, June 1998.

37. E.Y. Wu, J. Suné, W. Lai, A. Vayshenker, E. Nowak, and D. Harmon, "Critical Reliability Challenges in Scaling SiO_2-based Dielectric to Its Limit," *Microelectron. Reliab.* **43**, 1175–1184, Sept./Nov. 2003.

38. R.H. Fowler and L.W. Nordheim, "Electron Emission in Intense Electric Fields," *Proc. Royal Soc. Lond. A*, **119**, 173–181, 1928; M. Lenzlinger and E.H. Snow, "Fowler-Nordheim Tunneling into Thermally Grown SiO_2," *J. Appl. Phys.* **40**, 278–283, Jan. 1969; Z. Weinberg, "On Tunneling in Metal-Oxide-Structures," *J. Appl. Phys.* **53**, 5052–5056, July 1982.

39. See, e.g., M. Depas, B. Vermeire, P.W. Mertens, R.L. Van Meirhaeghe, and M.M. Heyns, "Determination of Tunneling Parameters in Ultra-Thin Oxide Layer Poly-Si/SiO$_2$/Si Structures," *Solid-State Electron.* **38**, 1465–1471, Aug. 1995; N. Matsuo, Y. Takami, and Y. Kitagawa, "Modeling of Direct Tunneling for Thin SiO_2 Film on n-Type Si(100) by WKB Method Considering the Quantum Effect in the Accumulation Layer," *Solid-State Electron.* **46**, 577–579, April 2002; N. Matsuo, Y. Takami, and H. Kihara, "Analysis of Direct Tunneling for Thin SiO_2 Film by Wentzel, Kramers, Brillouin Method—Considering Tail of Distribution Function," *Solid-State Electron.* **47**, 161–163, Jan. 2003; B. Govoreanu, P. Blomme, K. Henson, J. Van Houdt, and K. De Meyer, "An Effective Model for Analysing Tunneling Gate Leakage Currents Through Ultrathin Oxides and High-k Gate Stacks from Si Inversion Layers," *Solid-State Electron.* **48**, 617–625, April 2004.

40. Y-C Yeo, T-J King and C.M. Hu, "MOSFET Gate Leakage Modeling and Selection Guide for Alternative Gate Dielectrics Based on Leakage Considerations," *IEEE Trans Electron Dev.* **50**, 1027–1035, April 2003.

41. N.S. Saks, P.L. Heremans, L. van den Hove, H.E. Maes, R.F. De Keersmaecker, and G.J. Declerck, "Observation of Hot-Hole Injection in NMOS Transistors Using a Floating-Gate Technique," *IEEE Trans. Electron Dev.* **ED-33**, 1529–1534, Oct. 1986; B. Fishbein, D. Krakauer, and B. Doyle, "Measurement of Very Low Tunneling Current Density in SiO_2 Using the Floating-Gate Technique," *IEEE Electron Dev. Lett.* **12**, 713–715, Dec. 1991; B. De Salvo, G. Ghibaudo, G. Pananakakis, and B. Guillaumot, "Investigation of Low Field and High Temperature SiO_2 and ONO Leakage Currents Using the Floating Gate Technique," *J. Non-Cryst. Sol.* **245**, 104–109, April 1999.

42. A. Berman, "Time-Zero Dielectric Reliability Test by a Ramp Method," *IEEE Int. Reliab. Phys. Symp.*, IEEE, New York, 1981, 204–209.

43. K. Yoneda, K. Okuma, K. Hagiwara, and Y. Todokoro, "The Reliability Evaluation of Thin Silicon Dioxide Using the Stepped Current TDDB Technique," *J. Electrochem. Soc.* **142**, 596–600, Feb. 1995.

44. P.A. Heimann, "An Operational Definition of Breakdown of Thin Thermal Oxides of Silicon," *IEEE Trans. Electron Dev.* **ED-30**, 1366–1368, Oct. 1983; E.A. Sprangle, J.M. Andrews, and M.C. Peckerar, "Dielectric Breakdown Strength of SiO_2 Using a Stepped-Field Method," *J. Electrochem. Soc.* **139**, 2617–1620, Sept. 1992.

45. R. Degraeve, G. Groeseneken, R. Bellens, M. Depas, and H.E. Maes, "A Consistent Model for the Thickness Dependence of Intrinsic Breakdown in Ultra-Thin Oxides," *IEEE IEDM Tech. Digest*, 863–866, 1995.

46. C.M. Hu, "AC Effects in IC Reliability," *Microelectron. Reliab.* **36**, 1611–1617, Nov./Dec. 1996.

47. W. Kern, R.B. Comizzoli, and G.L. Schnable, "Fluorescent Tracers—Powerful Tools for Studying Corrosion Phenomena and Defects in Dielectrics," *RCA Rev.* **43**, 310–338, June 1982.

48. M. Itsumi, H. Akiya, M. Tomita, T. Ueki, and M. Yamawaki, "Observation of Defects in Thermal Oxides of Polysilicon by Transmission Electron Microscopy Using Copper Decoration," *J. Electrochem. Soc.* **144**, 600–605, Feb. 1997.

49. D.R. Wolters and J.F. Verwey, "Breakdown and Wear-Out Phenomena in SiO_2 Films," in *Instabilities in Silicon Devices* (B. Barbottin and A. Vapaille, eds.), Vol. 1, North-Holland, Amsterdam, 1986, 315–362; D.R. Wolters, "Breakdown and Wearout Phenomena in SiO_2," in *Insulating Films on Semiconductors* (M. Schulz and G. Pensl, Eds.), Springer Verlag, Berlin, 180–194, 1981.

50. D.R. Wolters and J.J. van der Schoot, "Dielectric Breakdown in MOS Devices," *Phil. J. Res.* **40**, 115–192, 1985.

51. J.H. Stathis, "Reliability Limits for the Gate Insulator in CMOS Technology," *IBM J. Res. Dev.* **46**, 265–286, March/May 2002.

52. J.H. Stathis, "Percolation Models for Gate Oxide Breakdown," *J. Appl. Phys.* **86**, 5757–5766, Nov. 1999.

53. B.E. Deal, M. Sklar, A.S. Grove, and E.H. Snow, "Characteristics of the Surface-State Charge (Q_{SS}) of Thermally Oxidized Silicon," *J. Electrochem. Soc.* **114**, 266–274, March 1967; A. Goetzberger, A.D. Lopez, and R.J. Strain, "On the Formation of Surface States During Stress Aging of Thermal Si-SiO_2 Interfaces," *J. Electrochem. Soc.* **120**, 90–96, Jan. 1973.

54. D.K. Schroder and J.A. Babcock, "Negative Bias Temperature Instability: A Road to Cross in Deep Submicron CMOS Manufacturing," *J. Appl. Phys.* **94**, 1–18, July 2003.

55. M. Makabe, T. Kubota, and T. Kitano, "Bias-temperature Degradation of pMOSFETs: Mechanism and Suppression," *IEEE Int. Reliability Phys. Symp.* **38**, 205–209, 2000.

56. K. Onishi, C.S. Kang, R. Choi, H.J. Cho, S. Gopalan, R. Nieh, E. Dharmarajan, and J.C. Lee, "Reliability Characteristics, Including NBTI, of Polysilicon Gate HfO_2 MOSFET's," *IEEE IEDM Tech. Digest*, 659–662, 2001.

57. N. Kimizuka, K. Yamaguchi, K. Imai, T. Iizuka, C.T. Liu, R.C. Keller, and T. Horiuchi, "NBTI Enhancement by Nitrogen Incorporation Into Ultrathin Gate Oxide for 0.10 μm Gate CMOS Generation," *IEEE VLSI Symp.* 92–93, 2000.

58. J. Maserijian and N. Zamani, "Behavior of the Si/SiO$_2$ Interface Observed by Fowler-Nordheim Tunneling," *J. Appl. Phys.* **53**, 559–567, Jan. 1982.

59. D.J. DiMaria and E. Cartier, "Mechanism for Stress-induced Leakage Currents in Thin Silicon Dioxide Films," *J. Appl. Phys.* **78**, 3883–3894, Sept. 1995.

60. A. Amerasekera and C. Duvvury, *ESD in Silicon Integrated Circuits*, Wiley, Chichester, 1995.

61. J. Colvin, "ESD Failure Analysis Methodology," *Microelectron. Reliab.* **38**, 1705–1714, Nov. 1998.

62. A.J. Walker, K.Y. Le, J. Shearer, and M. Mahajani, "Analysis of Tungsten and Titanium Migration During ESD Contact Burnout," *IEEE Trans. Electron Dev.* **50**, 1617–1622, July 2003.

63. R. Rajsuman, "Iddq Testing for CMOS VLSI," *Proc. IEEE*, **88**, 544–566, April 2000.

64. M. Rasras, I. De Wolf, H. Bender, G. Groeseneken, H.E. Maes, S. Vanhaeverbeke, and P. De Pauw, "Analysis of I$_{ddq}$ Failures by Spectral Photon Emission Microscopy," *Microelectron. Reliab.* **38**. 877–882, June/Aug. 1998.

65. S. Ito and H. Monma, "Failure Analysis of Wafer Using Backside OBIC Method," *Microelectron. Reliab.* **38**, 993–996, June/Aug. 1998.

66. E. Sabin, "High Temperature I$_{DDQ}$ Testing for Detection of Sodium and Potassium," *IEEE Int. Reliability Phys. Symp.* **34**, 355–359, 1996.

67. K.M. Wallquist, "Instrumentation for I$_{DDQ}$ Measurement," in S. Chakravarty and P.J. Thadikaran, *Introduction to I$_{DDQ}$ Testing*, Kluwer, Boston, 1997.

68. T. Schweinböck, S. Schömann, D. Alvarez, M. Buzzo, W. Frammelsberger, P. Breitschopf, and G. Benstetter, "New Trends in the Application of Scanning Probe Techniques in Failure Analysis," *Microelectron. Reliab.* **44**, 1541–1546, Sept./Nov. 2004; G. Benstetter, P. Breitschopf, W. Frammelsberger, H. Ranzinger, P. Reislhuber, and T. Schweinböck, "AFM-based Scanning Capacitance Techniques for Deep-submicron Semiconductor Failure Analysis," *Microelectron. Reliab.* **44**, 1615–1619, Sept./Nov. 2004.

69. N. Khurana and C.L. Chiang, "Dynamic Imaging of Current Conduction in Dielectric Films by Emission Spectroscopy," *IEEE Proc. 25th Int. Reliability Phys. Symp.*, San Diego, 1987, 72–76; J. Kölzer, C. Boit, A. Dallmann, G. Deboy, J. Otto, and D. Weinmann, "Quantitative Emission Microscopy," *J. Appl. Phys.* **71**, R23–R41, June 1992; C. Leroux and D. Blachier, "Light Emission Microscopy for Reliability Studies," *Microelectron. Eng.* **49**, 169–180, Nov.1999.

70. C.G.C. de Kort, "Integrated Circuit Diagnostic Tools: Underlying Physics and Applications," *Philips J. Res.* **44**, 295–327, 1989; F. Stellari, P. Song, M.K. McManus, A.J. Weger, R. Gauthier, K.V. Chatty, M. Muhammad, P. Sanda, P. Wu, and S. Wilson, "Latchup Analysis Using Emission Microscopy," *Microelectron. Reliab.* **43**, 1603–1608, Sept/Nov. 2003.

71. T. Kessler, F.W. Wulfert, and T. Adams, "Diagnosing Latch-up with Backside Emission Microscopy," *Semicond. Int.* **23**, 313–316, July 2000.

72. L. Liebert, "Failure Analysis from the Back Side of a Die," *Microelectron. Reliab.* **41**, 1193–1201, Aug. 2001.

73. F. Beaudoin, J. Lopez, M. Faucon, R. Desplats, and P. Perdu, "Femtosecond Laser Ablation for Backside Silicon Thinning," *Microelectron. Reliab.* **44**, 1605–1609, Sept/Nov. 2004.

74. M. Rasras, I. De Wolf, H. Bender, G. Groeseneken, H.E. Maes, S. Vanhaeverbeke, and P. De Pauw, "Analysis of I$_{ddq}$ Failures by Spectral Photon Emission Microscopy," *Microelectron. Reliab.* **38**. 877–882, June/Aug. 1998; I. De Wolf and M. Rasras, "Spectroscopic Photon Emission Microscopy: A Unique Tool for Failure Analysis of Microelectronics Devices," *Microelectron. Reliab.* **41**, 1161–1169, Aug. 2001.

75. G. Romano and M. Sampietro, "CMOS-Circuit Degradation Analysis Using Optical Measurement of the Substrate Current," *IEEE Trans. Electron Dev.* **44**, 910–912, May 1997.

76. J.C. Tsang and J.A. Kash, "Picosecond Hot Electron Light Emission from Submicron Complementary Metal-oxide-semiconductor Circuits," *Appl. Phys. Lett.* **70**, 889–891, Feb. 1997; J.A. Kash and J.C. Tsang, "Dynamic Internal Testing of CMOS Circuits Using Hot Luminescence," *IEEE Electron Dev. Lett.* **18**, 330–332, July 1997.

77. J.C. Tsang, J.A. Kash, and D.P. Vallett, "Time-Resolved Optical Characterization of Electrical Activity in Integrated Circuits," *Proc. IEEE* **88**, 1440–1459, Sept. 2000; F. Stellari, P. Song, J.C. Tsang, M.K. McManus, and M.B. Ketchen, "Testing and Diagnostics of CMOS Circuits Using Light Emission from Off-State Leakage Current," *IEEE Trans. Electron Dev.* **51**, 1455–1462, Sept. 2004.

78. P. Kolodner and J.A. Tyson, "Microscopic Fluorescent Imaging of Surface Temperature Profiles with 0.01°C Resolution," *Appl. Phys. Lett.* **40**, 782–784, May 1982.

79. C. Herzum, C. Boit, J. Kölzer, J. Otto, and R. Weiland, "High Resolution Temperature Mapping of Microelectronic Structures Using Quantitative Fluorescence Microthermography," *Microelectron. J.* **29**, 163–170, April/May 1998.

80. J. Kölzer, E. Oesterschulze, and G. Deboy, "Thermal Imaging and Measurement Techniques for Electronic Materials and Devices," *Microelectron. Eng.* **31**, 251–270, Feb. 1996.

81. G. Busse, D. Wu, and W. Karpen, "Thermal Wave Imaging with Phase Sensitive Modulated Thermography," *J. Appl. Phys.* **71**, 3962–3965, April 1992.

82. J.T.L. Thong, *Electron Beam Testing*, Plenum Press, New York, 1993.

83. M. Vallet and P. Sardin, "Electrical Testing for Failure Analysis: E-Beam Testing", *Microelectron. Eng.* **49**, 157–167, Nov. 1999.

84. H.K. Heinrich, "Picosecond Noninvasive Optical Detection of Internal Electrical Signals in Flip-chip-mounted Silicon Integrated Circuits," *IBM J. Res. Dev.* **34**, 162–172, March/May 1990.

85. M. Paniccia, R.M. Rao, and W.M. Yee, "Optical Probing of Flip Chip Packaged Microprocessors," *J. Vac. Sci. Technol.* **B16**, 3625–3630, Nov./Dec. 1998.

86. J.M. Keen, "Nondestructive Optical Technique for Electrically Testing Insulated-gate Integrated Circuits," *Electron. Lett.*, **7**, 432–433, July 1971.

87. C.E. Stephens and I.N. Sinnadurai, "A Surface Temperature Limit Detector Using Nematic Liquid Crystals With an Application to Microcircuits," *J. Phys. E* **7**, 641–643, Aug. 1974; D.J. Channin, "Liquid-Crystal Technique for Observing Integrated-Circuit Operation," *IEEE Trans. Electron Dev.* **21**, 650–652, Oct. 1974.

88. J. Hiatt, "A Method of Detecting Hot Spots on Semiconductors Using Liquid Crystals," *IEEE Int. Reliability. Phys. Symp.* **19**, 130–133, 1981.

89. D. Burgess and P. Tang, "Improved Sensitivity for Hot Spot Detection Using Liquid Crystals," *IEEE Int. Reliability. Phys. Symp.* **22**, 119–121, 1984.

90. K. Nikawa, C. Matsumoto, and S. Inoue, "Novel Method for Defect Detection in Al Stripes by Means of Laser Beam Heating and Detection of Changes in Electrical Resistance," *Japan. J. Appl. Phys.*, **34**, 2260–2265, May 1995.

91. K. Nikawa, T. Saiki, S. Inoue, and M. Ohtsu, "Imaging of Current Paths and Defects in Al and TiSi Interconnects on Very-large-scale Integrated-circuit Chips Using Near-field Optical-probe Stimulation and Resulting Resistance Change," *Appl. Phys. Lett.* **74**, 1048–1050, Feb. 1999.

92. K. Nikawa, "Failure Analysis Case Studies Using the IR-OBIRCH (Infrared Optical Beam Induced Resistance Change) Method," *Photonics Failure Analysis Workshop*, Boston, Oct. 1999.

93. E.I. Cole Jr., P. Tangyunyong, and D.L. Barton, "Backside Localization of Open and Shorted IC Interconnections," *IEEE Int. Reliability. Phys. Symp.* **36**, 129–136, 1998.

94. W.R. Runyan and T.J. Shaffner, *Semiconductor Measurements and Instrumentation*, 2nd ed., McGraw-Hill, New York, 1998.

95. H. Wong, "Low-frequency Noise Study in Electron Devices: Review and Update," *Microelectron. Reliab.* **43**, 585–599, April 2003.

96. C. Claeys, A. Mercha, and E. Simoen, "Low-Frequency Noise Assessment for Deep Submicrometer CMOS Technology Nodes," *J. Electrochem. Soc.* **151**, G307–G318, May 2004.

97. A. van der Ziel, *Noise: Sources, Characterization, Measurement*, Prentice-Hall, Englewood Cliffs, NJ, 1970; *Noise*, Prentice-Hall, Englewood Cliffs, NJ, 1954.

98. F.N.H. Robinson, *Noise and Fluctuations in Electronic Devices and Circuits*, Clarendon Press, Oxford, 1974.

99. C.D. Motchenbacher and F.C. Fitchen, *Low-Noise Electronic Design*, Wiley, New York, 1973.

100. A. Einstein, "A New Determination of the Molecular Dimensions (in German)," *Ann. Phys.* **19**, 289–306, Feb. 1906.

101. J.B. Johnson, "Thermal Agitation of Electricity in Conductors," *Phys. Rev.* **29**, 367–368, Feb. 1927; *Phys. Rev.* **32**, 97–109, July 1928.

102. H. Nyquist, "Thermal Agitation of Electric Charge in Conductors," *Phys Rev.* **32**, 110–113, July 1928.

103. R.J.T. Bunyan, M.J. Uren, J.C. Alderman, and W. Eccleston, "Use of Noise Thermometry to Study the Effects of Self-heating in Submicrometer SOI MOSFETs," *IEEE Electron Dev. Lett.* **13**, 279–281, May 1992.

104. W. Schottky, "On Spontaneous Current Fluctuations in Various Electricity Conductors (in German)," *Ann. Phys.* **57**, 541–567, Dec. 1918.

105. J.B. Johnson, "The Schottky Effect in Low Frequency Circuits," *Phys. Rev.* **26**, 71–85, July 1925.

106. A.L. McWhorter, "1/f Noise and Germanium Surface Properties," in *Semiconductor Surface Physics* (R.H. Kingston, ed.), University of Pennsylvania Press, Philadelphia, 1957, 207–228.

107. F.N. Hooge, "1/f Noise is No Surface Effect," *Phys. Lett.* **29A**, 139–140, April 1969; _____ Discussion of Recent Experiments on 1/f Noise," *Physica* **60**, 130–144, 1976; _____ "1/f Noise," *Physica* **83B**, 14–23, May 1976; F.N. Hooge and L.K.J. Vandamme, "Lattice Scattering Causes 1/f Noise," *Phys. Lett.* **66A**, 315–316, May 1978; F.N. Hooge, "1/f Noise Sources," *IEEE Trans. Electron Dev.*, **41**, 1926–1935, Nov. 1994.

108. S. Christensson, I. Lundström, and C. Svensson, "Low Frequency Noise in MOS Transistors I. Theory," *Solid-State Electron.* **11**, 797–812, Sept. 1968; S. Christensson and I. Lundström, "Low Frequency Noise in MOS Transistors II. Experiments," *Solid-State Electron.* **11**, 813–820, Sept. 1968.

109. M.J. Kirton and M.J. Uren, "Noise in Solid-State Microstructures: A New Perspective on Individual Defects, Interface States and Low-Frequency (1/f) Noise," *Advan. Phys.* **38**, 367–468, July/Aug. 1989.

110. K.K. Hung, P.K. Ko, C. Hu and Y.C. Cheng, "A Unified Model for Flicker Noise in Metal Oxide Semiconductor Field Effect Transistors," *IEEE Trans. Electron Dev.* **37**, 654–665, March 1990; K.K. Hung, P.K. Ko, C. Hu and Y.C. Cheng, "A Physics Based MOSFET Noise Model for Circuit Simulators," *IEEE Trans. Electron Dev.* **37**, 1323–1333, May 1990.

111. S. Christensson, I. Lundstrom, and C. Svensson, "Low-frequency Noise in MOS Transistors: I-Theory," *Solid-State Electron.* **11**, 797–812, Sept. 1968.

112. AKM Ahsan and D.K. Schroder. "Impact of Post-Oxidation Annealing on Low-Frequency Noise, Threshold Voltage, and Subthreshold Swing of p-Channel MOSFETs," *IEEE Electron Dev. Lett.* **25**, 211–213, April 2004.

113. F. Scholz, J.M. Hwang and D.K. Schroder, "Low Frequency Noise and DLTS as Semiconductor Characterization Tools," *Solid-State Electron.* **31**, 205–217, Febr. 1988; T. Hardy, M.J. Deen, and R.M. Murowinski, "Low Frequency Noise in Proton Damaged LDD MOSFETs," *IEEE Trans. Electron Dev.* **46**, 1339–1346, July 1999.

114. E. Simoen, A. Mercha, and C. Claeys, "Noise Diagnostics of Advanced Silicon Substrates and Deep Submicron Process Modules," in *Analytical and Diagnostic Techniques for Semiconductor Materials, Devices, and Processes* (B.O. Kolbesen et. al, eds.), Electrochem. Soc. **ECS PV 2003–03**, 420–439, 2003; G. Härtler, U. Golze, and K. Paschke, "Extended Noise

Analysis—A Novel Tool for Reliability Screening," *Microelectron. Reliab.* **38**, 1193–1198, June/Aug. 1998.

115. S. Zafar, Q. Liu, and E.A. Irene, "Study of Tunneling Current Oscillation Dependence on SiO_2 Thickness and Si Roughness at the Si/SiO_2 Interface," *J. Vac. Sci. Technol.* **A13**, 47–53, Jan./Feb. 1995; K.J. Hebert and E.A. Irene, "Fowler-Nordheim Current Oscillations at Metal/Oxide/Si Interfaces," *J. Appl. Phys.* **82**, 291–296, July 1997; L. Mao, C. Tan, and M. Xu, "Thickness Measurements for Ultrathin-Film Insulator Metal-Oxide-Semiconductor Structures Using Fowler-Nordheim Tunneling Current Oscillations," *J. Appl. Phys.* **88**, 6560–6563, Dec. 2000.

116. Y-C Yeo, T-J King and C.M. Hu, "MOSFET Gate Leakage Modeling and Selection Guide for Alternative Gate Dielectrics Based on Leakage Considerations," *IEEE Trans Electron Dev.* **50**, 1027–1035, April 2003.

PROBLEMS

12.1 For a particular failure mechanism two expressions for the mean time to failure are

$$\text{MTTF} = AF^{-n} \exp\left(\frac{E}{kT}\right); \text{MTTF} = B \exp\left(\frac{E - aF}{kT}\right)$$

where F is the driving force. For A, B, E, n, and a as positive constants independent of temperature and force:

Give expressions for the acceleration factor in each case when the temperature is raised from T_1 to T_2 at constant F.

Give expressions for the acceleration factor in each case when the force is increased from F_1 to F_2 at constant T.

Which of the two MTTF equations gives a higher value for AF for temperature acceleration?

Which of the two MTTF equations gives a higher value for AF for force acceleration?

12.2 An acceleration factor of two governs a chemical reaction rates when the temperature is raised by $10°C$. What activation energy is required for this to happen at room temperature?

12.3 High local electric fields promote avalanche breakdown in semiconductor junctions and dielectrics. Such breakdown is usually harmless in junction devices, but can be destructive in dielectrics. Explain why.

12.4 The extreme value distribution has a probability density function

$$f(t) = \frac{1}{b} \exp\left(\frac{t - \tau}{b}\right) \exp\left(-\exp\left(\frac{t - \tau}{b}\right)\right)$$

where b and τ are constants. What are the mathematical expressions for $F(t)$ and $\lambda(t)$?

12.5 This problem deals with electromigration. A number of metal lines are stressed and the median times to failure, t_{50}, are determined. They are plotted in Fig. P12.5. From these plots determine the activation energy E_A and the exponent n in the equation

$$t_{50} = AJ^{-n} \exp(E_A/kT)$$

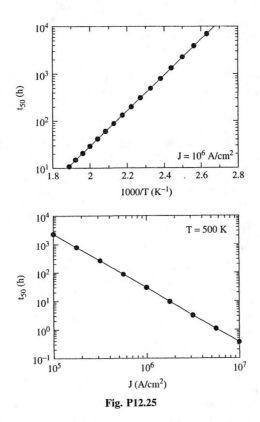

$J = 10^6$ A/cm^2

$T = 500$ K

Fig. P12.25

Then determine t_{50} at $T = 400$ K and $J = 10^5$ A/cm^2.

12.6 The oxide Fowler-Nordheim tunneling current density J_{FN} versus $\mathscr{E}_{ox}$ plot of an MOS capacitor is shown in Fig. P12.6. This is not Si/SiO$_2$. Determine the barrier height Φ_B (in eV) and m_{ox}/m.

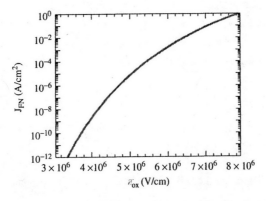

Fig. P12.26

12.7 Determine the charge-to-breakdown Q_{BD} in both plots in Fig. P12.7.

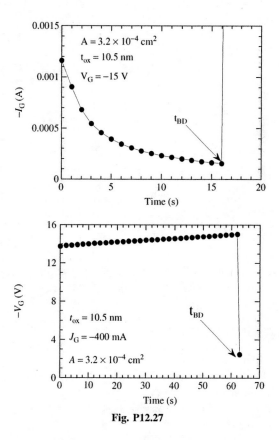

Fig. P12.27

12.8 From Fig. P12.8 determine the oxide defect density at $\mathscr{E}_{ox} = 8$ MV/cm. $A = 0.01$ cm^2.

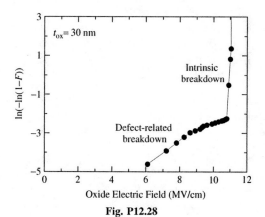

Fig. P12.28

12.9 Determine the activation energies for the three lines in Fig. P12.9.

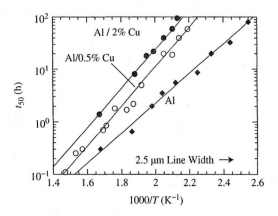

Fig. P12.29

Some of these problems were taken from M. Ohring, *Reliability and Failure of Electronic Materials and Devices*, Academic Press, San Diego, 1998.

REVIEW QUESTIONS

- What are acceleration factors?
- What are probability density and cumulative distribution functions?
- Name three distribution functions.
- What is "six sigma" and how many defects are allowed?
- What causes electromigration?
- Why do narrow metal lines have better electromigration resistance than wider lines?
- What is the "Blech" length?
- How are MOSFET hot carriers characterized?
- How is gate oxide integrity characterized?
- What distinguishes FN from direct tunnel currents?
- Below which oxide voltage is direct tunneling dominant?
- What is NBTI?
- What does electrostatic discharge do?
- How is I_{DDQ} implemented?
- What is emission microscopy?
- What is voltage contrast and how is it measured?
- How does OBIRCH work?
- Name and briefly describe three noise sources.

APPENDIX 1

LIST OF SYMBOLS

A	area (cm^2)
A^*	Richardson's constant (A/cm$^2 \cdot$ K^2)
A_c	contact area (cm^2)
A_G	gate area (cm^2)
A_J	junction area (cm^2)
B	magnetic field strength (G or T)
B	radiative recombination coefficient (cm^3/s)
b	mobility ratio μ_n/μ_p
C	capacitance (F)
c	velocity of light (2.998 $\times$ 10^{10} cm/s)
C_b	bulk capacitance (F/cm^2)
C_{ch}	channel capacitance
C_{dd}	deep-depletion capacitance (F/cm^2)
C_{hf}	high-frequency capacitance (F)
C_{hf}	high-frequency capacitance (F/cm^2)
C_{inv}	minimum (strong inversion) capacitance (F)
C_{inv}	minimum (strong inversion) capacitance (F/cm^2)
C_{it}	interface trap capacitance (F/cm^2)
C_{lf}	low-frequency capacitance (F)
C_{lf}	low-frequency capacitance (F/cm^2)
C_n	inversion (electron) capacitance (F/cm^2)
C_n	Auger recombination coefficient for n-type (cm^6/s)
c_n	electron capture coefficient (cm^3/s)
C_{ox}	oxide capacitance (F)
C_{ox}	oxide capacitance/unit area (F/cm^2)

Semiconductor Material and Device Characterization, Third Edition, by Dieter K. Schroder
Copyright © 2006 John Wiley & Sons, Inc.

C_P	parallel capacitance (F)
C_p	accumulation (hole) capacitance (F/cm^2)
C_p	Auger recombination coefficient for p-type (cm^6/s)
c_p	hole capture coefficient (cm^3/s)
C_S	semiconductor capacitance (F)
C_S	series capacitance (F)
$C_{S,dd}$	deep-depletion semiconductor capacitance (F/cm^2)
$C_{S,hf}$	high-frequency semiconductor capacitance (F/cm^2)
$C_{S,lf}$	low-frequency semiconductor capacitance (F/cm^2)
d	contact spacing (cm)
d	crystal plane spacing (cm)
d	distance (cm)
d	thickness (cm)
d	wafer diameter (cm)
D	defect density (cm^{-2})
D	diameter (cm)
D	diffusion constant (cm^2/s)
D	dissipation factor
D_{it}	interface trapped charge density (cm$^{-2} \cdot$ eV^{-1})
D_n	electron diffusion constant (cm^2/s)
E	energy (eV)
$\mathscr{E}$	electric field (V/cm)
E_A	acceptor energy level (eV)
E_A	activation energy (eV)
E_c	conduction band edge (eV)
E_D	donor energy level (eV)
$\mathscr{E}_{eff}$	effective electric field (V/cm)
E_{ehp}	mean energy to generate one electron-hole pair (eV)
E_F	Fermi energy (eV)
E_G	band gap (eV)
E_{it}	interface trapped charge energy (eV)
e_n	electron emission coefficient (s^{-1})
$\mathscr{E}_{ox}$	oxide electric field (V/cm)
E_p	phonon energy (eV)
e_p	hole emission coefficient (s^{-1})
E_T	trap energy (eV)
E_v	valence band edge (eV)
f	frequency (Hz)
f	probability density function
f	spatial frequency (cycles/cm)
F	cumulative distribution function
F	dimensionless electric field
F	Faraday constant (9.64×10^4 C)
F	van der Pauw F-function
G	bulk generation rate (cm$^{-3} \cdot$ s^{-1})
G	conductance (S)
G	Gibbs free energy (eV)
g	conductance (S)
g	degeneracy factor

g_d	drain conductance (S)
g_d	diode conductance (S)
g_{dk}	dark conductance (S)
g_m	transconductance (S)
G_P	parallel conductance (S)
g_{ph}	photoconductance (S)
G_S	surface generation rate ($cm^{-2} \cdot s^{-1}$)
G_{sh}	sheet conductance (1/ohms/square)
H	enthalpy (eV)
h	Planck's constant (6.626×10^{-34} $J \cdot s$)
I	current (A)
I_B	base current (A)
I_b	electron beam current (A)
I_C	collector current (A)
I_{cp}	charge pumping current (A)
I_D	drain current (A)
I_d	displacement current (A)
I_{dk}	dark current (A)
I_{EBIC}	electron beam induced current (A)
I_E	emitter current (A)
I_e	emission current (A)
I_G	gate current (A)
I_J	junction current (A)
I_{GIJ}	gate-induced junction current (A)
I_{ph}	photocurrent (A)
I_S	surface current (A)
I_{sc}	short-circuit current (A)
I_{sub}	substrate current (A)
J	current density (A/cm^2)
J_{dir}	direct tunnel current density (A/cm^2)
J_{FN}	Fowler-Nordheim tunnel current density (A/cm^2)
J_G	gate current density (A/cm^2)
J_{sc}	short-circuit current density (A/cm^2)
J_{scr}	space-charge region current density ($A)/cm^2$)
K	kinematic factor
k	Boltzmann's constant (8.617×10^{-5} eV/K)
k	extinction coefficient
k	spring constant
K_{ox}	oxide dielectric constant
K_s	semiconductor dielectric constant
L	channel length (cm)
L	contact or sample length (cm)
L	minority carrier diffusion length (cm)
L_D	Debye length (cm)
L_{Di}	intrinsic Debye length (cm)
L_{eff}	effective channel length (cm)
L_m	mask-defined channel length (cm)
L_n	electron diffusion length (cm)
L_p	hole diffusion length (cm)

L_T	$= \sqrt{\rho_c/R_{sh}}$, transfer length (cm)
L_{Tk}	$= \sqrt{\rho_c/R_{sk}}$, transfer length (cm)
L_{Tm}	$= \sqrt{\rho_c/(R_{sm} + R_{sk})}$, transfer length (cm)
M	elemental mass (kg)
M	molecular weight (g)
m	electron mass (9.11×10^{-31} kg)
m^*	effective mass (kg)
m_n	electron effective mass (kg)
m_{ox}	oxide electron effective mass (kg)
N	electron density (cm^{-2})
n	diode ideality factor
n	electron density (cm^{-3})
n	index of refraction
n	sub-threshold slope parameter
N_A	acceptor doping density (cm^{-3})
NA	numerical aperture
N_c	effective density of states in the conduction band (cm^{-3})
N_D	donor doping density (cm^{-3})
N_f	fixed oxide charge density (cm^{-2})
n_i	intrinsic carrier density (cm^{-3})
N_{it}	interface trapped charge density (cm^{-2})
N_m	mobile oxide charge density (cm^{-2})
n_o	equilibrium electron density (cm^{-3})
N_{ox}	oxide trapped charge density (cm^{-2})
n_{po}	equilibrium minority electron density (cm^{-3})
n_s	electron density at surface (cm^{-3})
N_T	deep-level impurity density (cm^{-3})
n_T	deep-level impurity density occupied by electrons (cm^{-3})
N_v	effective density of states in the valence band (cm^{-3})
n_1	electron density (cm^{-3})
P	power (W)
p	hole density (cm^{-3})
p	momentum (nt $\cdot$ s)
p	pressure (nt/cm^2)
$\mathcal{P}$	differential thermoelectric power (V/K)
p_o	equilibrium hole density (cm^{-3})
p_s	hole density at surface (cm^{-3})
p_T	deep-level impurity density occupied by holes (cm^{-3})
p_1	hole density (cm^{-3})
q	magnitude of electron charge (1.6×10^{-19} C)
Q	charge (C)
Q	charge density (C/cm^2)
Q	quality factor
Q_b	bulk charge density (C/cm^2)
Q_{BD}	charge-to-breakdown (C/cm^2)
Q_{cp}	charge pumping charge (C)
Q_f	fixed oxide charge density (C/cm^2)
Q_G	gate charge density (C/cm^2)
Q_i	interfacial charge density (C/cm^2)

Q_{it}	interface state charge density (C/cm^2)
Q_m	mobile oxide charge density (C/cm^2)
Q_N	electron charge density (C/cm^2)
Q_n	electron charge density (C/cm^2)
Q_n	inversion charge density (C/cm^2)
Q_{ot}	oxide trapped charge density (C/cm^2)
Q_p	hole charge density (C/cm^2)
Q_S	semiconductor charge density (C/cm^2)
Q_S	semiconductor charge (C)
R	recombination rate (cm$^{-3} \cdot$ s^{-1})
R	reflectivity, reflectance
R	reliability function
R	resistance (ohms)
r	contact radius (cm)
r	distance (cm)
r	Hall scattering factor
r	wafer radius (cm)
r_{dk}	dark resistance (ohms)
r_{ph}	photo resistance (ohms)
r_s	series resistance (ohms)
r_{sh}	shunt resistance (ohms)
R	bulk recombination rate (cm^{-3} s^{-1})
R_B	base resistance (ohms)
R_{Bi}	internal base resistance (ohms)
R_{Bx}	external base resistance (ohms)
R_c	contact resistance (ohms)
R_{ce}	contact end resistance (ohms)
R_{cf}	contact front resistance (ohms)
R_C	collector resistance (ohms)
R_{ch}	channel resistance (ohms)
R_D	drain resistance (ohms)
R_e	electron range (cm)
R_e	end resistance (ohms)
R_E	emitter resistance (ohms)
R_G	gate resistance (ohms)
R_{geom}	geometry-dependent resistance (ohms)
R_H	Hall coefficient (cm^3/C)
R_{Hs}	sheet Hall coefficient (cm^2/C)
R_k	measured contact resistance (ohms)
R_m	metal or poly-silicon resistance (ohms)
R_m	measured resistance (ohms)
R_p	probe resistance (ohms)
R_S	source resistance (ohms)
R_S	semiconductor resistance (ohms)
R_S	surface recombination rate (cm$^{-2} \cdot$ s^{-1})
R_{SD}	source-drain resistance (ohms)
R_{sh}	sheet resistance (ohms/square)
R_{sk}	sheet resistance under a contact (ohms/square)
R_{sm}	metal or poly-silicon sheet resistance (ohms/square)

R_{sp}	spreading resistance (ohms)
R_T	total resistance (ohms)
s_c	surface generation velocity (cm/s)
s, s_r	surface recombination velocity (cm/s)
s_g	surface generation velocity (cm/s)
$s_{g,eff}$	effective surface generation velocity (cm/s)
S	entropy (eV/K)
S	MOSFET sub-threshold swing (V/decade)
S	strain (cm/cm)
S_o	backscattering energy loss factor (eV/Å)
t	time (s)
t	wafer thickness (cm)
t_{BD}	time-to-breakdown (s)
t_d	drift time (s)
t_f	filling pulse width (s)
t_{ox}	oxide thickness (cm)
t_s	storage time (s)
t_t	transit time (s)
T	temperature (K)
T	transmissivity, transmittance
t_{50}	median time to failure (s)
U	$= q\phi/kT$
U_F	$= q\phi_F/kT$
U_S	surface recombination rate ($cm^{-2} \cdot s^{-1}$)
U_S	$= q\phi_S/kT$
v	velocity (cm/s)
v_d	drift velocity (cm/s)
v_n	electron velocity (cm/s)
v_{th}	thermal velocity (cm/s)
V	voltage (V)
V	volume (cm^3)
V_{air}	voltage across air gap (V)
V_0	defined in Eq. (6.14)
V_B	substrate voltage (V)
V_b	Dember potential (V)
V_{bi}	built-in potential (V)
V_{BS}	$V_B - V_S$ (V)
V_{cpd}	contact potential difference (V)
V_{CE}	collector-emitter voltage (V)
V_D	diode voltage (V)
V_D	drain voltage (V)
V_{DS}	$V_D - V_S$ (V)
V_{BE}	base-emitter voltage (V)
V_{FB}	flatband voltage (V)
V_G	gate voltage (V)
V_{GS}	$V_G - V_S$ (V)
V_H	Hall voltage (V)
V_j	junction voltage (V)
V_{oc}	open-circuit voltage (V)

V_{ox}	oxide voltage (V)
V_P	probe voltage (V)
V_S	source voltage (V)
V_S	surface voltage (V)
V_{SPV}	surface photovoltage (V)
V_T	threshold voltage (V)
w	width (cm)
W	channel width (cm)
W	diffusion window width (cm)
W	line width (cm)
W	space-charge region width (cm)
W_{eff}	effective channel width (cm)
W_{inv}	inversion space-charge region width (cm)
W_{inv}	$= (2K_s\varepsilon_0\phi_{s,inv}/qN_A)^{1/2}$ minimum (strong inversion) space-charge region width (cm)
x_{ch}	channel thickness (cm)
x_j	junction depth (cm)
Y	conductance (S)
Y	ratio of photocurrent to absorbed photon flux
z	dissolution valency
Z	atomic number
Z	contact or sample width (cm)
Z	impedance (ohms)
α	absorption coefficient (cm^{-1})
α	common-base current gain
α_F	forward common-base current gain
α_R	reverse common-base current gain
β	common-emitter current gain
β_F	forward common-emitter current gain
β_R	reverse common-emitter current gain
χ	semiconductor electron affinity (eV)
δ	skin depth (cm)
δ	$= W - Z$ (cm)
Δn	excess electron density (cm^{-3})
Δp	excess hole density (cm^{-3})
ε_0	permittivity of free space (8.854×10^{-14} F/cm)
Φ	photon flux density (photons/s · cm^2)
ϕ	work function (V)
Φ_B	barrier height (eV)
ϕ_B	Schottky diode barrier height (V)
ϕ_F	Fermi potential (V)
Φ_M	metal work function (eV)
ϕ_M	metal work function (V)
ϕ_{MS}	metal-semiconductor work function (V)
Φ_S	semiconductor work function (eV)
ϕ_S	semiconductor work function (V)
ϕ_s	surface potential (V)
γ	voltage acceleration factor (V^{-1})

λ	wavelength (cm)
λ	tunneling parameter
λ_e	electron wavelength (cm)
λ_p	plasma resonance wavelength (cm)
μ	mobility (cm^2/V $\cdot$ s)
μ/ρ	mass absorption coefficient (cm^2/g)
μ_{eff}	effective mobility (cm^2/V $\cdot$ s)
μ_{FE}	field-effect mobility (cm^2/V $\cdot$ s)
μ_{GMNR}	geometric magnetoresistance mobility (cm^2/V $\cdot$ s)
μ_H	Hall mobility (cm^2/V $\cdot$ s)
μ_n	electron mobility (cm^2/V $\cdot$ s)
μ_o	low-field mobility (cm^2/V $\cdot$ s)
μ_o	permeability of free space ($4\pi \times 10^{-9}$ H/cm)
μ_p	hole mobility (cm^2/V $\cdot$ s)
μ_{sat}	saturation mobility (cm^2/V $\cdot$ s)
ν	frequency of light (Hz)
ρ	density (g/cm^3)
ρ	resistivity (ohm $\cdot$ cm)
ρ_c	specific contact resistance (ohm $\cdot$ cm^2)
ρ_i	specific interface resistance (ohm $\cdot$ cm^2)
σ	conductivity (ohm^{-1} $\cdot$ cm^{-1} or S/cm)
σ_n	electron capture cross section (cm^2)
σ_{ns}	surface state electron capture cross-section (cm^2)
σ_p	hole capture cross section (cm^2)
σ_{ps}	surface state hole capture cross-section (cm^2)
τ	lifetime (s)
τ	time constant (s)
τ_{Auger}	Auger lifetime (s)
τ_B	bulk lifetime (s)
τ_c	capture time constant (s)
τ_e	$= 1/e_n$, electron emission time constant (s)
τ_{eff}	effective recombination lifetime (s)
τ_g	generation lifetime (s)
$\tau_{g,eff}$	effective generation lifetime (s)
τ_n	electron lifetime (s)
$\tau_{non-rad}$	non-radiative lifetime (s)
τ_p	hole lifetime (s)
τ_r	recombination lifetime (s)
τ_{rad}	radiative lifetime (s)
τ_s	surface recombination lifetime (s)
τ_{SRH}	Shockley-Read-Hall or multi-phonon lifetime (s)
ω	radial frequency (s^{-1})
ξ	magnetoresistance scattering factor
Ψ	ellipsometric angle
Δ	ellipsometric angle
θ	mobility degradation factor (V^{-1})

APPENDIX 2

ABBREVIATIONS AND ACRONYMS

AAS	atomic absorption spectroscopy
AEM	analytical transmission electron microscope (microscopy)
AES	Auger electron spectroscopy
AF	acceleration factor
AFM	atomic force microscope (microscopy)
ASTM	American Society for Testing of Materials
BE	bound exciton
BEEM	ballistic electron emission microscopy
BJT	bipolar junction transistor
BTS	bias temperature stress
cw	continuous wave
CAFM	conducting AFM
CBKR	cross-bridge Kelvin resistor
CC-DLTS	constant-capacitance DLTS
CCD	charge-coupled device
CD	critical dimension
CDM	charged device model
CER	contact end resistance
CFM	chemical force microscopy
CFR	contact front resistance
CL	cathodoluminescence
CMA	cylindrical mirror analyzer
CMOS	complementary MOS
COS	corona oxide semiconductor
CP	charge pumping

Semiconductor Material and Device Characterization, Third Edition, by Dieter K. Schroder
Copyright © 2006 John Wiley & Sons, Inc.

CRT	cathode ray oscilloscope
$C-V$	capacitance-voltage
CVD	chemical vapor deposition
dd	deep depletion
DC-IV	direct current-current voltage
DHE	differential Hall effect
DIBL	drain-induced barrier lowering
DLTS	deep-level transient spectroscopy
D-DLTS	double correlation DLTS
DUT	device under test
ehp	electron-hole pair
EBIC	electron beam induced current
EBS	elastic backscattering spectrometry
ECV	electrochemical CV
EDS	energy dispersive spectroscopy
EELS	electron energy loss spectroscopy
EEPROM	electrically-erasable programmable read-only memory
EFM	electrostatic force microscopy
ELYMAT	electrolytical metal tracer
EM	electromigration
EMMI	emission microscopy
EMP	electron microprobe
EPM	electron probe microanalysis
ESD	electrostatic discharge
ESCA	electron spectroscopy for chemical analysis
ESR	electron spin resonance
FA	failure analysis
FB	flatband
$F-D$	Fermi-Dirac
FE	field emission
FE	free exciton
FET	field-effect transistor
FIB	focused ion beam
FIT	failure unit (1 failure/10^9 hours)
FMT	fluorescent microthermography
FN	Fowler-Nordheim
FTIR	Fourier transform infrared spectroscopy
GIXXR	grazing incidence X-ray reflectometry
GMR	geometrical magnetoresistance
GOI	gate oxide integrity
$G-R$	generation-recombination
hf	high frequency
HBM	human body model
HEIS	high-energy ion backscattering spectrometry
HIBS	heavy ion backscattering spectrometry
HREM	high-resolution transmission electron microscopy
IC	integrated circuit
ICP-MS	inductively coupled plasma mass spectroscopy
IFM	interfacial force microscopy

$I-T$	current-temperature
$I-V$	current-voltage
IR	infrared
IRT	infrared thermography
JFET	junction field-effect transistor
lf	low frequency
L-DLTS	Laplace DLTS
LAMMA	laser microprobe mass spectrometer
LBIC	light beam induced current
LC	liquid crystal
LDD	lightly-doped drain
LEED	low energy electron diffraction
LIMS	laser ionization mass spectrometry
LOCOS	local oxidation of silicon
LVP	laser voltage probe
M	magnification
MBE	molecular beam epitaxy
MCA	multichannel analyzer
MEIS	medium energy ion scattering spectrometry
MESFET	metal-semiconductor field effect transistor
MFM	magnetic force microscopy
MM	machine model
MOCVD	metalorganic vapor deposition
MODFET	modulation-doped field effect transistor
MOS	metal oxide semiconductor
MOS-C	MOS capacitor
MOSFET	MOS field effect transistor
MRFM	magnetic resonance force microscopy
MSMS	micromagnetic scanning microprobe system
MTBF	mean time between failure
MTF	median time to failure
MTTF	mean time to failure
Nano-Field	nanometer electric field gradient
Nano-MNR	nanometer nuclear magnetic resonance
NA	numerical aperture
NAA	neutron activation analysis
NBTI	negative bias temperature instability
OBIRCH	optical beam induced resistance change
NDP	neutron depth profiling
NFOM	near field optical microscope (microscopy)
NRA	nuclear reaction analysis
opd	optical path difference
O-DLTS	optical DLTS
OCVD	open circuit voltage decay
PAS	positron annihilation spectroscopy
PC	photoconductance or photocurrent
PCD	photoconductance decay
PCSA	polarizer-compensator-sample-analyzer
PICA	picosecond imaging circuit analysis

PICTS	photoinduced current transient spectroscopy
PITS	photoinduced current transient spectroscopy
PL	photoluminescence
PMR	physical magnetoresistance
PSI	phase shift interferometry
PTIS	photothermal ionization spectroscopy
qn	quasi-neutral
qnr	quasi-neutral region
$Q-V$	charge-voltage
QSSPC	quasi steady-state photoconductance
RBS	Rutherford backscattering spectrometry
RHEED	reflection high energy electron diffraction
RIS	resonance ionization spectroscopy
RR	reverse recovery
scr	space-charge region
S-DLTS	scanning DLTS
SAM	scanning Auger microscopy
SCA	surface charge analyzer
SCCD	short circuit current decay
SCM	scanning capacitance microscopy
SCPM	scanning chemical potential microscopy
SE	spectroscopic ellipsometry
SEcM	scanning electrochemical microscopy
SEM	scanning electron microscope (microscopy)
SF	stacking fault
SI	semi-insulating
SICM	scanning ion-conductance microscopy
SILC	stress induced leakage current
SIMS	secondary ion mass spectrometry
SIP	surface impedance profiling
SKPM	scanning Kelvin probe microscopy
SM	stress migration
SNMS	secondary neutral mass spectrometry
SPM	scanning probe microscope (microscopy)
SPV	surface photovoltage
SRA	surface resistance analyzer
SRH	Shockley-Read-Hall
SRP	spreading resistance probe or profiling
SSRM	scanning SRP
STEM	scanning transmission electron microscope (microscopy)
SThM	scanning thermal microscopy
STM	scanning tunneling microscope (microscopy)
STOS	scanning tunneling optical microscopy
SV	surface voltage
SWEAT	standard wafer-level electromigration acceleration test
TCR	temperature coefficient of resistance
TE	thermionic emission
TEM	transmission electron microscope (microscopy)
TFE	thermionic-field emission

TIVA	thermally-activated voltage alteration
TLM	transmission line model *or* transfer length method
TOF-SIMS	time of flight SIMS
TSC	thermally stimulated current
TSCAP	thermally stimulated capacitance
TUNA	tunneling AFM
TVS	triangular voltage sweep
TXRF	total reflection XRF
UHV	ultra-high vacuum
UPS	ultraviolet photoelectron spectroscopy
UV	ultraviolet
VAMFO	variable-angle monochromator fringe observation
VPD	vapor phase decomposition
WDS	wavelength dispersive spectroscopy
WN	wave number
XPS	X-ray photoelectron spectroscopy
XRF	X-ray fluorescence
XRFA	X-ray fluorescence analysis
XRFS	X-ray fluorescence spectroscopy
XRT	X-ray topography

INDEX

A*, 133, 158, 190
Absorption coefficient:
  carbon in Si, 592
  EL2 in GaAs, 592
  free carrier, 413
  GaAs, 409, 614
  GaN, 614
  GaP, 614
  Ge, 614
  InP, 409, 614
  oxygen in Si, 592
  Si, 409, 613
Acceleration factors, 690
Activation energy, 293
  GaAs, 299
  Si, 298, 300
Airy disk, 565
Analytical techniques:
  analysis time, 677
  analyzed volume, 677
  depth resolution, 677
  detectable elements, 677
  detection limits, 677
  matrix effect, 677
Analytical transmission electron microscopy
  (AEM), 645
Anodic oxidation, 29, 472
  anodic solutions, 29

constant voltage method, 29
constant current method, 29
Antenna structure, 702
Arrhenius plot, 263, 288–289, 298, 299, 691
Atomic absorption spectroscopy (AAS), 667
Atomic force microscopy (AFM), 544, 715
  cantilever, 545
  contact mode, 546
  instrument, 545
  non-contact mode, 546
  piezoelectric scanner, 545
  tapping mode, 546
  van der Waal force, 546
Auger electron spectroscopy (AES), 634
  applications, 638
  Auger electron emission, 634
  Auger electron energy, 636
  band diagram, 635
  chemical information, 639
  cylindrical mirror analyzer, 636
  detection limit, 638
  depth profiling, 638
  energy band diagram, 635
  energy resolution, 636
  differentiated signal, 637
  electron energy analyzer, 636
  hemispherical analyzer, 636
  instrumentation, 636

Semiconductor Material and Device Characterization, Third Edition, by Dieter K. Schroder
Copyright © 2006 John Wiley & Sons, Inc.

Selected Properties of Some Semiconductors at $T = 300$ K.

Semicond	Band Gap (eV)	Electron Mobility* (cm²/V·s)	Hole Mobility* (cm²/V·s)	Static Dielectric Constant	Lattice Constant (Å)	Density (g/cm³)	Melting Point (K)
Si	1.12	1,500	470	11.7	5.43095	2.328	1685
Ge	0.67	3,900	1,900	16	5.64613	5.327	1231
Diamond	5.45	1,900	1,600	5.5	3.57	3.5	~4000
3C-SiC	2.3	800	40	9.7	4.36	3.2	sublimes
6H-SiC	3.03	400	100	9.7	$a = 3.081$ $c = 15.17$	3.2	>2100
GaAs (Hexagonal)	1.42	8,500	400	12.8	5.6533	5.32	1510
GaN (Wurtzite)	3.39	1,500	30	9	$a = 3.189$ $c = 5.185$	6.10	1500
GaP	2.26	110	75	11.2	5.4512	4.13	1750
GaSb	0.72	5,000	1,000	15.7	6.0959	5.619	980
InAs	0.36	33,000	460	15.1	6.0584	5.66	1215
InP	1.35	4,600	150	12.4	5.8693	4.787	1330
InSb	0.17	77,000	1,000	17.9	6.4794	5.775	798
AlAs	2.16	1,200	400	10.1	5.6622	3.81	1870
AlSb	1.6	200	420	14.4	6.1355	4.218	1330
AlP	3.0			9.8	5.4510	2.85	1770
CdS	2.5	300	50	11.6	5.8320	4.82	1750
CdTe	1.5	1,000	100	10.8	6.482	5.86	1365
PbS	0.41	600	700	175	5.9362	7.61	1390
PbSe	0.26	1,000	600	250	6.1243	8.15	1340
PbTe	0.32	1,800	900	400	6.4620	8.16	1180
ZnO	3.35	200	180	8.5	$a = 3.252$ $c = 5.213$	5.66	—
ZnS	3.66	165	5	8.3	5.410	4.079	2100
ZnSe	2.67	540	30	9.25	5.6676	5.42	1790
ZnTe	2.26	340	100	9.7	6.101	5.72	1568

* Drift mobilities in the purest materials.

Powers of Ten

10^{24}	yotta	Y
10^{21}	zetta	Z
10^{18}	exa	E
10^{15}	peta	P
10^{12}	tera	T
10^{9}	giga	G
10^{6}	mega	M
10^{3}	kilo	K
10^{2}	hecto	h
10^{1}	deka	da
10^{-1}	deci	d
10^{-2}	centi	c
10^{-3}	milli	m
10^{-6}	micro	μ
10^{-9}	nano	n
10^{-12}	pico	p
10^{-15}	femto	f
10^{-18}	atto	a
10^{-21}	zepto	z
10^{-24}	yocto	y